Computational Complexity

Mathematics and Its Applications (*East European Series*)

Computational Complexity

by

K. Wagner and G. Wechsung

D. Reidel Publishing Company

A MEMBER OF THE KLUWER ACADEMIC PUBLISHERS GROUP

Dordrecht / Boston / Lancaster / Tokyo

Library of Congress Cataloging-in-Publication Data

CIP

Wagner, K. (Klaus)
Computational complexity.

(Mathematics and its applications (East European series))
Bibliography: p.
Includes indexes.
1. Computational complexity. I. Wechsung, Gerd.
II. Title. III. Series: Mathematics and its applications (D. Reidel Publishing Company). East European series.
QA267.W34 1985 511.3 85-25587
ISBN 14-020-0313-7

Distributors for the Socialist Countries
VEB Deutscher Verlag der Wissenschaften, Berlin

Distributors for the U.S.A. and Canada
Kluwer Academic Publishers,
190 Old Derby Street, Hingham, MA 02043, U.S.A.

Distributors for all remaining countries
Kluwer Academic Publishers Group,
P.O. Box 322, 3300 AH Dordrecht, Holland

Published by VEB Deutscher Verlag der Wissenschaften, Berlin
in co-edition with
D. Reidel Publishing Company, Dordrecht, Holland.

Printed in the German Democratic Republic.

SERIES EDITOR'S PREFACE

Approach your problems from the right end and begin with the answers. Then one day, perhaps you will find the final question.

'The Hermit Clad in Crane Feathers' in R. van Gulik's *The Chinese Maze Murders*.

It isn't that they can't see the solution. It is that they can't see the problem.

G.K. Chesterton. *The Scandal of Father Brown* 'The point of a Pin'.

Growing specialization and diversification have brought a host of monographs and textbooks on increasingly specialized topics. However, the "tree" of knowledge of mathematics and related fields does not grow only by putting forth new branches. It also happens, quite often in fact, that branches which were thought to be completely disparate are suddenly seen to be related.

Further, the kind and level of sophistication of mathematics applied in various sciences has changed drastically in recent years: measure theory is used (non-trivially) in regional and theoretical economics; algebraic geometry interacts with physics; the Minkowsky lemma, coding theory and the structure of water meet one another in packing and covering theory; quantum fields, crystal defects and mathematical programming profit from homotopy theory; Lie algebras are relevant to filtering; and prediction and electrical engineering can use Stein spaces. And in addition to this there are such new emerging subdisciplines as "experimental mathematics", "CFD", "completely integrable systems", "chaos, synergetics and large-scale order", which are almost impossible to fit into the existing classification schemes. They draw upon widely different sections of mathematics. This programme, Mathematics and Its Applications, is devoted to new emerging (sub)disciplines and to such (new) interrelations as exempla gratia:

- a central concept which plays an important role in several different mathematical and/or scientific specialized areas;
- new applications of the results and ideas from one area of scientific endeavour into another;
- influences which the results, problems and concepts of one field of enquiry have and have had on the development of another.

The Mathematics and Its Applications programme tries to make available a careful selection of books which fit the philosophy outlined above. With such books, which are stimulating rather than definitive, intriguing rather than encyclopaedic, we hope to contribute something towards better communication among the practitioners in diversified fields.

Because of the wealth of scholarly research being undertaken in the Soviet Union, Eastern Europe, and Japan, it was decided to devote special attention to the work emanating from these particular regions. Thus it was decided to start three regional series under the umbrella of the main MIA programme.

New observational tools generate new discoveries. Often totally unexpected ones, pointing in directions unknown before. This is an observation in the history of science. New and different questions result for mathematical apparatus involved in the science concerned. New techniques, for instance new possibilities of calculation have similar effects. Thus it is no accident that questions like "what is computable" have taken on a new life and new meaning these days.

It is one thing to know that a quantity is precisely measurable (in principle), but if obtaining the 20-th decimal takes the energy-equivalent of 30 metric tons the fact is at best of marginal interest; it is one thing to know that a series is convergent, but if it would take a couple thousand years to actually do the summation up to a given accuracy the theoretical result loses much of its power. Similarly it is one thing to know that a certain function or number is computable in an abstract sense, that is, to have an algorithm for it (or an abstract result guranteeing its existence). It is another problem, and one of absorbing interest, to ask how well it is computable. This is what computational complexity is all about. It is a rather young field which generates new and different mathematical questions and answers. There are few comprehensive, good books on it. This seems to be one.

The unreasonable effectiveness of mathematics in science ...

Eugene Wigner

Well, if you know of a better 'ole, go to it.

Bruce Bairnsfather

What is now proved was once only imagined.

William Blake

As long as algebra and geometry proceeded along separate paths, their advance was slow and their applications limited.

But when these sciences joined company they drew from each other fresh vitality and thence forward marched on at a rapid pace towards perfection.

Joseph Louis Lagrange.

Bussum, August 1985 **Michiel Hazewinkel**

Preface

To Maria and Uta

This is no textbook but a monograph which is primarily intended for specialists doing research on computational complexity. It can be used by everybody who has a certain basic knowledge concerning effective computability, recursive functions and formal languages, for which the reader is referred to books like [Rog 67], [Schn 74a], [HoUl 79], [Gin 75], [Sal 73].

The suggestion for writing this book came from Professor Dr. Günter Asser (Greifswald). Our work has been considerably supported by the director of the Department of Mathematics of the Friedrich Schiller University at Jena, Professor Dr. Kurt Nawrotzki.

It would not have been possible to write this book without the help of many colleagues who provided us with conference proceedings and books on our topic. We express our deepest gratitude to Albert R. Meyer and to Ronald V. Book for sending us several STOC and FOCS proceedings, Giorgio Ausiello, Franz-Josef Brandenburg, Robert P. Daley, the late Calvin C. Elgot, Peter von Emde Boas, Jozef Gruska, Edward L. Robertson, Dirk Siefkes, Anatoli O. Slisenko, Volker Strassen, Boris A. Trakhtenbrot, and Klaus Weihrauch.

We owe numerous corrections and suggestions concerning the mathematical contents of the book to Franz-Josef Brandenburg, Michal Chytil, Robert Daley, Jozef Gruska, Hans Heller, Burkhard Monien, Werner Nagel, Werner Nehrlich, Rošal Nigmatullin, Peter Ružička, I. Hal Sudborough, and Ondrej Sykora. Especially gratefully we acknowledge the permanent constructive criticism of the colleagues of our group, Andreas Brandstädt, Gerhard Lischke, Jörg Vogel, Ludwig Staiger, Dietrich Meinhardt, and Holger Harz.

We are very much indebted to Lorna Froom (Liverpool) and Christine Conlin (Manchester) for correcting the English style of our book. Note that we have made many changes after the style correction procedure so that we have to take sole responsibility for any remaining offences against English style and grammar. Knowing that such a monograph cannot be written without mathematical errors we are grateful in advance to all readers who will give us hints as to errors and other suggestions for improving the book.

We thank Christine Heuschild for typing the whole manuscript and Maria for technical help.

Finally, we would like to thank Prof. Dr. H. Reichardt for accepting this book for the series „Mathematische Monographien", Mrs. B. Mai from the staff of the publishing house VEB Deutscher Verlag der Wissenschaften for the cooperation, and the staff of the printing office VEB Druckhaus „Maxim Gorki" Altenburg for the accurate work.

We have supplementary revised the finished manuscript for including the most important results published since 1982, where in most cases proofs and detailed explanations had to be omitted.

Jena, January 1985

Klaus Wagner
Gerd Wechsung

Contents

Introduction

Nearly everything in everyday life, in sciences and in arts has the aspect of complexity. From the very beginning we restrict our attention to complexity in computer science, and even here we restrict ourselves to a certain sub-area. Depending on the objects whose complexity is considered, two different areas can be distinguished, namely descriptional (structural) complexity and computational complexity. Without going into details we simply state that the object of descriptional complexity is the investigation of the complexity (in a sense to be made precise) of constructive objects like words, sequences, networks, automata, grammars, programs etc. Axiomatic foundations of different directions in this field have been given in [Blu 67b] and in [Kol 65]. For surveys see [Lup 74], [Sav 76], [Gru 76], [ZvLe 70], and [Aga 75].

Computational complexity is devoted to the study of the complexity of effective computations. With every program (machine) M a "complexity function" Φ_M can be associated in many different ways, where $\Phi_M(x)$ is interpreted as the complexity of the computation process of M on an input x. Whereas the descriptional complexity of a program reflects the structure of the program itself and is always expressed by a single value (a natural number), the computational complexity associated with a program is a function, whose values can be understood as the descriptional complexities of the computations described by the program (these being viewed as "finite objects").

This book deals with computational complexity of algorithms having the full power of Turing machines. In other words, apart from single exceptions we do not include results on computational complexity concerning restricted notions of algorithms, such as automata, straight line programs, or Boolean networks. In these restricted cases very restricted methods are used, which usually cannot be generalized to more powerful algorithms. A detailed treatment of these areas would require much more space than available here. Many results concerning restricted classes of algorithms can be found in the Soviet periodical "Problemy kibernetiki" and in [Sav 76].

When in the thirties of this century the notion of effective computability had been made precise, the question of what is computable had been answered, but evidently people were not immediately interested in the further question of how good the computability of a given function might be. Apparently the more practical needs of an ever further developing computing technique were necessary to stimulate researchers to investigate the question of which functions and computations are actually of human concern. In this situation, at the end of the fifties — note that complexity investigation for logical networks became popular somewhat earlier — a few papers (B. A. Trakhtenbrot 1956, see [Tra 67], M. O. Rabin [Rab 60] and H. Yamada [Yam 62] were among the very first) sufficed to initiate an enormously broad research activity on this topic, which led to the writing of more than 2500 papers by the end of 1984. Very often parts of the theory have been presented in books, indeed, this began to be done relatively early. Examples of such books are [Tra 67], [HoUl 69a], [AhHoUl 74], [Schn 74a], [Aus 75], [BoMu 75], [SpSt 76], [Meh 77], [BaLiRo 78], [HoSa 78], [MaYo 78], [Pau 78], [FeRa 79], [GaJo 79], [HoUl 79], [Kro 79]. All of them deal with well-defined restricted sub-areas of the theory of computational complexity.

Our book aspires to cover all the disciplines of this theory (as restricted above), which have emerged up to 1984. We try to develop the main features of each discipline, to state, as its core, the most important results with or without proofs and to give additional results and hints to the literature. Nevertheless, there are still many results which had to be omitted. Even a complete bibliography is far beyond the scope of this book. Note that applicational areas such as cryptography and VLSI systems are not included. Likewise, it was impossible to include the huge variety of results concerning parallel algorithms, distributed computation and communication complexity. Some aspects of parallelity are considered in §§ 20.3.1., 20.4.1. and 21.2., but hints to parallel algorithms for specific problems are given only sparsely.

The book is organized as follows. Part 1 presents the basic notions and results of the theory of computing and formal language theory (chapter I) and the basic notions of the theory of computational complexity (chapter II). The theory of computational complexity is developed in the remaining two parts. Part 2 studies problems concerning one complexity measure only, in particular the complexity of single problems (upper bounds, lower bounds and approximations of hard problems), it investigates the properties of complexity classes from various points of views (recursion theory, algebra, formal languages, automata, reducibilities) and it investigates the properties of complexity measures (naming and union theorems, hierarchies, speed-up and the problem of "natural" complexity measures). Part 3 deals with the relationships between different measures, in particular with time efficient simulations between different types of machines, with the time and space measures of "powerful machines", with the problem of determinism versus nondeterminism and with the problem of time versus space. The theory of relativized computational complexity is included in Part 3 although not all of its results concern relationships between different measures.

Although the material of Part 1, in particular that of § 5, is used throughout the book, the reader should use this part only for reference. There exist various interconnections between the sections of this book. Thus the sections form a "dependency graph" which is not acyclic. Besides the sections of chapter I, the sections 8, 13, 19 and 20 have a special position as "main sources". But nevertheless it should be possible to start reading the book with an arbitrary section.

We do not emphasize a difference between abstract and concrete complexity theory. For specific machine-oriented measures the results of the abstract theory can be completed by further results which can be proved only for these measures. This point of view is perhaps most evident in chapter V, especially in § 17. But it is natural that the greater part of the book is devoted to the study of machine-oriented measures. In particular, time and space problems which are undoubtedly of greatest practical importance have of course been treated in an especially detailed way.

Part 1
Basic Concepts

Chapter I
Preliminaries

The reader should be advised that this chapter is not intended to be read through, but only to be referred to as a source of references for the basic notions concerning various machine and automata models, recursive functions, reducibility, grammars, L-systems and formal languages which are necessary for this book. It is assumed that these topics are known, and therefore we confine ourselves to summarizing definitions of the basic notions and the basic results in a unified and systematic way. Proofs and historical references are omitted in most cases. For more details the reader is referred to the literature.

§ 0. Mathematical Preliminaries

The *set theoretical inclusion* is denoted by $\subseteq$, the *proper inclusion* is denoted by $\subset$ i.e. $A \subset B$ if and only if $A \subseteq B$ and $A \neq B$. The *empty set* is denoted by $\emptyset$, and $\mathfrak{P}(A)$ denotes the *power set* of the set A, i.e. $\mathfrak{P}(A) = \{B : B \subseteq A\}$. The *cartesian product* of the sets $A_1, A_2, \ldots, A_n$ is denoted by $A_1 \times A_2 \times \ldots \times A_n$, and we set $A^n = \underbrace{A \times A \times \ldots \times A}_{n \text{ times}}$. By card A we denote the *cardinality* of a set A.

The *existential* (*universal*) *quantifier* is denoted by $\bigvee$ ($\bigwedge$). Universal quantifiers on the left-hand side of a proposition can be omitted. Furthermore, for propositions $P(x)$ and $H(x)$ with the unquantified variable x,

$$\bigwedge_{P(x)} H(x) \text{ stands for } \bigwedge_{x} \big(P(x) \rightarrow H(x)\big),$$

$$\bigvee_{P(x)} H(x) \text{ stands for } \bigvee_{x} \big(P(x) \wedge H(x)\big),$$

$$\bigwedge_{x}^{\text{ae}} H(x) \text{ stands for } \bigvee_{B} \big(B \text{ finite} \wedge \bigwedge_{x \notin B} H(x)\big)$$

and

$$\bigvee_{x}^{\text{io}} H(x) \text{ stands for } \bigvee_{B} \big(B \text{ infinite} \wedge \bigwedge_{x \in B} H(x)\big).$$

From propositional calculus we use $\sim$ (*non*), $\wedge$ (*and*), $\vee$ (*or*), $\oplus$ (*exclusive or*), $\rightarrow$ (*if ... then ...*) and $\leftrightarrow$ (*if and only if*). Instead of "if and only if" we sometimes use "iff". $a =_{\text{df}} b$ means: a is defined by the value b, whereas $p \leftrightarrow_{\text{df}} q$ means: p is defined by the proposition q.

For the description of algorithms we use **if** … **then** … **else** and **for** … **do** … in a selfexplanatory way.

The end of a proof is marked by ∎ (or by □ for proofs of claims or lemmas within other proofs). If a theorem or corollary is not proved, then its end is marked by ∎.

The *algebraic closure operator defined by the operations* $\omega_1, \ldots, \omega_k$ is denoted by $\Gamma_{\omega_1,\ldots,\omega_k}$. Sometimes we write $\Gamma_{\omega_1,\ldots,\omega_k}(x_1, \ldots, x_m)$ instead of $\Gamma_{\omega_1,\ldots,\omega_k}(\{x_1, \ldots, x_m\})$.

A *graph* (or *directed graph* or *digraph*) G is a pair (V, E) where V is the set of *vertices* (or *nodes*) and $E \subseteq V \times V$ is the set of *edges*. We set $V(G) = V$ and $E(G) = E$. A graph $G = (V, E)$ is said to be *undirected* iff $(x, y) \in E$ implies $(y, x) \in E$ for all $x, y \in V$. The *indegree* (*outdegree*) of a graph $G = (V, E)$ is defined by $\operatorname{indeg} G = \max_{x \in V} \operatorname{card}(E \cap V \times \{x\})$ $\left(\operatorname{outdeg} G = \max_{x \in V} \operatorname{card}(E \cap \{x\} \times V)\right)$.

Let Σ be an arbitrary set of symbols. By Σ^* (Σ^ω) we denote the set of finite strings (infinite strings) of symbols from Σ. The set Σ is called the *alphabet*, and the elements of Σ^* are called *words over the alphabet* Σ. In particular, the *empty word* e is an element of Σ^*. Every subset of Σ^* is called a *language*. If Σ consists of one symbol only, then the subsets of Σ^* are called *single letter alphabet languages* or *SLA languages*. For $x, y \in \Sigma^*$, the word xy is called the *concatenation* of x and y. For $V, W \subseteq \Sigma^*$, we define $V \cdot W = \{vw : v \in V \wedge w \in W\}$, $V^0 = \{\mathrm{e}\}$, $V^{k+1} = V^k \cdot V$, $V^+ = \bigcup_{k=1}^{\infty} V^k$ and $V^* = V^0 \cup V^+$. The only element of $\{w\}^k$ is denoted by w^k. The *length* $|w|$ of $w \in \Sigma^*$ is defined by $|\mathrm{e}| = 0$ and $|wa| = |w| + 1$ for $w \in \Sigma^*$ and $a \in \Sigma$. The *reverse* w^{-1} of $w \in \Sigma^*$ is defined by $\mathrm{e}^{-1} = \mathrm{e}$ and $(wa)^{-1} = aw^{-1}$ for $w \in \Sigma^*$ and $a \in \Sigma$. By $w(i)$ we denote the ith symbol of $w \in \Sigma^*$, $1 \leqq i \leqq |w|$. The *initial word relation* $\sqsubseteq$ is defined by $v \sqsubseteq w \leftrightarrow \bigvee_u (vu = w)$. Furthermore, $v \sqsubset w$ iff $v \sqsubseteq w$ and $v \neq w$.

The set $\{0, 1, 2, \ldots\}$ of *natural numbers* is denoted by $\mathbb{N}$. Natural numbers can be presented as words in different ways. The *binary presentation* is defined by $\operatorname{bin} 0 = 0$, $\operatorname{bin} 1 = 1$, $\operatorname{bin}(2i) = (\operatorname{bin} i)0$ and $\operatorname{bin}(2i + 1) = (\operatorname{bin} i)1$ for $i \geqq 1$. The *r-ary presentation* can be defined in an analogous way. For $w \in \{0, 1\}^*$ we define $\operatorname{bin}^{-1}(w)$ as that $k \in \mathbb{N}$ for which there exists an $m \in \mathbb{N}$ such that $w = 0^m \operatorname{bin} k$. The *dyadic presentation* is defined by $\operatorname{dya} 0 = \mathrm{e}$, $\operatorname{dya}(2i + 1) = (\operatorname{dya} i)1$ and $\operatorname{dya}(2i + 2) = (\operatorname{dya} i)2$ for $i \geqq 0$. Note that dya is a one-one mapping from $\mathbb{N}$ onto $\{1, 2\}^*$.

The following functions will be used frequently ($x, y \in \mathbb{N}$):

$$x \dot{-} y = \begin{cases} x - y & \text{if } x \geqq y, \\ 0 & \text{otherwise,} \end{cases}$$

$$\operatorname{sgn} x = x \dot{-} (x \dot{-} 1) \quad \text{and} \quad \overline{\operatorname{sgn}}\, x = 1 \dot{-} \operatorname{sgn} x.$$

By $\mathbb{G}$ and $\mathbb{Q}$ we denote the set of all *integers* and *rational numbers*, resp. For $\alpha \in \mathbb{Q}$ we denote by $\lfloor \alpha \rfloor$ ($\lceil \alpha \rceil$) the *greatest* (*least*) *integer not greater* (*smaller*) *than* α. The logarithm with base k is denoted by $\log_k$. Instead of $\log_2$ we usually write log. Furthermore, for $n \in \mathbb{N}$,

$$\log^* n = \min \left\{k : \left.2^{2^{\cdot^{\cdot^{2}}}}\right\}k \geqq n\right\}.$$

We write $f: A \to B$ if f is a function such that $f \subseteq A \times B$. We define $D_f = \{x: \bigvee_y (x, y) \in f\}$ and $R_f = \left\{y: \bigvee_x (x, y) \in f\right\}$, and we write

$$f: A \mapsto B \quad \text{iff} \quad D_f = A,$$

$$f: A \mapsto\!\!\!\to B \quad \text{iff} \quad D_f = A \quad \text{and} \quad R_f = B,$$

$$f: A \overset{1\text{-}1}{\mapsto} B \quad \text{iff} \quad D_f = A \quad \text{and } f \text{ is } \textit{one-one} \text{ (i.e. } f(x) = f(y) \text{ implies } x = y),$$

$$f: A \overset{1\text{-}1}{\mapsto\!\!\!\to} B \quad \text{iff} \quad D_f = A, \quad R_f = B \text{ and } f \text{ is one-one}.$$

If $f: A \overset{1\text{-}1}{\mapsto\!\!\!\to} B$, then the *inverse* $f^{-1}: B \overset{1\text{-}1}{\mapsto\!\!\!\to} A$ is defined by $f^{-1} = \{(y, x): (x, y) \in f\}$. If $f: A \to B$ and $g: B \to C$, then the *product* $f \circ g: A \to C$ is defined by $(f \circ g)(x) = f\big(g(x)\big)$. For $f: A \times A \to A$ and $g: A \to A$, we define $f \,\square\, g: A \mapsto A$ by $(f \,\square\, g)(x) = f\big(x, g(x)\big)$. For $f: A \to A$ we define $f^{[k]} = \underbrace{f \circ f \circ \ldots \circ f}_{k \text{ times}}$. The set of all functions $f: A \mapsto B$ is denoted by B^A.

In general, we do not use the λ-calculus for the notation of functions, i.e. if no confusion is possible we use $f(x)$ instead of $\lambda x f(x)$. In particular, for constant functions we write a instead of $\lambda x a$.

Let Σ be a finite alphabet and $A \subseteq \Sigma^*$. Usually we use greek letters $\alpha, \beta, \gamma, \delta, \ldots, A, B, \Gamma, \Delta, \ldots$ for partial functions (i.e. for functions: $(\Sigma^*)^m \to \Sigma^*$), and we use latin letters $a, b, c, d, \ldots, A, B, C, D, \ldots$ for functions defined everywhere (i.e. for functions $(\Sigma^*)^m \mapsto \Sigma^*$). The *characteristic function* c_A of A and the *partial characteristic function* χ_A of A are defined by

$$c_A(w) = \begin{cases} 1 & \text{if} \quad w \in A, \\ 0 & \text{otherwise} \end{cases} \quad \text{and} \quad \chi_A(w) = \begin{cases} 1 & \text{if} \quad w \in A, \\ \text{undefined otherwise}, \end{cases}$$

resp. for all $w \in \Sigma^*$.

A function $f: \mathbb{N} \mapsto \mathbb{N}$ is said to be *increasing* (*strictly increasing, decreasing*) iff $f(x) \leqq f(y)$ $\big(f(x) < f(y), f(x) \geqq f(y)$, resp.$\big)$ for all $x < y$.

For a function $f: \mathbb{N} \mapsto \mathbb{N}$ and $s \geqq 1$, the function f^k is defined by $f^k(n) = \big(f(n)\big)^k$

Now let $f: \mathbb{N} \mapsto \mathbb{N}$ and $g: \mathbb{N} \mapsto \mathbb{N}$. We define

$$f \leqq g \quad \text{iff} \quad \bigwedge_x \big(f(x) \leqq g(x)\big),$$

$$f \leqq_{\text{ae}} g \quad \text{iff} \quad \bigwedge_x^{\text{ae}} \big(f(x) \leqq g(x)\big),$$

$$f \leqq_{\text{io}} g \quad \text{iff} \quad \bigvee_x^{\text{io}} \big(f(x) \leqq g(x)\big),$$

$$f \lesssim g \quad \text{iff} \quad \varliminf_{n \to \infty} \frac{g(n) + 1}{f(n) + 1} > 0,$$

$$f \lesssim_{\text{io}} g \quad \text{iff} \quad \varlimsup_{n \to \infty} \frac{g(n) + 1}{f(n) + 1} > 0,$$

$$f \prec g \quad \text{iff} \quad \overline{\lim_{n\to\infty}} \frac{f(n)+1}{g(n)+1} = 0,$$

$$f \prec_{io} g \quad \text{iff} \quad \underline{\lim}_{n\to\infty} \frac{f(n)+1}{g(n)+1} = 0$$

and

$$f \asymp g \quad \text{iff} \quad f \preceq g \quad \text{and} \quad f \succeq g.$$

Analogous definitions are made for $<$, $\geqq$, $>$, $<_{ae}$, $\geqq_{ae}$, $>_{ae}$, $<_{io}$, $\geqq_{io}$, $>_{io}$, $\succeq$, $\succeq_{io}$, $\succ$ and $\succ_{io}$.

§ 1. Machine Models

Here we introduce several types of Turing machines, random access machines and iterative arrays. Most of this book will deal with the problem of what these models of computation can do in general or with some restrictions on the complexity of their computations. Therefore the reader will very often refer to this section.

One of the most important preconditions for writing a monograph on computational complexity is a uniform terminology for machines and measures. We present such a uniform notational system for the various types of Turing machines and their restrictions: deterministic, nondeterministic, alternating and probabilistic Turing machines with and without one-way or two-way input tapes, with different numbers of worktapes of different dimensions, with different numbers of heads per tape, with tapes having restricted size of the tape alphabet, and with restricted Turing tapes such as stack, nonerasing stack, checking stack, pushdown store or counter. Our system is not so bad as it looks at first glance. It is not too difficult to keep it in mind because it is designed in a systematic and self-evident way, and, when possible, respects notations familar from the literature.

Once the reader has become accustomed to it, it will surely be found that it proves useful within our book.

1.1. Turing Machines

1.1.1. Ordinary Turing Machines

A *deterministic Turing machine with k tapes having $h_1, \ldots, h_k$ heads*, resp. (for short: $\mathrm{Th_1\text{-}Th_2\text{-}\ldots\text{-}Th_k\text{-}DM}$) is a device consisting of

1. a *finite control* which is at any time in one of finitely many *states*, and
2. *k tapes* (called *Turing tapes*) which are infinite to the left and to the right (Fig. 1.1) and which are divided into infinitely many *cells* (or *tape squares*). At any time each tape square holds exactly one of finitely many *tape symbols*. The *i*th tape has h_i *heads* each scanning one cell of the tape at a time ($i = 1, \ldots, k$).

The prefix $\mathrm{Th_1\text{-}Th_2\text{-}\ldots\text{-}Th_k}$ is said to be the type of this Turing machine.

A Turing machine performs one *move* (or *step*) per time unit. In such a move the Turing machine, depending on the state of the finite control and the $h_1 + h_2 + \ldots + h_k$ scanned tape symbols can

1. change the state,
2. change the $h_1 + h_2 + \ldots + h_k$ scanned tape symbols,
3. move the $h_1 + h_2 + \ldots + h_k$ heads: each head can either move one cell to the left, or one cell to the right, or remain stationary.

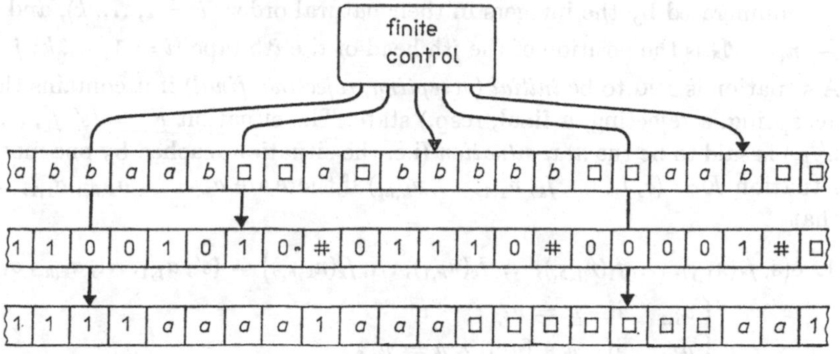

Figure 1.1

Formally, a Th_1-…-Th_k-DM M is a $(k+5)$-tuple $(S, \Delta_1, \ldots, \Delta_k, \delta, s_0, S_a, S_r)$ where

- S is the finite set of states.
- Δ_i is the set of tape symbols (*tape alphabet*) of the ith tape. The symbol $\square$ (the *blank*) is in Δ_i $(i = 1, \ldots, k)$.
- $\delta\colon S \times \Delta_1^{h_1} \times \ldots \times \Delta_k^{h_k} \mapsto S \times \Delta_1^{h_1} \times \ldots \times \Delta_k^{h_k} \times \{-1, +1, 0\}^{h_1+\ldots+h_k}$ is the *next move function*, where -1, $+1$ and 0 denote the *direction* of the head move (left, right, stationary, resp.). For $s \in S_a \cup S_r$ we require $\delta(s, a_{1,1}, \ldots, a_{k,h_k}) = (s, a_{1,1}, \ldots, a_{k,h_k}, 0, \ldots, 0)$, and we say that M *stops* or *halts*.
- $s_0 \in S$ is the *initial state*.
- $S_a \subseteq S$ is the set of *accepting states*.
- $S_r \subseteq S$ is the set of *rejecting states*. We require $S_a \cap S_r = \emptyset$. The states from $S_a \cup S_r$ are said to be *final*.

We agree upon some conventions. First we note that, except for the first tape which in any case has to hold the input, the order in which the tapes of a machine are mentioned in its type is not essential; for example, the Turing machine in Figure 1.1 can also be referred to as a T2-T2-T1-DM. The term $\underbrace{\text{Th-Th-}\ldots\text{-Th}}_{m \text{ times}}$ can be replaced by mTh, and T1 can be replaced by T. Thus, a T2-T1-T2-DM can also be referred to as a 2T2-T-DM. Furthermore, instead of T-DM we write DTM. For a DTM $M = (S, \Delta, \delta, s_0, S_a, S_r)$, instead of $\delta(s, a) = (s', a, \sigma)$ we also write the instruction $sa \to s'a\sigma'$,

where

$$\sigma' = \begin{cases} + & \text{if} \quad \sigma = +1, \\ - & \text{if} \quad \sigma = -1, \\ \mathrm{e} & \text{if} \quad \sigma = 0. \end{cases}$$

A *situation* of such a Turing machine is a $(h_1 + \ldots + h_k + k + 1)$-tuple $(s, f_1, \ldots, f_k, n_{1,1}, \ldots, n_{1,h_1}, \ldots, n_{k,1}, \ldots, n_{k,h_k})$ where

— s is the actual state,

— $f_i : \mathbf{G} \mapsto \Delta_i$ is the inscription of the ith tape, whose cells are considered to be enumerated by the integers in their natural order ($i = 1, \ldots, k$), and

— $n_{i,j} \in \mathbf{G}$ is the position of the jth head of the ith tape ($i = 1, \ldots, k$; $j = 1, \ldots, h_i$).

A situation is said to be *initial* (*accepting, rejecting, final*) if it contains the initial (an accepting, a rejecting, a final, resp.) state. The situation $K' = (s', f'_1, \ldots, f'_k, n'_{1,1}, \ldots, n'_{k,h_k})$ is said to be the *next situation* (i.e. the situation reached by one move M) to the situation $K = (s, f_1, \ldots, f_k, n_{1,1}, \ldots, n_{k,h_k})$ if there are $a_{1,1}, \ldots, a_{k,h_k}, \sigma_{1,1}, \ldots, \sigma_{k,h_k}$ such that

1. $\delta(s, f_1(n_{1,1}), \ldots, f_1(n_{1,h_1}), \ldots, f_k(n_{k,1}), \ldots, f_k(n_{k,h_k})) = (s', a_{1,1}, \ldots, a_{k,h_k}, \sigma_{1,1}, \ldots, \sigma_{k,h_k})$,

2. $$f'_i(n) = \begin{cases} a_{i,1} & \text{if} \quad n = n_{i,1}, \\ a_{i,2} & \text{if} \quad n \notin \{n_{i,1}\} \wedge n = n_{i,2}, \\ \vdots & \\ a_{i,h_i} & \text{if} \quad n \notin \{n_{i,1}, \ldots, n_{i,h_{i-1}}\} \wedge n = n_{i,h_i}, \\ f_i(n) & \text{if} \quad n \notin \{n_{i,1}, \ldots, n_{i,h_i}\}, \end{cases} \qquad (i = 1, \ldots, k).$$

(Note that in the case where a tape square is scanned by more than one head, its contents can be changed only by the head with the smallest number).

3. $n'_{i,j} = n_{i,j} + \sigma_{i,j}$ ($i = 1, \ldots, k$; $j = 1, \ldots, h_i$).

In this case we write $K \vdash_{\overline{M}} K'$. Note that $K \vdash_{\overline{M}} K$ for any final situation K. We write $K \overset{t}{\vdash_{\overline{M}}} K'$ if there are situations $K_0, K_1, \ldots, K_t$ such that

$$K = K_0 \vdash_{\overline{M}} K_1 \vdash_{\overline{M}} \ldots \vdash_{\overline{M}} K_t = K' \quad \text{for} \quad t \geqq 0.$$

The relation $\overset{*}{\vdash_{\overline{M}}}$ is the reflexive and transitive closure of $\vdash_{\overline{M}}$, i.e. $K \overset{*}{\vdash_{\overline{M}}} K'$ if $K \overset{t}{\vdash_{\overline{M}}} K'$ for some $t \geqq 0$. By $\mathrm{K}_{M,w}$ we denote the *initial situation of M induced by the input* $w \in \Delta_1^*$: $\mathrm{K}_{M,w} = (s_0, g_w, g_e, \ldots, g_e, 1, 1, \ldots, 1)$ where for $v \in \Delta_1^*$ we define $g_v(1)\, g_v(2) \ldots g_v(|v|) = v$ and $g_v(n) = \square$ for $n \leqq 0$ and $n > |v|$. A sequence $K_0, K_1, K_2, \ldots$ of situations of M is said to be a *computation of M* if $K_0 \vdash_{\overline{M}} K_1 \vdash_{\overline{M}} K_2 \vdash_{\overline{M}} \ldots$ Such a computation is said to be *halting* (*accepting, rejecting*) if there is a $t \geqq 0$ such that K_t is final (accepting, rejecting, resp.). In this case we have $K_t = K_{t+1} = K_{t+2} = \ldots$, and the number $\max \{t : K_t \text{ is not final}\}$ is said to be the *length of this computation.* The computation $K_0, K_1, K_2, \ldots$ of M is said to be the *computation of M on w* if $K_0 = \mathrm{K}_{M,w}$.

Let $M = (S, \Delta_1, \ldots, \Delta_k, \delta, s_0, S_a, S_r)$ be a kT-DM such that $\#, | \notin S \cup \Delta_1 \cup \ldots \cup \Delta_k$ and $S \cap (\Delta_1 \cup \ldots \cup \Delta_k) = \emptyset$, and let K be a situation of M in which all but finitely many tape squares are empty (i.e. have the contents $\square$). Then the word $w_1 s w_1' \# w_2 \mid w_2' \dots \# w_k \mid w_k'$ is said to be an *instantaneous description* (for short: *ID*) of K if $\dots \square\square\square w_i w_i' \square\square\square \dots$ is the content of the ith tape the head of which is scanning the first symbol of w_i' ($i = 1, \dots, k$). Note that a word can be the ID of several situations which differ from each other by translations of the tape inscriptions. On the other hand, each situation has several ID's. For example, if w is an ID of K, then $w\square$ is also an ID of K. Among the ID's of K there is exactly one shortest, which we shall refer to as *the* ID of K. An ID of an initial (accepting, rejecting, final) situation is said to be *initial* (*accepting, rejecting, final*, resp.). Let I_1 and I_2 be ID's of the situations K_1 and K_2, resp. We write $I_1 \vdash_M I_2$ $\left(I_1 \vdash_M^t I_2, I_1 \vdash_M^* I_2\right)$ if and only if $K_1 \vdash_M K_2$ $\left(K_1 \vdash_M^t K_2, K_1 \vdash_M^* K_2\text{, resp.}\right)$. Note that this definition is independent of which situations K_1 and K_2 are chosen from those described by I_1 and I_2, resp. A Th_1-...-Th_k-DM $M = (S, \Delta_1, \dots, \Delta_k, \delta, s_0, S_a, S_r)$ *decides* a language $L \subseteq \Sigma^*$ if $\Sigma \subseteq \Delta_1 \setminus \{\square\}$ and if for all $w \in \Sigma^*$: if $w \in L$ ($w \notin L$), then $K_{M,w} \vdash_M^* K$ for some accepting (rejecting) situation K. In this case we set $L(M) = L$. Note that not every deterministic Turing machine decides a language. A Th_1-...-Th_k-DM M decides the set $L \subseteq \mathbb{N}$ if $L(M) = \{\text{bin } n : n \in L\}$, where the digits 0 and 1 can be replaced by any two different symbols of the alphabet of the first tape.

Note that the input is given on the first tape described in the type of the machine. Note also that this tape is not always an input tape in the sense of the following definition.

A *two-way input tape* (or a *two-way read-only tape*) is a Turing tape with the following restrictions:

1. no symbol can be changed by the head, and
2. a tape square with a blank can be left by a head only in that direction from which it has been entered by this head.

A *one-way input tape* is a two-way input tape whose heads cannot move to the left. A *one*-way input tape* is a one-way input tape with only one head whose reading a blank causes a final state in the finite control. Machines with such an input tape are called *on-line machines*. They obviously require a new input symbol only after having computed the result for the input which has been read so far. A Th_1-Th_2-...-Th_k-DM whose first tape is a two-way (one-way, one*-way) input tape is said to be a $2{:}h_1$-Th_2-...-Th_k-DM ($1{:}h_1$-Th_2-...-Th_k-DM, 1^*-Th_2-...-Th_k-DM, resp.). In the same way as for full Turing tapes, if $h_1 = 1$, then h_1 can be omitted, i.e. we write 2-...(1-...) instead of 2:1-...(1:1-...). To conform to this pattern we sometimes write 0-Z, where Z is any type of Turing machine having no input tape at all.

An output tape (or write-only tape) is a Turing tape with only one head and with he following restrictions:

1. only the blank can be changed by the head, and
2. the head cannot move to the left.

A tape which is neither an input tape nor an output tape is called a *worktape*. A Th_1-Th_2-...-Th_k-DM with output tape is a Th_1-Th_2-...-Th_k-T-DM whose last tape is an output tape (i.e. the output tape is not mentioned in the description of the type). Let $m \in \mathbb{N}$ and let $M = (S, \Delta_1, \ldots, \Delta_k, \Delta, \delta, s_0, S_a, S_r)$ be a T^{h_1}-T^{h_2}-...-T^{h_k}-DM with output tape M *computes* the function $\varphi: (\Sigma^*)^m \to \Delta^*$ if $\Sigma \subseteq \Delta_1 \setminus \{\square\}$ and if for all $w = (w_1, \ldots, w_m) \in (\Sigma^*)^m$

- if $w \in \mathrm{D}_\varphi$, then $\mathrm{K}_{M,w_1\square w_2\square\ldots\square w_m} \mathrel{\vdash_M^*} (s_1, f_1, f_2, \ldots, f_k, g_{\varphi(w)}, n_{1,1}, \ldots, n_{k,h_k}, |\varphi(w)|)$ for some $s_1 \in S_a \cup S_r$ and some $f_1, \ldots, f_k, n_{1,1}, \ldots, n_{k,h_k}$, where $g_v(1)\, g_v(2) \ldots g_v(|v|) = v$ and $g_v(n) = \square$ for all $n \leq 0$ and $n > |v|$,

- if $w \notin \mathrm{D}_\varphi$, then there is no final situation K such that $\mathrm{K}_{M,w_1\square w_2\square\ldots\square w_m} \mathrel{\vdash_M^*} K$.

By $\varphi_M^{(m)}$ we denote the m-ary function computed by M. We set $\varphi_M = \varphi_M^{(1)}$. Note that every deterministic Turing machine computes some function. M computes the function $\varphi: \mathbb{N}^m \to \mathbb{N}$ if $\varphi_M^{(m)}$ (bin $n_1, \ldots,$ bin n_m) = bin $\varphi(n_1, \ldots, n_m)$.

If we want to emphasize that a certain type of a Turing machine has only m tape symbols, we use m as a subscript to the corresponding T in the type of the machine. For example, a $\mathrm{T_m}$-$\mathrm{T_k}$-DM is a deterministic Turing machine with two tapes, the first with at most m tape symbols and the second with at most k tape symbols. All definitions given above also apply to those types of Turing machines which have $\mathrm{T_m}$-tapes.

A *d-dimensional Turing tape* ($d \geq 1$) is a d-dimensional storage medium divided into infinitely many congruent d-dimensional cubes (also called *cells*). A d-dimensional Turing tape is indicated by $\mathrm{T^d}$ in our notational system. For example, a $\mathrm{T^3}$-$\mathrm{T^2}$-DM is a deterministic Turing machine with a 3-dimensional tape and a 2-dimensional tape. All definitions given above apply to those types of Turing machines which have multi-dimensional tapes, with the following minor modifications. The directions of the head moves are denoted by $0, -1, +1, -2, +2, \ldots, -d, +d$. The cells of a d-dimensional tape are considered to be enumerated by d-tuples of integers in the usual way. The inscription of a d-dimensional tape with tape alphabet Δ is a function $f: \mathbb{G}^d \mapsto \Delta$. The function g_v (used in the definition of $\mathrm{K}_{M,w}$) is defined for d-dimensional tapes as follows: $g_v(1, 0, \ldots, 0)\; g_v(2, 0, \ldots, 0,) \ldots g_v(|v|, 0, \ldots, 0) = v$ and $g_v(n_1, n_2, \ldots, n_d) = \square$ if $(n_1, n_2, \ldots, n_d) \notin \{(1, 0, \ldots, 0), \ldots, (|v|, 0, \ldots, 0)\}$.

A *nondeterministic Turing machine* M with k tapes having $h_1, \ldots, h_k$ heads resp., (for short: T^{h_1}-T^{h_2}-...-T^{h_k}-NM), is a $(k+5)$-tuple $(S, \Delta_1, \ldots, \Delta_k, \delta, s_0, S_a, S_r)$ whose components are specified in the same manner as for deterministic Turing machines, with the only difference that the next move function δ maps from $S \times \Delta_1^{h_1} \times \ldots \times \Delta_k^{h_k}$ into $\mathfrak{P}(S \times \Delta_1^{h_1} \times \ldots \times \Delta_k^{h_k} \times \{-1, +1, 0\}^{h_1+\ldots+h_k})$, and we require $\delta(s, a_{1,1}, \ldots, a_{k,h_k}) = \{(s, a_{1,1}, \ldots, a_{k,h_k}, 0, \ldots, 0)\}$ for $s \in S_a$. Intuitively, it can be seen that a nondeterministic Turing machine may have in any step several possibilities for the next move. We may imagine that duplicate copies of the machine are made in any step, one for each possible next move. Contrary to deterministic Turing machines, here there can be two different situations K_1, K_2 such that $K \mathrel{\vdash_M} K_1$ and $K \mathrel{\vdash_M} K_2$ for some K. Thus a nondeterministic Turing machine may have several computations on an input w. All definitions and conventions made above, except for the definition of the decision of a language and the computation of a function, also apply to nondeterministic machines.

A Th_1-Th_2-...-Th_k-NM $M = (S, \Delta_1, \ldots, \Delta_k, \delta, s_0, S_a, S_r)$ *accepts* a set $L \subseteq \Sigma^*$ if $\Sigma \subseteq \Delta_1 \setminus \{\square\}$ and if for all $w \in \Delta^*$: $w \in L$ if and only if there is an accepting situation K such that $K_{M,w} \overset{*}{\underset{M}{\vdash}} K$. By $L(M)$ we denote the language accepted by M. Note that every nondeterministic Turing machine accepts some language. The machine M accepts the set $L \subseteq \mathbb{N}$ if $L(M) = \{\text{bin}\, n : n \in L\}$, i.e. natural numbers are considered to be given in binary notation.

Since deterministic Turing machines are special cases of nondeterministic Turing machines, the above definitions apply to them also. However, in contrast to several authors we usually do not use the notion "a deterministic machine accepts a language". Thus, if M is specified as a deterministic machine, $L(M)$ is (if it is defined) the language decided by M. Therefore, in some theorems of this book, which state the existence of some deterministic Turing machine deciding a certain language, we need honesty suppositions for the bounding functions which do not appear in the original papers, where the deterministic machine accepts the language. For example, take Theorem 20.13.2 and Theorem 23.1.2.

Finally, let M be a nondeterministic TM with output tape, and let d be the maximum number of next situations for a given situation of M. Then a computation of M on w can be described by an infinite sequence $\pi \in \{1, \ldots, d\}^\omega$, and a halting computation of M on w can be described by a finite sequence $p \in \{1, \ldots, d\}^*$. We define

$$\varphi_M(w \mid \pi) = \begin{cases} \text{the final contents of the output tape obtained on the} \\ \text{path } \pi \text{ by } M \text{ on } w \text{ if } \pi \text{ describes an accepting computation} \\ \text{of } M \text{ on } w, \\ \text{undefined otherwise,} \end{cases}$$

$$\varphi_M(w \mid p) = \begin{cases} \varphi_M(w \mid p1^\omega) \text{ if } p \text{ describes an accepting computation of } M \\ \text{on } w, \\ \text{undefined otherwise.} \end{cases}$$

1.1.2. Restricted Models

In this subsection we introduce some restricted versions of the Turing tape called stack, nonerasing stack, checking stack, pushdown store, and counter, resp. In our notational system these restricted Turing tapes are denoted by S, NES, CS, PD, and C, resp. Note that all definitions and conventions made in § 1.1.1, except for the definition of d-dimensional tapes ($d > 1$), also apply (with the following minor modification) to those types of Turing machines whose first tape is of type 1:, 1*:, 2:, T, T^2, T^3, ... and whose other tapes are of type T, T^2, T^3, ..., S, NES, CS, PC and C. Since a tape of type S, NES, CS, PD or C can never be empty (we shall see that cell 1 always holds the symbol ⊡), we change the definition of $K_{M,w}$ as follows: If a tape of M is of type S, NES, CS, PD or C, then the corresponding function g_e in $K_{M,w}$ must be replaced by $g_{⊡}$.

Turing machines having only an input tape and at most one restricted Turing tape of type S, NES, CS, PD or C are said to be *automata*. For $i = 1, 2$ and $k \geqq 1$,

instead of	we write	and call this an
i:k-S-DM	i:k-DSA	*i-way k-head deterministic stack automaton*
i:k-NES-DM	i:k-DNESA	*i-way k-head deterministic nonerasing stack automaton*
i:k-CS-DM	i:k-DCSA	*i-way k-head deterministic checking stack automaton*
i:k-PD-DM	i:k-DPDA	*i-way k-head deterministic pushdown automaton*
i:k-C-DM	i:k-DCA	*i-way k-head deterministic counter automaton*
i:k-DM	i:k-DFA	*i-way k-head deterministic finite automaton*

This convention also applies to nondeterministic Turing machines, i.e. if "D" and "deterministic" are replaced by "N" and "nondeterministic". In the same way as for Turing machines, the reference to one input head can be omitted. For example, instead of 1:1-NPDA we can also write 1-NPDA.

A *stack* is a Turing tape (tape alphabet Δ), having only one head, such that

1. $\boxdot \in \Delta$; the symbol $\boxdot$, which is distinct from $\square$, is called the *bottom symbol* of the stack,
2. the inscription of the tape always has the form $\ldots \square\square\square \boxdot v \square\square\square \ldots$, where $v \in (\Delta \setminus \{\square, \boxdot\})^*$ and $\boxdot$ is stored in tape square 1. The word $\boxdot v$ is called the *stack word* and its rightmost symbol is called the *top symbol,*
3. the head can only move between tape square 1 (holding the bottom symbol) and the tape square to the right of the top symbol, and
4. the head can only change the top symbol or the blank to the right of the top symbol.

A *nonerasing stack* is a stack whose head can never change the top symbol. In particular, it cannot *erase* the top symbol, i.e. it cannot replace the top symbol by the blank.

A *checking stack* is a nonerasing stack whose head cannot change any symbol after the first move to the left. The time before (after) this first move to the left is called the *writing* (*checking*) *phase* of the checking stack.

A *pushdown store* is a stack whose head cannot leave the top symbol to the left. Hence, the head can essentially only read the top symbol and then

a) change the top symbol,

b) erase the top symbol, or

c) add a new top symbol.

For 2-NPDA's (and thus also for 1-NPDA's, 2-DPDA's and 1-DPDA's) we introduce new kinds of instructions for these three activities:

Ia) $sab \to s'b'\sigma'$ stands for $(s', a, b', \sigma, 0) \in \delta(s, a, b)$, where $b, b' \neq \square$ (the top symbol b is replaced by b') and

$$\sigma' = \begin{cases} + & \text{if} \quad \sigma = +1, \\ - & \text{if} \quad \sigma = -1, \\ \mathrm{e} & \text{if} \quad \sigma = 0, \end{cases}$$

Ib) $sab \to s' \uparrow \sigma'$ stands for $(s', a, \Box, \sigma, -1) \in \delta(s, a, b)$, where $b \neq \Box$ (the top symbol is erased or popped), and

Ic) $sab \to s''b' \downarrow \sigma'$ stands for $(s', a, b, 0, +1) \in \delta(s, a, b)$ and $(s'', a, b', \sigma, 0) \in \delta(s', a, \Box)$, where $b, b' \neq \Box$ (a new top symbol is added or pushed down).

If the situation K' of a 2-NPDA M is reached from a situation K by the application of an instruction of the form Ia), Ib) or Ic), then we write $K \vdash_{M,I} K'$. For $t \geq 0$ we write $K \vdash^{t}_{M,I} K'$ if there are situations $K_0, K_1, \ldots, K_t$ such that

$$K = K_0 \vdash_{M,I} K_1 \vdash_{M,I} \cdots \vdash_{M,I} K_t = K'.$$

Furthermore, $K \vdash^{*}_{M,I} K'$ if $K \vdash^{t}_{M,I} K'$ for some $t \geq 0$.

1.1. Lemma. Let M be a 2-NPDA (1-NPDA, 2-DPDA, 1-DPDA).

1. There is a 2-NPDA (1-NPDA, 2-DPDA, 1-DPDA, resp.) M' such that $L(M') = L(M)$ and the following property holds: if $K_{M,w} \vdash^{t}_{M} K$ for some accepting (rejecting) situation K, then $K_{M',w} \vdash^{t}_{M',I} K'$ for some accepting (rejecting) situation K'.
2. There is a 2-NPDA (1-NPDA, 2-DPDA, 1-DPDA) M' such that $L(M') = L(M)$ and the following property holds: if $K_{M,w} \vdash^{t}_{M,I} K$ for some accepting (rejecting) situation K, then $K_{M',w} \vdash^{t}_{M'} K'$ for some accepting (rejecting) situation K'. ▮

Consequently, if we want to construct a 2-NPDA (1-NPDA, 2-DPDA, 1-DPDA) M which accepts or decides a certain language in a given number of steps, then it is not important what we mean by a step: a move (in the sense of a Turing machine) or the execution of an instruction (of the form Ia), Ib) or Ic)).

Now we state the fact that 1-NPDA languages can be accepted by 1-NPDA's which are of a special form or which accept in a special manner. For a 1-NPDA $M = (S, \Delta_1, \Delta_2, \delta, s_0, \{s_1\}, S_r)$, we define $L'(M) = \{w : s_0 w \# | \boxdot \vdash^{*}_{M} w s_1 \# | \boxdot\}$ as the set of words accepted by M with an empty pushdown store.

1.2. Lemma. For a language $L \subseteq \Sigma^*$ the following statements are equivalent:

1. $L = L(M)$ for some 1-NPDA M.
2. $L = L'(M)$ for some 1-NPDA M.
3. $L = L'(M)$ for some 1-NPDA M having the following property: if it has the instruction $sab \to s'B0$ (where $B \in \{b', \uparrow, b'\downarrow\}$ for some stack symbol b'), then it has also the instruction $sa'b \to s'B0$ for each $a' \in \Sigma \cup \{\Box\}$, i.e. M does not read the input symbol. In this case we write $seb \to s'B0 \ldots$ ▮

A *counter* is a pushdown store with the tape alphabet $\{\boxdot, \Box, c\}$, i.e. whose stack word is of the form $\boxdot c^m$ for some $m \geq 0$. Thus the stack word can be considered as a natural number. The execution of an instruction $sab \to s'c\sigma$ ($sac \to s' \uparrow \sigma$, $sab \to s'c \downarrow \sigma$), $b \in \{c, \boxdot\}$, effects no alteration (subtraction of 1, addition of 1, resp.) in the counter. Note that sometimes in the literature a counter is assumed to be able to hold an arbitrary integer. However, by storing the sign of the integer in the finite control of the machine, our counters have the same ability.

For any type τ-XM of Turing machines we define

$$\boldsymbol{\tau}\textbf{-XM} = \{L(M) : M \text{ is a } \tau\text{-XM}\}.$$

All conventions made for the types of the machines carry over to the corresponding classes of languages, for example, **2:1-T1-T1-DM** = **2-2T-DM** and **1:1-PD-NM** = **1-NPDA**.

1.1.3. Alternating and Probabilistic Turing Machines

An *alternating Turing machine* is in fact the same thing as a nondeterministic Turing machine. Only the mode of acceptance (which in our understanding does not belong to the notion of "machine") differs from that used for nondeterministic machines. Let $M = (S, \Delta_1, \ldots, \Delta_k, \delta, s_0, S_a, S_r)$ be a nondeterministic Turing machine with k tapes (full or restricted Turing tapes). In contrast to our former terminology the states from $S_u =_{\mathrm{df}} S_r$ $\left(S_e =_{\mathrm{df}} S \setminus (S_a \cup S_r)\right)$ are called *universal* (*existential*). A situation of M is called *universal* (*existential*) if the state of this situation is universal (existential). An ID of a universal (existential) situation is called *universal* (*existential*).

A finite nonempty tree β is called an *accepting computation tree of* M *on* w if it is labelled by situations of M, l being the labelling function, such that

1. $l(\text{root of } \beta) = K_{M,w}$.
2. If b is an internal node of β (i.e. b is not a leaf), $l(b)$ is universal and $\{K_1, \ldots, K_r\} = \left\{K : l(b) \vdash_{\overline{M}} K\right\}$, then b has exactly the r successors $b_1, \ldots, b_r$ in β, and $l(b_i) = K_i$ for $i = 1, \ldots, r$.
3. If b is an internal node of β and $l(b)$ is existential, then b has exactly one successor b' in β, and $l(b) \vdash_{\overline{M}} l(b')$.
4. If b is a leaf of β, then $l(b)$ is accepting.

Now, $L^A(M) = \{w : \text{there is an accepting computation tree of } M \text{ on } w\}$ is the set of all w accepted by the alternating Turing machine M, or, in our understanding, $L^A(M)$ is the set of all w accepted in the alternating manner by the nondeterministic Turing machine M. Note that $S_u = \emptyset$ implies $L^A(M) = L(M)$.

For any type τ-NM of nondeterministic Turing machines we define

$$\boldsymbol{\tau}\textbf{-AM} = \{L^A(M) : M \text{ is a } \tau\text{-NM}\}.$$

All conventions of our notational system for the machines carry over to the notation of the classes of languages.

A *probabilistic Turing machine* is also the same thing as a nondeterministic Turing machine, but again another mode of acceptance is used. The idea is as follows: if the situation K of a nondeterministic Turing machine M has the possible next situations $K_1, \ldots, K_m$, then M goes from K to each of them with the *probability* $p_K = \frac{1}{m}$. The probability $p(K_0, K_1, K_2, \ldots)$ of a computation of M is given by

$\prod_{i=0}^{\infty} p_{K_i}$, the probability of some set B of computations by $p(B) = \sum_{\gamma \in B} p(\gamma)$. Now let $0 < \varepsilon \leqq \frac{1}{2}$. A machine M *accepts* (*rejects*) some input word w with *error probability* ε if $p(\{\gamma : \gamma$ is an accepting (rejecting) computation of M on $w\}) > 1 - \varepsilon$. The machine M *decides* the language $L \subseteq \Sigma^*$ *with error probability* ε if M accepts (rejects) all inputs $w \in L$ ($w \notin L$) with error probability ε. In this case we write $\mathrm{L}^\varepsilon(M) = L$. Note that not every nondeterministic Turing machine decides some language with error probability ε. The machine M *accepts* the language $L \subseteq \Sigma^*$ *with error probability* ε if M accepts exactly the inputs from L with error probability ε. In this case we write $\mathrm{L}^\varepsilon(M) = L$.

For any type τ-NM of nondeterministic Turing machines and any ε, $0 < \varepsilon \leqq \frac{1}{2}$, we define

$$\boldsymbol{\tau}\textbf{-R}^\varepsilon\textbf{M} = \{\mathrm{L}^\varepsilon(M) : M \text{ is a } \tau\text{-NM}\} \quad \text{and} \quad \boldsymbol{\tau}\textbf{-RM} = \tau\text{-R}^{1/2}\text{M}.$$

The letter R stands for "random".

1.2. Random Access Machines

A *Random access machine* (for short: *RAM*) is a computation model which is more related to real computers than a Turing machine. A random access machine M consists of

- a one-way input tape as defined in § 1.1.1 for TM's.
- an infinite number of *registers*, called register 0, 1, 2, ..., each of which is able to hold an arbitrary word from $\{0, 1\}^*$. By $\mathbf{R}_i$ we denote the current contents of the register i, i.e. $\mathbf{R}_i \in \{0, 1\}^*$. In many cases we shall also interpret $\mathbf{R}_i$ as a natural number. This number is denoted by $\langle \mathbf{R}_i \rangle$ and defined by $\mathbf{R}_i = 00 \ldots 0 \operatorname{bin} \langle \mathbf{R}_i \rangle$.
- an *instruction counter* which is able to hold an arbitrary natural number. By **IC** we denote the current contents of the instruction counter.
- a *program* operating on the registers and the instruction counter. A program is a finite set of *instructions* enumerated by 1, 2, ..., m, where the *final instruction* (i.e. the instruction m) is a stop instruction (defined below). We require that $\mathbf{IC} \in \{1, 2, \ldots, m\}$ (**IC** gives the number of that instruction which is to be executed next).

The sequence $K = (\mathbf{IC}, f, H, \mathbf{R}_0, \mathbf{R}_1, \mathbf{R}_2, \ldots)$ is said to be a *situation* of M where $f: \mathbb{G} \mapsto \Delta^*$ is the inscription of the input tape (having alphabet Δ) and H is the number of the tape square scanned by the input head. The sequence $K' = (\mathbf{IC}', f', H', \mathbf{R}'_0, \mathbf{R}'_1, \mathbf{R}'_2, \ldots)$ is said to be the *next situation* of K (for short: $K \vdash_M K'$) if it originates from K by the application of the instruction **IC**. The possible instructions and their effects are now listed. (If H' ($\mathbf{R}'_i$) is not mentioned, then the instruction effects no change of H ($\mathbf{R}_i$), i.e. $H' = H$ ($\mathbf{R}'_i = \mathbf{R}_i$). The contents of the input tape cannot be changed by an instruction, i.e. $f' = f$ in all cases.)

1. *Transport instructions*

a) $\mathbf{R}_i \leftarrow \mathbf{R}_j$ effects $\mathbf{R}'_i = \mathbf{R}_j$, $\mathbf{IC}' = \mathbf{IC} + 1$,

b) $\mathbf{R}_i \leftarrow \mathbf{R}_{\mathbf{R}_j}$ effects $\mathbf{R}'_i = \mathbf{R}_{\langle \mathbf{R}_j \rangle}$, $\mathbf{IC}' = \mathbf{IC} + 1$ (indirect addressing).

c) $\mathbf{R}_{\mathbf{R}_i} \leftarrow \mathbf{R}_j$ effects $\mathbf{R}'_{\langle \mathbf{R}_i \rangle} = \mathbf{R}_j$, $\mathbf{IC}' = \mathbf{IC} + 1$ (indirect addressing).

2. *Operational instructions*

a) $\mathbf{R}_i \leftarrow w$ effects $\mathbf{R}'_i = w$, $\mathbf{IC}' = \mathbf{IC} + 1$ $(w \in \{0, 1\}^*)$
($\mathbf{R}_i \leftarrow k$ stands for $\mathbf{R}_i \leftarrow \text{bin } k$, $k \in \mathbb{N}$),

b) $\mathbf{R}_i \leftarrow \mathbf{R}_j + 1$ effects $\mathbf{R}'_i = \text{bin}\,(\langle \mathbf{R}_j \rangle + 1)$, $\mathbf{IC}' = \mathbf{IC} + 1$,

c) $\mathbf{R}_i \leftarrow \mathbf{R}_j \dot{-} 1$ effects $\mathbf{R}'_i = \text{bin}\,(\langle \mathbf{R}_j \rangle \dot{-} 1)$, $\mathbf{IC}' = \mathbf{IC} + 1$,

d) $\mathbf{R}_i \leftarrow \mathbf{R}_j + \mathbf{R}_k$ effects $\mathbf{R}'_i = \text{bin}\,(\langle \mathbf{R}_j \rangle + \langle \mathbf{R}_k \rangle)$, $\mathbf{IC}' = \mathbf{IC} + 1$,

e) $\mathbf{R}_i \leftarrow \mathbf{R}_j \dot{-} \mathbf{R}_k$ effects $\mathbf{R}'_i = \text{bin}\,(\langle \mathbf{R}_j \rangle \dot{-} \langle \mathbf{R}_k \rangle)$, $\mathbf{IC}' = \mathbf{IC} + 1$,

f) $\mathbf{R}_i \leftarrow \mathbf{R}_j \cdot \mathbf{R}_k$ effects $\mathbf{R}'_i = \text{bin}\,(\langle \mathbf{R}_j \rangle \cdot \langle \mathbf{R}_k \rangle)$, $\mathbf{IC}' = \mathbf{IC} + 1$,

g) $\mathbf{R}_i \leftarrow \mathbf{R}_j\mathbf{R}_k$ effects $\mathbf{R}'_i = \mathbf{R}_j\mathbf{R}_k$, $\mathbf{IC}' = \mathbf{IC} + 1$ (concatenation),

h) $\mathbf{R}_i \leftarrow \overline{\mathbf{R}}_j$ effects $\mathbf{R}'_i =$ bitwise (letterwise) negation of $\mathbf{R}_j$, $\mathbf{IC}' = \mathbf{IC} + 1$,

i) $\mathbf{R}_i \leftarrow \mathbf{R}_j \wedge \mathbf{R}_k$ effects $\mathbf{R}'_i =$ bitwise (letterwise) conjunction of $\mathbf{R}_j 0^{|\mathbf{R}_k| \dot{-} |\mathbf{R}_j|}$ and $\mathbf{R}_k 0^{|\mathbf{R}_j| \dot{-} |\mathbf{R}_k|}$ $\mathbf{IC}' = \mathbf{IC} + 1$.

3. *Jump instructions*

a) GOTO k effects $\mathbf{IC}' = \min\,\{k, m\}$,

b) IF ($\mathbf{R}_i = 0$) THEN GOTO k effects $\mathbf{IC}' = \begin{cases} \min\,\{k, m\} \text{ if } \mathbf{R}_i = \text{e}, \\ \mathbf{IC} + 1 \text{ otherwise,} \end{cases}$

c) IF ($\mathbf{R}_i \varrho \mathbf{R}_j$) THEN GOTO k effects $\mathbf{IC}' = \begin{cases} \min\,\{k, m\} \text{ if } \langle \mathbf{R}_i \rangle\, \varrho \langle \mathbf{R}_j \rangle, \\ \mathbf{IC} + 1 \text{ otherwise} \end{cases}$
$(\varrho \in \{=, \leqq, <, \geqq, >\})$.

4. *Stop instructions*

a) STOP effects $\mathbf{IC}' = \mathbf{IC}$,

b) ACCEPT effects $\mathbf{IC}' = \mathbf{IC}$,

c) REJECT effects $\mathbf{IC}' = \mathbf{IC}$.

5. *Read instructions*

Let code: $\Delta \xrightarrow{1\text{-}1} \{0, 1\}^*$ be a fixed encoding of Δ.

$$\text{READ } \mathbf{R}_i \text{ effects } \quad \mathbf{R}'_i = \begin{cases} \text{code } f(H) \text{ if } f(H) \neq \Box, \\ \mathbf{R}_i \text{ otherwise,} \end{cases}$$

$$H' = H + 1, \quad \text{and}$$

$$\mathbf{IC}' = \begin{cases} \mathbf{IC} + 1 \text{ if } f(H) \neq \Box, \\ \mathbf{IC} + 2 \text{ otherwise.} \end{cases}$$

Now let $\vdash_M^*$ be the reflexive and transitive closure of the relation $\vdash_M$, i.e. $K \vdash_M^* K'$ if and only if there is a $t \geq 0$ and situations $K_0, K_1, \ldots, K_t$ such that $K = K_0 \vdash_M K_1 \vdash_M \ldots \vdash_M K_t = K'$. A situation $(\mathbf{IC}, f, H, \mathbf{R}_0, \mathbf{R}_1, \mathbf{R}_2, \ldots)$ is said to be *accepting* (*rejecting, final*) if the instruction with the number $\mathbf{IC}$ is ACCEPT (REJECT, STOP, resp.). The *initial situation of M induced by* $w = (w_1, \ldots, w_m)$ is defined as $K_{M,w} = (1, g_{w_1\Box w_2\Box\ldots\Box w_m}, 1, e, e, e, \ldots)$, where $g_v(1)\, g_v(2) \ldots g_v(|v|) = v$ and $g_v(n) = \Box$ for $n \leq 0$ and $n > |v|$. The notions *computation of M, computation of M on w, halting (accepting, rejecting) computation* and *length of a computation* are defined as for Turing machines.

The deterministic RAM M *decides* a language $L \in \Sigma^*$, $\Sigma \subseteq \Delta \setminus \{\Box\}$, if for all $w \in \Sigma^*$: if $w \in L$ ($w \notin L$), then $K_{M,w} \vdash_M^* K$ for some accepting (rejecting) situation K. In this case we define $\mathrm{L}(M) = L$. Note that not every deterministic RAM decides some language.

A *deterministic RAM M with an output tape* is an RAM having in addition an output tape as described in § 1.1.1 for TM's. Thus a situation of M is a sequence $(\mathbf{IC}, f, H, f_1, H_1, \mathbf{R}_0, \mathbf{R}_1, \mathbf{R}_2, \ldots)$ where $f_1 : \mathbb{G} \to \Delta_1$ is the inscription of the output tape and H_1 is the number of the tape square scanned by the input head. In order to write symbols on the output tape we need instruction of a new type which will change the situation $K = (\mathbf{IC}, f, H, f_1, H_1, \mathbf{R}_0, \mathbf{R}_1, \mathbf{R}_2, \ldots)$ to the next situation $K' = (\mathbf{IC}', f', H', f_1', H_1', \mathbf{R}_0, \mathbf{R}_1, \mathbf{R}_2, \ldots)$ as follows:

6. *Write instruction*

$$\text{WRITE } \mathbf{R}_i \text{ effects } f_1'(n) = \begin{cases} 0 & \text{if } n = H_1 \wedge \mathbf{R}_i = e, \\ 1 & \text{if } n = H_1 \wedge \mathbf{R}_i \neq e, \\ f_1(n) & \text{otherwise,} \end{cases}$$

$$H_1' = H_1 + 1,$$

$$\mathbf{IC}' = \mathbf{IC} + 1.$$

We define $\vdash_M$ and $\vdash_M^*$ as above. The initial situation of M induced by $w = (w_1, \ldots w_m)$ is defined as $K_{M,w} = (1, g_{w_1\Box\ldots\Box w_m}, 1, g_e, 1, e, e, \ldots)$. M computes a function $\varphi : (\Sigma^*)^m \mapsto \Sigma^*$ if for all $w \in (\Sigma^*)^m$:

– if $w \in \mathrm{D}_\varphi$, then $K_{M,w} \vdash_M^* (k, g_{w_1\Box\ldots\Box w_m}, |w_1 \Box \ldots \Box w_m| + 1, g_{\varphi(w)}, |\varphi(w)| + 1, v_1, v_2, v_3, \ldots)$ for some $k \geq 1, v_1, v_2, v_3, \ldots \in \{0, 1\}^*$, where the instruction k is STOP,

– if $w \notin \mathrm{D}_\varphi$, then there is no final situation K such that $K_{M,w} \vdash_M^* K$.

Note that every deterministic RAM M with output tape computes some function. By $\varphi_M^{(m)}$ we denote the m-ary function computed by M.

Now we distinguish between different types of RAM's. A deterministic RAM

using only instructions of type	is said to be a	for short
1, 2a, 2b, 2c, 3, 4, 5	*successor* RAM	DSRAM
1, 2a, 2d, 2e, 3, 4, 5	*ordinary* RAM	DRAM
1, 2a, 2d, 2e, 2h, 2i, 3, 4, 5	*boolean* RAM	DBRAM
1, 2a, 2d, 2e, 2f, 2h, 2i, 3, 4, 5	*multiplication* RAM	DMRAM
1, 2a, 2g, 2h, 2i, 3, 4, 5	*concatenation* RAM	DCRAM

If a DSRAM (DRAM, DBRAM, DMRAM, DCRAM) does not have instructions of type 1b and 1c, then it is said to have no *indirect addressing* and is called a DSRAM′ (DRAM′, DBRAM′, DMRAM′, DCRAM′, resp.).

A *nondeterministic RAM* differs from a deterministic RAM only in that there can be several instructions with the same number. If $\mathbf{IC} = k$, then the RAM chooses nondeterministically which one of the instructions with the number k will be executed next. All definitions and conventions made for deterministic RAM's, except for the definitions "M decides a language" and "M computes a function", also apply to nondeterministic RAM's. In the notation of the special types of RAM's, the letter D is replaced by N.

Let M be a nondeterministic RAM. We define the language accepted by M as $\mathrm{L}(M) = \{w \colon \mathrm{K}_{M,w} \vdash_M^* K \text{ for some accepting situation } K\}$.

1.3. *Iterative Arrays*

A *deterministic d-dimensional iterative array* (for short: $\mathrm{DIA^d}$) is a device which consists only of a d-dimensional storage which is essentially a d-dimensional Turing tape without any head ($d \geqq 1$). In one move a $\mathrm{DIA^d}$ can change the contents of all cells. The new contents of a cell depend on the old contents of this cell and its $2d$ adjacent cells.

Formally, a $\mathrm{DIA^d}$ M is a quadruple $(\Delta, \delta, S_a, S_r)$ where

— Δ is the finite set of *tape symbols*, $\square \in \Delta$,

— $\delta \colon \Delta^{2d+1} \mapsto \Delta$ is the *next move function*, $\delta(\square, \square, \ldots, \square) = \square$,

— $S_a \subseteqq \Delta$ is the set of *accepting symbols*, and

— $S_r \subseteqq \Delta$ is the set of *rejecting symbols*, where $S_a \cap S_r = \emptyset$.

A *situation* of M is an inscription $K \colon \mathbb{G}^d \mapsto \Delta$ of the d-dimensional tape of M. It is said to be *accepting* (*rejecting*, *final*) if $K(n_1, \ldots, n_d) \in S_a$ (if it is not accepting and $K(n_1, \ldots, n_d) \in S_r$; if $K(n_1, \ldots, n_d) \in S_a \cup S_r$; resp.) for some $(n_1, \ldots, n_d) \in \mathbb{G}^d$. The situation K' is said to be the *next situation* of K (for short: $K \vdash_M K'$) if $K'(n_1, \ldots, n_d) = \delta\big(K(n_1, \ldots, n_d),\ K(n_1 + 1, n_2, \ldots, n_d),\ K(n_1 - 1, n_2, \ldots, n_d), \ldots, K(n_1, \ldots, n_{d-1}, n_d + 1),\ K(n_1, \ldots, n_{d-1}, n_d - 1)\big)$. The relation $\vdash_M^*$ is the reflexive and transitive closure of $\vdash_M$, i.e. $K \vdash_M^* K'$ if there are a $t \geqq 0$ and situations $K_0, K_1, \ldots, K_t$ such that $K = K_0 \vdash_M K_1 \vdash_M \cdots \vdash_M K_t = K'$. The *initial situation* $\mathrm{K}_{M,w}$ *of M induced by* $w \in (\Delta \setminus \{\square\})^*$ is defined by

$$\mathrm{K}_{M,w}(1, 0, \ldots, 0)\, \mathrm{K}_{M,w}(2, 0, \ldots, 0) \ldots \mathrm{K}_{M,w}(|w|, 0, \ldots, 0) = w$$

and

$$\mathrm{K}_{M,w}(n_1, n_2, \ldots, n_d) = \square$$

$$\text{for all } (n_1, n_2, \ldots, n_d) \notin \{(1, 0, \ldots, 0), \ldots, (|w|, 0, \ldots, 0)\}.$$

The definitions of *computation*, *accepting* (*rejecting*, *halting*) *computation*, *computation of M on w*, and *decision of a language* are made as for Turing machines.

A *nondeterministic d-dimensional iterative array* (for short: NIA^d) is a quadruple $(\Delta, \delta, S_a, S_r)$ whose components are specified in the same way as for a DIA^d with the only difference that $\delta: \mathfrak{G}^d \to \mathfrak{P}(\Delta)$. All definitions made above except for the definition of decision of a language, also apply to NIA^d's. The definition of acceptance of a language is made as for nondeterministic Turing machines.

The cells $(n_1 + 1, n_2, \ldots, n_d)$, $(n_1 - 1, n_2, \ldots, n_d)$, $\ldots$, $(n_1, \ldots, n_{d-1}, n_d + 1)$, $(n_1, \ldots, n_{d-1}, n_d - 1)$ are said to be the *neighbourhood* of the cell $(n_1, \ldots, n_d)$.

§ 2. Recursive Functions

2.1. *Definitions*

Let Σ be a finite alphabet. Let $\mathbb{F}_i(\Sigma)$ be the set of all *partial i-place word functions* over Σ^*, i.e. all partial functions $\varphi: (\Sigma^*)^i \to \Sigma^*$ and $\mathbb{F}(\Sigma) = \bigcup_{i \in \mathbb{N}} \mathbb{F}_i(\Sigma)$. For $i \geqq 1$ we define the class $\mathbb{T}(\Sigma) = \left\{\varphi: \bigvee_i \left(\varphi \in \mathbb{F}_i(\Sigma) \wedge \mathrm{D}_\varphi = (\Sigma^*)^i\right)\right\}$ of *total word functions* over Σ^*.

If no confusion is possible we omit for all classes $\mathscr{C}(\Sigma) \subseteqq \mathbb{F}(\Sigma)$ defined above and below the reference to Σ. For $\mathscr{C} \subseteqq \mathbb{F}$ and $i \in \mathbb{N}$ we define $\mathscr{C}_i = \mathscr{C} \cap \mathbb{F}_i$.

Now we define the following operations on $\mathbb{F}$:

Substitution: For $h_1, \ldots, h_n \in \mathbb{F}_m$ and $g \in \mathbb{F}_n$ we define $f = \mathrm{SUB}_n(g, h_1, \ldots, h_n)$ by

$$f(w_1, \ldots, w_m) = g\big(h_1(w_1, \ldots, w_m), \ldots, h_n(w_1, \ldots, w_m)\big).$$

For algebraic closure operators $\Gamma_{\omega_1, \ldots \omega_r, \mathrm{SUB}_1, \mathrm{SUB}_2, \ldots}$ we shall also write $\Gamma_{\omega_1, \ldots, \omega_r, \mathrm{SUB}}$.

Primitive recursion: For $g \in \mathbb{F}_m$, $h_{x_1}, \ldots, h_{x_k} \in \mathbb{F}_{m+2}$ $(\Sigma = \{x_1, \ldots, x_k\})$ we define $f = \mathrm{PR}(g, h_{x_1}, \ldots, h_{x_k})$ by

$$f(w_1, \ldots, w_m, \mathrm{e}) = g(w_1, \ldots, w_m),$$
$$f(w_1, \ldots, w_m, vx_i) = h_{x_i}\big(w_1, \ldots, w_m, v, f(w_1, \ldots, w_m, v)\big), \quad i = 1, \ldots, k.$$

Let $\prec$ denote the fixed irreflexible ordering on Σ^* according to growing length where words of equal length are ordered lexicographically with respect to a fixed order on Σ.

Minimization: For $g \in \mathbb{F}_{m+1}$ we define $f = \mathrm{MIN}(g)$ by

$$f(w_1, \ldots, w_m) = \mu u\big(g(w_1, \ldots, w_m, u) = \mathrm{e}\big)$$
$$= \min \Big\{u: g(w_1, \ldots, w_m, u) = \mathrm{e} \wedge \bigwedge_{z \prec u} z \in \mathrm{D}_g\Big\}.$$

Note that $f(w_1, \ldots, w_m)$ is undefined if the set under min is empty.

Let e denote the 0-ary function with the (constant) value e, let s_x be the *x-successor function* defined by $\mathrm{s}_x(w) = wx$, and let $\mathscr{I} = \{\mathrm{I}_m^n: n \in \mathbb{N} \wedge 1 \leqq m \leqq n\}$ be the set of the identity (projection) functions I_m^n defined by $\mathrm{I}_m^n(w_1, \ldots, w_n) = w_m$. For $\mathscr{C} \subseteqq \mathbb{F}$

we define $\Gamma_\Omega(\mathscr{C}) = \Gamma_\Omega(\mathscr{C} \cup \{e\} \cup \mathscr{J})$. Now we introduce

$$\mathbb{P}\mathrm{r}(\Sigma) =_{\mathrm{df}} \Gamma_{\mathrm{SUB,PR}}(\{\mathrm{s}_x : x \in \Sigma\}),$$

the set of the *primitive recursive functions* over Σ^*,

$$\mathbb{P}(\Sigma) =_{\mathrm{df}} \Gamma_{\mathrm{SUB,PR,MIN}}(\{\mathrm{s}_x : x \in \Sigma\}),$$

the set of the *partial recursive functions* over Σ^*, and

$$\mathbb{R}(\Sigma) =_{\mathrm{df}} \mathbb{P}(\Sigma) \cap \mathbb{T}(\Sigma),$$

the set of the *recursive* (*total recursive*) *functions* over Σ^*. The importance of the partial recursive functions in the theory of computing is reflected in the following key result.

2.1. Theorem. Let τ be any type of deterministic machine defined in § 1.1.1, § 1.2 or § 1.3 and let $\varphi \in \mathbb{F}(\Sigma)$. Then $\varphi \in \mathbb{P}(\Sigma)$ iff φ is computable by some machine of type τ. ∎

The special case card $\Sigma = 1$ (we take $\Sigma = \{1\}$) where Σ^* is interpreted by $\mathbb{N}$ yields the number theoretic partial recursive functions. Then e becomes the constant function 0 (zero), and there is only one successor function $\mathrm{s} = \mathrm{s}_1$. In all parts of the book dealing with abstract complexity measures we use the number theoretic partial recursive functions, whereas in the other parts we usually require partial recursive functions over alphabets having at least two elements.

A relation between Σ^* and $(\Sigma^*)^2$ (and between unary functions and functions with two arguments) can be established by a (Cantor) *pairing function* $\mathrm{c}: (\Sigma^*)^2 \xrightarrow{1\text{-}1} \Sigma^*$, and its inverses $\mathrm{l}: \Sigma^* \mapsto \Sigma^*$ and $\mathrm{r}: \Sigma^* \mapsto \Sigma^*$ satisfying the equations

$$\mathrm{l}\big(\mathrm{c}(x, y)\big) = x, \qquad \mathrm{r}\big(\mathrm{c}(x, y)\big) = y \quad \text{and} \quad \mathrm{c}\big(\mathrm{l}(x), \mathrm{r}(x)\big) = x.$$

By $\langle x, y\rangle$ we denote a fixed pairing function. In the case of natural numbers take for instance $\langle x, y\rangle =_{\mathrm{df}} 2^x(2y + 1) - 1$. Enumeration functions for n-tuples ($n > 2$) are introduced inductively by $\langle x_1, \ldots, x_{n+1}\rangle =_{\mathrm{df}} \big\langle\langle x_1, \ldots, x_n\rangle, x_{n+1}\big\rangle$ and their inverses by $\mathrm{c}_m^{n+1}(x) = \mathrm{c}_m^n\big(\mathrm{l}(x)\big)$ for $m = 1, \ldots, n$ and $\mathrm{c}_{n+1}^{n+1}(x) = \mathrm{r}(x)$, where $\mathrm{c}_1^2 = \mathrm{l}$ and $\mathrm{c}_2^2 = \mathrm{r}$.

Using such enumeration functions a 1-1-correspondence between unary and n-ary functions can be defined by

$$\bar{\varphi}(x_1, \ldots, x_n) =_{\mathrm{df}} \varphi(\langle x_1, \ldots, x_n\rangle)$$

and it holds

2.2. Proposition. For all $n \geqq 2$,

$$\varphi \in \mathbb{P}_1 \leftrightarrow \bar{\varphi} \in \mathbb{P}_n, \quad f \in \mathbb{R}_1 \leftrightarrow \bar{f} \in \mathbb{R}_n, \quad f \in \mathbb{P}\mathrm{r}_1 \leftrightarrow \bar{f} \in \mathbb{P}\mathrm{r}_n. \ ∎$$

It is also possible to generate $\mathbb{P}\mathrm{r}_1(\{1\})$ using only 1-place functions.

Definition. (*Iteration*). For $f, g \in \mathbb{F}_1(\{1\})$,

$$f = \mathrm{IT}(g) \leftrightarrow f(0) = 0, \ f(x + 1) = g\big(f(x)\big).$$

Let $\mathrm{q}(x) =_{\mathrm{df}} x \dot{-} [\sqrt{x}]^2$.

2.3. Theorem. [Robi 47] $\mathbb{P}\mathrm{r}_1(\{1\}) = \Gamma_{\mathrm{SUB},+,\mathrm{IT}}(\mathrm{s}, \mathrm{q})$. ∎

2.2. *Gödel Numberings*

It is easy to construct an enumeration

$$M_0, M_1, M_2, \ldots$$

of all DTM's (or all machines of a given other type) in such a way that there exist two algorithms A and B with the properties: A computes for given i the (program of the) machine M_i, and B computes for given (program of the) machine M a number i such that $M = M_i$. Such an enumeration is called a *standard Gödel numbering*. Now, we choose once and for all a fixed standard Gödel numbering. It induces an enumeration $\varphi\colon \mathbb{N} \mapsto \mathbb{P}_1$ if we define $\varphi(i) = \varphi_{M_i} = \varphi_i =$ that function which is computed by M_i. We call this enumeration a standard Gödel numbering of $\mathbb{P}_1$. This enumeration carries over to an enumeration of $\mathbb{P}_n$ by $\varphi_i^{(n)}(x_1, \ldots, x_n) =_{\mathrm{df}} \varphi_i(\langle x_1, \ldots, x_n\rangle)$ (cf. 2.2).

2.4. Universal Function Theorem. For every $n \geq 1$ there exists a $u \in \mathbb{P}_{n+1}$ such that

$$u(i, x_1, \ldots, x_n) = \varphi_i(x_1, \ldots, x_n). \blacksquare$$

2.5. *s-m-n*-Theorem. For every m and n there exists an $s \in \mathbb{R}_{m+1}$ such that

$$\varphi_{s(k,x_1,\ldots,x_m)}(y_1, \ldots, y_n) = \varphi_k(x_1, \ldots, x_m, y_1, \ldots, y_n). \blacksquare$$

2.6. Padding Theorem. The function s in the *s-m-n*-Theorem can always be chosen as an arbitrarily large strictly increasing 1-1-function. ∎

We now briefly sketch the distinguished position of our standard Gödel numbering of $\mathbb{P}_1$ within a very large set of reasonable numberings of $\mathbb{P}_1$, namely that of the effective numberings.

Definition. A function $\alpha\colon \mathbb{N} \mapsto\!\!\!\!\!\!\to \mathbb{P}_1$ is called an *effective numbering* of $\mathbb{P}_1$ if it has a universal function, i.e. there exists an $a \in \mathbb{P}_2$ such that $\alpha_i(x) =_{\mathrm{df}} \big(\alpha(i)\big)(x) = a(i, x)$. This effective numbering α of $\mathbb{P}_1$ induces in the usual way an effective numbering of $\mathbb{P}_n$ by $\alpha_i^{(n)}(x_1, \ldots, x_n) = \alpha_i(\langle x_1, \ldots, x_n\rangle)$.

Numberings can be compared by the following reducibility.

Definition. Let α, β be effective numberings of $\mathbb{P}_1$.

$$\alpha \leq \beta \leftrightarrow \bigvee_{h \in \mathbb{R}_1} \alpha = \beta \circ h \qquad (\alpha \text{ is } \textit{reducible} \text{ to } \beta),$$

$$\alpha \equiv \beta \leftrightarrow \alpha \leq \beta \wedge \beta \leq \alpha \qquad (\alpha \text{ is } \textit{equivalent} \text{ to } \beta).$$

If numberings are understood as highly abstracted programming languages (numbers are programs), then $\alpha \leq \beta$ means that there is a compiler from α to β and hence β is at least as expressible as α.

2.7. Theorem. Let α, β be effective numberings of $\mathbb{P}_1$.

1. β satisfies the *s-m-n*-Theorem $\to \alpha \leq \beta$.
2. $\alpha \leq \beta \wedge \alpha$ satisfies the *s-m-n*-Theorem $\to \beta$ satisfies the *s-m-n*-Theorem. ∎

Definition. An effective numbering ψ of $\mathbb{P}_1$ is called *Gödel numbering* (*acceptable numbering*) if ψ satisfies the *s-m-n*-Theorem.

2.8. Corollary. α is a Gödel numbering iff $\alpha \equiv \varphi$. ∎

2.9. Theorem. For any two Gödel numberings φ and ψ there exists a recursive $h: \mathbb{N} \xrightarrow{1\text{-}1} \mathbb{N}$ such that $\varphi = \psi \circ h$. ∎

Remark. 2.8 and 2.9 show that φ, although apparently chosen with a high degree of arbitrariness, is in fact no worse and no better than any other Gödel numbering. This is no longer valid if complexity is taken into consideration. There are "easy" Gödel numberings α with the property that any other Gödel numbering can be reduced to α with a "low" complexity, and not every Gödel numbering is easy in this sense. Details are found in [Hart 74], [Schn 74a], [Schn 74b], [HaBa 75], [MaWiYo 78].

2.10. Recursion Theorem. For any $n \geqq 0$ and any $f \in \mathbb{R}_{n+1}$ there exists an $s \in \mathbb{R}_n$ such that

$$\varphi_{f(x_1,\ldots,x_n,s(x_1,\ldots,x_n))} = \varphi_{s(x_1,\ldots,x_n)}. \; ∎$$

2.3. *Decidability and Enumerability*

A set $A \subseteqq \Sigma^*$ is called *recursively decidable* (*recursive*) if $c_A \in \mathbb{R}_1$. A is called *recursively enumerable* (r.e.) if $A = \emptyset \vee \bigvee_{f \in \mathbb{R}_1} (A = \mathrm{R}_f)$. We define $\mathbf{REC}(\Sigma) = \{A : A \subseteqq \Sigma^* \wedge A \text{ is recursive}\}$ and $\mathbf{RE}(\Sigma) = \{A : A \subseteqq \Sigma^* \wedge A \text{ is r.e.}\}$.

2.11. Theorem. $\mathbf{REC} = \mathbf{RE} \cap \mathbf{coRE} \subset \mathbf{RE}$. ∎

A Gödel numbering of $\mathbb{P}_1$ implies two canonical enumerations of $\mathbf{RE}$. Define $\mathrm{D}_i = \mathrm{D}_{\varphi_i}$ and $\mathrm{R}_i = \mathrm{R}_{\varphi_i}$. Then

2.12. Theorem. $\mathbf{RE} = \{\mathrm{D}_i : i \in \mathbb{N}\} = \{\mathrm{R}_i : i \in \mathbb{N}\}$. ∎

2.13. Projection Theorem. $A \in \mathbf{RE} \leftrightarrow \bigvee_{B \in \mathbf{REC}} \left(A = \left\{x : \bigvee_y \langle x, y\rangle \in B\right\}\right)$. ∎

Example. The Turing predicate T defined by $\mathrm{T}(i, x, y, t) = 1 \leftrightarrow$ "M_i on input x stops after t steps with output y" is recursive. Therefore, the (special) *halting problem*

$$\mathrm{K} =_{\mathrm{df}} \{i : i \in \mathrm{D}_i\} \quad \text{is r.e., because} \quad i \in \mathrm{K} \leftrightarrow \bigvee_y \bigvee_t \mathrm{T}(i, i, y, t) = 1,$$

which allows us to apply the projection theorem.

2.14. Theorem. $\overline{\mathrm{K}} \notin \mathbf{RE}$, $\mathrm{K} \notin \mathbf{REC}$. ∎

Repeated application of projection and complementation leads to the *arithmetical sets*. They can be expressed in terms of multiply quantified recursive sets. For $n \geqq 0$,

$$A \text{ is a } \Sigma_n\text{-set} \leftrightarrow_{\mathrm{df}} \bigvee_{B \in \mathbf{REC}} \left(A = \left\{x : \bigvee_{y_1} \bigwedge_{y_2} \cdots \mathrm{Q}_{y_n} \langle y_1, \ldots, y_n, x\rangle \in B\right\}\right),$$

where $\mathrm{Q} = \bigwedge (\bigvee)$ if n is even (odd),

A is called a Π_n-set $\leftrightarrow_{df}$

$$\bigvee_{B \in \mathrm{REC}} \left(A = \left\{ x : \bigwedge_{y_1} \bigvee_{y_2} \dots \mathrm{Q}_{y_n} \langle y_1, \dots, y_n, x \rangle \in B \right\} \right),$$

where $\mathrm{Q} = \bigvee\ (\bigwedge)$ if n is even (odd),

$$\Sigma_n =_{df} \{A : A \text{ is a } \Sigma_n\text{-set}\}, \qquad \Pi_n =_{df} \{A : A \text{ is a } \Pi_n\text{-set}\}.$$

Note that $\Sigma_0 = \Pi_0 = \mathbf{REC}$, $\Sigma_1 = \mathbf{RE}$, $\Pi_1 = \mathbf{coRE}$ and $\Sigma_0 = \Pi_1 \cap \Sigma_1$.

2.15. Theorem. For every $n \geqq 1$,

$$\Sigma_n, \Pi_n \subset \Sigma_n \cup \Pi_n \subset \Sigma_{n+1} \cap \Pi_{n+1} \subset \Sigma_{n+1}, \Pi_{n+1}. \blacksquare$$

The sets Σ_n, Π_n form the *arithmetical hierarchy*.

The following theorem gives examples of $\leqq_1$-complete sets (see § 3) for some classes of the arithmetical hierarchy.

2.16. Theorem.

Θ Empty $=_{df} \{i : \mathrm{D}_i = \emptyset\}$ is $\leqq_1$-complete in Π_1.

Θ Finite $=_{df} \{i : \mathrm{D}_i \text{ is finite}\}$ is $\leqq_1$-complete in Σ_2.

Ω Total $=_{df} \{i : \varphi_i \in \mathbb{R}_1\}$ is $\leqq_1$-complete in Π_2.

Ω Bound $=_{df} \{i : \varphi_i \in \mathbb{R}_1 \wedge \mathrm{R}_i \text{ is finite}\} \in \Sigma_3 \cap \Pi_3$.

Θ Finite $\leqq_1$ Ω Bound (whence Ω Total $\not\equiv_1$ Ω Bound). $\blacksquare$

2.4. *Subrecursive Hierarchies*

The intensive study of $\mathbb{R}$ and $\mathbb{P}\mathrm{r}$ has led to a large number of classifications of the recursive functions by means of hierarchies which are very often based on different types of recursion. Multiple recursion leads to a hierarchy between $\mathbb{P}\mathrm{r}$ and $\mathbb{R}$ (see [Pét 51]). However, we are more interested in hierarchies below $\mathbb{P}\mathrm{r}$. Various complexity characterizations of such hierarchies are known, and, strictly speaking, not all of these complexity results belong to the scope of our book. However, the connections between computational and structural complexity are here so close and intimate that we give at least a brief sketch of this field. It is completed by § 10 as regards complexity characterizations of such hierarchies.

2.4.1. The Grzegorczyk Hierarchy

Various restrictions of primitive recursion have been used to define hierarchies within $\mathbb{P}\mathrm{r}$. Considering the case $\Sigma = \{1\}$ we define here bounded primitive recursion, delaying the definition of further restricted versions of primitive recursion, and a more detailed study, to § 10.

Definition. (*Bounded primitive recursion*). Let $n \geqq 0$, $g \in \mathbb{F}_{n+2}$, $h \in \mathbb{F}_n$ and $k \in \mathbb{F}_{n+1}$.

$$f = \mathrm{BPR}(g, h, k) \leftrightarrow f = \mathrm{PR}(g, h) \wedge f \leqq k.$$

Let

$$\mathrm{g}_0(x, y) = y + 1, \quad \mathrm{g}_1(x, y) = x + y, \quad \mathrm{g}_2(x, y) = x \cdot y,$$

and, for $n \geqq 2$,

$$\mathrm{g}_{n+1}(x, 0) = 1, \quad \mathrm{g}_{n+1}(x, y + 1) = \mathrm{g}_n(x, \mathrm{g}_{n+1}(x, y)).$$

Following [Ritr 63b], we define the Grzegorczyk classes

$$\mathscr{E}^n =_{\mathrm{df}} \Gamma_{\mathrm{SUB,BPR}}(\mathrm{s}, \mathrm{g}_n)$$

thus slightly modifying the original definition in [Grz 53].

These classes form the *Grzegorczyk hierarchy* (see 2.19.3). It is also possible to choose unary functions instead of the g_n. Let

$$\mathrm{a}_0(x) = x + 1, \ \mathrm{a}_1(x) = 2x, \ \mathrm{a}_2(x) = x^2, \ \mathrm{a}_3(x) = 2^x,$$

and, for $n \geqq 3$,

$$\mathrm{a}_{n+1}(0) = 1, \quad \mathrm{a}_{n+1}(x + 1) = a_{n+1}(\mathrm{a}_n(x)).$$

Furthermore, $\mathrm{m}(x, y) = \max(x, y)$.

2.17. Theorem.

$$\mathscr{E}^0 = \Gamma_{\mathrm{SUB,BPR}}(\mathrm{a}_0),$$

$$\mathscr{E}^n = \Gamma_{\mathrm{SUB,BPR}}(\mathrm{s}, \mathrm{m}, \mathrm{a}_n), \quad \text{for} \quad n \geqq 1. \ ▮$$

The function a_{n+1}, responsible for the growth of the functions of $\mathscr{E}^{n+1}$ is generated by n unrestricted primitive recursions. This turns out to be characteristic for all functions of $\mathscr{E}^{n+1}$.

2.18. Theorem. [Axt 65], [Schw 69], [Mül 73] $\mathscr{E}^{n+1}$ coincides with the class of all functions which can be generated from 0, s, $\mathscr{I}$ by substitution and at most n primitive recursions, provided $n \geqq 2$. ▮

2.19. Theorem. [Grz 53]

1. For $n \geqq 0$, $\bigwedge_{f \in \mathscr{E}_1{}^n} (f <_{\mathrm{ae}} \mathrm{a}_{n+1})$.
2. For $n \geqq 0$, $\bigwedge_{f \in \mathscr{E}_1{}^n} \bigvee_{m \in \mathbb{N}} (f \leqq_{\mathrm{ae}} \mathrm{a}_n^{[m]})$.
3. $\mathscr{E}^0 \subset \mathscr{E}^1 \subset \ldots$
4. $\bigcup_{n \in \mathbb{N}} \mathscr{E}^n = \mathbb{P}\mathrm{r}$.
5. $\{A : \mathrm{c}_A \in \mathscr{E}^2\} \subset \{A : \mathrm{c}_A \in \mathscr{E}^3\} \subset \ldots$ ▮

Remark. It is unknown whether the hierarchy $\{A : \mathrm{c}_A \in \mathscr{E}^0\} \subseteqq \{A : \mathrm{c}_A \in \mathscr{E}^1\} \subseteqq \{A : \mathrm{c}_A \in \mathscr{E}^2\}$ is proper. However, $\{A : \mathrm{c}_A \in \mathscr{E}^2\} = \{A : \mathrm{c}_A \in \mathscr{E}^1\}$ implies $\{A : \mathrm{c}_A \in \mathscr{E}^0\} = \{A : \mathrm{c}_A \in \mathscr{E}^1\}$ ([Bel 79]). See also [Muc 76] and [Kle 82] for further results.

It is of special interest that $\mathscr{E}^3$ coincides with the class $\mathscr{E}$ of the *elementary functions* introduced in [Kal 43]. It is defined by

$$\mathscr{E} = \Gamma_{\text{SUB,SUM,PRD}}(\text{s}, +, \dot{-})$$

where SUM and PRD are the following restricted versions of primitive recursion. For $n \in \mathbb{N}$ and $f, g \in \mathbb{F}_{n+1}$,

$$f = \text{SUM}(g) \leftrightarrow f(x_1, \ldots, x_n, y) = \sum_{z=0}^{y} g(x_1, \ldots, x_n, z),$$

$$f = \text{PRD}(g) \leftrightarrow f(x_1, \ldots, x_n, y) = \prod_{z=0}^{y} g(x_1, \ldots, x_n, z).$$

$\mathscr{E}$ contains all usual number theoretic functions like multiplication, division, square root, exponentiation etc. (cf. [Pét 51]). SUM and PRD are strictly weaker than PR. This follows from $\mathscr{E} \subset \mathbb{P}\text{r}$, which was first shown in [Ber 50].

2.20. Theorem. [Grz 53] $\mathscr{E} = \mathscr{E}^3$. ∎

As in the case of $\mathbb{P}\text{r}_1$ the classes $\mathscr{E}_1^n$ can also be generated using only 1-place functions.

Definitions. (*Bounded iteration*) For $f, g \in \mathbb{F}_1$,

$$f = \text{BIT}(g, h) \leftrightarrow f = \text{IT}(g) \wedge f \leqq h.$$

2.21. Theorem.

1. [Aus 75] For $n \geqq 2$, $\mathscr{E}_1^n = \Gamma_{\text{SUB},+,\cdot,\text{BIT}}\left(0, \text{s}, \text{q}, \left\lfloor\sqrt{x}\right\rfloor, \text{g}_n(x, x)\right)$.
2. [Geo 76] For $n \geqq 3$, $\mathscr{E}_1^n = \Gamma_{\text{SUB},\dot{-},\text{BIT}}(\text{s}, \text{a}_n)$. ∎

Characterizations of the $\mathscr{E}^n$ in terms of other restricted versions of primitive recursion are given in § 10.6.

A further characterization by bounded minimization can be found in [Har 75]. For further interesting relationships to various other classes of languages see [Har 78] and [Har 79].

In [MeRi 67] a subrecursive programming language for $\mathbb{P}\text{r}$ has been developed which has the nice property that the Grzegorczyk classes can be characterized by a structural property of these programs, namely by the depth of their loops. The primitive instructions of the LOOP language are $X = 0$, $X = Y$, $X = X + 1$, LOOP X ... END. A *LOOP program* is either one of the first three instructions, or it is of the form $P_1; P_2$ where P_1, P_2 are programs, or it is of the form LOOP X P END where P is a program. LOOP programs can be identified with register machines in a self-explanatory way (variables are understood as register names). Executing LOOP X P END means n-fold execution of P where n is the contents of register X.

The observation that both the structural complexity of a LOOP program and its running time are essentially dependent on the number of nestings of loops leads to the definition

$$\text{depth } P = 0 \text{ if } P \text{ is a program consisting of a single instruction},$$

$$\text{depth } (P_1; P_2) = \max (\text{depth } P_1, \text{depth } P_2),$$

$$\text{depth } (\text{LOOP } X\ P \text{ END}) = 1 + \text{depth } P.$$

$$\mathcal{L}_k =_{\text{df}} \{f : f \text{ is computable by a program } P \text{ such that depth } P \leqq k\}.$$

2.22. Theorem. [MeRi 67] $\mathcal{E}^{n+1} = \mathcal{L}_n$, for $n \geqq 2$. ∎

Remark. The classes of multiply recursive functions can be characterized in a similar way by suitably generalized LOOP programs (see [Ritd 66] and [Con 71 b]).

The material reviewed so far, with the exception of K. Harrow's and A. P. Beltjukov's results, is elaborated in a detailed way in [Aus 75].

2.4.2. Further Hierarchies

The following refinement of depth for LOOP programs has been considered in [GoNe 78]. A program P of depth n has length $k \leftrightarrow_{\text{df}} P$ has the form $P_1; P_2; \ldots; P_k$ where $P_1, \ldots, P_k$ have depth n.

$$\mathcal{L}_n^k =_{\text{df}} \{f : f \text{ is computable by a program of depth } n \text{ not being of length } k + 1\}.$$

For $n \geqq 3$ and $k \geqq 1$ these classes coincide with the classes of a hierarchy introduced as complexity classes in [Cle 63] (see [GoNe 78]).

A further refinement of the measure "length" leads to a refinement of this hierarchy which is of order type ω^ω ([GoNe 78]).

§ 3. Reducibility

The reader who wants more information about reducibility and operators is referred to [Rog 67]. The proofs of all theorems without references can be found there.

3.1. Basic Notions and Results

Reducibility notions are widely used in computability and complexity theory. The intuitive meaning of a statement like "A set A is reducible to a set B" is that a decision procedure of B can effectively be carried over to a decision procedure of A. We describe four standard ways to make this notion precise:

1-reducibility (reduction by a *1*-1 function)

$$A \leqq_1 B \leftrightarrow_{\text{df}} \bigvee_{f \in \mathbb{R}_1} (f \text{ is injective} \wedge c_A = c_B \circ f).$$

m-reducibility (reduction by a *many-one* function)

$$A \leqq_m B \leftrightarrow_{df} \bigvee_{f \in \mathbb{R}_1} c_A = c_B \circ f.$$

When $c_A = c_B \circ f$, then we say that A is reducible to B via f.

tt-reducibility (tt comes from *t*ruth *t*able). Let $\alpha_0, \alpha_1, \ldots$ be an effective enumeration of all Boolean functions. For given $f \in \mathbb{R}_1$ we interprete $f(x)$ as a pair $\langle m, \langle x_1, \ldots, x_n \rangle \rangle$ where n is the arity of α_m. We say

"B satisfies the *tt-condition* $f(x) = \langle m, \langle x_1, \ldots, x_n \rangle \rangle$"

if $\alpha_m(c_B(x_1), \ldots, c_B(x_n)) = 1$. Then we define

$$A \leqq_{tt} B \leftrightarrow_{df} \bigvee_{f \in \mathbb{R}_1} \bigwedge_x (x \in A \leftrightarrow B \text{ satisfies the tt-condition } f(x)).$$

T-reducibility (*T*uring reducibility):

$$A \leqq_T B \leftrightarrow_{df} c_A \in \Gamma_{\mathrm{SUB,PR,MIN}}(s, c_B).$$

3.1. Theorem.

1. $\leqq_1$, $\leqq_m$, $\leqq_{tt}$ and $\leqq_T$ are reflexive and transitive relations.
2. $\leqq_1 \subset \leqq_m \subset \leqq_{tt} \subset \leqq_T$.
3. Let $\leqq$ be any of these reducibility notions. Then

$$B \in \mathbf{REC} \land A \leqq B \to A \in \mathbf{REC}. \blacksquare$$

Statement 3 shows that these notions meet the intuitive meaning of reducibility stated at the beginning of this section. By $A \equiv_\alpha B =_{df} A \leqq_\alpha B \land B \leqq_\alpha A$ we have defined an equivalence relation whose equivalence classes $[A]_\alpha$ are called $\leqq_\alpha$-degrees ($\alpha \in \{1, \mathrm{m}, \mathrm{tt}, \mathrm{T}\}$). The degrees are partially ordered by $[A]_\alpha \leqq_\alpha [B]_\alpha \leftrightarrow_{df} A \leqq_\alpha B$. Statement 2 shows that the finest decomposition of $\mathfrak{P}(\mathbb{N})$ in degrees is given by $\leqq_1$.

Let $\mathscr{C}$ be a class of sets, and let $\leqq$ be a reflexive and transitive relation.

A is called *universal* or *hard* for $\mathscr{C}$ with respect to $\leqq$ ($\leqq$-hard for $\mathscr{C}$) iff $\bigwedge_{X \in \mathscr{C}} X \leqq A$.

A is called *complete* in $\mathscr{C}$ with respect to $\leqq$ ($\leqq$-complete in $\mathscr{C}$) iff $A \in \mathscr{C}$ and A is universal for $\mathscr{C}$ with respect to $\leqq$.

$\mathscr{C}$ is called *closed* under $\leqq$ iff $\bigwedge_A \bigwedge_B (A \leqq B \land B \in \mathscr{C} \to A \in \mathscr{C})$.

Theorem 2.16 illustrates these notions. In particular, K is $\leqq_1$-complete, and hence $\leqq_m$-complete, in $\mathbf{RE} = \Sigma_1$.

These definitions and the following simple but frequently used lemmas are also valid for complexity-bounded reducibilities as introduced in § 6.

3.2. Lemma. If A is $\leqq$-complete in $\mathscr{A}$ ($\leqq$-hard for $\mathscr{A}$) and $\mathscr{B}$ is closed with respect to $\leqq$, then

$$A \in \mathscr{B} \leftrightarrow \mathscr{A} \subseteq \mathscr{B} \quad (A \in \mathscr{B} \to \mathscr{A} \subseteq \mathscr{B}). \blacksquare$$

3.3. Lemma. If A is complete in $\mathscr{C}$ with respect to a relation $\leqq$ satisfying the condition $X \leqq Y \to \overline{X} \leqq \overline{Y}$ and if $\mathscr{C}$ is closed with respect to $\leqq$, then

$$\mathrm{co}\,\mathscr{C} = \mathscr{C} \leftrightarrow \bar{A} \in \mathscr{C}. \blacksquare$$

3.4. Lemma. If $\mathscr{B} = \bigcup_{i \geq 0} \mathscr{B}_i$ and $\bigwedge_i \mathscr{B}_i \neq \mathscr{B}$ and every $\mathscr{B}_i$ is closed with respect to $\leqq$, then $\mathscr{B}$ cannot have a $\leqq$-complete set. $\blacksquare$

3.5. Lemma. If $\mathscr{B} = \bigcup_{i \geq 0} \mathscr{B}_i$ and every $\mathscr{B}_i$ is closed with respect to $\leqq$, and if C is $\leqq$-complete in $\mathscr{A}$, then the following statements are mutually equivalent:

1. $\bigvee_i C \in \mathscr{B}_i$, 2. $\bigvee_i \mathscr{A} \subseteqq \mathscr{B}_i$, 3. $\mathscr{A} \subseteqq \mathscr{B}$, 4. $C \in \mathscr{B}$. $\blacksquare$

3.2. *Relativized Computability, Oracle Machines, and Recursive Operators*

Definition.

1. A function f is called *recursive in a function g* (*relative to g*) $\leftrightarrow f \in \Gamma_{\mathrm{SUB,PR,MIN}}(s, g)$.
2. A set A is called *recursive in the set B* (*relative to B*) $\leftrightarrow c_A$ is recursive in c_B.

It is possible to give a machine characterization of relative computability by *oracle machines*. Let τ be any type of deterministic or nondeterministic Turing machine as defined in § 1.1.1. An *oracle machine of type* τ (τ-OM) differs from an ordinary machine of type τ by having in addition a write-only tape, the *oracle tape* or *query tape*. Using this tape, the oracle machine M^B has access to the oracle B as follows (if the oracle is not specified we write $M^{(\,)}$): The set of states contains the special states s_q, s_y, and s_n (the query state, the yes-state and the no-state, resp.). If M^B enters the state s_q, then the following is done within one single step:

1. The machine enters the state s_y or s_n depending on whether the content of the query tape belongs to B or not.
2. The query tape is erased.

An oracle machine M^ψ with a function ψ as oracle differs from the oracle machine described above by having an additional write-only tape, the answer tape. If M^ψ enters the state s_q, then, if the contents x of the query tape does not belong to D_ψ, then the computation does not stop. Otherwise within one single step, the machine enters the state s_y, erases the query tape, and prints $\psi(x)$ onto the answer tape.

A formal definition of ID's and the notion of computation on OM's can be omitted here because it obviously follows the general scheme given in § 1.

Note that the idea of a standard enumeration (Gödel numering) of the Turing machines of type τ is also applicable to τ-OM's with function oracle, which leads us to a standard enumeration (Gödel numbering) $M_0^{(\,)}, M_1^{(\,)}, \ldots$ of all τ-OM's.

The functional of one function variable and one word variable computed by $M_i^{(\,)}$ is denoted by $\varphi_i^{(\,)}$. The standard enumeration $M_0^{(\,)}, M_1^{(\,)}, \ldots$ provides a standard enumeration (Gödel numbering) $\varphi_0^{(\,)}, \varphi_1^{(\,)}, \ldots$ of all functionals computable by τ-OM's. For given $\psi \in \mathbb{P}_1$ we obtain a standard enumeration $\varphi_0^\psi, \varphi_1^\psi, \ldots$ of all functions computable by τ-OM's with oracle ψ.

Using OM's with set oracle we obtain in the same way standard enumerations (Gödel numberings) $\varphi_0^{()}, \varphi_1^{()}, \dots$ of the set $\mathbb{P}_1^{()}(\Sigma)$ of all functionals with one set variable and one word variable computable by τ-OM's and $\varphi_0^A, \varphi_1^A, \dots$ of the set $\mathbb{P}_1^A(\Sigma)$ of all one-place function computable by τ-OM's with oracle $A \subseteq \Sigma^*$. Note that $\mathbb{P}_1^A(\Sigma) = \{\varphi_0^{c_A}, \varphi_1^{c_A}, \dots\}$.

The functional $\varphi_i^{()}$ can be considered as an operator mapping from $\mathbb{F}_1$ into $\mathbb{F}_1$.

A mapping $\mathsf{F}: \mathbb{F}_1 \to \mathbb{F}_1$ is called a *recursive operator* if $\mathsf{F} = \varphi_i^{()}$ for some i.

3.6. Theorem. F is a recursive operator if and only if $\mathsf{F} \in \Gamma_{\mathrm{SUB,PR,MIN}}(s, \ ) \cap \mathbb{F}_1^{\mathbb{F}_1}$, where the empty place indicates a function variable. ∎

3.7. Corollary. A set A is recursive in a set B if and only if there exists a recursive operator F such that $c_A = \mathsf{F}(c_B)$. ∎

3.8. Theorem. (J. Myhill/J. C. Shepherdson)

1. If F is recursive, then there exists an $s \in \mathbb{R}_1$ such that $\mathsf{F}(\varphi_i) = \varphi_{s(i)}$.
2. If $s \in \mathbb{R}_1$ is extensional, i.e. $\bigwedge_{i,j} \left(\varphi_i = \varphi_j \to s(i) = s(j)\right)$, then there exists a uniquely determined recursive operator F such that $\mathsf{F}(\varphi_i) = \varphi_{s(i)}$ for all $i \in \mathbb{N}$. ∎

Definition.

1. A recursive operator F is called *general recursive* $\leftrightarrow$ $\mathsf{F}(\mathbb{T}) \subseteq \mathbb{T}$.
2. A recursive operator F is called *total effective* $\leftrightarrow$ $\mathsf{F}(\mathbb{R}) \subseteq \mathbb{R}$.

In [Helm 71] it has been proved that there exist total effective operators which are not general recursive.

3.3. A General Setting of Reducibility

Any class $\mathfrak{R}$ of recursive operators defines a relation $\leqq_{\mathfrak{R}}$ on $\mathfrak{P}(\mathbb{N})$ by

$$A \leqq_{\mathfrak{R}} B \leftrightarrow \bigvee_{\mathsf{F} \in \mathfrak{R}} c_A = \mathsf{F}(c_B).$$

If $\mathfrak{R}$ contains the identity operator and is closed under composition, then $\leqq_{\mathfrak{R}}$ is reflexive and transitive and can be understood as a reducibility relation. K. Mehlhorn proposes some axioms to select reasonable reducibility notions with nice degree structures ([Meh 76]).

It is easy to determine the classes of operators belonging to the reducibilities $\leqq_1$, $\leqq_m$, $\leqq_{tt}$, $\leqq_T$. For instance, the class of operators inducing $\leqq_m$ is $\mathscr{C}_m = \{\mathsf{F}_f: f \in \mathbb{R}_1\}$ where $\mathsf{F}_f(g) =_{df} g \circ f$. The reducibility $\leqq_T$ is induced by the class of the recursive operators. This is one part of the following theorem:

3.9. Theorem.

1. $A \leqq_T B \leftrightarrow c_A = \mathsf{F}(c_B)$ for some recursive operator F.
2. $A \leqq_{tt} B \leftrightarrow c_A = \mathsf{F}(c_B)$ for some general recursive operator F. ∎

Similar notions of reducibility are $\leqq_{\mathscr{E}}$ (*elementary recursive in*), $\leqq_{\mathbf{Pr}}$ (*primitive recursive in*), which are induced by the operator classes $\Gamma_{\mathrm{SUB,SUM,PRD}}(s, +, \dot{-}, \ ) \cap \mathbb{F}_1^{\mathbb{F}_1}$,

$\Gamma_{\text{SUB,PR}}(\mathrm{s},\ ) \cap \mathbb{F}_1^{\mathbb{F}_1}$, respectively. In [Mac 72] the reader finds more about these reducibilities.

Finally we consider nondeterministic reducibilities. One approach is to use (deterministic) recursive operators having relations ("nondeterministic functions") as values instead of functions.

Let ϱ_M be the relation which is computed by the nondeterministic TM M as follows:

$$(x, y) \in \varrho_M \leftrightarrow \bigvee_p \varphi_M(x \mid p) = y$$

(i.e. M outputs y on input x on some path p).

Analogously, for the nondeterministic OM M^f we define ϱ_M^f. Now, a *relation-operator* $\mathsf{F}: \mathbb{T} \to \mathfrak{P}(\mathbb{N} \times \mathbb{N})$ is said to be *recursive* if there exists a nondeterministic OM $M^{()}$ such that $\mathsf{F}(f) = \varrho_M^f$ for all f. Equivalently, F is recursive if there exists a deterministic OM $M^{()}$ such that

$$(x, y) \in \mathsf{F}(f) \leftrightarrow \bigvee_p \varphi_M^{(f)}(x \mid p) = y.$$

From the class $\mathfrak{R}$ of recursive operators we derive the class

$$\mathfrak{R}' = \{\mathsf{F}_\sigma : \mathsf{F} \in \mathfrak{R} \wedge \sigma \text{ computable relation}\}$$

of relation-operators, where $\mathsf{F}_\sigma(g) = \mathsf{F}(g) \circ \sigma$. Equivalently, $\mathfrak{R}' = \{\mathsf{F}' : \mathsf{F} \in \mathfrak{R}\}$ where $(x, y) \in \mathsf{F}'(g) \leftrightarrow \bigvee_p \mathsf{F}(g)\,(x \mid p) = y$.

Now we are ready to define *nondeterministic reducibilities*. Any class $\mathfrak{R}'$ of recursive relation-operators defines a relation $\leqq_{\mathfrak{R}'}$ on $\mathfrak{P}(\mathbb{N})$ by

$$A \leqq_{\mathfrak{R}'} B \leftrightarrow \bigvee_{\mathsf{F} \in \mathfrak{R}'} \bigwedge_x \big(x \in A \leftrightarrow (x, 1) \in \mathsf{F}(\mathrm{c}_B)\big).$$

On the other hand, any class $\mathfrak{R}$ of recursive operators defines a relation $\leqq_{\mathfrak{R}}^{\mathrm{ND}}$ on $\mathfrak{P}(\mathbb{N})$ by

$$A \leqq_{\mathfrak{R}}^{\mathrm{ND}} B \leftrightarrow \bigvee_{\mathsf{F} \in \mathfrak{R}} \bigwedge_x \Big(x \in A \leftrightarrow \bigvee_y \mathsf{F}(\mathrm{c}_B)\,(x, y) = 1\Big).$$

It is easy to see that for any class $\mathfrak{R}$ of recursive operators we have the identity $\leqq_{\mathfrak{R}}^{\mathrm{ND}} = \leqq_{\mathfrak{R}'}$. Now we consider special cases.

T-reducibility. $A \leqq_{\mathrm{T}}^{\mathrm{ND}} B \leftrightarrow_{\mathrm{df}}$ there exists a recursive relation-operator F such that $x \in A \leftrightarrow (x, 1) \in \mathsf{F}(\mathrm{c}_B)$. Equivalently, $A \leqq_{\mathrm{T}}^{\mathrm{ND}} B \leftrightarrow$ there exists a recursive operator F such that $\Big(x \in A \leftrightarrow \bigvee_y \varphi_i^{(B)}(x, y) = 1\Big)$. It is easy to verify the following result.

3.10. Theorem.

1. The following statements are pairwise equivalent:
   (1) $A \leqq_{\mathrm{T}}^{\mathrm{ND}} B$.
   (2) $\chi_A = \mathsf{F}(\mathrm{c}_B)$ for some recursive operator F.
   (3) A recursive enumerable in B.
2. The relation $\leqq_{\mathrm{T}}^{\mathrm{ND}}$ is reflexive but not transitive. ∎

tt-reducibility. According to our general scheme we define

$$A \leqq_{\mathrm{tt}}^{\mathrm{ND}} B \leftrightarrow \text{there exists a computable relation } \varrho \text{ such that}$$
$$x \in A \leftrightarrow \bigvee_y ((x, y) \in \varrho \wedge B \text{ satisfies the tt-condition } y)$$
$$\leftrightarrow \text{there exists an } f \in \mathbb{R}_1 \text{ such that}$$
$$x \in A \leftrightarrow \bigvee_y B \text{ satisfies the tt-condition } f(x, y).$$

A comparison of 3.9.1 and 3.10 shows that the difference between deterministic and nondeterministic Turing reducibilities is the difference between c_A and χ_A. A similar result for tt-reducibilities is impossible because a general recursive operator cannot transform the total function c_B into the partial function χ_A. However, studying $\leqq_{\mathrm{tt}}^{\mathrm{ND}}$ separately is superfluous because of

3.11. Theorem. $\leqq_{\mathrm{tt}}^{\mathrm{ND}} = \leqq_{\mathrm{T}}^{\mathrm{ND}}$. ∎

m-reducibility. The nondeterministic m-reducibility $\leqq_{\mathrm{m}}^{\mathrm{ND}}$ is defined as $\leqq_{\mathscr{C}'_{\mathrm{m}}} = \leqq_{\mathscr{C}_{\mathrm{m}}}^{\mathrm{ND}}$ where $\mathscr{C}_{\mathrm{m}} = \{\mathsf{F}_g : g \in \mathbb{R}_1\}$ and $\mathsf{F}_g(f) = f \circ g$. It is easy to see that $\mathscr{C}'_{\mathrm{m}} = \{\mathsf{F}_\varrho : \varrho \text{ recursive relation}\}$, where $\mathsf{F}_\varrho(f) = f \circ \varrho$. More clearly we obtain

$$A \leqq_{\mathrm{m}}^{\mathrm{ND}} B \leftrightarrow \text{there exists a computable relation } \varrho \text{ such that}$$
$$x \in A \leftrightarrow \bigvee_y \big((x, y) \in \varrho \wedge y \in B\big)$$
$$\leftrightarrow \text{there exists a } g \in \mathbb{R}_2 \text{ such that}$$
$$x \in A \leftrightarrow \bigvee_y g(x, y) \in B.$$

3.12. Theorem. The relation $\leqq_{\mathrm{m}}^{\mathrm{ND}}$ is reflexive and transitive. ∎

§ 4. Formal Languages

In this section we make available some definitions and results from formal language theory which are needed in this book. In the first subsection we deal with some concepts concerning the generation of formal languages, especially with some types of grammars and L-systems. In the second subsection we define some important algebraic operations on languages, and we state some results on the dependence between these operations and on the closure and generation of some classes defined in the first subsection.

For a more detailed treatment of formal language theory, see [Gin 66], [Gin 75] and [Sal 73]. All results in this section can be found in these books unless otherwise stated.

4.1. Grammars and L-Systems

4.1.1. Grammars

A (unrestricted) *grammar* G is a quadruple (Σ, N, S, P) where Σ is the finite set of *terminals*, N is the finite set of *nonterminals* with $\Sigma \cap N = \emptyset$, $S \in N$ is the *start symbol*, and $P \subseteq (N \cup \Sigma)^+ \times (N \cup \Sigma)^*$ is the finite set of *productions*. A production

(x, y) is usually written in the form $(x \to y)$. For $(x \to y) \in P$ we write $u_1 x u_2 \underset{G}{\Longrightarrow} u_1 y u_2$ for all $u_1, u_2 \in (N \cup \Sigma)^*$. For $t \in \mathbb{N}$ we write $w \overset{t}{\underset{G}{\Longrightarrow}} v$ if there exist $w_0, w_1, \ldots, w_t$ such that

$$w = w_0 \underset{G}{\Longrightarrow} w_1 \underset{G}{\Longrightarrow} \ldots \underset{G}{\Longrightarrow} w_t = v,$$

where the latter is called a *derivation* of v from w. For grammars whose productions have no terminals on the left side we define a *leftmost derivation* as a derivation in which in any step the leftmost nonterminal is replaced. The relation $\overset{*}{\underset{G}{\Longrightarrow}}$ is the reflexive and transitive closure of the relation $\underset{G}{\Longrightarrow}$, i.e. $w \overset{*}{\underset{G}{\Longrightarrow}} v$ if and only if $s \overset{t}{\underset{G}{\Longrightarrow}} v$ for some $t \geqq 0$. Then $\mathrm{L}(G) = \left\{w: w \in \Sigma^* \wedge S \overset{*}{\underset{G}{\Longrightarrow}} w\right\}$ is the language generated by the grammar G.

4.1. Theorem. $\mathbf{RE} = \{\mathrm{L}(G): G$ is an unrestricted grammar$\}$. ∎

Thus we are interested in restricted types of grammars. A grammar $G = (\Sigma, N, S, P)$ is said to be X if all productions of G are of the form Y in Table 4.1. We say that these types of grammars (including unrestricted grammars), as well as the corresponding classes of languages defined below, form the Chomsky hierarchy.

Table 4.1

X	Y
context-sensitive	$(w \to v)$ where $\lvert w\rvert \leqq \lvert v\rvert$
context-free (or, for short, a *cfg*)	$(A \to v)$ where $A \in N$
linear	$(A \to vBu)$ or $(A \to v)$ where $A, B \in N$ and $v, u \in \Sigma^*$
right-linear	$(A \to vB)$ or $(A \to v)$ where $A, B \in N$ and $v \in \Sigma^*$
left-linear	$(A \to Bv)$ or $(A \to v)$ where $A, B \in N$ and $v \in \Sigma^*$

A language L is said to be *context-sensitive* if and only if $L - \{e\} = \mathrm{L}(G)$ for some context-sensitive grammar G. A language L is said to be *context-free* (*linear*) if and only if $L = \mathrm{L}(G)$ for some context-free (linear) grammar G. A language L is said to be *regular* if and only if $L = \mathrm{L}(G)$ for some right-linear or left-linear grammar G. By **CS** (**CF**, **LIN**, **REG**) we denote the class of all context-sensitive (context-free, linear, regular, resp.) languages.

Now we give some normal form theorems.

4.2. Theorem. A language $L \in \Sigma^+$ is context-sensitive if and only if $L = \mathrm{L}(G)$ for some grammar $G = (\Sigma, N, S, P)$ which has only productions of the form $(uAv \to uwv)$ where $A \in N$, $u, v \in (N \cup \Sigma)^*$ and $w \in \Sigma^+$. ∎

4.3. Theorem. The following statements are equivalent for $L \subseteqq \Sigma^+$:

1. $L \in \mathbf{CF}$.
2. $L \cup \{e\} \in \mathbf{CF}$.
3. $L = \mathrm{L}(G)$ for some grammar $G = (\Sigma, N, S, P)$ which has only productions of the form $(A \to BC)$ or $(A \to a)$ where $A, B, C \in N$ and $a \in \Sigma$ (*Chomsky normal form*).

4. $L = \mathrm{L}(G)$ for some grammar $G = (\Sigma, N, S, P)$ which has only productions of the form $(A \to aBC)$, $(A \to aB)$ or $(A \to a)$ where $A, B, C \in N$ and $a \in \Sigma$ (*Greibach normal form*). ∎

4.4. Theorem. The following statements are equivalent for $L \subseteq \Sigma^+$:

1. $L \in$ **LIN**.
2. $L \cup \{e\} \in$ **LIN**.
3. $L = \mathrm{L}(G)$ for some grammar $G = (\Sigma, N, S, P)$ which has only productions of the form $(A \to aBb)$, $(A \to aB)$, $(A \to Ba)$ or $(A \to a)$ where $A, B \in N$ and $a, b \in \Sigma$. ∎

4.5. Theorem. The following statements are equivalent for $L \subseteq \Sigma^+$:

1. $L \in$ **REG**.
2. $L \cup \{e\} \in$ **REG**.
3. $L = \mathrm{L}(G)$ for some grammar $G = (\Sigma, N, S, P)$ which has only productions of the form $(A \to aB)$ or $(A \to a)$ where $A, B \in N$ and $a \in \Sigma$.
4. $L = \mathrm{L}(G)$ for some grammar $G = (\Sigma, N, S, P)$ which has only productions of the form $(A \to Ba)$ or $(A \to a)$ where $A, B \in N$ and $a \in \Sigma$. ∎

4.6. Theorem. **REG** $\subset$ **LIN** $\subset$ **CF** $\subset$ **CS** $\subset$ **REC**. ∎

Remark. For witnesses of the proper inclusions take: $C =_{df} \{a^n b^n : n \geqq 0\} \in$ **LIN** $\setminus$ **REG**, $\{a^n b^n a^m b^m : n, m \geqq 0\}$, $D_1 \in$ **CF** $\setminus$ **LIN***) and $\{a^n b^n a^n : n \geqq 0\} \in$ **CS** $\setminus$ **CF**.

Now we mention some connections between grammar-defined and automata-defined classes of languages which are standard knowledge in the theory of formal languages.

4.7. Theorem. **REG** = **1-DFA** = **2-DFA** = **1-NFA** = **2-NFA**. ∎

4.8. Theorem. **CF** = **1-NPDA**. ∎

To conform with the pattern of Theorem 4.8 we set **DCF** = **1-DPDA**.

Now we define the *Szilardlanguage* $\mathrm{Sz}(G)$ of a grammar $G = (\Sigma, N, S, P)$. Let $P = \{r_1, r_2, \ldots, r_k\}$ and $w \in \Sigma^*$:

$$\mathrm{Sz}(w) = \{i_1 i_2 \ldots i_t : \text{there is a derivation } S = w_0 \underset{G}{\Longrightarrow} w_1 \underset{G}{\Longrightarrow} \ldots \underset{G}{\Longrightarrow} w_t = w \text{ such that } w_{\tau-1} \underset{G}{\Longrightarrow} w_\tau \text{ via } r_{i_\tau} \text{ for } \tau = 1, \ldots, t\},$$

$$\mathrm{Sz}_{\mathrm{left}}(w) = \{i_1 i_2 \ldots i_t : \text{there is a leftmost derivation } S = w_0 \underset{G}{\Longrightarrow} w_1 \underset{G}{\Longrightarrow} \ldots \underset{G}{\Longrightarrow} w_t = w \text{ such that } w_{\tau-1} \underset{G}{\Longrightarrow} w_\tau \text{ via } r_{i_\tau} \text{ for } \tau = 1, \ldots, t\},$$

$$\mathrm{Sz}(G) = \bigcup_{w \in \mathrm{L}(G)} \mathrm{Sz}(w) \text{ and } \mathrm{Sz}_{\mathrm{left}}(G) = \bigcup_{w \in \mathrm{L}(G)} \mathrm{Sz}_{\mathrm{left}}(w).$$

Furthermore we define

SZ = $\{\mathrm{Sz}(G) : G$ is an unrestricted grammar$\}$,

SZ(CF) = $\{\mathrm{Sz}(G) : G$ is a context-free grammar$\}$,

---

*) The *Dyck language* D_k over the alphabet $\{a_1, a_1', a_2, a_2', \ldots, a_k, a_k'\}$ is defined as the language generated by the grammar with the productions $S \to SS$, $S \to a_1 S a_1'$, $S \to a_2 S a_2'$, …, $S \to a_k S a_k'$, $S \to e$.

$\mathbf{SZ(REG)} = \{\mathrm{Sz}(G)\colon G \text{ is a right-linear or left-linear grammar}\}$,
$\mathbf{SZL} = \{\mathrm{Sz}_{\mathrm{left}}(G)\colon G \text{ is an unrestricted grammar}\}$,
$\mathbf{SZL(CF)} = \{\mathrm{Sz}_{\mathrm{left}}(G)\colon G \text{ is an context-free grammar}\}$.

4.9. Theorem.
1. $\mathbf{SZ} \subsetneqq \mathbf{CS}$.
2. $\mathbf{SZ(CF)} \not\subseteq \mathbf{CF}$.
3. $\mathbf{SZ(REG)} \subsetneqq \mathbf{REG}$.
4. $\mathbf{SZL} \subsetneqq \mathbf{CF}$. ∎

Let $G = (\Sigma, N, S, P)$ be a grammar and L a language. The language

$$\mathrm{L}(G, L) = \{w\colon w \in \mathrm{L}(G) \wedge \mathrm{Sz}(w) \cap L \neq \emptyset\}$$

is said to be the *language generated by G with the control set L*. We define

$$\mathbf{CS[CF]} = \{\mathrm{L}(G, L)\colon G \text{ is a context-sensitive grammar and } L \in \mathbf{CF}\}.$$

4.1.2. L-Systems

An *L-system* (*Lindenmayer system*) G is a triple (Σ, s, P) where Σ is the finite set of variables, $s \in \Sigma^*$ is the initial string and $P \subsetneqq \Sigma^* \times \Sigma \times \Sigma^* \times \Sigma^*$ is the finite set of productions. A production (x, a, y, z) is usually written in the form $\big((x, a, y) \to z\big)$. We write $a_1 a_2 \ldots a_k \underset{G}{\Longrightarrow} z_1 z_2 \ldots z_k$ if and only if for every $i \in \{1, \ldots, k\}$ there are $x, y \in \Sigma^+$ such that $x^{-1} \sqsubseteq a_{i-1} a_{i-2} \ldots a_1$, $y \sqsubseteq a_{i+1} a_{i+2} \ldots a_k$ and $\big((x, a_i, y) \to z_i\big)$ is a production. The relation $\overset{*}{\underset{G}{\Longrightarrow}}$ is the reflexive and transitive closure of the relation $\underset{G}{\Longrightarrow}$. Then $\mathrm{L}(G) = \big\{w\colon s \overset{*}{\underset{G}{\Longrightarrow}} w\big\}$ is the language generated by the L-system G.

An *EL-system* (*extended L-system*) G is a quadruple (Σ_1, Σ, s, P) where (Σ, s, P) is an L-system and $\Sigma_1 \subsetneqq \Sigma$ is the set of terminals. Then

$$\mathrm{L}(G) = \big\{w\colon s \overset{*}{\underset{G}{\Longrightarrow}} w \wedge w \in \Sigma_1^*\big\}$$

is the language generated by the EL-system G. Let $\mathbf{EL} = \{\mathrm{L}(G)\colon G \text{ is an EL-system}\}$.

Now we consider some restricted types of EL-systems. We add the symbols 2, 1, 0, D and P to the name EL, these indicating that only such EL-systems $G = (\Sigma_1, \Sigma, s, P)$ are permitted as satisfy the corresponding conditions:

2: $P \subsetneqq \Sigma \times \Sigma \times \Sigma \times \Sigma^*$	(G is two-sided context-sensitive),
1: $P \subsetneqq \Sigma \times \Sigma \times \{\mathrm{e}\} \times \Sigma^*$ or $P \subsetneqq \{\mathrm{e}\} \times \Sigma \times \Sigma \times \Sigma^*$	(G is one-sided-context-sensitive),
0: $P \subsetneqq \{\mathrm{e}\} \times \Sigma \times \{\mathrm{e}\} \times \Sigma^*$	(G is context-free),
D: $P\colon \Sigma^* \times \Sigma \times \Sigma^* \to \Sigma^*$	(G is deterministic),
P: $P \subsetneqq \Sigma^* \times \Sigma \times \Sigma^* \times \Sigma^+$	(G is propagating or e-free).

Thus we have the classes EiL, EDiL, EPiL and EPDiL for $i = \mathrm{e}, 2, 1, 0$. Now we generalize the notion of an EL-system. An ETL-system (extended table L-system)

G is a quadruple $(\Sigma_1, \Sigma, s, \mathscr{P})$ where $\mathscr{P} = (P_1, \ldots, P_r)$ and $G_i = (\Sigma, s, P_i)$ are EL-systems for $i = 1, \ldots, r$. We write $w \underset{G}{\Longrightarrow} v$ if $w \underset{G_i}{\Longrightarrow} v$ for some $i = 1, \ldots, r$.

The relation $\overset{*}{\underset{G}{\Longrightarrow}}$ is the reflexive and transitive closure of the relation $\underset{G}{\Longrightarrow}$, $L(G) = \{w: s \overset{*}{\underset{G}{\Longrightarrow}} w \wedge w \in \Sigma_1^*\}$ is the language generated by the ETL-system G.

Let **ETL** $= \{L(G): G$ is an ETL-system$\}$. Now we can make the same restrictions for ETL-systems as we have made for EL-systems. Thus we have the classes **ETiL**, **EDTiL**, **EPTiL** and **EPDTiL** of ETL-systems.

4.10. Theorem. [Vit 76], [Vit 77]

1. There is no difference between e and 2, i.e. **EL** = **E2L**, **EDL** = **ED2L**, **EPL** = **EP2L**, **EPDL** = **EPD2L**, **ETL** = **ET2L**, **EDTL** = **EDT2L**, **EPTL** = **EPT2L** and **EPDTL** = **EPDT2L**.
2. **E2L** = **ET2L** = **ED2L** = **EDT2L** = **E1L** = **ET1L** = **EDT1L** = **RE**.
3. **ED1L** $\subset$ **RE**. ∎

4.2. Algebraic Operations on Languages

In contrast to all other sections (except for § 11), this section deals with classes of languages that do not have a common finite alphabet. We fix an arbitrary countable alphabet Σ. A class $\mathscr{L} \subseteq P(\Sigma^*)$ is said to be a *family of languages* if and only if $\mathscr{L} \neq \emptyset$, $\mathscr{L} \neq \{\emptyset\}$ and, for every $L \in \mathscr{L}$, there is a finite $\Sigma_0 \subseteq \Sigma$ such that $L \subseteq \Sigma_0^*$. All classes of languages mentioned in the remainder of this subsection are families of languages. For example, the class of all context-free languages over Σ is a family of languages, since any context-free language is generated by a grammar having only finitely many terminals.

The following algebraic operations will be of special interest. Note that except for $\cup$, $\cap$, $^-$, $\cdot$ and $*$ all operations are actually families of operations. For such a family Ω of operations we define $\Omega(L) = \{f(L): f \in \Omega\}$.

$\cup$	— union.
$\cap$	— intersection.
$^-$	— complementation.
$\cap$ R	— intersection with regular sets: This is the set of all operations $L \to L \cap R_0$ where $R_0 \in$ **REG**.
$\cdot$	— concatenation.
$*$	— iteration.
h	— homomorphisms: This is the set of all operations $L \to h_0(L)$ where h_0 is a *homomorphism*. A homomorphism $h_0: \Sigma^* \mapsto \Sigma^*$ is characterized by the property $h_0(w \cdot v) = h_0(w) \cdot h_0(v)$ for $w, v \in \Sigma^*$. For $L \subseteq \Sigma^*$, $h_0(L) = \{h_0(w): w \in L\}$.
eh	— e-free homomorphism: This is the set of all operations $L \to h_0(L)$ where h_0 is an *e-free homomorphism*. A homomorphism h is said to be e-free if $h(a) \neq$ e for all $a \in \Sigma$.

h^{-1} — inverse homomorphism: This is the set of all operations $L \to h_0^{-1}(L)$ where h_0 is a homomorphism.

hr — *homomorphic replication:* This is the set of all operations $L \to \{h_1(w)\, h_2(w)^{-1} : w \in L\}$ where h_1, h_2 are homomorphisms.

hd — *homomorphic duplication:* This is the set of all operations $L \to \{h_1(w)\, h_2(w) : w \in L\}$ where h_1, h_2 are homomorphisms.

sub — *substitution:* This is the set of all operations sub_{Σ_0} which are defined as follows: For $\Sigma_0 = \{a_1, \ldots, a_k\} \subseteq \Sigma$ such that $a_1 < \ldots < a_k$ with respect to a given order "<" on Σ, let

$$\mathrm{sub}_{\Sigma_0}(L_0, L_1, \ldots, L_k) = \bigcup_{a_{i_1} \ldots a_{i_n} \in L_0 \cap \Sigma_0^*} L_{i_1} \cdot \ldots \cdot L_{i_n}.$$

Note that $h_0(L_0) = \mathrm{sub}_{\Sigma_0}\left(L_0, \{h_0(a_1)\}, \ldots, \{h_0(a_k)\}\right)$ for homomorphisms $h_0 : \Sigma_0^* \mapsto \Sigma_0^*$.

esub — *e-free substitution:* This is the set of all operations esub_{Σ_0} which are defined as follows: For $\Sigma_0 = \{a_1, \ldots, a_k\}$ as above, let

$$\mathrm{esub}_{\Sigma_0}(L_0, L_1, \ldots, L_k) = \begin{cases} \mathrm{sub}_{\Sigma_0}(L_0, L_1, \ldots, L_k) & \text{if} \quad e \notin L_1, \ldots, L_k \\ L_0 \quad \text{otherwise}. \end{cases}$$

sub^{-1} — *inverse substitution:* This is the set of all operations $\mathrm{sub}_{\Sigma_0}^{-1}$ which are defined as follows: For $\Sigma_0 = \{a_1, \ldots, a_k\}$ as above, let

$$\mathrm{sub}_{\Sigma_0}^{-1}(L_0, L_1, \ldots, L_k) = \{a_{i_1} \ldots a_{i_n} : L_{i_1} \cdot \ldots \cdot L_{i_n} \subseteq L_0\}.$$

gsm — *generalized sequential machine:* This is the set of all operations f which are the output function of a 1-DFA with output.

egsm — *e-free generalized sequential machine:* This is the set of all operations f which are the output function of a 1-DFA with output which never outputs e.

gsm^{-1} — *inverse generalized sequential machine:* This is the set of all operations f^{-1} where f is the output function of a 1-DFA with output.

All these operations have been defined for languages. Nevertheless, we shall also use some of them for families of languages. Let $\mathcal{L}$, $\mathcal{L}_1$ and $\mathcal{L}_2$ be families of languages. We define

$$\mathrm{f}(\mathcal{L}) = \bigcup_{L \in \mathcal{L}} \mathrm{f}(L), \quad \text{for} \quad \mathrm{f} \in \{\mathrm{h}, \mathrm{eh}, \mathrm{h}^{-1}, \mathrm{hr}, \mathrm{hd}\},$$

$$\mathcal{L}_1 \vee \mathcal{L}_2 = \{L_1 \cup L_2 : L_1 \in \mathcal{L}_1 \wedge L_2 \in \mathcal{L}_2\},$$

$$\mathcal{L}_1 \wedge \mathcal{L}_2 = \{L_1 \cap L_2 : L_1 \in \mathcal{L}_1 \wedge L_2 \in \mathcal{L}_2\},$$

$$\mathbf{co}\, \mathcal{L} = \{\bar{L} : L \in \mathcal{L}\}.$$

By $\Gamma_{\mathrm{f}_1, \mathrm{f}_2, \ldots, \mathrm{f}_k}(\mathcal{L})$ we denote the algebraic closure of the family of languages $\mathcal{L}$ with respect to the operations, or families of operations, $f_1, f_2, \ldots, f_k$.

A family of languages which is closed under eh, h^{-1}, $\cap$ R, $\cup$, $\cdot$, * (h, h^{-1}, $\cap$ R, $\cup$, $\cdot$, *) is said to be an *abstract family of languages* (*full abstract family of languages*), for short

AFL (full AFL). We define $\Gamma_{\mathrm{AFL}} = \Gamma_{\mathrm{eh,h^{-1},\cap R,\cup,\cdot,*}}$. Thus $\Gamma_{\mathrm{AFL}}(\mathscr{L})$ is the smallest AFL containing $\mathscr{L}$.

The operations defined above are not independent of each other, i.e. the closure of a family of languages under certain operations may imply the closure of this family under other operations.

4.11. Theorem. Let $\mathscr{L}$ be a family of languages.

If $\mathscr{L}$ is closed under	then $\mathscr{L}$ is closed under
eh, h^{-1}, $\cap$ R, $\cup$, *	$\cdot$
eh, h^{-1}, $\cap$ R, esub	$\cup$, $\cdot$, *
eh, h^{-1}, $\cdot$, $\setminus \{e\}$, $\cup \{e\}$	$\cup$, $\cap$ R
eh, h^{-1}, $\cap$ R, $\cup$, *, $\cap$	esub
h, h^{-1}, $\cap$ R, $\cup$, *, $\cap$	sub
eh, h^{-1}, $\cap$ R	gsm^{-1}, egsm
h, h^{-1}, $\cap$ R	gsm

(the nature of the operations $\setminus \{e\}$ and $\cup \{e\}$ is selfevident). ∎

The next theorem gives some closure properties of the most important grammar-defined families of languages.

4.12. Theorem. In Table 4.2 the class $\mathscr{C}$ is closed (not closed, not known to be closed) under the operation ω if + (—, ?, resp.) stands in row $\mathscr{C}$ and column ω. ∎

Table 4.2

	eh	h^{-1}	R	$\cup$	$\cdot$	*	AFL	esub	egsm	gsm^{-1}	h	full AFL	sub	gsm	$\cap$	—
RE	+	+	+	+	+	+	+	+	+	+	+	+	+	+	+	—
CS	+	+	+	+	+	+	+	+	+	+	—	—	—	—	+	?
CF	+	+	+	+	+	+	+	+	+	+	+	+	+	+	—	—
LIN	+	+	+	+	—	—	—	—	+	+	+	—	—	+	—	—
REG	+	+	+	+	+	+	+	+	+	+	+	+	+	+	+	+

An AFL $\mathscr{L}$ is said to be *principal* if there is a language L such that $\Gamma_{\mathrm{AFL}}(L) = \mathscr{L}$. In this case L is called a *generator* of $\mathscr{L}$. Next we give a method which can in some cases help to prove that an AFL is not principal.

4.13. Lemma. Let $\mathscr{L}_1, \mathscr{L}_2, \ldots$ be AFL's such that $\mathscr{L}_1 \subset \mathscr{L}_2 \subset \ldots$ Then $\mathscr{L} = \bigcup_{i=1}^{\infty} \mathscr{L}_i$ is an AFL which cannot be principal. ∎

In the subsequent theorems we state some interesting algebraic representations of some important families of languages.

4.14. Theorem.

$$\begin{aligned}\mathbf{RE} &= \mathrm{h}(\mathbf{LIN} \wedge \mathbf{LIN}) = \mathrm{h}\big(\mathrm{h}^{-1}(\mathrm{S}) \wedge \mathrm{h}^{-1}(\mathrm{S}) \wedge \mathbf{REG}\big) \text{ [BaBo 74]*)}\\ &= \Gamma_{\mathrm{h},\mathrm{h}^{-1},\cap\mathrm{R},\cap}(\mathrm{S}) = \Gamma_{\mathrm{h},\mathrm{h}^{-1},\cap\mathrm{R},\cap}(\mathrm{S}\#)\text{**)}\\ &= \Gamma_{\mathrm{h},\mathrm{h}^{-1},\cap\mathrm{R},\cap}(\mathrm{C}^*) \text{ [HaHo 71 b]***)}\\ &= \mathrm{h}(\mathbf{1\text{-}DCA} \wedge \mathbf{1\text{-}DCA}) = \Gamma_{\mathrm{h},\mathrm{h}^{-1},\cap\mathrm{R},\cap}(\mathrm{D}_1)\\ &= \Gamma_{\mathrm{hr},\cap}(\mathbf{REG}) \text{ [Boo 77]}\\ &= \mathrm{h}\big(\mathbf{EQ}^2 \wedge \{\mathrm{h}^{-1}(\{0\})\}\big) = \mathrm{h}(\mathbf{EQ}^2 \wedge \mathbf{REG}) \text{ [Bra 84]}\end{aligned}$$

(for $\mathbf{EQ}^2$ see p. 204)
(This implies $\mathbf{RE} = \mathrm{gsm}(\mathbf{EQ}^2)$ [Sal 78])

$$= \mathrm{h}(\mathbf{MINEQ}^2) \text{ [Čul 79]}$$

($\mathbf{MINEQ}^2 = \{\mathrm{Min}\, L : L \in \mathbf{EQ}^2\}$ and

$$\mathrm{Min}\, L = \Big\{u : u \in L \wedge \sim \bigvee_{v \in L} (v \neq \mathrm{e} \wedge v \sqsubset u)\Big\}). \quad ∎$$

For further characterizations of **RE** using only homomorphisms see [Bra 82].

4.15. Theorem. $\mathbf{CF} = \mathrm{eh}\big(\mathrm{h}^{-1}(\mathrm{D}_2) \wedge \mathbf{REG}\big) = \Gamma_{\mathrm{eh},\mathrm{h}^{-1},\cap\mathrm{R}}(\mathrm{D}_2)$. ∎

Because of **CF** = **1-NPDA** (cf. 4.8) the next theorem is analogous to Theorem 4.15.

4.16. Theorem. $\mathbf{1\text{-}NCA} = \mathrm{eh}\big(\mathrm{h}^{-1}(\mathrm{D}_1) \wedge \mathbf{REG}\big) = \Gamma_{\mathrm{eh},\mathrm{h}^{-1},\cap\mathrm{R}}(\mathrm{D}_1)$. ∎

4.17. Theorem.

1. $\mathbf{LIN} = \mathrm{eh}\big(\mathrm{h}^{-1}(\mathrm{S}\# \cup \{\mathrm{e}\}) \wedge \mathbf{REG}\big) = \Gamma_{\mathrm{eh},\mathrm{h}^{-1},\cap\mathrm{R}}(\mathrm{S})$.
2. $\mathbf{LIN} = \{\{h_1(w)\, h_2(w)^{-1} : w \in R\} : h_1, h_2$ homorphisms $\wedge\, R \in \mathbf{REG}\}$. ∎

Because of **LIN** = **1-PD-NREV**(1) (cf. 12.2) the next theorem is analogous to Theorem 4.17.

4.18. Theorem. $\mathbf{1\text{-}C\text{-}NREV}(1) = \mathrm{eh}\big(\mathrm{h}^{-1}(\mathrm{C}) \wedge \mathbf{REG}\big) = \Gamma_{\mathrm{eh},\mathrm{h}^{-1},\cap\mathrm{R}}(\mathrm{C})$. ∎

Finally we introduce the class **RUD** of rudimentary languages and the class **EXRUD** of extended rudimentary languages.

First we define three operations (families of operations) on relations over Σ^*.

Explicit transformation: Let $m \geqq 1$ and $t_1(x_1, \ldots, x_m), \ldots, t_k(x_1, \ldots, x_m) \in (\Sigma \cup \{x_1, \ldots, x_m\})^*$. For $R \subseteqq (\Sigma^*)^k$,

$$\mathrm{et}_{m,t_1,\ldots,t_k}(R) = \big\{(x_1, \ldots, x_m) : \big(t_1(x_1, \ldots, x_m), \ldots, t_k(x_1, \ldots, x_m)\big) \in R\big\},$$
$$\mathrm{et}\,(R) = \big\{\mathrm{et}_{m,t_1,\ldots,t_k}(R) : m \geqq 1 \wedge t_1, \ldots, t_k \in (\Sigma \cup \{x_1, \ldots, x_m\})^*\big\}.$$

Bounded existential quantification: For $R \subseteqq (\Sigma^*)^{k+1}$,

$$\mathrm{beq}\,(R) = \Big\{(x_1, \ldots, x_k, y) : \bigvee_z \big(|z| \leqq |y| \wedge (x_1, \ldots, x_k, z) \in R\big)\Big\}.$$

---

*) $\mathrm{S} =_{\mathrm{df}} \{w : w \in \{0, 1\}^* \wedge w = w^{-1}\}$.
**) $\mathrm{S}\# =_{\mathrm{df}} \{w \# w^{-1} : w \in \{0, 1\}^*\}$.
***) $\mathrm{C} =_{\mathrm{df}} \{0^n 1^n : n \in \mathbb{N}\}$.

Polynomial bounded existential quantification: Let $m \geqq 0$. For $R \subseteq (\Sigma^*)^{k+1}$,

$$\mathrm{pbeq}_m(R) = \left\{(x_1, \ldots, x_k, y) : \bigvee_z (|z| \leqq |y|^m \wedge (x_1, \ldots, x_k, z) \in R\right\}$$

$$\mathrm{pbeq}(R) = \{\mathrm{pbeq}_m(R) : m \geqq 0\}.$$

Let $\mathrm{CON}_{\Sigma_0} = \{(x, y, z) : x, y, z \in \Sigma_0^* \wedge x \cdot y = z\}$ be the concatenation predicate. For $R \subseteq (\Sigma^*)^k$, $\# \in \Sigma$, let $l(R) = \{x_1 \# x_2 \# \ldots \# x_k : (x_1, x_2, \ldots, x_k) \in R\}$. We define

$$\mathrm{RUD} = \bigcup_{\substack{\Sigma_0 \subseteq \Sigma \\ \Sigma_0 \text{ finite}}} \Gamma_{\cup,-,\mathrm{et},\mathrm{beq}}(\mathrm{CON}_{\Sigma_0}) \quad \text{and} \quad \mathbf{RUD} = \{l(R) : R \in \mathrm{RUD}\},$$

$$\mathrm{EXRUD} = \bigcup_{\substack{\Sigma_0 \subseteq \Sigma \\ \Sigma_0 \text{ finite}}} \Gamma_{\cup,-,\mathrm{et},\mathrm{pbeq}}(\mathrm{CON}_{\Sigma_0}) \quad \text{and}$$

$$\mathbf{EXRUD} = \{l(R) : R \in \mathrm{EXRUD}\}.$$

An arithmetical characterization of **RUD** is given by

4.19. Theorem. [Har 78] $A \in \mathbf{RUD}$ if and only if there exist $k \in \mathbb{N}$, $Q_1, \ldots, Q_k \in \{\bigwedge, \bigvee\}$ and a polynomial p such that

$$w \in A \leftrightarrow \underset{y_1 \leqq \mathrm{dya}^{-1} w}{Q_1} \ldots \underset{y_k \leqq \mathrm{dya}^{-1} w}{Q_k} \; p(\mathrm{dya}^{-1} w, y_1, \ldots, y_k) = 0. \;\blacksquare$$

Chapter II
The Notion of Computational Complexity

In § 5 the fundamental notions of complexity are defined (e.g. measures of computational complexity, complexity classes, honesty). Precise definitions of the time and space measures of all the kinds of machines dealt with in the book are given and further concrete measures for Turing machines such as reversal, crossing, return and dual return are introduced. Elementary relationships between complexity mesures are derived.

A general time measure (general space measure) is defined based on a large class of polynomially related time measures (linearly related space measures). Finally, the general concept of a Blum measure is dealt with, and the most elementary facts about them are proved.

Complexity bounded reducibility is a useful tool in various fields of complexity theory. § 6 gives a systematic definition of the time and space bounded versions of the notions defined in § 3. Elementary properties of these reducibilities are stated, among others the frequently used translational lemmas.

§ 5. Measures of Computational Complexity

The main interest of the theory of computational complexity is the question of how complex the computation of a recursive function, the decision of a recursive set or the acceptance of a recursively enumerable set is. To answer such questions we have to measure the complexity of a computation. There are various possibilities for doing so. For example, for computations of Turing machines one can count the length of the computation (i.e. the number of steps before the stop) or the number of tape squares used during the computation. We introduce these and other complexity measures and give some simple relationships between them.

It was M. Blum (cf. [Blu 67a]) who first observed that most of the familiar measures of computational complexity have a common core of properties (for example the Compression Theorem, the Gap Theorem, and the Speed-Up Theorem, cf. § 17 and § 18) and that these properties can be derived using only two very simple suppositions about a measure. A measure of computational complexity which satisfies these two axioms is called a Blum measure. We prove some simple properties of Blum measures and show that the previously introduced complexity measures for deterministic machines are Blum measures.

5.1. Specific Measures of Computational Complexity

5.1.1. Measures for Deterministic Machines

Let τ be any fixed type of deterministic machine (i.e. any type of deterministic Turing machine, random access machine or iterative array). A *measure of computational complexity* or *complexity measure* for machines of type τ is a function Φ assigning to each computation $K_0, K_1, K_2, \ldots$ of a machine of type τ a certain natural number which is interpreted as the *complexity of this computation.* If M is a machine of type τ and $K_0, K_1, K_2, \ldots$ is the computation of M on an input w, then we define

$$\Phi_M(w) = \Phi(K_0, K_1, K_2, \ldots),$$

and

$$\Phi_M(n) = \max_{|w|=n} \Phi_M(w) \quad \text{for} \quad n \in \mathbb{N}.$$

The function Φ_M is said to be the *complexity function* or the *step counting function of M with respect to Φ.* If $\Phi_M(n) \leqq_{\text{ae}} t(n)$, then we say that M *works within Φ-complexity t.*

Now we fix some finite alphabet Σ having at least two elements. For $t \in \mathbb{R}_1$, we define the *complexity class of languages* over Σ

$$\mathbf{\Phi}(t) = \{\mathrm{L}(M)\colon M \text{ is of type } \tau,\ \mathrm{L}(M) \subseteqq \Sigma^* \text{ and } \Phi_M(w) \leqq_{\text{ae}} t(|w|)\}^{*,**)}$$

or, equivalently,

$$\mathbf{\Phi}(t) = \{\mathrm{L}(M)\colon M \text{ is of type } \tau,\ \mathrm{L}(M) \subseteqq \Sigma^* \text{ and } \Phi_M(n) \leqq_{\text{ae}} t(n)\},$$

and the *complexity class of functions*

$$\Phi(t) = \{\varphi_M\colon M \text{ is of type } \tau \text{ with output tape, } \varphi_M\colon \Sigma^* \mapsto \Sigma^* \text{ and } \Phi_M(w) \leqq_{\text{ae}} t(|w|)\}$$

or, equivalently,

$$\Phi(t) = \{\varphi_M\colon M \text{ is of type } \tau \text{ with output tape, } \varphi_M\colon \Sigma^* \mapsto \Sigma^* \text{ and } \Phi_M(n) \leqq_{\text{ae}} t(n)\}.$$

The function t is said to be a *name* or a *bounding function* of the complexity classes $\mathbf{\Phi}(t)$ and $\Phi(t)$.

Let $L \subseteqq \mathbb{N}$ and $f\colon \mathbb{N} \mapsto \mathbb{N}$. By $L \in \mathbf{\Phi}(t)$ and $f \in \Phi(t)$ we mean $\{\text{bin}\, n\colon n \in L\} \in \mathbf{\Phi}(t)$ and $\bigvee_{f'} \left(f' \in \Phi(t) \wedge \bigwedge_n f'(\text{bin}\, n) = \text{bin}\, f(n)\right)$, resp., where the digits 0 and 1 which are necessary for the binary representation can be replaced by any two different symbols

---

*) The relation $\leqq_{\text{ae}}$ is used for the sake of convenience. For the time and space measures, the relation $\leqq$ can be used equivalently.

**) For § 11 and § 12 we use the slightly modified notion of a complexity class where the condition "$\mathrm{L}(M) \subseteqq \Sigma^*$" is omitted.

of the alphabet Σ. That is why we can also have sets of natural numbers or number-theoretical functions in a complexity class over an arbitrary alphabet Σ.

For $\mathfrak{K} \subseteq \mathbb{R}_1$ we define $\boldsymbol{\Phi}(\mathfrak{K}) = \bigcup_{t \in \mathfrak{K}} \boldsymbol{\Phi}(t)$ and $\Phi(\mathfrak{K}) = \bigcup_{t \in \mathfrak{K}} \Phi(t)$. Frequently used examples of such classes of functions are $t(\mathrm{Lin}) = \{t(k \cdot n) : k \in \mathbb{N}\}$, $t(\mathrm{Pol}) = \{t(n^k) : k \in \mathbb{N}\}$, $\mathrm{Lin}\, t = \{k \cdot t(n) : k \in \mathbb{N}\}$ and $\mathrm{Pol}\, t = \{t(n)^k : k \in \mathbb{N}\}$. We set Lin = Lin id, Pol = Pol id, and Const = Lin 1.

Let Φ and Φ' be two complexity measures for machines of type τ. For $t_1, t_2 \in \mathbb{R}_1$, we define the *double complexity classes*

$$\boldsymbol{\Phi}\text{-}\boldsymbol{\Phi}'(t_1, t_2) = \{\mathrm{L}(M) : M \text{ is of type } \tau,\ \mathrm{L}(M) \subseteq \Sigma^*,\ \Phi_M \leqq_{\mathrm{ae}} t_1 \text{ and } \Phi'_M \leqq_{\mathrm{ae}} t_2\}$$

$$\Phi\text{-}\Phi'(t_1, t_2) = \{\varphi_M : M \text{ is of type } \tau,\ \varphi_M : \Sigma^* \mapsto \Sigma^*,\ \Phi_M \leqq_{\mathrm{ae}} t_1 \text{ and } \Phi'_M \leqq_{\mathrm{ae}} t_2\}.$$

Now we introduce some abbreviations. Let Φ be any complexity measure of type τ, and let $f \in \mathbb{R}_1$, $A \in \mathbf{REC}$ and $t \geqq 0$. We define

$$\Phi\text{-Comp}(f) \geqq_{\mathrm{ae}} t \leftrightarrow \bigwedge_{M \text{ of type } \tau} (\varphi_M = f \to \Phi_M \geqq_{\mathrm{ae}} t),$$

$$\Phi\text{-Comp}(A) \geqq_{\mathrm{ae}} t \leftrightarrow \Phi\text{-Comp}(\mathrm{c}_A) \geqq_{\mathrm{ae}} t.$$

Analogous definitions are made for $>_{\mathrm{ae}}$, $\geqq_{\mathrm{io}}$, $>_{\mathrm{io}}$, $\geqq$, $>$, $\succeq$, $\succ$, $\succeq_{\mathrm{io}}$, and $\succ_{\mathrm{io}}$.

Further we define

$$\Phi\text{-Comp}(f) \leqq_{\mathrm{ae}} t \leftrightarrow \bigvee_{M \text{ of type } \tau} (\varphi_M = f \wedge \Phi_M \leqq_{\mathrm{ae}} t),$$

$$\Phi\text{-Comp}(A) \leqq_{\mathrm{ae}} t \leftrightarrow \Phi\text{-Comp}(\mathrm{c}_A) \leqq_{\mathrm{ae}} t.$$

Analogous definitions are made for $<_{\mathrm{ae}}$, $\leqq_{\mathrm{io}}$, $<_{\mathrm{io}}$, $\leqq$, $<$, $\preceq$, $\prec$, $\preceq_{\mathrm{io}}$, and $\prec_{\mathrm{io}}$.

Finally we set

$$\Phi\text{-Comp}(f) \asymp t \leftrightarrow \Phi\text{-Comp}(f) \preceq t \text{ and } \Phi\text{-Comp}(f) \succeq t.$$

5.1.1.1. *Time Measures*

Let τ-DM be any type of deterministic Turing machine. The complexity measure $\Phi = \tau$-DTIME assigns to each computation of a τ-DM the length of this computation, i.e. $\tau\text{-DTIME}(K_0, K_1, K_2, \ldots) = \max\{t : K_t \text{ is not final}\}$.

For d-dimensional iterative arrays ($d \geqq 1$), the complexity measure $\Phi = \mathrm{IA}^{\mathrm{d}}$-DTIME assigns to each computation $K_0, K_1, K_2, \ldots$ of a $\mathrm{DIA}^{\mathrm{d}}$ the length of this computation, i.e. $\mathrm{IA}^{\mathrm{d}}\text{-DTIME}(K_0, K_1, K_2, \ldots) = \max\{t : K_t \text{ is not final}\}$.

Now let DXZ be any type of deterministic RAM, i.e. X ∈ {S, e, B, M, C} and Z ∈ {RAM, RAM′}. The function Φ = XZ-DTIME assigns to each computation $\mathrm{K}_0, \mathrm{K}_1, \mathrm{K}_2, \ldots$ of an DXZ the length of this computation, i.e. XZ-DTIME(K_0, K_1,

$K_2, \ldots) = \max\{t: K_t$ is not final$\}$. This measure is said to be the time measure with unit cost criterion, since one step of the machine costs one unit of complexity. A possibly more realistic time measure for RAM's is the time measure with logarithmic cost criterion XZ-DTIME$^{\log}$. It assigns to each computation of length t the sum $\sum_{j=0}^{t-1} r_j$ where r_j is the sum of the lengths of all registers involved in that instruction which converts the situation K_j to the situation K_{j+1}.

A machine is called *t-time bounded* if it works within time t. We say that a machine M of type τ works in *realtime* if τ-DTIME$_M(w) = |w|$ for all inputs w. We say that a machine M of type τ having a one-way input tape works in *quasi-realtime with delay* d ($d \geqq 1$) iff it moves its input head in each d-th step, i.e. in the steps $d, 2d, 3d, \ldots$ of its work.

The next theorem shows that the time measures of various types of machines are polynomially related. Two measures Φ and Ψ are *polynomially related* if $\Phi(\text{Pol } t) = \Psi(\text{Pol } t)$ for all $t \geqq \text{id}$.

5.1. Theorem. The following complexity classes coincide for $t \geqq \text{id}$:

1. **i-T-DTIME**(Pol t), for $i = 0, 1, 2$,
2. **1-X$_1$-X$_2$-DTIME**(Pol t), for X$_1$, X$_2$ $\in$ {S, NES, PD},
3. **multiT^{multi}multi-DTIME**(Pol t)*),
4. **IA^d-DTIME**(Pol t), for $d \geqq 1$,
5. **XRAM-DTIME**(Pol t) and **XRAM-DTIME$^{\log}$**(Pol t), for X $\in$ {S, e, B},
6. **XRAM′-DTIME**(Pol t) and **XRAM′-DTIME$^{\log}$**(Pol t), for X $\in$ {e, B}.

Proof. It is sufficient to prove the inclusions represented in Figure 5.1, in which the term **-DTIME**(Pol t) is omitted. We restrict ourselves to the inclusions numbered by 1, 2, ..., 8. We give only the ideas of how the simulating machines work. The details of their construction are left to the reader.

Ad 1. The work of a Turing tape can be simulated in realtime (i.e. one step of the original machine can be simulated by one step) by two pushdown stores as follows: if $\square\, w$ and $\square\, v$ are the stack words of the two pushdown stores at a given moment, then $\ldots \square\square\square\, wv^{-1} \square\square\square \ldots$ is the inscription on the Turing tape at this moment where the head of the Turing tape scans either the last symbol of w or the first symbol of v^{-1}.

Ad 2. Let $\ldots \square\square\square\, w_i v_i \square\square\square \ldots$ be the inscription on the worktape of a 1-DTM M in the ith step of its work on an input w, where the worktape head scans

*) If the number of tapes, the number of heads and/or the number of dimensions of a Turing machine is not restricted, then we indicate this by *multi*. For example,

$$\textbf{multiT}^{\textbf{multi}}\textbf{multi-DTIME}(\text{Pol } t) = \bigcup_{k,h,d} \textbf{kT}^{\textbf{d}}\textbf{h-DTIME}(\text{Pol } t),$$

$$\textbf{multiT-DTIME}(t) = \bigcup_{k} \textbf{kT-DTIME}(t)$$

and

$$\textbf{Tmulti-DTIME}(\text{Lin } t) = \bigcup_{h} \textbf{Th-DTIME}(\text{Lin } t).$$

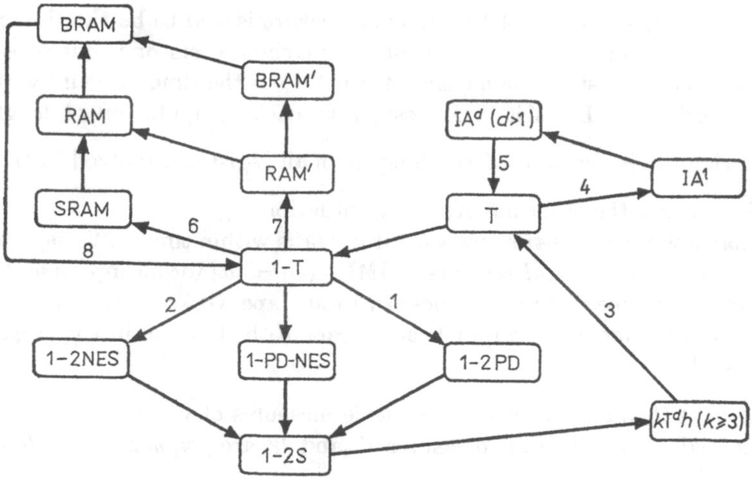

Figure 5.1

the first symbol of $v_i\square$ and the word w_iv_i does not contain the blank. A 1-2NES-DM M' can simulate the work of M by generating the sequence $w_1 \mid v_1 \# w_2 \mid v_2 \# w_3 \mid v_3 \# \ldots$ in each of the nonerasing stacks, where $\#$ and $\mid$ are symbols not used by M. If $w_1 \mid v_1 \# \ldots \# w_i \mid v_i$ is already constructed in the stacks, then $\# w_{i+1} \mid v_{i+1}$ can be produced on the first nonerasing stack by moving the head of the second nonerasing stack once back and forth on $\# w_i \mid v_i$. Then $\# w_{i+1} \mid v_{i+1}$ can be copied to the second stack by moving the head of the first nonerasing stack once back and forth on $\# w_{i+1} \mid v_{i+1}$. Consequently, if M works within time t, the length of w_iv_i does not exceed $t(|w|)$ for each i, and the generation of $\# w_{i+1} \mid v_{i+1}$ takes M' at most $\lesssim t(|w|)$ steps. Hence M' works within $\lesssim t(|w|)^2$ steps.

Ad 3. Let M be a kT^{d}h-DM working within time t. We describe the work of a DTM M' which simulates the work of M on an input w. After the simulation of a step of the work of M, the tape inscription of M' consists of some $(d+3)$-tuples which can be described as follows: if on completion of this step of M the square $(n_1, \ldots, n_d)$ of the ith tape of M has the contents $a \neq \square$ and is scanned by the heads $j_1, \ldots, j_m$ $(m \geqq 0)$, then the $(d+3)$-tuple $(n_1, \ldots, n_d, i, a, \{j_1, \ldots, j_m\})$ is written on the tape of M'. Since $n_1, \ldots, n_d \leqq t(|w|)$, the description of such a $(d+3)$-tuple is of length $\lesssim \log t(|w|)$, and since there are at most $\lesssim t(|w|)$ such $(d+3)$-tuples, the length of the description of all these $(d+3)$-tuples is $\lesssim t(|w|) \cdot \log t(|w|)$. Now it is obvious that M' can update this inscription, corresponding to the next step of M, within time $t(|w|) \cdot (\log t(|w|))^2$. Consequently, M works within time $\lesssim t(|w|)^2 (\log t(|w|))^2$.

Ad 4. Let M be a DTM working within time t. An IA^1 M' can simulate the work of M on an input w in realtime as follows: After the simulation of a step of M the tape inscriptions of M and M' coincide except for the inscription of the tape square scanned by the head of M. This square of M' holds, in addition, the actual state of M. In the next step of M' the symbols of the cells of M' which have this state in their neighbourhood are changed according to the next-move function of M, the symbols of all other cells remain unchanged.

Ad 5. This can be proved in the same way as for inclusion 3, with the exception that $(d+1)$-tuples $(n_1, \ldots, n_d, a)$ are stored on the tape of the DTM M' and that at most $\lesssim t(|w|)^d$ such tuples must be stored. Consequently, M' works within time $\lesssim t(|w|)^{d+1} (\log t(|w|))^2$.

Ad 6. Let M be a 1-DTM working within time t. First, M can be simulated in real-time by a 1-DTM M' which uses only the squares 1, 2, 3, ... on its worktape, storing in the tape square i the contents of the tape squares i and $-i+1$ of the worktape of M, for $i = 1, 2, 3, \ldots$ Now M' can be simulated by a DSRAM M'', where a suitable binary encoding of the contents of the square i of the worktape of M' is stored in the $(i+1)$th register of M'', for $i = 1, 2, 3, \ldots$ The position of the worktape head of M' is stored in register 0. It is evident that M'' can simulate one step of M' by a constant number of steps using register 1 as an auxiliary register. Thus M'' works within time $\lesssim t(n)$ on inputs of length n (unit cost criterion). The length of register 0 can be at most $\lesssim \log t(n)$, since the worktape head of M' can be at most $t(n) + 1$ tape squares far from tape square 0. The length of all other registers can be bounded by a constant. Thus $\text{SRAM-DTIME}_{M''}^{\log}(n) \lesssim t(n) \cdot \log t(n)$.

Ad 7. Let M be a 1-DTM which works within time t and which has the worktape symbols $a_1, \ldots, a_k$. Choose $m \in \mathbb{N}$ such that $k \leqq 2^m$. We encode the worktape symbols of M by a one-one mapping $b\colon \{a_1, \ldots, a_k\} \xrightarrow{1\text{-}1} \{0, 1\}^m$ (such a mapping is said to be a block encoding of length m), where $b(\square) = 0^m$. M can be simulated by a 1-DTM M' in such a way that the worktape inscription $\ldots c_{-2}c_{-1}c_0c_1c_2 \ldots$ $(c_i \in \{a_1, \ldots, a_k\})$ of M is represented by the worktape inscription $\ldots b(c_{-2})\, b(c_{-1})\, b(c_0)\, b(c_1)\, b(c_2) \ldots$ Thus M' needs only the worktape symbols 0 and 1 and works within time $\lesssim t$. Now we construct a DRAM' M'' which simulates the work of M' in such a way that the worktape inscription $\ldots 000w_1\overset{\downarrow}{a}w_2000\ldots$ $(w_1 \in 1\{0, 1\}^* \cup \{e\}$, $a \in \{0, 1\}$, $w_2 \in \{0, 1\}^* 1 \cup \{e\}$ and $\downarrow$ marks the head position) is represented by the registers 0, 1, and 2 in the following manner: $\mathbf{R}_0 = w_1$, $\mathbf{R}_1 = a$ and $\mathbf{R}_2 = w_2^{-1}$. Thus, in order to simulate a step of M' in which the worktape head moves to the left (moves to the right, does not move) M'' has to change the registers 0, 1 and 2 as follows:

$$
\begin{array}{ll}
\mathbf{R}_0 \leftarrow \left\lfloor \dfrac{\mathbf{R}_0}{2} \right\rfloor & \left(\begin{array}{ll} \mathbf{R}_0 \leftarrow 2 \cdot \mathbf{R}_0 + a' & \\[2ex] \mathbf{R}_1 \leftarrow \mathbf{R}_2 \mathbin{\dot{-}} 2 \cdot \left\lfloor \dfrac{\mathbf{R}_2}{2} \right\rfloor & \qquad \mathbf{R}_1 \leftarrow a' \\[2ex] \mathbf{R}_2 \leftarrow \left\lfloor \dfrac{\mathbf{R}_2}{2} \right\rfloor & \qquad\qquad \text{resp.} \end{array} \right) \\[2ex]
\mathbf{R}_1 \leftarrow \mathbf{R}_0 \mathbin{\dot{-}} 2 \left\lfloor \dfrac{\mathbf{R}_0}{2} \right\rfloor & \\[2ex]
\mathbf{R}_2 \leftarrow 2 \cdot \mathbf{R}_2 + a' &
\end{array}
$$

where $a' \in \{0, 1\}$ is determined by the instruction of M' which is executed in this step. For the first case this can be done by the following program part:

1 $\mathbf{R}_1 \leftarrow \mathbf{R}_0$
2 $\mathbf{R}_0 \leftarrow 0$
3 $\mathbf{R}_4 \leftarrow 0$
4 $\mathbf{R}_5 \leftarrow 1$
5 $\mathbf{R}_3 \leftarrow \mathbf{R}_4$
6 $\mathbf{R}_4 \leftarrow \mathbf{R}_5$
7 $\mathbf{R}_5 \leftarrow \mathbf{R}_5 + \mathbf{R}_5$
8 IF($\mathbf{R}_5 \leqq \mathbf{R}_1$)THEN GOTO 5
9 $\mathbf{R}_1 \leftarrow \mathbf{R}_1 \dot{-} \mathbf{R}_4$
10 $\mathbf{R}_0 \leftarrow \mathbf{R}_0 + \mathbf{R}_3$
11 IF($\mathbf{R}_1 > 1$)THEN GOTO 3
12 $\mathbf{R}_2 \leftarrow \mathbf{R}_2 + \mathbf{R}_2$
13 $\mathbf{R}_2 \leftarrow \mathbf{R}_2 + a'$.

In the second case we proceed analogously. Since the inner loop is executed at most $(|\mathbf{R}_1| + 1)^2$ times and $|\mathbf{R}_i| \precsim t(n)$ for inputs of length n ($i = 0, 1, 2, 3, 4, 5$), the DRAM' M'' works within time $\precsim t(n)^2$ (unit cost criterion) or $\precsim t(n)^3$ (logarithmic cost criterion).

Ad 8. Let M be a BRAM working within time t (unit cost criterion or logarithmic cost criterion). A 1-DTM M' can simulate M in such a way that it stores all pairs $(i, \mathbf{R}_i)$ such that $\mathbf{R}_i \neq \mathrm{e}$. For inputs of length n the 1-DTM M' must store at most $t(n)$ such pairs, each having a description of length $\precsim t(n)$. Consequently, the description of all these pairs is of length $\precsim t(n)^2$ and one step of M can be simulated by M' within $\precsim t(n)^3$ steps. Thus M' works within $\precsim t(n)^4$ steps. ∎

Remark. Theorem 5.1 also holds true for complexity classes of functions.

Since the time measures of the most frequently used types of machines are polynomially related, we introduce *general time complexity classes*, which do not depend on any special type of machine. For $t \geqq \mathrm{id}$ we define

$$\mathbf{DTIME}(\mathrm{Pol}\, t) = \mathbf{T\text{-}DTIME}(\mathrm{Pol}\, t)$$

and

$$\mathrm{DTIME}(\mathrm{Pol}\, t) = \mathrm{T\text{-}DTIME}(\mathrm{Pol}\, t).$$

A time measure which is polynomially related to the measure T-DTIME is said to be a *realistic time measure*.

Finally we introduce notations for frequently used time complexity classes:

$$\mathbf{REALTIME} = \mathbf{multiT\text{-}DTIME}(\mathrm{id}),$$

$$\mathbf{LINTIME} = \mathbf{multiT\text{-}DTIME}(\mathrm{Lin}),$$

$$\mathbf{P} = \mathbf{DTIME}(\mathrm{Pol}) \quad \text{and} \quad \mathrm{P} = \mathrm{DTIME}(\mathrm{Pol}),$$

$$\mathbf{DEXPTIME} = \mathbf{DTIME}(2^{\mathrm{Lin}}) \quad \text{and} \quad \mathrm{DEXPTIME} = \mathrm{DTIME}(2^{\mathrm{Lin}}).$$

5.1.1.2. Space Measures

Let $\mathrm{Z}_1\mathrm{h}_1\text{-}\mathrm{Z}_2\mathrm{h}_2\text{-}\ldots\text{-}\mathrm{Z}_k\mathrm{h}_k$-DM be any type of deterministic Turing machine with or without output tape (i.e. $\mathrm{Z}_1 \in \{1{:}, 1^*{:}, 2{:}, \mathrm{T}, \mathrm{T}^2, \mathrm{T}^3, \ldots\}$ and $\mathrm{Z}_2, \ldots, \mathrm{Z}_k \in \{\mathrm{T}, \mathrm{T}^2, \mathrm{T}^3, \ldots, \mathrm{S}, \mathrm{NES}, \mathrm{CS}, \mathrm{PD}, \mathrm{C}\}$). For a halting computation $K_0, K_1, K_2, \ldots$ of a machine M of this type, we define $\mathrm{Z}_1\mathrm{h}_1\text{-}\mathrm{Z}_2\mathrm{h}_2\text{-}\ldots\text{-}\mathrm{Z}_k\mathrm{h}_k\text{-DSPACE}(K_0, K_1, K_2, \ldots)$ as the number of worktape squares of M visited by one of the heads or holding a nonblank symbol in at least one of the situations $K_0, K_1, K_2, \ldots$ For nonhalting situations this measure remains undefined. Now we define another type of space measure, which disregards the tape squares used on certain specified worktapes, which are described as auxiliary tapes. Let $m \leqq k$ and $Z_2, Z_3, \ldots, Z_m \in \{S, NES, CS, PD, C\}$. We define Z_1h_1-auxZ_2h_2-auxZ_3h_3-…-auxZ_mh_m-$Z_{m+1}h_{m+1}$-…-Z_kh_k-DSPACE$(K_0, K_1, K_2, \ldots)$ as the number of squares of the nonauxiliary worktapes of M which are visited by one of the heads or which hold a nonblank symbol in at least one of the situations $K_0, K_1, K_2, \ldots$ For nonhalting computations this measure remains undefined.

For a halting computation $K_0, K_1, K_2, \ldots$ of a DIA^d $(d \geqq 1)$ we define IA^d-DSPACE$(K_0, K_1, K_2, \ldots)$ as the number of tape squares which hold a nonblank symbol in at least one of the situations $K_0, K_1, K_2, \ldots$ For nonhalting computations this measure remains undefined. Let DXZ be any type of deterministic RAM (i.e. $X \in \{S, e, B, M, C\}$ and $Z \in \{RAM, RAM'\}$). Further, let $K_0, K_1, K_2, \ldots$ be a halting computation of a DXZ. By $\mathbf{R}_{j,t}$ we denote the contents of the register j in the situation K_t. We define

$$\text{XZ-DSPACE}^1(K_0, K_1, K_2, \ldots) = \max_{t\in\mathbb{N}} \max_{j\in\mathbb{N}} |\mathbf{R}_{j,t}|,$$

$$\text{XZ-DSPACE}^2(K_0, K_1, K_2, \ldots) = \max_{t\in\mathbb{N}} \sum_{j=0}^{\infty} |\mathbf{R}_{j,t}|,$$

$$\text{XZ-DSPACE}^3(K_0, K_1, K_2, \ldots) = \sum_{j=0}^{\infty} \max_{t\in\mathbb{N}} |\mathbf{R}_{j,t}|$$

and

$$\text{XZ-DSPACE}(K_0, K_1, K_2, \ldots) = \sum_{j=0}^{\infty} \left(\max_{t\in\mathbb{N}} |\mathbf{R}_{j,t}| + |\text{bin}\, j| \cdot \text{sgn} \max_{t\in\mathbb{N}} |\mathbf{R}_{j,t}|\right).$$

For nonhalting computations these measures remain undefined.

A machine is called *s-space-bounded* if it works within space s. The next theorem shows that the space measures of various types of machines are linearly related. Two measures Φ and Ψ are *linearly related* if $\Phi(\text{Lin}\, t) = \Psi(\text{Lin}\, t)$ for all $t \geqq \text{id}$.

5.2. Theorem. The following complexity classes coincide for $s \geqq \text{id}$:

1. **i-T-DSPACE**(Lin s), for $i = 0, 1, 2$,
2. **1-2PD-DSPACE**(Lin s),
3. **multiT^{multi}multi-DSPACE**(Lin s),
4. **IA^d-DSPACE**(Lin s), for $d \geqq 1$,
5. **XRAM-DSPACE**(Lin s), for $X \in \{S, e, B, M, C\}$,
6. **XRAM′-DSPACE**(Lin s), for $X \in \{S, e, B, M, C\}$.

Proof. It is sufficient to prove the inclusions represented in Figure 5.2, in which the term **-DSPACE**(Lin s) is omitted. We restrict ourselves to the inclusions numbered by 1, 2, …, 6. We give only the idea of how the simulating machines work. The details of their construction are left to the reader.

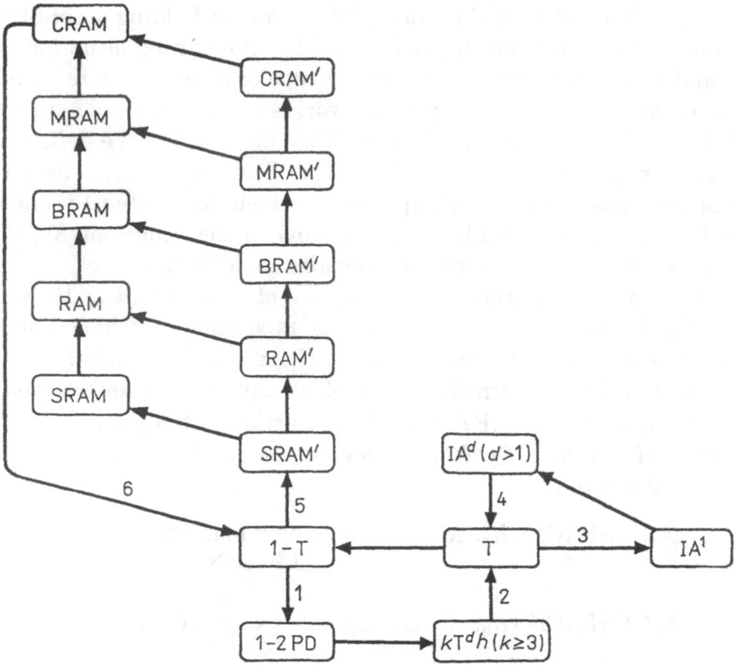

Figure 5.2

Ad 1. This inclusion can be shown in the same way as inclusion 1 in the proof of Theorem 5.1.

Ad 2. [Hem 79a] Let M be a kT^{d}h-DM which works within space s and, without loss of generality, whose heads do not print the blank. Consequently, at every moment of the work of M on an input of length n, we have for each tape the following property: the set of all tape squares not holding the blank is connected. (A set S of tape squares is said to be *connected* if for arbitrary $s, s' \in S$ there are $s_0, s_1, \ldots, s_r \in S$ $(r \geqq 0)$ such that $s = s_0$, $s' = s_r$ and s_{i-1} and s_i are neighbouring for $i = 1, \ldots, r$. Two tape squares $(n_1, \ldots, n_d)$ and $(n'_1, \ldots, n'_d)$ are said to be *neighbouring* if $\sum_{k=1}^{d} |n_k - n'_k| = 1$). There exists a simple algorithm which finds for a given connected set S of squares of a d-dimensional tape an *exhausting path*, i.e. a sequence $s_1, s_2, \ldots, s_r \in S$ such that

a) s_i and s_{i+1} are neighbouring, for $i = 1, \ldots, r - 1$,

b) $S = \{s_1, \ldots, s_r\}$,

c) $\text{card}\,\{i\colon 1 \leqq i \leqq r \wedge s_i = s\} \leqq 2d$, for all $s \in S$.

We describe such an algorithm, which determines in step i the tape square s_i:

Step 1. Choose $s_1 \in S$ arbitrarily.

Step $i + 1$. **If** there are $s \in S \setminus \{s_1, \ldots, s_i\}$ which are neighbouring to s_i
then choose s_{i+1} to be one of them. Go to step $i + 2$
else if $s_i = s_1$ **then** stop.
else set $s_{i+1} = s_{j-1}$ where j is the smallest index such that $s_j = s_i$. Go to step $i + 2$.

Let $f: \mathbb{G}^d \mapsto \Delta$ be the inscription of a tape of M at a given moment of its work and let the set S_f of all nonempty squares be represented by

$$f(s_1)\,(s_2 - s_1)\,f(s_2)\,(s_3 - s_2) \ldots (s_r - s_{r-1})\,f(s_r)$$

on the tape of a DTM M', where $s_1, s_2, \ldots, s_r$ is an exhausting path for S_f. Since card $S_f \leqq s(n)$, this takes space $\lesssim s(n)$. Now M' has enough information to find, with the help of the algorithm, a corresponding representation of the next tape inscription of M.

For the implementation of this algorithm, M' needs d counters of length $\lesssim s(n)$. In such a manner M' simulates the work of M within space $s(n)$.

Ad 3. This inclusion can be shown in the same way as inclusion 4 in the proof of Theorem 5.1.

Ad 4. This inclusion can be shown in the same way as inclusion 2 in the present proof.

Ad 5. We proceed as for inclusion 7 in the proof of Theorem 5.1, where the changing of the registers 0, 1 and 2 can be performed by the following program part:

1	$\mathbf{R}_3 \leftarrow \mathbf{R}_0$	9	$\mathbf{R}_3 \leftarrow \mathbf{R}_2$
2	$\mathbf{R}_0 \leftarrow 0$	10	$\mathbf{R}_2 \leftarrow 0$
3	IF ($\mathbf{R}_3 \leqq 1$)THEN GOTO 8	11	IF ($\mathbf{R}_3 = 0$)THEN GOTO 16
4	$\mathbf{R}_3 \leftarrow \mathbf{R}_3 \dot{-} 1$	12	$\mathbf{R}_3 \leftarrow \mathbf{R}_3 \dot{-} 1$
5	$\mathbf{R}_3 \leftarrow \mathbf{R}_3 \dot{-} 1$	13	$\mathbf{R}_2 \leftarrow \mathbf{R}_2 + 1$
6	$\mathbf{R}_0 \leftarrow \mathbf{R}_0 + 1$	14	$\mathbf{R}_2 \leftarrow \mathbf{R}_2 + 1$
7	GOTO 3	15	GOTO 11
8	$\mathbf{R}_1 \leftarrow \mathbf{R}_3$	16	$\mathbf{R}_2 \leftarrow \mathbf{R}_2 + a'$

In the second case we proceed analogously.

Ad 6. We use the simulation described in the proof of inclusion 8 of Theorem 5.1. The description of all pairs is of length

$$\sum_{|\mathbf{R}_j| > 0} (|\text{bin } j| + |\mathbf{R}_j|) = \sum_{j=0}^{\infty} (|\mathbf{R}_j| + |\text{bin } j| \cdot \text{sgn } |\mathbf{R}_j|) \leqq s(n). \blacksquare$$

Remark. Theorem 5.2 also holds true for complexity classes of functions.

Since the space measures of various types of machines are linearly related, we introduce *general space complexity classes* which do not depend on a special type of machine. For $s \geqq 0$ we define

$$\mathbf{DSPACE}(s) = \mathbf{2\text{-}T\text{-}DSPACE}(\mathrm{Lin}\ s)$$

and

$$\mathrm{DSPACE}(s) = \text{2-T-DSPACE}(\mathrm{Lin}\ s).$$

Remark. For all bounding functions $s > 0$ we have

$$\textbf{1-T-DSPACE}(\mathrm{Lin}\ s) = \textbf{1-multiT}^{\textbf{multi}}\textbf{multi-DSPACE}(\mathrm{Lin}\ s)$$

and

$$\textbf{2-T-DSPACE}(\mathrm{Lin}\ s) = \textbf{2-multiT}^{\textbf{multi}}\textbf{multi-DSPACE}(\mathrm{Lin}\ s).$$

The nontrivial inclusions can be proved in the same way as inclusion 2 in the proof of 5.2. However, **1-T-DSPACE**(Lin s) $\subset$ **2-T-DSPACE**(Lin s) for $\log\log \lesssim s <_{\mathrm{io}} \mathrm{id}$, because there is a language which is in **2-T-DSPACE**($\log\log n$) but not in **1-T-DSPACE**(s) for $s <_{\mathrm{io}} \mathrm{id}$ (cf. [Hart 70]). The set

$$\{ww \# \mathrm{bin}1 \# \mathrm{bin}2 \# \ldots \# \mathrm{bin}2^m \colon m \in \mathbb{N} \wedge |w| = m \wedge w \in \{0, 1\}^*\}$$

can serve as a witness. (The lower bound can be shown by the method of counting partial situations as described in § 8.1.1.) Since the measure 2-T-DSPACE seems to be more "flexible" we choose it to define the general space measure.

A space measure which is linearly related to the measure 2-T-DSPACE is said to be a *realistic space measure.*

Now we introduce notations for frequently used general space complexity classes.

L	= **DSPACE**(log)	and	L	= DSPACE(log),
DLINSPACE	= **DSPACE**(Lin)	and	DLINSPACE	= DSPACE(Lin),
PSPACE	= **DSPACE**(Pol)	and	PSPACE	= DSPACE(Pol),
EXPSPACE	= **DSPACE**(2^{Lin})	and	EXPSPACE	= DSPACE(2^{Lin}).

Remark. For the space measures for RAM's not mentioned in Theorem 5.2, we have the following results for $s \geqq \mathrm{id}$,

$$\textbf{XRAM'-DSPACE}^{j}(\mathrm{Lin}\ s) = \textbf{DSPACE}(s), \text{ for } X \in \{S, e, B\} \text{ and } j = 1, 2, 3$$
(see also [SlEm 83]),

$$\textbf{XRAM-DSPACE}^{1}(\mathrm{Lin}\ s) = \textbf{DSPACE}(2^{\mathrm{Lin}\ s}), \text{ for } X \in \{S, e, B\},$$

$$\textbf{XRAM-DSPACE}^{2}(\mathrm{Lin}\ s) = \textbf{DSPACE}(s^2), \text{ for } X \in \{S, e, B\}.$$

Since an analogue to the theorems 5.1 and 5.2 can also be proved for space-time double complexity classes, we introduce *general* "machine-independent" *space-time double complexity classes.* For $t \geqq \mathrm{id}$ and $s \geqq 0$ we define

$$\textbf{DSPACE-TIME}(s, \mathrm{Pol}\ t) = \textbf{2-T-DSPACE-TIME}(\mathrm{Lin}\ s, \mathrm{Pol}\ t)$$

and

$$\text{DSPACE-TIME}(s, \mathrm{Pol}\ t) = \text{2-T-DSPACE-TIME}(\mathrm{Lin}\ s, \mathrm{Pol}\ t).$$

5.1.1.3. *Further Measures*

1. Reversal measures. Let Z_1-Z_2-…-Z_k-DM be any type of deterministic Turing machine (i.e. $Z_1 \in \{1, 1^*, 2, T\}$ and $Z_2, \ldots, Z_k \in \{T, S, NES, CS, PD, C\}$). Let $K_0, K_1, K_2, \ldots$ be a halting computation of a Z_1-Z_2-…-Z_k-DM. By $\mathrm{rev}_m(K_0, K_1, K_2, \ldots)$ we denote the number of changes of the direction of the head moves on the mth tape ($m = 1, \ldots, k$).

$$Z_1\text{-}Z_2\text{-}\ldots\text{-}Z_k\text{-DREV}(K_0, K_1, K_2, \ldots) = \max_{m=1,\ldots,k} \text{rev}_m(K_0, K_1, K_2, \ldots);$$

for nonhalting computations this measure remains undefined.

2. Crossing measures. We restrict ourselves to Turing machines of type i-DTM ($i = 0, 1, 2$). Let $K_0, K_1, K_2, \ldots$ be a halting computation of an i-DTM. By $\text{cros}_j(K_0, K_1, K_2, \ldots)$ we denote the number of crossings of the boundary between the tape squares j and $j + 1$ ($j \in \mathbb{G}$), i.e. the number of t's such that $n_t = j$, $n_{t+1} = j + 1$ or $n_t = j + 1$ and $n_{t+1} = j$, where n_t denotes the position of the worktape head in the situation K_t. We define

$$\text{i-T-DCROS}(K_0, K_1, K_2, \ldots) = \max_{j \in \mathbb{G}} \text{cros}_j(K_0, K_1, K_2, \ldots);$$

for nonhalting computations this measure remains undefined.

3. Return measures. We restrict ourselves to Turing machines of type i-DTM ($i = 0, 1, 2$). Let $K_0, K_1, K_2, \ldots$ be a halting computation of an i-DTM. By $\text{ret}_j(K_0, K_1, K_2, \ldots)$ we denote the number of visits of the worktape head to the jth tape square after the first alteration of its contents (including the visit of the first alteration), i.e. the number of t's such that $n_t = j$ and $f_{t'}(j) \neq f_{t'+1}(j)$ for some $t' \leqq t$, where n_t denotes the position of the worktape head in the situation K_t, and $f_t(j)$ is the contents of the jth tape square in the situation K_t. We define

$$\text{i-T-DRET}(K_0, K_1, K_2, \ldots) = \max_{j \in \mathbb{G}} \text{ret}_j(K_0, K_1, K_2, \ldots);$$

for nonhalting computations this measure remains undefined.

4. Dual return measures. We restrict ourselves to Turing machines of type i-DTM ($i = 0, 1, 2$). Let $K_0, K_1, K_2, \ldots$ be a halting computation of an i-DTM. By $\text{dur}_j(K_0, K_1, K_2, \ldots)$ we denote the number of visits of the worktape head to the jth tape square before the final alteration on its contents (including the visit of the final alteration), i.e. the number of t's such that $n_t = j$ and $f_{t'}(j) \neq f_{t'+1}(j)$ for some $t' \geqq t$, where n_t denotes the position of the worktape head in the situation K_t. We define

$$\text{i-T-DDUR}(K_0, K_1, K_2, \ldots) = \max_{j \in \mathbb{G}} \text{dur}_j(K_0, K_1, K_2, \ldots);$$

for nonhalting computations this measure remains undefined.

Finally we state the following evident fact.

5.3. Lemma. Let Φ be any complexity measure defined in § 5.1.1. For $L \in \mathbf{REC}$ and $t \in \mathbb{R}_1$ we have $L \in \Phi(t) \leftrightarrow c_L \in \Phi(t)$. ∎

5.1.2. Measures for Nondeterministic Machines

We start with the observation that all complexity measures defined in § 5.1.1 for computations of deterministic machines also apply for computations of nondeterministic machines. Since a nondeterministic machine can have several computations on a given input w we define

$$\hat{\Phi}_M(w) = \begin{cases} \min\{\Phi(\gamma): \gamma \text{ is an accepting computation of } M \text{ on } w\} & \text{if } w \in \mathrm{L}(M), \\ 0 \quad \text{if} \quad w \notin \mathrm{L}(M), \end{cases}$$

$$\hat{\Phi}_M(n) = \max_{|w|=n} \hat{\Phi}_M(w),$$

where the denotation of $\hat{\Phi}$ is derived from the denotation of Φ by the replacement of "D" by "N". For example, for $\Phi = $ T-DTIME we get $\hat{\Phi} = $ T-NTIME.

Let d be the maximum number of next situations for a given situation of M. Then a computation of M on w can be described by an infinite sequence $\pi \in \{1, \ldots, \mathrm{d}\}^\omega$, and a halting computation can be described by a finite sequence $p \in \{1, \ldots, \mathrm{d}\}^*$. We define

$$\Phi_M(w \mid \pi) = \begin{cases} \Phi(\gamma) \text{ if } \gamma \text{ is that computation of } M \text{ on } w \text{ which is described} \\ \quad \text{by } \pi \text{ and if } \gamma \text{ is accepting,} \\ \text{undefined otherwise,} \end{cases}$$

$$\Phi_M(w \mid p) = \begin{cases} \Phi(w \mid p1^\omega) \text{ if } p \text{ describes an accepting computation of } M \\ \quad \text{on } w, \\ \text{undefined otherwise.} \end{cases}$$

Now let Φ be a measure for some type τ of nondeterministic machine. For $t \in \mathbb{R}_1$ we define

$$\Phi(t) = \{\mathrm{L}(M): M \text{ is of type } \tau,\ \mathrm{L}(M) \subseteq \Sigma^* \text{ and } \bigwedge_{w \in \mathrm{L}(M)}^{\mathrm{ae}} \Phi_M(w) \leqq t(|w|)\}.$$

If Φ and Φ' are two measures for the same type τ of nondeterministic machine, then we define for $t_1, t_2 \in \mathbb{R}_1$,

$$\Phi\text{-}\Phi'(t_1, t_2) = \Big\{\mathrm{L}(M): M \text{ is of type } \tau,\ \mathrm{L}(M) \subseteq \Sigma^*, \bigwedge_{w \in \mathrm{L}(M)}^{\mathrm{ae}} \big(\Phi_M(w) \leqq t_1(|w|) \wedge \Phi'_M(w) \leqq t_2(|w|)\big)\Big\}.$$

The definitions of $L \in \Phi(t)$ for $L \subseteq \mathbb{N}$ and of $\Phi(\mathfrak{R})$ for $\mathfrak{R} \subseteq \mathbb{R}_1$ are made in the same way as for measures for deterministic machines.

Since, as in the deterministic case, the time (space) measures of various types of nondeterministic machines are polynomially (linearly) related we can introduce general "machine-independent" complexity classes. For $t \geqq \mathrm{id}$ and $s \geqq 0$ we define

$$\mathbf{NTIME}(\mathrm{Pol}\, t) = \mathbf{T\text{-}NTIME}(\mathrm{Pol}\, t),$$

$$\mathbf{NSPACE}(s) = \mathbf{2\text{-}T\text{-}NSPACE}(\mathrm{Lin}\, s),$$

and

$$\mathbf{NSPACE\text{-}TIME}(s, \mathrm{Pol}\, t) = \mathbf{2\text{-}T\text{-}NSPACE\text{-}TIME}(\mathrm{Lin}\, s, \mathrm{Pol}\, t).$$

Finally, we introduce notations for some frequently used complexity classes:

$$\mathbf{Q} = \mathbf{multiT\text{-}NTIME}(\mathrm{id}),$$

$$\mathbf{RBQ} = \mathbf{multiT\text{-}NTIME\text{-}REV}(\mathrm{id}, \mathrm{Const}),$$

$$\mathbf{NP} = \mathbf{NTIME}(\mathrm{Pol}),$$

$$\mathbf{NEXPTIME} = \mathbf{NTIME}(2^{\mathrm{Lin}}),$$

$$\mathbf{NL} = \mathbf{NSPACE}(\log),$$

and

$$\mathbf{NLINSPACE} = \mathbf{NSPACE}(\mathrm{Lin}).$$

Now we deal with measures for alternating machines. Let Φ be any of the particular complexity measures defined in § 5.1.1 for computations of deterministic machines of type τ, and let M be a nondeterministic machine of type τ. We define for an input w,

$$\hat{\Phi}_M(w) = \min \{\max \{\Phi(\gamma) \colon \gamma \text{ is a path leading through a leaf of } \beta\} \colon \beta \text{ is an accepting computation tree of } M \text{ on } w\}$$

and for $t \in \mathbb{R}_1$,

$$\hat{\Phi}(t) = \{\mathrm{L}^{\mathrm{A}}(M) \colon M \text{ is of type } \tau \wedge \bigwedge_{w \in \mathrm{L}^{\mathrm{A}}(M)}^{\mathrm{ae}} \hat{\Phi}_M(w) \leqq_{\mathrm{ae}} t(|w|)\},$$

where the denotation of $\hat{\Phi}$ is derived from the denotation of Φ by the replacement of "D" by "A".

Finally we define measures for probabilistic machines. Let Φ be any of the particular complexity measures defined in § 5.1.1 for computations of deterministic machines of type τ, let M be a nondeterministic machine of type τ deciding some language with error probability ε $(0 < \varepsilon \leqq 1/2)$. We define for an input w

$$\hat{\Phi}_M(w) = \min \{t \colon p(\{\gamma \colon \gamma \text{ accepting computation of } M \text{ on } w \wedge \Phi(\gamma) \leqq t\}) > 1 - \varepsilon$$
$$\vee\, p(\{\gamma \colon \gamma \text{ rejecting computation of } M \text{ on } w \wedge \Phi(\gamma) \leqq t\}) > 1 - \varepsilon\}$$

and for $t \in \mathbb{R}_1$

$$\hat{\Phi}(t) = \{\mathrm{L}^{\varepsilon}(M) \colon M \text{ is of type } \tau \text{ and } \hat{\Phi}_M(w) \leqq_{\mathrm{ae}} t(|w|)\},$$

where the denotation of $\hat{\Phi}$ is derived from the denotation of Φ by the replacement of "D" by "R^ε". Furthermore, R stands for $\mathrm{R}^{1/2}$.

Let furthermore M be an arbitrary nondeterministic machine of type τ. We define for an input w,

$$\hat{\Phi}'_M(w) = \min \{t \colon p(\{\gamma \colon \gamma \text{ accepting computation of } M \text{ on } w \wedge \Phi(\gamma) \leqq t(|w|)\}) > 1 - \varepsilon\}$$

and for $t \in \mathbb{R}_1$,

$$\hat{\Phi}'(t) = \{L^e(M) : M \text{ is of type } \tau \text{ and } \hat{\Phi}'_M(w) \leqq_{ae} t(|w|)\},$$

where $\hat{\Phi}$ is defined as above.

5.1.3. Honest Functions

In many theorems (in particular in simulation theorems and hierarchy theorems) we shall need the assumption that bounding functions of complexity classes are in a sense "honest", i.e. that they have a small complexity relative to the size of their values. For example, $2^n \in$ multiT-DTIME$(2^n) \setminus$ multiT-DTIME$(2^n - 1)$ (since bin $2^n = \underbrace{10\ldots0}_{n\,\text{times}}$ must be computed from bin n and since $\max\limits_{|\text{bin}\, n| = m} |\text{bin}\, 2^n| = 2^m$, every machine computing the function 2^n needs at least 2^m steps on inputs of length m; but this amount of time is also sufficient), i.e. the absolute time complexity of 2^n is rather high. However, relative to the size of its values, the function 2^n has a small complexity.

There are several notions which make precise the intuitive idea of "honesty" of a function. Let τ be any type of deterministic machine and let Φ be a complexity measure for machines of type τ. A function $t: \mathbb{N} \mapsto \mathbb{N}$ is said to be

- *Φ-computable* if $t(|w|) \in \Phi(t)$,
- *Φ-constructible* if there is a machine M of type τ such that $\Phi_M(n) = t(n)$ for all $n \in \mathbb{N}$,
- *fully Φ-constructible* if there is a machine M of type τ such that $\Phi_M(w) = t(|w|)$ for all $w \in \Sigma^*$.

First we explain the difference between the properties "$t \in \Phi(t)$" and "$t(|w|) \in \Phi(t)$". If $t \in \Phi(t)$, then the Φ-complexity $t(|\text{bin}\, n|)$ is allowed for the computation of $t(n)$ (cf. the definition of "$f \in \Phi(t)$" for number-theoretic functions f on p. 61). On the other hand, $t(|w|)$ is a usual word function and therefore $t(|w|) \in \Phi(t)$ means that the Φ-complexity $t(n)$ is allowed for the computation of $t(n) = t(|w|)$ for $|w| = n$. The condition "$t(|w|) \in \Phi(t)$" can be reformulated as: the function $t': 1^* \mapsto \mathbb{N}$ defined by $t'(1^n) = t(n)$ belongs to $\Phi(t)$. Consequently, for increasing t, the condition "$t \in \Phi(t)$" is stronger than the condition "$t(|w|) \in \Phi(t)$". For example, $2^{|w|} \in$ multiT-DTIME$(n + 1)$ while we have seen that $2^n \notin$ multiT-DTIME$(2^n - 1)$.

Now we compare the three notions of "honesty" for some time and space measures. The following two theorems are not hard to prove.

5.4. Theorem. Let $Z \in \{T, 2T, 3T, \ldots, \text{multiT}\}$ and $t \geqq \text{id}$.

1. If t is fully Z-DTIME-constructible, then t is Z-DTIME-constructible.
2. If t is Z-DTIME-computable, then there exists a fully Z-DTIME-constructible function $t' \asymp t$.
3. If t is fully Z-DTIME-constructible, then $t(|w|) \in$ Z-DTIME$(t \cdot \log t)$. Moreover, if t is fully multiT-DTIME-constructible, then $t(|w|) \in$ multiT-DTIME(Lin t). ∎

5.5. Theorem.

1. If s is fully 2-T-DSPACE-constructible, then s is 2-T-DSPACE-constructible, for $s \geqq 0$.
2. If s is 2-T-DSPACE-constructible, then s is fully 2-T-DSPACE-constructible, for $s \geqq \text{id}$.
3. A function s is fully 2-T-DSPACE-constructible if and only if s is 2-T-DSPACE-computable, for $s \geqq 0$. ∎

5.6. Theorem.

1. [StHaLe 65] If s is 2-T-DSPACE-constructible, then $s \asymp 1$ or $s \succeq_{\text{io}} \log\log$ (see also 8.28).
2. [Sei 76] If s is fully 2-T-DSPACE-constructible, then $s \preceq_{\text{io}} 1$ or $s \succeq \log$.
3. [HaBe 75], [AlMe 75], [Sei 76] There are fully 2-T-DSPACE-constructible functions s such that $1 \prec_{\text{io}} s \preceq \log\log$. ∎

For every Blum measure Φ (see § 5.2), there is an $h \in \mathbb{R}_2$ such that every Φ-computable (Φ-constructible) function is *h-honest* in the following sense: f is called h-honest with respect to Φ if there exists a $\varphi_i = f$ such that $\Phi_i \leqq_{\text{ae}} h \square f$.

5.1.4. Simple Relationships between Specific Measures

In this subsection we deal with some simple relationships between measures for machines of type DTM, 1-DTM and 2-DTM. First we observe that deterministic complexity classes are included in the nondeterministic complexity classes with the same bounding function.

5.7. Theorem. For $i = 0, 1, 2$, Z ∈ {TIME, SPACE, REV, CROS, RET, DUR} and $t \in \mathbb{R}_1$

$$\textbf{i-T-DZ}(t) \subseteq \textbf{i-T-NZ}(t). \ ∎$$

5.8. Corollary.

1. **DTIME**(Pol t) ⊆ **NTIME**(Pol t), for $t \geqq \text{id}$,
2. **DSPACE**(s) ⊆ **NSPACE**(s), for $s \geqq 0$. ∎

Next we present some simple relationships between time and space measures.

5.9. Theorem. For X ∈ {D, N},

1. **i-T-XTIME**(t) ⊆ **i-T-XSPACE**(Lin t) for $i = 0, 1, 2$ and $t \geqq \text{id}$.
2. **i-T-XSPACE**(s) ⊆ **i-T-XSPACE-TIME**(s, $2^{\text{Lin}\,s}$) ⊆ **i-T-XTIME**($2^{\text{Lin}\,s}$), for ($i = 0$ and $s \geqq \text{id}$) or ($i = 1, 2$ and $s \geqq \log$).

Proof. Ad 1. This statement is obvious because within $t(n)$ steps of an i-DTM at most $t(n)$ tape squares can be scanned by the worktape head.

Ad 2. Let M be a DTM deciding a language L within space s. Hence M can be in at most $q \cdot k^{s(n)} \cdot s(n) \leqq 2^{c \cdot s(n)}$ different situations on an input of length n, where q is the number of states, k is the number of tape symbols, $k^{s(n)}$ is the number of possible tape inscriptions of length $s(n)$, and $s(n)$ is the number of possible head positions. Since in a halting computation no nonfinal situation can appear twice, M stops within $2^{c \cdot s(n)}$ steps. If M is a 1-DTM or a 2-DTM, then the number of possible situations is $q \cdot k^{s(n)} \cdot s(n) \cdot n$ where the factor n represents the number of possible positions of the input head.

For i-NTM's we argue as in the deterministic case. It is true that in a nondeterministic halting computation a nonfinal situation can appear twice, but one gets an equivalent (with respect to acceptance) computation if the part between two occurences of a nonfinal situation is cut out. ∎

5.10. Corollary. For $X \in \{D, N\}$,

1. **XTIME**(Pol t) $\subseteq$ **XSPACE**(Pol t), for $t \geqq$ id.
2. **XSPACE**(s) $\subseteq$ **XSPACE-TIME**(s, $2^{\text{Lin}\, s}$) $\subseteq$ **XTIME**($2^{\text{Lin}\, s}$), for $s \geqq$ log. ∎

Remark. The statements 5.9 and 5.10 (X = D) are also true for complexity classes of functions.

The following theorem shows how nondeterministic computations can be converted into deterministic computations.

5.11. Theorem.

1. For $i = 0, 1, 2$ and $t \geqq$ id: if $t(|w|) \in$ i-T-DSPACE(t), then

   i-T-NTIME(t) $\subseteq$ **i-T-DSPACE**(Lin t).

2. For ($i = 0$ and $s \geqq$ id) or ($i = 1, 2$ and $s \geqq$ log): if $s(|w|) \in$ i-T-DTIME($2^{\text{Lin}\, s}$), then

   i-T-NSPACE(s) $\subseteq$ **i-T-DTIME**($2^{\text{Lin}\, s}$).

3. For $i = 0, 1, 2$ and $t \geqq$ id: if $t(|w|) \in$ i-T-DTIME($2^{\text{Lin}\, t}$), then

   i-T-NTIME(t) $\subseteq$ **i-T-DTIME**($2^{\text{Lin}\, t}$).

Proof. Ad 1. Let M be an i-NTM working within time t. An i-DTM M' can simulate the work of M on an input w as follows: first M' computes $t(|w|)$ within space $t(|w|)$ and marks $t(|w|)$ tape squares. Then M' simulates $t(|w|)$ steps of every computation of M on w. This can be done by successively following all paths of length $t(|w|)$ (for $i = 1$ the input is copied onto the worktape of M') which can be stored within the marked space. The simulation itself can also be done within space $\lesssim t(|w|)$ since each of the first $t(|w|)$ situations of a computation of M on w has tape inscriptions of maximum length $t(|w|) + |w|$. If M accepts w, then M' finds an accepting situation of M in this way and also accepts w.

Ad 2. Let M be an i-NTM working within space s. There are at most $2^{c \cdot s(n)}$ situations using no more than $s(n)$ tape squares (cf. the proof of 5.9.2). An i-DTM M' can simulate the work of M on an input w as follows: First M' computes $s(|w|)$ within time $2^{c_1 \cdot s(|w|)}$ and marks $s(|w|)$ tape squares. Then M' writes down the ID of the initial situation and then for each ID which is already written down, all next ID's which use no more than $s(|w|)$ space and which are not already written down. In such a manner M' continues as long as new ID's can be found. Then M' accepts if and only if an accept-

ing ID is written down. Since at most $2^{c\cdot s(|w|)}$ ID's must be written down and writing down an ID takes M' at most $2^{c_2\cdot s(|w|)}$ steps, M' works within time $2^{c_3\cdot s(|w|)}$.
Ad 3. Similarly to Statement 1. ∎

5.12. Corollary.

1. If $t \geqq \mathrm{id}$ and $t(|w|) \in \mathrm{DSPACE}(\mathrm{Pol}\, t)$, then
$$\mathbf{NTIME}(\mathrm{Pol}\, t) \subseteq \mathbf{DSPACE}(\mathrm{Pol}\, t).$$
2. If $s \geqq \log$ and $s(|w|) \in \mathrm{DTIME}(2^{\mathrm{Lin}\, s})$, then
$$\mathbf{NSPACE}(s) \subseteq \mathbf{DTIME}(2^{\mathrm{Lin}\, s}).$$
3. If $t \geqq \mathrm{id}$ and $t(|w|) \in \mathrm{DTIME}(2^{\mathrm{Pol}\, t})$, then
$$\mathbf{NTIME}(\mathrm{Pol}\, t) \subseteq \mathbf{DTIME}(2^{\mathrm{Pol}\, t}). \;∎$$

From 5.8, 5.10 and 5.12 we obtain

5.13. Corollary.

1. **L** ⊆ **NL** ⊆ **P** ⊆ **NP** ⊆ **PSPACE.**
2. **DLINSPACE** ⊆ **NLINSPACE** ⊆ **DEXPTIME** ⊆ **NEXPTIME** ⊆ **EXPSPACE.**
3. **L** ⊊ **P** ⊊ **PSPACE.**
4. **DLINSPACE** ⊊ **DEXPTIME** ⊊ **EXPSPACE.** ∎

Now we deal with simple relationships between the time, reversal, crossing, return, and dual return measures. By the definition of these measures we immediately get

5.14. Theorem.

1. For $i = 0, 1, 2$, $\mathrm{X} \in \{\mathrm{D}, \mathrm{N}\}$ and $t \geqq \mathrm{id}$, **i-T-XTIME**$(t) \subseteq$ **i-T-XREV**(t).
2. For $i = 0, 1, 2$, $\mathrm{X} \in \{\mathrm{D}, \mathrm{N}\}$ and $t \geqq 0$,
$$\textbf{i-T-XREV}(t) \subseteq \textbf{i-T-XCROS}(t) \subseteq \textbf{i-T-XRET}(t) \subseteq \textbf{i-T-XDUR}(t).$$
3. For $t \geqq \log$, **2-T-DRET**$(t) \cup$ **2-T-DDUR**$(t) \subseteq$ **2-T-DSPACE**$(2^{\mathrm{Lin}\, t})$.

Proof. The statements 1 and 2 are obvious. We show Statement 3. Let M be a 2-DTM deciding some language L such that $\text{2-T-DRET}_M(w) \leqq t(|w|)$ for all w. Let w be an input of length n and let $f_0, f_1, \ldots, f_r \in \mathbb{N}$ be defined as follows: f_{i+1} is the first of the tape squares to the right of f_i (we set $f_0 = 0$) whose contents are altered, $i = 0, 1, \ldots, r - 1$; to the right of tape square f_r no symbol is changed. Further let $c_i = (s_{i,1}, a_{i,1}, s'_{i,1})\,(s_{i,2}, a_{i,2}, s'_{i,2}) \ldots (s_{i,m_i}, a_{i,m_i}, s'_{i,m_i})$ where m_i is the number of visits of the worktape head to tape square f_i, $s_{i,j}$ is the state in which the head enters tape square f_i at its jth visit, $a_{i,j}$ is the symbol read by the head in f_i at its jth visit and $s'_{i,j}$ is the state in which the head leaves tape square f_i at the jth visit. The essential part c'_i of c_i is defined as $(s_{i,k_i}, a_{i,k_i}, s'_{i,k_i}) \ldots (s_{i,m_i}, a_{i,m_i}, s'_{i,m_i})$ where k_i is the smallest number such that $a_{i,k_i-1} \neq a_{i,k_i}$, i.e. c'_i is that part of c_i which is important for the return measure. Consequently, $m_i - k_i + 1 \leqq \text{2-T-DRET}_M(n) \leqq t(n)$. For $i, i' \in \{1, \ldots, r\}$ such that $i \neq i'$ we have $c'_i \neq c'_{i'}$ because otherwise M would periodically enlarge the space used. Since there are at most $2^{d\cdot t(n)}$ different essential parts of length $t(n)$, we obtain $r \leqq 2^{d\cdot t(n)}$ (for suitable $d > 0$). Furthermore, we get $f_{i+1} - f_i \leqq d' \cdot n$ for suitable $d' > 0$ because otherwise M would periodically enlarge the space used. The same can be said for the distance between f_i and the rightmost tape square scanned by the worktape head. Consequently, M uses at most $\lesssim n \cdot 2^{d\cdot t(n)} \leqq 2^{d''\cdot t(n)}$ tape squares to the right of tape square 0. In the same manner we can show that M uses at most $2^{d''\cdot t(n)}$ tape squares to the left of tape square 0. The inclusion **2-T-DDUR**$(t) \subseteq$ **2-T-DSPACE**$(2^{\mathrm{Lin}\, t})$ can be shown in a similar way. ∎

In the next theorem we compare the reversal (crossing, return, dual return) measures for DTM's, 1-DTM's and 2-DTM's.

5.15. Theorem. For X ∈ {D, N} and $t \geqq 0$,

1. **T-XREV**$(t) \subseteq$ **1-T-XREV**$(t) \subseteq$ **2-T-XREV**(t),
2. **T-XCROS**$(t) \subseteq$ **1-T-XCROS**$(t) \subseteq$ **2-T-XCROS**(t),
3. **T-XRET**$(t) \cup$ **1-T-XRET**$(t) \subseteq$ **2-T-XRET**(t),
4. **T-XDUR**$(t) \subseteq$ **1-T-XDUR**$(t) \subseteq$ **2-T-XDUR**(t) ($t \geqq 1$ for the first inclusion).

Proof. The inclusions **1-T-XZ**$(t) \subseteq$ **2-T-XZ**(t) for Z ∈ {REV, CROS, RET, DUR} are obvious.

First we show the inclusions **T-XZ**$(t) \subseteq$ **1-T-XZ**(t) for Z ∈ {REV, CROS, DUR}. For a DTM M we construct an equivalent 1-DTM M' as follows. The worktape heads of M and M' move in the same way and write the same symbols. If the worktape head of M scans the tape square i ($1 \leqq i \leqq n$) for the first time, then the input head of M' also scans tape square i for the first time. Thus M' has at any time at least the same information as M. Consequently $\text{T-XZ}_M = \text{1-T-XZ}_{M'}$ for Z ∈ {REV, CROS}. Moreover, only during the first visit to a tape square it is possible for M' to alter the contents of this square while M does not. Hence, $\text{1-T-XZ}_{M'} = \max\{\text{T-XZ}_M, 1\}$.

Finally we show the inclusion **T-XRET**$(t) \subseteq$ **2-T-XRET**(t). For a DTM M we construct an equivalent 2-DTM M' as follows: The worktape heads of M and M' move in the same way and alter the symbols in the same way. If the worktape head of M scans the tape square i ($1 \leqq i \leqq n$), then the input head of M' also scans the tape square i. Thus M' has at any time the same information as M. Consequently, $\text{T-XRET}_M = \text{2-T-XRET}_{M'}$. ∎

Now we deal with measures for alternating machines. First we note that the time (space) measures for various types of Turing machines are polynomially (linearly) related. The proofs are analogous to those of 5.1 and 5.2.

5.16. Theorem.

1. **T-ATIME**(Pol t) = **multiT**$^{\mathbf{multi}}$**multi-ATIME**(Pol t), for $t \geqq \mathrm{id}$.
2. **T-ASPACE**(Lin s) = **multiT**$^{\mathbf{multi}}$**multi-ASPACE**(Lin s), for $s \geqq \mathrm{id}$. ∎

As in the deterministic and nondeterministic cases we define general "machine-independent" complexity classes for $t \geqq \mathrm{id}$ and $s \geqq 0$:

$$\mathbf{ATIME}(\mathrm{Pol}\, t) = \mathbf{T\text{-}ATIME}(\mathrm{Pol}\, t)$$

and

$$\mathbf{ASPACE}(s) = \mathbf{2\text{-}T\text{-}ASPACE}(\mathrm{Lin}\ s).$$

5.17. Theorem.

1. Let τ-NM be any type of Turing machine. For $t \geqq \mathrm{id}$

$$\tau\text{-}\mathbf{NTIME}(t) \subseteq \tau\text{-}\mathbf{ATIME}(t) \subseteq \tau\text{-}\mathbf{ASPACE}(t).$$

2. For $t \geqq \mathrm{id}$, $\mathbf{NTIME}(\mathrm{Pol}\ t) \subseteq \mathbf{ATIME}(\mathrm{Pol}\ t) \subseteq \mathbf{ASPACE}(\mathrm{Pol}\ t)$. ∎

Finally, we deal with measures for probabilistic Turing machines. Analogously to the alternating case we have

5.18. Theorem. For $0 < \varepsilon \leqq 1/2$, $t \geqq \mathrm{id}$ and $s \geqq 0$,

1. $\mathbf{T\text{-}R^{\varepsilon}TIME}(\mathrm{Pol}\ t) = \mathbf{multiT^{multi}multi\text{-}R^{\varepsilon}TIME}(\mathrm{Pol}\ t)$.
2. $\mathbf{T\text{-}R^{\varepsilon}SPACE}(\mathrm{Lin}\ s) = \mathbf{multiT^{multi}multi\text{-}R^{\varepsilon}SPACE}(\mathrm{Lin}\ s)$. ∎

For $0 < \varepsilon < 1/2$, $t \geqq \mathrm{id}$ and $s \geqq 0$, we define the "machine-independent" complexity classes

$$\mathbf{R^{\varepsilon}TIME}(\mathrm{Pol}\ t) = \mathbf{T\text{-}R^{\varepsilon}TIME}(\mathrm{Pol}\ t),$$
$$\mathbf{RTIME}(\mathrm{Pol}\ t) = \mathbf{T\text{-}RTIME}(\mathrm{Pol}\ t),$$
$$\mathbf{R^{\varepsilon}SPACE}(s) = \mathbf{2\text{-}T\text{-}R^{\varepsilon}SPACE}(\mathrm{Lin}\ s),$$
$$\mathbf{RSPACE}(s) = \mathbf{2\text{-}T\text{-}RSPACE}(\mathrm{Lin}\ s),$$

and similarly for the classes $\mathbf{R^{\varepsilon}TIME'}(\mathrm{Pol}\ t)$ etc.

5.19. Theorem. Let τ-NM be any type of nondeterministic Turing machines, $0 < \varepsilon \leqq 1/2$, $0 < \varepsilon_1 \leqq \varepsilon_2 \leqq 1/2$, $t \geqq \mathrm{id}$ and $s \geqq 0$.

1.1. $\tau\text{-}\mathbf{R^{\varepsilon}TIME}(t) \subseteq \tau\text{-}\mathbf{R^{\varepsilon}SPACE}(t)$.
1.2. $\tau\text{-}\mathbf{R^{\varepsilon_1}TIME}(t) \subseteq \tau\text{-}\mathbf{R^{\varepsilon_2}TIME}(t)$.
1.3. $\tau\text{-}\mathbf{R^{\varepsilon_1}SPACE}(s) \subseteq \tau\text{-}\mathbf{R^{\varepsilon_2}SPACE}(s)$.
1.4. $\tau\text{-}\mathbf{DTIME}(t) \subseteq \tau\text{-}\mathbf{R^{\varepsilon}TIME}(t)$.
1.5. $\tau\text{-}\mathbf{DSPACE}(s) \subseteq \tau\text{-}\mathbf{R^{\varepsilon}SPACE}(s)$.
2.1. $\mathbf{R^{\varepsilon}TIME}(\mathrm{Pol}\ t) \subseteq \mathbf{R^{\varepsilon}SPACE}(\mathrm{Pol}\ t)$.
2.2. $\mathbf{R^{\varepsilon_1}TIME}(\mathrm{Pol}\ t) \subseteq \mathbf{R^{\varepsilon_2}TIME}(\mathrm{Pol}\ t)$.
2.3. $\mathbf{R^{\varepsilon_1}SPACE}(s) \subseteq \mathbf{R^{\varepsilon_2}SPACE}(s)$.
2.4. $\mathbf{DTIME}(\mathrm{Pol}\ t) \subseteq \mathbf{R^{\varepsilon}TIME}(\mathrm{Pol}\ t) \subseteq \mathbf{R^{\varepsilon}TIME'}(\mathrm{Pol}\ t)$.
2.5. $\mathbf{DSPACE}(s) \subseteq \mathbf{R^{\varepsilon}SPACE}(s) \subseteq \mathbf{R^{\varepsilon}SPACE'}(s)$. ∎

5.2. *Abstract Measures of Computational Complexity*

Let $\{M_i : i \in \mathbb{N}\}$ be a Gödel numbering of all DTM's and let $\{\varphi_i : i \in \mathbb{N}\}$ be the corresponding Gödel numbering of $\mathbb{P}$ (i.e. $\varphi_i = \varphi_{M_i}$). Then we have for the complexity measures $\Phi = $ T-DTIME and $\Phi = $ T-DSPACE the following properties:

A1. $\mathrm{D}_{\varphi_i} = \mathrm{D}_{\Phi_i}$ for all $i \in \mathbb{N}$.

A2. $\{(i, w, m): \Phi_i(w) = m\}$ is a recursive set.

It is not quite evident that Property A2 holds true for Φ = T-DSPACE: Let M_i have q states and p worktape symbols. If M_i does not use more than m tape squares, then it can be in at most $q \cdot m \cdot p^m$ different situations. Thus after $q \cdot m \cdot q^m$ steps of the work of M_i on the input w, either at least one situation is repeated or more than m tape squares have been used or M_i has stopped. In the first case and the second case we have T-DSPACE$_i(w) \neq m$, in the third case the tape squares used during the work of M on w can be counted, and the question of T-DSPACE$_i(w) = m$ can be answered.

In the same manner one can show that the properties A1 and A2 hold true for all time and space measures mentioned in Theorem 5.1 and Theorem 5.2.

M. Blum has proposed to use just these two properties te define abstract measures of computational complexity [Blu 67a]. For another approach see [Rab 60].

Let $\varphi: \mathbb{N} \times \Sigma^* \to \Sigma^*$ be the universal function of a Gödel numbering of $\mathbb{P}(\Sigma)$ and let $\Phi: \mathbb{N} \times \Sigma^* \to \mathbb{N}$ be a partial recursive function. Set $\varphi_i(w) = \varphi(i, w)$ and $\Phi_i(w) = \Phi(i, w)$ for all $i \in \mathbb{N}$, $w \in \Sigma^*$. The pair (φ, Φ) is said to be an *abstract measure of computational complexity* or a *Blum measure* if (φ, Φ) satisfies the axioms A1 and A2. The reference to φ can be omitted if this Gödel numbering is understood.

5.20. Proposition. All time and space measures mentioned in Theorem 5.1 and Theorem 5.2 are Blum measures.

The above definition may astonish, since it is not evident that these two axioms actually make precise what we intuitively mean by a measure of computational complexity. Indeed, there are very unnatural, "pathological" measures which satisfy these axioms. Therefore, attempts have been made to use different axiom systems, as for instance in [Hav 71], or to exclude pathological measures by further restrictions (cf. § 16).

Nevertheless these two apparently weak axioms are powerful enough to yield some very strong theorems about measures (cf. § 15, § 17 and § 18). However, it turns out that even certain weakened versions of the axioms A1 and A2 have basically the same power (cf. [Aus 71] and [GiSi 76]).

Let (φ, Φ) be a Blum measure. In the same way as for the particular measures defined in § 5.1.1 we define for $t \in \mathbb{R}_1$ the complexity classes

$$\Phi(t) = \left\{A: \bigvee_i \left(\varphi_i = c_A \wedge \Phi_i(w) \leqq_{\mathrm{ae}} t(|w|)\right)\right\}$$

and

$$\Phi(t) = \{\varphi_i: \varphi_i \in \mathbb{R}_1(\Sigma) \wedge \Phi_i(w) \leqq_{\mathrm{ae}} t(|w|)\}.$$

The function t is said to be the name or the bounding function of these complexity classes.

For the more "theoretical" investigations of Blum measures in §§ 5, 15, 16, 17, 18, 27 we restrict ourselves for the sake of simplicity to the case $\Sigma = \{1\}$. Thus we can identify Σ^* with $\mathbb{N}$. However, all results are also valid for arbitrary finite alphabets Σ.

Note that for $\Sigma = \{1\}$ the complexity classes can be defined as

$$\Phi(t) = \left\{A : \bigvee_i (\varphi_i = c_A \wedge \Phi_i \leqq_{ae} t)\right\}$$

and

$$\Phi(t) = \{\varphi_i : \varphi_i \in \mathbb{R}_1 \wedge \Phi_i \leqq_{ae} t\}.$$

Now we deal with some simple properties of Blum measures. First we show that all Blum measures are recursively related.

5.21. Theorem. Let (φ, Φ) and (ψ, Ψ) be two Blum measures.

1. There exist a strictly increasing $r \in \mathbb{R}_1$ and an $s \in \mathbb{R}_1$ such that for all i,
   a) $\psi_{s(i)} = \varphi_i$,
   b) $\Psi_{s(i)} \leqq_{ae} r \circ \max\{\mathrm{id}, \Phi_i\}$ and $\Phi_i \leqq_{ae} r \circ \max\{\mathrm{id}, \Psi_{s(i)}\}$.*)
2. There exist an $r \in \mathbb{R}_2$ which is strictly increasing in the second argument and an $s \in \mathbb{R}_1$ such that for all i,
   a) $\psi_{s(i)} = \varphi_i$,
   b) $\Psi_{s(i)} \leqq_{ae} r \,\square\, \Phi_i$ and $\Phi_i \leqq_{ae} r \,\square\, \Psi_{s(i)}$.
3. There exist an $r \in \mathbb{R}_3$ which is strictly increasing in the third argument and an $s \in \mathbb{R}_1$ such that for all i,
   a) $\psi_{s(i)} = \varphi_i$,
   b) $\Psi_{s(i)}(x) \leqq r(i, x, \Phi_i(x))$ and $\Phi_i(x) \leqq r(i, x, \Psi_{s(i)}(x))$.

Proof. By Theorem 2.9 there exists an $s \in \mathbb{R}_1$ such that $\psi_{s(i)} = \varphi_i$. We define

$$g'(i, x, m) = \begin{cases} \Psi_{s(i)}(x) & \text{if } \Phi_i(x) = m, \\ 0 & \text{otherwise}, \end{cases}$$

and

$$g''(i, x, m) = \begin{cases} \Phi_i(x) & \text{if } \Psi_{s(i)}(x) = m, \\ 0 & \text{otherwise}. \end{cases}$$

Since it is decidable whether $\Phi_i(x) = m$, and $\Phi_i(x) = m$ implies $x \in D_{\Phi_i} = D_{\varphi_i} = D_{\psi_{s(i)}} = D_{\Psi_{s(i)}}$, the function g' is recursive. Analogously, g'' is recursive. Hence g is recursive where $g(i, x, m) = \max\{g'(i, x, m), g''(i, x, m)\}$. But $\Psi_{s(i)}(x) \leqq g(i, x, \Phi_i(x))$ and $\Phi_i(x) \leqq g(i, x, \Psi_{s(i)}(x))$.

Now we define $r(i, x, 0) = g(i, x, 0)$, $r(i, x, m + 1) = \max\{g(i, x, m + 1), r(i, x, m) + 1\}$ and obtain $\Psi_{s(i)}(x) \leqq_{ae} r(i, x, \Phi_i(x))$ and $\Phi_i(x) \leqq_{ae} r(i, x, \Psi_{s(i)}(x))$ for all i and $r(i, x, m) < r(i, x, m + 1)$ for all i, x, m. Defining $r'(x, m) = \max_{i \leqq x} r(i, x, m)$ we obtain $\Psi_{s(i)}(x) \leqq_{ae} r'(x, \Phi_i(x))$ and $\Phi_i(x) \leqq_{ae} r'(x, \Psi_{s(i)}(x))$ for all i and $r'(x, m) < r'(x, m + 1)$ for all x, m. Finally we set $r''(m) = \max_{x \leqq m} r'(x, m)$ and obtain $\Psi_{s(i)}(x) \leqq_{ae} r''(\Phi_i(x))$ and $\Phi_i(x) \leqq_{ae} r''(\Psi_{s(i)}(x))$ for all i such that $\Phi_i(x) \geqq x$. Furthermore, $r''(m) < r''(m + 1)$. ∎

Statement 5.21.3 can be in a sense converted.

---

*) Here and in what follows we define for $\alpha, \beta \in \mathbb{P}_1$: if $x \notin D_\beta$, then $\alpha(x) \leqq \beta(x)$. If in addition $x \in D_\alpha$, then $\alpha(x) < \beta(x)$.

5.22. Theorem. Let (ψ, Ψ) be a Blum measure, φ a Gödel numbering of $\mathbb{P}_1$ and $\Phi \in \mathbb{P}_2$ such that $D_{\varphi_i} = D_{\Phi_i}$ for all i, where $\Phi_i(x) =_{\text{df}} \Phi(i, x)$. Further, let $s \in \mathbb{R}_1$ be such that $\psi_{s(i)} = \varphi_i$. Then (φ, Φ) is a Blum measure if and only if there is a $g \in \mathbb{R}_3$ such that $\Psi_{s(i)}(x) \leqq g(i, x, \Phi_i(x))$ for all i.

Proof. We have to show that the question "$\Phi_i(x) = m$?" is decidable if and only if there is a $g \in \mathbb{R}_3$ such that $\Psi_{s(i)}(x) \leqq g(i, x, \Phi_i(x))$. The "only if" part is already proved by 5.21.3. Now let $g \in \mathbb{R}_3$ be such that $\Psi_{s(i)}(x) \leqq g(i, x, \Phi_i(x))$. The question "$\Phi_i(x) = m$?" can be decided as follows: if $\Psi_{s(i)} > g(i, x, m)$, then $\Phi_i(x) \neq m$. If $\Psi_{s(i)}(x) \leqq g(i, x, m)$, then $x \in D_{\Psi_{s(i)}} = D_{\psi_{s(i)}} = D_{\varphi_i} = D_{\Phi_i}$. Hence $\Phi_i(x)$ is defined and the question "$\Phi_i(x) = m$?" can easily be answered. ∎

The preceding theorem is a useful method of showing that specific complexity measures are Blum measures. By this theorem, 5.20, 5.14, 5.15 and because of the effectiveness of the proofs of 5.14 and 5.15, we obtain

5.23. Corollary. The measure i-T-DΩ is a Blum measure, for $i = 0, 1, 2$ and $\Omega \in$ {REV, CROS, RET, DUR}. ∎

By the same method as for Theorem 5.21 we can show that every partial recursive function can be uniformly recursively bounded by its complexities.

5.24. Theorem. Let (φ, Φ) be a Blum measure. There is a function $r \in \mathbb{R}_2$ such that $\varphi_i \leqq_{\text{ae}} r \,\square\, \Phi_i$. ∎

On the other hand, it is not possible to find an $r \in \mathbb{R}_2$ such that $\Phi_i \leqq_{\text{ae}} r \,\square\, \varphi_i$, since every recursive function has arbitrarily complex computations. This is shown by Theorem 5.25. The proof idea of Theorem 5.21 does not carry over to this case since $\{(i, x, m) : \varphi_i(x) = m\}$ is not decidable. However, if $s \in \mathbb{R}_1$ is chosen in such a way that $\{(i, x, m) : \varphi_{s(i)}(x) = m\}$ is decidable, then $\Phi_{s(i)} \leqq_{\text{ae}} r \,\square\, \varphi_{s(i)}$ for some $r \in \mathbb{R}_2$. This is stated by Theorem 5.26.

5.25. Theorem. Let (φ, Φ) be a Blum measure. For every $f, g \in \mathbb{R}_1$ there is an $i \in \mathbb{N}$ such that $\varphi_i = f$ and $\Phi_i \geqq_{\text{ae}} g$.

Proof. First we prove the theorem for Φ = 2-T-DTIME. Choose j and k such that $\varphi_j = g$ and $\varphi_k = f$. Let M_i be a machine which works as follows: First M_i works like M_j, but without using the output tape. Since M_j has to write $1^{g(n)}$ on its output tape, this takes M_i at least $g(n)$ steps. Now M_i erases the worktape inscription and works like M_k. Thus $\varphi_i = f$ and 2-T-DTIME$_i \geqq g$.

Now we assume that there is a Blum measure (φ, Φ) and functions $f, g \in \mathbb{R}_1$ such that $\Psi_i <_{\text{io}} g$ for all $\psi_i = f$. Because of the recursive relatedness of all Blum measures (cf. 5.21.1) there are $r, s \in \mathbb{R}_1$ such that r is increasing, $\varphi_i = \psi_{s(i)}$ and 2-T-DTIME$_i$ $\leqq_{\text{ae}} r \circ \max\{\text{id}, \Psi_{s(i)}\}$ for all i. If $\varphi_i = f$, then 2-T-DTIME$_i \leqq_{\text{ae}} r \circ \max\{\text{id}, \Psi_{s(i)}\}$ $\leqq_{\text{io}} r \circ \max\{\text{id}, g\}$. This contradicts the first part of the proof. ∎

Let $s \in \mathbb{R}_1$. The set $\{\varphi_{s(i)} : i \in \mathbb{N}\}$ is said to be *measured* if $\{(i, x, m) : \varphi_{s(i)}(x) = m\}$ is decidable.

5.26. Theorem. [McMe 69], [MoMe 74] Let (φ, Φ) be a Blum measure and $s \in \mathbb{R}_1$. If the set $\mathscr{C} = \{\varphi_{s(i)} : i \in \mathbb{N}\}$ is measured, then there is an $r \in \mathbb{R}_2$ such that $\Phi_{s(i)}$

$\leqq_{ae} r \circ \varphi_{s(i)}$ for all i, i.e. $\mathscr{C}$ consists only of functions which are r-honest with respect to Φ. Conversely, for every $r \in \mathbb{R}_2$, the set of all functions which are r-honest with respect to Φ is measured. ∎

Finally, we are interested in the question of whether the complexity of a combined computation can be recursively bounded by the complexities of the single computations. For example, by the s-m-n-Theorem there exists an $s \in \mathbb{R}_1$ such that $\varphi_{s(i,j)}(x) = \varphi_i(x) + \varphi_j(x)$. Is there an $r \in \mathbb{R}_3$ such that $\Phi_{s(i,j)}(x) \leqq_{ae} r(x, \Phi_i(x), \Phi_j(x))$ for all i and j?

5.27. Theorem. (Combining Lemma). Let (φ, Φ) be a Blum measure and let $s \in \mathbb{R}_2$ such that $D_i \cap D_j \subseteqq D_{s(i,j)}$ for all $i, j \in \mathbb{N}$. Then there is an $r \in \mathbb{R}_3$ which is increasing in its second and third arguments such that $\Phi_{s(i,j)}(x) \leqq_{ae} r(x, \Phi_i(x), \Phi_j(x))$ for all $i, j \in \mathbb{N}$.

Proof. Define

$$g(i, j, x, m, n) = \begin{cases} \Phi_{s(i,j)}(x) & \text{if } \Phi_i(x) \leqq m \text{ and } \Phi_j(x) \leqq n, \\ 0 & \text{otherwise.} \end{cases}$$

Since "$\Phi_i(x) \leqq m$" is decidable and since $\Phi_i(x) \leqq m \wedge \Phi_j(x) \leqq n$ implies $x \in D_i \cap D_j \subseteqq D_{s(i,j)}$, the function g is recursive, and we have $\Phi_{s(i,j)}(x) \leqq g(i, j, x, \Phi_i(x), \Phi_j(x))$. Defining $r(x, m, n) = \max_{i \leqq x} \max_{j \leqq x} g(i, j, x, m, n)$ we obtain $\Phi_{s(i,j)}(x) \leqq_{ae} r(x, \Phi_i(x), \Phi_j(x))$ for all i and j. Obviously, r is increasing in its last two arguments. ∎

5.3. *Complexity Measures of Relative Computations*

All complexity measures and notions introduced in § 5.1.1 and § 5.1.2 for the various types of machine can be introduced also for the corresponding types of oracle machine, where the oracle is referred to in the form of a superscript. For an oracle A we get, for example, the complexity functions Φ_M^A, the complexity classes $\Phi^A(t)$ and $\Phi^A(t)$, the specific measures T-DTIME^A and 2-T-DTIME^A, and the specific complexity classes $\mathbf{P}^A$ and $\mathbf{NP}^A$. For the space measures, however, it turns out that a canonical relativization does not seem to exist (cf. [Lyn 78], [Ang 80a]). We restrict ourselves to the following two possibilities: the measure Z-XSPACE^A, where also the length of the oracle questions is bounded by the bounding function, and the measure Z-XSPACE$^A_*$, where the length of the oracle questions is not restricted. For a more general approach see p. 492. In the nondeterministic case we additionally require for Z-NSPACE^A that all computations stop. Otherwise, it would be possible to accept all r.e. sets within constant space with a relatively simple oracle, as is shown by the following example. The set

$A = \{M \# w \# \gamma : \gamma$ describes an initial part of an accepting computation of M on w which includes an accepting situation$\}$

is obviously in **DLINSPACE** and we get $\mathbf{NSPACE}^A(0) = \mathbf{RE}$.

Analogously to § 5.2 we introduce the notion of an abstract measure of the complexity of relative computations. Let φ: $\mathfrak{P}(\Sigma^*) \times \mathbb{N} \times \Sigma^* \to \Sigma^*$ be the universal function of a Gödel numbering of $\mathbb{P}_1^{()}(\Sigma)$ and let $\Phi: \mathfrak{P}(\Sigma^*) \times \mathbb{N} \times \Sigma^* \to \mathbb{N}$ be a function computable by an oracle TM with set oracle. We define $\varphi_i^A(w) = \varphi^A(i, w) = \varphi(A, i, w)$, $\Phi_i^A(w) = \Phi^A(i, w) = \Phi(A, i, w)$. The pair $(\varphi^{()}, \Phi^{()})$ is said to be a *relative Blum measure* if φ and Φ satisfy the following two axioms:

A1. $D_{\varphi_i^A} = D_{\Phi_i^A}$ for all $i \in \mathbb{N}$ and $A \subseteq \Sigma^*$.

A2. There exists an oracle Turing machine computing for each oracle A the function

$$f^A(i, w, m) = \begin{cases} 1 & \text{if } \Phi_i^A(w) = m, \\ 0 & \text{otherwise}. \end{cases}$$

This definition can be found in [LyMeFi 76]. For other definitions see [Con 73] and [Sym 71].

The Gödel numbering φ can be omitted if it is understood.

5.28. Corollary. Let (φ, Φ) be a relative Blum measure and A a recursive set. Then (φ^A, Φ^A) is a Blum measure. ∎

5.29. Corollary. The measures T-DTIME$^{()}$, T-DSPACE$^{()}$, and T-DSPACE$_*^{()}$ are relative Blum measures. ∎

As for the "classical" Blum measures, we restrict ourselves to the case $\Sigma = \{1\}$ (i.e. $\Sigma^* = \mathbb{N}$) for the more "theoretical" investigations here and in § 27.

The following theorem states that all relative Blum measures are recursively related.

5.30. Theorem. [LyMeFi 76] Let $(\varphi^{()}, \Phi^{()})$ and $(\psi^{()}, \Psi^{()})$ be two relative Blum measures. Then there is an $s \in \mathbb{R}_1$ and an $r \in \mathbb{R}_2$ which is strictly increasing in the second argument such that for all $i \in \mathbb{N}$ and $A \subseteq \mathbb{N}$

a) $\psi_{s(i)}^A = \varphi_i^A$,

b) $\Psi_{s(i)}^A(x) \leqq r(x, \Phi_i^A(x))$ for all $x \geqq i$. ∎

For later use we state

5.31. Theorem. [LyMeFi 76] Let $(\varphi^{()}, \Phi^{()})$ be a relative Blum measure. Then there is an $h \in \mathbb{R}_3$ such that

$$\bigwedge_A \bigwedge_B \bigwedge_i \bigwedge_x \bigwedge_y \big(A \cap \{0, \ldots, h(i, x, y)\} = B \cap \{0, \ldots, h(i, x, y)\} \wedge \Phi_i^A(x) \leqq y$$
$$\to \big(\varphi_i^A(x) = \varphi_i^B(x) \wedge \Phi_i^A(x) = \Phi_i^B(x)\big)\big). \blacksquare$$

§ 6. Complexity-Bounded Reducibilities

Complexity-bounded reducibilities are the main tools for classifying problems according to their complexity (§ 14), for proving large lower bounds (§ 8) and for studying relationships between complexity classes (§§ 23, 24, 26, 28).

6.1. Basic Definitions

Using the operator approach to reducibility (§ 3.3), we study several reducibility notions induced by sets of operators which are complexity-bounded. Since recursive operators are given in the form $\varphi_i^{()}$ we can use the complexity measures for relativized computations as introduced in § 5.3.

6.1.1. Deterministic Reducibilities

For any class $\mathscr{C}$ of recursive operators, the corresponding reducibility notion $\leqq_{\mathscr{C}}$ defined by

$$A \leqq_{\mathscr{C}} B \leftrightarrow \bigvee_{\mathbf{F}\in\mathscr{C}} c_A = \mathbf{F}(c_B)$$

can be restricted by a relative Blum measure $\Phi^{()}$ and a set $K \subseteq \mathbb{R}_1$ of bounding functions as follows:

$$A \leqq_{\mathscr{C}}^{\Phi,K} B \leftrightarrow \bigvee_{\varphi_i{}^0\in\mathscr{C}} \bigvee_{h\in K} \Big(c_A = \varphi_i^B \wedge \bigwedge_x^{\mathrm{ae}} \Phi_i^B(x) \leqq h(|x|)\Big).$$

In the following we give some examples illustrating this general definition.

6.1.1.1. Turing Reducibilities

Let $\mathscr{C}_{\mathrm{T}}$ be the class of all recursive operators (see § 3.2).

Definition.

$\leqq_{\mathrm{T}}^{\mathrm{P}} =_{\mathrm{df}} \leqq_{\mathscr{C}_{\mathrm{T}}}^{\mathrm{DTIME,Pol}}$ [Coo 71a]

$\leqq_{\mathrm{T}}^{\mathrm{PSPACE}} =_{\mathrm{df}} \leqq_{\mathscr{C}_{\mathrm{T}}}^{\mathrm{DSPACE}_*,\mathrm{Pol}}$ (DSPACE$_*^{()}$ has been defined in § 5.3)

$\leqq_{\mathrm{T}}^{\log} =_{\mathrm{df}} \leqq_{\mathscr{C}_{\mathrm{T}}}^{\mathrm{DSPACE}_*,\log}$ [LaLy 76].

In addition, we introduce the following complexity classes of operators:

$$\mathrm{DTIME}^{()}(\mathrm{Pol}) =_{\mathrm{df}} \Big\{\varphi_i^{()} : \varphi_i^{()} \in \mathscr{C}_{\mathrm{T}} \wedge \bigvee_{k\in\mathbb{N}} \bigwedge_X \bigwedge_x^{\mathrm{ae}} \mathrm{DTIME}_i^X(x) \leqq |x|^k\Big\}.$$

$$\mathrm{DSPACE}_*^{()}(\mathrm{Pol}) =_{\mathrm{df}} \Big\{\varphi_i^{()} : \varphi_i^{()} \in \mathscr{C}_{\mathrm{T}} \wedge \bigvee_{k\in\mathbb{N}} \bigwedge_X \bigwedge_x^{\mathrm{ae}} \mathrm{DSPACE}_{*i}^X(x) \leqq |x|^k\Big\}.$$

$$\mathrm{DSPACE}_*^{()}(\log) =_{\mathrm{df}} \Big\{\varphi_i^{()} : \varphi_i^{()} \in \mathscr{C}_{\mathrm{T}} \wedge \bigwedge_X \bigwedge_x^{\mathrm{ae}} \mathrm{DSPACE}_{*i}^X(x) \leqq \log(|x|)\Big\}.$$

6.1. Lemma.

1. $\leqq_{\mathrm{T}}^{\mathrm{P}}, \leqq_{\mathrm{T}}^{\mathrm{PSPACE}}, \leqq_{\mathrm{T}}^{\log}$ are reflexive and transitive.
2. $A \leqq \bar{A}$ for $\leqq \in \{\leqq_{\mathrm{T}}^{\mathrm{P}}, \leqq_{\mathrm{T}}^{\mathrm{PSPACE}}, \leqq_{\mathrm{T}}^{\log}\}$ and all sets A.
3. $\leqq_{\mathrm{T}}^{\mathrm{P}} = \leqq_{\mathrm{DTIME}^{()}(\mathrm{Pol})}$.
4. $\leqq_{\mathrm{T}}^{\mathrm{PSPACE}} = \leqq_{\mathrm{DSPACE}_*^{()}(\mathrm{Pol})}$.

5. $\leqq_{\mathrm{T}}^{\log} = \leqq_{\mathrm{DSPACE}_*^{()}(\log)}$.
6. $\leqq_{\mathrm{T}}^{\log} \subseteq \leqq_{\mathrm{T}}^{\mathrm{P}} \subseteq \leqq_{\mathrm{T}}^{\mathrm{PSPACE}}$.

Statement 3 asserts that $A \leqq_{\mathrm{T}}^{\mathrm{P}} B$ can always be realized by an OM $M_i^{()}$ working within time $p \in \mathrm{Pol}$ not only for the oracle B but also for every oracle X. Statements 4 and 5 have a similar meaning.

Proof. Ad 1. The reflexivity is trivial. The transitivity for $\leqq_{\mathrm{T}}^{\mathrm{P}}$ can easily be verified. For $\leqq_{\mathrm{T}}^{\mathrm{PSPACE}}$ (and similarly for $\leqq_{\mathrm{T}}^{\log}$) the following difficulty arises: Let $c_A = \varphi_i^B$ and $c_B = \varphi_j^C$ and let M_i^B and M_j^C work within spaces p and q resp., $p, q \in \mathrm{Pol}$. We have to take into account the possibility that M_i^B produces queries of length $2^{kp(n)}$ on inputs of length n, and that therefore a direct combination of $M_i^{()}$ and $M_j^{()}$ may not result in a machine which works in polynomial space. We demonstrate how this difficulty is overcome for $\leqq_{\mathrm{T}}^{\mathrm{PSPACE}}$. For $\leqq_{\mathrm{T}}^{\log}$ the same idea is used.

Instead of a direct combination of $M_i^{()}$ and $M_j^{()}$ where $M_j^{()}$ uses the query tape of $M_i^{()}$ as input tape, we construct a combination $M^{()}$ of $M_i^{()}$ and $M_j^{()}$ which reduces A to C in polynomial space in the following manner: Instead of writing down a query of $M_i^{()}$ as in the direct combination of $M_i^{()}$ and $M_j^{()}$, it stores only that symbol of the query which is scanned by the input head of $M_j^{()}$ together with its position. If the input head of $M_j^{()}$ moves, then the new symbol is provided by the resimulation of the work of $M_i^{()}$ from the moment of the previous query up to the moment of printing this new symbol. Hence $M^{()}$ has to store the ID of the moment of the previous query. This, as well as counting up to at most $2^{kp(n)}$, can be done in polynomial space.

Ad 2. This statement is obvious.

Ad 3. Let $c_A = \varphi_i^B$ and let M_i^B work in time p. We construct a machine $M^{()}$ such that M^B computes c_A and M^X works in time q for every X and some $q \in \mathrm{Pol}$.

On input w ($|w| = n$) the machine $M^{()}$ marks within polynomial time $p(n)$ tape squares on an extra tape (the "clock") and the clock head moves to the leftmost square of the marked zone. Then $M^{()}$ works like $M_i^{()}$, and, in addition, at every step the clock head moves one square ahead. $M^{()}$ stops either when $M_i^{()}$ stops and the clock head has not yet left the marked zone (in this case the output $M^{()}$ is that of $M_i^{()}$), or when the clock head leaves the marked zone. In this case M outputs $2 \notin \{0, 1\}$. Obviously, $M^{()}$ satisfies the two properties stated initially.

Ad 4 and 5. The proof is analogous.

Statement 6 is obvious. ∎

6.1.1.2. *Truth Table Reducibilities*

Let $\mathscr{C}_{\mathrm{tt}}$ be the class of general recursive operators defining the tt-reducibility (see Theorem 3.9).

Definition. $\leqq_{\mathrm{tt}}^{\mathrm{P}} =_{\mathrm{df}} \leqq_{\mathscr{C}_{\mathrm{tt}}}^{\mathrm{DTIME,Pol}}$, $\leqq_{\mathrm{tt}}^{\log} =_{\mathrm{df}} \leqq_{\mathscr{C}_{\mathrm{tt}}}^{\mathrm{DSPACE*,log}}$.

Further complexity-restricted tt-reducibilities can be found in [LaLySe 75].

6.2. Lemma. [LaLySe 75]

1. $A \leqq_{tt}^{P} B \leftrightarrow \bigvee_{f \in P} \bigwedge_{x} (x \in A \leftrightarrow B$ satisfies the tt-condition $f(x))$.
2. $A \leqq_{tt}^{\log} B \leftrightarrow \bigvee_{f \in L} \bigwedge_{x} (x \in A \leftrightarrow B$ satisfies the tt-condition $f(x))$.
3. $\leqq_{tt}^{P}$, $\leqq_{tt}^{\log}$ are reflexive and transitive.
4. $A \leqq_{tt}^{P} \bar{A}$, $A \leqq_{tt}^{\log} \bar{A}$ for all A.
5. $\leqq_{tt}^{P} \subseteq \leqq_{T}^{P}$, $\leqq_{tt}^{\log} \subseteq \leqq_{T}^{\log}$.
6. $\leqq_{tt}^{\log} \subseteq \leqq_{tt}^{P}$.

Remark. The reader who is not interested in a systematic treatment of the complexity-bounded reducibilities may take Statements 1 and 2 of the foregoing lemma as definitions.

6.1.1.3. Many-one Reducibilities

Let $\mathscr{C}_m$ denote the class of operators defining m-reducibility, i.e. $\mathbf{F} \in \mathscr{C}_m \leftrightarrow \bigvee_{f \in \mathbb{R}_1} \bigwedge_{g \in \mathbb{R}_1} (\mathbf{F}(g) = g \circ f)$.

Definition. $\leqq_{m}^{P} =_{df} \leqq_{\mathscr{C}_m}^{\mathrm{DTIME,Pol}}$, $\leqq_{m}^{\log} =_{df} \leqq_{\mathscr{C}_m}^{\mathrm{DSPACE}_*,\log}$.

One-one reducibilities are defined analogously.

6.3. Lemma.

1. $A \leqq_{m}^{P} B \leftrightarrow \bigvee_{f \in P} \bigwedge_{x} \big(x \in A \leftrightarrow f(x) \in B\big)$.
2. $A \leqq_{m}^{\log} B \leftrightarrow \bigvee_{f \in L} \bigwedge_{x} \big(x \in A \leftrightarrow f(x) \in B\big)$.
3. $\leqq_{m}^{P}$, $\leqq_{m}^{\log}$ are reflexive and transitive.
4. $A \leqq B \leftrightarrow \bar{A} \leqq \bar{B}$ for $\leqq \in \{\leqq_{m}^{P}, \leqq_{m}^{\log}\}$.
5. $\leqq_{m}^{P} \subseteq \leqq_{tt}^{P}$, $\leqq_{m}^{\log} \subseteq \leqq_{tt}^{\log}$.
6. $\leqq_{m}^{\log} \subseteq \leqq_{m}^{P}$. ∎

Remark. Statements 1 and 2 can be considered as definitions, and for $\leqq_{m}^{P}$ this has actually been done in [Kar 72] where this notion was introduced.

A special case of one-one reducibility is the *padding*. For $h(n) \geqq n + 1$ we define

$$A \leqq_{\mathrm{pad},h} B \leftrightarrow_{df} x \in A \leftrightarrow x\, 0\, 1^{h(|x|)-1-|x|} \in B.$$

Polynomial padding is denoted by $\leqq_{\mathrm{pad}}$ and is defined by

$$A \leqq_{\mathrm{pad}} B \leftrightarrow \bigvee_{p \text{ polynomial}} (A \leqq_{\mathrm{pad},p} B).$$

Note that $\leqq_{\mathrm{pad}} \subseteq \leqq_{m}^{\log}$.

We get sharper translational results for the following refinement of $\leqq_{m}^{\log}$, introduced in [MeSt 72].

$$A \leqq_{m}^{\log\text{-lin}} B \leftrightarrow_{df} \bigvee_{f \in L} \bigvee_{c} \bigwedge_{x} \big[\big(x \in A \leftrightarrow f(x) \in B\big) \wedge |f(x)| \leqq c\,|x| + c\big].$$

Another refinement of $\leqq_{\mathrm{m}}^{\log}$ has been proposed in [HaImMa 78]:

$$A \leqq_{\mathrm{m}}^{\text{1-log}} B \leftrightarrow_{\mathrm{df}} \bigvee_{f \in \text{1-T-DSPACE(log)}} \bigwedge_{x} \big(x \in A \leftrightarrow f(x) \in B\big).$$

A special case of $\leqq_{\mathrm{m}}^{\text{log-lin}}$ and of $\leqq_{\mathrm{m}}^{\text{1-log}}$ is the reducibility defined by homomorphisms:

$$A \leqq_{\mathrm{hom}} B \leftrightarrow \text{there exists a homomorphism } h \text{ such that}$$
$$\bigwedge_{x \neq \mathrm{e}} \big(x \in A \leftrightarrow h(x) \in B\big).$$

The restriction $x \neq \mathrm{e}$ allows a certain freedom in the treatment of the empty word e. Without this restriction, $A \leqq_{\mathrm{hom}} B$ would be possible only if $\mathrm{e} \in A \leftrightarrow \mathrm{e} \in B$. But this is no natural restriction.

6.4. Lemma.
1. $\leqq_{\mathrm{m}}^{\text{log-lin}}$, $\leqq_{\mathrm{hom}}$, $\leqq_{\mathrm{m}}^{\text{1-log}}$, $\leqq_{\mathrm{pad}}$ are reflexive and transitive.
2. $A \leqq B \leftrightarrow \bar{A} \leqq \bar{B}$ for $\leqq \in \{\leqq_{\mathrm{m}}^{\text{log-lin}}, \leqq_{\mathrm{hom}}, \leqq_{\mathrm{m}}^{\text{1-log}}, \leqq_{\mathrm{pad}}\}$.
3. $\leqq_{\mathrm{hom}} \subseteq \leqq_{\mathrm{m}}^{\text{log-lin}} \subseteq \leqq_{\mathrm{m}}^{\log}$, $\leqq_{\mathrm{hom}} \subseteq \leqq_{\mathrm{m}}^{\text{1-log}} \subseteq \leqq_{\mathrm{m}}^{\log}$, $\leqq_{\mathrm{pad}} \subseteq \leqq_{\mathrm{m}}^{\text{1-log}}$. ∎

6.1.2. Nondeterministic Reducibilities

Based on the definition of $\leqq_{\mathrm{T}}^{\mathrm{ND}}$ and following the pattern of the preceding subsection we give the following

Definition. $A \leqq_{\mathscr{C}}^{\mathrm{N}\Phi,K} B \leftrightarrow \bigvee_{\varphi_i^{()} \in \mathscr{C}} \bigwedge_{h \in K} \bigwedge_{x} \Big(x \in A \leftrightarrow \bigvee_{y} \big(\varphi_i^B(x, y) = 1 \wedge \Phi_i^B(x, y) \leqq h(|x|)\big)\Big).$

Here $M_i^{()}$ is a deterministic machine, but $\varphi_i^B(x, y)$ can also be interpreted as being computed by a nondeterministic machine with oracle B on input x on path y (see § 3.3).

In the following we give some examples illustrating this general definition.

6.1.2.1. Turing Reducibilities

Let $\mathscr{C}_{\mathrm{T}}$ be the class of all recursive operators (see § 3.2). We define $\leqq_{\mathrm{T}}^{\mathrm{NP}} =_{\mathrm{df}} \leqq_{\mathscr{C}_{\mathrm{T}}}^{\mathrm{NTIME,Pol}}$.

Following [Lon 82a] we define a strong nondeterministic polynomial time reducibility $\leqq_{\mathrm{sT}}^{\mathrm{NP}}$ by $A \leqq_{\mathrm{sT}}^{\mathrm{NP}} B \leftrightarrow_{\mathrm{df}}$ there exists an oracle machine $M^{()}$ working within polynomial time p such that for every input x

1. there exists at least one computation of $M^{(B)}$ on input x that halts within $p(|x|)$ steps,
2. if $x \in A$, then every halting computation of $M^{(B)}$ on input x of length $\leqq p(|x|)$ is accepting,
3. if $x \notin A$, then every halting computation of $M^{(B)}$ on input x of length $\leqq p(|x|)$ is rejecting.

6.5. Lemma.

1. $\leqq_T^{NP}$ is reflexive but not transitive, $\leqq_{sT}^{NP}$ is reflexive and transitive.
2. $A \leqq_T^{NP} \bar{A}$ and $A \leqq_{sT}^{NP} \bar{A}$ for every A.
3. $\leqq_T^P \subseteq \leqq_{sT}^{NP} \subseteq \leqq_T^{NP} \subseteq \leqq_T^{PSPACE}$.

Proof. The only statement of this lemma which deserves a proof is the nontransitivity of $\leqq_T^{NP}$. Our proof goes back to an idea of [LaLySe 75]. We construct sets $A \subseteq \{0, 1\}^*$ and $B \subseteq \{0, 1\}^*$ such that $\bar{A} \leqq_T^{NP} B$ and $A \nleqq_T^{NP} B$. As $A \leqq_T^{NP} \bar{A}$ is automatically true (see Statement 2 of this lemma), this implies that $\leqq_T^{NP}$ is not transitive.

The sets A and B will be defined in such a way that we have

$$x \in \bar{A} \leftrightarrow \bigvee_y (|y| = |x| \wedge y \in B). \tag{1}$$

This obviously guarantees $\bar{A} \leqq_T^{NP} B$. To ensure $A \nleqq_T^{NP} B$ we construct by a diagonalization process two sequences $(A_m)_{m\in\mathbb{N}}$ and $(B_m)_{m\in\mathbb{N}}$ of finite sets and a sequence $(k_m)_{m\in\mathbb{N}}$ of natural numbers such that $A_{m+1} \subseteq A_m \cup \{0, 1\}^{k_m}$ and $B_{m+1} \subseteq B_m \cup \{0, 1\}^{k_m}$. Finally, we put $A = \bigcup_{m=0}^{\infty} A_m$ and $B = \bigcup_{m=0}^{\infty} B_m$.

Stage 0. $A_0 = \emptyset$, $B_0 = \{e\}$, $k_1 = 1$.

Stage m. Let $m = \langle i, j\rangle$.

Case 1. There is a path of length $< 2^{k_m}$ on which the nondeterministic oracle machine $M_i^{B_{m-1}}$ accepts 0^{k_m}. Let $y_1, \ldots, y_k$ be the queries on such a path. We choose $y \notin \{y_1, \ldots, y_k\}$ such that $|y| = k_m$ and define $A_m = A_{m-1}$ and $B_m = B_{m-1} \cup \{y\}$.

Case 2. There is no accepting path of length $< 2^{k_m}$ of $M_i^{B_{m-1}}$ on 0^{k_m}. We define $A_m = A_{m-1} \cup \{0, 1\}^{k_m}$ and $B_m = B_{m-1}$. For both cases, let $z_1, \ldots, z_s$ be the queries of $M_i^{B_{m-1}}$ on input 0^{k_m} used on any path during the first 2^{k_m-1} steps. We define $k_{m+1} = \max\{|z_1|, \ldots, |z_s|\} + 1$.

Now it can easily be seen that A and B satisfy (1). Assume that M_i^B accepts A within time $p(n)$ (p being a polynomial). Without loss of generality we can assume that each computation of M_i^B on input of length n is not longer than $p(n)$. Choose j such that $p(k_m) < 2^{k_m}$ for $m = \langle i, j\rangle$. We conclude

$$\begin{aligned} 0^{k_m} \in A &\leftrightarrow M_i^B \text{ accepts } 0^{k_m} \text{ within time } p(k_m)\\ &\leftrightarrow M_i^{B_{m-1}} \text{ accepts } 0^{k_m} \text{ within time } p(k_m)\\ &\leftrightarrow M_i^{B_{m-1}} \text{ accepts } 0^{k_m} \text{ within time } < 2^{k_m}\\ &\leftrightarrow 0^{k_m} \notin A_m\\ &\leftrightarrow 0^{k_m} \notin A. \end{aligned}$$

Consequently, there is no i such that M_i^B accepts A in polynomial time. ∎

6.1.2.2. Truth Table Reducibilities

Analogous definitions of $\leqq_{tt}^{NP}$ and $\leqq_{stt}^{NP}$ lead to $\leqq_{tt}^{NP} = \leqq_T^{NP}$ (cf. Theorem 3.11), but $\leqq_{stt}^{NP} \subset \leqq_{sT}^{NP}$ ([Lon 82a]).

6.1.2.3. *Many-one Reducibilities*

Let $\mathscr{C}_{\mathrm{m}}$ be the class of operators described on p. 87. Then

$$\leqq_{\mathrm{m}}^{\mathrm{NP}} =_{\mathrm{df}} \leqq_{\mathscr{C}_m}^{\mathrm{NTIME,Pol}}.$$

$A \leqq_{\mathrm{m}}^{\mathrm{NP}} B$ can be interpreted as follows: There exist an NTM M computing a relation ϱ (see p. 50) and a polynomial p such that

$$\bigwedge_x (x \in A \leftrightarrow \bigvee_y [(x, y) \in \varrho \wedge y \in B \wedge M \text{ outputs } y \text{ within } p(|x|) \text{ steps}]).$$

In [AdMa 77] a restriction $\leqq_{\mathrm{sm}}^{\mathrm{NP}}$ (called $\leqq_\gamma$ there) of $\leqq_{\mathrm{m}}^{\mathrm{NP}}$ has been introduced: $A \leqq_{\mathrm{sm}}^{\mathrm{NP}} B \leftrightarrow_{\mathrm{df}}$ there exists an NTM M computing a relation ϱ and a polynomial p such that, whenever M stops on input x and path y then it does so within $p(|x|)$ steps and

$$\bigwedge_x \bigvee_y (x, y) \in \varrho \wedge \bigwedge_x \bigwedge_y [(x, y) \in \varrho \rightarrow (x \in A \leftrightarrow y \in B)].$$

In [Lon 79] a necessary and sufficient condition for $\leqq_{\mathrm{sm}}^{\mathrm{NP}} = \leqq_{\mathrm{m}}^{\mathrm{P}}$ has been proved.

A nondeterministic logspace reducibility analogous to $\leqq_{\mathrm{m}}^{\mathrm{NP}}$ would lead to the undesirable result that every r.e. set is reducible to some set in **L** (by padding). However, a reasonable nondeterministic logspace reducibility has been defined in [Lag 84]: $A \leqq_{\mathrm{m}}^{\mathrm{Nlog}} B \leftrightarrow_{\mathrm{df}}$ there exist an NTM M computing a relation ϱ (see p. 50) and a polynomial p such that

$$\bigwedge_x (x \in A \leftrightarrow \bigvee_y [(x, y) \in \varrho \wedge y \in B \wedge M \text{ outputs } y \text{ within } p(|x|) \text{ steps}$$
$$\text{and using no more than } \log x \text{ squares on its work tape}]).$$

6.6. Lemma

1. $\leqq_{\mathrm{m}}^{\mathrm{NP}}, \leqq_{\mathrm{sm}}^{\mathrm{NP}}$ are reflexive and transitive, $\leqq_{\mathrm{m}}^{\mathrm{Nlog}}$ is not transitive unless **NL** = **NP**.
2. $\leqq_{\mathrm{m}}^{\mathrm{NP}} \subseteqq \leqq_{\mathrm{T}}^{\mathrm{NP}}$.
3. $\leqq_{\mathrm{m}}^{\mathrm{P}} \subseteqq \leqq_{\mathrm{sm}}^{\mathrm{NP}} \subseteqq \leqq_{\mathrm{m}}^{\mathrm{NP}}$.
4. $A \leqq_{\mathrm{sm}}^{\mathrm{NP}} B \leftrightarrow \bar{A} \leqq_{\mathrm{sm}}^{\mathrm{NP}} \bar{B}$. ∎

Proof. We prove only the $\leqq_{\mathrm{m}}^{\mathrm{Nlog}}$ part of Statement 1 [Lag 84]. Assume **NL** ⊂ **NP**. Then Remark 4 after Theorem **14.3** and Theorem **12.19.3.3** yield

$$\mathscr{R}_{\mathrm{m}}^{\mathrm{Nlog}}(\mathbf{LIN}) = \mathbf{NL} \subset \mathbf{NP} = \mathscr{R}_{\mathrm{m}}^{\mathrm{Nlog}}(\mathbf{EDTOL}) \subseteqq \mathscr{R}_{\mathrm{m}}^{\mathrm{Nlog}}(\mathbf{NL})$$
$$= \mathscr{R}_{\mathrm{m}}^{\mathrm{Nlog}}(\mathscr{R}_{\mathrm{m}}^{\mathrm{Nlog}}(\mathbf{LIN})). \ ∎$$

For further work on complexity-bounded reducibilities the reader should consult [Sel 78], [Sel 79], [Sel 80], [Lon 82a] and [ChStVi 82].

6.2. *Relationships between Different Reducibility Notions*

One of the most important questions in the field of complexity-bounded reducibilities concerns the comparison of their power. We start with

6.7. Theorem. [LaLySe 75] $\leqq_{\mathrm{m}}^{\mathrm{P}} \subset \leqq_{\mathrm{tt}}^{\mathrm{P}} \subset \leqq_{\mathrm{T}}^{\mathrm{P}}$.

Proof. From 6.3.5 and 6.2.5 we know the inclusions $\leqq_m^P \subseteq \leqq_{tt}^P \subseteq \leqq_T^P$. It remains to show $\leqq_m^P \neq \leqq_{tt}^P$ and $\leqq_{tt}^P \neq \leqq_T^P$. The first of these inequalities is proved by constructing a set A such that $A \nleqq_m^P \bar{A}$. By 6.2.4 it holds that $A \leqq_{tt}^P \bar{A}$, and this proves $\leqq_m^P \neq \leqq_{tt}^P$. Now we outline the construction of A. We make use of the auxiliary function h defined by $h(0) = 1$, $h(n+1) = 2^{h(n)}$. The set A will be defined by diagonalization in such a way that $A \subseteq \{0^{h(m)}: m \in \mathbb{N}\}$ and that no polynomial time machine can reduce A to $\bar{A}$: For $x = 0^n$ and $n = h(m)$ determine i, j such that $\langle i, j\rangle = m$. Then let M_i run $2^n - 1$ steps on input x. If M_i does not stop, then $x \in \bar{A}$. If it stops with output y, then

$$\text{if} \quad |y| > h(m-1), \quad \text{then} \quad x \in \bar{A},$$

$$\text{if} \quad |y| \leqq h(m-1), \quad \text{then} \quad x \in A \leftrightarrow y \in A.$$

We leave it to the reader to verify that $A \leqq_m^P \bar{A}$ is impossible.

The second inequality $\leqq_{tt}^P \neq \leqq_T^P$ can also be proved by diagonalization. Sets A, B can be constructed such that $A \equiv_T^P B$, $A \nleqq_{tt}^P B$ and $B \nleqq_{tt}^P A$. These sets can be chosen to be in **EXPTIME**. ∎

Remark 1. The set A constructed in the first part of the proof can be shown to belong to **EXPTIME**. If we were to succeed in finding such an A in $\mathbf{NP} \setminus \{\emptyset, \Sigma^*\}$, then, since A and $\bar{A}$ cannot belong to **P**, we would have $\mathbf{P} \neq \mathbf{NP}$. In [Sel 82] it is shown that if $\mathbf{DEXPTIME} \neq \mathbf{NEXPTIME}$, then $\leqq_m^P$ and $\leqq_{tt}^P$ differ even on **NP**, i.e. there exist sets $B \in \mathbf{NP}$ and $C \leqq_{tt}^P B$ such that $C \nleqq_m^P B$.

Remark 2. The proof of $\leqq_{tt}^P \nsubseteq \leqq_m^P$ can be adapted to show even $\leqq_{tt}^{\log} \nsubseteq \leqq_m^{NP}$.

6.8. Theorem. [Simi 76]

1. $\bigwedge_{A \notin \mathbf{P}} \bigvee_{B \in \mathbf{REC}} (A \leqq_T^P B \wedge A \nleqq_m^P B)$.

2. $\bigvee_{B \notin \mathbf{P}} \bigwedge_{A} (A \leqq_T^P B \leftrightarrow A \leqq_m^P B)$. ∎

Statement 1 of Theorem 6.8 shows that $\leqq_T^P$-hardness for **NP** differs from $\leqq_m^P$-hardness for **NP** if (and only if) $\mathbf{P} \neq \mathbf{NP}$. (For the "if" part, apply the statement to $A = \mathbf{SAT}$. The "only if" part is elementary.)

6.9. Theorem. (For an illustration see Figure 6.1.) [LaLySe 75], [LaLy 76], [Lon 82a].

1. $\leqq_m^{NP} \subset \leqq_T^{NP}$. 2. $\leqq_{sT}^{NP} \subset \leqq_T^{NP}$.
3. $\leqq_T^P \subset \leqq_{sT}^{NP}$. 4. $\leqq_{stt}^{NP} \subset \leqq_{sT}^{NP}$.
5. $\leqq_{sm}^{NP} \subset \leqq_{stt}^{NP}$. 6. $\leqq_{sm}^{NP} \subset \leqq_m^{NP}$.
7. $\leqq_m^{Nlog} \subset \leqq_{tt}^P$. 8. [Lag 84] $\leqq_m^{\log} \subset \leqq_m^{Nlog}$ unless $\mathbf{L} = \mathbf{NP}$.

Proof. Ad 1. $\leqq_m^{NP} \subseteq \leqq_T^{NP}$ is valid because of Theorem 6.6.2. As $\leqq_m^{NP}$ is transitive (Theorem 6.6.1) and $\leqq_T^{NP}$ is not (Theorem 6.5.1), they must be different.

Statement 2 can be proved in the same way.

Ad 3.—6. This follows from Theorems 6.10.3.—6.10.6.

The proof of Statement 7 is omitted.

Ad 8. In [Lag 84] it is shown that

$$\mathbf{NP} = \mathcal{R}_{\mathrm{m}}^{\mathrm{Nlog}}\left(\{(\#w)^n : n \in \mathbb{N} \wedge w \in \{0,1\}^*\right),$$

but

$$\mathbf{L} = \mathcal{R}_{\mathrm{m}}^{\mathrm{log}}\left(\{(\#w)^n : n \in \mathbb{N} \wedge w \in \{0,1\}^*\right). \blacksquare$$

Remark. The proof of 6.5.1 shows that the difference between $\leq_{\mathrm{m}}^{\mathrm{NP}}$ and $\leq_{\mathrm{T}}^{\mathrm{NP}}$ can be stated on $\mathbf{DTIME}(2^{2^n})$.

The next result is an analogue of 6.8.

6.10. Theorem.

1. [SiGi 77] $\bigwedge_{A \notin \mathbf{NP}} \bigvee_{B \in \mathbf{REC}} (A \leq_{\mathrm{T}}^{\mathrm{NP}} B \wedge A \not\leq_{\mathrm{m}}^{\mathrm{NP}} B)$.
2. [SiGi 77] $\bigvee_{B \notin \mathbf{P}} \bigwedge_{A} (A \leq_{\mathrm{T}}^{\mathrm{PSPACE}} B \leftrightarrow A \leq_{\mathrm{m}}^{\mathrm{P}} B)$.
3. [Lon 82a] $\bigwedge_{A \notin \mathbf{P}} \bigvee_{B \in \mathbf{REC}} (A \leq_{\mathrm{sT}}^{\mathrm{NP}} B \wedge A \not\leq_{\mathrm{T}}^{\mathrm{P}} B)$.
4. [Lon 82a] $\bigwedge_{A \notin \mathbf{NP} \cap \mathbf{coNP}} \bigvee_{B \in \mathbf{REC}} (A \leq_{\mathrm{T}}^{\mathrm{P}} B \wedge A \not\leq_{\mathrm{stt}}^{\mathrm{NP}} B)$.
5. [Lon 82a] $\bigwedge_{A \notin \mathbf{NP} \cap \mathbf{coNP}} \bigvee_{B \in \mathbf{REC}} (A \leq_{\mathrm{tt}}^{\mathrm{P}} B \wedge A \not\leq_{\mathrm{sm}}^{\mathrm{NP}} B)$.
6. [Lon 82a] $\bigwedge_{A \notin \mathbf{NP} \cap \mathbf{coNP}} \bigvee_{B \in \mathbf{REC}} (A \leq_{\mathrm{m}}^{\mathrm{NP}} B \wedge A \not\leq_{\mathrm{sm}}^{\mathrm{NP}} B)$.

Remark 1. Statement 4 of Theorem 6.10 shows that $\leq_{\mathrm{sT}}^{\mathrm{NP}}$-hardness for **NP** differs from $\leq_{\mathrm{stt}}^{\mathrm{NP}}$-hardness for **NP** if (and only if) $\mathbf{NP} \neq \mathbf{coNP}$. (For the "if" part, apply the statement to $A = \mathrm{SAT}$. The "only if" part is elementary.) Statements 1, 3, 5 and 6 can be interpreted similarly.

Remark 2. Statement 2 is stronger than 6.8.2 and stronger than $\bigvee_{B \notin \mathbf{P}} \bigwedge_{A} (A \leq_{\mathrm{T}}^{\mathrm{NP}} B \leftrightarrow A \leq_{\mathrm{m}}^{\mathrm{NP}} B)$ which is the direct analogue of 6.8.2. It is also stronger than $\bigvee_{B} (\mathrm{PSPACE}^B = \mathrm{P}^B)$ stated in 28.10.1.

In contrast to the polynomial time reducibilities, the logspace truth table and Turing reducibilities coincide:

6.11. Theorem. [LaLy 76] $\leq_{\mathrm{m}}^{\mathrm{log}} \subset \leq_{\mathrm{tt}}^{\mathrm{log}} = \leq_{\mathrm{T}}^{\mathrm{log}}$.

Proof. The first part is proved by a slight modification of the proof of 6.7. For the second part the reader is referred to [LaLy 76]. ∎

All results stated so far and concerning the comparison of the reducibility notions considered here are summarized in Figure 6.1.

Remark. The inclusions in Figure 6.1 follow from the definitions. The strict inclusions follow from 6.7, 6.9, and 6.11. Note that each of the inclusions $\leq_{\mathrm{m}}^{\mathrm{P}} \subseteq \leq_{\mathrm{m}}^{\mathrm{log}}$, $\leq_{\mathrm{m}}^{\mathrm{P}} \subseteq \leq_{\mathrm{tt}}^{\mathrm{log}}$ and $\leq_{\mathrm{tt}}^{\mathrm{P}} \subseteq \leq_{\mathrm{tt}}^{\mathrm{log}}$ would imply $\mathbf{P} = \mathbf{L}$.

It has been observed that $\leq_{\mathrm{T}}^{\mathrm{log}}$ is not invariant over different machine models. Whereas $\leq_{\mathrm{T}}^{\mathrm{P}}$ and $\leq_{\mathrm{T}}^{\mathrm{NP}}$ are not influenced by the number of query tapes, we get logspace Turing reducibilities of different powers when oracle machines with different numbers of query tapes are considered (see [Lyn 78]). This shows that there seems

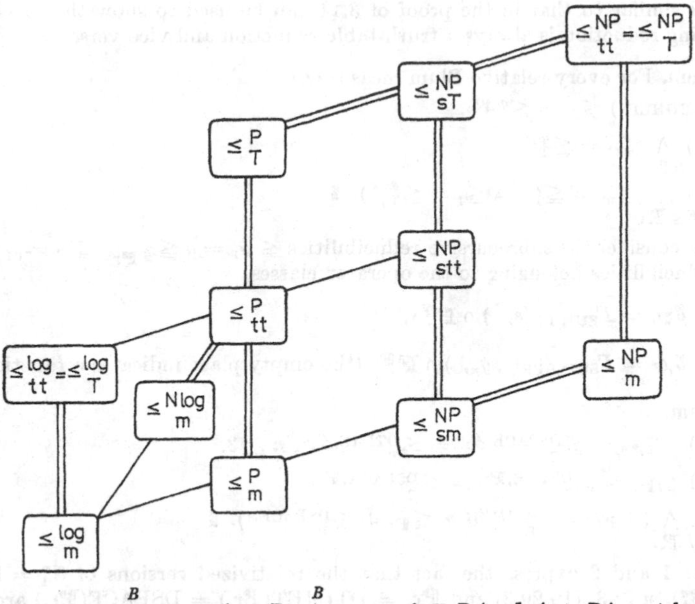

Figure 6.1. $\overset{B}{\underset{A}{\|}}$ means $A \subset B$, $\overset{B}{\underset{A}{|}}$ means $A \subseteq B$ (and $A \neq B$ is not known)

to be no uniform definition of space bounded relativized computations (see also p. 83). The relations between the restrictions of $\leqq_m^{\log}$ defined above are illustrated in Figure 6.2.

Remark. The inclusions in Figure 6.2 are clear by definition. Because of

$$D^{\#} \in \mathbf{L} \setminus \mathbf{1\text{-}T\text{-}DSPACE}(\log)$$

(cf. 8.2) we have $D^{\#} \leqq_m^{\text{log-lin}}\{1\}$ and $D^{\#} \not\leqq_m^{\text{1-log}}\{1\}$, and because of $\mathcal{R}_m^{\text{log-lin}}(\mathbf{NLINSPACE}) = \mathbf{NLINSPACE} \subset \mathbf{PSPACE} = \mathcal{R}_{\text{pad}}(\mathbf{NLINSPACE})$ (cf. 14.4 and 14.1) we have $\leqq_{\text{pad}} \not\subseteq \leqq_m^{\text{log-lin}}$. The relation $\leqq_{\text{pad}} \subset \leqq_m^{\text{1-log}}$ is immediately clear. This proves all statements contained in Figure 6.2.

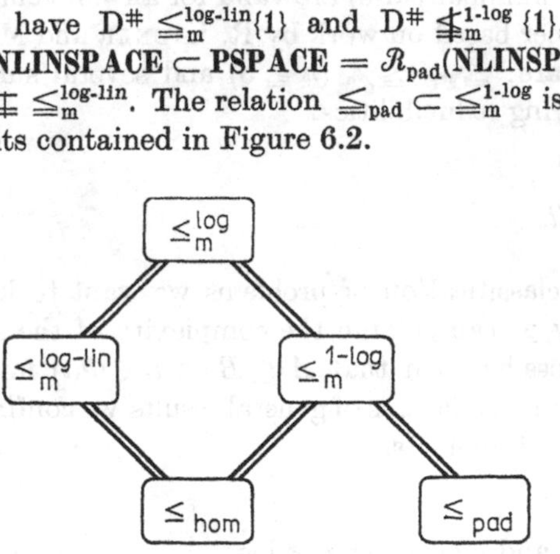

Figure 6.2. $\overset{B}{\underset{A}{\|}}$ means $A \subset B$

A method similar to that in the proof of 3.11 can be used to show that a complexity-restricted Turing reduction is always a truth-table reduction and vice versa:

6.12. Theorem. For every relative Blum measure Φ,

1. (T. McLaughlin) $\leqq_{tt} = \leqq_T^{\Phi,\mathbb{R}_1}$.
2. ([Lyn 72]) $\bigwedge_{f \in \mathbb{R}_1} \leqq_{tt} \neq \leqq_T^{\Phi,f}$.
3. ([Lyn 72]) $\bigwedge_{K \subseteq \mathbb{R}_1} (\leqq_m \neq \leqq_T^{\Phi,K} \wedge \leqq_1 \neq \leqq_T^{\Phi,K})$. ∎

Finally we consider the subrecursive reducibilities $\leqq_{\mathbb{P}r} =_{df} \leqq_{\mathscr{C}_{\mathbb{P}r}}$, $\leqq_{\mathscr{E}^n} =_{df} \leqq_{\mathscr{C}_{\mathscr{E}^n}}$ defined as the reducibilities belonging to the operator classes

$$\mathscr{C}_{\mathbb{P}r} = \Gamma_{SUB,PR}(s, \ ) \cap \mathbb{F}_1^{\mathbb{F}_1},$$

$$\mathscr{C}_{\mathscr{E}^n} = \Gamma_{SUB,BPR}(s, g_n, \ ) \cap \mathbb{F}_1^{\mathbb{F}_1} \text{ (the empty place indicates a function variable).}$$

6.13. Theorem.

1. ([Lad 75]) $\leqq_{\mathscr{E}^n} = \leqq_T^{DSPACE,\mathscr{E}_1^n} = \leqq_T^{DTIME,\mathscr{E}_1^n}$, $n \geqq 3$.
2. ([Lyn 72]) $\leqq_{\mathbb{P}r} = \leqq_T^{DTIME,\mathbb{P}r_1} = \leqq_T^{DSPACE,\mathbb{P}r_1}$.
3. ([Lyn 72]) $\bigwedge_{f \in \mathbb{R}_1} (\leqq_{\mathbb{P}r} \neq \leqq_T^{DTIME,f} \wedge \leqq_{\mathbb{P}r} \neq \leqq_T^{DSPACE,f})$. ∎

Statements 1 and 2 express the fact that the relativized versions of $\mathscr{E}_1^n = DSPACE(\mathscr{E}_1^n) = DTIME(\mathscr{E}_1^n)$, $n \geqq 3$, (10.20.3) and $\mathbb{P}r_1 = DTIMET(\mathbb{P}r_1) = DSPACE(\mathbb{P}r_1)$ are also valid.

The relativized version of $P = \Gamma_{SUB,BSR}(s_1, s_2, m, x^{|x|})$, (10.14.2) leads to a characterization of $\leqq_T^P$ which can be found in [Meh 76]. Further results of § 10 could be relativized.

6.3. *Degree Structure*

For every reducibility notion we can define, along the general lines of § 3, degrees which form a partially ordered set. The structure of the set of $\leqq_T^P$-degrees will be investigated in § 24.2.1.2.3. The results of that subsection (nonexistence of nontrivial minimal degrees, density, minimal pairs) are valid for all subrecursive reducibilities defined in [Meh 76] a paper based on work by R. Ladner and M. Machtey. Examples of such reducibilities are: $\leqq_{\mathbb{P}r}$, $\leqq_{\mathscr{E}^n}$ ($n \geqq 3$) and several suitably time-restricted or space-restricted Turing reducibilities.

6.4. *Translational Results*

For the purpose of complexity classification of problems we want to know how a complexity bounded reducibility $\leqq$ can enlarge the complexity of the sets, i.e. we want to know an h as small as possible such that $A \leqq B$ an $B \in \Phi(t)$ implies $A \in \Phi \times (t \circ h)$. Instead of cumbersome formulations of general results we confine ourselves to two very often applied sample statements.

6.14. Lemma.

1. Let $A \leqq_m^{\log} B$. For $X \in \{D, N\}$ and increasing $s \geqq \log$,

$$B \in \mathbf{XSPACE}(s) \rightarrow A \in \mathbf{XSPACE}(s(\mathrm{Pol})).$$

2. Let $A \leqq_m^P B$. For X $\in$ {D, N} and increasing $t \geqq \mathrm{id}$,

$$B \in \textbf{multiT-XTIME}(t) \to A \in \textbf{multiT-XTIME}(t(\mathrm{Pol})).$$

Proof. We only mention that for small s the technique of the proof of 6.1.1 should be used. ∎

In what follows the reducibility $\leqq_{\mathrm{pad},h}$, $h > \mathrm{id}$, will be of interest on account of the fact that for A and $A_h = \{w01^{h(|w|)-|w|-1} : w \in A\}$, besides the "downwards" translational lemma of the form $A_h \in \Phi(t) \to A \in \Phi(t \circ h)$, there is also an "upwards" translational lemma of the form $A \in \Phi(t \circ h) \to A_h \in \Phi(t)$. This idea has been developed in [RuFi 65].

We say that (Φ, h, t) satisfies Condition 1 if

a) $\Phi = \mathrm{DSPACE}$, $t \geqq \log$, $h(|w|) \in \mathrm{DSPACE}(t \circ h)$, or

b) $\Phi = \mathrm{NSPACE}$, $t \geqq \log$, or

c) $\Phi = \text{multiT-DTIME}$, $t(n) \geqq c \cdot n$ for some $c > 1$, $h(|w|) \in \text{multiT-DTIME}(t \circ h)$, or

d) $\Phi = \text{multiT-NTIME}$, $t \geqq \mathrm{id}$.

We say that (Φ, h, t) satisfies Condition 2 if

a) $\Phi \in \{\mathrm{DSPACE}, \mathrm{NSPACE}\}$, $t \geqq \log$, $\{w01^{h(|w|)-|w|-1} : w \in \{0, 1\}^*\} \in \textbf{DSPACE}(t)$ (the latter can be concluded from $h(|w|) \in \mathrm{DSPACE}(t \circ h) \wedge t(|w|) \in \mathrm{DSPACE}(t)$) or

b) $\Phi \in \{\text{multiT-DTIME}, \text{multiT-NTIME}\}$, $t(n) \geqq c \cdot n$ for some $c > 1$, $\{w01^{h(|w|)-|w|-1} : w \in \{0, 1\}^*\} \in \textbf{multiT-DTIME}(t)$. (The latter can be concluded from $h(|w|) \in \text{multiT-DTIME}(t \circ h) \wedge t(|w|) \in \text{multiT-DTIME}(t)$.)

6.15. Lemma. If (Φ, h, t) satisfies Condition 1, then

$$A_h \in \Phi(t) \to A \in \Phi(t \circ h). \;∎$$

6.16. Lemma. If (Φ, h, t) satisfies Condition 2, then

$$A \in \Phi(t \circ h) \to A_h, \overline{A}_h \in \Phi(t). \;∎$$

6.17. Corollary. If (Φ, h, t_1) satisfies Condition 2 and (Ψ, h, t_2) satisfies Condition 1, then

1. $\Phi(t_1) \subseteq \Psi(t_2) \to \Phi(t_1 \circ h) \subseteq \Psi(t_2 \circ h)$.
2. $\Phi(t_1 \circ h) \setminus \Psi(t_2 \circ h) \neq \emptyset \to \Phi(t_1) \setminus \Psi(t_2) \neq \emptyset$. ∎

6.18. Corollary. If (Φ, h, t_1) satisfies Condition 2 and (Φ, h, t_2) satisfies Condition 1, then

1. $t_1 \leqq_{\mathrm{ae}} t_2 \wedge \Phi(t_1 \circ h) \subset \Phi(t_2 \circ h) \to \Phi(t_1) \subset \Phi(t_2)$.
2. $\Phi(t_1) \subseteq \mathbf{co}\,\Phi(t_2) \to \Phi(t_1 \circ h) \subseteq \mathbf{co}\,\Phi(t_2 \circ h)$. ∎

In some cases the constructibility supposition for h in the conditions 1 and 2 can be replaced by such a supposition for t.

6.19. Lemma. If $s \geqq \log$ is increasing, $s(|w|) \in \mathrm{DSPACE}(t)$, $h > \mathrm{id}$, $A \in \mathbf{DSPACE}(s \circ h)$, then there exists an $h' \in \mathbb{R}_1$ such that

1. $s \circ h' \leqq_{ae} s \circ h$.
2. $A \in \mathbf{DSPACE}(s \circ h')$.
3. $h'(|w|) \in \mathrm{DSPACE}(s \circ h')$. ∎

6.20. Lemma. If $t \geqq \mathrm{id}$ is increasing, $t(|w|) \in \text{multiT-DTIME}(t)$, $h > \mathrm{id}$, $A \in \textbf{multiT-DTIME}(t \circ h)$, then there exists an $h' \in \mathbb{R}_1$ such that

1. $t \circ h' \leqq_{ae} t \circ h$.
2. $A \in \textbf{multiT-DTIME}(t \circ h')$.
3. $h'(|w|) \in \text{multiT-DTIME}\big((t \circ h')^2\big)$. ∎

The theorems 26.2 and 26.6 can be considered as corollaries of the preceding lemmas.

The conversion of a word into the unary notation of that number which it presents in binary is something like padding. Defining

$$\text{Tally } A = \{1^{\mathrm{bin}^{-1}(1w)} : w \in A\}$$

we obtain the following two lemmas which were first mentioned in [Savi 73b], and which allow a translation of equalities downwards, but at the cost of having "below" only languages over a single letter alphabet.

6.21. Lemma. For $\mathrm{X} \in \{\mathrm{D}, \mathrm{N}\}$ and $s \geqq \mathrm{id}$,

$$A \in \mathbf{XSPACE}\big(s(\mathrm{Lin})\big) \to \text{Tally } A \in \mathbf{XSPACE}\big(s(\mathrm{Lin}\log)\big). \blacksquare$$

6.22. Lemma. For $\mathrm{X} \in \{\mathrm{D}, \mathrm{N}\}$ and $t(n) \geqq 2^n$,

$$A \in \textbf{multiT-XTIME}\big(t(\mathrm{Lin})\big) \leftrightarrow \text{Tally } A \in \textbf{multiT-XTIME}\big(t(\mathrm{Lin}\log)\big). \blacksquare$$

Part 2

Complexity Classes and Complexity Measures

Chapter III
Complexity of Single Problems

In this chapter we mainly study the complexity of specific problems (deciding sets or computing functions). Given such a problem, it would be most desirable to have a feasible algorithm for solving it. The attempts to find efficient algorithms lead to results about upper bounds which are discussed in § 7. To show that improvements of upper bounds are impossible or at least limited requires proving lower bounds. This is the objective of § 8. From a practical point of view, intractable problems are not necessarily hopeless, because sometimes it is possible to slightly weaken or modify a problem, thus making it tractable. A survey of such possibilities is given in § 9.

§ 7. Upper Bounds

We call $t \in \mathbb{R}_1$ an *upper bound* for $A \subseteq \Sigma^*$ with respect to a measure (φ, Φ) if there exists an i such that $c_A = \varphi_i$ and $\Phi_i \leq_{\text{ae}} t$. Thus every algorithm solving A provides us with an upper bound for A. Of course, our interest lies in finding very efficient algorithms — one of the most important problems which initiated the study of complexity theory.

The state of the art is characterized by quite extreme types of results: In some cases the best possible algorithms are found without effort, sometimes clever tricks must be invented to design good algorithms. In some cases, lower bounds are known which teach us that further improvements of the existing upper bounds are impossible, and sometimes neither reasonable lower bounds nor essential improvements of the upper bounds can yet be proved.

We do not repeat the general techniques for designing good algorithms, such as divide and conquer, branch and bound, dynamic programming etc., which have been described in [AhHoUl 74], [HoSa 78], [Meh 77]. Our aim is to present some results which are used throughout the book and to gather some results which are either of particular attractiveness (linear programming, prime numbers, graph isomorphism) or have ingenious solutions (string matching and related problems). Because of restricted space we can neither give the proofs of these latter results nor can we describe the technique of using suitable data structures for developing particularly efficient algorithms, for example for graph theoretical problems. The only more general method for proving upper bounds described here, namely the method of quantifier elimination, is restricted to logical theories.

The material is arranged according to growing time bounds, but we have felt that close thematical connections may sometimes justify a deviation from this principle.

7.1. *Types of Problems*

The problems we are dealing with are decision, existence, search, optimization and counting problems. We start with a general framework and introduce a unified notation. Let prob $\subseteq (\Sigma^*)^2$ be a decision problem. It gives rise to four different types of problem.

The existence problem is also a decision problem. It is defined by

$$\mathrm{PROB} = \left\{x : \bigvee_y (x, y) \in \mathrm{prob}\right\}.\text{*)}$$

The search problem Prob can be stated like this: Given an *input* x, find a *solution* y, i.e. a y such that $(x, y) \in$ prob. To solve the search problem Prob means to compute a *search function f of* prob, i.e. a total function f with the property

$$\big(x, f(x)\big) \in \mathrm{prob} \quad \text{if} \quad \bigvee_y (x, y) \in \mathrm{prob},$$

$$f(x) = \mathrm{no} \quad \text{if} \quad \sim\bigvee_y (x, y) \in \mathrm{prob}.$$

If there exists a search function f of prob such that $f \in \Phi(t)$, for some measure Φ and $t \in \mathbb{R}_1$, we shall write Prob $\in \Phi(t)$.

Optimization problems are special types of search problems. To define them we start from a search problem Prob and a *valuation function*

$$b_{\mathrm{prob}} : \mathrm{prob} \mapsto \mathbb{N}.\text{**)}$$

The value $b_{\mathrm{prob}}(x, y)$ tells us how good the solution y for the input x is. If good solutions correspond to large (small) values of b_{prob}, we define

$$b^*_{\mathrm{prob}}(x) = \max_{y \in \Sigma^*} b_{\mathrm{prob}}(x, y) \qquad \Big(b^*_{\mathrm{prob}}(x) = \min_{y \in \Sigma^*} b_{\mathrm{prob}}(x, y)\Big).$$

The optimization problem Max prob is the search problem belonging to

$$\{(x, y) : (x, y) \in \mathrm{prob} \wedge b_{\mathrm{prob}}(x, y) = b^*_{\mathrm{prob}}(x)\}$$

in the case where b^*_{prob} is defined by the maximum operator. In the other case the optimization problem is called Min prob.

The counting problem corresponding to prob is the problem of computing the function

$$n_{\mathrm{prob}}(x) =_{\mathrm{df}} \mathrm{card}\,\{y : (x, y) \in \mathrm{prob}\}.$$

We introduce the convention of denoting existence problems by upper case letters and search (and optimization) problems by lower case letters with upper case initials.

---

*) In practice, both x and y may occur as tuples.

**) In many cases b_{prob} does not depend on its first argument. In such cases we omit the first argument.

Example. A clique of a graph is a complete subgraph, i.e. a subgraph which contains, for any two of its vertices u, v, also the edge (u, v).

$$\text{clique} =_{\text{df}} \{((G, k), G') : G \text{ is a finite graph} \wedge k \in \mathbb{N} \wedge G' \text{ is a clique of } G \text{ with at least } k \text{ vertices}\}.$$

$$\text{CLIQUE} = \left\{(G, k) : \bigvee_{G'} ((G, k), G') \in \text{clique}\right\}.$$

The search problem related to clique is denoted by Clique.

If $b_{\text{clique}}(G') =_{\text{df}} \text{card } V'$, then the corresponding optimization problem Max clique is the search problem belonging to the decision problem

$$\{((G, k), G') : ((G, k), G') \in \text{clique} \wedge b_{\text{clique}}(G') = b^*_{\text{clique}}(G, k)\}.$$

7.2. *Polynomial Time*

There is a point of view from which it is sufficient to know matching upper and lower bounds for given problems. From a more practically oriented point of view this cannot always be satisfying. Let us consider for instance two problems having the exact time complexities $\asymp n^2$ and $\asymp 2^n$, resp. The immense difference between the growth rates of a polynomial and an exponential function accounts for incredibly long computation times for the exponential time algorithm, even for relatively short inputs. This is most impressively demonstrated by a table in [GaJo 79], from which we quote one single illustrative example concerning the time bounds n^2 and 2^n. Assume we have a computer performing 10^6 steps per second. An algorithm with the time complexity n^2 stops on an input of length 60 after 0.0036 seconds, whereas an algorithm with the time complexity 2^n on the same input would require 36300 years!

This observation has led to the widely accepted consensus that feasible problems should have polynomial time complexity. This is reasonable insofar as polynomial time complexity does not depend on the machine model as long as realistic machines are considered (§ 19). On the other hand this notion does not adequately reflect what is necessary for practical computations. A polynomial with astronomical coefficients or a large degree would hardly be taken for a good running time, and on the contrary, there are exponential time algorithms with a good behaviour for small arguments. For instance, under the same assumptions as above, for inputs of length 60 an algorithm with time complexity n^{10} stops after 19140 years whereas an algorithm with time complexity $2^{\sqrt[4]{n}}$ stops after 0.000008 seconds.

7.2.1. Realtime Computations

If a machine has the identity function as running time, then we say that it works in realtime. Of course, we are mainly interested in realtime computations of realistic machines.

Apart from a very few exceptions the problems discussed in this subsection are related to *string matching*. This is a problem which is, for instance, of importance in retrieval systems where according to one or more key word(s) [pattern(s)] all documents [texts] have to be found which contain the given key(s). String matching can be formulated as the decision problem "Is a given pattern v a substring of a given text w?" More formally,

$$\mathrm{SM} =_{\mathrm{df}} \left\{v \# w : v, w \in \{0, 1\}^* \wedge \bigvee_{v_1, v_2} w = v_1 v v_2\right\}$$

or, not completely equivalently, as we shall see,

$$\mathrm{SM}' =_{\mathrm{df}} \left\{w \# v : v, w \in \{0, 1\}^* \wedge \bigvee_{v_1, v_2} w = v_1 v v_2\right\}.$$

String matching is one of the problems for which on-line solutions are especially desirable. By a result of S. A. Cook (13.20.2) we know **2-DPDA** $\subseteq$ **RAM-DTIME** (Lin). As SM $\in$ **2-DPDA**, there must exist a linear time string matching algorithm.

Such algorithms have been proposed in [Wein 73], [Gal 76c], [BoMo 77] (see also [Gal 78], [GuOd 80]), [KnMoPr 77], [Riv 77], [AhCo 75] and [Com 79] (the two latter for more than one pattern). For more details the reader is referred to the survey paper [Aho 80]. In [FimPa 74] the Knuth-Morris-Pratt algorithm has been implemented in linear time on a multiT-DM. This implementation can be modified in such a way that it works in realtime (see [Gal 76c], where a sufficient condition is developed for an on-line algorithm to be transformable into a realtime algorithm).

7.1. Theorem.

1. [Gal 76c] SM $\in$ **REALTIME**.
2. [Sli 77b] SM′ is acceptable in quasi-realtime by an on-line DRAM using logarithmic space. ∎

Unfortunately, the proofs are too complicated to be presented here.

Remark 1. Statement 1 strengthens a result from [Mati 71] which states SM $\in$ **T²-DTIME**(id).

Remark 2. Note that SM′ $\notin$ **REALTIME** because of 8.8.2.

Remark 3. For string matching in sublinear time by parallel algorithms see [Gal 84].

Differences like those between the behaviour of SM and SM′ can also be observed for

$$\mathrm{MS} =_{\mathrm{df}} \{u_1 \# \dots \# u_n \# u_i^{-1} : n \in \mathbb{N} \wedge 1 \leqq i \leqq n \wedge u_1, \dots, u_n \in \{0, 1\}^*\}$$

which is not in **REALTIME** (8.8.2) and

$$\mathrm{MS}' =_{\mathrm{df}} \{u_i^{-1} \# u_1 \# \dots \# u_n : n \in \mathbb{N} \wedge 1 \leqq i \leqq n \wedge u_1, \dots, u_n \in \{0, 1\}^*\}$$

for which we have

7.2. Theorem. [Nek 72] MS′ $\in$ **3T-DTIME**(id). ∎

The proof is not too hard and makes use of the given block structure of the words of MS'. The following string matching problem which contains MS' as a special case, is presumed to be in **REALTIME**;

$$\left\{v \# u : u, v \in \{0, 1\}^* \wedge \bigvee_{u_1, u_2} u = u_1 v^{-1} u_2\right\}.$$

A time-space trade-off within a whole family of string matching algorithms has been derived from the Knuth-Morris-Pratt algorithm by Z. Galil and J. I. Seiferas. We present only the extreme results in the Turing machine implementation, although the authors give RAM and 2:multi-DFA implementations as well. In the formulation of their results n is the length of the pattern string and m the length of the text string. The pattern string is read by two heads from a 2-way input tape and the text is read on-line.

7.3. Theorem. [GaSe 77]

1. For every $\varepsilon > 0$, SM $\in$ **1*-2:2-T-DSPACE-TIME**($\log n, n^\varepsilon \cdot m$).
2. SM $\in$ **1*-2:2-multiT-DSPACE-TIME**($\log^2 n, m$). ∎

Further papers on string matching: [AvMa 80], [BaDr 80], [GaSe 81a], [GaSe 81b] and [Sli 81].

By modifying the string matching algorithms it is possible to prove D $\in$ **REALTIME** for the set $D =_{df} \{ww : w \in \{0, 1\}^*\}$ of "double words". The same result is true for the symmetric words or palindromes

$$S =_{df} \{w : w = w^{-1} \wedge w \in \{0, 1\}^*\}.$$

7.4. Theorem. [Sli 77a] S $\in$ **REALTIME**. ∎

Remark 1. The first proof of this result goes back to A. O. Slisenko in 1973. His proof is about 170 pages long and has been shortened by Z. Galil (see [Gal 76b]) and by himself (see [Sli 77a]). This proof is, however, still too long to be included in this book. Note that $S_i =_{df} S \cap \{w : |w| \equiv i(2)\}$ is also in **REALTIME** for $i = 0, 1$.

Remark 2. Note that in [Col 69] it has already been observed that S, D $\in$ **IA^1-DTIME**(id).

We add two further highly nontrivial results concerning S^*.

7.5. Theorem.

1. [Gal 76b] $S^* \in$ **REALTIME**.
2. [GaSe 78] $S^* \in$ **1*-RAM-DTIME**(Lin). ∎

It is not known whether S^* can be recognized in realtime by a DRAM.

In [FimPa 74] a linear time algorithm on a multiT-DM for S_0^* has been given.

D and S_0 are special cases of sets of the type $M_v = \{w^v : w \in \{0, 1\}^*\}$, where $w^e =_{df} e$, $w^{v0} =_{df} w^v w$, $w^{v1} =_{df} w^v w^{-1}$, namely $D = M_{00}$ and $S_0 = M_{01}$. Moreover, they are special cases of sets of the type

$$\mathrm{M}^0_{k,l,m} =_{\mathrm{df}} \{xwywz\colon w, x, y, z \in \{0, 1\}^* \wedge |x| = k\,|w| \wedge |y| = l\,|w| \wedge |z| = m\,|w|\},$$

$$\mathrm{M}^1_{k,l,m} =_{\mathrm{df}} \{xwyw^{-1}z\colon w, x, y, z \in \{0, 1\}^* \wedge |x| = k\,|w| \wedge |y| = l\,|w| \wedge |z| = m\,|w|\},$$

namely $\mathrm{D} = \mathrm{M}^0_{0,0,0}$ and $\mathrm{S}_0 = \mathrm{M}^1_{0,0,0}$.

7.6. Theorem. [SeGa 77] For every $k, l, m \in \mathbb{N}$, $\mathrm{M}^0_{k,l,m}, \mathrm{M}^1_{k,l,m} \in$ **REALTIME**. ∎

7.7. Corollary. For every $v \in \{0, 1\}^*$, $\mathrm{M}_v \in$ **REALTIME**.

Proof. Let $v = x_1, \ldots, x_n$. Then evidently

$$M_v = M^{x_1 \oplus x_2}_{0,0,n-2} \cap M^{x_2 \oplus x_3}_{1,0,n-3} \cap \ldots \cap M^{x_{n-1} \oplus x_n}_{n-2,0,0},$$

(here the x_i are treated as Boolean values, and $\oplus$ means "exclusive or"), all constituents of the intersection belong to **REALTIME** (by 7.6), and **REALTIME** is closed under intersection (11.2). ∎

The following theorem also generalizes Corollary 7.7.

7.8. Theorem. [SeGa 77] For every n, every $w_1, \ldots, w_n \in \{0, 1\}^*$ and every $x_1, \ldots, x_n \in \{0, 1\}$,

$$\{w_1 w^{x_1} w_2 \ldots w_n w^{x_n}\colon w \in \{0, 1\}^*\} \in \textbf{REALTIME}. \ ∎$$

Further related results are contained in [Sli 78] and [Sli 79].

We close this subsection by mentioning a further realtime result. For $d \geqq 1$ let $A_d = \{a_1, \ldots, a_d, b_1, \ldots, b_d\}$ and $n_c(w)$ = number of c's in w, and

$$\text{ORIGIN CROSSING}_d = \{w\colon w \in A_d^* \wedge n_{a_1}(w) = n_{b_1}(w) \wedge \ldots \wedge n_{a_d}(w) = n_{b_d}(w)\}.$$

7.9. Theorem. [FimRo 68] For every $d \geqq 1$, ORIGIN CROSSING$_d \in$ **1-T-DTIME**(id). ∎

7.2.2. Selected Polynomial Time Problems

It is impossible to enumerate or even to analyze all known efficient programs for discrete problems, i.e. problems arising in graph theory, automaton theory, algebra, number theory, logic, programming, game theory, and other branches of discrete mathematics. Instead we refer the reader to books like [Knu 69], [Knu 73], [AhHoUl 74], [Chr 75], [BoMu 75], [Meh 77], [HoSa 78], [Even 79], [Kro 79], [Gob 80], [Kuč 83], and the survey paper [Tar 78], which at least partially cover the topic. We restrict ourselves to just a few representative sample results which are referred to elsewhere in this book.

7.2.2.1. Standard Examples

Besides the sets S, S_0, S_1, D the following sets are very often used as examples, counter-examples or witness sets:

$$S\# =_{df} \{w \# w^{-1} : w \in \{0, 1\}^*\}, \quad D\# =_{df} \{w \# w : w \in \{0, 1\}^*\},$$

$$C =_{df} \{0^n 1^n : n \in \mathbb{N}\}.$$

As is known from § 7.1.1, these are realtime problems. If one-tape Turing machines are considered, the situation changes slightly.

7.10. Theorem. An upper bound for the set A on the measure Φ is given in row A and column Φ of Table 7.1.

Table 7.1. $q_i(t)\,(n) = \begin{cases} t(n) & \text{if } n \equiv i(2) \\ n & \text{otherwise} \end{cases}$ for $i = 0, 1$.

The real number $\varepsilon > 0$ can be chosen arbitrarily

	T-DTIME	1-T-DTIME	2-T-DTIME	T-NTIME	1-T-NTIME	2-T-NTIME
S	εn^2	εn^2	$(1.5+\varepsilon)\,n$	εn^2	n	n
$\bar{S}$	εn^2	εn^2	$(1.5+\varepsilon)\,n$	$\varepsilon n \log n$	n	n
S_0	$q_0(\varepsilon n^2)$	$q_0(\varepsilon n^2)$	$q_0((1.5+\varepsilon)\,n)$	$q_0(\varepsilon n^2)$	n	n
S_1	$q_1(\varepsilon n^2)$	$q_1(\varepsilon n^2)$	$q_1((1.5+\varepsilon)\,n)$	$q_1(\varepsilon n^2)$	n	n
S#	$q_1(\varepsilon n^2)$	n	n	$q_1(\varepsilon n^2)$	n	n
D	$q_0(\varepsilon n^2)$	$q_0(\varepsilon n^2)$	$q_0((1.5+\varepsilon)\,n)$	$q_0(\varepsilon n^2)$	$q_0((1+\varepsilon)\,n)$	$q_0((1+\varepsilon)\,n)$
D#	$q_1(\varepsilon n^2)$	$q_1((1+\varepsilon)\,n)$	$q_1((1+\varepsilon)\,n)$	$q_1(\varepsilon n^2)$	$q_1((1+\varepsilon)\,n)$	$q_1((1+\varepsilon)\,n)$
C	$q_0(\varepsilon n \log n)$	n	n	$q_0(\varepsilon n \log n)$	n	n
$\bar{C}$	$q_0(\varepsilon n \log n)$	n	n	$q_0(\varepsilon n \log\log n)$	n	n

Proof. We confine ourselves to an outline of the idea for proving 1. $D \in$ **2-T-DTIME** $\left(q_0((1.5+\varepsilon)\,n)\right)$ and 2. $\bar{C} \in$ **T-NTIME**$\left(q_0(\varepsilon n \log\log n)\right)$, the remaining claims being obvious.

Ad 1. Let $\varepsilon > 0$ be given. Choose k such that $\frac{2}{k} < \varepsilon$. A 2-DTM M accepting D can work as follows:

Phase 1: The machine copies the whole input word onto the worktape using k-fold contraction (i.e. k input symbols are written into one square) and simultaneously checking its parity. Inputs of odd length are rejected.

Phase 2: (Finding the middle of the word on the worktape.) Both heads go to the left with equal speed until the worktape head has reached the left end. Then both heads go to the right until the input head has reached the last symbol, however the worktape head goes with half the speed of the input head.

Phase 3: Both heads go to the left and compare.

This takes altogether $n + \frac{2n}{k} + \frac{n}{2} + \text{const} \leqq_{ae} (1.5 + \varepsilon)\, n$ steps if n is even.

Ad 2. [AlMe 76] The following NTM M accepts $\overline{\mathrm{C}}$. Words which do not belong to the regular set 0^+1^+ are accepted. For words of the form $w = 0^i1^j$, $|w| = n$, the machine guesses a prime number $p \leqq c_1 \log n$ (c_1 will be determined in the remaining part of the proof) and accepts w if $i \not\equiv j(p)$.

We must prove that if $i \neq j$, then there exists always a $p \leqq c_1 \log n$ such that $i \not\equiv j(p)$. Let p_i be the ith prime number. Obviously, the smallest prime number p such that $i \not\equiv j(p)$ is equal to p_r where

$$r = \min \{k\colon p_k \text{ does not divide } |i - j|\}.$$

Consequently,

$$n \geqq |i - j| \geqq p_1 \ldots p_r \geqq (r - 1)! > \left(\frac{r-1}{2}\right)^{\frac{r-1}{2}}.$$

Hence $r \leqq c_0 \dfrac{\log n}{\log \log n}$.

By the prime number theorem, $p_r \leqq c_1 \log n$ for suitable c_1, and consequently it has a binary representation of length $\lesssim \log \log n$. The computation of the remainders of i and j takes time $\lesssim n \log \log n$, and their comparison is possible in time $(\log \log n)^2 \lesssim n \log \log n$. ∎

Remark 1. Using the same idea as used for D one easily proves

$$M_v \in \mathbf{2\text{-}T\text{-}DTIME}\left(\left(2 - \frac{1}{|v|} + \varepsilon\right) \mathrm{n}\right) \text{ for any } v \in \{0, 1\}^* \setminus \{\mathrm{e}, 0, 1\}.$$

Remark 2. Theorem 8.13 shows that S, S_0, S_1, $\mathrm{S}^\#$, D, $\mathrm{D}^\#$ and C cannot be recognized any more quickly by DTM's or NTM's.

7.11. Theorem. The sets S, S_0, S_1, $\mathrm{S}^\#$, D, $\mathrm{D}^\#$, C can be recognized by DTM's in linear average time. The same is true for 1-DTM's.

Proof. We show this for S and DTM's. Assume $n \equiv 0(2)$, the other case is similar. We define

$$\mathrm{S}_{n,m} = \left\{waubw^{-1}\colon waubw^{-1} \in \{0, 1\}^n \wedge |w| = m \wedge a \neq b\right\}$$

for $m = 0, 1, \ldots, \frac{n}{2} - 1$ and

$$\mathrm{S}_{n,\frac{n}{2}} = \left\{ww^{-1}\colon ww^{-1} \in \{0, 1\}^n\right\}.$$

Then the sets $\mathrm{S}_{n,1}, \ldots, \mathrm{S}_{n,\frac{n}{2}}$ are pairwise disjoint and $\{0, 1\}^n = \mathrm{S}_{n,0} \cup \ldots \cup \mathrm{S}_{n,\frac{n}{2}}$. Furthermore, $\operatorname{card} \mathrm{S}_{n,\frac{n}{2}} = 2^{n/2}$ and $\operatorname{card} \mathrm{S}_{n,m} = 2^{n-m-1}$ for $m = 0, \ldots, \frac{n}{2} - 1$. The usual strategy for accepting S (compare the first with the last letter, the second with the last but one etc.) works within $c \cdot m \cdot n$ steps for words from S_m. Hence the average time for input length n is not greater than

$$\frac{1}{2^n}\left(c \cdot \frac{n}{2} \cdot n \cdot 2^{n/2} + \sum_{m=0}^{\frac{n}{2}-1} c \cdot m \cdot n \cdot 2^{n-m-1}\right) \leqq c' \cdot n \cdot \sum_{m=0}^{\infty} \frac{m}{2^m} \leqq c''n. \blacksquare$$

Note also that probabilistic machines can accept the languages in question in less than quadratic time (cf. 9.20 for $D^{\#}$).

Upper space bounds for S, S_0, S_1, $\overline{S}$, $S^{\#}$, D, $D^{\#}$, C and $\overline{C}$ are given in the next theorem.

7.12. Theorem. An upper bound for the set A on the measure Φ is given in row A and column Φ of Table 7.2.

Table 7.2. $p_i(s)\,(n) = \begin{cases} s(n) & \text{if } n \equiv i(2) \\ & \text{for } i = 0, 1 \\ 0 & \text{otherwise} \end{cases}$

	1-T-DSPACE	2-T-DSPACE	1-T-NSPACE	2-T-NSPACE
S	n	$\log n$	n	$\log n$
$\overline{S}$	n	$\log n$	$\log n$	$\log n$
S_0, D	n	$p_0(\log n)$	$p_0(n)$	$p_0(\log n)$
S_1, $S^{\#}$, $D^{\#}$	n	$p_1(\log n)$	$p_1(n)$	$p_1(\log n)$
C	$\log n$	$p_0(\log n)$	$p_0(\log n)$	$p_0(\log n)$
$\overline{C}$	$\log n$	$p_0(\log n)$	$p_0(\log \log n)$	$p_0(\log \log n)$

Proof. The only statement which deserves a proof is $\overline{C} \in$ **1-T-NSPACE**$(p_0(\log \log n))$. The machine constructed in the proof of $\overline{C} \in$ **T-NTIME**$(q_0(\varepsilon n \log \log n))$ does work within space $p_0(\log \log n)$ if the input is given on a 1-way input tape. ∎

Remark. For lower space bounds of the sets dealt with in Theorem 7.12 see 8.2, 8.29.2, and 8.31.2.

7.13. Theorem. [Dek 69] The function $\text{un}(n) = 1^n$ can be computed by a **DTM** within time $2n + |\text{bin}\, n| + 2$.

Proof. A DTM can compute the function un according to the following algorithm. (We assume that the input n stands in binary notation in a first track of the tape.)

1. Mark the first symbol by #, then move to the right, mark the last symbol by * and, simultaneously, write a 0 into the second track of the square holding the last input symbol.
2. **If** # and * mark the same tape square holding x
   then if $x = 0$ **then** stop
   else print 1 onto the output tape. Stop.

3. Print 1 onto the output tape and move to the left replacing in the second track all symbols 1 by 0 until a square with 0 or □ is found.
4. Change this symbol into 1, print 1 onto the output tape and
   if the scanned square is unmarked
   then move to the square marked by $*$.
   Go to 3.
   else (the head scans # and reads 1 in track 2)
   if the input word to the right of # consists only of zeros
   then stop
   else move to the right, write # into the first square whose first track contains a 1 and move to $*$. Go to 2.

Let $\sigma(2^n)$ be the number of steps of a DTM executing this algorithm between printing the (2^{n-1})th (exclusively) and the (2^n)th (inclusively) output symbol. Evidently, $\sigma(2) = 1$ and $\sigma(2^{n+1}) = 2\sigma(2^n) + 1$, from which we conclude $\sigma(2^n) = 2^n - 1$. Let $\tau(n)$ be the number of steps which are used for printing n output symbols (without counting the steps of executing item 1 of the algorithm). Then

$$\tau(2^m) = 1 + \sigma(2) + \ldots + \sigma(2^m) = 2^{m+1} - (m + 1).$$

Claim. $\tau(x) \leqq 2x$ for all x.

This is true for all x of the form 2^m. For the remaining x we prove the claim by induction. For $x \leqq 2$ we evidently have $\tau(x) \leqq 2x$. Let $\tau(x) \leqq 2x$ for all $x \leqq 2^m$. We prove the claim for $2^m + 1 \leqq x < 2^{m+1}$. When the (2^m)th output symbol is printed, the head scans 1, and to the right of this square there are m zeros (in track 2). The next output symbol will be printed only after m steps. Hence

$$\begin{aligned}\tau(2^m + x) &\leqq \tau(2^m) + (m - 1) + \tau(x) \leqq 2^{m+1} - (m + 1) + m - 1 + 2x \\ &\leqq 2(2^m + x).\end{aligned}$$

This proves the claim.

On input n the DTM needs $|\text{bin}\, n| + 2$ steps for executing item 1 and then $\tau(n)$ steps for printing the n output symbols. Thus its time is bounded by $2n + |\text{bin}\, n| + 2$. ∎

7.14. Corollary.

1. $\{\text{bin}\, n \# 1^n : n \in \mathbb{N}\} \in$ **2T-DTIME**(Lin).
2. $|w| \in$ 2-T-DTIME(Lin). ∎

7.2.2.2. *Integer Multiplication*

The usual algorithm for multiplying binary numbers taught at school works within quadratic time. Iterative arrays can multiply in realtime. We define mult$(x, y) = x \cdot y$.

7.15. Theorem. [Atr 65] mult $\in$ **IA^1-DTIME**(id). ∎

The simulation results known at present (19.32.1) imply only a quadratic time bound for multiT-DM's. A much better result exploiting the fast Fourier transform has been proved by A. Schönhage and V. Strassen. Their algorithm can be realized on multiT-DM's, and a technique developed by M. J. Fischer and L. J. Stockmeyer allows us to construct an on-line multiT-DM for computing mult whose running time should be compared with the lower bound stated in 8.36.

7.16. Theorem. [ScStr 71], [FimSt 74]

1. mult $\in$ multiT-DTIME($n \log n \log \log n$).
2. mult $\in$ 1*-multiT-DTIME($n \log^2 n \log \log n$). ∎

For a detailed treatment of integer multiplication see [Knu 69].

It can readily be seen that integer multiplication can be done by RAM's in linear time (unit cost). A stronger result is given by

7.17. Theorem. [Schö 78] mult $\in$ SRAM-DTIME(Lin). ∎

7.2.2.3. *Graph Theoretical Problems*

First of all we have to agree upon how to understand problems about discrete objects. As algorithms can understand only words as inputs, it is necessary to encode the objects by words. It is clear that the complexity of a problem is dependent on the encoding of the inputs. The question of which encoding should be preferred cannot be solved in a general way. However, in any case, there always exist several very natural, very reasonable encodings which do not contain additional information, are not artificially padded, and where integers occuring as parts of the objects are not represented in unary. It is a general experience that any two of such reasonable encodings can be exchanged within polynomial time. Thus it is of no concern to feasibility which reasonable encoding is chosen. However, the encoding is of concern for concrete upper bounds, which is important in particular for small bounds. In the following examples we do not inform the reader about the chosen encodings although this, strictly speaking, would be necessary.

Let us consider as an example an encoding of a graph (V, E). We always define $v = \text{card } V$ and $e = \text{card } E$. Assume $V = \{x_1, \ldots, x_v\}$, $E = \{(x_{i_1}, x_{j_1}), \ldots, (x_{i_e}, x_{j_e})\} \subseteq V^2$. The vertices $x_1, \ldots, x_v$ can be identified with the numbers $0, \ldots, v - 1$ which we represent in binary notation. Then we can define

$$\text{Code}(V, E) =_{\text{df}} x_1 \# \ldots \# x_v \# \# x_{i_1} \# x_{j_1} \# \# \ldots \# \# x_{i_e} \# x_{j_e}.$$

Note that $n =_{\text{df}} |\text{Code}(V, E)| \lesssim v^2 \log v$. It is often more informative to express the complexity of the algorithms in terms of v and e instead of n. The latter can always be reached by an obvious calculation. Another possibility for encoding a graph is by an incidence matrix, which can be easily written as a word by simply concatenating the rows. This word is of length v^2, and a change from one to the other encoding can be carried out in quadratic time.

The design of efficient algorithms for specific problems usually requires detailed studies exploiting to a large extent the corresponding mathematical theories. Very often the complexity bounds of the algorithms found this way can be improved by a clever use of appropriate data structures. The following results about graph problems, quoted without proofs or explanations, can serve as examples for this statement.

Our first result concerns planarity testing. We define

$$\text{PLANAR} =_{\text{df}} \{\text{Code}(V, E) : (V, E) \text{ is a planar graph}\}.$$

7.18. Theorem. [HoTa 74] PLANAR $\in$ **RAM-DTIME**$\big(\text{Lin}(v)\big)$. ∎

Remark. In [JaSi 80] a deterministic parallel RAM is designed which decides PLANAR and runs within time $\text{Lin}(\log^2 v)$ (the input being stored in registers). This parallel algorithm is remarkable in that it is not obvious at all how parallelity could be used for planarity testing. The algorithm uses $c \cdot v^4$ processors and can be implemented on a DTM in space $\log^4 v$.

Definition. A *spanning tree* of a graph (V, E) whose edges are labelled by integers is a connected subgraph of (V, E) which is a tree and contains all vertices of (V, E). Let spanning tree $= \{(G, T) : T$ is a spanning tree of $G\}$ and let Min spanning tree denote the corresponding optimization problem, with respect to the valuation function "sum of labels of the edges".

7.19. Theorem. [Yaoa 75] Min spanning tree $\in$ RAM-DTIME$\big(\text{Lin}(e \log \log v)\big)$. ∎

Remark. Finding minimal spanning trees in planar graphs is possible in time Lin v ([ChTa 76]). In [SaJa 81] a deterministic parallel RAM is designed which solves Min spanning tree and runs within time $\text{Lin}(\log^2 v)$.

Definition. A *matching of a graph* (V, E) is a set $M \subseteq E$ such that no two edges in M have a common vertex. Let matching $= \{(G, M) : M$ is a matching of $G\}$ and let Max matching be the corresponding search problem with respect to the valuation function "cardinality of M".

7.20. Theorem. [MiVa 80] Max matching $\in$ RAM-DTIME$\big(\sqrt{v} \cdot e\big)$. ∎

Remark. Matching problems with restrictions are considered in [ItRo 77]. A restriction is given by a subset $E_1 \subseteq E$ and a number $r_1 \in \mathbb{N}$ and the demand card $(M \cap E_1) \leqq r_1$. (The time scheduling problem, see p. 270, can be reformulated as a restricted matching problem for bipartite graphs.) The existence problem for the matching problem with many restrictions is **NP**-complete. If only one restriction is given, then this problem is in **P** and can be decided within time $c \cdot v \cdot e$. Further matching problems have been considered in [ItRoTa 78].

Definition. Let (V, E, c, s, t) be a graph with distinguished nodes s (source) and t (target) and a function $c: E \mapsto \mathbb{Q}$ (capacity). A *flow* is a function $f: E \mapsto \mathbb{Q}$ with the following properties:

(1) $0 \leqq f \leqq c, c > 0$.

(2) Let $\text{in}\, x =_{\text{df}} E \cap (V \times \{x\})$, $\text{out}\, x =_{\text{df}} E \cap (\{x\} \times V)$.

For all $x \in V \setminus \{s, t\}$ the equation

$$\sum_{e \in \text{in}\, x} f(e) = \sum_{e \in \text{out}\, x} f(e)$$

is satisfied.

Let flow $= \{((V, E, c, s, t), f) : f$ is a flow of $(V, E, c, s, t)\}$ and let Max flow be the search problem with respect to the valuation function

$$b_{\text{flow}}((V, E, c, s, t), f) = \sum_{e \in \text{out}s} f(e) - \sum_{e \in \text{in}s} f(e).$$

7.21. Theorem. [GaNa 79], [Karz 74] Max flow $\in$ RAM-DTIME(Lin min($ev \log^2 v$, v^3)). ∎

For restricted versions of the flow problem see [ItSh 79], [KoŘi 81], [Kuč 81] and [Kuč 84]. A parallel algorithm for Max flow can be found in [JoVe 82].

7.2.2.4. Linear Programming

Linear Programming is the following (existence) problem:

$$\text{LP} =_{\text{df}} \Big\{(A, b, c, d) : A \text{ is an } (m, k) \text{ integer matrix} \wedge c \in \mathbb{G}^k \wedge b \in \mathbb{G}^m \wedge d, k, m \in \mathbb{N} \wedge \bigvee_{x \in \mathbb{Q}^k} (Ax \leqq b \wedge c'x \geqq d)\Big\}.$$

(Vectors are understood as columns, transposition is indicated by '.) In the literature "Linear Programming" usually means the optimization problem Max lp with the valuation function

$$b_{\text{lp}}((A, b, c, d), x) =_{\text{df}} c'x.$$

Max lp and LIQ($\mathbb{Q}$) which is defined below are known to be polynomially equivalent.

$$\text{LIQ}(\mathbb{Q}) =_{\text{df}} \Big\{(A, b) : A \text{ is an } (m, k) \text{ integer matrix } \wedge b \in \mathbb{G}^m \wedge k, m \in \mathbb{N} \wedge \bigvee_{x \in \mathbb{Q}^k} (Ax \leqq b)\Big\}.$$

LIQ($\mathbb{Q}$) stands for linear inequalities with intended solutions in $\mathbb{Q}$. The following special cases of LIQ($\mathbb{Q}$) are also of interest:

a) $\mathbb{Q}$ can be replaced by $\mathbb{G}$, $\mathbb{N}$ or {0, 1}. These problem will be denoted by LIQ($\mathbb{G}$), LIQ($\mathbb{N}$), and LIQ({0, 1}), respectively.

b) Instead of inequalities we can consider equalities. The corresponding problems will be denoted by LEQ( ).

c) The systems can consist of only one single (in)equality. The corresponding problems will be denoted by SLEQ( ), (SLIQ( )).

This terminology and the following slightly modified table are taken from [Emd 79].

7.22. Theorem. In the following table **P** [**NP**] in row A and in column B means $B(A) \in$ **P** [$B(A)$ is **NP**-complete]:

	LIQ	LEQ	SLEQ	SLIQ
$\mathbb{Q}$	P	P	P	P
$\mathbb{G}$	NP	P	P	P
$\mathbb{N}$	NP	NP	NP	P
{0, 1}	NP	NP	NP	P

Proof. We restrict ourselves to presenting L. G. KHACHIAN's proof [Kha 79] of $LIQ(\mathbb{Q}) \in \mathbb{P}$, which was a long standing open question. That the special case of $LIQ(\mathbb{Q})$ where at most two variables per inequality occur is in $\mathbb{P}$ has already been shown in [AsSh 79]. For the proofs of the remaining statements the reader is referred to [Emd 79].

To prove $LIQ(\mathbb{Q}) \in \mathbb{P}$ it is sufficient to consider the case of strict inequalities (see [Kha 79] and [GaLo 79]). Let the system

$$a_i'x < b_i, \qquad i = 1, \ldots, m, \tag{1}$$

where $a_i \in \mathbb{G}^k$, $b_i \in \mathbb{G}$ and $a_i' = (a_{i1}, \ldots, a_{ik})$.

Let n denote the length of the input (A, b) where all integers are represented in binary. Then using methods of linear algebra it is easy to prove the following result (cf. [GaLo 79]).

Fact 1. If the system (1) has a rational solution, then it has solutions inside the cube $|x_i| \leqq 2^n$, $i = 1, \ldots, k$, and the volume of the solution set S inside this cube satisfies the inequality $\mathrm{vol}\,(S) \geqq 2^{-kn}$. □

KHACHIAN's algorithm consists of the construction of a sequence of ellipsoids of strictly decreasing volume, each containing the set S. An ellipsoid is represented by its centre z and its matrix C in the form $(x - z)'C^{-1}(x - z) = 1$. We use the notation $(z, C) =_{\mathrm{df}} \{x : (x - z)'C^{-1}(x - z) \leqq 1\}$.

Fact 2. Let $E = (z, C)$ be an ellipsoid in the k-dimensional Euclidian space and let $a'(x - z) = 0$ be a hyperplane. Then the following statements hold true:

1. The smallest (with respect to volume) ellipsoid F containing $E \cap \{x : a'(x - z) \leqq 0\}$ is given by $F = (y, B)$ where

$$y = z - \frac{1}{k+1} \cdot \frac{Ca}{\sqrt{a'Ca}}, \qquad B = \frac{k^2}{k^2 - 1} \cdot \left(C - \frac{2}{k+1} \cdot \frac{(Ca)\,(Ca)'}{a'Ca}\right).$$

2. $\mathrm{vol}(F) \leqq 2^{-\frac{1}{2(k+1)}}\,\mathrm{vol}(E)$. □

Both statements are first proved for the unit sphere and the hyperplane $x_k = 0$, and then the results are carried over to the general case by an affine transformation which leaves the volume ratio invariant and which transforms the smallest ellipsoid containing the lower half ball into F (provided the unit sphere is transformed into E).

In what follows we need the ellipsoid $E_0 = (x_0, C_0)$ with $x_0 = 0$,

$$C_0 = \begin{pmatrix} 2^{2n} & 0 \dots & 0 \\ 0 & \ddots & \vdots \\ & & 0 \\ 0 & \dots & 0 \; 2^{2n} \end{pmatrix}.$$

This contains the cube $|x_i| \leqq 2^n$ and hence also S.

KHACHIAN's algorithm:

1. $j = 0$.
2. If all given inequalities are satisfied by x_j then return x_j as a solution else let i_j be an index such that $a'_{i_j} x_j \geqq b_{i_j}$. The hyperplane $a'_{i_j}(x - x_j) = 0$ divides E_j into two parts.
3. Define the minimum ellipsoid $E_{j+1} = (x_{j+1}, C_{j+1})$ containing $E_j \cap \{x: a'_{i_j}(x - x_j) \leqq 0\}$, i.e. that half of E_j which contains S. [x_{j+1} and C_{j+1} can be computed from x_j, C_j and a'_{i_j} according to the formulas in Fact 2.]
4. $j = j + 1$.
   If $j < 6k(k + 1)n$ then goto 2 else return the answer "The system has no solution".

End.

Analysis of the algorithm: Statement 2 of Fact 2 shows that for the sequence $E_0, E_1, \dots$ of ellipsoids constructed by the algorithm it holds that

$$\dots < \operatorname{vol}(E_1) < \operatorname{vol}(E_0) \leqq 2^{2kn}.$$

More precisely,

$$\operatorname{vol}(E_j) \leqq 2^{-\frac{j}{2(k+1)}} \cdot 2^{2kn} < 2^{2kn - \frac{j}{2(k+1)}}.$$

Hence for $j = 6k(k + 1)n$ we get $\operatorname{vol}(E_j) < 2^{-kn}$.

On the other hand, each of these ellipsoids contains the set S ($S \subseteqq E_0$; if $S \subseteqq E_j$, then $S \subseteqq \{x: a'_{i_j}(x - x_j) \leqq 0\} \cap E_j \subseteqq E_{j+1}$ because the points of S satisfy $a'_{i_j} x < b_{i_j} \leqq a'_{i_j} x_j$) and, by Fact 1, $S = \emptyset$ or $2^{-kn} \leqq \operatorname{vol} S$. This means that if after $6k(k + 1)n$ steps no solution has been found, then no solution exists at all. ∎

Remark 1. Although the well-known simplex method has an exponential time complexity [KlMi 72] it runs in practically relevant cases often faster than KHACHIAN's algorithm (for a linear expected time simplex algorithm see [AdMe 84]). The ellipsoid method of the above proof has been improved in [Karm 84] to yield a faster algorithm for LIQ($\mathbb{Q}$) which is, moreover, numerically stable. A polynomial time algorithm for LIQ($\mathbb{Q}$) using simplexes instead of ellipsoids is described in [YaLe 82].

Remark 2. In [KaPa 80] connections between LIQ($\mathbb{Q}$) and certain optimization problems are investigated.

Remark 3. LIQ($\mathbb{G}$) restricted to a bounded number of variables is in **P** ([Len 81], see also [Emd 81] and [Kan 83a]).

7.3. *Nondeterministic Polynomial Time*

7.3.1. Examples

We know of many problems from various mathematical disciplines, which can easily be solved nondeterministically in polynomial time but for which no deterministic polynomial time algorithm is as yet known. We present some of them which have the property of being NP-complete. Their significance for complexity theory will become clear later (§ 14, § 24).

Nondeterministic polynomial time algorithms for these problems always work according to the following pattern: Guess a solution and verify whether the guess is correct.

As nondeterministic algorithms are not satisfactory in practice, we ask whether nondeterminism can be eliminated without more than polynomial loss of time. 5.11.3 and 5.12.3 state our present knowledge: We are not yet able to avoid exponential time in general, but in some cases subexponential bounds can be found. For instance, for

$$\mathrm{SAT} =_{\mathrm{df}} \{H : H \text{ is a Boolean formula in conjunctive normal form} \wedge H \text{ is satisfiable}\}$$

we have $\mathrm{SAT} \in \mathrm{DTIME}\left(2^{\mathrm{Lin}\frac{n}{\log n}}\right)$ because there can be only $c\,\frac{n}{\log n}$ variables in an input of length n. Hence, only $2^{c\frac{n}{\log n}}$ assignments of the variables have to be checked, and a checking does not take more than $c'n^2$ steps. For better results see [MoSp 79].

Let SLP denote the problem LP restricted to a 1-row matrix and let KNAPSACK $=$ SLP($\mathbb{N}$) and (0, 1)-KNAPSACK $=$ SLP($\{0, 1\}$). An interpretation for (0, 1)-KNAPSACK is: Given n objects of size $a_1, \ldots, a_k$ and of value $c_1, \ldots, c_k$ and a knapsack of size b, find a subset of these objects fitting into the knapsack and having a total value of at least d, where d is also given.

The result (0, 1)-KNAPSACK $\in \mathrm{DTIME}(2^{\mathrm{Lin}\sqrt{n}})$ has been communicated to the authors by B. Monien. A similar result for graph colouring is contained in [Law 76], for the clique problem see [TaTr 77], for the exact cover problem see [MoSpVo 80].

7.3.2. Pseudo-Polynomial Time

For $x = (w_1, \ldots, w_m, z_1, \ldots, z_k) \in \bigcup_{m,k\geq 1} (\Sigma^*)^m \times \mathbb{N}^k$ we define

$$|x| = \sum_{i=1}^{m} |w_i| + \sum_{i=1}^{k} \log z_i + m + k \quad \text{and} \quad \max x = \max\,(z_1, \ldots, z_k).$$

Note that an indexed variable x_i is understood as the word $x \# \mathrm{bin}\; i \in \Sigma^*$. A problem $A \subseteq \bigcup_{m,k\geq 1} (\Sigma^*)^m \times \mathbb{N}^k$ is said to be a pseudo-polynomial time problem iff there exists a

DTM M deciding A and a polynomial p such that $t_M(x) \leq p(|x|, \max x)$. For a detailed discussion of such problems see [GaJo 79].

We give the following example. Let SOS = SLEQ({0, 1}) be the "sum of subset" problem.

7.23. Theorem. SOS is a pseudo-polynomial time problem.

Proof. For the input $(a_1, \ldots, a_k, b) \in \mathbb{N}^{k+1}$ we construct a list L which initially contains 0. At stage i the list contains all values $\leq b$ which can be represented in the form $a_{i_1} + \ldots + a_{i_r}$ with $i_1, \ldots, i_r \leq i$. At stage $i + 1$ we form $a_{i+1} + x$ for all entries x of L and update L by including all values $a_{i+1} + x \leq b$ which are not already in L. The value b is in L after stage k if and only if $(a_1, \ldots, a_k, b) \in$ SOS. This algorithm needs no more than kb additions. ∎

This algorithm is not a polynomial time algorithm because it is possible that $b \geq 2^{c \cdot |(a_1, \ldots, a_k, b)|}$ for some $c > 0$ and for infinitely many inputs. If however SOS is restricted to instances where b is bounded by a polynomial in the input length, then this is polynomially decidable. The fact that SOS is **NP**-complete (14.21) is hence caused by the occurence of very large integers in the instances.

In [Mon 80] the notion of pseudo-polynomiality is extended and the rich structure in the class of pseudo-polynomial time problems with respect to a suitable reducibility notion is described.

As an example for a problem which is not a pseudo-polynomial time problem (unless **P** = **NP**) we mention TRAVELLING SALESMAN defined on p. 166 (see [GaJo 79]).

7.3.3. Prime Numbers

Let PRIME be the set of all prime numbers and $\text{PRIME}^1 = \{0^n : n \text{ is a prime number}\}$. PRIME plays a role in the study of the **P-NP** problem (cf. § 24.2.1.2.3).

It is trivial that $\overline{\text{PRIME}}$ belongs to **NP**: Given n, guess two nontrivial factors k and m, compute $k \cdot m$ and verify $n = km$. This algorithm runs within polynomial time. It is not so evident that PRIME belongs to **NP**.

7.24. Theorem. [Pra 75] PRIME $\in$ **NP**.

Proof. The proof is based on the following number theoretic fact.

Lemma. m is a prime $\leftrightarrow \bigvee\limits_{x \leq m} \left(x^{m-1} \equiv 1(m) \wedge \bigwedge\limits_{\substack{q \text{ divides } m-1 \\ q \neq 1}} x^{\frac{m-1}{q}} \not\equiv 1(m) \right)$, i.e. if and only if there exists a primitive root modulo m.

A nondeterministic algorithm for accepting PRIME works as follows:

Test (m):

(1) Guess $x \leq m$ such that $x^{m-1} \equiv 1(m)$.
$a = 1$.

(2) **If** $a = m - 1$ **then** accept **else** goto (3).

(3) Guess a divisor q of $m - 1$.

(4) Test (q).

(5) **If** $x^{\frac{m-1}{q}} \not\equiv 1(m)$ **then** $a = aq$ **and goto** 2 **else reject.**

By the lemma it is clear that a prime p is accepted when a primitive root x of p and all prime factors of $p - 1$ are guessed correctly. On the other hand, nonprimes cannot be accepted.

Let $t(m)$ be the number of steps the algorithm has to perform on a successful path on m. Let $m - 1 = p_1^{\alpha_1} \dots p_k^{\alpha_k}$. To accept m the program Test has to be called $\alpha_1 + \dots + \alpha_k$ times as a subroutine, i.e. at most $\log m$ times. As the remaining steps of Test (m) can be performed in time polynomial in $\log m$ the algorithm runs within polynomial time (in the input length). ∎

The usual deterministic algorithm for deciding PRIME, which consists of checking for all numbers $m \leqq \sqrt{n}$ whether m divides n, runs in time 2^{cn}, but it is easily seen that it can be implemented in linear space.

7.25. Theorem. PRIME $\in$ **DLINSPACE.** ∎

7.26. Corollary. PRIME^1 $\in$ **L.** ∎

Remark 1. In [HsYe 72] it is shown that PRIME^1 can be accepted by a finite deterministic automaton with three markers.

Remark 2. The preceding corollary should be compared with the result PRIME^1 $\notin$ **2-T-DSPACE**(s) for $s \prec \log$ (see 8.33).

From Corollary 7.26 we obtain PRIME^1 $\in$ **P**. In particular, for iterative arrays one achieves real-time:

7.27. Theorem. [Fip 65] PRIME^1 $\in$ **IA^1-DTIME**(id). ∎

This result leaves unresolved the question of whether PRIME belongs to **P**. On probabilistic machines polynomial time algorithms for PRIME are possible (9.24). G. L. Miller has shown that the *extended Riemann hypothesis* implies PRIME $\in$ **P**. We now explain this hypothesis. The Jacobi symbol $\left(\frac{n}{d}\right)$ for integers n and odd natural numbers d is defined inductively as follows: If d is an odd prime, then $\left(\frac{n}{d}\right) \in \{0, 1, -1\}$ and $\left(\frac{n}{d}\right) \equiv n^{\frac{d-1}{2}} (d)$.

If $d = d_1 d_2$, then $\left(\frac{n}{d}\right) = \left(\frac{n}{d_1}\right) \cdot \left(\frac{n}{d_2}\right)$.

The extended Riemann hypothesis (ERH) is the following statement: For every $d \equiv 1(4)$ such that d is either a prime or a product of two primes the Dirichlet L-function $L_d(s) =_{\mathrm{df}} \sum_{n=1}^{\infty} \left(\frac{n}{d}\right) \cdot n^{-s}$ with complex s has in the strip $0 \leqq \mathrm{Re}\,(s) \leqq 1$ zeros only on the line $\mathrm{Re}\,(s) = \frac{1}{2}$. G. L. Miller's approach has been simplified by H. W. Lenstra (see also [Paj 80]). We first state his result.

7.28. Lemma. [Len 79] Assume the ERH. Let $m > 1$ be an odd integer, $m - 1 = 2^t \cdot u$, $u \equiv 1(2)$. Then there exists a constant c independent of m such that

$$m \in \mathrm{PRIME} \leftrightarrow \bigwedge_{\substack{a \leq c \cdot \log^2 m \\ a \in \mathrm{PRIME} \\ a \neq m}} \left(a^u \equiv 1(m) \vee \bigvee_{0 \leq j < t} a^{2^j u} \equiv -1(m)\right). \blacksquare$$

From this purely number theoretic result, in which the ERH is responsible for the small bound $c \cdot \log^2 m$, G. L. MILLER's result follows immediately.

7.29. Theorem. [Mil 75] If the ERH is true, then PRIME $\in$ **P**. ∎

The constant c in Theorem 7.28 is specified in [Bac 82] and thus a fast primality test (under the assumption of the ERH) can actually be implemented.

Remark. It is still unknown whether polynomial time algorithms for PRIME can be found, which are not based on the assumption of the ERH. A first step towards this goal might be [Adl 80] where a deterministic algorithm is proposed, which does not use the ERH and which is conjectered to run in time $n^{\mathrm{Lin} \log \log n}$.

7.3.4. Graph Isomorphism

The graph isomorphism problem is defined by

$$\mathrm{GRAPHISOM} = \{(G, G') : G \text{ and } G' \text{ are finite isomorphic graphs}\}.$$

Two graphs (V, E) and (V', E') are called isomorphic iff there exists a bijection $f: V \mapsto V'$ such that $(x, y) \in E \leftrightarrow (f(x), f(y)) \in E'$. For an introduction to GRAPHISOM the reader is referred to the book [Hfm 82a]. The graph isomorphism problem is of similar importance for the **P**-**NP**-problem as PRIME (cf. § 24.2.1.2.3) in that it is so far neither proved to be **NP**-complete nor to be in **P**.

For **NP**-complete problems similar to GRAPHISOM see [Yaof 79] and [Lub 81].

In [Bab 81] an algorithm for GRAPHISOM is given which runs in time $2^{\mathrm{Lin}\, n^{2/3}}$. Furthermore, in [Mil 78] some graph classes have been described for which isomorphism can be decided within time $n^{\mathrm{Lin} \log n}$, and many similar results can be found in [Bab 81]. More, namely the existence of polynomial time algorithms, is known for several restricted graph classes.

7.30. Theorem. The isomorphism of graphs can be decided in polynomial time within the following classes:

1. [Mil 80], [FiMa 80] Graphs with bounded genus.
2. [Luk 80], [GaHoLuScWe 82] Graphs with bounded valence. ∎
3. [BaGrMo 82] Graphs with bounded eigenvalue multiplicity. ∎

Remark. For further results on graph isomorphism see [Gri 79a], [LuBo 79], [Scht 79], [BaErSe 80], [CrKi 80], [BaLu 83], [FüScSp 83] and [Schi 83]. Parallel algorithms for graph isomorphism have been studied in [Ref 83].

7.4. *Larger Upper Bounds: Quantifier Elimination*

Many upper bounds beyond **NP** will be given together with the corresponding lower bound results in § 8. Most of them concern logical theories, and they are usually proved using *quantifier elimination*. This method rests upon the fact that if a formula is satisfiable, then it is satisfiable in a model of predictable cardinality. As an example we consider the set MON of all satisfiable first order formulas having only monadic (i.e. one place) predicate symbols and no function symbols.

The decidability of MON is a direct consequence of the following theorem which is due to L. LÖWENHEIM.

7.31. Theorem. Let F be a formula in the language of MON containing k predicate symbols. Then $F \in$ MON iff F has a model of cardinality at most 2^k. ∎

7.32. Theorem. [Lewh 78a] MON $\in \mathbf{NTIME}\left(2^{c\frac{n}{\log n}}\right)$.

Proof. We construct an NTM M to accept MON. Given a formula F in the language of MON containing k predicate symbols $P_1, \ldots, P_k$ and q quantifiers, if $|F| = n$, then $k, q \leqq c_1 \cdot \dfrac{n}{\log n}$ for some c_1.

From Löwenheim's theorem we know that F is satisfiable over a set of 2^k elements iff $F \in$ MON. So we guess a $p \leqq 2^k$ and a model of cardinality p for F, i.e. a matrix $(\varepsilon_{ij})_{k,p}$ over $\{0, 1\}$ where $\varepsilon_{ij} = 1$ means that P_i is satisfied by a_j.

Now the quantifiers of F are successively eliminated. Let $\underset{x}{\mathsf{Q}} G$ be a subformula of F such that G is quantifier free. We consider the case $\mathsf{Q} = \bigwedge$, the case $\mathsf{Q} = \bigvee$ being similar. The following steps are carried out by M:

1. G is transformed into conjunctive normal form

$$G \leftrightarrow G_1 \wedge \ldots \wedge G_l \wedge G_{l+1} \wedge \ldots \wedge G_s .$$

We assume that $G_1, \ldots, G_l$ contain x and $G_{l+1}, \ldots, G_s$ do not contain x. For $i = 1, \ldots, l$ we write $G_i = G_i' \vee G_i''$ where G_i' is that part of the disjunction G_i, which depends on x. Let G_i' have the form

$$P_{i_1}^{\delta_{i1}}(x) \vee \ldots \vee P_{i_{m_i}}^{\delta_{im_i}}(x)$$

where $\delta_{ij} = 0$ (1) indicates the negated (unnegated) occurence of P_{i_j}. Then we have

$$\bigwedge_x G \leftrightarrow \left(\bigwedge_x G_1' \vee G_1''\right) \wedge \ldots \wedge \left(\bigwedge_x G_l' \vee G_l''\right) \wedge G_{l+1} \wedge \ldots \wedge G_s .$$

Comment. G has length bounded by n. Hence at most $c_1 \cdot \dfrac{n}{\log n}$ "variables" of the form $P_i(y)$ occur in G. To find the conjunctive normal form of G simply guess one and check the $2^{c_1 \cdot \frac{n}{\log n}}$ possible assignments. This takes altogether time $\tau_1 \leqq 2^{c_2 \cdot \frac{\log n}{n}}$ for some c_2.

2. For $i = 1, \ldots, l$ it is checked whether $\bigwedge_x G_i'$ is valid in the model previously guessed. If yes, it is replaced by 1, and otherwise by 0, and the formula is simplified. This takes altogether time $\tau_2 \leq l \cdot k \cdot 2^k$.

When all quantifiers are eliminated we get the truth value 1 iff F is satisfiable.

What is the running time of M? It needs $\leq kp \leq k \cdot 2^k$ steps to write down the model. The steps 1 and 2 are carried out q times (once for each quantifier). This gives a time bound of

$$k \cdot 2^k + q(\tau_1 + \tau_2) \leq 2^{2c_1 \cdot \frac{n}{\log n}} + \frac{n}{\log n} \left(2^{c_2 \cdot \frac{n}{\log n}} + 2^{c_1 \cdot \frac{n}{\log n}} \cdot \frac{n}{\log n} \cdot 2^{c_1 \cdot \frac{n}{\log n}} \right)$$

$$\leq 2^{c \cdot \frac{n}{\log n}} \quad \text{for some} \quad c > 0. \blacksquare$$

The reader should consult [FeRa 79] for further results and reference to original papers.

§ 8. Lower Bounds

M. O. Rabin's result (17.1) shows the existence of problems which cannot be solved with arbitrarily small complexity. Hence, the endeavours to improve upper bounds of given problems must principially be of limited success. The limits on the improvability of upper bounds can be quantitatively made precise in terms of lower bounds.

Ideally, the best lower and upper bounds of a problem coincide (up to order). Theorem 18.1 shows that this is not always achievable.

Although ae lower bounds (i.e. lower bounds which are valid for almost all inputs) are in principle preferable, there are many natural problems (for example logical theories) where we get io lower bounds (i.e. lower bounds which are valid for infinitely many inputs) which cannot be improved to ae bounds (cf. 18.27).

It is very difficult to find lower bounds for completely unrestricted general models of non-parallel computation, and so far only a few examples are known (see [BoCo 80]). In general, the lower bounds are proved with respect to a certain machine model, and at least the small lower bounds can usually not be maintained for all reasonable models. Typical examples are the quadratic lower time bounds of 8.13 which are, for instance, not valid for multiT-DM's.

In general, it is much more difficult to prove good lower bounds than good upper bounds. An upper bound on the complexity of a given problem may be deduced simply from a careful analysis of any algorithm solving this problem, whereas lower bounds always have to take into consideration all (infinitely many) algorithms that solve the problem. Note, however, that there are cases in which a careful analysis of algorithms with the given means must necessarily give bad results. The following result is proved in [HaHo 76]: If $\mathcal{F}$ is an arbitrary axiomatizable and consistent theory, then there exists an algorithm with running time n^2 such that no upper bound better than 2^n can be proved within $\mathcal{F}$.

At present only two essentially different methods to prove lower bounds are known, namely counting arguments and the reducibility method. Usually, in the applications of the latter method diagonalization is involved.

Whereas up to now counting arguments have yielded at most quadratic lower bounds, diagonalization allows us to prove arbitrarily large lower bounds. On the other hand, counting arguments may be used even for machine classes with such a small computational power that the reducibility method fails. In particular, there are many lower bound results for restricted classes of algorithms such as straight line programs, search tree algorithms, logical networks etc. Surveys on these topics can be found in [Sav 76], [Lup 74] and [Knu 73].

8.1. Counting Arguments

The principle which is basic to all counting arguments can be described as follows: Let M be a machine deciding (accepting) a language or computing a function whose complexity we are interested in. For every computation we choose one or several characteristic parts, called "characteristics". In order to accept L a machine M must be able to produce at least a certain number $l(n)$ of different characteristics on inputs of length n. If the number of different characteristics which M is able to produce within complexity t is bounded above by $f(n, t(n))$ (f being increasing in its second argument), then we get the inequality $l(n) \leqq f(n, t(n))$ which yields a lower bound for t.

In most of the applications given below (namely in § 8.1.1 and § 8.1.2) a characteristic describes the information flow across the boundary between two squares of the (a) tape of a Turing machine, i.e. it is the sequence of certain parts of those situations in which the boundary is crossed by the head. Such a sequence is called a crossing sequence. If the boundary in question belongs to a one-way tape, then the crossing sequence degenerates to consist of one element only, i.e. only of the part of one situation. This case is treated in § 8.1.1. In § 8.1.2 we present lower bounds by counting nontrivial crossing sequences (belonging to boundaries of a two-way tape).

8.1.1. Counting of Partial Situations

8.1.1.1. 1-T-DSPACE and 1-T-NSPACE

Lower space bounds for one-way Turing machine computations are very often provided by counting partial situations described by words of the form $x_1 \ldots x_i s x_{i+1} \ldots x_k$, where $x_1 \ldots x_i x_{i+1} \ldots x_k$ is the worktape inscription, $x_1, x_k \neq \square$, the remaining part of the tape being empty and the head scans x_{i+1} in state s.

The following lemma yields lower bounds for many standard languages. For $A \subseteq \Sigma^*$ we fix for every n sets $A_n, A' \subseteq (\Sigma \cup \{e\})^n$ with the properties

$$\bigwedge_{u \in A_n} \bigwedge_{v \in A_n} \left(u \neq v \rightarrow \bigvee_p (up \in A \leftrightarrow vp \notin A)\right)$$

and

$$\bigwedge_{u \in A_n'} \bigvee_{p} \left(|up| = n \wedge up \in A \wedge \bigwedge_{v \in A_n'} (u \neq v \rightarrow vp \notin A) \right).$$

8.1. Lemma.

1. For every 1-DTM M deciding A,

$$\text{1-T-DSPACE}_M(n) \geq \log \operatorname{card} A_n .$$

2. For every 1-NTM M accepting A,

$$\text{1-T-NSPACE}_M(n) \geq \log \operatorname{card} A_n' .$$

Proof. Ad 1. Let M be a 1-DTM deciding A and having q states and p worktape symbols. For $u \in \Sigma^*$ let $k(u)$ be the description of that partial situation in which the input head leaves the word u or, if it does not, in which u is accepted or rejected. Let further

$$\hat{s}(n) = \max_{u \in A_n} |k(u)| - 1 \quad \text{for} \quad n \in \mathbb{N}.$$

Evidently, if $u, v \in A_n$ and $u \neq v$, then $k(u) \neq k(v)$. Consequently,

$$\operatorname{card} A_n \leq q \cdot \hat{s}(n) \cdot p^{\hat{s}(n)} \leq 2^{c \cdot \hat{s}(n)}$$

for some $c > 0$. But $\text{1-T-DSPACE}_M(n) \geq \hat{s}(n)$.

Ad 2. Let M be a 1-NTM accepting A and having q states and p worktape symbols. For $u \in A_n'$ we fix a p_u such that $|up_u| = n$, $up_u \in A$ and $\bigwedge_{v \in A_n'} (u \neq v \rightarrow vp_u \notin A)$. The set $K(u) = \{w$: there exists an accepting computation of M on up_u such that w is the description of that partial situation in which the input head leaves the word u or, if it does not, in which up_u is accepted$\}$ is not empty. Since for different $u, v \in A_n'$ the sets $K(u)$ and $K(v)$ must be disjoint, we obtain for $\hat{s}(n) =_{\text{df}} \max_{u \in A_n'} \min_{w \in K(u)} |w| - 1$ the inequality

$$\operatorname{card} A_n' \leq q \cdot \hat{s}(n) \cdot p^{\hat{s}(n)} \leq 2^{c \cdot \hat{s}(n)}$$

for some $c > 0$. But $\text{1-T-NSPACE}(n) \geq \hat{s}(n)$. ∎

8.2. Theorem. A lower bound for the set A on the measure Φ is given in row A and column Φ of Table 8.1.

Table 8.1. $p_i(s)(n) = \begin{cases} s(n) & \text{if} \quad n \equiv i(2) \\ 0 & \text{otherwise} \end{cases} \quad \text{for } i = 0, 1$

	1-T-DSPACE	1-T-NSPACE
S	$\geq n$	$\geq n$
S_0, D	$\geq n$	$\geq p_0(n)$
S_1, S#, D#	$\geq n$	$\geq p_1(n)$
C	$\geq \log n$	$\geq p_0(\log n)$

Proof. All statements are consequences of Lemma 8.1, where $A_n = A'_n = \Sigma^{\lfloor \frac{n}{2} \rfloor}$ is chosen for S, S_0, D, S_1, S#, D#, and where $A_n = A'_n = \left\{0^1, 0^2, \ldots, 0^{\lfloor \frac{n}{2} \rfloor}\right\}$ is choosen for C. ∎

Remark. The proof of the following statement is analogous to that of Lemma 8.1: For every 1-kC-XM M deciding A we have

$$\text{1-kC-DSPACE}_M(n) \succeq \sqrt[k]{\text{card } A_n} \quad \text{and} \quad \text{1-kC-NSPACE}_M(n) \succeq \sqrt[k]{\text{card } A'_n}.$$

Thus we obtain in the same way as in 8.2 for example 1-kC-NSPACE-Comp(S) $\succeq 2^{\frac{n}{2k}}$ and (see [FipMeRo 68]) 1-kC-DSPACE-Comp(C) $\succeq \sqrt[k]{n}$.

The next result has the double character of being a lower bound result and a gap result. The latter becomes clear in the formulation 8.4.

8.3. Theorem. [StHaLe 65] For every 1-DTM M,

$$\text{1-T-DSPACE}_M \preceq 1 \quad \text{or} \quad \text{1-T-DSPACE}_M \succeq_{\text{io}} \log.$$

Proof. Let M be a 1-DTM and define $s = \text{1-T-DSPACE}_M$. If s is not bounded, then for every $k \in \mathbb{N}$ there exists a word w_k with minimal length n_k such that 1-T-DSPACE$_M(w_k) = s(n_k) \geqq k$.

For different initial words u_1, u_2, u_3 of w_k we cannot have $k(u_1) = k(u_2) = k(u_3)$, k being defined in the same way as in the proof of Lemma 8.1. Otherwise w_k could be shortened, and the shortened word would still require k squares. Consequently, we must have $\frac{1}{2}(n_k + 1) \leqq a^{s(n_k)}$ for some $a > 0$ and hence $c \log n \leqq s(n)$ for some $c > 0$ and for infinitely many n. ∎

Remark. Theorem 8.3 cannot be maintained for 1-NTM as follows from 7.12.

Using 4.7 and 8.29 we obtain

8.4. Corollary. For $s \prec \log$,

1. **1-T-DSPACE**(0) = **1-T-DSPACE**(s).
2. $A \notin$ **REG** $\to A \notin$ **1-T-DSPACE**(s).

   For $s \prec \log\log$,

3. **1-T-NSPACE**(0) = **1-T-NSPACE**(s).
4. $A \notin$ **REG** $\to A \notin$ **1-T-NSPACE**(s). ∎

Remark. Theorem 7.12 shows that these gaps cannot be amplified.

8.1.1.2. 1-auxPD-T-DSPACE and 1-auxPD-T-NSPACE

The counting arguments of § 8.1.1.1 used parts of situations consisting of state, worktape inscription and worktape head position. For 1-auxPD-T-XM's we add the top symbol of the pushdown store. Let

$$B = \{u \# v \# u^{-1} \# v^{-1} \colon |u| = |v| \wedge u, v \in \{0, 1\}^*\}.$$

8.5. Theorem. [Bra 77]
1. 1-auxPD-T-DSPACE-Comp B $\succeq$ id.
2. 1-auxPD-T-NSPACE-Comp B $\succeq_{io}$ id. (io refers to all input lengths of the form $4n + 3$.) ∎

The following two results are of the same type as 8.3 and 8.4.

8.6. Theorem. [Bra 77] If a 1-auxPD-T-DM works with space complexity s, then $s \preceq 1$ or $s \succeq_{io}$ log log. ∎

8.7. Corollary. For $s \prec$ log log,
1. **1-auxPD-T-DSPACE**(0) = **1-auxPD-T-DSPACE**(s).
2. $A \notin$ **DCF** $\to A \notin$ **1-auxPD-T-DSPACE**(s). ∎

Remark 1. Theorem 8.6 and Corollary 8.7 cannot be improved because of

$$\{0^{2^{2^k}} : k \geq 0\} \in \textbf{1-auxPD-T-DSPACE}(\log\log).$$

Remark 2. The remarks at the end of § 17.2.2 show that Corollary 8.7 is not valid for **1-auxPD-T-NSPACE** (see also [Chy 82] and [Chy 84]).

8.1.1.3. multiT-DTIME

Let M be a kT-DM with q states and p symbols working in realtime. We consider the computation of M on uv where $|v| = m$ and in particular that part of the worktapes that, after reading of u, can still have influence on the following m steps. This is on every tape a zone of $2m + 1$ squares whose centres are those squares which are scanned by the corresponding heads after having read u. In this subsection we prove lower bounds by counting these "m-parts". As examples we consider the sets

$$\mathrm{R} = \{0^{n_1} \# 0^{n_2} \# \ldots \# 0^{n_s} \#^j 0^{n_{s-j+1}} : s, n_1, \ldots, n_s \geq 1 \wedge 1 \leq j \leq s\},$$

$$\mathrm{MS} = \{w_1 \# \ldots \# w_j \# w_i^{-1} : 1 \leq i \leq j \wedge w_1, \ldots, w_j \in \{0, 1\}^*\},$$

$$\mathrm{MD} = \{w_1 \# \ldots \# w_j \# w_i : 1 \leq i \leq j \wedge w_1, \ldots, w_j \in \{0, 1\}^*\}.$$

8.8. Theorem.
1. [Ros 67] R $\notin$ **multiT-DTIME**(id).
2. [HaSt 65] MS, MD $\notin$ **multiT-DTIME**(id).

Proof. We restrict ourselves to the second statement. Let M be a 1-kT-DM deciding MS and having q states and p worktape symbols. We consider inputs of the form $w = w_1 \# w_2 \# \ldots \# w_j \# w_i^{-1}$, where $1 \leq i \leq j \leq 2^{m-1}$ and $w_1, \ldots, w_j \in \{0, 1\}^m$.

We define the partial situation $k(w_1, \ldots, w_j) = (s, f_1, \ldots, f_k)$, where s is the state of the moment at which the input head enters w_i^{-1} and f_l is the contents of those $2m + 1$ squares of tape l, whose distances from the head do not exceed m at this moment. If M works in realtime its further work depends only on $k(w_1, \ldots, w_j)$ and the further input. Hence, if $w_1' \# \ldots \# w_{j'}' \# w_{i_1}'^{-1}$ is also of the described form and

$\{w_1, \ldots, w_j\} \neq \{w_1', \ldots, w_{j'}'\}$, then we must have $k(w_1, \ldots, w_j) \neq k(w_1', \ldots, w_{j'}')$. This means, the number of the different partial situations cannot be smaller than the number of different nonempty subsets of $\{0, 1\}^m$. In formulas

$$q \cdot p^{k(2m+1)} \geqq 2^{2^m} - 1,$$

which is a contradiction for sufficiently large m. ∎

Remark 1. The proof can be maintained to show MS, MD ∉ **kTl-DTIME**(id). The only change would be to replace $qp^{k(2m+1)}$ by $qp^{k(2m+1)'}$.

Remark 2. It is an interesting contrast to Statement 2 that

$$\mathrm{MS}' = \left\{w_i^{-1} \# w_1 \# \ldots \# w_j : i, j \in \mathbb{N} \wedge i \leqq j \wedge w_1, \ldots, w_j \in \{0, 1\}^*\right\}$$

belongs to **3T-DTIME**(id) (see 7.2).

In combination with other ideas this counting argument can also be used to derive nonlinear lower time bounds for on-line machines. We define

$$\mathrm{A} = \left\{w_1 \# \ldots \# w_{2^s} \# v_1 \# \ldots \# v_t : s \geqq 0 \wedge t \geqq 1 \wedge w_i, v_i \in \{0, 1\}^* \wedge |w_1| = \ldots = |v_t| = s \wedge v_t \in \{w_1, \ldots, w_{2^s}\}\right\}.$$

8.9. Theorem. [Hen 66] 1*-kT-DTIME-Comp(A) $\geq \dfrac{n^2}{\log^2 n}$. ∎

Remark. This lower bound is tight.

In [Strn 68] F. C. Hennie's method has been generalized to prove for a variety of similarly constructed sets lower time bounds between n and $\dfrac{n^2}{\log^2 n}$ for one-line recognition.

The largest lower time bound for on-line recognition known so far has been proved for

$$\mathrm{G} = \left\{xyz \# u_1 \# \ldots \# u_k : k \in \mathbb{N} \wedge \bigvee_{i \leq k} y = u_i^{-1} \wedge x, y, z, u_1, \ldots, u_k \in \{0, 1\}^*\right\}.$$

8.10. Theorem. [Gall 69] 1*-multiT-DTIME-CompG $\geq \dfrac{n^2}{\log n}$. ∎

8.1.2. Counting of Crossing Sequences

A crossing sequence can be understood as a non-numerical measure for the amount of information flowing across the boundary between two tape segments during a computation. The idea of using crossing sequences as characteristics for a counting method goes back to M. O. Rabin (see [Rab 63], cf. 8.24), who used them (in a slightly modified form) in the case of 1-DTM's. We first present the method in the easier case of DTM's and NTM's.

8.1.2.1. *T-DTIME, T-NTIME and T-DCROS*

A sequence $(s_1, s_2, \ldots, s_k)$ is called the *crossing sequence* of the computation of a DTM or NTM on input w of the boundary g between two squares of the worktape if the head crosses g exactly k times during this computation, being in state s_i when crossing g for the ith time ($i = 1, \ldots, k$).

Let M be a DTM. For an input word uv we denote the crossing sequence arising in the computation of M on the input uv at the boundary between u and v by $\mathrm{CS}_M(u:v)$. The following simple lemma is basic.

8.11. Lemma. If $\mathrm{CS}_M(u:v) = \mathrm{CS}_M(u':v')$, then $\mathrm{CS}_M(u:v) = \mathrm{CS}_M(u:v') = \mathrm{CS}_M(u':v)$. ∎

Figure 8.1 illustrates Lemma 8.11.

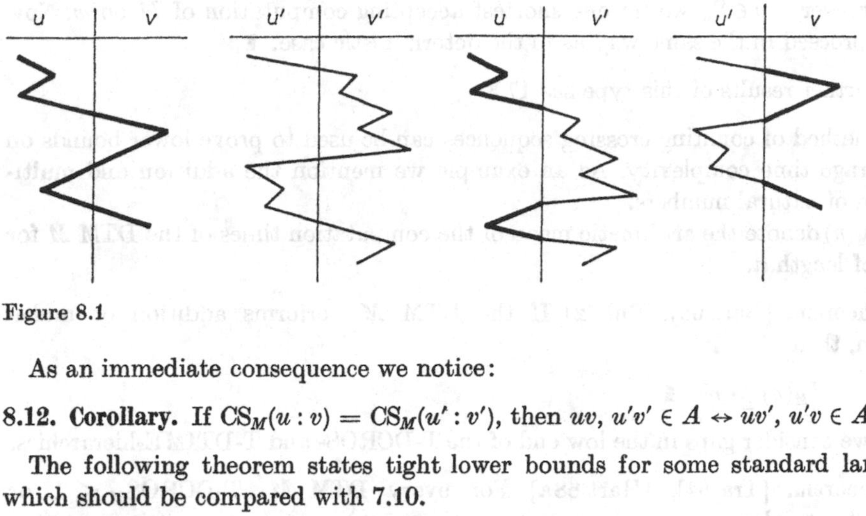

Figure 8.1

As an immediate consequence we notice:

8.12. Corollary. If $\mathrm{CS}_M(u:v) = \mathrm{CS}_M(u':v')$, then $uv, u'v' \in A \leftrightarrow uv', u'v \in A$. ∎

The following theorem states tight lower bounds for some standard languages which should be compared with 7.10.

8.13. Theorem. [Hen 65], [Barz 65] For $\mathrm{X} \in \{\mathrm{D}, \mathrm{N}\}$,

1. T-XTIME-CompS $\geq n^2$.

2. For $A \in \{\mathrm{S}_0, \mathrm{D}\}$, T-XTIME-Comp$A \geq \begin{cases} n^2 & \text{if } n \equiv 0(2), \\ n & \text{otherwise}. \end{cases}$

3. For $A \in \{\mathrm{S}_1, \mathrm{S}^\#, \mathrm{D}^\#\}$, T-XTIME-Comp$A \geq \begin{cases} n^2 & \text{if } n \equiv 1(2), \\ n & \text{otherwise}. \end{cases}$

4. T-XTIME-CompC $\geq \begin{cases} n \log n & \text{if } n \equiv 0(2), \\ n & \text{otherwise}. \end{cases}$

Proof. We restrict ourselves to the case $A = \mathrm{S}$. First let M be a DTM accepting S and let M have q states. For

$$\mathrm{S}_n = \left\{ w0^k w^{-1} : w \in \{0, 1\}^{\left\lfloor \frac{n}{4} \right\rfloor} \wedge k = n - 2 \cdot \left\lfloor \frac{n}{4} \right\rfloor \right\} \subseteq \mathrm{S},$$

Corollary 8.12 implies:

For any two different words $uv, u'v' \in S_n$ with $\left\lfloor \frac{n}{4} \right\rfloor \leqq |u|, |u'| \leqq n - \left\lfloor \frac{n}{4} \right\rfloor$ we must have $CS_M(u:v) \neq CS_M(u':v')$.

The number of crossing sequences whose length is not greater than $\frac{\left\lfloor \frac{n}{4} \right\rfloor}{\log q} - 1$ is less than $2^{\left\lfloor \frac{n}{4} \right\rfloor}$, the number of words in S_n. Therefore, there must exist a $w \in S_n$ having between $\left\lfloor \frac{n}{4} \right\rfloor$ and $n - \left\lfloor \frac{n}{4} \right\rfloor$ only crossing sequences of length at least $\frac{\left\lfloor \frac{n}{4} \right\rfloor}{\log q}$. This leads to a computation time of at least $\frac{n}{2} \cdot \frac{\left\lfloor \frac{n}{4} \right\rfloor}{\log q} \geqq_{\text{ae}} \frac{n^2}{9 \log q}$ steps. If M is an NTM, then for every $w \in S_n$ we fix one shortest accepting computation of M on w. Now we can proceed in the same way as in the deterministic case. ∎

For further results of this type see 17.38.

The method of counting crossing sequences can be used to prove lower bounds on the average time complexity. As an example we mention the addition and multiplication of natural numbers.

Let $\bar{t}_M(n)$ denote the arithmetic mean of the computation times of the DTM M for inputs of length n.

8.14. Theorem. [Barz 65], [Bul 72] If the DTM M performs addition or multiplication, then

$$\bar{t}_M(n) \succeq n^2. \blacksquare$$

Now we consider gaps in the low end of the T-DCROS- and T-DTIME-hierarchies.

8.15. Theorem. [Tra 64], [Hart 68a] For every DTM M, T-DCROS$_M \preceq 1$ or T-DCROS$_M \succeq_{\text{io}} \log$.

Proof. If T-DCROS$_M$ is unbounded, then for every $k \in \mathbb{N}$ there exists a shortest word w_k — let $n_k = |w_k|$ — and an m_k such that T-DCROS$_M(w_k) = m_k \geqq k$. Every crossing sequence of M in the input zone of w_k can occur at most twice. Assume to the contrary that w_k has the same crossing sequence at the boundaries g_1, g_2, and g_3 (assume g_2 to lie between g_1 and g_3) and that these boundaries define the decomposition $w_k = w_{1k}w_{2k}w_{3k}w_{4k}$. Then

$$w'_k =_{\text{df}} \begin{cases} w_{1k}w_{3k}w_{4k} & \text{if to the left of } g_2 \text{ there does not exist a crossing sequence of length } m_k, \\ w_{1k}w_{2k}w_{4k} & \text{otherwise} \end{cases}$$

defines a word which is shorter than w_k but which still requires crossing complexity m_k. This contradicts the choice of w_k. Consequently, w_k has at least $\left\lfloor \frac{n_k - 1}{2} \right\rfloor$ different crossing sequences. If M has q states, this requires $\left\lfloor \frac{n_k - 1}{2} \right\rfloor \leqq q^{m_k+1}$. This yields $\log n \lesssim_{io} \text{T-DCROS}_M(n)$. ∎

Remark. The proof is not valid for NTM's because we are not sure that the word w_k' still requires k crossings. There could exist another path on which w_k' is accepted with fewer crossings.

Because of **T-DCROS**(Const) = **T-DCROS**(1) = **REG** (see 12.1.3) we obtain:

8.16. Corollary. For $t \prec \log$,

1. **T-DCROS**(1) = **T-DCROS**(t),
2. $A \notin$ **REG** $\rightarrow A \notin$ **T-DROS**(t). ∎

Remark. $C = \{0^n 1^n : n \in \mathbb{N}\} \in$ **T-DCROS**(log), which shows that the gap cannot be amplified.

The following theorem is a simple consequence from 8.15.

8.17. Theorem. For every DTM M,

$$\text{T-DTIME}_M \lesssim n \quad \text{or} \quad \text{T-DTIME}_M \gtrsim_{io} n \log n .$$

Proof. If $t = \text{DTIME}_M$ is not linearly bounded, then T-DCROS_M is not bounded by a constant as follows from the trivial inclusion **T-DCROS**(Const) $\subseteq$ **T-DTIME**(Lin). Hence, according to the proof of 8.15, there exists a sequence $w_1, w_2, \ldots$ with $|w_1| < |w_2| < \ldots$ such that w_i has at least $\left\lfloor \frac{|w_i| - 1}{2} \right\rfloor$ different crossing sequences. However, w_i has at most $\left\lfloor \frac{|w_i| + 1}{4} \right\rfloor$ different crossing sequences of length $\leqq \frac{\log \left\lfloor \frac{|w_i|+1}{4} \right\rfloor}{\log q} - 1$ (q being the number of states of M). Consequently, w_i has at least $\left\lfloor \frac{|w_i| - 1}{4} \right\rfloor$ crossing sequences of length $\geqq \frac{\log \left\lfloor \frac{|w_i|+1}{4} \right\rfloor}{\log q}$, i.e. M works at least $\left\lfloor \frac{|w_i| - 1}{4} \right\rfloor \cdot \frac{\log \left\lfloor \frac{|w_i|+1}{4} \right\rfloor}{\log q} \gtrsim |w_i| \log |w_i|$ steps on w_i. ∎

The preceding proof shows

8.18. Corollary. For every DTM M,

$$\text{T-DTIME}_M \prec n \log n \text{ implies } \text{T-DCROS}_M \lesssim 1. \ ∎$$

From this corollary and **T-DCROS**(Const) = **REG** = **T-DTIME**(n) (12.1.3 and 12.1.2) we obtain

8.19. Corollary. For $t \prec n \log n$,

1. **T-DTIME**(n) = **T-DTIME**(t),
2. $A \notin$ **REG** $\rightarrow A \notin$ **T-DTIME**(t). ∎

Remark 1. Corollary 8.19 is not valid for NTM's and RTM's as is shown by 7.10 and 9.22, resp.

Remark 2. $C \in$ **T-DTIME**$(n \log n)$ (7.10), which shows that 8.19 cannot be improved.

8.1.2.2. 1-auxPD-T-NSPACE

An interesting version of the method of counting crossing sequences, which allows lower bound proofs on 1-auxPD-T-NSPACE goes back to M. Chytil. He uses the following characterization of 1-auxPD-T-NSPACE:

Let $A \subseteq \Sigma^*$. We say $A \in$ **CTX**(s) ("A is of *context-sensitivity* bounded by s") iff there exist an alphabet $\Sigma' \supseteq \Sigma$, a context-free language $L \subseteq \Sigma'^*$ and a DTM M such that

$$v \in A \leftrightarrow \bigvee_{w} \left(h(w) = v \wedge w \in L \cap \mathrm{L}(M) \wedge \text{T-DCROS}_M(w) \leq s(|v|)\right),$$

where h denotes the projection from Σ'^* onto Σ^*.

8.20. Theorem. [Chy 77] **CTX**$(s) =$ **1-auxPD-T-NSPACE**(s) for all $s \in \mathbb{R}_1$. ▮

Furthermore, the following technical result is necessary:

8.21. Lemma. Let $G = (\Sigma', N, S, R)$ be a context-free grammar in Chomsky normal form, let $\Sigma \subseteq \Sigma'$ and $h: \Sigma'^* \mapsto \Sigma^*$ be the homorphism defined by

$$h(x) = \begin{cases} x \text{ if } x \in \Sigma, \\ \text{e otherwise}. \end{cases}$$

Then for every $w \in \mathrm{L}(G)$ with $|h(w)| \geq 2$ there exist a decomposition $w = w_1w_2w_3$ and a $B \in N$ such that $w = w_1w_2w_3$, $S \overset{*}{\underset{G}{\Longrightarrow}} w_1Bw_3$, $B \overset{*}{\underset{G}{\Longrightarrow}} w_2$ and

$$\frac{1}{3}\,|h(w)| \leq |h(w_2)| \leq \frac{2}{3}\,|h(w)|. \ ▮$$

8.22. Theorem. [Chy 76]

1. $D^{\#} \notin$ **1-auxPD-T-NSPACE**(s) for $s \prec$ id.
2. $\{0^n1^n0^n : n \in \mathbb{N}\} \notin$ **1-auxPD-T-NSPACE**(s) for $s \prec \log$.
3. $\{0^{2^n} : n \in \mathbb{N}\} \notin$ **1-auxPD-T-NSPACE**(s) for $s \prec \log\log$.

Proof. We restrict ourselves to a proof of Statement 1. In view of 8.20 it suffices to prove $D^{\#} \notin$ **CTX**(s) for $s \prec$ id. For $i \geq 1$ we consider the sets

$$D^{\#}_i = \left\{u01^{4i-2}0u \# u01^{4i-2}0u : u \in \{0,1\}^* \wedge |u| = i\right\}$$

and define $n = 6i + 1$. Evidently, $D^{\#}_i \subseteq D^{\#}$ has the following property:

(1) If an arbitrary subword of $p_u =_{\mathrm{df}} u01^{4i-2}0u \# u01^{4i-2}0u$ of length λ, $\frac{n}{3} \leq \lambda \leq \frac{2n}{3}$, is replaced by an arbitrary subword of p_v of length λ and $u \neq v$, then the new word does not belong to $D^{\#}$.

Assume now $D^{\#} \in \mathbf{CTX}(s)$. Then there exist a $\Sigma' \supseteq \{0, 1, \#\}$, a DTM with alphabet Σ' and an $L \subseteq \Sigma'^*$ such that $L \in \mathbf{CF}$ and L satisfies the definition of CTX given above. Thus, for every $p_u \in D_i^{\#}$ there exists a $w_u \in L \cap \mathrm{L}(M)$ such that $h(w_u) = p_u$ and $\text{T-DCROS}_M(w_u) \leqq s(|p_u|)$. According to Lemma 8.21, for every $p_u \in D_i^{\#}$ there exist a decomposition $w_u = w_u'\hat{w}_u w_u''$ and a nonterminal A such that $S \overset{*}{\underset{G}{\Longrightarrow}} w_u' A w_u''$ and $A \overset{*}{\underset{G}{\Longrightarrow}} \hat{w}_u$ and

$$(2) \qquad \frac{n}{3} \leqq |h(\hat{w}_u)| \leqq \frac{2n}{3}.$$

Each of these $\hat{w}_u$ determines a triple (A, S_1, S_2), where A is that nonterminal generating $\hat{w}_u$ and S_1 and S_2 are the crossing sequences at the right and left boundaries of $\hat{w}_u$ during the accepting of w_u by M. If $\hat{w}_u$ and $\hat{w}_v$ have the same triple (A, S_1, S_2), then $\hat{w}_u$ can be replaced by $\hat{w}_v$, and the resulting word will also be accepted. For $u \neq v$ this would be a wrong acceptance because of (2) and (1). Consequently, the number of words in $D_i^{\#}$ cannot be larger than the number of different triples. If r is the number of nonterminals in a grammar generating L and q is the number of states of M, then we conclude

$$2^{\frac{n}{13}} \leqq \text{card}\, D_i^{\#} \leqq r \cdot q^{2(s(n)+1)} \quad \text{for} \quad n \equiv 1(12)$$

and hence $n \lesssim_{\text{io}} s(n)$. ∎

8.1.2.3. *1-T-DTIME, 1-T-DRET, 1-T-DCROS and 1-T-DREV*

As in the case of DTM's, crossing sequences on the worktape are also considered in the case of 1-DTM's. But here the information brought across a boundary consists not only of a state but also of an input head position. This gives rise to a slight modification of the notion of a crossing sequence. A crossing sequence of a 1-DTM with input alphabet Σ and state set S is a sequence $(a_1, \ldots, a_r)$ where $a_1, \ldots, a_r \in S \cup (\Sigma \times S)$.

Let $(a_1, \ldots, a_r)$ be the crossing sequence of the boundary g of the worktape of the 1-T-DTM M. If $a_i \in S$, then the ith crossing of g is made with state a_i and with simultaneously moving input head. If $a_i = (x_i, s_i) \in \Sigma \times S$, then during the ith crossing of g the input head rests and scans x_i in state s_i. A boundary g defines a crossing sequence $(a_1, \ldots, a_r)$ and a decomposition $w_1 w_2 \ldots w_{r+1}$ of the input word:

If $a_i \in S$, then w_i ends at that boundary which is crossed by the input head when the worktape head makes its ith crossing. If $a_i = (x_i, s_i)$, then w_i is empty if between the $(i-1)$th and the ith crossing of the worktape head the input head was quiescent. Otherwise, w_i ends with that x_i which is scanned during the ith crossing of the worktape head.

If during the computation of M on input w a crossing sequence $(a_1, \ldots, a_r)$ determines the decomposition $w = w_1 w_2 \ldots w_{r+1}$, we write $(a_1, \ldots, a_r) = \text{CS}_M(w_1 : w_2 : \ldots : w_{r+1})$. Corollary 8.12 has to be modified as follows:

8.23. Lemma. If $\mathrm{CS}_M(w_1 : w_2 : \ldots : w_{r+1}) = \mathrm{CS}_M(v_1 : v_2 : \ldots : v_{r+1})$, then

$$w_1 w_2 \ldots w_{r+1},\ v_1 v_2 \ldots v_{r+1} \in \mathrm{L}(M) \leftrightarrow w_1 v_2 w_3 v_4 \ldots,\ v_1 w_2 v_3 w_4 \ldots \in \mathrm{L}(M). \blacksquare$$

Using this lemma to construct words not belonging to the set under consideration requires much more effort than in the comparable case of Lemma 8.12 (as illustrated for example in the proof of 8.13). For the proof of the following theorem where the method of crossing sequences has been used for the first time, M. O. Rabin has introduced the "bottleneck squares" to make the lemma applicable with the desired effect.

We define

$$\mathrm{T} = \{u \# v \# u^{-1} : u, v \in \{0, 1\}^*\} \cup \{u \# v * v^{-1} : u, v \in \{0, 1\}^*\}.$$

8.24. Theorem. [Rab 63] $\mathrm{T} \notin$ **1-T-DTIME**(id). ∎

Remark. One easily proves

$$\mathrm{T} \in \textbf{1-T-DTIME}((1 + \varepsilon)n) \cap \textbf{1-2T-DTIME}(\mathrm{id}) \text{ for all } \varepsilon > 0.$$

The following result has a very complex proof.

8.25. Theorem. [Frv 65] If the 1-DTM M decides S in such a way that within the first n steps the input word (of length n) is read completely, then $\text{1-T-DTIME}_M \geq_{\mathrm{io}} n^2$. ∎

Note that so far it has not been possible to state the same result without the special input assumption.

Finally some results concerning return, reversal and crossing are reported. Define for $k \geq 2$

$$\mathrm{L}_k = \{w_1 \# w_2 \# \ldots \# w_l : l \leq k \wedge w_1 \in \{0, 1\}^* \wedge w_1 = w_3 = w_5 = \ldots = w_4^{-1} = w_2^{-1}\}$$

and

$$\mathrm{M}_k = \{(0^m \#)^l : l \leq k \wedge k, m \in \mathbb{N}\}.$$

8.26. Theorem. [BrSa 77]

1. $\mathrm{L}_{k+1} \notin$ **1-T-DRET**(k) for $k \geq 1$.
2. $\mathrm{M}_{k+1} \notin$ **1-T-DCROS**(k) for $k \geq 1$. ∎

Remark. For all $k \geq 2$,

$$\mathrm{L}_k \in \textbf{1-T-DRET}(k) \quad \text{and} \quad \mathrm{M}_k \in \textbf{1-T-DCROS}(k) \cap \textbf{1-T-DRET}(2).$$

Define for $k \geq 1$

$$\mathrm{H}_k = \{0^{n_1} \# 0^{n_1} \# 0^{n_2} \# 0^{n_2} \# \ldots \# 0^{n_l} \# 0^{n_l} : 1 \leq j < k \wedge n_1, \ldots, n_l \in \mathbb{N}\},$$

$$\mathrm{H}(g) = \{0^{g(m)} \# w_1 \# w_1^{-1} \# \ldots \# w_m \# w_m^{-1} : m \geq 1 \wedge |w_1| = \ldots = |w_m| = g(m)\}$$

and

$$t_g(n) = \mu m\big(n \leq (2m + 1)\,(g(m) + 1)\big).$$

8.27. Theorem. [Hart 68b]
1. $\mathrm{H}_{k+1} \notin$ **1-T-DREV**(k) for $k \in \mathbb{N}$.
2. $\mathrm{H}(g) \notin$ **1-T-DREV**$(t_g - 1)$ provided $g(n) \geq 3n \log n$. ∎

Remark 1. $\mathrm{H}_{k+1} \in$ **1-T-DREV**$(k + 1)$ for all $k \in \mathbb{N}$.

Remark 2. If g satisfies suitable honesty conditions, then $\mathrm{H}(g) \in$ **1-T-DREV**(t_g).

8.1.2.4. 2-T-DSPACE and 2-T-NSPACE

The crossing sequence method can be applied to two-way TM's to prove lower space bounds. Here crossing sequences describe the information flow across the boundary between two squares of the input tape. More formally, for a 2-DTM M, we call $\mathrm{CS}_M(w_1 : w_2) = (K_1, \ldots, K_r)$ the crossing sequence at the boundary g between w_1 and w_2 in the computation of M on input $w_1 w_2$ iff the input head crosses g exactly r times and K_i consists of the state, the worktape contents and the position of the worktape head of M corresponding to the moment of the ith crossing.

Note that 8.12 is also valid for this type of crossing sequence. In a finite computation every K_i can occur at most once in every crossing sequence, otherwise the computation would cycle. If there are altogether k different K_i, there exist at most k^{k+1} different crossing sequences. If M works within space s, this number can be bounded above by $2^{2^{cs(n)}}$ for some $c > 0$.

The following theorem is also a gap result (cf. 8.29).

8.28. Theorem. [HoUl 69b] For every 2-NTM M,

$$\text{2-T-NSPACE}_M \leq 1 \quad \text{or} \quad \text{2-T-NSPACE}_M \geq_{\text{io}} \log\log .$$

Proof. The proof of the deterministic case which has already been given in [StHaLe 65] is nearly word for word the same as that of 8.15. It does not work in the nondeterministic case for the same reason as that mentioned in the remark after 8.15. But with the help of the method of transition matrices the result can also be proved in the nondeterministic case. ∎

Using 4.7 we obtain

8.29. Corollary. For $s < \log\log$,
1. **2-T-NSPACE**$(0) =$ **2-T-NSPACE**(s),
2. $A \notin$ **REG** $\to A \notin$ **2-T-NSPACE**(s). ∎

Remark 1. In [StHaLe 65] it has been proved that the result cannot be improved:

$$\{\text{bin } 0 \# \text{bin } 1 \# \ldots \# \text{bin } k : k \in \mathbb{N}\} \in \textbf{2-T-DSPACE}(\log\log).$$

Remark 2. It is shown in [Hon 84] that for $s \cdot r < \mathrm{id}$ we have **2-T-DSPACE-REV**$(s, r) =$ **2-T-DSPACE-REV**$(0, 0)$ and, consequently, $A \notin$ **REG** implies $A \notin$ **2-T-DSPACE-REV**(s, r).

The next result goes back to [Sei 77a], but a very similar result has been obtained in [HoUl 69b].

Let $f(n) \leq \log \frac{n}{2}$ and define $S_f = \{u2^k u^{-1}: |u| = 2^{f(|u2^ku^{-1}|)} \wedge k \in \mathbb{N} \wedge u \in \{0, 1\}^*\}$.

8.30. Theorem. 2-T-NSPACE-Comp(S_f) $\succeq f$. Moreover, if f is fully 2-T-DSPACE-constructible, then $S_f \in$ **2-T-DSPACE**(f).

Proof. We prove only the first part. Different words $u_1 2^k u_1^{-1}$, $u_2 2^k u_2^{-1} \in S_f$ must have different crossing sequences on their middle parts; otherwise $u_1 2^k u_2^{-1}$ would also be accepted. This shows: If the NTM M accepts S_f within space s, then

$$\begin{aligned} 2^{2^{f(n)}} &= \text{number of different words of length } n \text{ in } S_f \\ &\leqq \text{number of crossing sequences of } M \\ &\leqq 2^{2^{cs(n)}}. \end{aligned}$$

Hence, $s \succeq f$. ∎

Combining this method with algebraic results one can prove the subsequent result concerning **SEMILIN** which is defined as follows.

Subsets of $\mathbb{N}^m$ of the form $\{\mathfrak{a} + n_1\mathfrak{b}_1 + \ldots + n_k\mathfrak{b}_k : n_1, \ldots, n_k \in \mathbb{N}\}$ where $n \in \mathbb{N}$ and $\mathfrak{a}, \mathfrak{b}_1, \ldots, \mathfrak{b}_k \in \mathbb{N}^m$, are called *linear* sets. Finite unions of linear sets are called *semilinear*. A language L is called a *semilinear language* iff there exists an m and an alphabet $\{a_1, \ldots, a_m\}$ such that $L \subseteqq a_1^* \ldots a_m^*$ and the set $\{(n_1, \ldots, n_m): a_1^{n_1} \ldots a_m^{n_m} \in L\}$ is semilinear.

SEMILIN denotes the set of all semilinear languages (over some countable alphabet).

8.31. Theorem. [Alt 77]

1. If $L \in$ **SEMILIN** $\setminus$ **REG** (for example, L bounded context-free and nonregular) and $s \prec \log$, then

$$L \notin \textbf{2-T-DSPACE}(s).$$

2. [AlMe 76] If $L \in$ **DCF** $\setminus$ **REG** and $s \prec \log$, then $L \notin$ **2-T-DSPACE**(s).

Remark. The bound of Statement 1 is tight for nonregular bounded context-free languages (see also Remark 4 after 12.14). Thus we have a subclass of **CF** for which the known lower and upper space bounds (8.29, 12.14) can be considerably improved.

Counting partial situations instead of crossing sequences as used in this subsection we get the following result which should be compared with 8.28.

8.32. Theorem. [HaBe 75] If L is an infinite SLA-language not containing an infinite regular subset, then for $s \prec \log$

$$L \notin \textbf{2-T-DSPACE}(s). \ ∎$$

8.33. Corollary. PRIME[1] $\notin$ **2-T-DSPACE**(s) for $s \prec \log$. ∎

Remark. For PRIME the same statement is valid (see [HaSh 69]).

8.1.3. Counting of Reaction Sequences

Let M be a DTM with state set S and card $S = q$. Every word $u \in \Sigma^*$ defines a mapping $R_u: S \times \{l, r\} \mapsto S \times \{l, r\}$ in the following way (l and r stand for "left" and "right", resp.): $R_u(s, \sigma) = (s', \sigma') \leftrightarrow$ if the head enters the word u in the state s from the σ end, then it leaves u from the σ' end in state s' and u is not rewritten. R_u gives a description of all reactions of u to possible visits which leave u unaltered.

If the interval containing initially the word u is rewritten k times during some computation, and if the ith rewriting takes place when the head enters the interval in state s_i from direction ε_i and leaves it in state s_i' in direction ε_i' and yields the new contents w_i in the interval, then we say that u has the *reaction sequence* (of length $k + 1$) $\left(R_u, (s_1, \varepsilon_1, s_1', \varepsilon_1'), R_{w_1}, \ldots, (s_k, \varepsilon_k, s_k', \varepsilon_k'), R_{w_k}\right)$. The number of different reaction sequences of length k is bounded by $(2q)^{2qk} \cdot (4q^2)^{k-1} \leqq 2^{ck}$ for some $c > 0$.

In [Pec 77] it is proved that $\mathrm{L} = \{0^n 1^{n+m} 0^m : n, m \in \mathbb{N}\}$ is not acceptable by a DTM within constant return. A modification of this proof allows us to state a stronger result:

8.34. Theorem. T-DRET-Comp(L) $\geq_{\mathrm{io}}$ log.

Proof. Let M be a DTM deciding L with q states and let $t =$ T-DRET$_M$. For an input word of the special kind $0^r 1^{2r} 0^r$ we observe the middle part 1^{2r} during the computation. Assume $(2q)^{2q} < 2r$. Then in the middle part at least one rewriting has to occur. Otherwise, as there exist only $(2q)^{2q}$ different reaction sequences of length 1, some subword could be replaced by another, shorter one with the same reaction sequence, and thus a word would be accepted that does not belong to L.

Another important observation is the following: If the head enters a subword of the form 1^k, then either it leaves it unaltered or it performs an alteration within the first q steps. Let us now assume that the first alteration in the middle part 1^{2r} takes place in step τ in square i_1 and $r < i_1 \leqq r + q$. (The other case $3r - q < i_1 \leqq 3r$ is treated analogously.)

Let us assume that M needs the squares $-l, -l_{+1}, \ldots, 4r + l'$.

For $i < j$ we define w_{ij} to be that part of $\ldots \square\square\square w \square\square\square \ldots$ that stands in the tape segment consisting of the squares $i + 1, \ldots, j - 1$. Let

$$j_1 = \begin{cases} \text{that square to the right of } i_1 \text{ in which the last alteration has} \\ \text{taken place in the time interval } [1, \tau - 1] \text{ if such a tape square} \\ \text{exists,} \\ 4r + l' + 1 \quad \text{otherwise.} \end{cases}$$

We define a sequence of nested words $v_1, v_2, \ldots$

$$v_1 =_{\mathrm{df}} w_{i_1 j_1}.$$

If v_m is defined, let k be the tape square of the first alteration in v_m.

$$i_{m+1} =_{\mathrm{df}} \begin{cases} k & \text{if} \quad k - i_m \leqq q, \\ i_m & \text{otherwise,} \end{cases} \qquad j_{m+1} =_{\mathrm{df}} \begin{cases} j_m & \text{if} \quad k - i_m \leqq q, \\ k & \text{otherwise,} \end{cases}$$

$$v_{m+1} =_{\mathrm{df}} w_{i_{m+1} j_{m+1}}.$$

Because of the observations mentioned above, the sequence $(v_1, v_2, \ldots)$ contains $c' \cdot n$ elements for some $c' > 0$. We observe

1. v_m is not changed before the contents of the squares i_m and, if $j_m \neq 4r + l' + 1$, j_m are altered.
2. After this moment v_m is visited from outside at most $2t(n)$ times. Otherwise i_m or j_m would have a return number greater than $t(n)$. From this we conclude that every v_m has a reaction sequence of length at most $2t(n)$. The number of reaction sequences of length $2t(n)$ is bounded by $2^{2ct(n)}$.

It can be shown that there exists a $c'' > 0$ and a subsequence $(v_{m_1}, v_{m_2}, \ldots)$ of $(v_1, v_2, \ldots)$ containing $c''n$ elements such that

$$i_{m_{\nu+1}} - i_{m_\nu} > j_{m_\nu} - j_{m_{\nu+1}}.$$

This inequality ensures that a replacement of v_{m_μ} by v_{m_ν} $(\nu \neq \mu)$ leads to a word which is not in L. Therefore, all v_{m_ν} must have pairwise different reaction sequences. Consequently, $c''n \leqq 2^{2ct(n)}$ for sufficiently large $n \equiv 0(4)$ and hence $t \geq_{\text{io}} \log$. ∎

For further examples consult [Pec 77].

8.1.4. Overlap Arguments

This method will here be used for 1*-kTl-DM's computing length preserving functions, i.e. machines printing the ith output symbol after reading the ith and before reading the $(i+1)$th input symbol, and halting when reading □. More general machine types are considered in [CoAa 69] and [PaFiMe 74].

Let M be a 1*-kTl-DM working within time t and having q states and p tape symbols. (t_1, t_2) is called an *overlapping pair* with respect to w iff during the computation of M on input w some square is visited by some head in step t_1 and in step $t_2 > t_1$, but in no step t where $t_1 < t < t_2$.

Let $|w| = n$ and I be a subinterval of $[1, \ldots, t(n)]$.

$$\Omega_w(I) =_{\text{df}} \{(t_1, t_2) : t_1, t_2 \in I \wedge (t_1, t_2) \text{ is an overlapping pair with respect to } w\}.$$

$$\omega_w(I) =_{\text{df}} \text{card}\, \Omega_w(I).$$

Every t_2 may occur in at most kl overlapping pairs. Consequently, we have

$$\omega_w(I) \leqq k \cdot l \cdot t(n). \tag{1}$$

Let I, J be neighbouring subintervals of $[1, \ldots, t(n)]$,

$$\Omega_w(I, J) =_{\text{df}} \{(t_1, t_2) : t_1 \in I \wedge t_2 \in J \wedge (t_1, t_2) \in \Omega_w(I \cup J)\},$$

$$\omega_w(I, J) =_{\text{df}} \text{card}\, \Omega_w(I, J).$$

Let $w = x_1 \ldots x_n$, let $U = [g + 1, g + R]$ be a subintervall of $[1, n]$ and define $w(U) = x_{g+1} \ldots x_{g+R}$. Let V be a right neighbour interval of U. Let f be a function to be computed. We consider all $w = w_1 x w_2$ where $x = w(U)$ varies arbitrarily, but U, w_1 and w_2 are fixed. How sensitive is the reaction of $y = f(w)\,(V)$ when x is changed?

An answer which is sufficient for our aims is provided by the index of the following equivalence relation on Σ^R:

$$x \sim_{f,U,V} x' \leftrightarrow_{\text{df}} f(w_1 x w_2)\,(V) = f(w_1 x' w_2)\,(V).$$

Let I_U, I_V be the time interval during which the input head is in U, V, resp.

The influence of x on y depends only on what information about x can be stored during I_U in such a way that it is accessible during I_V. We consider the following items to be the defining constituents of the characteristics of the overlap method:

1. state of M at the beginning of I_V,
2. head positions at the beginning of I_V,
3. the contents and the initial and end points of those intervals which have been visited during I_U by the worktape heads.

The number of characteristics of a 1*-kTl-DM working within time t and with overlap $\omega_w(I, J) \leqq s$ can be bounded by

$$q \cdot \big(2t(n)\big)^{kl} \cdot \big(4t^2(n)\big)^{kl} \cdot p^s \leqq t(n)^c \cdot p^s \quad \text{for some} \quad c > 0. \tag{2}$$

Now the basic notions of the overlap method are ready for use. The next lemma links different overlap-values with each other. We assume the input word to be of the form $w = x_1 \ldots x_n$, and $n = 2^r$. For every i with $0 \leqq i \leqq r$ we decompose $[1, \ldots, n]$ into 2^{r-i} subintervals U_{ij} of length 2^i, i.e. $U_{ij} = [j \cdot 2^i + 1, \ldots, (j + 1) \cdot 2^i]$ for $j = 0, \ldots, 2^{r-i} - 1$.

Furthermore we define $I_{ij} = I_{U_{ij}}$ and $t_{ij} = |I_{ij}|$. Then evidently

$$1^*\text{-}k\text{T}l\text{-DTIME}_M(w) = \sum_{j=0}^{2^{r-i}-1} t_{ij} \quad \text{for every} \quad i = 0, \ldots, r.$$

We define $\Omega_{ij} = \Omega(I_{ij}, I_{ij+1})$, $\Omega = \Omega(I_{r0})$, $\omega_{ij} = \text{card}\,\Omega_{ij}$ and $\omega = \text{card}\,\Omega$.

Lemma 1. $\sum_{i=1}^{r} \sum_{j=0(2)} \omega_{ij} = \omega$. ∎

We now apply the overlap method to the multiplication of binary numbers. As a special case we consider multiplication by a fixed number.

Define $\alpha_m = \sum_{2^{2^i} < m} 2^{2^i}$. We investigate the function f defined by $f(a_1, \ldots, a_n) =$ the initial word of length n of $\big(\text{bin}\,(m \cdot \alpha_m)\big)^{-1}$ where $a_1 \ldots a_n = (\text{bin}\,m)^{-1}$. Let g and $R = 2^d$ be chosen in such a way that $g + 2^{d+1} \leqq n$.

Define $U = [g + 1, \ldots, g + 2^d]$ and $V = [g + 2^d + 1, \ldots, g + 2^{d+1}]$. With the help of elementary number theoretic facts one proves

Lemma 2. Every equivalence class of $\sim_{f,U,V}$ may have at most two members. ∎

8.35. Theorem. [PaFiMe 74] If M is a 1*-kTl-DM computing the function f with average time $\bar{t}$, then

$$\bar{t}(n) \succeq n \log n.$$

Remark. In [CoAa 69], for a more general class of computers and a wider class of functions, the worst case lower bound $c \cdot \dfrac{n \log n}{(\log \log n)^2}$ is proved.

Proof. First a notational convention. If ξ depends on w, and w varies in some set W, then we write

$$\text{average}_W\ \xi(w) =_{\text{df}} \frac{1}{\text{card } W} \sum_{w \in W} \xi(w).$$

Let $U = [g+1, g+R]$ and V be neighbouring intervals of length R and let w_1, w_2 be fixed words with $|w_1| = g$, $|w_2| = n - g - R$ and $n = 2^r$ for some $r \in \mathbb{N}$.

$$W =_{\text{df}} \left\{ w_1 x w_2 : x \in \{0,1\}^R \wedge \omega_{w_1 x w_2}(I_U, I_V) \leqq \frac{\frac{R}{2} - c \log t(n)}{\log p} \right\}.$$

Because of (2) the number of different characteristics that can be generated by words from W is bounded by

$$t(n)^c \cdot p^{\frac{\frac{R}{2} - c \cdot \log t(n)}{\log p}} \leqq 2^{\frac{R}{2}}.$$

Hence the index of $\sim_{f,U,V}$ restricted to W is bounded above by $2^{\frac{R}{2}}$. Lemma 2 implies card $W \leqq 2 \cdot 2^{\frac{R}{2}}$. Consequently, at least $2^R - 2 \cdot 2^{\frac{R}{2}} \geqq 2^{\frac{R}{2}}$ (for $R \geqq 4$) words $w \in w_1\{0,1\}^R w_2 = W'$ satisfy the inequality $\omega_w(I_U, I_V) \cdot \log p + c \cdot \log t(|w|) > \dfrac{R}{2}$. This yields

$$\text{average}_{W'}\left(\omega_w(I_U, I_V) \cdot \log p + c \cdot \log t(|w|)\right) \geqq \frac{1}{2^R}\left(2^R - 2 \cdot 2^{\frac{R}{2}}\right) \cdot \frac{R}{2}$$
$$\geqq \frac{R}{3}, \quad (R \geqq 6).$$

As this is independent of w_1 and w_2 we get

$$\text{average}_{\{0,1\}^n}\ \omega_w(I_U, I_V) \cdot \log p + c \cdot \log t(|w|) \geqq \frac{R}{3}.$$

We use the fact that an upper bound for our problem is known, namely n^2, and thus $\log t < 2 \log n$. Furthermore, we choose $R \geqq 24c \log n$. Then we get

$$\text{average}_{\{0,1\}^n}\ \omega_w(I_U, I_V) \geqq \frac{R}{4 \log p}.$$

This result is applied to $U = U_{i,2j}$ and $V = U_{i,2j+1}$. In this case $\omega_w(I_U, I_V) = \omega_{i,2j}$ and $R = 2^i$, and summing up all inequalities

$$\text{average}_{\{0,1\}^n}\ \omega_{i,2j} \geqq \frac{2^i}{4 \log p}$$

for fixed i, we get

$$\text{average}_{\{0,1\}^n} \sum \omega_{i,2j} \geqq \frac{2^i}{4 \log p} \cdot \frac{2^{r-i}}{2} = \frac{n}{8 \log p}.$$

This is valid for all i such that $n \geqq 2^i = R \geqq 24c \log n$. Summation using inequality (1) and Lemma 1 yields

$$\begin{aligned} k \cdot l \cdot \text{average}_{\{0,1\}^n} t(n) &= k \cdot l \cdot \bar{t}(n) \\ &\geqq \text{average}_{\{0,1\}^n} \omega = \text{average}_{\{0,1\}^n} \sum_i \sum_j \omega_{i,2j} \\ &\geqq \text{average}_{\{0,1\}^n} \sum_{i > \log(24c \log n)} \sum_j \omega_{i,2j} \\ &\geqq \left(\log n - \log (24c \log n)\right) \frac{n}{8 \log p} \\ &\geqq \frac{1}{9 \log p} n \log n \end{aligned}$$

for sufficiently large n. ∎

Remark. Using Kolmogorov complexity the proof of Theorem 8.35 can be simplified [ReSc 82].

The proof of this result can easily be modified to give the following more general result.

Let $\text{prod}\left(\begin{pmatrix} a_1 \\ b_1 \end{pmatrix} \cdots \begin{pmatrix} a_n \\ b_n \end{pmatrix}\right)$ be the initial word of length n of $(\text{bin}\,(a \cdot b))^{-1}$ where $a_1 \ldots a_n = (\text{bin}\, a)^{-1}$ and $b_1 \ldots b_n = (\text{bin}\, b)^{-1}$.

8.36. Theorem. [PaFiMe 74] If the 1*-kTl-DM M computes the function prod, then the average time $\bar{t}(n)$ of M satisfies the inequality $n \log n \lesssim \bar{t}(n)$. ∎

The next theorem is S. O. Aanderaa's result that realtime 1-kT-DM's are more powerful than realtime 1-$(k-1)$T-DM's. We choose alphabets $A_k = \{a_1, b_1, \ldots, a_k, b_k\}$ and $C_k = \{c_1, \ldots, c_k\}$ for every $k \in \mathbb{N}'$ such that $A_k \cap C_k = \emptyset$. For every $i = 1, \ldots, k$ the homomorphism σ_i is defined by

$$\sigma_i(x) = \begin{cases} x & \text{if } x \in \{a_i, b_i\}, \\ e & \text{otherwise}. \end{cases}$$

Furthermore, we define

$$\begin{aligned} \text{AA}_k = \Big\{ wc_i^m : w \in A_k^* \wedge i \in \{1, \ldots, k\} \wedge m \in \mathbb{N} \\ \wedge \bigvee_v \bigvee_{w'} (w'v = w \wedge v \in a_i A_k^* \wedge |\sigma_i(v)| = m) \Big\}. \end{aligned}$$

8.37. Theorem. [Aan 73] For every $k \geqq 1$,

$$\text{AA}_{k+1} \in \textbf{1-(k + 1)T-DTIME}(\text{id}) \setminus \textbf{1-kT-DTIME}(\text{id}).\ ∎$$

The proof is too lengthy to be reproduced here. Simplified proofs have been published in [Per 79] and [PaSeSi 80].

Remark. The question of what time bound t is required for an inclusion **1-(k + 1)T-DTIME**(id) $\subseteq$ **1-kT-DTIME**(t) is answered for on-line machines in [Pau 82]: For

$t \prec_{io} n(\log n)^{\frac{1}{k+1}}$,

$$\mathbf{1\text{-}(k+1)T\text{-}DTIME}(\mathrm{id}) \not\subseteq \mathbf{1^*\text{-}kT\text{-}DTIME}(t).$$

A generalization to probabilistic on-line machines is contained in [PaSi 83]. Furthermore, for $t \prec_{io} n(\log n)^{\frac{1}{k+1}}$,

$$\mathbf{1\text{-}kT\text{-}DTIME}(\mathrm{id}) \not\subseteq \mathbf{1^*\text{-}kPD\text{-}DTIME}(t) \quad \text{([DuGaPaRe 83])}.$$

Define $H_2 = AA_2$ and $H_{k+1} = H_k \cup H_k \# AA_{k+1}$.

8.38. Theorem. [Vit 80] For every $k \geqq 1$,

$$H_{k+1} \in \mathbf{1\text{-}T(k+1)\text{-}DTIME}(\mathrm{id}) \setminus \mathbf{1\text{-}Tk\text{-}DTIME}(\mathrm{id}). \blacksquare$$

Theorems 8.37 and 8.38 are special cases of a very general result on real-time hierarchies in [Vit 84].

8.1.5. Further Results

A method to prove lower bounds using the Kolmogorov complexity has been described in [Pau 79]. It is applied for proving lower time bounds for a sorting problem on TM's. In [Imm 79] Ehrenfeucht games are used for proving lower space bounds.

Finally we mention a result on lower bounds on the execution time of Markov algorithms. In [Zej 71b] it is shown that every Markov algorithm computing the function $r(w) = w^{-1}$ needs quadratic time.

8.2. The Reducibility Method

8.2.1. Description of the Method

The reducibility method for proving lower bounds differs completely from the counting method dealt with in § 8.1. It can be described as follows: Let A be a set for which we want to have a lower bound and let B be another set for which a lower bound t is already known. If B can be reduced to A by a certain reducibility, then we obtain a lower bound t' for A where t' is smaller than t by that amount by which the used reducibility can enlarge the complexity of a set. In particular, if the reducibility preserves the complexity, then $t = t'$. Often we do not know a specific set B with a suitable lower bound t', but we know (for example by diagonalization) that there exist such sets in a class $\mathcal{A}$. Then the reduction of all sets of A to $\mathcal{A}$ ensures the lower bound t' for A. This idea is made precise by the following obvious lemma which is a modification of Lemma 3.2.

8.39. Lemma. If A is hard for $\mathcal{A}$ with respect to $\leqq$, then

$$\mathcal{A} \not\subseteq \mathcal{R}_{\leqq}(\mathcal{B}) \to A \notin \mathcal{B}. \blacksquare$$

In applications we usually choose $\mathcal{A} = \Phi(F)$ and $\mathcal{B} = \Phi(F')$ for some complexity measure Φ and $F, F' \subseteq \mathbb{R}_1$.

There are two kinds of applying this method.

1. It is unknown whether $\mathcal{R}_{\leq}(\mathcal{B}) \subset \mathcal{A}$, but there is a strong belief that $\mathcal{A} \not\subseteq \mathcal{R}_{\leq}(\mathcal{B})$. Then we use this belief to give evidence for $A \notin \mathcal{B}$. (For example, let $\mathcal{B} = \mathbf{P}$, $\mathcal{A} = \mathbf{NP}$ and $\leq\, = \,\leq_m^P$. Then the belief that $\mathbf{P} \subset \mathbf{NP}$ gives evidence for $A \notin \mathbf{P}$ for sets A which are $\leq_m^P$-hard for $\mathbf{NP}$. Note that the degree of evidence for $A \notin \mathbf{P}$ depends on the reducibility used. Reducibilities $\leq\, \supset\, \leq_m^P$ admit easier proofs of A being hard for $\mathbf{NP}$ but give weaker evidence for $A \notin \mathbf{P}$. For example, if A is $\leq_{sm}^{NP}$-hard for $\mathbf{NP}$, then we have only the evidence $(\mathbf{NP} \cap \mathbf{coNP} =)\ \mathcal{R}_{sm}^{NP}(\mathbf{P}) \subset \mathbf{NP}$ for $A \notin \mathbf{P}$. For related results see [AdMa 77], [AdMa 79] and [Schg 83a].)

Examples concerning the most important complexity classes like **NP**, **P**, **NL** and **L** are given in § 8.2.2.

Note that this kind of application of the reducibility method does not really provide lower bounds for hard problems. In many cases, however, it is possible to prove lower bounds with respect to subrecursive computational models. See for instance [CoSe 76], [Koz 77a], [CoRa 78], [Imm 79], [BeSi 83], [Ukk 83b].

2. It is known that $\mathcal{R}_{\leq}(\mathcal{B}) \subset \mathcal{A}$, for instance by hierarchy results (diagonalization). Then we really obtain $A \notin \mathcal{B}$, i.e. a lower bound for A (see § 8.2.3).

In many cases we shall use the following theorem which is a consequence of 8.39.

8.40. Theorem.

1. Let $\Phi \in$ {DSPACE, NSPACE, T-DTIME, T-NTIME}, and let t be an increasing function such that $\log t > \log$. If $\Phi(t(n)) \subset \Phi(t(2n))$ and A is $\leq_m^{P-lin}$-hard for $\Phi(t(\mathrm{Lin}))$, then $A \notin \Phi(t(c \cdot n))$ for some $c > 0$.
2. Let Φ and t be as above. If $\Phi(t(n)) \subset \Phi(t(n^2))$ and A is $\leq_m^P$-hard for $\Phi(t(\mathrm{Pol}))$, then $A \notin \Phi(t(n^c))$ for some $c > 0$.
3. Let $\Phi \in$ {DSPACE, NSPACE}, and let t be an increasing function such that $t > \log$. If $\Phi(t(n)) \subset \Phi(t(n^2))$ and A is $\leq_m^{\log}$-hard for $\Phi(t(\mathrm{Pol}))$, then $A \notin \Phi(t(n^c))$ for some $c > 0$.

Proof. We prove Statement 1 for Φ = DSPACE. The other statements can be proved in the same way.

Let $B \in \mathbf{DSPACE}(s(2n)) \setminus \mathbf{DSPACE}(s(n))$. Since A is $\leq_m^{P-lin}$-hard for **DSPACE** $(s(\mathrm{Lin}))$, we have $B \leq_m^{\mathrm{DSPACE},\log,k\cdot n} A$ for some $k \geq 1$. Now we set $\mathcal{A} = \{B\}$, $\mathcal{B} = \mathrm{DSPACE}\left(s\left(\left\lfloor \frac{n}{k} \right\rfloor\right)\right)$ and $\leq\, = \,\leq_m^{\mathrm{DSPACE},\log,k\cdot n}$ and apply Lemma 8.39. ∎

In [CoHe 84] and [MaYo 81] other methods for proving lower bounds are developed.

It has to be noticed that hardness is not the only possible reason for a set to be complex. There exist sets which are not hard but which are nevertheless complex, as is shown by the next theorem. For these sets, new techniques for lower bound proofs are required.

8.41. Theorem. [Lyn 76] Let

$$t \in \textbf{T-DTIME}(\text{Pol } t)$$

and

$$t > n^k \text{ for all } k \in \mathbb{N}.$$

Then there exists a polynomial p such that

$$\bigwedge_{s \in \mathbb{R}_1} \bigvee_{A \in \mathbf{REC}} (s <_{\text{ae}} \text{T-DTIME-Comp } A \wedge A \text{ not } \leqq_{\text{T}}^{\text{P}}\text{-hard for } \textbf{T-DTIME}(p \circ t)). \blacksquare$$

8.2.2. Hard Sets for some Important Complexity Classes

At first we present a $\leqq_{\text{m}}^{\log}$-complete set for **NP**.

8.42. Theorem. [Coo 71a] SAT is $\leqq_{\text{m}}^{\log}$-complete for **NP**.

Proof. 1. SAT $\in$ **NP**. An NTM is constructed which tests deterministically, in linear time, whether a given input H is a Boolean formula in conjunctive normal form. If it is, an assignment of the variables is guessed (linear time) and the value of the formula with respect to this assignment is computed (quadratic time).
2. $B \in \mathbf{NP} \to B \leqq_{\text{m}}^{\text{P}}$ SAT. Let $B \in \mathbf{NP}$ be accepted by the NTM M working in time p where p is a polynomial. Let $M = (S, \Sigma, \delta, s_0, s_1, \emptyset)$ where s_0 and s_1 are the initial and the w.l.o.g. only accepting state, respectively. Assume that $S \cap \Sigma = \emptyset$, that M uses the tape squares $1, 2, \ldots, p(n)$ only and that M stops only when the head scans tape square 1.

Let w be an input of length n. For describing the work of M on w it is sufficient to consider those ID's of the situations of M which represent the state and the tape contents of the squares $0, 1, 2, \ldots, p(n) + 1$. These ID's are of length $p(n) + 3$.

The idea for the reduction is to map the word w to the Boolean formula Comp(w) describing the behaviour of M on input w and having the following properties:
1. $w \in B \leftrightarrow \text{Comp}(w) \in \text{SAT}$.
2. Comp(w) can be computed from w in polynomial time.

Comp(w) will depend on the Boolean variables $x_{t,i,j}$ ($t = 0, 1, \ldots, p(n)$, $i = 0, 1, \ldots, p(n) + 2$, $j = 1, \ldots, m$, where $m = \lceil \log \text{card } (\Sigma \cup S) \rceil$). The variables $x_{t,i,1}, \ldots, x_{t,i,m}$ will describe the ith symbol of the ID of the tth step of M. Encoding each $c \in \Sigma \cup S$ by a $(c_1, \ldots, c_m) \in \{0, 1\}^m$ we shall write

$$x_{t,i} = c \quad \text{instead of} \quad x_{t,i,1}^{c_1} \wedge x_{t,i,2}^{c_2} \wedge \ldots \wedge x_{t,i,m}^{c_m}$$

and

$$xyz = abc \quad \text{instead of} \quad (x = a) \wedge (y = b) \wedge (z = c).$$

An assignment φ of the variables $x_{t,i,j}$ will satisfy the formula Comp(w) if and only if $\varphi(x_{001})\, \varphi(x_{002}) \ldots \varphi(x_{p(n),p(n+2),m})$ describes the sequence of ID's of an accepting computation of M on w which takes at most $p(n)$ steps, uses at most the tape squares

1, 2, ..., $p(n)$ and stops in tape square 1. The formula has the form $\text{Comp}(w) = A \wedge B \wedge C$ where

A describes the correct start of M on input $w = a_1 \dots a_n$,
B describes a correct move of M, and
C describes the correct stop of M.

$$A \equiv (x_{00} = \square) \wedge (x_{01} = s_0) \wedge (x_{02} = a_1) \wedge \dots \wedge (x_{0,n+1} = a_n)$$
$$\wedge\, (x_{0,n+2} = \square) \wedge \dots \wedge (x_{0,p(n)+1} = \square) \wedge (x_{0,p(n)+2} = \square),$$

$$B \equiv \bigwedge_{t=0}^{p(n)-1} \left[\bigwedge_{i=0}^{p(n)} T(x_{t,i}, x_{t,i+1}, x_{t,i+2}, x_{t+1,i}, x_{t+1,i+1}, x_{t+1,i+2}) \wedge (x_{t0} = \square) \wedge (x_{t,p(n)+2} = \square) \right],$$

$T(y_1, y_2, y_3, y_1', y_2', y_3')$ is a formula in conjunctive normal form which is equivalent to

$$\bigwedge_{b_1,b_2,b_3 \in \Sigma} (y_1 y_2 y_3 = b_1 b_2 b_3 \to y_2' = b_2) \wedge \bigwedge_{b_1,b_3 \in \Sigma, s \in S} \Big(y_1 y_2 y_3 = b_1 s b_3$$
$$\to \Big(\bigvee_{(s,'b_2,0) \in \delta(s,b_3)} y_1' y_2' y_3' = b_1 s' b_2 \vee \bigvee_{(s',b_2,+1) \in \delta(s,b_3)} y_1' y_2' y_3' = b_1 b_2 s'$$
$$\vee \bigvee_{(s',b_2,-1) \in \delta(s,b_3)} y_1' y_2' y_3' = s' b_1 b_2 \Big) \Big),$$

$$C \equiv (x_{p(n),1} = s_1).$$

Computing Comp(w) is simply a question of writing down $A_1 \wedge A_2 \wedge A_3$, and this can be done in logarithmic space. ∎

Logical implications of Theorem 8.42, related topics and some historical comments are presented in [Boe 83].

The next theorem concerns some problems related to the "graph accessibility problem", which is the set of all graphs for which a distinguished goal node is accessible (reachable by a path) from a distinguished source node. It can be made precise as follows. Let

$$E = \{(1, i_{11}), \dots, (1, i_{1k_1}), \dots, (n, i_{n1}), \dots, (n, i_{nk_n})\} \subseteq \{1, \dots, n\}^2.$$

Then we define

$$\text{Code } E = [0 \# 0^{i_{11}} \# \dots \# 0^{i_{1k_1}}] \dots [0^n \# 0^{i_{n1}} \# \dots \# 0^{i_{nk_n}}].\text{*)}$$

The graph accessiblity problem is then defined by

$$\text{GAP} = \{\text{Code } E : \bigvee_n (E \subseteq \{1, \dots, n\}^2 \wedge (1, n) \text{ belongs to the transitive closure of } E)\}.$$

*) Note that using binary representation we would get a code which is polynomially length related to our code. So binary representation could be used as well. The only reason that we prefer unary representation is that in this representation GAP_{ao} (see below) becomes acceptable by a 1:2-NCA.

This problem is also considered for undirected graphs and graphs with outdegree bounded by 1:

$$\mathrm{UGAP} = \{\mathrm{Code}\, E : \mathrm{Code}\, E \in \mathrm{GAP} \wedge E \text{ is symmetric}\},$$

$$1\mathrm{GAP} = \Big\{\mathrm{Code}\, E : \mathrm{Code}\, E \in \mathrm{GAP} \wedge \bigwedge_i \bigwedge_j \bigwedge_k \big((i, j) \in E \wedge (i, k) \in E \to j = k\big)\Big\}.$$

Some versions of GAP acceptable by simple types of automata will be of interest in § 23.

$$\mathrm{GAP}_o = \Big\{\mathrm{Code}\, E : \bigvee_n \bigvee_k \bigvee_{i_1} \cdots \bigvee_{i_k} \Big[E \subseteq \{1, \ldots, n\}^2 \wedge i_1 = 1 \wedge i_k = n \wedge \bigwedge_{j<k} \big((i_j, i_{j+1}) \in E \wedge i_j < i_{j+1}\big)\Big]\Big\}.$$

The index o comes from "*o*rder preserving".

$$\mathrm{GAP}_{ao} = \Big\{\mathrm{Code}\, E : \bigvee_n \bigvee_k \bigvee_{i_1} \cdots \bigvee_{i_k} \Big[E \subseteq \{1, \ldots, n\}^2 \wedge i_1 = 1 \wedge i_k = n \wedge \bigwedge_{j<k} (i_j, i_{j+1}) \in E \wedge \bigwedge_{j+1<k} i_j < i_{j+2}\Big]\Big\}.$$

The index ao comes from "*a*lmost *o*rder preserving".

That the following language is closely related to GAP_o becomes clear by the proof of Statement 5 of the subsequent theorem.

Let $\mathrm{S} \subseteq \{0, 1\}^*$ be the set of all symmetrical words

$$\mathrm{L(S)} = \{[u_1 v_1 w_1] \ldots [u_k v_k w_k] : k \geqq 1 \wedge v_1 \ldots v_k \in \mathrm{S} \wedge u_1, \ldots, u_k \in (\{0, 1\}^* \#)^* \wedge w_1, \ldots, w_k \in (\# \{0, 1\}^*)^*\}.$$

The value $\max_{(i,j) \in E} |i - j|$ is called the *bandwidth* of $E \subseteq \{1, \ldots, n\}^2$.

$$\mathrm{GAP}(f) = \{\mathrm{Code}\, E : \mathrm{Code}\, E \in \mathrm{GAP} \wedge E \text{ has bandwidth bounded by } f(|\mathrm{Code}\, E|)\}.$$

GAP can also be considered for "alternating graphs", i.e. for graphs whose vertices are labelled by $\wedge$ or $\vee$. For such a graph $(\{1, \ldots, n\}, E)$ we define

0. i is always reachable from i.
1. n is reachable from the $\wedge$-vertex i if n is reachable from all j such that $(i, j) \in E$.
2. n is reachable from the $\vee$-vertex i if there exists a j with $(i, j) \in E$ such that n is reachable from j.

$$\mathrm{AGAP} = \{\mathrm{Code}\, G : G \text{ is an alternating graph with vertex set } \{1, \ldots, n\} \text{ such that } n \text{ is reachable from } 1\}.$$

By a slight modification of the interpretation of the alternating graphs we obtain the path systems introduced in [Coo 73a]. A *path system* is a quadruple (X, R, S, T) where $R \subseteq X^3$, $S \subseteq X$ (the set of *source* nodes) and $T \subseteq X$ (the set of *terminal* nodes). For $A \subseteq X$ let $\Gamma_R(A)$ be the least set M including A and having the property $\bigwedge_i \bigwedge_j \bigwedge_k \big((i, j, k) \in R \wedge j, k \in M \to i \in M\big)$. A path system (X, R, S, T) is called *solvable* if $\Gamma_R(T) \cap S \neq \emptyset$.

$$\mathrm{SP} = \{P : P \text{ is a solvable path system}\}.$$

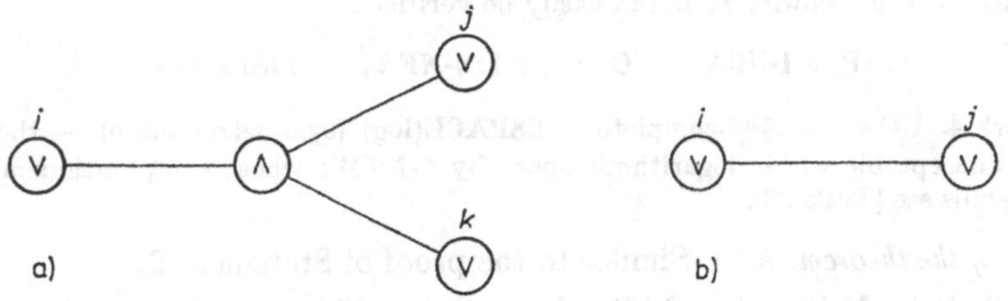

Figure 8.2

To show that AGAP and SP are closely related, let a path system (X, R, S, T) be given such that $1, n \notin X$. We construct an alternating graph G by representing each $(i, j, k) \in R$ by a subgraph of the form shown in Figure 8.2a. For $j = k$ this can be simplified to a subgraph of the form shown in Figure 8.2b. Furthermore, we add a vertex 1 labelled by $\vee$ and for every $s \in S$ the edge $(1, s)$, and we add a node n and for every $t \in T$ the edge (t, n). G belongs to AGAP if and only if (X, R, S, T) belongs to SP.

Conversely, if an alternating graph G is given, we construct a path system $P = (X, R, S, T)$ where X is the vertex set of G, $S = \{1\}$, $T = \{n\}$ and R is defined as follows:

1. If i is a vertex of G labelled with $\vee$, then for all j such that $(i, j) \in E$ the triple (i, j, j) belongs to R.
2. If i is vertex of G labelled with $\wedge$ and $(i, j_1), \ldots, (i, j_k)$ are all edges arising from i, then the triples
$$(i, j_1, (i, j_2)),\ \big((i, j_2), j_2, (i, j_3)\big),\ \ldots,\ \big((i, j_{k-2}), j_{k-2}, (i, j_{k-1})\big),\ \big((i, j_{k-1}), j_{k-1}, j_k\big)$$
belong to R where $(i, j_2), \ldots, (i, j_{k-1})$ do not belong to X.
3. Further triples do not belong to R.

It is obvious that $P \in \mathrm{SP}$ if and only if $G \in \mathrm{AGAP}$.

8.43. Theorem.

1. [HaImMa 78] 1GAP is $\leq_m^{1-\log}$-complete in **L**.
2. [Savi 73a] GAP is $\leq_m^{\log}$-complete in **NL**.
3. [Sud 73] GAP_o is $\leq_m^{\log}$-complete in **NL**.
4. [Sud 73] GAP_{ao} is $\leq_m^{\log}$-complete in **NL**.
5. [Sud 75] L(S) is $\leq_m^{\log}$-complete in **NL**.
6. [MoSu 79] $\{\mathrm{GAP}(2^{kf}) : k \in \mathbb{N}\}$ is $\leq_m^{\log}$-complete in **NSPACE**(f)*), provided $f(2^{2^n})$ is T-DSPACE-constructible and $\log\log n \leq f(n) \leq \log n$.

*) i.e. for every $A \in$ **NSPACE**(f) there exists a $k \in \mathbb{N}$ such that $A \leq_m^{\log} \mathrm{GAP}(2^{kf})$.

7. [Imm 79] AGAP is $\leqq_m^{\log}$-complete in **ASPACE**(log) = **P**.
8. [Coo 73a] SP is $\leqq_m^{\log}$-complete in **P**.

Remark 1. 1GAP and similar problems were considered in [Jon 75], where, however, not $\leqq_m^{1-\log}$-reducibility was used.

Remark 2. See [Imm 79] for further results concerning GAP.

Remark 3. The following facts can easily be verified:

$$\mathrm{GAP}_o \in \mathbf{1\text{-}NCA}, \qquad \mathrm{GAP}_{ao} \in \mathbf{1{:}2\text{-}NFA}, \qquad \mathrm{L(S)} \in \mathbf{LIN}.$$

Remark 4. UGAP is $\leqq_m^{\log}$-complete in **SSPACE**(log) (symmetric space) — the class of languages acceptable within logarithmic space by 1-NTM's whose computations are reversable. For details see [LePa 80].

Proof of the theorem. Ad 1. Similar to the proof of Statement 2.

Ad 2. It is obvious that GAP belongs to **NL**. To prove the completeness of GAP, let $A \in \mathbf{NL}$. W.l.o.g. we choose a 2-NTM M accepting A in logarithmic space in such a way that there is only one distinguished accepting situation. The desired reduction $A \leqq_m^{\log}$ GAP will be done by a machine computing for an input w the computation graph Comp(w) of M on input w. (The vertices of Comp(w) are those situations of M in which the input tape contains w and at most $\log |w|$ worktape squares are used. The edges of Comp(w) are given by $\vdash_M$. The role of $1(n)$ is played by the initial (accepting) situation of M on w.) This is possible within logarithmic space, and evidently $w \in A \leftrightarrow \mathrm{Comp}(w) \in \mathrm{GAP}$.

For 3 and 4 the reader is referred to [Sud 73].

Ad 5. The reader may check that L(S) ∈ **NL**. We prove $\mathrm{GAP}_o \leqq_m^{\log} \mathrm{L(S)}$. The function

$$\begin{aligned} &f([1 \# i_{11} \# \ldots \# i_{1k_1}] \ldots [n \# i_{n1} \# \ldots \# i_{nk_n}]) \\ &= [01]\left[\# 010^{i_{11}}1 \# \ldots \# 010^{i_{1k_n}}1\right] \ldots \left[\# 0^n10^{i_{n1}}1 \# \ldots \# 0^n10^{i_{nk_n}}1\right] \\ &\quad [0^n1]\,[10^n10^n \#]\,[10^{n-1}10^{n-1} \#] \ldots [10^210^2 \#]\,[1010 \#] \end{aligned}$$

is computable within logarithmic space, and $\mathrm{GAP}_o \leqq_m^{\log} \mathrm{L(S)}$ via f.

Ad 6. See [MoSu 79].

Ad 7. This is proved by adapting the proof of Statement 2 to alternating TM's. For **ASPACE**(log) = **P** see 20.36.

Ad 8. This follows from the considerations before this theorem showing the $\leqq_m^{\log}$-equivalence of AGAP and SP. ∎

8.2.3. Lower Bounds for Selected Classes of Problems

In this subsection lower bounds for a number of problems from various branches of mathematics serve as illustration of the reducibility method. They are arranged according to their topic rather than their complexity. Many further lower bounds immediately follow from an application of 8.40 or 8.39 to the problems listed in § 14.2.2.

8.2.3.1. Regular-Like Expressions

Let X be a finite alphabet with at least two elements, and let $\mathfrak{X} = \{\boldsymbol{x} : x \in X\}$. In what follows Ω denotes a set satisfying $\mathfrak{X} \subseteq \Omega$ and $\Omega \setminus \mathfrak{X} \subseteq \{\cup, \cdot, {}^*, {}^2, \cap, \overline{\ }\}$. Let $\mathbf{E}(\Omega)$ be the set of all *Ω-expressions* i.e. the smallest set M such that

1. $\boldsymbol{\emptyset}, \mathbf{e} \in M, \mathfrak{X} \subseteq M$,
2. ${}^*({}^2, \overline{\ }) \in \Omega \wedge t \in M \to t^*(t^2, \overline{t}, \text{resp.}) \in M$,
3. $\cdot\ (\cup, \cap) \in \Omega \wedge t_1, t_2 \in M \to t_1 \cdot t_2 (t_1 \cup t_2, t_1 \cap t_2, \text{resp.}) \in M$.

The Ω-expression t describes the language $\mathrm{L}(t) \subseteq X^*$ which is defined as follows:

$$\mathrm{L}(\boldsymbol{\emptyset}) = \emptyset, \quad \mathrm{L}(\mathbf{e}) = \{e\}, \quad \mathrm{L}(\mathbf{x}) = \{x\} \quad \text{for} \quad x \in \Sigma,$$

$$\mathrm{L}(t^*) = \mathrm{L}(t)^*, \quad \mathrm{L}(t^2) = \mathrm{L}(t) \cdot \mathrm{L}(t), \quad \mathrm{L}(\overline{t}) = \overline{\mathrm{L}(t)}, \quad \mathrm{L}(t_1 \cdot t_2) = \mathrm{L}(t_1) \cdot \mathrm{L}(t_2),$$

$$\mathrm{L}(t_1 \cup t_2) = \mathrm{L}(t_1) \cup \mathrm{L}(t_2), \quad \mathrm{L}(t_1 \cap t_2) = \mathrm{L}(t_1) \cap \mathrm{L}(t_2).$$

Since we have $\mathbf{REG} \cap \mathfrak{P}(X^*) = \{\mathrm{L}(t) : t \in \mathbf{E}(\Omega)\}$ if $\{\cup, \cdot, {}^*\} \in \Omega$, the Ω-expressions are also called *regular-like expressions*.

The following problems concerning regular-like expressions are considered:

$$\mathrm{INEQ}(\Omega) =_{\mathrm{df}} \{(t, t') : t, t' \in \mathbf{E}(\Omega) \wedge \mathrm{L}(t) \neq \mathrm{L}(t')\},$$

$$\mathrm{NEC}(\Omega) =_{\mathrm{df}} \{t : t \in \mathbf{E}(\Omega) \wedge \mathrm{L}(t) \neq X^*\},$$

$$\mathrm{NE}(\Omega) =_{\mathrm{df}} \{t : t \in \mathbf{E}(\Omega) \wedge \mathrm{L}(t) \neq \emptyset\}.$$

(INEQ, NEC, NE are abbreviations of "inequivalence", "nonempty complement", "nonempty", resp.)

INEQ(Ω), NEC(Ω) and NE(Ω) are equivalent with respect to $\leq_{\mathrm{m}}^{\text{log-lin}}$ provided $\{\cap, \overline{\ }\} \subseteq \Omega$ or $\{\cup, \overline{\ }\} \subseteq \Omega$.

NEC(Ω) $\leq_{\mathrm{m}}^{\text{log-lin}}$ INEQ(Ω) provided $\overline{\ } \in \Omega$ or $\{\cup, {}^*\} \subseteq \Omega$.

The following theorem gives a quantitative statement of the intuitively clear insight that growing succinctness of a term language leads to a shortening of the expressions and thus to more complex decision problems. The most exciting fact is that the use of complementation makes the nonempty complement problem nonelementary.

8.44. Theorem.

A	lower bound		upper bound	
	$A \notin$	author(s)	$A \in$	authors
NEC(0, 1, ∪, ·, *)	**NSPACE**(s) for every $s \prec n$	[MeSt 72]	**LINSPACE**	A. R. Meyer (see [Sto 74])
NEC(0, 1, ∪, ·, ²)	**NTIME**(2^{cn})	[StMe 73]	**NEXPTIME**	[StMe 73]
NE(0, 1, ∪, ·, *, ∩)	**NSPACE**(s) for every $s \prec n$	[Für 80]	**LINSPACE**	J. E. Hopcroft
NEC(0, 1, ∪, ·, *, ∩),	**DSPACE**(2^{cn})	[Für 80]	**EXPSPACE**	H. B. Hunt III/ J. E. Hopcroft
INEQ(0, 1, ∪, ·, *, ∩)	**DSPACE**(2^{cn})	[Für 80]	**EXPSPACE**	H. B. Hunt III/ J. E. Hopcroft
NEC(0, 1, ∪, ·, ², *)	**DSPACE**(2^{cn})	[MeSt 72]	**EXPSPACE**	[MeSt 72]
NEC(0, 1, ∪, ·, ¯)	**DSPACE** $\left(2^{2^{\cdot^{\cdot^{2}}}}\right\}\lfloor\log n\rfloor\Big)$	[Sto 74]	**DSPACE** $\left(2^{2^{\cdot^{\cdot^{2}}}}\right\} n\Big)$	[Sto 74]

($c > 0$ is a suitable constant.)

Remark. In [Hun 73a], [Hun 73b], [HuRo 74], [Sto 74], [Hun 79] the reader finds lower and upper bounds for many further problems concerning regular-like expressions.

Proof. We do not prove all statements of 8.44. In order to explain the method it is sufficient to describe it by a few examples. For the proofs of the remaining statements the reader is referred to the literature.

Starting with NEC(0, 1, ∪, ·, ², *) we first show:

(*) If $A \in$ **EXPSPACE**, then there exists a finite alphabet X such that

$$A \leqq_{\mathrm{m}}^{\text{log-lin}} \mathrm{NEC}(\mathfrak{X}, \cup, \cdot, {}^2, {}^*), \quad \text{where} \quad \mathfrak{X} = \{\mathbf{x} : x \in X\}.$$

Let $M = (S, \Sigma, \delta, q_0, \{q_1\}, \emptyset)$ be an NTM accepting A within space 2^{kn} for some $k \geqq 1$. Assume $\Sigma \cap S = \emptyset$ and $\# \notin \Sigma \cup S$. W.l.o.g. we assume that M uses only the tape squares $1, 2, \ldots, 2^{kn}$ on inputs of length n. Define $X = S \cup \Sigma \cup \{\#\}$.

Thus an accepting computation of M on input of length n can be described by the word $\# K_0 \# K_1 \# \ldots \# K_m$ where K_i is that ID of the ith situation of this computation which describes the contents of the tape squares $1, 2, \ldots, 2^{kn} + n + 1$. Consequently, each K_i is of length $2^{kn} + n + 2$.

$$\mathrm{NAC}_M(x) =_{\mathrm{df}} \{w : w \in X^* \wedge w \text{ does not describe in the way explained above an accepting computation of } M \text{ on input } x\}.$$

The idea of the proof rests on the equivalence

$$x \in A \leftrightarrow \mathrm{NAC}_M(x) \neq X^*$$

and the possibility of describing $\mathrm{NAC}_M(x)$ by an $\mathfrak{X} \cup \{\cup, \cdot, {}^2, {}^*\}$-expression $\mathrm{E}_M(x)$: $\mathrm{L}(\mathrm{E}_M(x)) = \mathrm{NAC}_M(x)$. This yields the desired reducibility

$$x \in A \leftrightarrow \mathrm{E}_M(x) \in \mathrm{NEC}(0, 1, \cup, \cdot, {}^2, {}^*)$$

as soon as the log-lin-computability of $\mathrm{E}_M(x)$ as a function of x is demonstrated.

All proofs of results of the form "$\Phi(t) \leqq A$" (with some reducibility notion $\leqq$) have as their central part a description by means of A of the work of all machines running within Φ-complexity t (later on we will refer to such descriptions as *encodings of the computations of the machines running within Φ-complexity t by means of A*). We outline this description in the following in order to establish the equation $\mathrm{L}(\mathrm{E}_M(x)) = \mathrm{NAC}_M(x)$. The present proof (and also the proof of 8.42) can serve as a basic pattern for all proofs of the above-mentioned type of results.

A word $w \in X^*$ belongs to $\mathrm{NAC}_M(x)$ if and only if

(a) it starts incorrectly, i.e. w does not begin with $\# q_0 x \square^{2kn+1} \#$, or

(b) it "moves" incorrectly, i.e. w contains a subword $\# K \# K' \#$, where K' is not possible next ID of K, or

(c) it ends incorrectly, i.e. w does not contain q_1 or does not end with $\#$.

Let B_1, B_2 and B_3 be the set of words satisfying (a), (b), or (c), resp. Then $\mathrm{NAC}_M(x) = B_1 \cup B_2 \cup B_3$, and it is sufficient to describe regular-like expressions for these three sets. B_1, the set of words starting incorrectly, is described by the expression (let $x = x_1 \ldots x_n$ and let $\bar{\bar{a}}$ and $\mathbf{X}$ be abbreviations for expressions describing $\mathrm{X} \setminus \{a\}$, for $a \in X$, and X, resp.).

$$\mathrm{E}_M^{(1)}(x) =_{\mathrm{df}} \Big(\bar{\bar{\#}} \cup \# \big(\bar{\bar{q}}_0 \cup \mathbf{q}_0(\bar{\bar{x}}_1 \cup \mathbf{x}_1(\bar{\bar{x}}_2 \cup \mathbf{x}_2(\ldots \cup \mathbf{x}_{n-1}\bar{\bar{x}}_n) \ldots\big) \mathbf{X}^*$$
$$\cup \# \mathbf{q}_0\mathbf{x}_1 \ldots \mathbf{x}_n(\mathbf{X} \cup \mathrm{e})^{2kn} \bar{\bar{\square}} \mathbf{X}^* \cup \mathbf{X}^{2kn+n+3} \bar{\bar{\#}} \mathbf{X}^*.$$

The three subexpressions describe words that

— do not start with $\# q_0 x_1 \ldots x_n$,

— do not have enough blanks in the first ID,

— do not have a $\#$ at the position $2^{kn} + n + 4$.

To describe the set B_2 of words moving incorrectly let $\mathrm{N}(a, b, c)$ be the set of words $a'b'c'$ with the property: If a, b, c occupy the positions $j, j+1, j+2$ in some ID K of M, then a', b', c' may occupy the positions $j, j+1, j+2$ in some next ID of K. Furthermore, if $a_1'a_2'a_3' \in \mathrm{N}(a_1, a_2, a_3)$ and $a_i = \#$, then $a_i' = \#$. Then, if $t(a, b, c)$ is a term describing $X^3 \setminus \mathrm{N}(a, b, c)$,

$$\mathrm{E}_M^{(2)}(x) =_{\mathrm{df}} \bigcup_{a,b,c \in X} \mathbf{X}^* \, \mathbf{abc} \, \mathbf{X}^{2kn+n} \cdot t(a, b, c) \cdot \mathbf{X}^*$$

describes B_2.

Finally

$$\mathrm{E}_M^{(3)}(x) =_{\mathrm{df}} (\bar{\bar{q}}_1)^* \cup \mathbf{X}^* \, \bar{\bar{\#}}$$

describes the set B_3 of words ending incorrectly.

To complete the proof of (*) it remains to show that

$$\mathrm{E}_M(x) =_{\mathrm{df}} \mathrm{E}_M^{(1)}(x) \cup \mathrm{E}_M^{(2)}(x) \cup \mathrm{E}_M^{(3)}(x)$$

can be computed from x in logarithmic space and that $|\mathrm{E}_M(x)| = d\,|x|$ for some constant d. To see the latter we first notice that the length of $\bar{a}$ and of $t(a, b, c)$ is independent of x. The crucial fact is that $\mathbf{X}^{2^{kn}}$ can be described by an expression using kn squarings and thus having length $d' \cdot n$ for some $d' \in \mathbb{N}$. The remaining details, in particular the logspace computability of the mapping E_M, are left to the reader.

(*) is proved. A simple coding argument can be used to get rid of the alphabet X depending on the individual set $A \in$ **EXPSPACE** and to show NEC($\mathfrak{X}$, ∪, ·, ², *) $\leqq_{\mathrm{m}}^{\text{log-lin}}$ NEC(0, 1, ∪, ·, ², *). Consequently,

$$\mathrm{NEC}(0, 1, \cup, \cdot, {}^2, {}^*) \quad \text{is} \quad \leqq_{\mathrm{m}}^{\text{log-lin}}\text{-hard for } \mathbf{EXPSPACE}.$$

From this result the lower bound indicated in the theorem follows by an application of 8.40.1.

The lower bound result for NEC(0, 1, ∪, ·, *) is proved in the same way. The essential difference is that 2^{kn} has to be replaced by kn. Although the squaring operation is now missing, the resulting expression has the length $d \cdot n$ (for some d) because the subexpressions $\mathbf{X}^{2^{kn}}$ are replaced with $\mathbf{X}^{kn}$.

The same idea is used to prove the lower bound for NEC(0, 1, ∪, ·, ²). ∎

The application of this method to the case of NEC(0, 1, ∪, ·, ¯) is much more complicated (cf. [AhHoUl 74]). Instead of repeating a proof here, we give an intuitive explanation of why the complementation operator makes the decision of "nonempty complement" so difficult. NEC(0, 1, ∪, ·, ¯) can be decided by inductively constructing from a given expression E a DFA or an NFA A such that $\mathrm{L}(A) = \mathrm{L}(E)$ and by checking whether A accepts the whole set $\{0, 1\}^*$. The obvious construction of an automaton A_3 such that $\mathrm{L}(A_3) = \mathrm{L}(A_1) \cdot \mathrm{L}(A_2)$ yields an NFA A_3. On the other hand, the obvious construction of an automaton A_5 such that $\mathrm{L}(A_5) = \overline{\mathrm{L}(A_4)}$ requires a DFA A_4. If the concatenation is followed by a complementation, then the NFA A_3 must be converted into an equivalent DFA A_4 which possibly causes an exponential increase in the number of states. Thus, alternating concatenations and complementations lead to iterated exponentiations in the number of states and hence in the complexity. In fact, [Sto 74] contains a result that closely relates the depth of nesting of concatenations and complementations to the number of iterated exponentiations in a lower complexity bound.

8.2.3.2. *Logical Theories*

Logical theories are understood by their semantic approach: Let $\mathfrak{K}$ be a set of models of a given signature. This signature determines in a well-known way a first order and a second order predicate calculus language of this signature. The *first order theory of* $\mathfrak{K}$ is the set of all sentences of the first order language of the signature of $\mathfrak{K}$ that are valid in all models of $\mathfrak{K}$. It is denoted by $\mathcal{T}(\mathfrak{K})$. For further details the reader is referred to [Sho 67].

If $\mathfrak{K}$ is a model class containing only one model (K, Ω), then we write $\mathcal{T}(K, \Omega)$ instead of $\mathcal{T}\left(\{(K, \Omega)\}\right)$.

We start with several order theories. Let $\mathfrak{W}$ be the class of all well-ordered sets and $\mathfrak{O}$ the class of all totally ordered sets. We use the notations WELLORD $=_{\text{df}} \mathcal{T}(\mathfrak{W})$, LINORD $=_{\text{df}} \mathcal{T}(\mathfrak{O})$, LEXORD $=_{\text{df}} \mathcal{T}(\{0, 1\}^*, \prec)$, where $\prec$ is the usual lexicographic order on $\{0, 1\}^*$. Note that we have in mind the order e, 0, 00, 000, ..., 1, 10, 100, ..., ... and not the quasilexicographic order e, 0, 1, 00, 01, 10, 11, ... as described on p. 39. We also consider the theories $\mathcal{T}(K, <)$ and $\mathcal{T}_M(K, <) =_{\text{df}} \mathcal{T}\left(\{(K, <, P) : P \subseteq K\}\right)$, where $K \in \{\mathbb{N}, \mathbb{G}, \mathbb{Q}\}$, ABA denotes the theory of atomless Boolean algebras.

Together with or instead of $\mathcal{T}(\mathfrak{K})$ we are sometimes interested in the set SAT($\mathfrak{K}$) of all sentences in the language of $\mathfrak{K}$'s signature that are satisfiable by at least one model in $\mathfrak{K}$. The obvious relation

$$p \in \text{SAT}(\mathfrak{K}) \leftrightarrow \sim p \notin \mathcal{T}(\mathfrak{K})$$

yields

$$\text{SAT}(\mathfrak{K}) \leqq_{\text{m}}^{\text{log-lin}} \overline{\mathcal{T}(\mathfrak{K})}.$$

8.45. Theorem.

A	lower bound		upper bound	
	$A \notin$	authors	$A \in$	authors
$\mathcal{T}(\mathbb{Q}, <)$	**NSPACE**(s) for every $s \prec \sqrt{n}$	[Sto 74]	**DSPACE**$(n \log n)$	[Fer 74]
$\mathcal{T}(\mathbb{N}, <)$	**NSPACE**(s) for every $s \prec \text{id}$	[Fer 74]	**DSPACE**(n^2)	[Fer 74]
$\mathcal{T}(\mathbb{G}, <)$	**NSPACE**(s) for every $s \prec \text{id}$	[Fer 74]		
LEXORD	**NSPACE**(s) for every $s \prec \text{id}$	[Fer 74]	**DSPACE**(n^2)	[Fer 74]
WELLORD	**NSPACE**(s) for every $s \prec \text{id}$	[Fer 74]	**DSPACE**(n^3)	[Fer 74]
LINORD	**DSPACE** $\left(2^{2^{\cdot^{\cdot^{2}}}}\Big\}\, cn\right)$	[Mey 74]	**DSPACE** $\left(2^{2^{\cdot^{\cdot^{2}}}}\Big\}\, dn\right)$	M. O. Rabin (see [Mey 74])
$\mathcal{T}_M(K, <)$ for arbitrary infinite K	**DSPACE** $\left(2^{2^{\cdot^{\cdot^{2}}}}\Big\}\, cn\right)$	[Sto 74] [Mey 74] [Rob 74b]	need not always be decidable	
ABA	**NTIME** $\left(2^{c \cdot \frac{n}{\log n}}\right)$	[Berp 79]	**DSPACE** $\left(2^{d \frac{n}{\log n}}\right)$	[Berp 79]

($c, d > 0$ are suitable constants.) ∎

Remark. Let BA be the theory of all Boolean algebras. Compare the last statement of Theorem 8.45 with the result that BA is $\leqq_{\mathrm{m}}^{\log}$-complete in $\mathbf{ATIME\text{-}ALT}(2^{\mathrm{Lin}}, n)$ ([Koz 80b]).

The next theorem concerns several successor theories.

$$\mathrm{SUC}_1 =_{\mathrm{df}} \mathcal{T}(\mathbb{N}, \mathrm{s}),$$

$$\mathrm{SUC}_1\mathrm{M} =_{\mathrm{df}} \mathcal{T}\big(\{(\mathbb{N}, \mathrm{s}, P) : P \subseteq \mathbb{N}\}\big), \text{ M comes from } monadic,$$

$$\mathrm{SUC}_2 =_{\mathrm{df}} \mathcal{T}(\{0, 1\}^*, \mathrm{s}_0, \mathrm{s}_1),$$

$$\mathrm{SUC}_2(\sqsubseteq) =_{\mathrm{df}} \mathcal{T}(\{0, 1\}^*, \mathrm{s}_0, \mathrm{s}_1, \sqsubseteq).$$

WS1S denotes the weak second order theory of one successor. Here

$$\mathrm{s}(n) =_{\mathrm{df}} n + 1 \quad \text{for} \quad n \in \mathbb{N},$$

$$\mathrm{s}_0(w) =_{\mathrm{df}} w0 \quad \text{and} \quad \mathrm{s}_1(w) =_{\mathrm{df}} w1 \quad \text{for} \quad w \in \{0, 1\}^*.$$

8.46. Theorem.

A	lower bound		upper bound	
	$A \notin$	author	$A \in$	author
SUC_1	$\mathbf{NSPACE}(s)$ for every $s \prec \mathrm{id}$	[Fer 74]	$\mathbf{DSPACE}(n^2)$	[Fer 74]
$\mathrm{SUC}_1\mathrm{M}$	$\mathbf{NTIME}(2^{2^{cn}})$	[Fer 74]	$\mathbf{DSPACE}(2^{2^n})$	[Fer 74]
SUC_2	$\mathbf{NTIME}(2^{cn})$	[Fer 74]	$\mathbf{DSPACE}(2^n)$	[Fer 74]
$\mathrm{SUC}_2(\sqsubseteq)$	$\mathbf{DTIME}\Big(2^{2^{\cdot^{\cdot^{2}}}}\Big\}\, c\lfloor\log n\rfloor\Big)$	A. R. MEYER see [Sto 74]	$\mathbf{DTIME}\Big(2^{2^{\cdot^{\cdot^{2}}}}\Big\}\, dn\Big)$	[ElRa 66]
WS1S	$\mathbf{DTIME}\Big(2^{2^{\cdot^{\cdot^{2}}}}\Big\}\, cn\Big)$	[Mey 73] M. O. RABIN	$\mathbf{DTIME}\Big(2^{2^{\cdot^{\cdot^{2}}}}\Big\}\, dn\Big)$	[Büc 60]

($c, d > 0$ are suitable constants.) ∎

Remark. In [Vol 83a] it has been shown that SUC_2 is $\leqq_{\mathrm{m}}^{\text{log-lin}}$-complete in $\mathbf{ATIME\text{-}ALT}(2^{\mathrm{Lin}}, \mathrm{id})$ (for **ATIME-ALT** see Remark 1 after Theorem 8.47). This result remains true if the equal length predicate is added to SUC_2.

The next theorem deals with the following arithmetical theories:

$$\mathrm{PA} =_{\mathrm{df}} \mathcal{T}(\mathbb{N}, 0, +, \leqq) \quad \text{(Presburger arithmetic)},$$

$$\mathrm{RA} =_{\mathrm{df}} \mathcal{T}(\mathcal{R}, 0, +, \leqq) \quad \text{(reals with addition)},$$

$$\mathrm{PM} =_{\mathrm{df}} \mathcal{T}(\mathbb{N}', 1, \cdot, \leqq) \quad \text{(positive natural numbers with multiplication)}.$$

Sometimes the Presburger arithmetic is considered with subtraction, and the sentences are interpreted over the integers $\mathbb{G}$. If this theory is denoted by $\mathrm{PA}(\mathbb{G})$, then we can state

$$\mathrm{PA} \leqq_{\mathrm{m}}^{\text{log-lin}} \mathrm{PA}(\mathbb{G}).$$

This can be done simply by adding to every variable x the restriction $x \geqq 0$ that forces every interpretation over $\mathbb{G}$ to work actually over $\mathbb{N}$. We state the lower bound for PA and the upper bound for PA($\mathbb{G}$).

8.47. Theorem.

A	lower bound		upper bound	
	$A \notin$	authors	$A \in$	authors
RA	**NTIME**(2^{cn})	[FeRa 75]	**EXPSPACE**	[FeRa 75]
PA	**NTIME**$(2^{2^{cn}})$	[FimRa 74]	**DSPACE**$(2^{2^{dn}})$	[FeRa 75]
PM	**NTIME**$\left(2^{2^{2^{cn}}}\right)$	[FimRa 74]	**DSPACE**$\left(2^{2^{2^{dn}}}\right)$	[Rac 76]

($c, d > 0$ are suitable constants.) ∎

Remark 1. The gap between lower and upper bounds for RA and PA can be explained by the fact that these theories are complete in complexity classes which are between the time and space complexity classes describing the lower and upper bounds, resp. Thus NTIME and DSPACE do not seem to be adequate measures for describing the complexity of RA and PA. More precisely, let **ATIME-ALT**(Pol t, s) be the class of all languages which can be accepted by alternating TM's working both within time Pol t and with at most s alternations. By 20.38.2 we have **NTIME**(Pol t) $\subseteqq$ **ATIME-ALT**(Pol t, 0) $\subseteqq$ **ATIME-ALT**(Pol t, Pol t) $=$ **DSPACE**(Pol t) for "well-behaved" bounding functions t.

In [Berl 78] it is shown that RA is $\leqq_{\mathrm{m}}^{\text{P-lin}}$-complete in **ATIME-ALT**$(2^{\mathrm{Lin}}, n)$ and PA is $\leqq_{\mathrm{m}}^{\text{P-lin}}$-complete in **ATIME-ALT**$(2^{2^{\mathrm{Lin}}}, n)$.

Remark 2. In order to accept RA alternating TM's need either $2^{c_1 \cdot n^2}$ time or $2^{c_2 \cdot n}$ space or $c_3 \cdot n$ alternations for some $c_1, c_2, c_3 > 0$ (see [Berl 78]). This improves a result of [BrMe 78].

Remark 3. In [ScKl 77] it has been proved that there exists a subset S of RA and a $c > 0$ such that, for every nondeterministic TM deciding RA, almost all sentences of S require $2^{c \cdot n}$ steps on that machine.

Remark 4. In [Schö 80] the lower bound for PA is extended to nondeterministic RAM's with multiplication (but without boolean operations).

Remark 5. If only one single monadic predicate is added to PA, the theory becomes undecidable (see [Dow 72]).

Remark 6. In [BeKoRe 84] it is shown that the first-order theory of real numbers with addition, multiplication and equality is complete in **EXPSPACE**.

We now deal with the following theories.

MON denotes the set of all satisfiable sentences of monadic predicate calculus without function symbols.

MON(=) denotes the set of all satisfiable sentences of monadic predicate calculus (without function symbols) with equality.

FUN denotes the set of all satisfiable sentences in a (first order) language having one function symbol and the equality predicate.

INJ denotes the theory of one injection, i.e. the theory of the model class $\{(A, f)$: A is set and f is an injection from A to $A\}$.

INJM denotes the theory of one injection with one monadic predicate, i.e. the theory of the model class $\{(A, f, P): A$ is a set $\wedge f$ is an injection from A to $A \wedge P \subseteq A\}$.

EQUALITY denotes the set of all formulas in prefix form which are true in all models $(A, \{(x, x): x \in A\})$.

Our definition of INJ is slightly different from that given in [Fer 74]. One can, however, easily verify that the results of this paper are also valid for our version.

8.48. Theorem.

A	lower bounds		upper bounds	
	$A \notin$	authors	$A \in$	authors
MON	$\mathbf{NTIME}\left(2^{c\frac{n}{\log n}}\right)$	[Mey 74] [Rac 75]	$\mathbf{NTIME}\left(2^{d\frac{n}{\log n}}\right)$	[Lewh 78a]
EQUALITY	$\mathbf{NSPACE}(s)$ for every $s < \sqrt{n}$	[Sto 77]	$\mathbf{DSPACE}(n)$	[Sto 77]
INJ	$\mathbf{NTIME}(2^{cn})$	[Fer 74]	$\mathbf{NSPACE\text{-}TIME}$ $(2^n, 2^{dn^2})$	[Fer 74]
INJM	$\mathbf{NTIME}(2^{2^{cn}})$	[Fer 74]	$\mathbf{NSPACE\text{-}TIME}$ $(2^{2^n}, 2^{2^{dn}})$	[Fer 74]
MON(=)	$\mathbf{NTIME}\left(2^{2^{c\frac{n}{\log n}}}\right)$	[Rac 75]		
FUN	$\mathbf{DSPACE}$ $\left(2^{2^{\cdot^{\cdot^{2}}}}\Big\}\, c\lfloor\log n\rfloor\right)$	[Mey 74]	$\mathbf{DSPACE}\left(2^{2^{\cdot^{\cdot^{2}}}}\Big\}\, dn\right)$	M. J. Fischer/ A. R. Meyer

($c, d > 0$ are suitable constants.) ∎

The next theorem deals with purely logical theories. Our first example is B_ω, which could be viewed as a first order propositional calculus, and it is defined by $\mathrm{B}_\omega = \bigcup_{k=1}^{\infty} \mathrm{B}_k$ and

$$\mathrm{B}_k = \{F(X_1, \dots, X_k): F(X_1, \dots, X_k) \text{ is a Boolean formula and}$$

$$\bigvee_{X_1} \bigwedge_{X_2} \bigvee_{X_3} \dots \mathsf{Q}_{X_k} F(X_1, X_2, \dots, X_k) = 1\},$$

where $X_1, \ldots, X_k$ are pairwise disjoint sets of propositional variables and a quantification $\underset{X_i}{\mathsf{Q}}$ means

$$\underset{x_{i1}}{\mathsf{Q}}\ \underset{x_{i2}}{\mathsf{Q}} \ldots \underset{x_{in_i}}{\mathsf{Q}} \text{ where } X_i = \{x_{i1}, \ldots, x_{in_i}\}.$$

IIC $=_{\mathrm{df}}$ set of all satisfiable formulas of the intuitionistic implicational calculus.

S4 $=_{\mathrm{df}}$ set of all satisfiable formulas of the modal propositional calculus S4.

PMLP $=_{\mathrm{df}}$ set of all satisfiable formulas of the following *p*ropositional *m*odal *l*ogic of *p*rograms, which is a logical framework for reasoning about correctness, termination and equivalence of programs: The essential difference from the propositional calculus is that there exists a set of well-formed programs, and every program A defines an operator $\langle A \rangle$ that can be applied to assertions p. The interpretation of $\langle A \rangle\, p$ is "A can terminate with p holding on termination".

8.49. Theorem.

A	lower bounds		upper bounds	
	$A \notin$	authors	$A \in$	authors
B_ω	**DSPACE**(n^c)	[StMe 73]	**PSPACE**	[StMe 73]
IIC	**DSPACE**(n^c)	[Sta 79]	**PSPACE**	[Sta 79]
S4	**DSPACE**(n^c)	[Lad 77]	**PSPACE**	[Lad 77]
PMLP	**DTIME**$\left(2^{c\frac{n}{\log n}}\right)$	[FimLa 77]	**DTIME**(2^{dn})	V. R. Pratt (see [Val 80])

($c, d > 0$ are suitable constants.) ∎

Remark. Further systems of modal logic are dealt with in [Lad 77]. In [Vav 80] a result similar to Statement 4 has been proved. In [Chl 81] it is shown that the satisfiability problem of the propositional algorithmic (dynamic) logic is complete in **PSPACE**(**DTIME**(2^{Pol})). Satifiability problems for propositional calculi have also been discussed in [Lewh 78b]. In [Lewh 78a], [Für 83] and [Pla 84] lower and upper bounds for the decidable fragments of the first order predicate calculus have been proved.

We now consider the set theoretical calculus of pure finite types and the set PFT of all true sentences of this calculus. Its language $\mathscr{L}$ contains two constants **0**, **1** of type 0. Every variable of $\mathscr{L}$ has a natural number type indicated by an upper index.

Prime formulas have the form $\mathbf{0} \in x^1$, $\mathbf{1} \in x^1$ or $x^n \in y^{n+1}$, and formulas are built up from them in the usual way. The interpretation is as follows: Variables of type n range over D_n, where $D_0 =_{\mathrm{df}} \{0, 1\}$ and $D_{n+1} =_{\mathrm{df}} \mathfrak{P}(D_n)$, the interpretation of **0** (**1**) is 0 (1).

8.50. Theorem. [Mey 74] PFT $\notin$ **DSPACE**$\left(2^{2^{\cdot^{\cdot^{2}}}}\Big\}\, c\lfloor \log n \rfloor\right)$. ∎

The following restricted versions of the concatenation theory are of interest for two different reasons. First, they provide examples of theories with a desired built-in complexity. Second, as concatenation is very elementary and easy to handle one can use these theories as standard sets to be reduced to the ories of unknown complexity. Many of our lower bounds stated so far can actually be proved using these theories.

The concatenation theory $\mathrm{CONC}(\Sigma)$ for a finite alphabet Σ is defined by $\mathrm{CONC}(\Sigma) = \mathscr{T}(\Sigma^*, \mathrm{CON}_\Sigma)$, where CON_Σ means the ternary predicate "$xy = z$". As all Turing machine computations can be encoded in $\mathrm{CONC} =_{\mathrm{df}} \mathrm{CONC}\,\{0, 1\}$, CONC is undecidable. Define $t\text{-BCT}(\Sigma) = \mathscr{T}(\Sigma^*, t\mathrm{CON}_\Sigma)$, where $t\mathrm{CON}_\Sigma$ denotes the 4-place predicate "$xy = z \wedge |z| < t(|u|)$". Let $t\text{-BCT} = t\text{-BCT}(\{0, 1\})$.

8.51. Theorem. [FlMaSi 76] If t is an increasing T-DTIME-computable function with $t(n) \succ n^k$ for all $k \in \mathbb{N}$, then there exists a $c > 0$ such that $t\text{-BCT} \notin \mathbf{NTIME}\big(t(cn)\big)$. ∎

8.2.3.3. *Games*

The 2-person perfect-information games like chess, checkers, go, hex etc. (extended to larger boards) have been studied intensively. They are of particular interest since they are closely related to the working mode of alternating TM's. A survey of complexity results for games is given in [Joh 81–], 4 (1983), 397—411. See also 14.14 and 14.24.

Informally, such a game consists of a set of positions (for instance positions of stones on a board or of pebbles on a graph or of assignments of variables). The two players move alternately, a move consisting of changing one position into another one (for instance by adding, moving or removing a stone on a board or by changing the position of a pebble on a graph or by setting further variables to constants). The loser is that player who is no longer able to move. The main problem for a given game consists of deciding whether the player moving first has a forced win. Our main concern is the complexity of such decision problems.

We make precise our notions.

Definition. $G = (P_1, P_2, R)$ is a *game* iff P_1 and P_2 are disjoint finite sets (of positions) and $R \subseteq (P_1 \times P_2) \cup (P_2 \times P_1)$.

The intended interpretation is: P_i is the set of positions in which player i has to move ($i = 1, 2$). Player 1 may move from $p \in P_1$ to $p' \in P_2$ if and only if $(p, p') \in R$. Player 2 may move from $q' \in P_2$ to $q \in P_1$ if and only if $(q', q) \in R$.

We define the set $W(G)$ of positions ($\in P_1 \cup P_2$) for which player 1 has a forced win:

$$W_0(G) =_{\mathrm{df}} \Big\{p_2 : p_2 \in P_2 \wedge \sim \bigvee_{p_1 \in P_1} (p_2, p_1) \in R\Big\},$$

$$W_{i+1}(G) =_{\mathrm{df}} W_i(G) \cup \Big\{p_1 : \bigvee_{p_2 \in P_2} \big(p_2 \in W_i(G) \wedge (p_1, p_2) \in R\big)\Big\}$$

$$\cup \Big\{p_2 : \bigwedge_{p_1 \in P_1} \big((p_2, p_1) \in R \to p_1 \in W_i(G)\big)\Big\}.$$

Evidently, $W_i(G)$ is the set of positions for which player 1, if playing correctly, can always be sure to win within no more than i steps and hence, $W(G) =_{df} \bigcup_{i=0}^{\infty} W_i(G)$ is the desired definition for $W(G)$.

What are the connections between alternating TM's and games? One can interpret the work of an alternating TM M as a game G by viewing the situations as positions and the next-move function of M as the rule of the game. The existential situations are understood to be the positions in which player 1 has to move and the universal ones are those in which player 2 has to move. (To be in accordance with our game definition we assume without loss of generality that every next situation of an existential situation is universal or accepting and that every universal situation does have next situations, which are all existential.) Accepting situations are losing positions for player 2. Existential situations without next situation are losing positions for player 1. It is quite obvious (cf. the definition of acceptance by alternating TM's) that $w \in \mathrm{L}(M) \leftrightarrow K_M(w) \in W(G)$, where $K_M(w)$ is the initial situation with input w.

From this point of view the relationships 20.36 and 20.38

$$\mathbf{ASPACE}(s) = \mathbf{DTIME}(2^{\mathrm{Lin}\,s}),$$

$$\mathbf{ATIME}(\mathrm{Pol}\,t) = \mathbf{DSPACE}(\mathrm{Pol}\,t)$$

suggest the construction of **DSPACE**(Pol t)-complete [**DTIME**($2^{\mathrm{Lin}\,s}$)-complete] games from **DTIME**(Pol t)-complete [**DSPACE**(s)-complete] problems by alternating quantification of variables (which have to be found out in a suitable way). The construction of **PSPACE**-complete games from **NP**-complete problems has been investigated in [Sch 76]. The simplest example to illustrate this construction is B_ω (cf. the definition before 8.49) which can be viewed as an alternating quantified version of the **NP**-complete problem SAT. The interpretation as a game is obvious: The variables of a formula $F(X_1, \ldots, X_k)$ (which plays the role of the "board" of the game) are divided into two disjoint sets $X_1 \cup X_3 \cup \ldots$ and $X_2 \cup X_4 \cup \ldots$

The ith move of player 1 (2) consists of choosing assignments to the variables from $X_{2i-1}(X_{2i})$. Player 1 (2) wins if F has the value 1 (0) as soon as assignments to all variables have been chosen. Player 1 has a forced win if and only if $F \in \mathrm{B}^\omega$.

From the large number of games known we choose only a few typical ones.

$$\mathrm{DEC} = \{X, X \times X, R\}, \text{ where } (X, \cdot) \text{ is a gruppoid, } T \subseteq X \text{ and}$$

$$R = \{(p, (q, r)) : p = qr\} \cup \{((p, q), p) : p \notin T\} \cup \{((p, q), q) : q \notin T\}.$$

Player 1 offers decompositions of elements of X and player 2 picks an element of the decompositions that does not belong to T. Player 2 loses if player 1 offers him a decomposition $t_1 \cdot t_2$ with $t_1, t_2 \in T$.

GHEX is played on graphs with two distinguished nodes a and b. The players alternately choose nodes which have not been chosen so far. Player 1 wins as soon as some of his nodes form a path from a to b.

NODE KAYLES: The players alternately put a marker on a node that is unoccupied and not adjacent to an occupied one. The loser is the first player unable to move.

GEOGRAPHY is played on directed graphs with a starting node s. The players select in every move one node x such that there exists a directed arc from the most recently selected node (from s in the first move) to x. The loser is the first player unable to move.

SC is played on Boolean formulas $F(X, Y)$ in conjunctive normal form, where X and Y are disjoint sets of Boolean variables. Positions are pairs (i, α), where $i \in \{1, 2\}$ is only an index indicating which player has to move and α is a complete assignment of the variables from $X \cup Y$, i.e. $\alpha: X \cup Y \mapsto \{0, 1\}$. A move of player 1 (2) consists of changing the value of at most one variable of X (Y). Player 1 wins if and only if F becomes true.

8.52. Theorem.

1. [JoLa 77] DEC is $\leq_m^{\log}$-complete in **P**.
2. [EvTa 76], [Sch 76] GHEX, NODE KAYLES, GEOGRAPHY are ${}_m^{\log}\leq$-complete in **PSPACE**. Consequently, GHEX, NODE KAYLES, GEOGRAPHY are in **DSPACE**(n^d) but not in **DSPACE**(n^c) for some $c, d > 0$.
3. [StCh 79] SC $\notin$ **DTIME**$\left(2^{c\sqrt[3]{\frac{n}{\log n}}}\right)$ and SC $\in$ **DTIME**$\left(2^{d\frac{n}{\log n}}\right)$ for some $c, d > 0$. ∎

Finally we note that in [AdIwKa 81] lower time bounds of the form n^k for certain games G_k, $k \geqq 1$, have been proved.

8.2.3.4. *Some Hints to Literature*

Further lower bound results concerning automata and grammars can be found in [Gal 76a], [Hun 76], [HuRo 77], [HuRoSz 76], [HuSz 76], [HuSzUl 75] and [HuRo 80].

Results on algebraic problems can be found in [Koz 77a], [Koz 77b], [CaLiMe 76], [AdMa 75] and [MaMe 82].

Petri net problems are dealt with in [LaRo 78], [JoLaLi 77], [Rac 78a], [BöKl 80] and [MaMe 81]. In the latter paper an uncontrived decidable problem is exhibited which is not primitive recursive.

§ 9. Overcoming Lower Bounds

There are cases where, from a practical point of view, the lower bounds need not be taken too seriously. Many attitudes are possible which help to solve hard problems in a way which is satisfiable in practice. So far the following have been considered in the literature:

1. Not to insist on getting exact solutions. For instance, in solving optimization problems we are content with near optimal solutions (§ 9.1).

2. Not to insist on getting always correct solutions, but only almost everywhere with respect to a randomization of the inputs (§ 9.2.1), or with a certain probability using probabilistic algorithms (§ 9.2.2), or with a certain frequency (§ 9.2.3).

It is of theoretical importance to learn that equally hard problems may have quite different behaviour with respect to approximations. This is discussed in particular in § 9.1.3.

9.1. Approximate Solutions

In this subsection we can neither develop our topic in full detail nor give an exhaustive collection of the relevant results. Instead we explain the basic problems and illustrate them by sample results. Further material and bibliographic references the reader will find in Chapter 6 of [GaJo 79] and Chapter 12 of [HoSa 78].

9.1.1. Preliminaries

It is possible to extend the Turing reducibility, originally defined only for decision problems, to search problems too.

Definition. Let A and B be decision or search problems. $A \leq_T^P B$ iff there exists an oracle machine M with function oracle working in polynomial time such that

Case 1. A, B decision problems: $c_A = \varphi_M^{(c_B)}$.

Case 2. A decision problem, B search problem: if g search function for B, then $c_A = \varphi_M^{(g)}$.

Case 3. A search problem, B decision problem: $\varphi_M^{(c_B)}$ is search function for A.

Case 4. A, B search problems: if g is search function for B, then $\varphi_M^{(g)}$ is search function for A.

Note that for decision problems A and B this definition of $\leq_T^P$ is equivalent to that given in § 6. If an NP-complete problem A is $\leq_T^P$-reducible to a search problem B, we employ the word *NP-hard* for B. Thus an NP-hard search problem cannot be solved within polynomial time unless $\mathbf{P} = \mathbf{NP}$.

Examples. 1. CLIQUE $\leq_T^P$ Clique. This is true because the ability to produce a k-clique for the graph G implies the ability to decide the question of whether G has a k-clique.

2. Clique $\equiv_T^P$ Max clique. If a maximum clique for G can be found, then a k-clique for G can be constructed if k is not greater than the size of the maximum clique. Conversely, if Clique is solvable, then a maximum clique of G is the first clique which is found if Clique is applied to $(G, n), (G, n-1), \ldots$ in this order, where $n = \text{card } V(G)$.

Similar statements are true for all NP-complete problems and their corresponding search and optimization problems.

In the following Opt (opt) stands for Min (min) or Max (max). Let $h: \mathbb{N} \mapsto \mathbb{N}$. As defined in § 7.1 the optimization problem Opt prob is the search problem of the decision problem opt prob $=_{df} \{(x, y): (x, y) \in \text{prob} \wedge b_{\text{prob}}(x, y) = b^*_{\text{prob}}(x)\}$ with respect to the valuation function b_{prob}. We define two relaxations of Opt prob, namely the search problem Opt prob$^{h,\mathrm{abs}}$ of the decision problem

$$\text{opt prob}^{h,\mathrm{abs}} =_{\mathrm{df}} \{(x, y): (x, y) \in \text{prob} \wedge |b_{\mathrm{prob}}(x, y) - b^*_{\mathrm{prob}}(x)| \leqq h(|x|)\},$$

and the search problem Opt prob$^{h,\mathrm{rel}}$ of the decision problem

$$\text{opt prob}^{h,\mathrm{rel}} =_{\mathrm{df}} \{(x, y): (x, y) \in \text{prob} \wedge |b_{\mathrm{prob}}(x, y) - b^*_{\mathrm{prob}}(x)| \leqq h(|x|) \cdot b^*_{\mathrm{prob}}(x)\}.$$

Opt prob$^{h,\mathrm{abs}}$(Opt prob$^{h,\mathrm{rel}}$) is called the *absolute* (*relative*) *h-approximation* of Opt prob.

Note that the difference between these two relaxations lies only in the way of measuring the deviations of the approximate solutions from the optimal ones.

9.1.2. Absolute Approximations

As an illustrative example we consider the following results which show that admitting a deviation by 1 from the minimum makes certain NP-hard problems polynomially solvable. The problems in question are the optimization problems corresponding to

$$\text{PROGRAM STORE} =_{\mathrm{df}} \{(a_1, \ldots, a_n, l, k): n, l, k, a_1, \ldots, a_n \in \mathbb{N} \text{ and there exist two subsets } I_1, I_2 \subseteq \{1, \ldots, n\} \text{ with the properties } I_1 \cap I_2 = \emptyset,\ k = \mathrm{card}(I_1 \cup I_2),\ \sum_{i \in I_1} a_i \leqq l,\ \sum_{i \in I_2} a_i \leqq l\}$$

(Is it possible to store k of n programs having lengths $a_1, \ldots, a_n$ on two storage devices of capacity l?) and, for $m \geqq 1$,

$$\text{SAFEOPEN}_m =_{\mathrm{df}} \{(A, b, k): A = (a_{ij}) \text{ is an } (m, n)\text{-matrix of natural numbers, and } b \text{ is an } m\text{-vector of natural numbers, and } n \in \mathbb{N}, \text{ and there exists an } I \subseteq \{1, \ldots, n\} \text{ with } k = \mathrm{card}\, I \text{ and } \bigwedge_{1 \leqq i \leqq m} \sum_{j \in I} a_{ij} \geqq b_i\}$$

(Do there exist k of n safes each containing known amounts of m currencies so that opening them guarantees desired amounts of the currencies in question?) and to the valuation functions $b(I_1, I_2) = \mathrm{card}(I_1 \cup I_2)$ and $b(I) = \mathrm{card}\, I$, resp.

PROGRAM STORE is NP-complete because its special case PARTITION $\left(k = n \wedge l = \frac{1}{2} \sum_{i=1}^{n} a_i\right)$ is NP-complete (14.21). Hence Max program store is also NP-hard. Min safeopen$_m$ is NP-hard for $m \geqq 2$ (see [DiKa 78]).

9.1. Theorem.

1. [HoSa 78] Max program store$^{1,\mathrm{abs}} \in$ P.
2. [DiKa 78] Min safeopen$_m^{1,\mathrm{abs}} \in$ P.

Proof. We prove only Statement 1. First order $a_1, \ldots, a_n$ according to increasing length: $b_1, \ldots, b_n$. Next determine the largest k' such that $b_1 + \ldots + b_{k'} \leqq 2l$. Evidently $k' \geqq b^*(a_1, \ldots, a_n, l, k)$. Then find the maximum k'' such that $b_1 + \ldots + b_{k''} \leqq l$. Storing $b_1, \ldots, b_{k''}$ on the first device and $b_{k''+2}, \ldots, b_{k'}$ on the second is a solution where the number of stored programs differs from the optimum value by at most 1.

The proof of Statement 2 is by far not so easy. ∎

9.1.2.1. *Negative Results*

Unfortunately, most of the NP-complete problems do not allow absolute h-approximations with constant h. To explain this in more detail we need as a technical tool the following m-reducibility variant of the Turing reducibility for search problems Prob_1, Prob_2 belonging to the decision problems prob_1, prob_2, resp.

$$\mathrm{Prob}_1 \leqq_m^P \mathrm{Prob}_2 \leftrightarrow_{\mathrm{df}} \bigvee_{f,g \in P} \Big(\mathrm{PROB}_1 \leqq_m^P \mathrm{PROB}_2 \text{via } f \wedge \bigwedge_{x,y} \big((f(x), y) \in \mathrm{prob}_2 \rightarrow (x, g(y)) \in \mathrm{prob}_1\big)\Big).$$

We write $\mathrm{Prob}_1 \leqq_m^P \mathrm{Prob}_2$ via (f, g) for short.

If $\mathrm{Prob}_1 \leqq_m^P \mathrm{Prob}_2$ via (f, g) and s is a search function for Prob_2, then obviously $s' =_{\mathrm{df}} g \circ s \circ f$ is a search function for Prob_1 which, moreover, belongs to P if $s \in$ P.

As a first example we consider

$$\mathrm{KNAPSACK} = \{(a_1, \ldots, a_n, a, b_1, \ldots, b_n, b): n, a, b, a_i, b_j \in \mathbb{N} \text{ and there exist } \delta_1, \ldots, \delta_n \in \mathbb{N} \text{ such that } \sum_{i=1}^n \delta_i a_i \leqq a \wedge \sum_{i=1}^n \delta_i b_i \geqq b\},$$

the evaluation function $b((a_1, \ldots, a_n, a, b_1, \ldots, b_n, b), (\delta_1, \ldots, \delta_n)) = \sum_{i=1}^n \delta_i b_i$ and the corresponding optimization problem Max knapsack.

9.2. Theorem. Max knapsack $\leqq_m^P$ Max knapsack$^{h,\mathrm{abs}}$ for every $h \in \mathbb{N}$.

Proof. We define $f(a_1, \ldots, a_m, a, b_1, \ldots, b_m, b) = (a_1, \ldots, a_m, a, (h+1) b_1, \ldots, (h+1) b_m, (h+1) b)$ and $g = \mathrm{id}$. Assume $(f(a_1, \ldots, a_m, a, b_1, \ldots, b_m, b), (\delta_1, \ldots, \delta_m)) \in$ max knapsack$^{h,\mathrm{abs}}$. This means

$$b^*(a_1, \ldots, a_m, a, (h+1) b_1, \ldots, (h+1) b_m, (h+1) \cdot b) - \sum_{i=1}^m \delta_i (h+1) b_i \leqq h$$

or

$$(h+1)\left(b^*(a_1, \ldots, a_m, a, b_1, \ldots, b_m, b) - \sum_{i=1}^m \delta_i b_i\right) \leqq h.$$

Since all expressions are integer valued we conclude

$$\sum_{i=1}^{m} \delta_i b_i = b^*(a_1, \ldots, a_m, a, b_1, \ldots, b_m, b),$$

i.e. $((a_1, \ldots, a_m, a, b_1, \ldots, b_m, b), (\delta_1, \ldots, \delta_m)) \in$ max knapsack. ∎

Theorem 9.2 shows Max knapsack$^{h,\text{abs}}$ to be NP-hard.

Many results of this type can be proved using the following technical lemma which generalizes the idea of the preceding proof. We formulate it for the Min case, the Max case being analogous.

9.3. Lemma. [Nig 75] Let Prob$_1$, Prob$_2$ be search problems with the valuation functions b_1, b_2, resp., and let $b_1, b_2 \in$ P. If, for some $h \in \mathbb{N}$, there exist $f, g_1, \ldots, g_{h+1} \in$ P such that Prob$_1 \leqq_m^P$ Prob$_2$ via (f, g_i) for $i = 1, \ldots, h+1$, $b_2^*(f(x)) \leqq (h+1)\, b_1^*(x)$ and $\sum_{i=1}^{h+1} b_1(x, g_i(y)) \leqq b_2(f(x), y)$, then

$$\text{Min prob}_1 \leqq_m^P \text{Min prob}_2^{h,\text{abs}}.$$

Proof. Since always PROB = MIN PROB = MIN PROB$^{h,\text{abs}}$ for search problems Prob, f actually reduces MIN PROB$_1$ to MIN PROB$_2^{h,\text{abs}}$. It remains to give a $g \in$ P such that

$$(f(x), y) \in \text{min prob}_2^{h,\text{abs}} \to (x, g(y)) \in \text{min prob}_1$$

or, equivalently,

$$(f(x), y) \in \text{prob}_2 \wedge b_2(f(x), y) - b_2^*(f(x)) \leqq h \to (x, g(y)) \in \text{prob}_1 \wedge b_1(x, g(y)) = b_1^*(x).$$

Let $x \in$ PROB$_1$, and consider a y such that $(f(x), y) \in$ prob$_2$ and $b_2(f(x), y) - b_2^*(f(x)) \leqq h$. By hypothesis,

$$\sum_{i=1}^{h+1} b_1(x, g_i(y)) - (h+1)\, b_1^*(x) \leqq b_2(f(x), y) - b_2^*(f(x)) \leqq h,$$

and hence

$$\sum_{i=1}^{h+1} \left(b_1(x, g_i(y)) - b_1^*(x)\right) \leqq h.$$

As the differences $b_1(x, g_i(y)) - b_1^*(x)$ are natural numbers, at least one of them must vanish. We define $g(y) = g_{i_0}(y)$, where i_0 is the least number $j \leqq h+1$ such that $b_1(x, g_j(y)) = b_1^*(x)$. This latter is equivalent to $b_1(x, g_j(y)) = \min_{1 \leqq i \leqq h+1} b_1(x, g_i(y))$. This shows that g is computable in polynomial time. ∎

To give applications of this lemma we introduce the following problems, which are all known to be NP-complete (14.21). Together with each of these problems we define a valuation function, which allows us to consider associated optimization problems.

$$\text{SET COVER} = \{(S, \mathfrak{S}, k): S \text{ is a finite set} \wedge k \in \mathbb{N} \wedge \mathfrak{S} \subseteq \mathfrak{P}(S) \wedge \bigvee_{\mathfrak{S}' \subseteq \mathfrak{S}} (\text{card } \mathfrak{S}' \leq k \wedge \bigcup \mathfrak{S}' = \mathfrak{S})\},$$

$b_{sc}(\mathfrak{S}') = \text{card } \mathfrak{S}'$.

$$\text{EXACT COVER} = \{\mathfrak{S}: \mathfrak{S} \text{ is a system of finite sets} \wedge \bigvee_{\mathfrak{S}'} (\mathfrak{S}' \subseteq \mathfrak{S} \wedge \mathfrak{S}' \text{ disjoint} \wedge \bigcup \mathfrak{S}' = \bigcup \mathfrak{S})\}.$$

$b_{ec}(\mathfrak{S}') = \text{card } \mathfrak{S}'$.

$$\text{NODE COVER} = \Big\{(G, k): G \text{ is a graph} \wedge k \in \mathbb{N} \wedge \bigvee_{V' \subseteq V(G)} \Big(\text{card } V' \leq k \wedge \bigwedge_{x,y} \big((x, y) \in E(G) \to \{x, y\} \cap V' \neq \emptyset\big)\Big)\Big\}.$$

$b_{n\varrho}(V') = \text{card } V'$.

$$\text{COLOUR} = \{(G, k): G \text{ is a graph which is colourable with } k \text{ colours, i.e. there is a mapping } c: V(G) \mapsto \{1, \ldots, k\} \text{ such that } \bigwedge_{(u,v) \in E(G)} c(u) \neq c(v)\}.$$

$b_c(c) = \text{card } R_c$ (R_c is the range of c).

9.4. Theorem. [Nig 75], [Nig 78] For every $h \in \mathbb{N}$ the following problems are NP-hard:

Max clique$^{h,\text{abs}}$, Min exact cover$^{h,\text{abs}}$, Min set cover$^{h,\text{abs}}$,

Min colour$^{h,\text{abs}}$, Min node cover$^{h,\text{abs}}$.

Proof. We consider only one sample problem, namely Max clique. Let $G_i = (V_i, E_i)$ be an ith copy of the graph $G = (V, E)$ and define

$$G^m = (V_1 \cup \ldots \cup V_m, E_1 \cup \ldots \cup E_m \cup \bigcup_{i \neq j} V_i \times V_j.$$

We define $f(G, k) = \big(G^{h+1}, (h + 1) k\big)$ and for cliques C of G^{h+1}

$$g_i(C) = \text{that clique which is induced by } C \text{ in the subgraph } G_i \text{ of } G^{h+1}.$$

It can easily be seen that

$$b^*(G^{h+1}) \geq (h + 1)\, b^*(G) \quad \text{and} \quad \sum_{i=1}^{h+1} b\big(g_i(C)\big) = b(C)$$

for cliques C of G^{h+1}. This shows that the conditions of 9.3 (applied to the Max case) are satisfied, and we can conclude Max clique $\leq_m^P$ Max clique$^{h,\text{abs}}$. As Max clique is NP-hard (it is $\leq_T^P$-equivalent to the NP-complete problem CLIQUE) we conclude that Max clique$^{h,\text{abs}}$ is also NP-hard. ∎

Remark 1. Further problems could be included in Theorem 9.4 (see [Nig 78]). We have chosen only those problems which will occur again in this section in other connections.

Remark 2. Theorem 9.4 holds true in the stronger form where h is replaced by $|x|^{\frac{1}{2}-\varepsilon}$, $\varepsilon > 0$.

The fact that for many problems there do not exist feasible absolute approximations does not render us hopeless, because it is possible that at least good relative approximations exist. This question is studied in the next subsection.

9.1.3. Relative Approximations

9.1.3.1. Polynomial and Fully Polynomial Approximation

In this subsection we deal with relative ε-approximability where $\varepsilon > 0$ is a rational number. Instead of constructing for every $\varepsilon > 0$ a separate algorithm solving Opt prob$^{\varepsilon,\text{rel}}$ we are interested in *one* algorithm solving Opt prob$^{\varepsilon,\text{rel}}$ for every rational $\varepsilon > 0$.

Definition. An optimization problem Opt prob is called *p-approximable* iff there is an algorithm with two inputs (problem instance and rational $\varepsilon > 0$) solving Opt prob$^{\varepsilon,\text{rel}}$ within time polynomial in n (length of problem instance).

The disadvantage of this notion is that the running time can grow in a non-polynomial way in $\frac{1}{\varepsilon}$. For instance,

$$\text{Max } (0, 1)\text{-knapsack}^{\varepsilon,\text{rel}} \in \text{RAM-DTIME}(n^{1/\varepsilon}),$$

as shown in [Sah 75].

A stronger notion of approximability is given by the following

Definition. An optimization problem Opt prob is called *fully p-approximable* iff there is an algorithm with two inputs (problem instance and rational $\varepsilon > 0$) solving Opt prob$^{\varepsilon,\text{rel}}$ within time polynomial in n (length of problem instance) and $\frac{1}{\varepsilon}$. We can also say that Opt prob is fully p-approximable iff the search problem Opt prob^{rel} of the decision problem opt prob$^{\text{rel}} = \{((x, \varepsilon), y): (x, y) \in \text{prob} \wedge |b(x, y) - b^*(x)| \leq \varepsilon \cdot b^*(x) \wedge \varepsilon > 0\}$ is solvable within time polynomial in $|x|$ and $\frac{1}{\varepsilon}$.

9.1.3.1.1. Examples of Fully p-Approximable Problems

The first examples of fully p-approximable problems have been discovered in [IbKi 75a]. Their results have been improved as stated in the next theorem. [GeLe 79] contains many further results concerning several versions of KNAPSACK.

Modifying the decision problem for SOS we define

$$\mathrm{sos}' = \left\{\big((a_1, \ldots, a_n, b), (\delta_1, \ldots, \delta_n)\big) : a_1, \ldots, a_n, b, n \in \mathbb{N} \right.$$
$$\left. \wedge\ \delta_1, \ldots, \delta_n \in \{0, 1\} \wedge \sum_{i=1}^{n} a_i\delta_i \leqq b\right\}.$$

A maximization problem Max sos′ can be derived using the following valuation function

$$\mathrm{b}_{\mathrm{sos}'}\big((a_1, \ldots, a_n, b), (\delta_1, \ldots, \delta_n)\big) = \sum_{i=1}^{n} \delta_i a_i .$$

9.5. Theorem.

1. [Law 77] Max knapsack^{rel} $\in$ RAM-DSPACE-TIME $\left(\mathrm{Lin}\left(n + \frac{1}{\varepsilon^3}\right), \mathrm{Lin}\left(n + \frac{1}{\varepsilon^3}\right)\right)$.
2. [Law 77] Max (0, 1)-knapsack^{rel} $\in$ RAM-DSPACE-TIME $\left(\mathrm{Lin}\left(n + \frac{1}{\varepsilon^3}\right), \mathrm{Lin}\left(n \log \frac{1}{\varepsilon} + \frac{1}{\varepsilon^4}\right)\right)$.
3. [GeLe 79] Max sos′^{rel} $\in$ RAM-DSPACE-TIME $\left(\mathrm{Lin}\left(n + \frac{1}{\varepsilon^2}\right), \mathrm{Lin}\left(n + \frac{1}{\varepsilon^2}\right)\right)$ and Max sos′^{rel} $\in$ RAM-DSPACE-TIME $\left(\mathrm{Lin}\left(n + \frac{1}{\varepsilon}\right), \mathrm{Lin}\left(\frac{n}{\varepsilon} + \frac{1}{\varepsilon^3}\right)\right)$. ∎

The proof is too lengthy to be given here, but basically similar methods are used as in the proof of the next result, which is about scheduling problems, these problems being defined below:

JOB SEQUENCING WITH DEADLINES

$$= \{(p_1, \ldots, p_n, t_1, \ldots, t_n, d_1, \ldots, d_n, p) : n \in \mathbb{N} \wedge p_1, \ldots, p \in \mathbb{N}$$
$$\wedge \bigvee_{\pi} (\pi \text{ permutation of } \{1, \ldots, n\} \wedge \sum_{i=1}^{n} p_{\pi(i)}\delta_i \geqq p \text{ where } \delta_i = 1$$
$$\text{if } t_{\pi(1)} + \cdots + t_{\pi(i)} \leqq d_{\pi(i)} \text{ and } = 0 \text{ otherwise}\}$$

(n independent jobs with processing time t_i and deadline d_i have to be scheduled on one processor in such a way that the profit p is guaranteed. Profit p_i is earned if job i is completed by its deadline),

m-FINISH TIME

$$= \{(t_1, \ldots, t_n, t) : n, t_1, \ldots, t_n, t \in \mathbb{N} \wedge \text{there exists a partition}$$
$$I_1 \cup \ldots \cup I_m = \{1, \ldots, n\} \text{ such that } \sum_{j \in I_i} t_j \leqq t \text{ for all } i = 1, \ldots, m\}$$

(n independent jobs of processing time t_i have to be scheduled on m processors in such a way that all processors finish their work by time t).

It should be obvious how to define the corresponding valuation functions. Note that Max job sequencing with deadlines and Min m-finish time, for $m \geqq 2$, are NP-hard.

9.6. Theorem. [Sah 76]

1. Max job sequencing with deadlines^{rel} $\in$ RAM-DTIME $\left(\text{Lin}\,\frac{n^2}{\varepsilon}\right)$.

2. Min 2-finish time^{rel} $\in$ RAM-DTIME $\left(\text{Lin}\left(n\log n + \frac{n}{\varepsilon^2}\right)\right) \cap$ RAM-DTIME $\left(\text{Lin}\,\frac{n^2}{\varepsilon}\right)$.

3. Min m-finish time^{rel} $\in$ RAM-DTIME $\left(\text{Lin}\,\frac{n^{2m-1}}{\varepsilon^{m-1}}\right)$ for $m \geqq 3$.

Proof. We confine ourselves to one single sample statement and show Min 2-finish time^{rel} $\in$ RAM-DTIME $\left(\text{Lin}\,\frac{n^2}{\varepsilon}\right)$.

Let $x = (t_1, \ldots, t_n)$ be a problem instance and $M =_{\text{df}} \sum_{i=1}^{n} t_i$. The minimum finish time $b^*(x)$ satisfies the inequality $b^*(x) \geqq \frac{M}{2}$. To determine a minimum solution we can use the following dynamic programming algorithm:

1. $S_0 = \{0\}$
   For $i = 1, \ldots, n$ **do**
   $S_i = S_{i-1} \cup \left\{s + t_i : s \in S_{i-1} \wedge s + t_i \leqq \frac{M}{2}\right\}$.
2. $t = \max S_n$ (Obviously, $M - t$ is the minimum finish time)
   $I_1 = \emptyset$
   For $i = n, \ldots, 1$ **do**
     If $t - t_i \in S_{i-1}$ **then** $I_1 = I_1 \cup \{i\}$ and $t = t - t_i$.

The above algorithm can be implemented on a DRAM which runs within time

$$c \cdot n \cdot \max_i \text{card}\, S_i \lesssim n \cdot M.$$

As M is not polynomially bounded in n, the given algorithm is only a pseudo-polynomial time algorithm (§ 7.3.2).

To come to an ε-approximate algorithm, we modify the above algorithm in such a way that $\max_i \text{card}\, S_i$ is linear in $\frac{n}{\varepsilon}$. This can be achieved by dividing the interval $[0, M]$ into intervals J_ν of length $\left\lfloor \frac{M\varepsilon}{2n} \right\rfloor$ each — there are about $\frac{2n}{\varepsilon}$ such intervals — and replacing in our algorithm S_i by $\overline{S}_i$ and the third line by

"$V = \overline{S}_{i-1}$
For $s \in \overline{S}_{i-1}$ **do**

if $s + t_i \leqq \frac{M}{2}$ and the interval J_ν to which $s + t_i$ belongs contains no element of V

then $V = V \cup \{s + t_i\}$

$\bar{S}_i = V$"

Thus every $\bar{S}_i$ contains at most one element from every interval. Hence, card $\bar{S}_i \leq \frac{n}{\varepsilon}$.

The maximum element of $\bar{S}_i$ may be less than the maximum element of S_i by at most $i\,|J_\nu| \leqq n \cdot \frac{M\varepsilon}{2n} \leqq \varepsilon \cdot b^*(x)$. Hence, if $f(x)$ denotes the finish time computed by the modified algorithm, it holds $|f(x) - b^*(x)| \leqq \varepsilon b^*(x)$, and the modified algorithm realizes an ε-approximation. As its computation time is bounded by $c \cdot n \cdot \max_i$ card $\bar{S}_i \leq n \cdot \frac{n}{\varepsilon}$, i.e. a polynomial in n and $\frac{1}{\varepsilon}$, Min 2-finish time is shown to be fully p-approximable. ∎

Remark 1. To prove the remaining statements of this and the preceding theorem still finer methods are required. Such methods can be found in [HoSa 78].

Remark 2. The full p-approximability of further scheduling problems is dealt with in [Sah 76].

9.1.3.1.2. Negative Results

Not too many fully p-approximable problems are known. The next two theorems provide tools for proving that the optimization problems for many NP-complete problems cannot be fully p-approximable unless **P** = **NP**.

As defined in connection with the introduction of pseudo-polynomial time algorithms (§ 7.3.2) $\max(x)$ is the magnitude of the largest integer occuring in the problem instance x. Let prob be a decision problem, and let b be a corresponding valuation function.

9.7. Theorem. [GaJo 79] If there is a polynomial q such that $b^*(x) \leqq q(|x|, \max(x))$, then it holds: If Opt prob is fully p-approximable, then Opt prob is solvable in pseudo-polynomial time.

Proof. By definition a search function f can be computed within time $r\left(|x|, \frac{1}{\varepsilon}\right)$ where r is a polynomial, and f satisfies $|b(f(x)) - b^*(x)| \leqq \varepsilon \cdot b^*(x)$. For $\varepsilon = \frac{1}{q(|x|, \max(x)) + 1}$ this yields

$$|b(x, f(x)) - b^*(x)| \leqq \frac{b^*(x)}{q(|x|, \max(x)) + 1} < 1.$$

As b, and consequently also b^*, are integer valued, we get $b(x, f(x)) = b^*(x)$, and f is computable in time $r(|x|, q(|x|, \max(x)) + 1)$, i.e. in pseudo-polynomial time. ∎

For formulating a consequence of this result we need the notion of strongly NP-complete problems. An NP-complete problem M is called *strongly NP-complete* iff there is a polynomial p such that $M_p = \{x: x \in M \wedge \max(x) \leq p(|x|)\}$ is also NP-complete.

Examples. CLIQUE and COLOUR are strongly NP-complete because of CLIQUE $=$ CLIQUE$_{\mathrm{id}}$ and COLOUR $=$ COLOUR$_{\mathrm{id}}$. TRAVELLING SALESMAN (see below) is strongly NP-complete (see [GaJo 75]).

It is evident that strongly NP-complete problems cannot be pseudo-polynomial time problems unless $\mathbf{P} = \mathbf{NP}$ (see § 7.3.2).

9.8. Corollary. If PROB is strongly NP-complete and $b^*(x) \leq q(|x|, \max(x))$ for some polynomial q, then Opt prob is not fully p-approximable unless $\mathbf{P} = \mathbf{NP}$.

Proof. If Opt prob is fully p-approximable, then, by the preceding theorem, it is solvable in pseudo-polynomial time. Consequently, PROB is decidable in pseudo-polynomial time. Since PROB is strongly NP-complete this means PROB $\in$ P. ∎

To illustrate the preceding theorem we introduce some further problems:

$$\text{TRAVELLING SALESMAN} = \{(V, f, k): f\colon V^2 \to \mathbb{N} \wedge k \in \mathbb{N} \wedge \text{there exists a permutation } \pi \text{ of } V \text{ of order card } V \text{ such that } \sum_{v \in V} f(v, \pi(v)) \leq k\}.$$

If V is interpreted as a set of cities and $f(i, j)$ as the distance between cities i and j, then the problem is whether there is a closed way no longer than k visiting every city exactly once. Note that (V, f) is not required to be embeddable into a metric space. For later use we also define the subproblems

$$\text{METRIC TRAVELLING SALESMAN} = \{(V, f, k): (V, f, k) \in \text{TRAVELLING SALESMAN} \wedge (V, f) \text{ is a metric space}\}$$

and

$$\text{EUCLIDEAN TRAVELLING SALESMAN} = \{(V, f, k): (V, f, k) \in \text{TRAVELLING SALESMAN} \wedge V \text{ is a point set in the plane} \wedge f(i, j) \text{ is the integer part of the euclidean distance between } i \text{ and } j\}.$$

The corresponding minimization problems are all defined with the same function b_{ts}

$$b_{\mathrm{ts}}\big((V, f, k), \pi\big) = \sum_{v \in V} f\big(v, \pi(v)\big).$$

An independent set of a graph $G(V, E)$ is a subset V' of V such that $(V' \times V') \cap E = \emptyset$:

$$\text{INDEPENDENT SET} =_{\mathrm{df}} \{(G, k): G \text{ graph} \wedge k \in \mathbb{N} \wedge G \text{ has an independent set of at least } k \text{ vertices}\}.$$

The maximization function b_{is} is defined by $b_{is}\big((V, E), V'\big) = \text{card } V'$.

For the next theorem we need a valuation function for SAT:

$b_{sat}(F, f)$ = number of clauses of F satisfied by the assignment f.

9.9. Theorem. The following problems are not fully p-approximable unless $\mathbf{P} = \mathbf{NP}$:

Min travelling salesman	Min set cover
Max clique	Min exact cover
Max independent set	Min colour
Min node cover	Max sat

Proof. The existence problems corresponding to the optimization problems listed in the theorem are strongly NP-complete and hence, by 9.8, cannot be fully p-approximable if P = NP. ∎

A problem which is not fully p-approximable could still be p-approximable. Sometimes this possibility can be ruled out. This can be done, for example, by showing that the necessary condition for p-approximability stated in the next theorem is not satisfiable.

Following [PaMo 77] we define for given PROB and $f: \mathbb{N} \mapsto \mathbb{N}$ the existence problem f-PROB $= \{x: x \in \text{PROB} \wedge b^*(x) \leq f(|x|)\}$. For example, k-CLIQUE $= \{(G, m): G \text{ finite graph} \wedge m \in \mathbb{N} \wedge m \leq b^*_{\text{CLIQUE}}(G) \leq k\}$.

Definition. Opt prob is called *simple* iff k-PROB $\in \mathbf{P}$ for every $k \in \mathbb{N}$. Opt prob is called *p-simple* iff $\{(x, k): x \in k\text{-PROB}\} \in \mathbf{P}$.

We consider the further problems

$$\text{BIN PACKING} = \{(x_1, \ldots, x_m, l, k): k, l, m, x_1, \ldots, x_m \in \mathbb{N} \wedge x_1, \ldots, x_m \leq l \wedge \text{there exist an } r \leq k \text{ and a partition } X_1 \cup \ldots \cup X_r = \{x_1, \ldots, x_m\} \text{ such that } \bigwedge_{i \leq r} \sum_{x \in X_i} x \leq l\},$$

$$b_{bp}((X_1, \ldots, X_r)) = r.$$

Now we illustrate the notions "simple" and "p-simple" by some examples.

Example 1. If $\mathbf{P} \neq \mathbf{NP}$, then Min colour and Min bin packing are not simple because 3-COLOUR and 2-BIN PACKING are known to be NP-complete. This follows from Remark 6 and Remark 4 after 14.21. On the other hand, Max sat, Max clique, Min node cover, Min set cover, Max independent set and Min exact cover are examples which are obviously simple.

Note that for a p-simple set Opt prob, it must hold r-PROB $\in \mathbf{P}$ for any polynomial r. Taking $r(n) = n$ we state as an obvious consequence

Example 2. If $\mathbf{P} \neq \mathbf{NP}$, then Max sat, Max clique, Min node cover, Min set cover, Max independent set, and Min exact cover are not p-simple.

9.10. Theorem. [PaMo 77] Let Opt prob be an optimization problem with respect to the valuation function $b \in \mathrm{P}$.

1. If Opt prob is p-approximable, then Opt prob is simple. More precisely:

   a) If Max $\mathrm{prob}^{\varepsilon,\mathrm{rel}} \in \mathbf{P}$ for some $\varepsilon < \frac{1}{k}$, then $(k-1)$-Prob $\in \mathbf{P}$.

   b) If Min $\mathrm{prob}^{\varepsilon,\mathrm{rel}} \in \mathbf{P}$ for some $\varepsilon < \frac{1}{k}$, then k-PROB $\in \mathbf{P}$.

2. If Opt prob is fully p-approximable, then Opt prob is p-simple.

Proof. Ad 1. We consider the case Opt = Max. If Max prob is p-approximable, then, for any $\varepsilon > 0$, there exists an algorithm A_ε solving Max $\mathrm{prob}^{\varepsilon,\mathrm{rel}}$ in polynomial time. This is valid in particular for $\varepsilon < \frac{1}{k}$. For the search function f computed by the algorithm A_ε it holds that

$$b^*(x) - b(x, f(x)) \leqq \varepsilon b^*(x) < \frac{1}{k}\, b^*(x).$$

This yields

$$b(x, f(x)) \leqq b^*(x) < b(x, f(x)) + \frac{1}{k-1} \cdot b(x, f(x)).$$

For $b(x, f(x)) \leqq k-1$ we get $b(x, f(x)) = b^*(x)$ and thus

$$\begin{aligned} x \in (k-1)\text{-PROB} &\leftrightarrow x \in \text{PROB} \wedge b^*(x) \leqq k-1 \\ &\leftrightarrow f(x) \neq \text{no} \wedge b(x, f(x)) \leqq k-1, \end{aligned}$$

which shows that A_ε provides us with a polynomial decision algorithm for $(k-1)$-PROB.

The case Opt = Min and Statement 2 are proved analogously. ∎

An application of the preceding theorem is provided by the next theorem.

9.11. Theorem. Min bin packing and Min colour are not p-approximable unless $\mathrm{P} = \mathrm{NP}$. ∎

Remark 1. Theorem 9.10 shows that Min bin $\mathrm{packing}^{\varepsilon,\mathrm{rel}}$ for $\varepsilon < \frac{1}{2}$ and Min $\mathrm{colour}^{\varepsilon,\mathrm{rel}}$ for $\varepsilon < \frac{1}{3}$ are NP-hard.

Remark 2. That even Min $\mathrm{colour}^{1,\mathrm{rel}}$ is NP-hard has been proved in [GaJo 76].

An even stronger result is true for Min travelling salesman, but this does not follow from 9.10.1 because Min travelling salesman is simple. To get the result we first introduce the problem

HAMILTONIAN CIRCUIT $=_{\mathrm{df}}$ $\{G$: G is a finite graph such that there exists a circuit in G containing each vertex exactly once (a ***Hamiltonian circuit***)$\}$.

9.12. Theorem. [SaGo 76] Min travelling salesman$^{k,\mathrm{rel}}$ is NP-hard for all $k \in \mathbb{N}$.

Proof. The result follows from

$$\text{HAMILTONIAN CIRCUIT} \leq_T^p \text{Min travelling salesman}^{k,\mathrm{rel}}$$

(for every $k \in \mathbb{N}$) and the NP-completeness of HAMILTONIAN CIRCUIT (cf. 14.21). To prove the reduction let G be a graph and define an instance $H =_{\mathrm{df}} (V, f, m)$ of Min travelling salesman by

$$V =_{\mathrm{df}} V(G), \quad f(i,j) = \begin{cases} 1 & \text{if } (i,j) \in E(G), \\ 2 + kn & \text{otherwise} \end{cases} \quad m = n(2 + kn),$$

where $n = \operatorname{card} V(G)$. Let A be an algorithm finding tours $t(H)$ with

$$b_{\mathrm{ts}}\big(H, t(H)\big) - b_{\mathrm{ts}}^*(H) \leq k \cdot b_{\mathrm{ts}}^*(H).$$

Then $G \in$ HAMILTONIAN CIRCUIT iff $b_{\mathrm{ts}}\big(H, t(H)\big) = n$. ∎

Remark. [SaGo 76] and [GeLe 79] contain many other results of this type.

9.1.3.2. *Bounded Polynomial Approximation*

Though we have seen that problems like Min bin packing are not p-approximable one could hope that Min bin packing$^{c,\mathrm{rel}} \in$ P at least for certain c. For Min bin packing a slight modification of this question can be answered in the affirmative as shown in the next theorem. For Min travelling salesman there is no such hope because of 9.12.

9.13. Theorem. [Joh 74b] There exists a search function $f \in$ P for Min bin packing such that for any x

$$b_{\mathrm{bp}}\big(f(x)\big) - b_{\mathrm{bp}}^*(x) \leq \frac{2}{9}\, b_{\mathrm{bp}}^*(x) + 4. \ ∎$$

The proof is too lengthy to be reproduced here. The chosen algorithm does not provide a better bound. Note that 9.13 does not contradict Remark 1 after 9.11. Further results concerning or akin to bin packing will be found in [KoMa 77], [GaJo 79], [Yaoa 80], [FeLu 81], [Lia 80], [CoLe 77], and [CoLeSl 78]. Similar results for postman problems are contained in [Fre 79].

Whereas Min travelling salesman$^{k,\mathrm{rel}}$ is NP-hard for all $k > 0$, the two restricted subproblems Min metric travelling salesman and Min euclidean travelling salesman allow polynomial approximation.

9.14. Theorem. [Chr 76] Min metric travelling salesman$^{1/2,\mathrm{rel}} \in$ RAM-DTIME $(\mathrm{Lin}\,n^3)$. ∎

For related results see [HoSh 84].

Similar results are known for the following subproblems of SAT, SET COVER and NODE COVER:

$k\text{SAT} =_{\text{df}} \{H : H \in \text{SAT} \wedge \text{each clause of } H \text{ has at least } k \text{ distinct literals}\}$,

$k\text{SET COVER} =_{\text{df}} \{(S, \mathfrak{S}, m) : (S, \mathfrak{S}, m) \in \text{SET COVER} \wedge \text{each set of } \mathfrak{S} \text{ has at most } k \text{ elements}\}$,

$k\text{NODE COVER} =_{\text{df}} \{(G, l) : (G, l) \in \text{NODE COVER} \wedge G \text{ has degree bounded by } k\}$.

9.15. Theorem. [Joh 74a] For every $k \geqq 1$,

$$\text{Max } k\text{sat}^{2^{\frac{1}{k-1}},\text{rel}} \in \text{RAM-DTIME}(\text{Lin}(n \log n)),$$

$$\text{Min } k\text{set cover}^{\ln k,\text{rel}} \in \text{RAM-DTIME}(\text{Lin}(n \log n)),$$

$$\text{Min } k\text{node cover}^{\ln k,\text{rel}} \in \text{RAM-DTIME}(\text{Lin}(n \log n)). \blacksquare$$

Remark 1. For more recent results see [Hoc 82], [BaEv 82], [MoSp 83] and [MoSp 84]. Similar results for further problems can be found in [SaGo 76].

Remark 2. Investigations of Max clique and Min colour are contained in [Joh 74a], but so far even results like Max $\text{clique}^{1-n^{-\varepsilon},\text{rel}} \in \text{P}$ or Min $\text{colour}^{n^{\varepsilon},\text{rel}} \in \text{P}$ cannot be guaranteed for all $\varepsilon > 0$. For Min colour see also [Wig 82].

We have seen that NP-complete problems may have very different behaviour with respect to approximation. Some of them remain NP-hard like Min travelling salesman, some afford full p-approximations like Max knapsack and some of them are very difficult to tackle like Max clique. Note also that very closely related problems may have completely different approximation behaviour because near optimal solutions of the one problem need not be translatable into near optimal solutions of the other. See for example NODE COVER and INDEPENDENT SET discussed in [GaJo 79].

An approximation algorithm working in time $\text{Lin}(n \log n)$ for Max independent set on planar graphs is described in [ChNiSa 82].

9.2. *Probability and Frequency*

9.2.1. Random Inputs

The amazingly good behaviour of several heuristic algorithms applied to hard problems leads to the idea that there might be problems which have a large worst-case complexity but are easy "on average". One possible approach to a deeper study of such phenomena is to enrich our problems with probability distributions on the inputs. We provide some examples to demonstrate how much simpler such problems with random inputs can be in comparison with the original problems.

We start with an existence problem $\text{PROB} = \left\{x : \bigvee_{y} (x, y) \in \text{prob}\right\}$ as discussed in § 7.1 and fix for it a measure "size"

$$\text{size} : \Sigma^* \mapsto \mathbb{N}$$

which is intended to characterize the size of the inputs. For example, for problems connected with graphs one might choose $\text{size}(G) =_{\text{df}} \text{card}\, V(G)$. Let $I_n =_{\text{df}} \{x: \text{size}(x) = n\}$ and $\pi = (p_n)_{n\in\mathbb{N}}$, where p_n is a probability distribution on I_n.

We are primarily interested in asymptotic probabilistic statements (with respect to π), and therefore we consider $\prod I_n$ with the product measure P induced on $\prod I_n$ by π. We recall the following facts from the theory of probability.

Definition. The property E on $\bigcup I_n$ is called *asymptotically almost sure* with respect to π iff $P\left(\left\{(x_1, x_2, \ldots): \bigvee_n \bigwedge_{m>n} E(x_m)\right\}\right) = 1$ or, equivalently, $P(\{(x_1, x_2, \ldots): \{m: \sim E(x_m)\}$ is finite$\}) = 1$. Putting $E_m =_{\text{df}} \{(x_1, x_2, \ldots): E(x_m)\}$ we get

E is asymptotically almost sure with respect to $\pi \leftrightarrow P\left(\bigcup_n \bigcap_{m>n} E_m\right) = 1$,

i.e. if and only if $\bigcup_n \bigcap_{m>n} E_m$ is almost sure with respect to the product measure.

A sufficient (and for completely independent E_m also necessary) condition for E to be almost sure is provided by the

Lemma. (Borel/Cantelli)

1. $\sum_{m\in\mathbb{N}} P(\bar{E}_m) < \infty \rightarrow P\left(\bigcup_n \bigcap_{m>n} E_m\right) = 1.$
2. If the E_m are completely independent, then

$$\sum_{m\in\mathbb{N}} P(\bar{E}_m) = \infty \rightarrow P\left(\bigcup_n \bigcap_{m>n} E_m\right) = 0. \blacksquare$$

We will refer to the condition $\sum_{m\in\mathbb{N}} P(\bar{E}_m) < \infty$ by saying "E is valid almost everywhere with respect to π".

Now we are ready to give our definitional set-up.

Definition. Let Prob be a search problem and let π be a probability distribution on the input set. $\text{Prob}(\pi) \in_{\text{ae}} \Phi(t)$ iff there exists an $f \in \mathbb{R}_1$ such that

1. $f \in \Phi(t)$,
2. $\sum_n P\left(\left\{(x_1, x_2, \ldots): \left(x_n \in \text{PROB} \wedge (x_n, f(x_n)) \notin \text{prob}\right)\right.\right.$
$\left.\left.\vee \left(x_n \notin \text{PROB} \wedge f(x_n) \neq \text{no}\right)\right\}\right) < \infty,$

i.e. f is almost everywhere a search function of Prob with respect to π. We say also, f solves the problem $\text{Prob}(\pi)$.

Unlike the weakened versions of Prob considered in the preceding subsection $\text{Prob}(\pi)$ should better be considered as an independent problem which, however, can yield some insight into the nature of Prob. Note, for instance, that a good algorithm for $\text{Prob}(\pi)$ need not be good for $\text{Prob}(\pi')$ with $\pi' \neq \pi$. Whether statements about $\text{Prob}(\pi)$ are practically useful statements about Prob depends greatly on whether π is chosen in such a way as to reflect essential features of the real situation.

We start with some graph problems. The distribution on the set I_n of all loop-free graphs with n vertices is generated as follows: A fixed probability q_n is chosen, and a graph with the vertices $1, \ldots, n$ is constructed by selecting independently the $\binom{n}{2}$ edges with probability q_n. As before, we denote the distributions on I_n by p_n. Let $\pi_0 = (p_n)_{n\in\mathbb{N}}$, where p_n is defined by $q_n = p = \text{const}$ (this is called the *constant density model*). Let π_1 be the family of distributions belonging to the *constant average model* $q_n = \dfrac{c}{n-1}$.

The first two statements of the following theorem should be compared with Remark 2 after 9.15, and Statement 2 with Remark 2 after 9.11.

Similar results for BIN PACKING with random inputs can be found in [Sha 77], [Fre 80], [Hof 80], [Knö 81] and [BeJoLeMcMc 84]. For further related results see [Blo 80], [KaTa 80] and [Weid 80].

9.16. Theorem.

1. [GrDi 75], [Kar 75] Max clique$^{1/2,\text{rel}}(\pi_0) \in_{\text{ae}}$ RAM-DTIME(Lin n^2).
2. [GrDi 75] Min colour$^{1+\varepsilon,\text{rel}}(\pi_0) \in_{\text{ae}}$ RAM-DTIME(Lin n^2) for all $\varepsilon > 0$.
3. [Kar 76] Max independent set$^{1/2+\varepsilon,\text{rel}}(\pi_0) \in_{\text{ae}}$ RAM-DTIME(Lin n^2) for all $\varepsilon > 0$.

Proof. We restrict ourselves to outline the proof of Statement 2. We achieve the claimed result with the following algorithm.

1. Vertex 1 is coloured with colour 1.
2. For i = 2, ..., n, vertex i is coloured with colour j, where j is the smallest number which is compatible with the colouring reached for the subgraph containing the vertices $1, \ldots, i-1$.

To analyse the performance of this algorithm which obviously runs in quadratic time we introduce the random variables

$$\chi_{n,p}(G) = \text{chromatic number of } G(\in I_n)\ \big(= b^*(G)\big).$$

$$c_{n,p}(G) = \text{number of colours which the algorithm needs to colour } G.$$

In [GrDi 75] the following result has been proved: For every $\varepsilon > 0$,

1. $\chi_{n,p} \geqq (1-\varepsilon) \cdot \dfrac{n}{2 \log_{\frac{1}{1-p}} n}$ almost everywhere with respect to π_0,

2. $c_{n,p} \leqq (1+\varepsilon) \cdot \dfrac{n}{\log_{\frac{1}{1-p}} n}$ almost everywhere with respect to π_0.

For $\varepsilon \leqq 1/2$ this implies $c_{n,p} - \chi_{n,p} \leqq (1 + 8\varepsilon) \cdot \chi_{n,p}$ almost everywhere with respect to π_0. ∎

Our next example is Min euclidean travelling salesman. Let E be the unit square, $I_n =_{\text{df}} E^n$, and $\pi = (p_n)_{n\in\mathbb{N}}$ where p_n is the uniform distribution over E^n. The following result should be compared with 9.14.

9.17. Theorem. [Kar 76] For every $\varepsilon > 0$,
Min euclidean travelling salesman$^{\varepsilon,\mathrm{rel}}(\pi) \in_{\mathrm{ae}}$ RAM-DTIME(Lin($n \log n$)). ▮

Our last problem in this subsection affords an exact solution ae in contrast to the near optimal solutions stated in 9.16 and 9.17. We consider once more our basic model of random graphs and define ϱ_α to be the family of distributions belonging to $q_n = \alpha \cdot \dfrac{\ln n}{n}$. For existence problems we define PROB(π) $\in_{\mathrm{ae}} \Phi(t)$ by the definition on p. 171 where $P(\{(x_1, x_2, \ldots): f(x_n) \neq c_{\mathrm{PROB}}(x_n)\})$ is used.

9.18. Theorem. [Pos 76], [Kar 76]

1. For $\alpha < 1$ a random graph does not possess a Hamiltonian circuit ae.
2. For $\alpha > 1$, HAMILTONIAN CIRCUIT(ϱ_α) $\in_{\mathrm{ae}}$ P. ▮

In [AnVa 77] fast probabilistic algorithms for the directed and undirected Hamiltonian circuit problem with random inputs are given which can also be converted to deterministic polynomial algorithms with random inputs.

Algorithms for solving SAT with random inputs have been considered in [Gol 77], [GoPuBr 82], [ApDi 82] and [FrPa 83]. Linear time algorithms for SOS with uniform instance distribution can be found in [AtPu 82].

A survey on algorithms with random inputs is given in [Wel 83].

9.2.2. Probabilistic Machines

A relaxation of a problem can also be achieved by weakening the reliability of the results, for instance by using probabilistic machines or frequency computations.

Are probabilistic machines superior to deterministic ones with respect to (time) complexity? As early as 1958 S. V. Jablonskiĭ ([Jab 58]) dealt with a problem where exhaustive search, which otherwise was provably inevitable, could be avoided by using probabilistic techniques. In [Barz 69], [Tra 74], and [Gil 74] further examples are given showing the stronger power of probabilistic machines in comparison with deterministic ones.

We restrict ourselves to the presentation of some uncontrived problems. First we consider the special case where an RTM computes a correct solution on all its paths. In [Rab 76] it is proved that the problem of finding a nearest pair of n given points of the k-dimensional space can be solved with a probabilistic algorithm whose expected number of distance computations is $\preceq n$ whereas the best-known deterministic algorithm requires $\asymp n \log \log n$ distance computations (cf. [FoHo 79]). In [ItRo 78] a similar result for the minimum length circuit cover problem is proved, but their algorithm may not stop with probability 0. A *circuit cover* is a set C of circuits covering all the edges of a graph. Its length is the sum of the lengths of all circuits belonging to C. In [ItRo 78] a probabilistic cn^2 expected time algorithm is given which, with probability 1, stops on (bridge-free) graphs and yields a minimum length circuit cover. The best known deterministic algorithm to solve this problem requires $t \succeq n^3$ steps.

Results concerning probabilistic machines which may stop with incorrect values are reported in greater detail. The following two theorems should be noted in connection with 8.13.

9.19. Theorem. [Vai 76] $D^{\#} \in$ **T-RTIME**(Lin). ∎

Higher reliability can be gained at the expense of larger computation time (see 9.20).

We define: A is called *asymptotically reliably acceptable* within time t iff there exists an RTM M such that for any $\varepsilon > 0$ there exists an $n(\varepsilon)$, such that for all x with $|x| \geqq n(\varepsilon)$,

$$\mathrm{p}(M \text{ on input } x \text{ halts with output } c_A(x) \text{ within } t(x) \text{ steps}) > 1 - \varepsilon.$$

Remark. If A is asymptotically reliably acceptable within time t, then it belongs to **T-R$^\varepsilon$TIME**(t) for all $\varepsilon > 0$.

9.20. Theorem. [Frv 75], [Frv 79a]

1. $D^{\#} \in$ **T-R$^\varepsilon$TIME**$(\mathrm{Lin}(n \log n))$ for all $\varepsilon > 0$.
2. $D^{\#} \notin$ **T-R$^\varepsilon$TIME**(t) for $t <_{\mathrm{io}} n \log n$ and $\varepsilon < \frac{1}{2}$.
3. $D^{\#}$ can be accepted asymptotically reliably within time $n \log^2 n \log \log n$.

Proof. We restrict ourselves to Statement 1.

We need some preliminary facts. Let $\pi(x)$ be the number of primes not exceeding x. The prime number theorem states, that there exist a, b such that $0 < a < b$ and for all $x \geqq 2$, $a \cdot \frac{x}{\log x} < \pi(x) < b \frac{x}{\log x}$. Let $P_2(l, n', n'')$ be the number of primes $p \leqq l$ with $n' - n'' \equiv 0(p)$ and $P_3(l, n) = \max \{P_2(l, n', n'') : n'' < n' \leqq 2^n\}$.

Lemma. For any $\varepsilon > 0$ there exists a c such that

$$\lim_{n \to \infty} \frac{P_3(cn, n)}{\pi(cn)} < \varepsilon.$$

Proof. On the one hand, there exist $n', n'' \leqq 2^n$ such that $n' > n''$ and $P_3(cn, n) = P_2(cn, n', n'')$. If $p_1^{\alpha_1} \cdot \ldots \cdot p_k^{\alpha_k}$ is the prime number decomposition of $n' - n''$ then there exists a $c_1 > 0$ such that $k \leqq \frac{c_1 n}{\log n}$. This follows from $\left(\frac{k}{2}\right)^{k/2} \leqq k! \leqq p_1^{\alpha_1} \cdot \ldots \cdot p_k^{\alpha_k} = n' - n'' \leqq 2^n$. Consequently $P_3(cn, n) = P_2(cn, n', n'') \leqq k \leqq \frac{c_1 n}{\log n}$. On the other hand, by the prime number theorem, it holds that $\pi(cn) > a \frac{cn}{\log cn}$ for some $a > 0$. Then we get the desired relation for $c > \frac{c_1}{a\varepsilon}$. □

We construct an RTM M accepting $D^{\#}$ within time $n \log n$ with given error probability ε. On inputs of length n the RTM M works as follows.

Phase 1 (deterministic). It is checked whether the input is of the form $u \# v$ with $u, v \in \{0, 1\}^*$ and $|u| = |v|$. If not, the input is rejected. Otherwise u and v are interpreted as binary presentation of n' and n'', resp. (possibly with leading zeros). Note that $n', n'' \leqq 2^n$.

Phase 2 (deterministic). Let c be chosen according to the lemma, applied for $\frac{\varepsilon}{2}$. The machine lays off a tape zone of $\lceil \log cn \rceil$ squares.
Phase 3 (stochastic). M guesses a word of length $\lceil \log cn \rceil$ which is interpreted as the binary presentation of the natural number m ($m \leq 2^{\lceil \log cn \rceil}$). The words of equal length are considered to be uniformly distributed.
Phase 4 (deterministic). M tests primality of m by checking whether one of the numbers $1 < x \leq \sqrt{m}$ divides m. If m is a prime, M works according to phase 5, otherwise phase 3 is repeated.
Phase 5 (deterministic). M computes the smallest nonnegative remainders of n' and n'' modulo m. The input is accepted if and only if $n' \equiv n''(m)$.

It is easy to see that all deterministic steps require $\precsim n \log n$ steps, the primality test requires only $\precsim \sqrt{n} \log^2 n$ steps. Therefore, even a $\log^2 cn$-fold repetition of phases 3 and 4 will lead to an overall running time of $\precsim n \log n$. (Thus many repetitions of phases 3 and 4 are necessary to achieve the desired error probability.)

In phase 3 a prime is generated with probability $\frac{\pi(2^{\lceil \log cn \rceil})}{2^{\lceil \log cn \rceil}} > \frac{a}{\lceil \log cn \rceil}$ and hence a composite with probability $\leq 1 - \frac{a}{\lceil \log cn \rceil}$. The probability that during $\log^2 cn$ repetitions of phase 3 only composite numbers are generated is $\leq \left(1 - \frac{a}{\lceil \log cn \rceil}\right)^{\log^2 cn}$. Consequently, the probability that during $\log^2 cn$ repetitions of phase 3 a prime number is generated, is $> 1 - \left(1 - \frac{a}{\lceil \log cn \rceil}\right)^{\log^2 cn}$. This tends to 1 for large n. Therefore there exists a number n_1 such that for $n > n_1$ the probability that during $\log^2 cn$ repetitions of phase 3 a prime number is generated is $> 1 - \frac{\varepsilon}{2}$.

The decision criterion of phase 5 is not reliable only for the prime divisors of $n' - n''$. But by the lemma the number of these primes is small in comparison with the number of all primes of length $\leq \lceil \log cn \rceil$. Hence there exists a number n_2 such that for $n > n_2$ we have $\frac{P_3(cn, n)}{\pi(2^{\lceil \log cn \rceil})} \leq \frac{\varepsilon}{2}$ and therefore the probability for a correct answer is $> 1 - \frac{\varepsilon}{2}$. As phase 5 is applied only with probability $> 1 - \frac{\varepsilon}{2}$ (for paths on which in phase 3 a prime has been generated) the probability for a correct decision within the given time is $> \left(1 - \frac{\varepsilon}{2}\right)\left(1 - \frac{\varepsilon}{2}\right) > 1 - \varepsilon$. M can be constructed in such a way that for inputs of length $\leq \max(n_1, n_2)$ an absolute correct decision is guaranteed. Note that inputs $w \in \mathrm{D}^{\#}$ are always accepted correctly. ∎

The same idea can be used to prove

9.21. Theorem. [Frv 79c] For every $\varepsilon > 0$,

$$\{(k, l, m) : k \cdot l = m\} \in \mathbf{T\text{-}R^{\varepsilon}TIME}(n \log n). \blacksquare$$

A comparison with 8.14 shows again the superiority of probabilistic machines over deterministic ones. The next result, which should be compared with 8.13.4 and 7.10, shows that for RTM's there can exist at most an $n \log \log n$ time gap in the "low end" of the hierarchy, instead of an $n \log n$ gap as known for DTM's (cf. 8.17).

9.22. Theorem. [Frv 75], [Frv 79a]

1. C is asymptotically reliably acceptable within time $\preceq n \log \log n$.

2. $C \notin$ **T-R$^\varepsilon$TIME**(t) for $t \prec n \log \log n$ and $\varepsilon < \frac{1}{2}$. ∎

The next result concerns realtime computation. Let $\tilde{\mathrm{S}} =_{\mathrm{df}} \{(w\,\#)^k : k = \lceil \log |w| \rceil \wedge w \in \mathrm{S}\}$.

9.23. Theorem. [Frv 79d]

1. For every $\varepsilon > 0$, $\tilde{\mathrm{S}} \in$ **T-R$^\varepsilon$TIME**(id).

2. **T-DTIME-Comp**$(\tilde{\mathrm{S}}) >_{\mathrm{io}} \dfrac{n^2}{\log n}$. ∎

Further related results are contained in [Frv 74], [Frv 79a], [Frv 79d] and, concerning several types of automata, in [Frv 78], and [Frv 79b].

The next result is of interest in connection with the still open question of whether PRIME belongs to P (cf. § 7.3.3 and § 24.2.1.2.3).

9.24. Theorem. [SoSt 77], [Rab 76], [Rab 80] For every $\varepsilon > 0$,

$$\mathrm{PRIME} \in \textbf{RAM-R}^\varepsilon\textbf{TIME}(\mathrm{Lin}\, n^3).$$

Proof. We reproduce the elegant proof given by R. Solovay and V. Strassen. It extensively exploits algebraic and number-theoretic facts, which are considered to be familiar.

If m is an odd prime and $a \leq m - 1$, then $\left(\frac{a}{m}\right) \equiv a^{\frac{m-1}{2}} (m)$, where $\left(\frac{a}{m}\right)$ is the Jacobi symbol (see p. 116). The converse is not valid. However, defining

$$G_m = \left\{a : a \in \{1, \ldots, m-1\} \wedge (a, m) = 1 \wedge \left(\frac{a}{m}\right) \equiv a^{\frac{m-1}{2}} (m)\right\}$$

we have for odd composite m

$$\operatorname{card} G_m \leq \frac{m-1}{2}. \tag{1}$$

This allows the following primality test of m: If a number a is chosen at random from a uniform distribution over $\{1, \ldots, m-1\}$, then m is accepted as a prime if and only if $a \in G_m$. For $m \in \mathrm{PRIME}$ it is a correct decision, and for composite m the error probability is less than $\frac{1}{2}$. For given $\varepsilon > 0$ determine k such that $\frac{1}{2^k} < \varepsilon$. A k-fold repetition of the test guarantees an error probability of less than ε.

To estimate the computation time, take into consideration that the Euclidean algorithm for testing relative primality of a and m takes $c \log m$ arithmetic operations, including division with remainder. The same is true for computing $a^{\frac{m-1}{2}}$ modulo m and for computing $\left(\frac{a}{m}\right)$ with the help of the reciprocity law. For $n = |\text{bin } m| = \lfloor \log m \rfloor + 1$, this is an overall computation time of $c \cdot n \cdot n^2$ for a DRAM.

It remains to prove (1). $H_m =_{\text{df}} \{a : a \in \{1, \ldots, m-1\} \wedge (a, m) = 1\}$ is a multiplicative group (a is understood as the congruence class of a modulo m). G_m is a subgroup of H_m. It remains to show that G_m is a proper subgroup of H_m. This implies

$$\text{card } G_m \leqq \frac{1}{2} \cdot \text{card } H_m \leqq \frac{m-1}{2}.$$

Assume on the contrary that $G_m = H_m$, i.e. for $(a, m) = 1$ we have automatically $\left(\frac{a}{m}\right) \equiv a^{\frac{m-1}{2}} (m)$.

Case 1: $m = p^i$ where p is a prime and $i \geqq 2$. Then for all a not divisible by p we get $a^{p^i - 1} \equiv 1(p^i)$. (Note that $\left(\frac{a}{m}\right)^2 = 1$.) As the multiplicative group of all congruence classes modulo p^i has order $p^{i-1}(p-1)$ we conclude $p^{i-1}(p-1)/p^i - 1$, hence $p^{i-1}/p^{i-1} + p^{i-2} + \ldots + 1$, which is possible only for $i \leqq 1$, a controdiction.

Case 2: $m = r \cdot s$, $(r, s) = 1$. This assumption implies the controdiction $\left(\frac{a}{m}\right) = 1$ for all a with $(a, m) = 1$. This can be seen as follows: Assume $a^{\frac{m-1}{2}} \equiv -1(m)$. We determine x and y such that $rx + 1 = sy + a$. This is possible because of $(r, s) = 1$. Let b be the common value. Then $b \equiv 1(r)$ and $b \equiv a(s)$. Hence $b^{\frac{m-1}{2}} \equiv 1(r)$ and $b^{\frac{m-1}{2}} \equiv a^{\frac{m-1}{2}} (s)$ i.e. $b^{\frac{m-1}{2}} \equiv -1(s)$. b is prime relative to m: We know already $b \equiv 1(r)$, and a common divisor of b and s would also divide a because of $b \equiv a(s)$, which is impossible because of $(a, m) = 1$. Hence $b \in H_m$, and by assumption $b^{\frac{m-1}{2}} \equiv \left(\frac{b}{m}\right) (m)$. But $\left(\frac{b}{m}\right)^2 = 1$ and therefore $b^{\frac{m-1}{2}} \equiv 1(m)$ or $b^{\frac{m-1}{2}} \equiv -1(m)$. Both possibilities are incompatible with $b^{\frac{m-1}{2}} \equiv 1(r) \wedge b^{\frac{m-1}{2}} \equiv -1(s)$. Consequently, $a^{\frac{m-1}{2}} \equiv 1(m)$ and thus $\left(\frac{a}{m}\right) = 1$ for all a with $(a, m) = 1$. ∎

Remark. A comparison between the primality tests in [SoSt 77] and [Rab 76] has been carried out in [Monr 80], for related results see [FiLi 84].

In [AnVa 77] fast probabilistic algorithms for the directed and undirected Hamiltonian circuit problem are presented where, additionally, the inputs are randomized.

No NP-complete problem is known to be acceptable within probabilistic polynomial time with small error bound. The same question for the problems in the polynomial time hierarchy is dealt with in [Ko 82].

9.2.3. Frequency Computations

In [McN 61] the following notion has been suggested.

Definition. The function $f\colon \Sigma^* \to \Sigma^*$ is called *computed with frequency* $\frac{m}{n}$ or (m, n)-*computed* by a deterministic machine M of type τ iff M computes a function g: $(\Sigma^* \#)^n \to (\Sigma^* \#)^n$ such that for all $x = x_1 \# \ldots \# x_n$ and $y = y_1 \# \ldots \# y_n$ with $g(x) = y$ there are m valid equations among the n equations $f(x_1) = y_1, \ldots, f(x_n) = y_n$.

In [Tra 63] it is shown: Every (m, n)-computable function for $\frac{m}{n} > \frac{1}{2}$ is recursive, but there exists an algorithm (according to [Kinb 76], a finite automaton) which (1, 2)-computes $2^{\aleph_0}$ predicates. Consequently, for $\frac{m}{n} \leqq \frac{1}{2}$, among the (m, n)-computable functions there exist noncomputable ones (see also [Tra 77]).

A set $A \subseteqq \Sigma^*$ is called accepted with frequence $\frac{m}{n}$ or (m, n)-accepted by a deterministic machine M of type τ iff c_A is (m, n)-computed by M. For $t \geqq \mathrm{id}$ we define τ-**(m, n)-FTIME**(t) as the class of all sets (m, n)-accepted by machines of type τ which run within time t.

9.25. Theorem. [Kinb 75] For every $t \in \mathbb{R}_1$ and every $k \geqq 2$ there exists a set $A \notin T$-**DTIME**(t) with $A \in$ **multiT-(k − 1, k)-FTIME**(id). ∎

Chapter IV
Properties of Complexity Classes

In this chapter we have in mind properties of single complexity classes and not properties of the family of all complexity classes of a measure. The latter are considered to be measure properties and are dealt with in Chapter V.

Complexity classes describe in a very exact way the computational power of complexity bounded machines, and therefore many important problems of complexity theory (determinism versus nondeterminism, space versus time) are actually questions about complexity classes. That is why a profound knowledge of the properties of complexity classes could be helpful for solving such problems, and this motivates the study of complexity classes from as many points of views as possible.

In § 10 recursion theoretic characterizations of complexity classes are given. Algebraic closure properties, in particular AFL properties, are studied in § 11. § 12 relates complexity classes to various types of grammars and § 13 relates them to various types of automata. § 14 summarizes the closure properties of complexity classes with respect to complexity bounded reducibilities and states some facts about complete sets in complexity classes with respect to such reducibilities.

Since the results in [Fag 73], [JoSe 74] (see § 24.2.2.2.2.) [Sto 77] (end of § 24.3.) and [Imm 80] the possibility of characterizing complexity classes by logical means was known. But only quite recently ([Imm 83]) it became obvious to what extent complexity classes can be characterized by expressibility in suitable first and second order languages. Further papers belonging to this field which is too new to be presented in an extra section are [Grd 83] and [Sca 83].

§ 10. Complexitiy Classes and Recursion

The concept of defining restricted classes of recursive functions in terms of complexity classes (i.e. by bounding the complexity of their computation) is relatively new — it dates from the sixties. However, as early as in the thirties another method of defining classes of functions was created, namely by certain types of recursion which cannot generate all recursive functions. An important example is the primitive recursion, but it is still too powerful to generate natural classes of recursive functions which are in a sense simple, since it can generate functions of exorbitant growth from very simple functions. Therefore there has been a search for suitable restrictions of the operation of primitive recursion or of equivalent operations. It has been found that such restrictions can be obtained by "bounding" the operation in the following sense: an operation is applicable only if the generated function can be bounded above by some function which is already generated in this way.

In this chapter we investigate in particular the relationships between complexity classes and classes of recursive functions defined by several types of bounded primitive recursion. We show that many space and time complexity classes can be characterized recursion-theoretically, i.e. in terms of such types of recursion.

10.1. *Various Types of Primitive Recursion and Their Bounded Versions*

In this section it is useful to identify words over $\{1, 2\}$ with natural numbers by the following bijection (dyadic presentation): $e = 0$, $x1 = 2x + 1$, $x2 = 2x + 2$. Therefore, every function $f: (\{1,2\}^*)^k \mapsto \{1,2\}^*$ is a number-theoretic function and vice versa.

The primitive recursion defined in § 2 gives rise to two versions:

Syntactic recursion. For $s \in \mathbb{N}$, $f: \mathbb{N}^s \mapsto \mathbb{N}$, $g_1, g_2: \mathbb{N}^{s+2} \mapsto \mathbb{N}$ and $h: \mathbb{N}^{s+1} \mapsto \mathbb{N}$,

$$h = \mathrm{SR}(f, g_1, g_2) \leftrightarrow 1.\ h(\bar{x}, e) = f(\bar{x}), ^{*)}$$
$$2.\ h(\bar{x}, y1) = g_1(\bar{x}, y, h(\bar{x}, y)),$$
$$h(\bar{x}, y2) = g_2(\bar{x}, y, h(\bar{x}, y)).$$

Thus the primitive recursion of § 2 on $\{1, 2\}^*$ is henceforth called syntactic recursion whereas the term *primitive recursion* is understood in the following sense (suggested by the primitive recursion of § 2 based on $\{1\}^*$).

Primitive recursion. For $s \in \mathbb{N}$, $f: \mathbb{N}^s \mapsto \mathbb{N}$, $g: \mathbb{N}^{s+2} \mapsto \mathbb{N}$ and $h: \mathbb{N}^{s+1} \mapsto \mathbb{N}$,

$$h = \mathrm{PR}(f, g) \leftrightarrow 1.\ h(\bar{x}, 0) = f(\bar{x}), \quad 2.\ h(\bar{x}, y + 1) = g(\bar{x}, y, h(\bar{x}, y)).$$

Note that the computation of $h(\bar{x}, y)$ by syntactic (primitive) recursion requires $|y|$**) (y) recursion steps.

The *recursion by two values* allows us to use in the recursion step for $h(\bar{x}, y + 1)$ not only the value $h(\bar{x}, y)$ but also a further value $h(\bar{x}, z)$ where $z \leqq y$.

For $s \in \mathbb{N}$, $f: \mathbb{N}^s \mapsto \mathbb{N}$, $g: \mathbb{N}^{s+3} \mapsto \mathbb{N}$ and $r, h: \mathbb{N}^{s+1} \mapsto \mathbb{N}$ such that $r(\bar{x}, y) \leqq y$,

$$h = \mathrm{VR}(f, g, r) \leftrightarrow 1.\ h(\bar{x}, 0) = f(\bar{x}),$$
$$2.\ h(\bar{x}, y + 1) = g(\bar{x}, y, h(\bar{x}, y), h(\bar{x}, r(\bar{x}, y))).$$

Now we introduce bounded versions of SR, PR and VR. Let $s \in \mathbb{N}$, $f: \mathbb{N}^s \mapsto \mathbb{N}$, $g_1, g_2: \mathbb{N}^{s+2} \mapsto \mathbb{N}$, $g: \mathbb{N}^{s+3} \mapsto \mathbb{N}$ and $k, r, h: \mathbb{N}^{s+1} \mapsto \mathbb{N}$

$$h = \mathrm{BSR}(f, g_1, g_2, k) \leftrightarrow h = \mathrm{SR}(f, g_1, g_2) \wedge h \leqq k,$$

$$h = \mathrm{BPR}(f, g_1, k) \quad \leftrightarrow h = \mathrm{PR}(f, g_1) \wedge h \leqq k,$$

$$h = \mathrm{BVR}(f, g, r, k) \leftrightarrow h = \mathrm{VR}(f, g, r) \wedge h \leqq k.$$

---

*) $\bar{x}$ is an abbreviation for $x_1, \ldots, x_s$.

**) Remember that $|y|$ is the length of the word y. Thus we have $|2^x| = x$, $2^{|x|} \leqq x + 1$ and $x \leqq 2^{|x|+1}$.

Furthermore the operations BPR and BVR can be modified in such a way that only $|y|$ recursion steps are permitted for the computation of $h(\bar{x}, y)$ (similar to the case of syntactic recursion):

Weak bounded primitive recursion. For $s \in \mathbb{N}$, $f: \mathbb{N}^s \mapsto \mathbb{N}$, $g: \mathbb{N}^{s+2} \mapsto \mathbb{N}$ and $k, h: \mathbb{N}^{s+1} \mapsto \mathbb{N}$,

$$h = \mathrm{WBPR}(f, g, k) \leftrightarrow h(\bar{x}, y) = h'(\bar{x}, |y|) \text{ where } h' = \mathrm{BPR}(f, g, k).$$

Finally we define the corresponding modifications of the operations SUM and PRD, which are not as powerful as primitive recursion (see before 2.20).

Weak sum operator and *weak product operator*. For $f, g: \mathbb{N}^{s+1} \mapsto \mathbb{N}$,

$$\mathrm{WSUM}(f) = g \leftrightarrow g(\bar{x}, y) = \sum_{z=0}^{|y|} f(\bar{x}, z),$$

$$\mathrm{WPRD}(f) = g \leftrightarrow g(\bar{x}, y) = \prod_{z=0}^{|y|} f(\bar{x}, z).$$

Now we deal with some relationships between the operators defined above. First we state that the operators PR, SR and VR are equivalent.

10.1. Theorem.

1. If $s \in A$, then $\Gamma_{\mathrm{SUB,PR}}(A) = \Gamma_{\mathrm{SUB,SR}}(A) = \Gamma_{\mathrm{SUB,VR}}(A)$.
2. $\Gamma_{\mathrm{SUB,SR}}(s) = \Gamma_{\mathrm{SUB,VR}}(s) = \mathbb{P}\mathrm{r}$. ∎

It is not so easy to relate the bounded versions of the operators PR, SR and VR, since some relationships between them are connected with some open questions about the relationships between time and space complexity classes. This will become clear in § 10.2. However, some simple relationships can be stated. Let $\mathrm{m} =_{\mathrm{df}} \max$.

10.2. Lemma.

1. If $s \in \Gamma_{\mathrm{SUB,BSR}}(A)$, then $\Gamma_{\mathrm{SUB,WBPR}}(A) \subseteq \Gamma_{\mathrm{SUB,BSR}}(A)$.
2. If s, m, $2 \cdot x$, $\left\lfloor \frac{x}{2} \right\rfloor$, $\dot{-} \in \Gamma_{\mathrm{SUB,WBPR}}(A)$, then $\Gamma_{\mathrm{SUB,BSR}}(A) \subseteq \Gamma_{\mathrm{SUB,WBPR}}(A)$. ∎

10.3. Lemma.

1. If s, m, $2 \cdot x \in \Gamma_{\mathrm{SUB,BPR}}(A)$, then $\Gamma_{\mathrm{SUB,WBPR}}(A) \subseteq \Gamma_{\mathrm{SUB,BPR}}(A)$.
2. If $2^x \in \Gamma_{\mathrm{SUB,WBPR}}(A)$, then $\Gamma_{\mathrm{SUB,BPR}}(A) \subseteq \Gamma_{\mathrm{SUB,WBPR}}(A)$. ∎

10.4. Lemma.

1. $\Gamma_{\mathrm{SUB,BPR}}(A) \subseteq \Gamma_{\mathrm{SUB,BVR}}(A)$.
2. If s, m, $2^x \in \Gamma_{\mathrm{SUB,BPR}}(A)$ and every function from A can be bounded above by an increasing function from $\Gamma_{\mathrm{SUB,BPR}}(A)$, then $\Gamma_{\mathrm{SUB,BVR}}(A) \subseteq \Gamma_{\mathrm{SUB,BPR}}(A)$. ∎

From 10.2, 10.3 and 10.4 we can conclude

10.5. Lemma. If s, m, $2^x \in \Gamma_{\mathrm{SUB,BPR}}(A) \cap \Gamma_{\mathrm{SUB,BVR}}(A) \cap \Gamma_{\mathrm{SUB,BSR}}(A) \cap \Gamma_{\mathrm{SUB,WBPR}}(A)$ and every function from A can be bounded above by an increasing function from

$\Gamma_{\mathrm{SUB,BPR}}(A)$, then

$$\Gamma_{\mathrm{SUB,BPR}}(A) = \Gamma_{\mathrm{SUB,BVR}}(A) = \Gamma_{\mathrm{SUB,BSR}}(A) = \Gamma_{\mathrm{SUB,WBPR}}(A). \blacksquare$$

10.6. Corollary. For $k \geqq 3$,

$$\Gamma_{\mathrm{SUB,BVR}}(\mathrm{s}, \mathrm{m}, \mathrm{a}_k) = \Gamma_{\mathrm{SUB,BSR}}(\mathrm{s}, \mathrm{m}, \mathrm{a}_k) = \Gamma_{\mathrm{SUB,WBPR}}(\mathrm{s}, \mathrm{m}, \mathrm{a}_k) = \mathcal{E}^k. \blacksquare$$

Finally we introduce some simple functions which will be used in § 10.2.

For $x, y, z, x_1, x_2, \ldots, y_1, y_2, \ldots \in \{1, 2\}^* = \mathbb{N}$,

$$\mathrm{s}_1(x) =_{\mathrm{df}} x1, \quad \mathrm{s}_2(x) = x2, \quad \mathrm{c}(x, y) =_{\mathrm{df}} xy,^{*)}$$

$$\tilde{\mathrm{e}} = \mathrm{e}, \quad \widetilde{x1} = \tilde{x}112, \quad \widetilde{x2} = \tilde{x}221,$$

$$\mathrm{b}_1(x) =_{\mathrm{df}} \begin{cases} y & \text{if} \quad x = y1 \quad \text{or} \quad x = y2, \\ \mathrm{e} & \text{if} \quad x = \mathrm{e}, \end{cases}$$

$$\mathrm{b}_2(x) =_{\mathrm{df}} \begin{cases} y & \text{if} \quad x = 1y \quad \text{or} \quad x = 2y, \\ \mathrm{e} & \text{if} \quad x = \mathrm{e}, \end{cases}$$

$$\mathrm{b}_3(x) =_{\mathrm{df}} \begin{cases} 1 & \text{if} \quad x = y1 \quad \text{for some} \quad y, \\ 2 & \text{if} \quad x = y2 \quad \text{for some} \quad y, \\ 1 & \text{if} \quad x = \mathrm{e}, \end{cases}$$

$$\mathrm{b}_4(x) =_{\mathrm{df}} \begin{cases} 1 & \text{if} \quad x = 1y \quad \text{for some} \quad y, \\ 2 & \text{if} \quad x = 2y \quad \text{for some} \quad y, \\ 1 & \text{if} \quad x = \mathrm{e}, \end{cases}$$

$$\mathrm{b}_5(x, y) =_{\mathrm{df}} \begin{cases} \mathrm{e} & \text{if} \quad x = y, \\ 1 & \text{otherwise}, \end{cases}$$

$$\mathrm{b}_6(x_1, y_1, x_2, y_2) =_{\mathrm{df}} \begin{cases} \mathrm{e} & \text{if} \quad x_1 = y_1 \quad \text{and} \quad x_2 = y_2, \\ 1 & \text{otherwise}, \end{cases}$$

$$\mathrm{b}_7(x) =_{\mathrm{df}} \begin{cases} y & \text{if} \quad x = \tilde{y}21z \quad \text{or} \quad x = \tilde{y}12z \quad \text{for some} \quad z, \\ \mathrm{e} & \text{otherwise}, \end{cases}$$

$$\mathrm{b}_8(x) =_{\mathrm{df}} \begin{cases} z & \text{if} \quad x = \tilde{y}21z \quad \text{or} \quad x = \tilde{y}12z \quad \text{for some} \quad y, \\ \mathrm{e} & \text{otherwise}, \end{cases}$$

$$\mathrm{b}_9(x, y) =_{\mathrm{df}} \begin{cases} |y|\text{th} & \text{symbol of } x \text{ if } 1 \leqq |y| \leqq |x|, \\ \mathrm{e} & \text{otherwise}, \end{cases}$$

$$\mathrm{b}_9'(x, y) =_{\mathrm{df}} \begin{cases} y\text{th} & \text{symbol of } x \text{ if } 1 \leqq y \leqq |x|, \\ \mathrm{e} & \text{otherwise}, \end{cases}$$

*) We have to distinguish carefully between the concatenation xy of the words (numbers) x and y and their product $x \cdot y$. Especially we distinguish between $2x$ and $2 \cdot x$. The term x^l stands for $\underbrace{x \cdot x \cdot \ldots \cdot x}_{l \text{ times}}$.

for $s \geq 1$,

$$\mathrm{v}_s(x_1, \ldots, x_s, y_1, \ldots, y_s, z)$$

$$=_{\mathrm{df}} \begin{cases} x_1 & \text{if} \quad y_1 = \mathrm{e}, \\ x_2 & \text{if} \quad y_2 = \mathrm{e} \quad \text{and} \quad y_1 \neq \mathrm{e}, \\ x_3 & \text{if} \quad y_3 = \mathrm{e}, \quad y_2 \neq \mathrm{e} \quad \text{and} \quad y_1 \neq \mathrm{e}, \\ \vdots & \\ x_s & \text{if} \quad y_s = \mathrm{e}, \; y_{s-1} \neq \mathrm{e}, \ldots, y_2 \neq \mathrm{e} \; \text{and} \; y_1 \neq \mathrm{e}, \\ z & \text{if} \quad y_s \neq \mathrm{e}, \ldots, y_2 \neq \mathrm{e} \quad \text{and} \quad y_1 \neq \mathrm{e} \end{cases}$$

$$\mathrm{abs}(x - y) =_{\mathrm{df}} (x \dot{-} y) + (y \dot{-} x).$$

10.7. Lemma.

1. $\mathrm{s}_1, \mathrm{s}_2 \in \Gamma_{\mathrm{SUB}}(\mathrm{s}, 2 \cdot x) \subseteq \Gamma_{\mathrm{SUB}}(\mathrm{s}, +)$.
2. $\mathrm{c} \in \Gamma_{\mathrm{SUB,BSR}}(\mathrm{s}, +, x^2) \subseteq \Gamma_{\mathrm{SUB,BSR}}(\mathrm{s}, \mathrm{m}, 2 \cdot x, x^2)$.
3. $\Gamma_{\mathrm{SUB}}(\mathrm{m}, 2 \cdot x, \dot{-}) = \Gamma_{\mathrm{SUB}}(+, \dot{-})$.
4. $\tilde{\ }, \mathrm{b}_1, \mathrm{b}_2, \ldots, \mathrm{b}_9, \mathrm{v}_1, \mathrm{v}_2, \ldots, \mathrm{sgn}, \overline{\mathrm{sgn}} \in \Gamma_{\mathrm{SUB,BSR}}(\mathrm{s}_1, \mathrm{s}_2, \mathrm{c})$.
5. $\mathrm{s}_1, \mathrm{s}_2, \mathrm{c}, \tilde{\ }, \mathrm{b}_1, \mathrm{b}_2, \ldots, \mathrm{b}_9, \mathrm{b}_9', \mathrm{v}_1, \mathrm{v}_2, \ldots, \mathrm{sgn}, \overline{\mathrm{sgn}}, 2 \cdot x, \left\lfloor \frac{x}{2} \right\rfloor, +, \dot{-}, \mathrm{abs}(x \dot{-} y)$, $|x|, x^3, x^4, \ldots, 2^{|x|}, \lfloor \sqrt{x} \rfloor \in \Gamma_{\mathrm{SUB,BPR}}(\mathrm{s}, \mathrm{m}, x^2) \subseteq \Gamma_{\mathrm{SUB,BVR}}(\mathrm{s}, \mathrm{m}, x^2)$. ∎

10.2. Recursion-Theoretic Characterizations of Complexitiy Classes

10.2.1. Bounded Syntactic Recursion and Weak Bounded Primitive Recursion

We show in this subsection that some complexity classes of type DSPACE-TIME (Lin t, Pol t)*) can be characterized recursion-theoretically using bounded syntactic recursion or weak bounded primitive recursion. We start with the characterization of the class DSPACE-TIME(Lin, Pol) which will be the base for all other characterizations in this subsection.

To this end it is necessary to describe the local behaviour of a Turing machine, i.e. the step from a given situation to a next situation, by means of bounded syntactic recursion. Let M be a DTM with the worktape alphabet $\{1, 2\}$ and the states $1, 2, \ldots, q$. (Note that our definition of a worktape alphabet requires the symbol □. But for the reason of encoding we take 1 instead of □.) Further, let S be a situation of M

---

*) In this section we assume (in contrast to all other sections) that the Turing machines which compute functions do not have a separate output tape. Thus the length of the computed value cannot exceed the given bound for the space on the worktape. Furthermore, an s-ary function f is said to belong to a complexity class if the function $\tilde{f}$ belongs to this complexity class where

$$\tilde{f}(w) = \begin{cases} f(x_1, \ldots, x_s) & \text{if} \quad w = \tilde{x}_1 * \tilde{x}_2 * \ldots * \tilde{x}_s, \\ w & \text{otherwise.} \end{cases}$$

Here and in the remainder of this section $*$ stands for 21.

with the worktape contents $\ldots 111 w_1 \overset{\downarrow}{a} w_2 111 \ldots$ ($w_1, w_2 \in \{1,2\}^*$, $a \in \{1,2\}$), with the head position marked by the arrow, and with the state s. As $*$ stands for 21 this situation S can be described by the word $\tilde{s} * \tilde{w}_1 * \tilde{a} * \tilde{w}_2 *$ and we set $\beta(\tilde{s} * \tilde{w}_1 * \tilde{a} * \tilde{w}_2 *) =_{\mathrm{df}} S$. If w has not the form $\tilde{s} * \tilde{w}_1 * \tilde{a} * \tilde{w}_2 *$ for some $s \in \{1, \ldots, q\}$, w_1, $w_2 \in \{1,2\}^*$ and $a \in \{1,2\}$, then $\beta(\mathrm{w})$ is not defined.

10.8. Lemma. There is a $\mathrm{k}\colon \{1,2\}^* \mapsto \{1,2\}^*$ such that

1. $\mathrm{k} \in \Gamma_{\mathrm{SUB,BSR}}(\mathrm{s}_1, \mathrm{s}_2, \mathrm{c})$.
2. If $\beta(w)$ is defined, then $\beta(\mathrm{k}(w))$ is defined and $\beta(w) \vdash_{\overline{M}} \beta(\mathrm{k}(w))$.

Proof. Let $ji \to s_{ji} a_{ji} \sigma_{ji}$ ($j = 1, \ldots, q$; $i = 1, 2$) be the instructions of M. We define for $j, i, w_1, w_2 \in \{1,2\}^*$

$$\mathrm{k}_1'(j, w_1, i, w_2) =_{\mathrm{df}} \begin{cases} s_{ji} & \text{if } j \in \{1, \ldots, q\}, \text{ and } i \in \{1,2\}, \\ \mathrm{e} & \text{otherwise}, \end{cases}$$

$$\mathrm{k}_2'(j, w_1, i, w_2) =_{\mathrm{df}} \begin{cases} w_1 a_{ji} & \text{if } j \in \{1, \ldots, q\}, i \in \{1,2\} \text{ and } \sigma_{ji} = +, \\ w_1 & \text{if } j \in \{1, \ldots, q\}, i \in \{1,2\} \text{ and } \sigma_{ji} = \mathrm{e}, \\ \mathrm{b}_1(w_1) & \text{if } j \in \{1, \ldots, q\}, i \in \{1,2\} \text{ and } \sigma_{ji} = -, \\ \mathrm{e} & \text{otherwise}, \end{cases}$$

$$\mathrm{k}_3'(j, w_1, i, w_2) =_{\mathrm{df}} \begin{cases} \mathrm{b}_4(w_2) & \text{if } j \in \{1, \ldots, q\}, i \in \{1,2\} \text{ and } \sigma_{ji} = +, \\ a_{ji} & \text{if } j \in \{1, \ldots, q\}, i \in \{1,2\} \text{ and } \sigma_{ji} = \mathrm{e}, \\ \mathrm{b}_3(w_1) & \text{if } j \in \{1, \ldots, q\}, i \in (1,2\} \text{ and } \sigma_{ji} = -, \\ \mathrm{e} & \text{otherwise}, \end{cases}$$

$$\mathrm{k}_4'(j, w_1, i, w_2) =_{\mathrm{df}} \begin{cases} \mathrm{b}_2(w_2) & \text{if } j \in \{1, \ldots, q\}, i \in \{1,2\} \text{ and } \sigma_{ji} = +, \\ w_2 & \text{if } j \in \{1, \ldots, q\}, i \in \{1,2\} \text{ and } \sigma_{ji} = \mathrm{e}, \\ a_{ji} w_2 & \text{if } j \in \{1, \ldots, q\}, i \in \{1,2\} \text{ and } \sigma_{ji} = -, \\ \mathrm{e} & \text{otherwise}. \end{cases}$$

Each of the functions k_1', k_2', k_3' and k_4' is defined by $6q + 1$ mutually exclusive cases. Therefore, these functions can be generated by substitution using only functions mentioned in Lemma 10.7.4, especially b_6 and v_{6q}. $\underbrace{\text{Consequently}}$ $\underbrace{\mathrm{k}_1', \mathrm{k}_2', \mathrm{k}_3', \mathrm{k}_4' \in \Gamma_{\mathrm{SUB,BSR}}(\mathrm{s}_1, \mathrm{s}_2, \mathrm{c})}$. $\underbrace{\text{Defining}}$ $\mathrm{k}'(j, w_1, i, w_2) =_{\mathrm{df}} \mathrm{k}_1'(j, w_1, i, w_2) * \mathrm{k}_2'(j, w_1, i, w_2) * \mathrm{k}_3'(j, w_1, i, w_2) * \mathrm{k}_4'(j, w_1, i, w_2) *$ we get $\beta(\tilde{j} * \tilde{w}_1 * \tilde{i} * \tilde{w}_2 *) \vdash_{\overline{M}} \beta(\mathrm{k}'(j, w_1, i, w_2))$ if $\beta(\tilde{j} * \tilde{w}_1 * \tilde{i} * \tilde{w}_2 *)$ is defined. Now we set

$$\mathrm{k}(w) =_{\mathrm{df}} \mathrm{k}'\Big(\mathrm{b}_7(w), \mathrm{b}_7(\mathrm{b}_8(w)), \mathrm{b}_7\big(\mathrm{b}_8(\mathrm{b}_8(w))\big), \mathrm{b}_7\big(\mathrm{b}_8(\mathrm{b}_8(\mathrm{b}_8(w)))\big)\Big)$$

and get $\mathrm{k}(\tilde{j} * \tilde{w}_1 * \tilde{i} * \tilde{w}_2 *) = \mathrm{k}'(j, w_1, i, w_2)$ and consequently $\beta(w) \vdash_{\overline{M}} \beta(\mathrm{k}(w))$ if $\beta(w)$ is defined. Because of Lemma 10.7.4 the functions k' and k are in $\Gamma_{\mathrm{SUB,BSR}}(\mathrm{s}_1, \mathrm{s}_2, \mathrm{c})$. ∎

Now let $\mathrm{k}_l(x, y) =_{\mathrm{df}} \mathrm{k}^{[|y|^l]}(\mathrm{x})$ for $l \in \mathbb{N}$.

10.9. Lemma. If there is a $d \in \mathbb{N}$ such that $|\mathrm{k}^{[z]}(x)| \leqq_{\mathrm{ae}} d \cdot |x|$ for all $z \in \mathbb{N}$, then $\mathrm{k}_l \in \Gamma_{\mathrm{SUB,BSR}}(\mathrm{s}_1, \mathrm{s}_2, \mathrm{c})$ for all $l \in \mathbb{N}$.

Proof. By induction on l.

For $l = 0$ we have $\mathrm{k}_0(x, y) = \mathrm{k}(x)$. By the preceding lemma we get

$$\mathrm{k}_0 \in \Gamma_{\mathrm{SUB,BSR}}(\mathrm{s}_1, \mathrm{s}_2, \mathrm{c}).$$

Induction step. Let $\mathrm{t}_l(x, y, z) =_{\mathrm{df}} \mathrm{k}^{[|z|\cdot|y|^l]}(x)$. Then

$$\mathrm{t}_l(x, y, \mathrm{e}) = x,$$

$$\mathrm{t}_l(x, y, zi) = \mathrm{k}^{[|y|^l]}\big(\mathrm{k}^{[|z|\cdot|y|^l]}(x)\big) = \mathrm{k}_l\big(\mathrm{t}_l(x, y, z)\big),$$

and from $|\mathrm{t}_l(x, y, z)| = |\mathrm{k}^{[|z|\cdot|x|^l]}(x)| \leqq d \cdot |x| + a$ for some $a \in \mathbb{N}$ we conclude

$$|\mathrm{t}_l(x, y, z)| \leqq |\underbrace{xx\ldots x}_{d\text{ times}}\ \underbrace{11\ldots1}_{a\text{ times}}| \quad \text{and} \quad \mathrm{t}_l(x, y, z) \leqq \underbrace{xx\ldots x}_{d\text{ times}}\ \underbrace{11\ldots1}_{(a+1)\text{times}}.$$

Therefore t_l and (because of $\mathrm{k}_{l+1}(x, y) = \mathrm{t}_l(x, y, y)$) k_{l+1} are in $\Gamma_{\mathrm{SUB,BSR}}(\mathrm{s}_1, \mathrm{s}_2, \mathrm{c})$. ∎

The following theorem is (like the preceding two lemmas) due to D. B. Thompson.

10.10. Theorem. [Tho 72] DSPACE-TIME(Lin, Pol) $= \Gamma_{\mathrm{SUB,BSR}}(\mathrm{s}_1, \mathrm{s}_2, \mathrm{c})$.

Proof. 1. Let M be a DTM which computes the function $f\colon \mathbb{N}^s \mapsto \mathbb{N}$ within space $d \cdot n$ and time n^l for some $d, l \geqq 1$. Then there exists a DTM M' such that

- M' has the worktape alphabet $\{1, 2\}$ and the states $\{1, 2, \ldots, q\}$ where 1 is the initial state and 2 is the (only) final state,
- M' has the instruction $22 \to 220$,
- M' converts the initial situation with the worktape contents $\ldots 11\overset{\downarrow}{1}w\,111\ldots$ ($w \in \{1, 2\}^*$) into the final situation $\ldots 111\,\overset{\downarrow}{2}\,\overbrace{f(x_1, \ldots, x_s)}\,21v111\ldots$ by simulating M if $w = \tilde{x}_1 * \ldots * \tilde{x}_s * v$ for some $v \in \{1, 2\}^*$, and into the final situation $\ldots 111\overset{\downarrow}{2}w111\ldots$ otherwise (the head positions are marked by an arrow).

If $w = \tilde{x}_1 * \ldots * \tilde{x}_s * \mathrm{v}$, then M' has to encode the worktape alphabet of M and needs therefore $d_1 n + a_1$ tape squares and $d_2 \cdot n^2 + a_2$ steps for the encoding of the initial situation and $d_3 \cdot n + a_3$ tape squares and $d_4 \cdot n^{l_1} + a_4$ steps for the simulation (for some $d_1, d_2, d_3, d_4, a_1, a_2, a_3, a_4, l_1 \in \mathbb{N}$ and $n = |w|$). Otherwise M' needs $d_5 \cdot n + a_5$ tape squares and $2(d_5 \cdot n + a_5)$ steps. Consequently, M' needs $d_6 \cdot n + a_6$ tape squares and $d_7 \cdot n^{l_2} + a_7$ steps for all w such that $|w| = n$ (for some $d_6, a_6, d_7, l_2 \in \mathbb{N}$). If k describes the local behaviour of M, we obtain

$$\begin{aligned} f(x_1, \ldots, x_s) &= \mathrm{b}_7^{[2]}\Big(\mathrm{b}_8^{[3]}\Big(\mathrm{k}^{[n^{l_2}+a_7]}(\tilde{1} * \tilde{1} * \tilde{1} * \overbrace{\tilde{x}_1 * \ldots * \tilde{x}_s * *})\Big)\Big) \\ &= \mathrm{b}_7^{[2]}\Big(\mathrm{b}_8^{[3]}\Big(\mathrm{k}^{[a_7]}\Big(\mathrm{k}^{[|x_1*\ldots*x_s*|^{l_2}]}(\tilde{1} * \tilde{1} * \tilde{1} * \overbrace{\tilde{x}_1 * \ldots * \tilde{x}_s * *})\Big)\Big)\Big) \\ &= \mathrm{b}_7^{[2]}\Big(\mathrm{b}_8^{[3]}\Big(\mathrm{k}^{[a_7]}\Big(\mathrm{k}_{l_2}(\tilde{1} * \tilde{1} * \tilde{1} * \overbrace{\tilde{x}_1 * \ldots * \tilde{x}_s * *}, \tilde{x}_1 * \ldots * \tilde{x}_s *)\Big)\Big)\Big). \end{aligned}$$

Since M' works within space $d_6 \cdot n + a_6$ we have $|\mathrm{k}^{[z]}(x)| \leqq_{\mathrm{ae}} (d_6 + 1) \cdot |x|$ for all $z \in \mathbb{N}$ whence $\mathrm{k}_{l_2} \in \Gamma_{\mathrm{SUB,BSR}}(\mathrm{s}_1, \mathrm{s}_2, \mathrm{c})$ follows (Lemma 10.9).

By Lemma 10.8 and Lemma 10.7.4 we have $f \in \Gamma_{\mathrm{SUB,BSR}}(\mathrm{s}_1, \mathrm{s}_2, \mathrm{c})$.

2. The other inclusion can easily be shown by induction on the stages of the generation of the functions in $\Gamma_{\mathrm{SUB,BSR}}(\mathrm{s}_1, \mathrm{s}_2, \mathrm{c})$. ∎

Now we can characterize for all "well-behaved" functions t the class DSPACE-TIME(Lin t, Pol t) by means of bounded syntactic recursion. The Statement 10.11.1 can be found in a similar form in [Matr 76].

10.11. Theorem.

1. For $t \geqq \mathrm{id}$ such that $t \in$ DSPACE-TIME(Lin $|t(2^n)|$, Pol $|t(2^n)|$),

$$\text{DSPACE-TIME}\big(\text{Lin } |t(2^n)|, \text{Pol } |t(2^n)|\big) = \big\{g\big(\bar{x}, t(2^{|\bar{x}|})\big) : g \in \Gamma_{\mathrm{SUB,BSR}}(\mathrm{s}_1, \mathrm{s}_2, \mathrm{c})\big\}.^{*)}$$

2. For $t \geqq \mathrm{id}$ such that $t \in$ DSPACE-TIME(Lin $|t(2^{\mathrm{Lin}})|$, Pol $|t(2^{\mathrm{Lin}})|$),

$$\text{DSPACE-TIME}\big(\text{Lin } |t(2^{\mathrm{Lin}})|, \text{Pol } |t(2^{\mathrm{Lin}})|\big) = \Big\{g\Big(\bar{x}, t\big(h(\bar{x})\big)\Big) : g, h \in \Gamma_{\mathrm{SUB,BSR}}(\mathrm{s}_1, \mathrm{s}_2, \mathrm{c})\Big\}.$$

Proof. Because of 10.10 it is sufficient to show the equalities with DSPACE-TIME (Lin, Pol) instead of $\Gamma_{\mathrm{SUB,BSR}}(\mathrm{s}_1, \mathrm{s}_2, \mathrm{c})$. We show the inclusions "$\subseteq$". The other inclusions are obvious.

Let $f: \mathbb{N}^s \mapsto \mathbb{N}$ be a function which is computed by a DTM M within space $d \cdot |t(2^n)| + a$ and time $|t(2^n)|^l + a$ (space $d \cdot |t(2^{d' \cdot n})| + a$ and time $|t(2^{d' \cdot n})|^l + a$ for Statement 2). Define

$$g(\bar{x}, y) =_{\mathrm{df}} \begin{cases} f(\bar{x}) & \text{if } \text{T-DSPACE}_M(x_1 \square x_2 \square \dots \square x_s) \leqq d \cdot |y| + a \\ & \text{and } \text{T-DTIME}_M(x_1 \square x_2 \square \dots \square x_s) \leqq |y|^l + a, \\ 0 & \text{otherwise}. \end{cases}$$

Now it is easy to see that $g \in$ DSPACE-TIME(Lin, Pol) and $f(\bar{x}) = g\big(\bar{x}, t(2^{|\bar{x}|})\big)$ $\Big(f(\bar{x}) = g\Big(\bar{x}, t\big(h(\bar{x})\big)\Big)$ for Statement 2 where

$$h(x_1, x_2, \dots, x_s) = \underbrace{zz\dots z}_{d' \text{ times}} \quad \text{and} \quad z = 1x_1 1x_2 1 \dots 1x_s\Big). \ ∎$$

Now we give two applications of 10.11.

10.12. Corollary. For $k \geqq 1$,

1. DSPACE-TIME(Lin n^k, Pol) $= \{g(\bar{x}, 2^{|\bar{x}|^k}) : g \in \Gamma_{\mathrm{SUB,BSR}}(\mathrm{s}_1, \mathrm{s}_2, \mathrm{c})\}$.

---

*) $|\bar{x}|$ is an abbrevation of $|x_1 \square x_2 \square \dots \square x_s|$.

$$2.\ \mathrm{DTIME}\Big(2^{2^{\cdot^{\cdot^{2^{\mathrm{Lin}}}}}}\Big\}k\Big) = \mathrm{DSPACE\text{-}TIME}\Big(\mathrm{Lin}\ 2^{2^{\cdot^{\cdot^{2^{\mathrm{Lin}}}}}}\Big\}k,\ \mathrm{Pol}\ 2^{2^{\cdot^{\cdot^{2^{\mathrm{Lin}}}}}}\Big\}k\Big)$$
$$= \Big\{g\Big(\bar{x}, 2^{2^{\cdot^{\cdot^{2^{h(|\bar{x}|)}}}}}\Big\}k\Big) : g, h \in \Gamma_{\mathrm{SUB,BSR}}(\mathrm{s}_1, \mathrm{s}_2, \mathrm{c})\Big\}$$
$$= \Big\{g\Big(x, 2^{2^{\cdot^{\cdot^{2^{r\cdot|\bar{x}|}}}}}\Big\}k+1\Big) : r \in \mathbb{N} \wedge g \in \Gamma_{\mathrm{SUB,BSR}}(\mathrm{s}_1, \mathrm{s}_2, \mathrm{c})\Big\}. \ \blacksquare$$

The characterization of the complexity classes in 10.11 is possible because the substitution of the function t, which is responsible for the growth of all functions in the generated class, is allowed only once. If we allow the unrestricted use of the substitution, then, in general, complexity classes are generated which are bounded by a family of the form $\Gamma_{\mathrm{SUB}}(t)$. This is shown by the next theorem.

10.13. Theorem. [Wag 79] If t is an increasing function such that $t(x) \geqq x^r$ for some $r > 1$ and $t \in \mathrm{DSPACE\text{-}TIME}\big(\Gamma_{\mathrm{SUB}}(|t(2^n)|), \mathrm{Pol}\ \Gamma_{\mathrm{SUB}}(|t(2^n)|)\big)$, then

$$\begin{aligned}
&\mathrm{DSPACE\text{-}TIME}\big(\Gamma_{\mathrm{SUB}}(|t(2^n)|), \mathrm{Pol}\ \Gamma_{\mathrm{SUB}}(|t(2^n)|)\big)\\
&\qquad = \Gamma_{\mathrm{SUB,BSR}}(\mathrm{s}_1, \mathrm{s}_2, \mathrm{m}, t)\\
&\qquad = \Gamma_{\mathrm{SUB,BSR}}(\mathrm{s}, \mathrm{m}, 2\cdot\mathrm{x}, t)\\
&\qquad = \Gamma_{\mathrm{SUB,BSR}}(\mathrm{s}, +, t)\\
&\qquad = \Gamma_{\mathrm{SUB,WBPR}}\Big(\mathrm{s}, \mathrm{m}, 2\cdot x, \Big\lfloor\frac{x}{2}\Big\rfloor, \dot{-}, t\Big)\\
&\qquad = \Gamma_{\mathrm{SUB,WBPR}}\Big(\mathrm{s}, \Big\lfloor\frac{x}{2}\Big\rfloor, x, \dot{-}, t\Big).
\end{aligned}$$

Proof. We conclude

$$\begin{aligned}
&\mathrm{DSPACE\text{-}TIME}\big(\Gamma_{\mathrm{SUB}}(|t(2^n)|), \mathrm{Pol}\ \Gamma_{\mathrm{SUB}}(|t(2^n)|)\big)\\
&= \bigcup_{k\geqq 0} \mathrm{DSPACE\text{-}TIME}\big(|\hat{\mathrm{t}}^{[k]}(2^n)|, \mathrm{Pol}\ |\hat{\mathrm{t}}^{[k]}(2^n)|\big)\\
&\quad \text{(where } \hat{\mathrm{t}}(n) = \mathrm{t}(n) + 1, \text{ cf. second footnote on p. 180)}\\
&\subseteqq \bigcup_{k\geqq 0} \big\{g\big(\bar{x}, \hat{\mathrm{t}}^{[k]}(2^{|\bar{x}|})\big) : g \in \Gamma_{\mathrm{SUB,BSR}}(\mathrm{s}_1, \mathrm{s}_2, \mathrm{c})\big\} \text{ (by 10.11.1)}\\
&\subseteqq \Gamma_{\mathrm{SUB,BSR}}(\mathrm{s}_1, \mathrm{s}_2, \mathrm{c}, 2^{|\bar{x}|}, \mathrm{t}, \mathrm{s})\\
&\subseteqq \Gamma_{\mathrm{SUB,BSR}}(\mathrm{s}_1, \mathrm{s}_2, \mathrm{c}, t) \quad \text{(by 10.10)}\\
&\subseteqq \Gamma_{\mathrm{SUB,BSR}}(\mathrm{s}, +, \mathrm{x}^2, t) \quad \text{(by 10.7.1 and 10.7.2)}\\
&\subseteqq \Gamma_{\mathrm{SUB,BSR}}(\mathrm{s}, +, t)\\
&\quad \text{(since } \mathrm{e}^2 = 0,\ (x1)^2 = 4x^2 + 4x + 1,\ (x2)^2 = 4x^2 + 8x + 4\\
&\quad \text{and since } \mathrm{t}^{[k]}(x) \geqq x^2 \text{ for some } k \geqq 1)\\
&\subseteqq \Gamma_{\mathrm{SUB,WBPR}}\Big(\mathrm{s}, \mathrm{m}, 2\cdot x, \Big\lfloor\frac{x}{2}\Big\rfloor, \dot{-}, t\Big) \quad \text{(by 10.2.2 and 10.7.3)}\\
&\subseteqq \Gamma_{\mathrm{SUB,WBPR}}\Big(\mathrm{s}, \Big\lfloor\frac{x}{2}\Big\rfloor, +, \dot{-}, t\Big) \quad \text{(by 10.7.3)}
\end{aligned}$$

and

$$\begin{aligned}\Gamma_{\mathrm{SUB,BSR}}(\mathrm{s}_1, \mathrm{s}_2, \mathrm{c}, t) &\subseteq \Gamma_{\mathrm{SUB,BSR}}(\mathrm{s}_1, \mathrm{s}_2, \mathrm{m}, t)\\ &\quad (\text{since } \mathrm{c}(x, \mathrm{e}) = x,\ \mathrm{c}(x, y1) = \mathrm{s}_1(\mathrm{c}(x, y)),\\ &\quad \mathrm{c}(x, y2) = \mathrm{s}_2(\mathrm{c}(x, y)),\\ &\quad \mathrm{c}(x, y) = 2^{|y|} \cdot x + y \leqq (y + 1) \cdot x + y\\ &\quad \leqq (y + 1) \cdot (x + 1) \leqq (y1) \cdot (x1) \leqq \mathrm{m}(x1, y1)^2\\ &\quad \text{and since } x^2 \leqq t^{[k]}(x) \text{ for some } k \geqq 1)\\ &\subseteq \Gamma_{\mathrm{SUB,BSR}}(\mathrm{s}, \mathrm{m}, 2 \cdot x, t) \quad (\text{by 10.7.1})\\ &\subseteq \Gamma_{\mathrm{SUB,WBPR}}\left(\mathrm{s}, \mathrm{m}, 2 \cdot x, \left\lfloor \frac{x}{2} \right\rfloor, \dot{-}, t\right) \quad (\text{by 10.2.2}).\end{aligned}$$

The remaining inclusion

$$\Gamma_{\mathrm{SUB,WBPR}}\left(\mathrm{s}, \left\lfloor \frac{x}{2} \right\rfloor, x, \dot{-}, t\right)$$
$$\subseteq \text{DSPACE-TIME}\left(\Gamma_{\mathrm{SUB}}(|t(2^n)|), \text{Pol } \Gamma_{\mathrm{SUB}}(|t(2^n)|)\right)$$

can easily be shown by induction on the stages of the generation of the functions in $\Gamma_{\mathrm{SUB,WBPR}}\left(\mathrm{s}, \left\lfloor \frac{x}{2} \right\rfloor, +, \dot{-}, t\right)$. ∎

We now give some applications of 10.13.

10.14. Corollary.

1. (Supplement to 10.10)

$$\begin{aligned}\text{DSPACE-TIME(Lin, Pol)} &= \Gamma_{\mathrm{SUB,BSR}}(\mathrm{s}_1, \mathrm{s}_2, \mathrm{m}, x^2)\\ &= \Gamma_{\mathrm{SUB,BSR}}(\mathrm{s}, \mathrm{m}, 2 \cdot x, x^2)\\ &= \Gamma_{\mathrm{SUB,BSR}}(\mathrm{s}, +, x^2)\\ &= \Gamma_{\mathrm{SUB,WBPR}}\left(\mathrm{s}, \mathrm{m}, 2 \cdot x, \left\lfloor \frac{x}{2} \right\rfloor, \dot{-}, x^2\right)\\ &= \Gamma_{\mathrm{SUB,WBPR}}\left(\mathrm{s}, \left\lfloor \frac{x}{2} \right\rfloor, +, \dot{-}, x^2\right)\end{aligned}$$

2. (includes a result of [Cob 64])

$$\begin{aligned}\mathrm{P} = \text{DSPACE-TIME(Pol, Pol)} &= \Gamma_{\mathrm{SUB,BSR}}(\mathrm{s}_1, \mathrm{s}_2, \mathrm{m}, x^{|x|})\\ &= \Gamma_{\mathrm{SUB,BSR}}(\mathrm{s}, \mathrm{m}, 2 \cdot x, x^{|x|})\\ &= \Gamma_{\mathrm{SUB,BSR}}(\mathrm{s}, +, x^{|x|})\\ &= \Gamma_{\mathrm{SUB,WBPR}}\left(\mathrm{s}, \mathrm{m}, 2 \cdot x, \left\lfloor \frac{x}{2} \right\rfloor, \dot{-}, x^{|x|}\right)\\ &= \Gamma_{\mathrm{SUB,WBPR}}\left(\mathrm{s}, \left\lfloor \frac{x}{2} \right\rfloor, +, \dot{-}, x^{|x|}\right)\end{aligned}$$

3. Let $|x|_k =_{\mathrm{df}} \underbrace{\|\ldots |x| \ldots \|}_{k \text{ times}}$. For $k \geqq 0$,

$$\begin{aligned}
&\mathrm{DSPACE\text{-}TIME}(n \cdot \mathrm{Pol}(\log^{[k]} n), \mathrm{Pol}) \\
&= \Gamma_{\mathrm{SUB,BSR}}(\mathrm{s}_1, \mathrm{s}_2, \mathrm{m}, x^{|x|_{k+1}}) \\
&= \Gamma_{\mathrm{SUB,BSR}}(\mathrm{s}, \mathrm{m}, 2 \cdot x, x^{|x|_{k+1}}) \\
&= \Gamma_{\mathrm{SUB,BSR}}(\mathrm{s}, +, x^{|x|_{k+1}}) \\
&= \Gamma_{\mathrm{SUB,WBPR}}\left(\mathrm{s}, \mathrm{m}, 2 \cdot x, \left\lfloor \frac{x}{2} \right\rfloor, \dot{-}, x^{|x|_{k+1}}\right) \\
&= \Gamma_{\mathrm{SUB,WBPR}}\left(\mathrm{s}, \left\lfloor \frac{x}{2} \right\rfloor, +, \dot{-}, x^{|x|_{k+1}}\right). \blacksquare
\end{aligned}$$

In [Pak 79] for suitably restricted loosely connected classes $S, T \subseteq \mathbb{R}_1$ a generalization of weak bounded primitive recursion has been considered which allows the characterization of DSPACE-TIME(S, T). The special cases $T = S$ and $T = \{2^s : s \in S\}$ give characterizations of DTIME(T) and DSPACE(S), resp.

10.2.2. Bounded Primitive Recursion

In this subsection we show that some complexity classes of type DSPACE(Lin t) can be characterized recursion-theoretically using bounded primitive recursion. We proceed as in the previous subsection. Thus we first characterize the class DSPACE (Lin). Again we use the function k of Lemma 10.8 which describes the local behaviour of a Turing machine. Defining $\hat{\mathrm{k}}(x, y) =_{\mathrm{df}} \mathrm{k}^{[y]}(x)$ we get

10.15. Lemma. If there is a $d \geqq 1$ such that $|\mathrm{k}^{[y]}(x)| \leqq_{\mathrm{ae}} d \cdot |x|$ for all $y \in \mathbb{N}$, then $\hat{\mathrm{k}} \in \Gamma_{\mathrm{SUB,BPR}}(\mathrm{s}, \mathrm{m}, x^2)$.

Proof. Because of $\hat{\mathrm{k}}(x, 0) = x$, $\hat{\mathrm{k}}(x, y + 1) = \mathrm{k}(\hat{\mathrm{k}}(x, y))$ and since $|\mathrm{k}^{[y]}(x)| \leqq_{\mathrm{ae}} d \cdot |x|$ implies $\mathrm{k}^{[y]}(x) \leqq 2 \cdot 2^{|\mathrm{k}^{[y]}(x)|} \leqq 2^a \cdot 2^{d \cdot |x|} \leqq 2^a \cdot (x + 1)^d$ we have $\hat{\mathrm{k}} \in \Gamma_{\mathrm{SUB,BPR}}(\mathrm{s}, \mathrm{m}, 2 \cdot x, x^2, \mathrm{k})$.

By 10.8, 10.10 and 10.14.1 we have $\mathrm{k} \in \Gamma_{\mathrm{SUB,BSR}}(\mathrm{s}, \mathrm{m}, 2 \cdot x, x^2)$, by 10.2.2 and 10.3.1 we get $\mathrm{k} \in \Gamma_{\mathrm{SUB,BPR}}\left(\mathrm{s}, \mathrm{m}, 2 \cdot x, \left\lfloor \frac{x}{2} \right\rfloor, \dot{-}, x^2\right)$. Consequently, $\hat{\mathrm{k}} \in \Gamma_{\mathrm{SUB,BPR}}\left(\mathrm{s}, \mathrm{m}, 2 \cdot x, \left\lfloor \frac{x}{2} \right\rfloor, \dot{-}, x^2\right)$ and by 10.7.5 we get the result. ∎

10.16. Theorem. [Ritr 63a] DLINSPACE $= \Gamma_{\mathrm{SUB,BPR}}(\mathrm{s}, \mathrm{m}, x^2)$.

Proof. We proceed as in the proof of Theorem 10.10 except that, in the estimation of the necessary number of steps, n must be replaced by 2^n. Then we have

$$\begin{aligned}
\mathrm{f}(x_1, \ldots, x_s) &= \mathrm{b}_7^{[2]}\left(\mathrm{b}_8^{[3]}\left(\mathrm{k}^{(2^n)^{l_2} + a_7}(\tilde{1} * \tilde{1} * \tilde{1} * \overbrace{\tilde{x}_1 * \ldots * \tilde{x}_s * *})\right)\right) \\
&= \mathrm{b}_7^{[2]}\left(\mathrm{b}_8^{[3]}\left(\mathrm{k}^{(\tilde{x}_1 * \cdots * \tilde{x}_s * + 1)^{l_2} + a_7}(\tilde{1} * \tilde{1} * \tilde{1} * \overbrace{\tilde{x}_1 * \ldots * \tilde{x}_s * *})\right)\right) \\
&= \mathrm{b}_7^{[2]}\left(\mathrm{b}_8^{[3]}\left(\hat{\mathrm{k}}(\tilde{1} * \tilde{1} * \tilde{1} * \overbrace{\tilde{x}_1 * \ldots * \tilde{x}_s * *}), (\tilde{x}_1 * \ldots * \tilde{x}_s * + 1)^{l_2} + a_7\right)\right)
\end{aligned}$$

Since M' works within space $d_6 \cdot n + a_6$ we have $|\mathrm{k}^{[y]}(x)| \leqq_{\mathrm{ae}} (d_6 + 1) \cdot |x|$ for all $y \in \mathbb{N}$. Therefore $\hat{\mathrm{k}} \in \Gamma_{\mathrm{SUB,BPR}}(\mathrm{s}, \mathrm{m}, x^2)$ (Lemma 10.15) and $f \in \Gamma_{\mathrm{SUB,BPR}}(\mathrm{s}, \mathrm{s}_1, \mathrm{s}_2, \mathrm{m}, \tilde{\ }, \mathrm{b}_7, \mathrm{b}_8, \mathrm{c}, x^2)$. By Lemma 10.7.5 we get $f \in \Gamma_{\mathrm{SUB,BPR}}(\mathrm{s}, \mathrm{m}, x^2)$. ∎

Remark. Remember that $\mathscr{E}^2 =_{\mathrm{df}} \Gamma_{\mathrm{SUB,BPR}}(\mathrm{s}, \mathrm{m}, x^2)$. Thus we have the characterization $\mathscr{E}^2 = \mathrm{DLINSPACE}$. The remaining theorems of this subsection can be shown in a similar way to the corresponding theorems of the preceding subsection.

10.17. Theorem.

1. For $t \geqq \mathrm{id}$ such that $t \in \mathrm{DSPACE}\big(\mathrm{Lin}\ |t(2^n)|\big)$,

$$\mathrm{DSPACE}\big(\mathrm{Lin}\ |t(2^n)|\big) = \big\{g\big(\bar{x}, t(2^{|\bar{x}|})\big) : g \in \mathscr{E}^2\big\}.$$

2. For $t \geqq \mathrm{id}$ such that $t \in \mathrm{DSPACE}\big(\mathrm{Lin}\ |t(2^{\mathrm{Lin}})|\big)$,

$$\mathrm{DSPACE}\big(\mathrm{Lin}\ |t(2^{\mathrm{Lin}})|\big) = \Big\{g\Big(\bar{x}, t\big(h(\bar{x})\big)\Big) : g, h \in \mathscr{E}^2\Big\}. \ ∎$$

Now we give some applications of 10.17.

10.18. Corollary. For $k \geqq 1$,

1. $\mathrm{DSPACE}(\mathrm{Lin}\ n^k) = \{g(\bar{x}, 2^{|\bar{x}|^k}) : g \in \mathscr{E}^2\}$.

2. $$\mathrm{DSPACE}\Big(\underbrace{2^{2^{\cdot^{\cdot^{\cdot^{2^{\mathrm{Lin}}}}}}}}_{k}\Big) = \Big\{g\Big(\bar{x}, \underbrace{2^{2^{\cdot^{\cdot^{\cdot^{2^{h(\bar{x})}}}}}}}_{k}\Big) : g, h \in \mathscr{E}^2\Big\}$$
$$= \Big\{g\Big(\bar{x}, \underbrace{2^{2^{\cdot^{\cdot^{\cdot^{2^{r \cdot |\bar{x}|}}}}}}}_{k+1}\Big) : g \in \mathscr{E}^2,\ r \in \mathbb{N}\Big\}$$

3. $$\mathrm{DSPACE}\Big(\underbrace{2^{2^{\cdot^{\cdot^{\cdot^{2^{n+\mathrm{const}}}}}}}}_{k}\Big) = \Big\{g\Big(\bar{x}, \underbrace{2^{2^{\cdot^{\cdot^{\cdot^{2^{|\bar{x}|+r}}}}}}}_{k+1}\Big) : g \in \mathscr{E}^2, r \in \mathbb{N}\Big\}. \ ∎$$

Remark. By 10.18.3 the *Ritchie classes* F_k are characterized, since

$$\mathrm{F}_k =_{\mathrm{df}} \mathrm{DSPACE}\Big(\underbrace{2^{2^{\cdot^{\cdot^{\cdot^{2^{n+\mathrm{const}}}}}}}}_{k}\Big)$$

(cf. [Ritr 63a] and [Her 71]).

10.19. Theorem. [Wag 79] If t is an increasing function such that $t(x) \geqq_{\mathrm{ae}} x^r$ for some $r > 1$ and $t \in \mathrm{DSPACE}\Big(\Gamma_{\mathrm{SUB}}\big(|t(2^n)|\big)\Big)$, then

$$\mathrm{DSPACE}\Big(\Gamma_{\mathrm{SUB}}\big(|t(2^n)|\big)\Big) = \Gamma_{\mathrm{SUB,BPR}}(\mathrm{s}, \mathrm{m}, t) = \Gamma_{\mathrm{SUB}}(\mathscr{E}^2 \cup \{t\}). \ ∎$$

Now we give some applications of 10.19.

10.20. Corollay.

1. [Tho 72] $\mathrm{PSPACE} = \Gamma_{\mathrm{SUB,BPR}}(\mathrm{s}, \mathrm{m}, x^{|x|})$.
2. $\mathrm{DSPACE}\big(n \cdot \mathrm{Pol}(\log^{[k]} n)\big) = \Gamma_{\mathrm{SUB,BPR}}(\mathrm{s}, \mathrm{m}, x^{|x|_{k+1}})$ for $k \geqq 1$.
3. $\mathrm{DSPACE}\big(\Gamma_{\mathrm{SUB}}(\mathrm{a}_k)\big) = \mathrm{DSPACE}(\mathscr{E}_1^k) = \mathscr{E}^k = \Gamma_{\mathrm{SUB}}(\mathscr{E}^2 \cup \{\mathrm{a}_k\})$ for $k \geqq 3$.

Proof of Statement 3. By 10.19 we have $\mathrm{DSPACE}\Big(\Gamma_{\mathrm{SUB}}\big(|\mathrm{a}_k(2^n)|\big)\Big) = \mathscr{E}^k = \Gamma_{\mathrm{SUB}}(\mathscr{E}^2 \cup \{\mathrm{a}_k\})$. The functions from $\Gamma_{\mathrm{SUB}}\big(|\mathrm{a}_k(2^n)|\big)$ can be bounded above by functions from $\Gamma_{\mathrm{SUB}}(\mathrm{a}_k)$ and vice versa. Consequently $\mathscr{E}^k = \mathrm{DSPACE}\big(\Gamma_{\mathrm{SUB}}(\mathrm{a}_k)\big)$. By 2.19.2 we get $\mathscr{E}_k = \mathrm{DSPACE}(\mathscr{E}_1^k)$. ∎

Remark. In [Mar 80] it is shown that $\mathscr{E} = \mathscr{E}^3 = \Gamma_{\mathrm{SUB}}\left(x+1, \left\lfloor \frac{x}{y} \right\rfloor, x^y, \varphi(x,y)\right)$ where $\varphi(x, y)$ is the greatest r such that x^r is a divisor of y if $x > 1$ and 0 otherwise. For $\tau(x) = \varphi(2, x)$ it is known $\mathscr{E} = \Gamma_{\mathrm{SUB}}\left(x+1, \left\lfloor \frac{x}{y} \right\rfloor, x^y, \tau(x)\right)$ ([Mar 84]).

The restriction of $\mathscr{E}$ to 0-1-valued functions coincides with the restriction of the following classes to 0-1-valued functions:

$$\Gamma_{\mathrm{SUB}}\left(x+1, x \mathbin{\dot-} y, \left\lfloor \frac{x}{y} \right\rfloor, x^y\right) \quad \text{[see Mar 84]),}$$

$$\Gamma_{\mathrm{SUB}}\left(x+y, x \mathbin{\dot-} 1, \left\lfloor \frac{x}{y} \right\rfloor, \lfloor\sqrt{x}\rfloor, 2^x\right), \qquad \Gamma_{\mathrm{SUB}}\left(x \mathbin{\dot-} y, \left\lfloor \frac{x}{y} \right\rfloor, \lfloor\sqrt{x}\rfloor, 2^{x+y}\right).$$

$$\Gamma_{\mathrm{SUB}}\left(x+y, x \mathbin{\dot-} y, \left\lfloor \frac{x}{y} \right\rfloor, x!, 2^x\right) \quad \text{(see [JoMa 82]).}$$

10.2.3. Bounded Recursion by Two Values

In this subsection we show that some complexity classes of the special type

$$\mathrm{DTIME}(2^{\mathrm{Lin}\,t}, \mathrm{Lin}\,t) =_{\mathrm{df}} \{f : f \in \mathrm{DTIME}(2^{\mathrm{Lin}\,t}) \wedge |f(x)| \lesssim t(|x|)\}$$

can be characterized recursion-theoretically using bounded recursion by two values. Again we proceed as in § 10.2.1. Thus we characterize first the class DTIME(2^{Lin}, Lin).

To this end it is necessary to describe the local behaviour of a Turing machine, i.e. the step from a given situation to the next situation by means of bounded recursion by two values (the function k of § 10.2.1 cannot be used here since it can give values which are too large).

Let M be a DTM whose head (starting in tape square 1 for the first step) moves in the special manner described in Figure 13.1, and which has the worktape alphabet $\{0, 1, \dots, l-1\}$ and the states $0, 1, \dots, q-1$. The input words are written on the worktape in the tape squares $1, 2, \dots, n$. The symbol 0 acts as the empty symbol. Let $h(t) =_{\mathrm{df}}$ the number of the tape square which is scanned in the tth step, $z(y, t) =_{\mathrm{df}}$ the state in which M will be on completion of the tth step of its work on the input y and $a(y, t) =_{\mathrm{df}}$ the symbol written by M (into the tape square $h(t)$) in the tth step of its work on the input y.

10.21. Lemma. $h, z, a \in \Gamma_{\mathrm{SUB,BVR}}(\mathrm{s}, \mathrm{m}, x^2)$.

Proof. It is not hard to see that $h(t) = \mathrm{abs}\left(\lfloor\sqrt{t \mathbin{\dot-} 1}\rfloor - \left((t \mathbin{\dot-} 1) \mathbin{\dot-} \left(\lfloor\sqrt{t \mathbin{\dot-} 1}\rfloor\right)^2\right)\right) + 1$. By Lemma 10.7.5 we have $h \in \Gamma_{\mathrm{SUB,BVR}}(\mathrm{s}, \mathrm{m}, x^2)$. Now we define the auxiliary functions p and p':

$$p(t) =_{\mathrm{df}} \sup\{t' : t' < t \wedge h(t') = h(t)\},$$
$$p'(t, t') =_{\mathrm{df}} \sup\{t'' : t'' \leqq t' \wedge h(t'') = h(t)\}.$$

Because of $p(t) = p'(t, t \dot{-} 1)$ and

$$p'(t, 0) = 0,$$
$$p'(t, t' + 1) = p'(t, t') \cdot \operatorname{sgn} \operatorname{abs} \big(h(t) - h(t' + 1)\big) + (t' + 1) \cdot \overline{\operatorname{sgn}} \operatorname{abs} \big(h(t) - h(t' + 1)\big),$$
$$p'(t, t') \leqq t',$$

and by Lemma 10.7.5 we get $p', p \in \Gamma_{\mathrm{SUB,BVR}}(\mathrm{s}, \mathrm{m}, x^2)$.

Now let 0 be the initial state of M and let $ji \to s_{ji}\mathrm{a}_{ji}\sigma_{ji}$ be the instructions of M ($j = 0, 1, \ldots, q - 1$, $i = 0, 1, \ldots, l - 1$). Then we have

$$z(y, 0) = 0, \qquad a(y, 0) = \mathrm{b}_9'(y, 1)$$
$$z(y, t + 1) = s_{ji} \quad \text{if} \quad \big(p(t + 1) > 0 \wedge j = z(y, t) \wedge i = a(y, p(t + 1))\big) \text{ or } \big(p(t + 1) = 0 \wedge j = z(y, t) \wedge i = \mathrm{b}_9'(y, h(t + 1))\big)$$
$$a(y, t + 1) = a_{ji} \quad \text{if} \quad \big(p(t + 1) > 0 \wedge j = z(y, t) \wedge i = a(y, p(t + 1))\big) \text{ or } \big(p(t + 1) = 0 \wedge j = z(y, t) \wedge i = \mathrm{b}_9'(y, h(t + 1))\big).$$

Thus $z(y, t + 1)$ and $a(y, t + 1)$ are expressed by $2 \cdot q \cdot l$ mutually exclusive cases $\big(\operatorname{sgn} p(t + 1) = 0, 1,\ j = 0, 1, \ldots, q - 1,\ i = 0, 1, \ldots, l - 1\big)$. Using the function $\mathrm{v}_{2 \cdot q \cdot l}$ we get, by Lemma 10.7.5, functions $g_1, g_2 \in \Gamma_{\mathrm{SUB,BVR}}(\mathrm{s}, \mathrm{m}, x^2)$ such that

$$z(y, t + 1) = g_1\big(y, t, z(y, t), a(y, p(y, t + 1))\big)$$

and

$$a(y, t + 1) = g_2\big(y, t, z(y, t), a(y, p(y, t + 1))\big).$$

In order to avoid this simultaneous recursion of two functions we introduce the function $r(y, t) =_{\mathrm{df}} \widetilde{z(y, t)} * \widetilde{a(y, t)} *$ for which we get $z(y, t) = \mathrm{b}_7\big(r(y, t)\big)$ and $a(y, t) = \mathrm{b}_7\Big(\mathrm{b}_8\big(r(y, t)\big)\Big)$. By Lemma 10.7.5 we have $r \in \Gamma_{\mathrm{SUB,BVR}}(\mathrm{s}, \mathrm{m}, x^2)$ and consequently $z, a \in \Gamma_{\mathrm{SUB,BVR}}(\mathrm{s}, \mathrm{m}, x^2)$. ∎

By the preceding lemma, which is due to B. Monien, we are able to characterize DTIME(2^{Lin}, Lin).

10.22. Theorem. [Mon 77a] $\mathrm{DTIME}(2^{\mathrm{Lin}}, \mathrm{Lin}) = \Gamma_{\mathrm{SUB,BVR}}(\mathrm{s}, \mathrm{m}, x^2)$.

Proof. Let $f: \mathbb{N}^s \mapsto \mathbb{N}$ be computed by a DTM within time $2^{d \cdot n}$ and let $|f(\bar{x})| \leqq d_1 \cdot |\bar{x}| + a_1$ for some $d_1, a_1 \in \mathbb{N}$. Then there is a DTM M such that

- M has the worktape alphabet $\{0, 1, \ldots, l - 1\}$, 0 being the empty symbol, and the states $0, 1, \ldots, q$, 0 being the initial state,
- the head of M moves ad infinitum in the special manner described in Figure 13.1.
- if the input $\tilde{x}_1 * \tilde{x}_2 * \ldots * \tilde{x}_s$ is written in the tape squares 1, 2, 3, ... and M starts in tape square 1, then after $2^{d_2 \cdot |\bar{x}|} + a_2$ steps (for some $a_2, d_2 \in \mathbb{N}$) $f(\bar{x})$ is written in the tape squares $1, 2, \ldots, |f(\bar{x})|$ and will never be changed during the further work of M. The other tape squares contain the symbol 0.

Since $h(t^2 + t + 1) = 1$ for all $t \in \mathbb{N}$, the head of M scans tape square 1 during the step $\left((2^{d_2 \cdot |x|} + a_2)^2 + 2^{d_2 \cdot |x|} + a_2 + 1\right)_{\mathrm{df}} = k$, and we get

$$f(\bar{x}) = a(\tilde{x}_1 * \ldots * \tilde{x}_s, k)\, a(\tilde{x}_1 * \ldots * \tilde{x}_s, k+1) \ldots$$
$$\ldots a(\tilde{x}_1 * \ldots * \tilde{x}_s, k + d_1 \cdot |\bar{x}| + a_1).$$

Defining $g(y, t, z) =_{\mathrm{df}} a(y, t)\, a(y, t+1) \ldots a(y, t + |z|)$ we get

$$f(\bar{x}) = g\left(\tilde{x}_1 * \ldots * \tilde{x}_s, \left((2^{|\bar{x}|})^{d_2} + a_2\right)^2 + (2^{|\bar{x}|})^{d_2} + a_2 + 1, (2^{|\bar{x}|})^{d_1} \cdot 2^{a_1}\right)$$

whence $f \in \Gamma_{\mathrm{SUB,BVR}}(\mathrm{s}, \mathrm{m}, x^2, g)$ follows by Lemma 10.7.5. Therefore it remains to show that $g \in \Gamma_{\mathrm{SUB,BVR}}(\mathrm{s}, \mathrm{m}, x^2)$. Because of

$$g(y, t, e) = a(y, t),$$
$$g(y, t, zi) = g(y, t, z)\, a(y, t + |z| + 1) \quad \text{for} \quad i = 1, 2$$

and since $|g(y, t, z)| \leqq |z| + 1$ implies $g(y, t, z) \leqq 4 \cdot (z + 1)$ we have $g \in \Gamma_{\mathrm{SUB,BSR}}(a, c, s, +, |z|)$. Finally by 10.2.2, 10.3.1, 10.4.1, 10.7.5 and 10.21 we get $g \in \Gamma_{\mathrm{SUB,BVR}}(\mathrm{s}, \mathrm{m}, x^2)$. The inclusion "$\supseteq$" can be shown by induction on the stages of the generation of the functions in $\Gamma_{\mathrm{SUB,BVR}}(\mathrm{s}, \mathrm{m}, x^2)$. ∎

The remaining theorems of this subsection can be shown in the same manner as the corresponding theorems of § 10.2.1.

10.23. Theorem.

1. For $t \geqq \mathrm{id}$ such that $t \in \mathrm{DTIME}(2^{\mathrm{Lin}\,|t(2^n)|}, \mathrm{Lin}\,|t(2^n)|)$,

$$\mathrm{DTIME}(2^{\mathrm{Lin}\,|t(2^n)|}, \mathrm{Lin}\,|t(2^n)|) = \left\{g(\bar{x}, t(2^{|\bar{x}|})) : g \in \Gamma_{\mathrm{SUB,BVR}}(\mathrm{s}, \mathrm{m}, x^2)\right\}.$$

2. For $t \geqq \mathrm{id}$ such that $t \in \mathrm{DTIME}(2^{\mathrm{Lin}|t(2^{\mathrm{Lin}})|}, \mathrm{Lin}\,|t(2^{\mathrm{Lin}})|)$,

$$\mathrm{DTIME}(2^{\mathrm{Lin}|t(2^{\mathrm{Lin}})|}, \mathrm{Lin}\,|t(2^{\mathrm{Lin}})|)$$
$$= \left\{g(\bar{x}, t(h(\bar{x}))) : g, h \in \Gamma_{\mathrm{SUB,BVR}}(\mathrm{s}, \mathrm{m}, x^2)\right\}. \ ∎$$

Now we give some applications of 10.23.

10.24. Corollary. For $k \geqq 1$,

1. $\mathrm{DTIME}(2^{\mathrm{Lin}\, n^k}, \mathrm{Lin}\, n^k) = \{g(\bar{x}, 2^{|\bar{x}|^k}) : g \in \Gamma_{\mathrm{SUB,BVR}}(\mathrm{s}, \mathrm{m}, x^2)\}$.
2. $\mathrm{DTIME}\left(2^{2^{\cdot^{\cdot^{2^{\mathrm{Lin}}}}}}\Big\}k+1, 2^{2^{\cdot^{\cdot^{2^{\mathrm{Lin}}}}}}\Big\}k\right) = \left\{g\left(\bar{x}, 2^{2^{\cdot^{\cdot^{2^{h(\bar{x})}}}}}\Big\}k\right) : g, h \in \Gamma_{\mathrm{SUB,BVR}}(\mathrm{s}, \mathrm{m}, x^2)\right\}$

$$= \left\{g\left(\bar{x}, 2^{2^{\cdot^{\cdot^{2^{r \cdot |\bar{x}|}}}}}\Big\}k+1\right) : g \in \Gamma_{\mathrm{SUB,BVR}}(\mathrm{s}, \mathrm{m}, x^2), r \in \mathbb{N}\right\}. \ ∎$$

10.25. Theorem. [Wag 79] If f is an increasing function such that $t(x) \geqq_{\mathrm{ae}} x^r$ for some $r > 1$ and $t \in \mathrm{DTIME}\left(2^{\Gamma_{\mathrm{SUB}}(|t(2^n)|)}, \Gamma_{\mathrm{SUB}}(|t(2^n)|)\right)$, then

$$\mathrm{DTIME}\left(2^{\Gamma_{\mathrm{SUB}}(|t(2^n)|)}, \Gamma_{\mathrm{SUB}}(|t(2^n)|)\right) = \Gamma_{\mathrm{SUB,BVR}}(\mathrm{s}, \mathrm{m}, t). \ ∎$$

Now we give two applications of 10.25.

10.26. Corollary.

1. $\mathrm{DTIME}(2^{n \cdot \mathrm{Pol} \log n}, n \cdot \mathrm{Pol} \log n) = \Gamma_{\mathrm{SUB,BVR}}(\mathrm{s}, \mathrm{m}, x^{\|x\|})$.
2. $\mathrm{DTIME}(2^{\mathrm{Pol}}, \mathrm{Pol}) = \Gamma_{\mathrm{SUB,BVR}}(\mathrm{s}, \mathrm{m}, x^{|x|})$. ∎

10.2.4. Further Results

First we deal with weak sum and weak product.

10.27. Theorem. [Con 73] $\mathrm{P} = \Gamma_{\mathrm{SUB,WSUM,WPRD}}\left(\mathrm{s}, +, \dot{-}, \cdot, \left\lfloor \frac{x}{y} \right\rfloor\right)$. ∎

Remark. In the light of Theorem 10.27 the class P appears as the "weak analogue" of the class of elementary functions. Further studies could show that the operators WSUM and WPRD are together as powerful as WBPR or BSR, i.e. that results analogous to those of § 10.2.1 could hold true. However this is an open problem. Further promising directions of research could be the investigation of WBVR and the application of S. V. Pakhomov's approach (cf. p. 189) to WBVR.

Finally we hint at some characterizations of time complexity classes Lin t. These complexity classes depend in general on the type of machines chosen. Therefore characterizations of these classes seem to be very complicated. Special types of recursion must be found which are adapted to the given types of machines. However, such results are possible. B. Monien shows in [Mon 74a] that certain classes of type multiT-DTIME(Lin t) and the corresponding time classes of a simple type of RAM's can be characterized recursion-theoretically.

§ 11. Complexity Classes and Algebraic Operations

Language theoretic aspects of complexity classes, i.e. their behaviour under certain algebraic operations, have been investigated from the very beginning of the theory of computational complexity. All those questions which are already familiar from the well-developed theory of formal languages (cf. [Gin 66], [Sal 73], [Gin 75]) have also been asked for complexity classes. As a result, the following observations have been made:

- Many of the complexity classes studied are "good" classes from the formal language theoretic point of view, i.e. they are closed under certain algebraic operations and can be generated by them from smaller well-known classes or even from a single language.
- Some open questions about the closure of complexity classes have been shown to be equivalent to well-known open complexity-theoretic problems, e.g. to the **P-NP** problem or to time-space problems.

In what follows we illustrate these observations, dealing especially with the algebraic properties of time and space complexity classes.

11.1. *Closure under Algebraic Operations*

In this subsection we investigate which complexity classes*) are closed under the algebraic operations $\cup$, $\cap$, $^-$, $\cap$R, $\cdot$, $*$, h, eh, h^{-1}, sub, esub, sub^{-1}, gsm, egsm, and gsm^{-1}.

First we list some conditions for the resource bound functions which are sufficient for the closure of space and time complexity classes under these operations. There are two kinds of such conditions for resource bound functions t: 1. to be *increasing* (for short: t incr.), to be *superadditive*, i.e. $t(n) + t(m) \leqq t(n + m)$ for all $n, m \in \mathbb{N}$ (for short: t supadd.), and/or to exceed some function, 2. to be of the form t(Lin), Pol t etc. instead of t.

Our first theorem can be found partially in [BoGrWe 70].

11.1. Theorem. Table 11.1 is to be interpreted as follows: The complexity class X is closed under the operation ω if the, in row X and column ω, resource bound Y satisfies the condition Z.

Proof. Most of the proofs are straightforward. Hint: In order to verify the eh, $\cdot$, $*$, esub, and egsm results for multiT-NTIME the linear speed-up for this measure (see 18.13.3) should be used. In order to verify the $\cap$R, $\cup$, and $\cap$ results for both multiT-DTIME and multiT-NTIME the theorem on the real-time simulation of multihead TM's by multitape TM's (see 19.11) should be used. ∎

Remark 1. It is not known whether the eh, esub, and egsm results for DSPACE, NSPACE, and multiT-DTIME are valid for smaller bounds (see Theorem 11.3).

Remark 2. The smallest multiT-NTIME complexity class which is known to be closed under complementation is

$$\mathscr{E} = \bigcup_{k \geqq 0} \textbf{multiT-NTIME}\left(2^{2^{\cdot^{\cdot^{2^n}}}}\Big\}\, k\right).$$

Remark 3. Theorem 11.1 does not say anything about closure under h, sub, and gsm. This is caused by the fact that only very small space and time complexity classes (for instance those which are equal to **REG**) can be closed under these operations (see 11.2.2).

The next theorem on the closure properties of some space and time classes includes many special cases of the previous theorem.

11.2. Theorem.

1. The class X is closed (not closed, not known to be closed) under the operation ω iff the corresponding field of Table 11.2 contains + (—, ?, resp.).
2. None of the classes of Table 11.2 is closed under h, sub or gsm.

Proof. Negative results:

1. The h, sub and gsm results (Statement 2): Since $S^{\#}$ is in every complexity class mentioned in the theorem and all these classes are closed under h^{-1}, $\cap$R and $\cap$, the

*) See Footnote ** on p. 61.

Table 11.1.

		eh	h^{-1}	$\cap R$	$\cup$
DSPACE(Y)	Y	s	s(Lin)	s	s
	Z	$s \geqq id$ s incr.	$s \geqq 0$ s incr.	$s \geqq 0$	$s \geqq 0$
NSPACE(Y)	Y	s	s(Lin)	s	s
	Z	$s \geqq id$ s incr.	$s \geqq 0$ s incr.	$s \geqq 0$	$s \geqq 0$
multiT-DTIME(Y)	Y	Pol t	t(Lin)	t	t
	Z	$t(n) \geqq 2^n$ t incr.	$t \geqq id$ t incr.	$t \geqq id$	$t \geqq id$
multiT-NTIME(Y)	Y	t	t(Lin)	t	t
	Z	$t \geqq id$ t incr.	$t \geqq id$ t incr.	$t \geqq id$	$t \geqq id$

equality $\mathbf{RE} = \Gamma_{h^{-1},h,\cap R,\cap}(S^{\#})$ (see 4.14) implies that these classes are not closed under h, sub and gsm.

2. **REALTIME** results (except of h, sub and gsm): It can readily be seen that $P = \{11w1100v2w^{-1}: v \in \{0, 1\}^*, w \in (0^+1)^*0^*\} \in$ **REALTIME**. However, using the method described in § 8.1.1.3, in [Ros 67] it is shown that $\{0, 1, 2\}^* \cdot P \notin$ **REALTIME**. Therefore this class is not closed under $\cdot$. Furthermore, $H =_{df} P \cdot \{3\} \cup \{0, 1, 2\}^* \in$ **REALTIME**. Assume that $H^* \in$ **REALTIME**. Then $H^* \cap \{0, 1, 2\}^* \cdot \{3\} \in$ **REALTIME**, but $H^* \cap \{0, 1, 2)^* \cdot \{3\} = \{0, 1, 2\}^* \cdot P \cdot \{3\}$, and this implies $\{0, 1, 2\}^* \cdot P \in$ **REALTIME**. With this contradiction we know that **REALTIME** is not closed under *. Finally, $\{0', 1', 2'\}^* \cdot P \in$ **REALTIME** but $h(\{0', 1', 2'\}^* \cdot P) = \{0, 1, 2\}^* \cdot P$ is not, where $h(i') = i$ for $i = 0, 1, 2$. Consequently, **REALTIME** is not closed under eh, esub, and egsm.

3. The **LINTIME** results follow from **LINTIME** $\subset$ **Q** $=$ **eh(LINTIME)** where the proper inclusion is stated in 24.6.10 and the equality is obvious.

Positive results:

1. The h^{-1} and gsm^{-1} results for **REALTIME**. Let A be a finite deterministic automaton with output, let L' be the set of images of L by A, and let L' be accepted by a 1-kT-DTM M_1 in realtime. In order to accept L a 1-kT-DTM M_2 is constructed which works on the output word of A (without writing it down) like M_1. This would imply linear time for M_2. However, by block encoding of the worktape contents M_1 can process a single output of A in one step only. Thus M_2 works in realtime.

2. All other positive results are special cases of Theorem 11.1. ∎

$\cdot$	$*$	AFL	esub	egsm	gsm^{-1}	$\cap$	$^{-}$
s	s	s(Lin)	s	s	s(Lin)	s	s
$s \geqq \log$ s incr.	$s \geqq \mathrm{id}$ s incr.	$s \geqq \mathrm{id}$ s incr.	$s \geqq \mathrm{id}$ s incr.	$s \geqq \mathrm{id}$ s incr.	$s \geqq \log$ s incr.	$s \geqq 0$	$s \geqq 0$
s	s	s(Lin)	s	s	s(Lin)	s	Pol s
$s \geqq \log$ s incr.	$s \geqq \log$ s incr.	$s \geqq \mathrm{id}$ s incr.	$s \geqq \mathrm{id}$ s incr.	$s \geqq \mathrm{id}$ s incr.	$s \geqq \log$ s incr.	$s \geqq 0$	$s \geqq \log$
Pol t	Pol t	Pol t(Lin)	Pol t	Pol t	t(Lin)	t	t
$t \geqq \mathrm{id}$ t incr.	$t \geqq \mathrm{id}$ t incr.	$t(n) \geqq 2^n$ t incr.	$t(n) \geqq 2^n$ t incr.	$t(n) \geqq 2^n$ t incr.	$t \geqq \mathrm{id}$ t incr.	$t \geqq \mathrm{id}$	$t \geqq \mathrm{id}$
t	t	t(Lin)	t	t	t(Lin)	t	
$t \geqq \mathrm{id}$ t incr.	$t \geqq \mathrm{id}$ t supadd.	$t \geqq \mathrm{id}$ t supadd.	$t \geqq \mathrm{id}$ t supadd.	$t \geqq \mathrm{id}$ t incr.	$t \geqq \mathrm{id}$ t incr.	$t \geqq \mathrm{id}$	

Except for the two question marks in the row **LINTIME** for $\cdot$ and $*$ all other question marks indicate that the corresponding closure problem is equivalent to some well-known open problem of complexity theory (see § 23, 24, 26).

Especially, the problems of whether the classes **NL**, **NLINSPACE**, **NP**, **NEXPTIME** are closed under complement are themselves open problems which will be dealt with in detail in § 23 and § 24. For the other question marks see the next theorem.

11.3. Theorem.

1. [Mon 75a] **L** = **NL** $\leftrightarrow$ **L** is closed under $*$.
2. **L** = **NP** $\leftrightarrow$ **L** is closed under eh
   $\leftrightarrow$ **L** is an AFL
   $\leftrightarrow$ **L** is closed under esub
   $\leftrightarrow$ **L** is closed under egsm
   $\leftrightarrow$ **L** = **RUD**.
3. **NL** = **NP** $\leftrightarrow$ **NL** is closed under eh
   $\leftrightarrow$ **NL** is an AFL
   $\leftrightarrow$ **NL** is closed under esub
   $\leftrightarrow$ **NL** is closed under egsm.
4. **P** = **NP** $\leftrightarrow$ **P** is closed under eh
   $\leftrightarrow$ **P** is an AFL
   $\leftrightarrow$ **P** is closed under esub
   $\leftrightarrow$ **P** is closed under egsm.
5. **Q** = **RUD** $\leftrightarrow$ **Q** is closed under $^{-}$.

Table 11.2.

	eh	h^{-1}	$\cap R$	$\cup$	$\cdot$	*	AFL	esub	egsm	gsm^{-1}	$\cap$	$-$
L	?	+	+	+	+	?	?	?	?	+	+	+
NL	?	+	+	+	+	+	?	?	?	+	+	?
DLINSPACE	+	+	+	+	+	+	+	+	+	+	+	+
NLINSPACE	+	+	+	+	+	+	+	+	+	+	+	?
PSPACE	+	+	+	+	+	+	+	+	+	+	+	+
EXPSPACE	+	+	+	+	+	+	+	+	+	+	+	+
REALTIME	—	+	+	+	—	—	—	—	—	+	+	+
LINTIME	—	+	+	+	?	?	—	—	—	+	+	+
Q	+	+	+	+	+	+	+	+	+	+	+	?
P	?	+	+	+	+	+	?	?	?	+	+	+
NP	+	+	+	+	+	+	+	+	+	+	+	?
DEXPTIME	+	+	+	+	+	+	+	+	+	+	+	+
NEXPTIME	+	+	+	+	+	+	+	+	+	+	+	?
$\{A: c_A \in \mathscr{E}^n\}$, $n \geqq 3$	+	+	+	+	+	+	+	+	+	+	+	+
$\{A: c_A \in \mathbb{P}r\}$	+	+	+	+	+	+	+	+	+	+	+	+

Proof. We omit the proof of Statement 1.

Ad 2. Assume that **L** is closed under eh. By 11.13 we would have $\mathbf{Q} \subseteq \mathbf{L}$. Since **Q** contains sets which are $\leqq_m^{\log}$-complete in **NP** (see 14.9.2) we can conclude **L** = **NP**. Assume **L** = **NP**. Then **L** is an AFL and is closed under esub and egsm (see 11.2). Assume that **L** is an AFL or is closed under esub or under egsm. Then **L** is closed under eh. Finally, because of 11.17, **L** = **RUD** is equivalent to the closure of **L** under eh.

Statements 3 and 4 can be proved in the same manner.

Ad 5. This equivalence follows from the equality $\mathbf{RUD} = \Gamma_{\text{eh},-}(\mathbf{Q})$ (see 11.14.3) and the fact that **Q** is closed under eh. ∎

Remark 1. In [FlSt 74a] it is shown that **L** ∩ **CF** is closed under * iff **L** = **NL**. For Statement 3 and for further results concerning **NL** see also [Gre 77].

Remark 2. It is not known whether **Q** = **RUD**, but because of 11.16.2 the equality seems to be unlikely.

Finally we state some closure properties of some classes which will play a role in what follows.

11.4. Theorem.

1. The class X is closed (not closed, not known to be closed) under the operation ω iff the corresponding field of Table 11.3 contains + (—, ?, resp.).
2. None of the classes of Table 11.3 is closed under h, sub or gsm. ∎

Table 11.3.

	eh	h^{-1}	$\cap R$	$\cup$	$\cdot$	*	AFL	esub	egsm	gsm^{-1}	$\cap$	$-$
1-multiC-DTIME(id)	−	+	+	+	−	−	−	−	−	+	+	−
1-multiC-NTIME(id)	+	+	+	+	+	+	+	+	+	+	+	?
1-multiC-NTIME-REV (id, Const)	+	+	+	+	+	+	+	+	+	+	+	?
RBQ = 1-multiT-NTIME-REV(Lin, Const)	+	+	+	+	+	?	?	?	+	+	+	?
$\mathcal{R}_{m}^{\log}$(**DCF**)	?	+	+	+	+	?	?	?	?	+	+	+
$\mathcal{R}_{m}^{\log}$(**CF**)	?	+	+	+	+	+	?	?	?	+	+	?
RUD	+	+	+	+	+	+	+	+	+	+	+	+
EXRUD	+	+	+	+	+	+	+	+	+	+	+	+
eh(L)	+	+	+	+	+	+	+	+	+	+	+	?
eh(NL)	+	+	+	+	+	+	+	+	+	+	+	?
eh(P)	+	+	+	+	+	+	+	+	+	+	+	?

Remark. The above results can be found partially in [FipMeRo 68] for **1-multiC-DTIME**(id), in [Gre 78c] for both **1-multiC-NTIME**(id) and **1-multiC-NTIME-REV**(id, Const), in [BoNiPa 74] for **RBQ**, in [Sud 77c] for both $\mathcal{R}_{m}^{\log}$(**DCF**) and for $\mathcal{R}_{m}^{\log}$(**CF**) (hint: to prove use 14.3), in [Wra 78] and [Boo 78] for **RUD** (hint: to prove use 11.14), in [Boo 78] for **EXRUD** (hint: to prove use 11.20), and in [KiWr 78] for both **eh(L)** and **eh(NL)**.

11.2. Algebraic Closure

In this section we investigate the problem of whether a given complexity class can be generated from a second class by certain algebraic operations. The question of whether a complexity class is a principal AFL (see p. 57) is treated in § 11.2.1, other results can be found in § 11.2.2.

11.2.1. Principal AFL's

In the following theorem conditions are given under which space and time complexity classes will be principal AFL's. First we define the sets which will serve as generators for certain space or time complexity classes. Let ^ be a simple encoding of the symbols of Σ over {0, 1} (note that Σ may be countably infinite), and let ^ also denote a simple encoding of the k-tape TM's, $k \geqq 1$, over {0, 1} which uses the encoding ^

of the symbols. Then, for $X \in \{D, N\}$ and $s \geqq 0$, we set

$$A_s^X =_{df} \{\hat{M}^k \# \hat{a}_1 \# \hat{M}^k \# \hat{a}_2 \# \ldots \# \hat{M}^k \# \hat{a}_n \#: M \text{ is a 2-XTM}, k, n \geqq 1, a_1, \ldots, a_n \in \Sigma, \text{ and } M \text{ accepts } a_1a_2\ldots a_n \text{ within space } s(k \cdot n)\},$$

for $t \geqq \mathrm{id}$ such that $t(n)^2 \leqq t(l \cdot n)$ for some l and all n we fix such an l and set

$$B_t^D =_{df} \{\hat{M}^{k\cdot l} \# \hat{a}_1 \# \hat{M}^{k\cdot l} \# \hat{a}_2 \# \ldots \# \hat{M}^{k\cdot l} \# \hat{a}_n \#: n, k \geqq 1, M \text{ is an mT-DM}, m \geqq 1, a_1, \ldots, a_n \in \Sigma \text{ and } M \text{ accepts } a_1a_2\ldots a_n \text{ within time } t(k \cdot n)\},$$

and, for $t \geqq \mathrm{id}$, we set

$$B_t^N =_{df} \{\hat{M}^k \# \hat{a}_1 \# \hat{M}^k \# \hat{a}_2 \# \ldots \# \hat{M}^k \# \hat{a}_n \#: M \text{ is a 3T-NM}, k, n \geqq 1, a_1, \ldots, a_n \in \Sigma, \text{ and } M \text{ accepts } a_1a_2\ldots a_n \text{ within time } t(k \cdot n)\}.$$

11.5. Theorem. [BoGrIbWe 70], [GiRo 70].

1. Let $s \geqq \mathrm{id}$ be a fully 2-T-DSPACE-constructible increasing function such that $k \cdot s(n) \leqq s(k \cdot n)$ for $k, n \in \mathbb{N}$. Then, for $X \in \{D, N\}$, $\mathbf{XSPACE}(s(\mathrm{Lin}))$ is a principal AFL, and $\mathbf{XSPACE}(s(\mathrm{Lin})) = h^{-1}(A_s^X) \vee \{\emptyset, \{e\}\}$.
2. Let $t(n) \geqq 2^n$ be a fully multiT-DTIME-constructible increasing function such that $k \cdot t(n) \leqq t(k \cdot n)$ for all $k, n \in \mathbb{N}$ and $t(n)^2 \leqq t(l \cdot n)$ for some l and all $n \in \mathbb{N}$. Then $\mathbf{multiT\text{-}DTIME}(t(\mathrm{Lin}))$ is a principal AFL and $\mathbf{multiT\text{-}DTIME}(t(\mathrm{Lin})) = h^{-1}(B_t^D) \vee \{\emptyset, \{e\}\}$.
3. Let $t(n) \geqq \mathrm{id}$ be a fully 3T-DTIME-constructible superadditive function. Then $\mathbf{multiT\text{-}NTIME}(t(\mathrm{Lin}))$ is a principal AFL, and $\mathbf{multiT\text{-}NTIME}(t(\mathrm{Lin})) = h^{-1}(B_t^N) \vee \{\emptyset, \{e\}\}$.

Proof. Ad 1. We restrict ourselves to the case $X = D$, the case $X = N$ can be treated analogously.

First we show $A_s^D \in \mathbf{DSPACE}(s)$. We describe how a 2-DTM M_0 can decide the set A_s^D within space s.

M_0 checks whether the input w has the form $\hat{M}^k \# \hat{a}_1 \# \ldots \# \hat{M}^k \# \hat{a}_n \#$ for some 2-DTM M, $k \geqq 1$, $n \geqq 1$, $a_1, \ldots, a_n \in \Sigma$, and whether $a_1, \ldots, a_n$ are input symbols of M. If this is not the case, M_0 rejects w. M_0 can perform this within space $\log |w| \leqq s(|w|)$.

M_0 lays off the space $|\hat{M}| \cdot s(k \cdot n)$ on its worktape. Because of the full 2-T-DSPACE-constructibility this can be done within space $|\hat{M}| \cdot s(k \cdot n) \leqq s(|\hat{M}| \cdot k \cdot n) \leqq s(|w|)$.

In order to avoid cycles in the simulation of M the machine M_0 constructs the 16-ary representation of $n \cdot$ (number of states of M) $\cdot\, s(k \cdot n) \cdot$ (number of worktape symbols of $M)^{s(k\cdot n)} \leqq 2^{4\cdot s(|w|)}$ which will serve as a clock. This clock has the length $s(|w|)$.

M_0 simulates the work of M on the input $a_1 \dots a_n$ as long as M needs at most $s(k \cdot n)$ tape squares (if M_0 encodes all symbols of M with code length $|\hat{M}|$, this can be checked by the marked tape squares) and at most $2^{4 \cdot s(|w|)}$ steps (this can be checked by the clock). If M stops within these bounds, M_0 accepts as M does, otherwise M_0 rejects w. Consequently, M_0 accepts A_s^D within space $s(|w|)$.

Now we note that by the suppositions and Theorem 11.1 $\mathbf{DSPACE}(s(\mathrm{Lin}))$ is an AFL. Thus the inclusion "$\supseteq$" is proved, and it remains to show that for every $L \in \mathbf{DSPACE}(s(\mathrm{Lin}))$ there is a homomorphism h such that $L - \{e\} = h^{-1}(A_s^D)$. If L is accepted by a 2-DTM M within space $s(k \cdot n)$, then we set $h(a) =_{\mathrm{df}} \hat{M}^k \# \hat{a} \#$ for all a which are input symbols of M.

Ad 2. First we show $B_t^D \in \mathbf{2T\text{-}DTIME}(t)$. We describe the work of a 2T-DTM M_0 accepting B_t^D within time t.

M_0 checks whether the input has the form $\hat{M}^{k \cdot l} \# \hat{a}_1 \# \dots \# \hat{M}^{k \cdot l} \# \hat{a}_n \#$ for some mT-DM M, $m \geqq 1$, $k \geqq 1$, $n \geqq 1$, $a_1, \dots, a_n \in \Sigma$, and whether $a_1, \dots, a_n$ are input symbols of M. M_0 can perform this within time $c \cdot |w| \leqq_{\mathrm{ae}} t(|w|)$.

M_0 marks $t(k \cdot n)$ tape squares on its second tape (which will serve as a clock for the simulation) in the following way: Because of the constructibility supposition there is an rT-DM M' working exactly $t(k \cdot n)$ steps on every input of length $k \cdot n$. M_0 produces first some input of length $k \cdot n$ and then M_0 simulates on its first tape the work of M' on this input within $t(k \cdot n)^2 \leqq t(l \cdot k \cdot n) \leqq t(|w|)$ steps. This is possible by Theorem 19.15. After simulating one step of M' the machine M_0 marks one tape squares on its second tape.

The machine M_0 simulates with its first tape $t(k \cdot n)$ steps of the work of M on the input $a_1 \dots a_n$. This is possible within $|\hat{M}| \cdot t(k \cdot n)^2 \leqq t(|\hat{M}| \cdot l \cdot k \cdot n) \leqq t(|w|)$ steps (again by Theorem 19.15). If M stops within the time $t(k \cdot n)$, then M_0 accepts as M does, otherwise M_0 rejects w.

The proof can be completed as in 1.

Ad 3. This proof can be carried out as in the deterministic case with the only exception that the simulations can be performed using $\mathbf{multiT\text{-}NTIME}(t) = \mathbf{3T\text{-}NTIME}(t)$ (19.20 instead of 19.15). Therefore we construct a 4T-NM M_0 accepting B_t^N within time t. ∎

Remark. The supposition "$k \cdot s(n) \leqq s(k \cdot n)$ for all $k, n \in \mathbb{N}$" excludes functions below id. By a slight modification of A_s^X (using $\hat{M}^{k \cdot |\hat{M}|^{r-1}}$ instead of $\hat{M}^k$) one can show that the representation of $\mathbf{XSPACE}(s(\mathrm{Lin}))$ in 11.5.1 remains valid under the weaker supposition "$k \cdot s(n) \leqq s(k^r \cdot n)$ for all $k, n \in \mathbb{N}$" (for instance $s(n) = n^{1/r}$) for $r \geqq 1$.

11.6. Corollary. The following complexity classes are principal AFLs:

1. **DLINSPACE**, $\mathbf{DSPACE}(n^k)$, **EXPSPACE**, $\mathbf{DSPACE}(2^{\mathrm{Lin}\, n^k})$, $\mathbf{DSPACE}\left(2^{2^{\cdot^{\cdot^{2^{\mathrm{Lin}}}}}}\right\}k\Big)$.
2. **NLINSPACE**, $\mathbf{NSPACE}(n^k)$.
3. **DEXPTIME**, $\mathbf{DTIME}(2^{\mathrm{Lin}\, n^k})$.
4. **Q**, $\mathbf{multiT\text{-}NTIME}(n^k)$, **NEXPTIME**, $\mathbf{NTIME}(2^{\mathrm{Lin}\, n^k})$. ∎

Finally we mention some negative results.

As an immediate consequence of Lemma 4.13, Theorem 11.1 and the corresponding hierarchy results (17.23, 17.35, 17.43) we obtain

11.7. Corollary. The following AFLs are not principal:
1. **PSPACE**, **DSPACE**(2^{Pol}), $\{A : c_A \in \mathcal{E}\}$, $\{A : c_A \in \mathbb{P}\mathrm{r}\}$.
2. **DTIME**(2^{Pol}).
3. **NP**, **NTIME**(2^{Pol}). ∎

11.2.2. Other Algebraic Closures

11.2.2.1. Realtime Languages

Some of the complexity classes included in the class $\mathbf{Q} =_{\mathrm{df}}$ **multiT-NTIME**(id) have pleasing properties with respect to their generation by some AFL operations combined with intersection. For the results in question we need the following lemma.

11.8. Lemma. [GrHo 69] For $k \geqq 1$,

$$\textbf{1-kPD-NTIME}(\mathrm{id}) \subseteqq \mathbf{eh}(\underbrace{\textbf{1-PD-DTIME}(\mathrm{id}) \wedge \ldots \wedge \textbf{1-PD-DTIME}(\mathrm{id})}_{k \text{ times}}).$$

Moreover, if one the pushdown stores of the 1-kPD-NM works with the reversal bound $r(n)$ (is actually a counter), then the corresponding 1-DPDA works with reversal bound $r(n)$ and in realtime (is a 1-DCA working in realtime).

Proof. Let M be a 1-kPD-NM accepting a language L in realtime, and let m be the maximum number of possible next situations. Thus a computation path of M is given by an element of $\{1, 2, \ldots, m\}^n$. Let s_0 be the initial state of M. We define

$L' =_{\mathrm{df}} \{(x_1, \mu_1, s_1, a_{1,1}, \ldots, a_{k,1}) \ldots (x_n, \mu_n, s_n, a_{1,n}, \ldots, a_{k,n}) : s_n$ is an accepting state and for the input $x_1x_2 \ldots x_n$ the machine M has on the computation path $(\mu_1, \ldots, \mu_n)$ the states $s_0, s_1, \ldots, s_n$ and reads the top symbols $(a_{1,1}, \ldots, a_{k,1}), (a_{1,2}, \ldots, a_{k,2}), \ldots, (a_{1,n}, \ldots, a_{k,n})\}$.

Evidently, $h(L') = L$ where $h((x, \mu, s, a_1, \ldots, a_k)) =_{\mathrm{df}} x$. Now, for $j = 1, \ldots, k$ let M_j be that 1-NPDA which acts in the following way: if the input head of M_j reads $(x, a_1, \ldots, a_{j-1}, a_{j+1}, \ldots, a_k)$ and its pushdown head reads a_j then M_j does the same as M, reading x with its input head and $a_1, \ldots, a_k$ with its pushdown heads. Defining

$L_j =_{\mathrm{df}} \{(x_1, \mu_1, s_1, a_{1,1}, \ldots, a_{k,1}) \ldots (x_n, \mu_n, s_n, a_{1,n}, \ldots, a_{k,n}) : s_n$ is an accepting state and for the input $(x_1, a_{1,1}, \ldots a_{j-1,1}a_{j+1,1}, \ldots, a_{k,1}) \ldots (x_n, a_{1,n}, \ldots, a_{j-1,n}, a_{j+1,n}, \ldots, a_{k,n})$ the machine M_j has on the computation path $(\mu_1, \ldots, \mu_n)$ the states $s_0, s_1, \ldots, s_n$ and reads the top symbols $a_{j,1}, \ldots, a_{j,n}\}$

we get $L' = L_1 \cap L_2 \cap \ldots \cap L_k$, and there are 1-DPDA $A_1, A_2, \ldots, A_n$ accepting the sets $L_1, L_2, \ldots, L_k$, resp. in realtime. ∎

Remark. This lemma can be formulated in a more general way in terms of so-called multitape AFA (abstract families of automata, cf. [Gin 75]).

11.9. Theorem. [BoGr 70]

$$\begin{aligned}\mathbf{Q} &= \mathbf{eh}(\mathbf{DCF} \wedge \mathbf{DCF} \wedge \mathbf{DCF})\\ &= \Gamma_{\mathrm{eh},\cap}(\mathbf{CF})\\ &= \Gamma_{\mathrm{eh},\mathrm{h}^{-1},\cap\mathrm{R},\cap}(\mathrm{D}_2)\\ &= \Gamma_{\mathrm{eh},\mathrm{h}^{-1},\cap}\big(\mathrm{L}_{\mathrm{N}}(\mathrm{D}_2)\big).\end{aligned}$$

Proof. Since $\mathbf{CF} \subseteq \mathbf{Q}$ (12.4) and $\mathbf{Q}$ is closed under $\cap$ and eh (11.2) we have $\Gamma_{\mathrm{eh},\cap}(\mathbf{CF}) \subseteq \mathbf{Q}$.

Conversely, we know $\mathbf{Q} \subseteq$ **1-3PD-NTIME**(id) $\subseteq \mathbf{eh}(\mathbf{DCF} \wedge \mathbf{DCF} \wedge \mathbf{DCF})$ by Theorem 22.2 and Lemma 11.8.

Finally, $\mathbf{CF} = \Gamma_{\mathrm{eh},\mathrm{h}^{-1},\cap\mathrm{R}}(\mathrm{D}_2) = \Gamma_{\mathrm{h}^{-1}}\big(\mathrm{L}_{\mathrm{N}}(\mathrm{D}_2)\big)$ (4.15 and 14.2, resp.) and the fact that $\mathbf{Q}$ is closed under eh, h^{-1}, $\cap$R and $\cap$ provides the other two equalities. ∎

Remark. Because of the remark after 22.2 every $L \in \mathbf{Q}$ can be accepted in realtime by a 1-3PD-NM which operates in such a way that on every accepting computation path two of the pushdown heads make at most one reversal. Therefore, by 11.8 and 12.2 we even have $\mathbf{Q} = \mathbf{eh}(\mathbf{DCF} \wedge \mathbf{LIN} \wedge \mathbf{LIN})$ (see [BoNiPa 74]).

Analogous results can also be established for some subclasses of $\mathbf{Q}$. For some relationships between these classes and for related results see § 22. We start with a representation theorem for the class $\mathbf{RBQ} =_{\mathrm{df}}$ **multiT-NTIME-REV**(id, Const).

11.10. Theorem. [BoNiPa 74]

$$\begin{aligned}\mathbf{RBQ} &= \mathbf{eh}(\mathbf{LIN} \wedge \mathbf{LIN} \wedge \mathbf{LIN})\\ &= \Gamma_{\mathrm{eh},\cap}(\mathbf{LIN})\\ &= \Gamma_{\mathrm{eh},\mathrm{h}^{-1},\cap\mathrm{R},\cap}(\mathrm{S})\\ &= \Gamma_{\mathrm{eh},\mathrm{h}^{-1},\cap\mathrm{R},\cap}(\mathrm{S}^{\#} \cup \{\mathrm{e}\}).\end{aligned}$$

Proof. Since $\mathbf{LIN} \subseteq \mathbf{RBQ}$ (12.2 and 22.8) and $\mathbf{RBQ}$ is closed under $\cap$ and eh (11.4) we have $\Gamma_{\mathrm{eh},\cap}(\mathbf{LIN}) \subseteq \mathbf{RBQ}$.

Conversely,

$$\begin{aligned}\mathbf{RBQ} &\subseteq \textbf{1-3PD-NTIME-REV}(\mathrm{id}, 1)\ (22.8)\\ &\subseteq \mathbf{eh}\big(\textbf{1-PD-DREV}(1) \wedge \textbf{1-PD-DREV}(1) \wedge \textbf{1-PD-DREV}(1)\big)\ (11.8)\\ &\subseteq \mathbf{eh}(\mathbf{LIN} \wedge \mathbf{LIN} \wedge \mathbf{LIN})\ (12.2).\end{aligned}$$

Finally, $\mathbf{LIN} = \Gamma_{\mathrm{eh},\mathrm{h}^{-1},\cap\mathrm{R}}(\mathrm{S}) = \Gamma_{\mathrm{eh},\mathrm{h}^{-1},\cap\mathrm{R}}(\mathrm{S}^{\#} \cup \{\mathrm{e}\})$ (4.17) and the fact that $\mathbf{RBQ}$ is closed under eh, h^{-1}, $\cap$R, and $\cap$ (11.4) provides the other two equalities. ∎

Remark 1. The class **RBQ** is not known to be closed under *, but it is closed under the remaining AFL operations and under intersection (11.4). The least intersection closed AFL containing **RBQ** can be represented as follows (cf. [BrWa 82]):

$$\Gamma_{\mathrm{AFL},\cap}(\mathbf{RBQ}) = \Gamma_{\mathrm{AFL},\cap}(\mathbf{LIN}) = \mathbf{eh}(\mathbf{LIN}^* \wedge \mathbf{LIN} \wedge \mathbf{LIN}) = \Gamma_{\mathrm{eh},\mathrm{h}^{-1},\cap\mathrm{R},\cap}(\mathrm{S}^*) = \mathbf{multiT\text{-}NTIME\text{-}CROS}(\mathrm{id}, \mathrm{Const}).$$

The equation $\Gamma_{\mathrm{AFL},\cap}(\mathbf{RBQ}) = \mathbf{eh}(\mathbf{LIN}^* \wedge \mathbf{LIN} \wedge \mathbf{LIN})$ rests on Remark 2 after 22.8 and the fact that

$$\mathbf{1\text{-}PD\text{-}NTIME\text{-}CROS}(\mathrm{Lin}, 3) = \mathbf{1\text{-}PD\text{-}NCROS}(3) = \mathrm{eh}(\mathbf{LIN}^* \wedge \mathbf{REG}).$$

Remark 2. In [Hul 79] it has been shown by a long proof that

$$\mathbf{eh}(\mathbf{LIN} \wedge \mathbf{LIN}) \subset \mathbf{eh}(\mathbf{LIN} \wedge \mathbf{LIN} \wedge \mathbf{LIN}).$$

Remark 3. An interesting restriction of the class RBQ has been introduced in [BoGrWr 78]. A *reset tape* (for short: RT) is an infinite (to the right) tape whose read-write head can move only left-to-right ecxept that the head can be reset to the left end of the tape within one step. The component RESET in the denotation of a complexity measure gives an upper bound for the number of reset moves (maximum over all tapes). The classes $\mathbf{DUP} =_{\mathrm{df}} \mathbf{1\text{-}RT\text{-}NTIME\text{-}RESET}(\mathrm{id}, 1)$, $\mathbf{RSBQ} =_{\mathrm{df}} \mathbf{1\text{-}multiRT\text{-}NTIME\text{-}RESET}(\mathrm{id}, \mathrm{Const})$, and $\mathbf{RSQ} =_{\mathrm{df}} \mathbf{1\text{-}multiRT\text{-}NTIME}(\mathrm{id})$ are the "duplication-based" analogues of the "replication-based" classes **LIN**, **RBQ** and **Q**, resp. The following results are proved in [BoGrWr 78].

$$\mathbf{DUP} = \Gamma_{\mathrm{eh},\mathrm{h}^{-1},\cap\mathrm{R}}(\mathrm{D}) = \{\{h_1(w)\, h_2(w) : w \in R\} : h_1, h_2 \text{ e-free homomorphisms}, R \in \mathbf{REG}\}$$

(compare with 4.17).

$$\mathbf{RSQ} = \mathbf{RSBQ} = \mathbf{1\text{-}multiRT\text{-}NTIME}(\mathrm{Lin}) = \mathbf{1\text{-}3RT\text{-}NTIME\text{-}REV}(\mathrm{id}, 1)$$

(compare with 22.8)

$$= \mathbf{eh}(\mathbf{DUP} \wedge \mathbf{DUP} \wedge \mathbf{DUP}) = \Gamma_{\mathrm{eh},\mathrm{h}^{-1},\cap\mathrm{R},\cap}(\mathrm{D}) = \Gamma_{\mathrm{AFL},\cap}(\mathrm{D}).$$

Further characterizations in terms of *Post machines* as well as *equality sets* can be found in [Bra 79a], [Bra 81b] and [Bra 84]. Let us deal with the latter. A *k-fold equality set* L is defined by k homomorphisms $h_1, h_2, \ldots, h_k$ as

$$L = \{w : h_1(w) = h_2(w) = \ldots = h_k(w)\}.$$

Let $\mathbf{EQ^k}$ be the class of all k-fold equality sets and $\mathbf{EQ^+} =_{\mathrm{df}} \bigcup_{k \geq 2} \mathbf{EQ^k}$. It holds that $\mathbf{EQ^2} \subset \mathbf{EQ^3} \subset \ldots \subset \mathbf{EQ^k} \ldots$ and $\mathbf{EQ^+} \subseteqq \mathbf{RSQ} \cap \mathbf{L}$. There are $\leqq^{\mathrm{hom}}$-complete sets L_k in $\mathbf{EQ^k}$, i.e. $\mathbf{EQ^k} = \mathbf{h^{-1}}(L_k)$.

RSQ can be generated from $\mathbf{EQ^3}$ in the following way ([Bra 81]):

$$\mathbf{RSQ} = \mathbf{eh}(\mathbf{EQ^3} \wedge \mathbf{REG}) = \Gamma_{\mathrm{eh},\mathrm{h}^{-1},\cap\mathrm{R}}(\mathrm{L}_3).$$

11.11. Theorem.

$$\mathbf{1\text{-}multiC\text{-}NTIME}(\mathrm{id}) = \mathbf{eh}\big(\Gamma_\cap(\mathbf{1\text{-}DCA})\big) = \Gamma_{\mathrm{eh},\cap}(\mathbf{1\text{-}NCA}) = \Gamma_{\mathrm{eh},\mathrm{h}^{-1},\cap\mathrm{R},\cap}(\mathrm{D}_1) = \Gamma_{\mathrm{AFL},\cap}(\mathrm{D}_1).$$

Proof. Since $\mathbf{1\text{-}NCA} \subseteqq \mathbf{1\text{-}C\text{-}NTIME}(\mathrm{id})$ (13.6.5) and $\mathbf{1\text{-}multiC\text{-}NTIME}(\mathrm{id})$ is closed under eh and $\cap$ (11.4) we have $\mathbf{eh}\big(\Gamma_\cap(\mathbf{1\text{-}NCA})\big) \subseteqq \mathbf{1\text{-}C\text{-}NTIME}(\mathrm{id})$. Conversely, $\mathbf{1\text{-}multiC\text{-}NTIME}(\mathrm{id}) \subseteqq \mathbf{eh}\big(\Gamma_\cap(\mathbf{1\text{-}DCA})\big)$ follows from 11.8.

The other equality is a consequence of **1-NCA** $= \Gamma_{\mathrm{eh},\mathrm{h}^{-1},\cap\mathrm{R}}(\mathbf{D}_1)$ (4.16) and the fact that **1-multiC-NTIME**(id) is closed under eh, h^{-1}, $\cap$R, and $\cap$ (11.4). ∎

11.12. Theorem. [Gre 78c]

$$\textbf{1-multiC-NTIME-REV}(\mathrm{id}, \mathrm{Const}) = \mathbf{eh}\big(\Gamma_\cap(\textbf{1-C-DREV}(1))\big)$$
$$= \Gamma_{\mathrm{eh},\cap}\big(\textbf{1-C-NREV}(1)\big) = \Gamma_{\mathrm{eh},\mathrm{h}^{-1},\cap\mathrm{R},\cap}(\mathbf{C}).$$

Proof. Since **1-C-NREV**(1) $\subseteq$ **1-multiC-NTIME-REV**(id, 1) (22.16) and **1-multiC-NTIME-REV**(id, Const) is closed under eh and $\cap$ (11.4) we have $\Gamma_{\mathrm{eh},\cap}(\textbf{1-C-NREV}(1))$ $\subseteq$ **1-multiC-NTIME-REV**(id, Const). Conversely,

$$\textbf{1-multiC-NTIME-REV}(\mathrm{id}, \mathrm{Const}) \subseteq \textbf{1-multiC-NTIME-REV}(\mathrm{id}, 1) \quad (22.16)$$
$$\subseteq \mathbf{eh}\big(\Gamma_\cap(\textbf{1-C-DREV}(1))\big) \quad (11.8).$$

The other equality is a consequence of **1-C-NREV**(1) $= \Gamma_{\mathrm{eh},\mathrm{h}^{-1},\cap\mathrm{R}}(\mathbf{C})$ (4.18) and the fact that **1-multiC-NTIME-REV**(id, Const) is closed under eh, h^{-1}, $\cap$R and $\cap$ (11.4). ∎

We conclude this subsection with some results concerning the relationships between **Q** and other classes of interest.

11.13. Theorem. [Spr 76] $\mathbf{Q} \subseteq \mathbf{eh}(\mathbf{L}) \subseteq \mathbf{NP}$.

Proof. Because of $\mathbf{Q} = \Gamma_{\mathrm{eh},\cap}(\mathbf{CF})$ (11.9) and the closure properties of **eh(L)** (11.4) and **NP** (11.2) it suffices to show that $\mathbf{CF} \subseteq \mathbf{eh}(\mathbf{L})$.

By 12.14 and 11.8 it is sufficient to show **1-PD-DTIME**(id) $\subseteq$ **eh(L)**. Let M be a 1-DPDA accepting a language L in realtime. We define

$L' =_{\mathrm{df}} \{(x_1, s_1, a_1, \sigma_1)\,(x_2, s_2, a_2, \sigma_2) \ldots (x_n, s_n, a_n, \sigma_n)$:
s_n is an accepting state, and for the input $x_1x_2 \ldots x_n$ the machine M has the states $s_0, s_1, \ldots, s_n$, reads the top symbols $a_1, a_2, \ldots, a_n$, and its pushdown head performs the moves $\sigma_1, \sigma_2, \ldots, \sigma_n \in \{\uparrow, \downarrow, \mathrm{e}\}$ (e stands for "no move")$\}$.

Evidently, $h(L') = L$ where $h\big((x, s, a, \sigma)\big) =_{\mathrm{df}} x$. The language L' can be accepted by a 2-DTM M' as follows: All steps in which M does not pop its top symbol can be simulated by M' without the use of its worktape. If M pops the top symbol, M' has to find the new top symbol of M. This can be done by moving the head to the left and counting the symbols $\uparrow$ and $\downarrow$. Consequently, M' can work within space $\log n$. ∎

Remark 1. The class $\Gamma_{\mathrm{eh}}(\mathbf{L})$ also contains **1-DSA**. Analogously, **1-NSA** $\subseteq \Gamma_{\mathrm{eh}}(\mathbf{NL})$ (cf. [KiWr 78]).

Remark 2. Theorem 11.13 shows that the class **eh(L)** has a position between four interesting classes (cf. Figure 11.1).

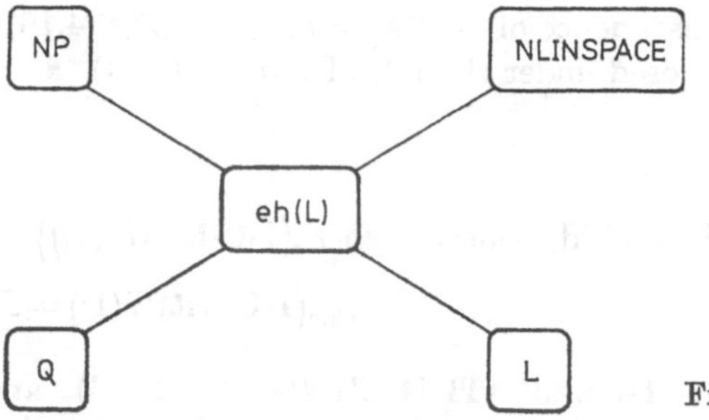

Figure 11.1

Starting with **LINTIME** one can define (analogously to the polynomial-time hierarchy) the *linear-time hierarchy*.

$$\Sigma_0^{\mathbf{Q}} =_{\mathrm{df}} \mathbf{LINTIME}, \quad \Sigma_{k+1}^{\mathbf{Q}} =_{\mathrm{df}} \mathcal{R}_{\mathrm{T}}^{\mathrm{NTIME,Lin}}(\Sigma_k^{\mathbf{Q}}),$$

$$\Pi_k^{\mathbf{Q}} =_{\mathrm{df}} \mathrm{co}\Sigma_k^{\mathbf{Q}}, \quad \text{for} \quad k \geqq 0.$$

11.14. Theorem. [Wra 78]

1. $\Sigma_1^{\mathbf{Q}} = \mathbf{Q}$.
2. $\Sigma_{k+1}^{\mathbf{Q}} = \mathrm{eh}(\Pi_k^{\mathbf{Q}})$, for $k \geqq 0$.
3. $\bigcup\limits_k \Sigma_k^{\mathbf{Q}} = \Gamma_{\mathrm{eh},-}(\mathbf{Q}) = \mathbf{RUD}$. ∎

The following corollary can be found partially in [Yu 70].

11.15. Corollary. $\mathbf{RUD} = \Gamma_{\mathrm{eh},-,\cap}(\mathbf{CF}) = \Gamma_{\mathrm{eh},\mathrm{h}^{-1},\cap,-}(\mathrm{L_N}(\mathrm{D_2}))$
$= \Gamma_{\mathrm{eh},\mathrm{h}^{-1},\cap\mathrm{R},\cap,-}(\mathrm{C})$. ∎

Remark 1. The latter equality is a consequence of the inclusion $\mathbf{CF} \subseteqq \Gamma_{\mathrm{eh},\mathrm{h}^{-1},\cap\mathrm{R},\cap,-}(\mathbf{C})$ which is shown in [Wra 77b].

Remark 2. In [Wot 78] it has been shown that $\mathbf{CF} \setminus \Gamma_{\cap,-}(\mathbf{DCF}) \neq \emptyset$ and $\Gamma_{\cap,-}(\mathbf{DCF}) \subset \Gamma_{\cap}(\mathbf{CF}) = \Gamma_{\cap,\mathrm{h}^{-1}}(\mathbf{CF}) \subset \Gamma_{\cap,-}(\mathbf{CF})$.

There are further interesting relationships between **RUD** and some complexity classes.

11.16. Theorem.

1. [Myh 60] **RUD** $\subseteqq$ **DLINSPACE**.
2. [Nep 66] $\bigcup\limits_{\alpha<1}$ **DSPACE-TIME**$(n^\alpha, \mathrm{Pol}) \subseteqq$ **RUD**. ∎

Remark. In [Bel 79] RUD has been characterized as a space complexity class of a certain tape of RAM's. In [Lip 78] it has been proved that ATIME-ALT(Lin, Const) = **RUD** (for ATIME-ALT see Remark 1 after Theorem 8.47). Theorem 11.16.2 is strengthened in [Vol 83b] where it is shown that

$$\bigcup_{\alpha<1} \text{ASPACE-TIME-ALT}(n^\alpha, \text{Pol}, \text{Const}) \subseteqq \text{RUD}.$$

From 11.16 one gets easily the following characterization of **RUD**.

11.17. Corollary. $\mathbf{RUD} = \Gamma_{\mathrm{eh},-}(\mathbf{L})$.

Proof. $\mathbf{RUD} \subseteq \Gamma_{\mathrm{eh},-}(\mathbf{Q})$ (11.14)

$\subseteq \Gamma_{\mathrm{eh},-}(\mathbf{L})$ (11.13)

$\subseteq \Gamma_{\mathrm{eh},-}(\mathbf{RUD})$ (11.16.2)

$\subseteq \mathbf{RUD}$ (11.14). ∎

11.2.2.2. *Bounded Erasing*

Now we state some results concerning closure with respect to homomorphisms, homomorphic replication and homomorphic duplication, all with bounded erasing.

A homomorphism (homomorphic replication, homomorphic duplication) h is said to be *t-erasing* on a language L iff there is a $k \geqq 1$ such that $|w| \leqq k \cdot t(|h(w)|)$ for all $w \in L$; h is said to be *linearly erasing* (*polynomially erasing*) iff h is id-erasing (n^l-erasing for some $l \geqq 1$). By h_t ($\mathrm{h}_{\mathrm{lin}}$, $\mathrm{h}_{\mathrm{pol}}$) we denote the operator of the application of a t-erasing (linearly erasing, polynomially erasing, resp.) homomorphism, and analogously hr_t, $\mathrm{hr}_{\mathrm{lin}}$, and $\mathrm{hr}_{\mathrm{pol}}$ for homomorphic replication, and hd_t, $\mathrm{hd}_{\mathrm{lin}}$, and $\mathrm{hd}_{\mathrm{pol}}$ for homomorphic duplication.

11.18. Lemma. [BoNi 78] For every family of languages $\mathscr{L} \supseteq \mathbf{RBQ}$,

1. If $\mathscr{L}$ is closed under h^{-1}, eh, ∩, ∪, then $\mathscr{L}$ is closed under $\mathrm{h}_{\mathrm{lin}}$.
2. If $\mathscr{L}$ is closed under h^{-1}, eh, ∩, ∪, ·, then $\mathscr{L}$ is closed under $\mathrm{hr}_{\mathrm{lin}}$. ∎

11.19. Theorem. [Boo 78], [Boo 79b]

1. $\mathbf{RBQ} = \Gamma_{\mathrm{hr}_{\mathrm{lin}},\cap}(\mathbf{REG})$.
2. $\mathbf{Q} = \Gamma_{\mathrm{hr}_{\mathrm{lin}},\cap}(\mathbf{CF})$.
3. $\mathbf{RUD} = \Gamma_{\mathrm{hr}_{\mathrm{lin}},\cap,-}(\mathbf{REG})$. ∎

For relativized versions see [Boo 79b] and [BoWr 78].

Remark. For the class **RSQ** (see Remark 3 after Theorem 11.10) one can prove a result analogous to that for **RBQ**, namely $\mathbf{RSQ} = \Gamma_{\mathrm{hd}_{\mathrm{lin}},\cap}(\mathbf{REG})$ (cf. [BoGrWr 78]).

11.20. Theorem. [Boo 78], [Wra 75], [BoWr 78], [BaBo 74], [BoBr 79], [Bran 80]

1. $\mathbf{NP} = \Gamma_{\mathrm{hr}_{\mathrm{pol}},\cap}(\mathbf{REG}) = \mathbf{h}_{\mathrm{pol}}(\mathbf{LIN} \wedge \mathbf{LIN}) = \Gamma_{\mathrm{h}_{\mathrm{pol}},\cap\mathrm{R}}(\mathbf{EQ}^2)$.
2. $\mathbf{EXRUD} = \Gamma_{\mathrm{hr}_{\mathrm{pol}},\cap,-}(\mathbf{REG}) = \Gamma_{\mathrm{h}_{\mathrm{pol}},-}(\mathbf{LIN} \wedge \mathbf{LIN}) = \bigcup_k \Sigma_k^{\mathbf{P}}$.
3. $\mathbf{1\text{-}T\text{-}NSPACE}(\log) = \mathbf{h}_{\mathrm{pol}}(\mathbf{1\text{-}DCA} \wedge \mathbf{1\text{-}DCA} \wedge \mathbf{1\text{-}DCA})$. ∎

For relativized versions see [BoWr 78].

11.21. Theorem.

1. [BaBo 74] $\mathrm{NTIME}(\mathrm{Pol}\, t) = \mathrm{h}_{\mathrm{pol}\, t}(\mathrm{LIN} \wedge \mathrm{LIN})$, for $t \geqq \mathrm{id}$.
2. [BoGrWe 70] $\mathbf{multiT\text{-}NTIME}(t) = \mathbf{h}_t(\mathbf{Q}) = \mathbf{h}_t(\mathbf{CF} \wedge \mathbf{CF} \wedge \mathbf{CF})$, for $t \geqq \mathrm{id}$.

3. [BoBr 79] **multiT-NTIME**$(\Gamma_{\mathrm{SUB}}(t)) = \mathrm{h}_{\Gamma_{\mathrm{SUB}}(t)}(\mathbf{EQ}^2 \wedge \mathbf{REG})$, for $t(n) \geqq n^2$.
4. [Boo 79 b] **multiT-NTIME**$(\Gamma_{\mathrm{SUB}}(t)) = \Gamma_{\mathrm{hr}_t,\cap}(\mathbf{REG})$, for superadditive t, $t(n) \geqq n^2$. ∎

11.22. Theorem.
1. [BoGrWe 70] **XSPACE**$(s) = \mathbf{h}_s(\mathbf{XLINSPACE})$, for $s \geqq \mathrm{id}$ and $\mathrm{X} \in \{\mathrm{D}, \mathrm{N}\}$.
2. [Bran 80] **1-T-NSPACE**$(\log s) = \mathbf{h}_{\mathrm{pol}\,s}(\mathbf{1\text{-}DCA} \wedge \mathbf{1\text{-}DCA} \wedge \mathbf{1\text{-}DCA})$, for $s \geqq \mathrm{id}$. ∎

For related results see [Čul 79] and [ČuDi 79], for relativized versions see [Boo 79 b]. In [Gre 78 a] grammar-like characterizations of $\Gamma_{\mathrm{hr(REG)}}$ and $\Gamma_{\mathrm{hd(REG)}}$ are given.

11.2.2.3. Weak Transitive Closure

Assume $\# \notin \Sigma$ and $A, B \subseteqq \Sigma^* \# \Sigma^*$. The set B is said to be the *weak transitive closure* of A (for short: $B = \mathrm{wtc}(A)$) iff for all $x, y \in \Sigma^*$: $x \# y \in B \leftrightarrow$ there are $z_0, z_1, \ldots, z_r \in \Sigma^*$ $(r \geqq 0)$ such that

a) $x = z_0, y = z_r$,
b) $|x| = |z_i|$, for $i = 0, 1, \ldots, r$,
c) $z_{i-1} \# z_i \in A$, for $i = 1, \ldots, r$.

It has been found that in some interesting cases the closure under wtc is precisely the difference between time and space (if any), i.e. some space classes are the wtc-closure (joint with other operations under which the time classes are closed) of the corresponding time classes (cf. 11.19.2, 11.20.2 and 11.21.4).

11.23. Theorem. [Boo 79 c], [Boo 79 d], [BoBr 79]
1. **NLINSPACE** $= \Gamma_{\mathrm{hr_{lin}},\cap,\mathrm{wtc}}(\mathbf{REG}) = \Gamma_{\mathrm{hd_{lin}},\cap,\mathrm{wtc}}(\mathbf{REG})$.
2. **PSPACE** $= \Gamma_{\mathrm{hr_{pol}},\cap,\mathrm{wtc}}(\mathbf{REG}) = \Gamma_{\mathrm{hd_{pol}},\cap,\mathrm{wtc}}(\mathbf{REG}) = \Gamma_{\mathrm{h_{pol}},\cap\mathrm{R},\mathrm{wtc}}(\mathbf{EQ}^2)$.
3. **NSPACE**$(\Gamma_{\mathrm{SUB}}(s)) = \Gamma_{\mathrm{hr}_s,\cap,\mathrm{wtc}}(\mathbf{REG}) = \Gamma_{\mathrm{hd}_s,\cap,\mathrm{wtc}}(\mathbf{REG})$, for superadditive $s \geqq \mathrm{id}$.

For related results see [Jon 68]; relativized versions can be found in [Boo 79 d].

§ 12. Complexitiy Classes and Grammars

Since some grammar-defined classes of languages are of great importance in formal language theory, it is quite natural to compare them with complexity classes.

Especially the time and tape bounds for the decision of context-free languages have been investigated from the very beginning of the study of computational complexity, and the problem of whether these bounds can be improved is one of the most important problems in complexity theory.

In this section we deal in detail with the complexity of regular, linear, context-free, and context-sensitive languages, and quote some results concerning the complexity of languages of other grammar-defined classes.

12.1. Right(Left)-Linear Grammars

The class of right-linear or left-linear languages (i.e. languages defined by regular grammars) coincides with various complexity classes with small bounds.

12.1. Theorem.

1. **REG** = **1-T-DSPACE**(s)	if $0 \leq s \prec \log$.*)
= **2-T-DSPACE**(s) = **2-T-NSPACE**(s)	if $0 \leq s \prec \log\log$.
2. **REG** = **T-DTIME**(t) = **T-NTIME**(id)	if $n \leq t(n) \prec n \cdot \log n$.
3. **REG** = **T-DCROS**(t) = **T-NCROS**(k)	if $1 \leq t \prec \log$, $k \geq 1$.
4. **REG** = **T-DREV**(t) = **T-NREV**(k)	if $0 \leq t \prec \log$, $k \geq 0$.
5. **REG** = **T-DRET**(k) = **T-NRET**(k)	for $k = 0, 1$.
6. **REG** = **T-DDUR**(k) = **T-NDUR**(k)	for $k = 0, 1$.

Proof. Ad 1. Corollary 8.4.2 (Corollary 8.29.2) yields **1-T-DSPACE**$(s) \subseteq$ **REG** for $s \prec \log$ (**2-T-NSPACE**$(s) \subseteq$ **REG** for $s \prec \log\log$). On the other hand, **REG** = **1-DFA** $\subseteq$ **1-T-DSPACE**(0) by 4.7.

Ad 2. Corollary 8.18 implies **T-DTIME**$(t) \subseteq$ **T-DCROS**(Const) for $t(n) \prec n \cdot \log n$. By Statement 3 we have **T-DCROS**(Const) $\subseteq$ **REG**. On the other hand, **REG** $\subseteq$ **1-DFA** $\subseteq$ **T-DTIME**(id) by 4.7.

Ad 3. Let M be an NTM accepting a language L such that **T-NCROS**$_M(n) \leq k$ for some $k \geq 1$. We describe the work of a 1-NFA simulating M on an input word $a_1 a_2 \ldots a_n$. First A guesses nondeterministically a crossing sequence of M (of length $\leq k$) for the boundary on the left-hand side of a_1. Since the work of M on the left of a_1 depends only on this crossing sequence, A can "know" whether this sequence is compatible with M's work on the left of a_1. Then A guesses nondeterministically a crossing sequence (of length $\leq k$) for the boundary between a_1 and a_2, and using a_1 and the preceding crossing sequence A checks whether it is compatible with M's work on the left of a_2. If not, then A rejects on this computation path. A continues in such a manner up to the boundary on the right-hand side of a_n. Finally, A guesses a crossing sequence and checks whether it can appear on the right-hand side of the input word.

Now, by 4.7 and 8.16.1 we have for $t \prec \log$, **T-DCROS**$(t) \subseteq$ **T-DCROS**(Const) $\subseteq$ **T-NCROS**(Const) $\subseteq$ **1-NFA** $\subseteq$ **REG** $\subseteq$ **1-DFA** $\subseteq$ **T-DCROS**(1).

Ad 4. By Statement 3 and 4.7 we get for $t \prec \log$: **T-NREV**(Const) $\subseteq$ **T-NCROS** (Const) $\subseteq$ **REG** $\subseteq$ **1-DFA** $\subseteq$ **T-DREV**(0) $\subseteq$ **T-DREV**(t) $\subseteq$ **T-DCROS**(t) $\subseteq$ **REG**.

Ad 5. An NTM M such that T-NRET$_M(n) \leq 1$ cannot return to a tape square after the first alteration of its contents. Therefore M need not alter any symbol. Since cycles can be cut off from an accepting computation path of M, we have T-NCROS$_M$ $(n) \preceq 1$.

By Statement 3 and 4.7 we get **T-NCROS**(Const) $\subseteq$ **REG** $\subseteq$ **1-DFA** $\subseteq$ **T-DRET**(0).

Ad 6. First we outline the proof of **T-NDUR**(1) $\subseteq$ **1-NFA**. Let M be an NTM such that T-NDUR$_M \leq 1$ and w.l.o.g. it halts on the rightmost input symbol. We simulate M by a 1-NFA A. Since M alters the contents of a tape square either during the first visit or it does not alter it at all the squares become "frozen" after their first visits and therefore it is possible to replace the movements of M's head to the left by table

*) See Footnote ** on p. 61.

look-ups. If A scans the ith input symbol, then its state contains the corresponding state of M and in a finite table the information about how the left part of the Turing tape with respect to its ith input symbol reacts when it is visited by the head of M. More precisely, this table answers the following four questions: In which state M can leave the nonempty zone left of square i to right (left, left or right) after entering it from the right (right, left or left, resp.). This table is updated in every step of A: If the head of M goes one step to the right, then so does the head of A, and knowing what M prints it can easily update the table. If M's head goes to the left, then A's head does not go to the left but uses its table to learn in which state M's head can come back. Furthermore, A is able to find out in its finite memory how the table must be changed after such visits which cause the head to replace blanks on the left end by nonblank symbols. When the head reaches the first blank on the right the work of M may not yet be finished. However, using in addition a similar table for entering the zone right of the input word, A is able to find out wether M can reach the accepting state.

Because of 4.7 we have **T-NDUR**(1) $\subseteq$ **REG**. On the other hand, we have **REG** $\subseteq$ **1-DFA** $\subseteq$ **T-DDUR**(0). ∎

12.2. Linear Grammars

The class of linear languages (i.e. languages generated by linear grammars) can be characterized in terms of reversal-bounded pushdown automata or Turing machines.

12.2. Theorem. LIN = 1-PD-NREV(1) = **1-T-NREV**(1).

Proof. 1. Let L be a linear language. By 4.17.2 there are homomorphisms h_1 and h_2 and a regular language R such that $L = \{h_1(w)h_2(w)^{-1} : w \in R\}$. A 1-NPDA A can accept L easily in the following manner. Let w be the input word. First A guesses nondeterministically a symbol a_1 and checks whether $w = h_1(a_1)w_1$ for some w_1. If this is so, then a_1 is pushed into the store, and A guesses a_2 and checks whether $w_1 = h_1(a_2)w_2$ for some w_2 etc. After having guessed, checked, and stored $a_1, a_2, \ldots, a_r$ in this way, and having stated (simultaneously) that $a_1a_2\ldots a_r \in R$, the 1-NPDA A decides nondeterministically to check whether

$$w_r = h_2(a_1a_2\ldots a_r)^{-1} = h_2(a_r)^{-1} \ldots h_2(a_2)^{-1}h_2(a_1)^{-1}.$$

During this process the pushdown store is emptied. If any check ends negatively, A rejects w, otherwise w is accepted.

2. Evidently, **1-PD-NREV**(1) = **1-T-NREV**(1).

3. Let L be accepted by a 1-NPDA A whose pushdown head makes at most one reversal and (w.l.o.g.) cannot simply change the top symbol (i.e. it can either push or pop or leave the top simple unchanged). Furthermore, A accepts (w.l.o.g.) with an empty pushdown store. Representing a step of A by (s, a, b, c) where s is the state, a is the symbol read by the input head if it moves and the empty word otherwise, b is the top symbol and c is of the form e, $x\downarrow$ or $\uparrow$ indicating that b is not changed, x is pushed or b is popped, respectively, we can describe a computation path of A

by a sequence of the form

$$\sigma = (s_1, a_1, b_1, c_1) \dots (s_m, a_m, b_m, c_m).$$

If during this computation altogether r symbols are stored, then there exist exactly $2r$ steps, say $i_1, \dots, i_{2r}$, such that $c_{i_j} = x_j\downarrow$ for $j = 1, \dots, r$ and some $x_1, \dots, x_r$, $c_{i_{r+1}} = \dots = c_{i_{2r}} = \uparrow$ and $b_{i_{r+1}} = x_r, \dots, b_{i_{2r}} = x_1$. Emphasizing these $2r$ quadruples, σ has the form

$$\sigma = u_1(s_{i_1}, a_{i_1}, \square, x_1\downarrow) \dots u_r(s_{i_r}, a_{i_r}, x_{r-1}, x_r\downarrow)\, v_{r+1}(s_{i_{r+1}}, a_{i_{r+1}}, x_r, \uparrow) \dots v_{2r}(s_{i_{2r}}, a_{i_{2r}}, x_1, \uparrow)\, v_{2r+1}.$$

Now we introduce for every $v = (s, a, b, c) \dots (s', a', b', c')$ its modified inverse $w = [s', a', b', c'] \dots [s, a, b, c]$ and define

$$\hat{\sigma} = w_{2r+1} u_1(s_{i_1}, a_{i_1}, x_1, a_{i_{2r}}, s_{i_{2r}})\, w_{2r} \dots u_r(s_{i_r}, a_{i_r}, x_r, a_{i_{r+1}}, s_{i_{r+1}})\, w_{r+1}$$

and the homomorphisms h_1, h_2 by

$$h_1\big((s, a, b, c)\big) = a, \quad h_1\big((s, a, x, b, t)\big) = a, \quad h_1([s, a, b, c]) = \mathrm{e},$$

$$h_2\big((s, a, b, c)\big) = \mathrm{e}, \quad h_2\big((s, a, x, b, t)\big) = b, \quad h_2([s, a, b, c]) = a.$$

The input word w can be obtained from $\hat{\sigma}$ by $w = h_1(\hat{\sigma})\,\big(h_2(\hat{\sigma})\big)^{-1}$. Now define

$$R = \{\hat{\sigma} : \sigma \text{ is the description of an accepting computation path of } A'\}.$$

The set R can be accepted by a 2-NFA, the head of which runs first from the left to the right and then back to the left. Consequently, R is regular.

Finally, we have: $w \in L \leftrightarrow$ there is a $\hat{\sigma} \in R$ such that $h_1(\hat{\sigma}) h_2(\hat{\sigma})^{-1} = w$. ∎

Remark. By the same method it is proved in [Kam 70b] that

$$\textbf{1-T-NREV}(k) = \{\{h_1(w)h_2(w)^{-1}h_3(w) \dots h_{k+1}(w)^\tau : w \in R\} : h_1, h_2, \dots, h_{k+1} \text{ homomorphisms}, R \in \textbf{REG}\}$$

for $k \geqq 1$ and $\tau = (-1)^k$. This class is said to be the class of k-unfolded regular languages. For related results see [Kui 71], [Sir 71] and [Iba 73b].

Now we deal with the time and space complexity of linear languages.

12.3. Theorem. [Kas 72]
1. $\textbf{LIN} \subseteq \textbf{T-DTIME}(n^2)$.
2. $\textbf{LIN} \subseteq \textbf{NL}$.

Proof. Ad 1. Let L be a language generated by the linear grammar G. Then there is an equivalent grammar $G' = [\Sigma, N, S, R]$ having only rules of the forms $A \to Ba$, $A \to aB$, and $A \to \mathrm{e}$ where $A, B \in N$, $a \in \Sigma$. Now we describe the work of a DTM M accepting L. On an input word $w = a_1 a_2 \dots a_n$ ($a_1, a_2, \dots, a_n \in \Sigma$), the work of M will consist of the stages $0, 1, \dots, n$. At the end of the ith stage the tape of M has the contents

$$a_1 a_2 \dots a_i (a_{i+1}, N_{i,i+1}) (a_{i+2}, N_{i,i+2}) \dots (a_n, N_{i,n}) N_{i,n+1},$$

where

$$N_{i,k} =_{\mathrm{df}} \left\{B : B \in N,\ S \overset{*}{\underset{G'}{\Longrightarrow}} a_1 a_2 \ldots a_i B a_k a_{k+1} \ldots a_n\right\}$$

for $k = i + 1, i + 2, \ldots, n + 1$. Evidently, each $(a_k, N_{i,k})$ can be written in one tape square, and w is accepted iff there is an $i \leqq n$ and $B \in N_{i,i+1}$ such that $(B \to \mathrm{e}) \in R$.

In the 0th stage M first determines $N_{0,n+1} = \{S\}$ and then successively $N_{0,k} = \{B$: there is an $A \in N_{0,k+1}$ such that $(A \to Ba_k) \in R\}$.

In the ith stage $(i > 0)$ M first replaces all $N_{i-1,k}$ for $k > i$ by $N'_{i,k} =_{\mathrm{df}} \{B$: there is an $A \in N_{i-1,k}$ such that $(A \to a_i B) \in R\}$ and then determines successively $N_{i,k} = N'_{i,k} \cup \{B$: there is an $A \in N'_{i,k+1}$ such that $(A \to Ba_k) \in R\}$.

Evidently, M can perform every stage within linear time. Thus M can accept L within time n^2 (applying linear speed-up, see 18.12.1).

Ad 2. A linear language L can be accepted by a 2-NTM M' working in principle in the same way as the DTM M in the above proof. The only difference is that M' nondeterministically chooses in the ith stage a number $k_i \in \{i + 1, i + 2, \ldots, n + 1\}$ such that $i < k_i \leqq k_{i-1}$, and determines in the 0th stage

$$\hat{N}_{0,k_0} =_{\mathrm{df}} \left\{B : S \overset{*}{\underset{G'}{\Longrightarrow}} Ba_{k_0} a_{k_0+1} \ldots a_n\right\}$$

and in the ith stage $(i > 0)$

$$\hat{N}_{i,k_i} =_{\mathrm{df}} \Big\{B\text{: there is an } A \in \hat{N}_{i-1,k_{i-1}} \text{ and a } C \in N \text{ such that } (A \to a_i C) \in R \text{ and } C \overset{*}{\underset{G'}{\Longrightarrow}} Ba_{k_i} a_{k_i+1} \ldots a_{k_{i-1}-1}\Big\}.$$

Again, w is accepted iff there is an $i \leqq n$ and a $B \in N_{i,i+1}$ such that $(B \to \mathrm{e}) \in R$. Thus M has to store, at any point of its computation, at most three numbers between 0 and $n + 1$ (namely i, k_{i-1}, k_i), which can be done in logarithmic space. ∎

Remark. It is not known whether **LIN** $\subseteq$ **L**. However **LIN** $\subseteq$ **L** is equivalent to **L** = **NL** (cf. 23.19).

12.3. Context-Free Grammars

The class of context-free languages (i.e. the languages generated by context-free grammars) coincides with the class of languages accepted by nondeterministic one-way pushdown automata and even with the realtime class of these automata. This is shown by the following theorem, whose first equality has already been proved in [Chom 62] and [Eve 63].

12.4. Theorem. CF = 1-NPDA = 1-PD-NTIME(id).

Proof. Since **CF = 1-NPDA** is known from 4.8 and **1-PD-NTIME**(id) $\subseteq$ **1-NPDA** is obvious it remains to show **CF** $\subseteq$ **1-PD-NTIME**(id).

Let $L \in$ **CF**. By 4.3, $L \setminus \{\mathrm{e}\} = \mathrm{L}(G)$ for some context-free grammar $G = (\Sigma, N, S, P)$ which is in Greibach normal form, i.e. having only rules of the form $A \to a$, $A \to aB$, and $A \to aBC$, where $A, B, C \in N$ and $a \in \Sigma$.

Consider the 1-NPDA M with the input alphabet Σ, the pushdown alphabet N, the set $N \cup \{E\}$ of states ($E \notin N$), the initial state S, the accepting state E, and the instructions given in Table 12.1. Evidently, M checks whether the input word can be generated by a leftmost derivation of G. Furthermore, M works in realtime. ∎

Table 12.1

the rule of G	induces the instructions	
$(A \to a)$	$AaD \to D \uparrow +$	for all $D \in N$
	$Aa\square \to E\square +$	
$(A \to aB)$	$AaD \to BD +$	for all $D \in N$
$(A \to aBC)$	$AaD \to BC \downarrow +$	for all $D \in N$

The class **CF** also appears as a complexity class of the measures T-NRET and T-NDUR.

12.5. Theorem. [Wec 76], [Hib 67] For $k \geqq 2$, $\mathbf{CF} = \mathbf{T\text{-}NRET}(k) = \mathbf{T\text{-}NDUR}(k)$.

Proof. Because of 21.9.1 and the remark after 21.12 it remains to show that $\mathbf{CF} \subseteq \mathbf{T\text{-}NRET}(2)$ and $\mathbf{CF} \subseteq \mathbf{T\text{-}NDUR}(2)$. We do the first, the latter can be done analogously.

Let $L \in \mathbf{CF}$ and $L = \mathrm{L}(G)$ where $G = (\Sigma, N, S, P)$ is a context-free grammar in Greibach normal form (4.3). We describe an NTM M which accepts L with $\mathrm{NRET}_M \leqq 2$. We assume that the tape of M has two tracks. The first one holds the input word $w = a_1 \ldots a_n$ within the squares $1, \ldots, n$. We let B_i denote the contents of the second track of square i. M tries to simulate the generation of w by G. If $B_i \in N$, $B_{i+1} = \ldots = B_k = \square$ and $B_{k+1} \neq \square$ or $k = n$, then M tries to verify $B_i \overset{*}{\underset{G}{\Longrightarrow}} a_1 \ldots a_k$. In detail, M works as follows on input w:

1. $B_1 \leftarrow S$. $i \leftarrow 1$.
2. Guess nondeterministically a production $B_i \to a_i V$.
3. **If** $V = \mathrm{e}$ **then if** $B_{i+1} = \square$ **then if** $i = n$ **then** accept.
   else reject.
   else if $i = n$ **then** reject.
   else $i \leftarrow i + 1$. Go to 2.
4. **If** $V = B$ **then if** $i = n$ or $B_{i+1} \neq \square$ **then** reject.
   else $B_{i+1} \leftarrow B$. $i \leftarrow i + 1$. Go to 2.
5. **If** $V = BC$ **then if** $i \in \{n - 1, n\}$ or $B_{i+1} \neq \square$ or $B_{i+2} \neq \square$
   then reject.
   else guess nondeterministically $j \in \{i + 2, \ldots, n\}$ such that $B_k = \square$ for $k \in \{i + 2, \ldots, j\}$. $B_j \leftarrow C$. Return to square $i + 1$. $B_{i+1} \leftarrow B$. $i \leftarrow i + 1$. Go to 2. ∎

Now we deal with the time and space requirements for the decision of context-free languages. We restrict ourselves to the case of context-free languages L which do not include the empty word, since an algorithm which decides L can easily be converted into an algorithm which decides $L \cup \{e\}$ and which does not require more time or space.

Let $G = (\Sigma, N, S, P)$ be a context-free grammar, which is without loss of generality in Chomsky normal form (cf. 4.3.3), and let $w = a_1 a_2 \ldots a_n$, where $a_1, a_2, \ldots, a_n \in \Sigma$. For $i = 1, \ldots, n$ and $j = i, \ldots, n$ define

$$N_{i,j} = \left\{A : A \overset{*}{\underset{G}{\Longrightarrow}} a_i a_{i+1} \ldots a_j\right\}.$$

Obviously, $w \in \mathrm{L}(G) \leftrightarrow S \in N_{1,n}$. Therefore we can decide the question of "$w \in \mathrm{L}(G)$?" by the successive computation of the sets $N_{i,j}$, according to the evident formulas

$$N_{i,i} = \{A : (A \to a_i) \in P\} \quad \text{for} \quad i = 1, \ldots, n,$$

$$N_{i,j} = \bigcup_{k=i}^{j-1} N_{i,k} \cdot N_{k+1,j} \quad \text{for} \quad i = 1, \ldots, n-1 \quad \text{and} \quad j = i+1, \ldots, n$$

where

$$N' \cdot N'' =_{\mathrm{df}} \left\{A : \bigvee_{B \in N'} \bigvee_{C \in N''} (A \to BC) \in P\right\} \quad \text{for} \quad N', N'' \subseteq N.$$

It is not hard to see that this can be done by a multitape Turing machine within time $\preceq n^3$.

12.6. Theorem. [Your 67] $\mathbf{CF} \subseteq \mathbf{multiT\text{-}DTIME}(n^3)$. ∎

Remark. In [IgHo 74] it is shown that $\mathbf{CF} \subseteq \mathbf{1\text{-}3PD\text{-}DTIME}(n^3)$.

However, this is not the best known result. Faster algorithms for the decision of context-free languages can be derived from fast algorithms for matrix multiplication.

With the context-free grammar $G = (\Sigma, N, S, P)$, which is in Chomsky normal form, we associate the G-multiplication for $m \times m$ matrices with elements from $\mathfrak{P}(N)$, which is defined as follows:

$$a \circ b = c \quad \text{where} \quad c_{i,j} = \bigcup_{k=1}^{m} a_{i,k} \cdot b_{k,j}$$

($\cdot$ being defined as above). Furthermore, we define the transitive G-closure "$^+$" for $m \times m$ matrices with elements from $\mathfrak{P}(N)$ as follows:

$$a^+ = \bigcup_{k=1}^{\infty} a^{(k)} \quad \text{where} \quad a^{(1)} =_{\mathrm{df}} a \quad \text{and} \quad a^{(k)} =_{\mathrm{df}} \bigcup_{j=1}^{k-1} a^{(j)} \circ a^{(k-j)}.$$

Now consider the $(m+1) \times (m+1)$ matrix b which is defined as follows

$$b_{i,j} = \begin{cases} N_{i,i} & \text{if} \quad j = i+1, \\ \emptyset & \text{otherwise,} \end{cases} \qquad i = 1, \ldots, m+1, \; j = 1, \ldots, m+1.$$

It is not hard to see that $N_{i,j} = b^+_{i,j}$. This leads us to the following lemma. For an $m \times m$ matrix a, we define its linear representation

$$g(a) = a_{11} a_{12} \ldots a_{1m} a_{21} a_{22} \ldots a_{2m} \ldots a_{m1} a_{m2} \ldots a_{mm}.$$

Furthermore, we define the function G-TC (transitive G-closure):

$$\text{G-TC}(w) = \begin{cases} g(a^+) & \text{if } w = g(a) \text{ and } a \text{ is an } m \times m\text{-matrix for some } m, \\ \mathrm{e} & \text{otherwise.} \end{cases}$$

The subsequent lemmas can be found in a similar form in [Val 75].

12.7. Lemma. For $t \geqq \mathrm{id}$,

$$\text{G-TC} \in \text{multiT-DTIME}(t) \to \mathbf{CF} \subseteq \textbf{multiT-DTIME}(t(n^2)). \ \blacksquare$$

The following lemma shows that transitive G-closure is no more difficult than G-multiplication of matrices, which is represented by the function

$$\text{G-MM}(w) = \begin{cases} g(a \circ b) & \text{if } w = g(a)\, g(b) \text{ and } a, b \text{ are } m \times m \text{ matrices for some } m \in \mathbf{N}, \\ \mathrm{e} & \text{otherwise.} \end{cases}$$

Note that this lemma is the crucial one for the proof of Theorem 12.10.

12.8. Lemma. For every rational $r \geqq 1$,

$$\text{G-MM} \in \text{multiT-DTIME}(n^r) \to \text{G-TC} \in \text{multiT-DTIME}(n^r). \ \blacksquare$$

The following lemma shows that G-multiplication of matrices is no more difficult than Boolean multiplication of matrices. We define

$$\text{BMM}(w) = \begin{cases} g(a \circ b) & \text{if } w = g(a)\, g(b) \text{ and } a, b \text{ are 0-1 matrices of size } m \times m \text{ for some } m \in \mathbf{N}, \\ \mathrm{e} & \text{otherwise,} \end{cases}$$

where here $a \circ b = c$ with $c_{i,j} = \bigvee_{k=1}^{m} a_{i,k} \wedge b_{k,j}$.

12.9. Lemma. For every rational $r \geqq 1$,

$$\text{BMM} \in \text{multiT-DTIME}(n^r) \to \text{G-MM} \in \text{multiT-DTIME}(n^r). \ \blacksquare$$

By the Lemmas 12.7, 12.8 and 12.9 we have

12.10. Theorem. [Val 75] For every rational $r \geqq 1$,

$$\text{BMM} \in \text{multi T-DTIME}(n^r) \to \mathbf{CF} \in \text{multiT-DTIME}(n^{2r}). \ \blacksquare$$

Finally, the multiplication of Boolean $m \times m$ matrices can be implemented by multitape TM's as fast as the multiplication of $m \times m$ matrices over the finite ring $\{0, 1, \ldots, m\}$ (the Boolean elements are treated as integers modulo $m + 1$, and the nonzero elements are reduced to one in the result).

At present (i.e. end of 1983), the best time bound for the matrix multiplication on multitape Turing machines is $\sqrt{n^{2.496}}$ (cf. [Tar 80] and [CoWi 81]).

12.11. Corollary. $\mathbf{CF} \subseteq \textbf{multiT-DTIME}(n^{2.496})$. $\blacksquare$

Next we deal with the time complexity of context-free languages on random access machines.

12.12. Theorem. [Weic 77]

1. **CF** $\subseteq$ **RAM-DTIME**(Lin $n^2 \cdot \log n$).
2. **CF** $\subseteq$ **BRAM-DTIME**(Lin n^2). ∎

Proof. Ad 1. Let $G = (\Sigma, N, S, P)$ be a context-free grammar, which is without loss of generality in Chomsky normal form (cf. 4.3.3), and let $w = a_1 a_2 \ldots a_n$ where $a_1, a_2, \ldots, a_n \in \Sigma$. For $A \in N$ and $i = 1, \ldots, n$ we define

$$A(i) = \left\{j\colon A \overset{*}{\underset{G}{\Longrightarrow}} a_{j+1} \ldots a_i\right\}.$$

Consequently, $w \in \mathrm{L}(G) \leftrightarrow 0 \in S(n)$. Furthermore we get

$$A(i) = \bigcup_{k=1}^{i} \left(A(k, i) \cap \{k - 1\}\right), \tag{1}$$

where

$$A(i, i) =_{\mathrm{df}} \begin{cases} \{i - 1\} & \text{if } (A \to a_i) \in P, \\ \emptyset & \text{otherwise}, \end{cases} \tag{2}$$

and for $k = 1, \ldots, i - 1$,

$$A(k, i) =_{\mathrm{df}} \left\{j\colon j < k \wedge \bigvee_{m=k}^{i-1} \bigvee_{B \in N} \bigvee_{C \in N} \left((A \to BC) \in P \wedge B \overset{*}{\underset{G}{\Longrightarrow}} a_{j+1} \ldots a_m \wedge C \overset{*}{\underset{G}{\Longrightarrow}} a_{m+1} \ldots a_i\right)\right\}.$$

For $k < i$ the set $A(k, i)$ can also be expressed as

$$\begin{aligned} A(k, i) &= \left\{j\colon j < k \wedge \bigvee_{B \in N} \bigvee_{C \in N} \left((A \to BC) \in P \wedge B \overset{*}{\underset{G}{\Longrightarrow}} a_{j+1} \ldots a_k \wedge C \overset{*}{\underset{G}{\Longrightarrow}} a_{k+1} \ldots a_i\right)\right\} \cup \left(A(k + 1, i) \setminus \{k\}\right) \\ &= \left\{j\colon j < k \wedge \bigvee_{B \in N} \bigvee_{C \in N} \left((A \to BC) \in P \wedge j \in B(k) \wedge k \in C(k + 1, i)\right)\right\} \cup \left(A(k + 1, i) \setminus \{k\}\right) \\ &= \bigcup_{\substack{(A \to BC) \in P \\ k \in C(k+1, i)}} \left(B(k) \cup \left(A(k + 1, i) \setminus \{k\}\right)\right). \end{aligned} \tag{3}$$

Based on the formulas (1), (2) and (3) we get the following algorithm, which computes the sets $A(k, i)$ and $A(i)$ and decides whether $w \in \mathrm{L}(G)$ $\left(\leftrightarrow 0 \in S(1, n)\right)$.

For $i = 1, \ldots, n$ **do**

  For $A \in N$ **do**

    $A(i) \leftarrow \emptyset$

    If $(A \to a_i) \in P$ **then** $A(i, i) \leftarrow \{i - 1\}$ **else** $A(i, i) \leftarrow \emptyset$

For $k = i - 1, i - 2, \ldots, 1$ **do**
  For $A \in N$ **do**
    $A(k, i) \leftarrow \emptyset$
    For $B, C \in N$ **do**
      if $(A \to BC) \in P$ **and** $k \in C(k + 1, i)$ **then** $A(k, i) \leftarrow A(k, i) \cup B(k)$
    $A(k, i) \leftarrow A(k, i) \cup \big(A(k + 1, i) \setminus \{k\}\big)$
    if $k - 1 \in A(k, i)$ **then** $A(i) \leftarrow A(i) \cup \{k - 1\}$

If $0 \in S(1, n)$ **then accept else reject.**

Now we explain how the sets $A(i)$ and $A(k, i)$ are represented in the registers of a random access machine and how the operations $\cup$ and $\setminus$ and the tests can be performed by a DRAM. The sets $A(i)$ and $A(k, i)$ are represented as binary numbers of length $n \cdot (c + \log n)$ (for a sufficiently large $c > 0$) consisting of n blocks each of length $c + \log n$:

block $n-1$ block $n-2$ ... block 1 block 0

For the set $A(i)$, block j has the contents 00 ... 00 (00 ... 01) if $j \notin A(i)$ $\big(j \in A(i)\big)$. For the set $A(k, i)$, block j has the contents 00 ... 00 if and only if $j \notin A(k, i)$, i.e. for $j \in A(k, i)$, block j has some contents not representing 0.

The Table 12.2 shows how to replace the instruction I of the above algorithm by programs P which can be performed by a DRAM. The details, for instance where the number $2^{(c+\log n)\cdot k}$ comes from, are left to the reader.

Table 12.2

	I	P
a	$R \leftarrow \emptyset$	$R \leftarrow 0$
b	$R \leftarrow \{k\}$	$R \leftarrow 2^{(c+\log n)\cdot k}$
c	$R \leftarrow R_1 \cup R_2$	$R \leftarrow R_1 + R_2$
d	$R \leftarrow A(k + 1, i) \setminus \{k\}$	$R \leftarrow A(k + 1, i)$ **For** $r = (c + \log n) \cdot k + c + \log n, \ldots,$ $(c + \log n) \cdot k$ **do** **if** $2^r \leq R$ **then** $R \leftarrow R - 2^r$
e	$k \in A(k + 1, i)$?	$A(k + 1, i) \geq 2^{(c+\log n)\cdot k}$?

Comments: Ad c. The instruction $R \leftarrow R_1 \cup R_2$ is used in the algorithm in such a way that each block of the representation of $A(i)$ $(i = 1, \ldots, n)$ has the contents 0 or 1, and registers representing $A(l)$ can be added to an initially empty register representing $A(k, i)$ at most $d \cdot n$ times where $d = \text{card}\,(N \times N)$. Thus, in the blocks, numbers of maximum length $\log (d \cdot n) \leq c + \log n$ are generated, which fit into their blocks and do not interfere with neighbouring blocks. Ad d and e. For the correctness of the corresponding programs P, it is important that $j \notin A(k+1, i)$ for all $j > k$, i.e. that all blocks left of the block k have the contents $00\ldots00$.

Time estimation: Each instruction of the program is executed at most $\preceq n^2$ times. The instruction $R \leftarrow A(k+1, i) \setminus \{k\}$ can be executed by a DRAM within $\preceq \log n$ steps, all other instructions can be executed by a DRAM within a constant number of steps.

Ad 2. In contrast to 1, the blocks of the representation of $A(k, i)$ also have the contents $00\ldots01$ or $00\ldots00$. Then the instruction $R \leftarrow R_1 \cup R_2$ can be replaced by $R \leftarrow R_1 \vee R_2$ and the instruction $R \leftarrow A(k+1, i) \setminus \{k\}$ can be replaced by $R \leftarrow \sim 2^{(c+\log n)\cdot k} \wedge A(k+1, i)$. All other instructions are treated as in 1. Thus every instruction of the algorithm can be executed by a DBRAM within a constant number of steps. ∎

Remark. In [Ryt 84] it is shown that the context-free languages can be decided on-line by deterministic RAM's within time Lin $n^3/\log^2 n$.

12.13. Theorem.

1. [Kos 75] $\mathbf{CF} \subseteq \mathbf{IA^1\text{-}DTIME}(n^2)$.
2. [Kos 75] $\mathbf{CF} \subseteq \mathbf{IA^2\text{-}DTIME}\big((1+\varepsilon)\cdot n\big)$, for all $\varepsilon > 0$.
3. [Smi 72] $\mathbf{CF} \subseteq \mathbf{IA^1\text{-}NTIME}(\mathrm{id})$. ∎

12.14. Theorem. [LeStHa 65]

$$\mathbf{CF} \subseteq \mathbf{DSPACE}\big((\log n)^2\big).$$

Proof. Because of 12.4 and 23.2 we have

$$\begin{aligned}\mathbf{CF} &= \mathbf{1\text{-}PD\text{-}NTIME}(\mathrm{id}) = \mathbf{1\text{-}auxPD\text{-}T\text{-}NSPACE\text{-}TIME}(0, \mathrm{id})\\ &\subseteq \mathbf{2\text{-}auxPD\text{-}T\text{-}NSPACE\text{-}TIME}(\log, \mathrm{Pol}) \subseteq \mathbf{DSPACE}\big((\log n)^2\big). \ ∎\end{aligned}$$

Remark 1. In [LeStHa 65] the above result has been proved via context-free grammars using Lemma 8.21 which says that a generation $A \overset{*}{\underset{G}{\Longrightarrow}} w$ in a Chomsky normal form context-free grammar G can always be split up into $A \overset{*}{\underset{G}{\Longrightarrow}} w_1 B w_3$ and $B \overset{*}{\underset{G}{\Longrightarrow}} w_2$ such that $w = w_1 w_2 w_3$ and $\frac{1}{3}|w| \leqq |w_2| \leqq \frac{2}{3}|w|$. All possible w_2's and B's are tried successively. To check $A \overset{*}{\underset{G}{\Longrightarrow}} w_1 B w_3$ and $B \overset{*}{\underset{G}{\Longrightarrow}} w_2$ the lemma is applied again. This leads to a recursive procedure of recursion depth $\asymp \log |w|$ (because of the length condition). Since in each recursion step w_2 has to be stored and this can be done by two numbers between 0 and $|w|$ this recursive procedure can be implemented within space $\log^2 |w|$. This outlines the proof given in [LeStHa 65].

Remark 2. For some consequences of $\mathbf{CF} \nsubseteq \mathbf{XL}$ or $\mathbf{CF} \subseteq \mathbf{XL}$, see the Remarks 3, 4 and 5 after 13.15.

Remark 3. Though the question dealt with in Remark 2 is still unresolved, a lot of subclasses of **CF** has been exhibited which are included in **NL** or even in **L**. Here are some of them:

1. **1-NCA** $\subsetneqq$ **NL** (cf. 13.6).
2. [Bar 74] **1-PD-NREV(Const)** $\subsetneqq$ **NL**. Note that this is a generalization of **LIN** $\subsetneqq$ **NL** (cf. 12.2, 12.3.2).
3. **1-DCA** $\subsetneqq$ **L** (cf. 13.6).
4. The bounded context-free languages belong to **L** where a set is called *bounded* if it is a subset of $a_1^* \dots a_m^*$ for some m and $a_1, \dots, a_m \in \Sigma$ (see [RiSp 72]).
5. Input-driven languages are those languages which can be accepted by **1-DPDA**'s (or, equivalently, by **1-NPDA**'s) in realtime in such a way that the moves of the pushdown head depend only on the symbols read on the input tape. The paranthesis languages (see [Meh 75], [Lyn 77]) and the leftmost Szilard-languages (see Theorems 4.9.4 and 12.18.3) are input-driven. Input-driven languages are in **L** (see [BrVe 84]). For further such subclasses see [RiSp 72].

Remark 4. Because of 5.10.2, Theorem 12.14 yields **CF** $\subsetneqq$ **DSPACE-TIME**$((\log n)^2, n^{\mathrm{Lin}(\log n)})$, and this is the best known result. For DCF, however, we have **DCF** $\subsetneqq$ **DSPACE-TIME**$((\log n)^2, \mathbf{Pol})$ (see [Coo 79]).

Finally, we mention **CF** $\in$ **NC** ([Ruz 79a]) (**NC** defined on p. 402).
Parallel algorithms for DCF-recognition of order $\log n$ are contained in [Ref 82].

12.4. *Context-Sensitive Grammars*

The class of context-sensitive languages (i.e. languages defined by context-sensitive grammars) can be characterized in terms of space-bounded computations.

12.15. Theorem. [Kur 64], [Lan 63] **CS** = **NLINSPACE**.

Proof. 1. **CS** $\subsetneqq$ **NLINSPACE**. Let $G = (\Sigma, N, S, R)$ be a context-sensitive grammar. For any $w \in \mathrm{L}(G)$ and any derivation $S \underset{G}{\Longrightarrow} v_1 \underset{G}{\Longrightarrow} v_2 \underset{G}{\Longrightarrow} \dots \underset{G}{\Longrightarrow} v_r = w$ we have $1 = |S| \leqq |v_1| \leqq |v_2| \leqq \dots \leqq |v_r| = |w|$ because of $|u| \leqq |v|$ for every $(u \to v) \in R$. Therefore an NTM M starting with w and using no more than $|w|$ tape squares can apply nondeterministically all possible rules in reverse order. If in this manner S is reached on some computation path, M accepts w on this path.

2. **NLINSPACE** $\subsetneqq$ **CS**. For **L** $\in$ **NLINSPACE** $\cap\, \Sigma^*$ we choose an NTM $M = (S, \Delta, \delta, s_0, \{s_1\}, \emptyset)$ accepting L in space $n + 1$, i.e. the head never leaves the input word to the left, but it is allowed to move one step to the right of the input word. This is possible because of the linear speed-up (cf. 18.11). W.l.o.g. we assume that a) s_0 never appears after the first step, b) the accepting state appears when the head scans the leftmost square of the workspace, c) a nonblank is never replaced by the blank, and d) whenever the head leaves the input word to the right it returns immediately without changing the blank. We consider the language $L_1 = L \setminus (\Sigma \cup \{e\})$ and construct a context-sensitive grammar $G = (\Sigma, N, S', P)$ such that $L(G) = L_1$. By adding the productions $S' \to a$ for $a \in L \cap \Sigma$ we obtain a grammar for $L \setminus \{e\}$.

Together with $\Delta' = \Delta \setminus \{\Box\}$ we use $\Delta_1 = \{\bar{a} : a \in \Delta'\}$ and $\Delta_2 = \{\bar{\bar{a}} : a \in \Delta'\}$ and define $N = \{S'\} \cup \Delta_1 \cup \Delta_2 \cup S \times (\Delta \cup \Delta_1 \cup \Delta_2)$. P contains three groups of productions. A first one, which will not be specified, enables us to generate the regular set $(\{s_1\} \times \Delta_1)\, \Delta^* \Delta_2$ from S'. Speaking in terms of M these productions allow to generate final ID's of M from the start symbol. The productions of the second group simulate the work of M backward and allow obtaining initial ID's of M from final ones. They are listed as follows:

$$\left.\begin{array}{l} y(s', z) \to (s, x)\, z \\ \bar{y}(s', z) \to (s, \bar{x})\, z \\ y(s', \bar{\bar{z}}) \to (s, x)\, \bar{\bar{z}} \end{array}\right\} \quad \text{for } z \in \Delta' \text{ and } (s', y, +1) \in \delta(s, x),$$

$$(s', \bar{y}) \to (s, \bar{\bar{x}}) \quad \text{for } x \in \Delta' \text{ and } (s'', y, +1) \in \delta(s, x) \text{ and } (s', \Box, -1) \in \delta(s'', \Box),$$

$$\left.\begin{array}{l} (s', z)\, y \to z(s, x) \\ (s', \bar{z})\, y \to \bar{z}(s, x) \\ (s', z)\, \bar{\bar{y}} \to z(s, \bar{\bar{x}}) \end{array}\right\} \quad \text{for } z \in \Delta' \text{ and } (s', y, -1) \in \delta(s, x),$$

$$\left.\begin{array}{l} (s', y) \to (s, x) \\ (s', \bar{y}) \to (s, \bar{x}) \\ (s', \bar{\bar{y}}) \to (s, \bar{\bar{x}}) \end{array}\right\} \quad \text{for } z \in \Delta' \text{ and } (s', y, 0) \in \delta(s, x).$$

The productions of the third group allow to eliminate s_0 and the bars on the first and last symbol:

$$\left.\begin{array}{l} (s_0, \bar{a})\, b \to a(s_0, b) \\ (s_0, a)\, b \to a(s_0, b) \\ (s_0, a)\, \bar{\bar{b}} \to ab \end{array}\right\} \quad a, b \in \Delta'. \blacksquare$$

From this, Theorem 21.3.1, and Theorem 21.6.1 we immediately get

12.16. Corollary. **CS** = **T-NCROS**(id) = **T-NREV**(id). ∎

12.5. Further Results

The first result in this subsection should be compared with the result **CS** = **NLINSPACE**. The inclusion **CS** ⊊ **DEXPTIME** can be made to be an equality by adding context-free control sets to context-sensitive grammars.

12.17. Theorem. [Mon 77a] **CS[CF]** = **DEXPTIME**. ∎

Remark. Further characterizations of **DEXPTIME** in terms of grammars can be found in [Rou 75] and [Bra 78]. The latter paper also includes such a characterization of **DTIME**(2^{Pol}). In [DeMe 77] the class **EXPSPACE** is shown to coincide with the class of languages generated by the *van Wijngaarden grammars* of type L or, equivalently, of type R.

Next we deal with the time and space complexity of Szilard languages.

12.18. Theorem. [Iga 77], [Pen 77]

1. $\mathbf{SZ} \subsetneq \mathbf{T\text{-}NTIME}(n^2) \subsetneq \mathbf{DLINSPACE}$.
2. $\mathbf{SZ(CF)} \subsetneq \mathbf{1\text{-}DSPACE}(\log)$.
3. $\mathbf{SZL} \subsetneq \mathbf{L} \cap \mathbf{1\text{-}T\text{-}NTIME}(\mathrm{id}) \cap \mathbf{multiT\text{-}DTIME}(n^{2.496})$.
4. $\mathbf{SZL(CF)} \not\subseteq \mathbf{1\text{-}NSPACE}(s)$, for $s \prec \mathrm{id}$.

Remark. The time results of 12.18.3 are a consequence of $\mathbf{SZL} \subsetneq \mathbf{CF}$ (cf. 4.9.4). The set S# can serve as a witness for 12.18.4.

Finally, we give some results on the time and space complexity of L-languages (i.e. languages generated by L-systems).

12.19. Theorem.

1.1. [Dal 71], [Woo 75] $\mathbf{EP2L} = \mathbf{EPDT2L} = \mathbf{EPT2L} = \mathbf{NLINSPACE}$.
1.2. [Vit 76] $\mathbf{EPD2L} = \mathbf{DLINSPACE}$.
2.1. [Vit 77] $\mathbf{EP1L} = \mathbf{EPDT1L} = \mathbf{EPT1L} = \mathbf{NLINSPACE}$.
2.2. [Vit 76] $\mathbf{EPD1L} \subset \mathbf{DLINSPACE}$.
3.1. [Lee 76] $\mathbf{ETOL} \subset \mathbf{DLINSPACE}$.
3.2. [JoSk 77a] $\mathbf{ETOL} \subsetneq \mathbf{NP}$.
3.3. [JoSk 77b], [Sud 77b] $\mathbf{EDTOL} \subsetneq \mathbf{NL}$.
3.4. [Lee 75b] $\Gamma_{\mathrm{AFL,h}}(\mathbf{EOL}) \subsetneq \mathbf{NSPACE\text{-}TIME}((\log n)^2, \mathrm{Pol})$.
3.5. [Sud 77b] $\mathbf{EOL} \subsetneq \mathbf{DSPACE}((\log n)^2)$.
3.6. [Gra 77] $\mathbf{EOL} \subsetneq \mathbf{multiT\text{-}DTIME}(n^{3.496})$.
3.7. [Sud 77b] $\mathbf{EDOL} \subsetneq \mathbf{L}$. ∎

Remark 1. Note that $\mathbf{E2L} = \mathbf{ET2L} = \mathbf{ED2L} = \mathbf{EDT2L} = \mathbf{E1L} = \mathbf{ET1L} = \mathbf{EDT1L} = \mathbf{RE}$ and $\mathbf{ED1L} \subset \mathbf{RE}$ (cf. 4.10.2 and 4.10.3).

Remark 2. $\mathbf{ETOL} \subsetneq \mathbf{P}$ would imply $\mathbf{P} = \mathbf{NP}$ because there are **ETOL** languages which are $\leq_{\mathrm{m}}^{\log}$-complete in **NP** (cf. [Lee 75a]).

Remark 3. Because of $\mathbf{CF} \subsetneq \mathbf{EOL} \subsetneq \mathcal{R}_{\mathrm{m}}^{\log}(\mathbf{CF})$ (cf. [Sud 77b]) we have $\mathbf{EOL} \subsetneq \mathbf{XL} \leftrightarrow \mathbf{CF} \subsetneq \mathbf{XL}$, for $\mathrm{X} \in \{\mathrm{D}, \mathrm{N}\}$.

§ 13. Complexity Classes and Automata

Although the declared intentions of this book do not include dealing with subrecursive computation models (automata), this section is one of the most important. From the very beginning of complexity theory, the investigation of the computational complexity of automata defined languages has played an important role. And this is still the case today. There are several reasons for this phenomenon:

1. Some classes of languages defined by 1-way automata (e.g. the context-free languages) are of great practical and theoretical importance. Therefore good algorithms for deciding these languages are of interest.
2. Some classes of languages defined by 2-way automata coincide with time or space complexity classes. Very often the corresponding proofs carry over to the

case where these automata have an infinite memory (Turing tape) instead of a finite one only (see § 20.2).

3. Most of the central problems in the theory of computational complexity can be formulated as problems regarding the relationships between different classes of automata (see chapters VII and VIII).

In what follows we investigate only the most important types of automata, i.e. finite, counter, pushdown, stack, nonerasing stack, and checking stack automata. A comparison of the time and tape complexities of the languages decided/accepted by these types of automata can be found at the end of this section.

13.1. Finite Automata

13.1.1. Relationships to Time and Space Complexity Classes

13.1. Theorem. For 1-way finite automata we have the following relationships to time and space complexity classes:

	automata	time	space
1	**1-DFA**	= **T-DTIME**(id)	= **DSPACE**(0)
2	**1:k-DFA**	⊆ **kT-DTIME**(Lin)	⊆ **L**
3	**1-NFA**	= **T-DTIME**(id)	= **DSPACE**(0)
4		⊆ **2T-DTIME**(n^{k+1})	⊆ **NL**
5	**1:k-NFA**	⊆ **RAM-DTIME**(Lin n^k)	
6		⊆ **1-2T-NTIME**(id)	

Proof. Time results.

Ad 1. This statement follows from 4.7 and 12.1.2.

Ad 2. A 1:k-DFA A can be simulated by a kT-DM M in the following way. First M makes copies of the input on the other $k-1$ tapes within $2n+2$ steps. Then the k heads of M work exactly as the k heads of A. Since A halts on any input, at least one head has to move after q steps if A has q states. Therefore A works at most $q \cdot k \cdot n$ steps. Consequently, M works at most $q \cdot k \cdot n + 2n + 2$ steps.

Ad 3. This statement is an immediate consequence of 4.7 and 12.1.2.

Ad 4. Let A be a 1:k-NFA with the set S of states, the initial state s_0, the (w.l.o.g. only) accepting state s_1 and the next move function $\delta: S \times \{0, 1, \square\}^k \mapsto \mathfrak{P}(S \times \{0, +1\}^k)$. W.l.o.g. we assume that A moves in one step exactly one head, i.e. $(s', \sigma_1, \ldots, \sigma_k) \in \delta(s, a_1, \ldots, a_k)$ implies $\sigma_1 + \ldots + \sigma_k = 1$, and that A halts in the case of acceptance with all heads on the first $\square$ to the right side of the input. For the input w with $|w| = n$, the heads of A can visit only the tape squares $1, 2, \ldots, n, n+1$ with the symbols $w(1), w(2), \ldots, w(n)$, $\square = w(n+1)$. Let $S(i_1, i_2, \ldots, i_k, t)$ be the set of all states which are possible after exactly t steps if the heads have the positions $i_1, \ldots, i_k$ ($i_j = 1, 2, \ldots, n+1$ and $j = 1, \ldots, k$). Since every computation path has length $k \cdot n$, we have: A accepts $w \leftrightarrow s_1 \in S(n+1, n+1, \ldots, n+1, k \cdot n)$. Now we describe the work of a 2T-DTM M computing $S(n+1, n+1, \ldots, n+1, k \cdot n)$. The work of M is subdivided into the stages $0, 1, \ldots, k \cdot n$. In the tth stage M computes the sets $S(i_1, i_2, \ldots, i_k, t)$ in such a way that at the end of this stage the word

$$u_t = \prod_{i_1=1}^{n+1} \prod_{i_2=2}^{n+1} \ldots \prod_{i_k=1}^{n+1} w(i_1)\, w(i_2) \ldots w(i_k)\; S(i_1, i_2, \ldots, i_k, t)$$

is written on the second tape, where every word $w(i_1) w(i_2) \ldots w(i_k)\; S(i_1, i_2, \ldots, i_k, t)$ is written in one tape square which is called the tape square $(i_1, i_2, \ldots, i_k)$.

For the 0th stage we have $S(1, 1, \ldots, 1, 0) = \{s_0\}$ and $S(i_1, i_2, \ldots, i_k, 0) = \emptyset$ for $(i_1, i_2, \ldots, i_k) \neq (1, 1, \ldots, 1)$. Therefore u_0 can be computed within $\lesssim n^k$ steps. In the $(t+1)$th stage, starting with u_t, the word u_{t+1} must be computed. To this end it is sufficient to replace the sets $S(i_1, i_2, \ldots, i_k, t)$ by $S(i_1, i_2, \ldots, i_k, t+1)$. We use the fact that $s \in S(i_1, i_2, \ldots, i_k, t)$ and $(s', \sigma_1, \ldots, \sigma_k) \in \delta(s, w(i_1), w(i_2), \ldots, w(i_k))$ imply $s' \in S(i_1 + \sigma_1, i_2 + \sigma_2, \ldots, i_k + \sigma_k, t+1)$. Remember that exactly one of the σ's is $+1$, the others are 0. Consequently, M can proceed successively for $j = 1, 2, \ldots, k$ as follows: The head of tape 2 moves over the word u_t and being in the tape square $(i_1, i_2, \ldots, i_k)$ it collects all states s', for which there is a state $s \in S(i_1, i_2, \ldots, i_k, t)$ such that $(s', \underbrace{0, \ldots, 0}_{j-1}, +1, 0, \ldots, 0) \in \delta\big(s, w(i_1), w(i_2), \ldots, w(i_k)\big)$. The head of tape 1 writes these states in the square $(i_1, \ldots, i_{j-1}, i_j + 1, i_{j+1}, \ldots, i_k)$ of tape 1, which is situated $(n+1)^{k-j}$ tape squares to the right of $(i_1, i_2, \ldots, i_k)$. After doing so for $j = 1, 2, \ldots, k$ the contents of the square $(i_1, i_2, \ldots, i_k)$ of tape 1 are $S(i_1, i_2, \ldots, i_k, t+1)$. Finally, M replaces the set $S(i_1, i_2, \ldots, i_k, t)$ in the square $(i_1, \ldots, i_k)$ of tape 2 by the set $S(i_1, i_2, \ldots, i_k, t+1)$ and erases the contents of tape 1. For the $(t+1)$th stage M needs $\lesssim n^k$ steps. After the $(k \cdot n)$th stage M checks whether $s_1 \in S(n+1, n+1, \ldots, n+1, k \cdot n)$ and accepts if and only if the answer is yes. For all stages M needs $\lesssim n^{k+1}$ steps, but **2T-DTIME**(Lin n^{k+1}) = **2T-DTIME**(n^{k+1}) for $k \geqq 1$.

Ad 5. This statement is a special case of 13.2.6.

Ad 6. The inclusion **1:k-NFA** $\subseteq$ **kT-NTIME**(Lin) can be proved as in the deterministic case (Statement 2). Actually, it is possible that the heads of a 1:k-NFA do not move q steps or more, but cutting out all cycles one gets an equivalent (with respect to acceptance) computation path of length at most $q \cdot k \cdot n$. By 22.2 we have **kT-NTIME**(Lin) $\subseteq$ **Q** = **1-2T-NTIME**(id).

Space results. The statements 1 and 3 follows from 4.7 and 12.1.1. The statements 2 and 4 are immediate consequences of 13.2.2 and 13.2.5 respectively. ∎

13.2. Theorem. For 2-way finite automata we have the following relationships to time and space complexity classes:

	automata	time	space
1	**2-DFA**	$=$ **T-DTIME**(id)	$=$ **DSPACE**(0)
2	**2:k-DFA**	$\subseteq$ **kT-DTIME**(n^k)	$\subseteq$ **L**
3	**2:multi-DFA**	$\subseteq$ **P**	$=$ **L**
4	**2-NFA**	$=$ **T-DTIME**(id)	$=$ **DSPACE**(0)
5		$\subseteq$ **2T-DTIME**(n^{2k})	$\subseteq$ **NL**
6	**2:k-NFA**	$\subseteq$ **RAM-DTIME**(Lin n^k)	
7		$\subseteq$ **kT-NTIME**(n^k)	
8	**2:multi-NFA**	$\subseteq$ **P**	$=$ **NL**

Proof. Time results. The statements 1 and 4 follow from 4.7 and 12.1.2. The statements 3 and 8 are immediate consequences of the corresponding space results and the inclusions $\mathbf{L} \subseteq \mathbf{NL} \subseteq \mathbf{P}$ (see 5.13.1).

Ad 2. Because of Statement 1 we can restrict ourselves to the case $k \geqq 2$. A 2:k-DFA A can be simulated by a kT-DM as described in the proof of 13.1.2. Since A has only $q(n+2)^k$ different situations (q being the number of states), any finite computation path has at most $q \cdot (n+2)^k$ steps and the simulating kT-DM works at most $q \cdot (n+2)^k + 2n + 2$ steps. The linear speed-up for multiT-DTIME completes the proof.

Ad 5. This statement can be proved as 13.1.4. Since it is necessary to perform $q \cdot (n+2)^k$ stages, the simulating 2T-DM works $\lesssim n^{2k}$ steps. The linear speed-up for 2T-DTIME completes the proof.

Ad 6. Let A be a 2:k-NFA. To an input word w we assign a graph $G(A, w)$ whose nodes are the $q \cdot (n+2)^k$ possible situations of A on the input word w and whose edges are the pairs (S_1, S_2) such that $S_1 \vdash_{\overline{A}} S_2$, i.e. such that S_2 is a possible next situation of S_1. Let m be the maximum number of possible next situations for any situation of A. The problem of whether $w \in \mathrm{L}(A)$ is equivalent to the problem of whether there is a path in $G(A, w)$ from the initial situation to an accepting situation.

For the situation S which is determined by the state $p \in \{0, 1, \ldots, q-1\}$ and the positions $i_1, \ldots, i_k \in \{0, 1, \ldots, n+1\}$ of the input heads we define $r_S =_{\mathrm{df}} \left(q \cdot \sum_{j=1}^{k} i_j(n+2)^{j-1} + p\right) \cdot (m+1) + N$ where N is a sufficiently large constant. Now a DRAM M can record for a given input word w the graph $G(A, w)$ in such a way that the elements of $\{r_{S_2} : S_1 \vdash_{\overline{A}} S_2\}$ are stored in the registers $r_{S_1} + 1, r_{S_1} + 2, \ldots, r_{S_1} + m$. The registers $0, 1, \ldots, N-1$ are used for intermediate computations. The registers $N' =_{\mathrm{df}} (m+1) \cdot q \cdot (n+2)^k + N$, $N'+1$, $N'+2, \ldots$ are used to store a push-down list. The DRAM M now checks according to the following algorithm whether there is a path in $G(A, w)$ from the initial situation to an accepting situation.

1. For the initial situation S, execute $R_{r_S} \leftarrow 1$ and place r_S on the list.
2. For every accepting situation S, execute $R_{r_S} \leftarrow 2$.
3. If the list is empty, reject.
4. If i is the top element of the list, then delete i from the list.

   For $j = 1, 2, \ldots, m$ **do**

   if $R_{R_{i+j}} = 2$ **then accept**

   else if $R_{R_{i+j}} = 0$ **then** $R_{R_{i+j}} \leftarrow 1$ and place R_{i+j} on the list.

   Go to 3.

Since every r_S can be placed on the list only once and since the deletion of an element from the list (Item 4 of the algorithm) causes only a constant number of steps, M executes $\lesssim n^k$ steps.

Ad 7. This statement can be proved as Statement 2. Actually, finite computation paths of **2:k-NFA** can have cycles, but cutting out all cycles one gets an equivalent (with respect to acceptance) computation path of length at most $q \cdot (n + 2)^k$ steps.

Space results. Statements 1 and 4 follows from 4.7 and 12.1.1.

Ad 2. The head positions of a **2:k-DFA** can be stored in binary presentation on the tape of a 2-DTM in logarithmic space. With this information the input head of this 2-DTM can find the corresponding squares of the input tape.

Ad 3. Because of Statement 2 we have only to show that $\mathbf{L} \subseteq$ **2:multi-DFA**. Let M be a 2-DTM which works within space $\log n$. As in Item 6 of the proof of Theorem 5.1 we construct a 2-DTM M' such that a) $L(M') = L(M)$, b) M' has only the worktape symbols 0 and 1 and c) for every worktape inscription $\ldots 000 w_1 \overset{\downarrow}{a} w_2 000 \ldots$ ($w_1, w_2^{-1} \in 1\{0, 1\}^* \cup \{e\}$, $a \in \{0, 1\}$ and $\downarrow$ marks the head position) during the work of M' on an input of length n we have $|w_1|, |w_2| \leqq m \cdot \log n \leqq n^m$ for some $m > 0$ and hence $\mathrm{bin}^{-1} w_1, \mathrm{bin}^{-1} w_2 < n^m$. Since every number $k < n^m$ can uniquely be represented in the form $k = \sum_{i=0}^{m-1} \alpha_i n^i$, where $\alpha_0, \ldots, \alpha_{m-1} \in \{0, \ldots, n-1\}$, it can be represented by m input heads having the positions $\alpha_0 + 1, \alpha_1 + 1, \ldots, \alpha_{m-1} + 1$. Thus, the contents of the worktape of M' can be represented by $2m$ input heads of a finite automaton M''. For the initial situation we have $w_1 = w_2 = e$ and therefore all input heads have a priori the correct position. In order to simulate one step of M', operations of the kind $k \to 2k$, $k \to 2k + 1$ and $k \to \left\lfloor \frac{k}{2} \right\rfloor$ must be applied to the two numbers represented by these $2m$ input heads. It can easily be seen that this can be managed by an additional input head. One further input head is necessary for the simulation of the input head of M'.

The statements 5 and 8 can be proved exactly as the statements 2 and 3. ∎

13.3. Theorem.

1. [ChSt 76] **2-AFA** = **T-DTIME**(id).

2. [ChSt76] **2:multi-AFA = P.**
3. [Kin 81a] **1:multi-AFA $\subseteq$ DLINSPACE.** ∎

13.1.2. Relationships between Different Classes of Finite Automata

Because of Theorem 4.7 all types of one-head finite automata are mutually equivalent.

For several types of finite automata it is known that $k + 1$ heads are better than k heads.

13.4. Theorem. For $k \geqq 1$ and X $\in$ {D, N},
1. [YaRi 76] **1:k-XFA $\subset$ 1:(k+1)-XFA.**
2. [Mon 78b] **2:k-XFA $\subset$ 2:(k+1)-XFA.**
3. [Jan 79] **2:k-XTIME**(id) $\subset$ **2:(k+1)-XTIME**(id).
4. [Kin 81a] **2:*k*-AFA $\subset$ 2:(*k*+1)-AFA.** ∎

Remark 1. (On the proof 13.4) Ad 1. In [YaRi 76] it is shown that

$$\left\{w_1 \# w_2 \# \ldots \# w_{\frac{k\cdot(k+1)}{2}} \# w_{\frac{k\cdot(k+1)}{2}} \# \ldots \# w_2 \# w_1 : \bigvee_{m \geqq 1} \bigwedge_i w_i \in \{0, 1\}^m\right\}$$
$$\in \mathbf{1:(k+1)\text{-}DFA} \setminus \mathbf{1:k\text{-}NFA}.$$

The proper inclusion **1-DFA $\subset$ 1:2-DFA** is obvious, **1:2-DFA $\subset$ 1:3-DFA** can be already found in [IbKi 75b].

Ad 2. These hierarchy results are proved by translational methods. Defining $T_k(w) = w^{|w|^{k-1}}$ for $k \geqq 1$, $\mathfrak{M} =_{\mathrm{df}} \mathfrak{P}(\{0^{2^n}; n \in \mathbb{N}\})$, and using the lemmas

a) $A \in \mathbf{XL} \cap \mathfrak{M}$ implies $\bigvee_k T_k(A) \in$ **2:3-XFA,**

b) for $k, j \geqq 1$, $T_k(A) \in$ **2:j-XFA** $\cap\ \mathfrak{M}$ implies $A \in$ **2:(k · j)-XFA,**

c) for $k > j \geqq 2$, $T_{k+1}(A) \in$ **2:j-XFA** $\cap\ \mathfrak{M}$ implies $T_k(A) \in$ **2:(j+1)-XFA,**

it is easy to show that **2:k-XFA = 2:(k+1)-XFA** for $k \geqq 2$ implies **XL** $\cap\ \mathfrak{M} \subseteq$ **2:(k(k+1))-XFA** which contradicts the following lemma:

d) For $k \geqq 1$, **2:k-XFA** $\cap\ \mathfrak{M} \subset$ **XL** $\cap\ \mathfrak{M}$.

Thus an SLA language can serve as a witness set for **2:k-XFA $\subset$ 2:(k+1)-XFA.**

Ad 3. In [Jan 79] it is shown that $P_k =_{\mathrm{df}} \{w_k \# w_k^{-1} \# \ldots \# w_1 \# w_1^{-1} : w_i \in \{0, 1\}^*\} \in$ **2:(k+1)-DTIME**(id) $\setminus$ **2:k-NTIME**(id). Furthermore, $\overline{P_k} \in$ **2:2-NTIME**(id) $\setminus$ **2:k-DTIME**(id) for all $k \geqq 2$.

Remark 2. It is easy to see that there is a **2:(k+2)-DFA** set which is universal for all 2:k-DFA sets, i.e. there is a $B \in$ **2:(k+2)-DFA** such that **2:k-DFA** $= \left\{C: \bigvee_v \bigwedge_w (w \in C \leftrightarrow w \# v \in B)\right\}$ (see [KoMč 69]). By a simple diagonalization argument one gets $\{w: (w \# w) \notin B\} \in$ **2:(k+2)-DFA $\setminus$ 2:k-DFA.**

Remark 3. For 1-way nondeterministic automata with writing input heads there is no infinite hierarchy with respect to the number of heads. Actually, it is obvious that two writing heads are better than one, and it is shown in [Sud 76] that three writing heads are better than two; but it is shown in [Bra 79b] that three writing heads are as good as k writing heads for $k > 3$. Consequently 1-way nondeterministic automata with only three writing heads can simulate all 1:k-NFA for $k \geqq 1$.

Furthermore, there are some proper inclusions between the different types of finite automata having the same number of heads.

13.5. Theorem. For $k \geqq 2$,

1. **1:k-DFA** $\subset$ **2:k-DFA**.
2. **1:k-DFA** $\subset$ **1:k-NFA** $\subset$ **1:k-AFA**.
3. **1:k-NFA** $\subset$ **2:k-NFA**.

Proof. Simple counting arguments show that S $\notin$ **1:k-NFA**. On the other hand it can be readily verified that S $\in$ **2:2-DFA** and $\bar{\text{S}}$ $\in$ **1:2-NFA**. Consequently, S $\in$ **2:k-DFA** $\setminus$ **1:k-DFA**, $\bar{\text{S}}$ $\in$ **1:k-NFA** $\setminus$ **1:k-DFA**, and S $\in$ **2:k-NFA** $\setminus$ **1:k-NFA** for $k \geqq 2$. The set $\{1^{2^n} : n \in \mathbb{N}\}$ ($\notin$ **REG**) belongs to **1:2-AFA** ([Kin 81a]), but because of Theorem 13.19 the SLA-languages in the class **1:multi-NFA** are regular. Therefore, **1:k-NFA** $\subset$ **1:k-AFA** ($k \geqq 2$). ∎

Remark 1. It is not known whether the inclusion **2:k-DFA** $\subseteqq$ **2:k-NFA** is proper for $k \geqq 2$. Theorem 23.18.1.1 shows that **2:k-DFA** = **2:k-NFA** for some $k \geqq 2$ implies **L** = **NL**.

Remark 2. There exists a set in **1:2-AFA** which is $\leqq_m^{\log}$-complete in **P** ([Kin 81a]). This together with Theorems 13.2 and 13.3.2 implies

$$\mathbf{1{:}2\text{-}AFA} \subseteqq \mathbf{2{:}multi\text{-}XFA} \leftrightarrow \mathbf{P} = \mathbf{XL} \qquad (\mathrm{X} = \mathrm{D}, \mathrm{N}).$$

Furthermore,

$$\mathbf{1{:}multiAFA} = \mathbf{2{:}multiAFA} \rightarrow \mathbf{P} \subseteqq \mathbf{DLINSPACE}$$

because of Theorem 13.3.3.

13.2. Counter Automata

13.2.1. Relationships to Time and Space Complexity Classes

13.6. Theorem. For 1-way counter automata we have the following relationships to time and space complexity classes (see table on p. 228).

Proof. Time results. The statements 2 and 7 can be proved in the same way as the corresponding results for 1-way pushdown automata, i.e. in the same way as the time results of 13.15.2 and 13.15.8, respectively. The statements 1 and 3 are immediate consequences of Statement 2.

Ad 5. The lengthy proof can be found in [Gre 75b].

Ad 4. [Gre 75a] Because of Statement 5 we can start with a 1-NCA A that moves its input head in each step. Let S be the set of states, s_0 the initial state, s_1 the

	automata	time	space
1	1-DCA	$\subseteq$ **1-T-DTIME**(Lin)	
2	1:k-DCA	= **1:k-C-DTIME**(Lin)	$\subseteq$ **L**
3		$\subseteq$ **2-kT-DTIME**(Lin)	$\nsubseteq$ **DSPACE**(s), $s \prec \log$
4	1-NCA	$\subseteq$ **1-T-DTIME**(n^2)	
5		= **1-C-NTIME**(id)	
6		$\subseteq$ **2T-DTIME**(n^{k+2})	$\subseteq$ **NL**
7	1:k-NCA	= **1:k-C-NTIME**(Lin)	$\nsubseteq$ **DSPACE**(s), $s \prec \log$
8		$\subseteq$ **1-2T-NTIME**(id)	

(w.l.o.g. only) accepting state and

$$\delta: S \times \{0, 1, \square\} \times \{0, 1\} \to \mathfrak{P}(S \times \{+1\} \times \{+1, 0, -1\})$$

the next move function of A. Further, let $S(j, t)$ be the set of all states which A can reach after exactly t steps having the contents j in the counter. We construct a 1-DTM M whose work is subdivided into the stages $0, 1, \ldots, n$. In the tth stage M computes $S(0, t), S(1, t), \ldots, S(n, t)$ where $S(j, t)$ is stored in the tape square j. Since, on all computation paths of A, in the tth step the input head is situated on the tth symbol of the input word, only that symbol is of interest for the tth stage of M. For the 0th stage we have $S(0, 0) = \{s_0\}$ and $S(j, 0) = \emptyset$ for all $j = 1, 2, \ldots, n$. For the $(t+1)$th stage each $S(j, t)$ must be replaced by $S(j, t+1)$. Since $S(j, t)$ is stored in the jth tape square this can be done by looking at the tape squares $j - 1$, j and $j + 1$. In addition, in the nth stage M checks whether $s_1 \in \bigcup_{j=0}^{n} S(j, n)$ and accepts correspondingly. Thus, M needs for the tth stage $\preceq n$ steps and for all stages $\preceq n^2$ steps. The equality **1-T-DTIME**(Lin n^2) = **1-T-DTIME**(n^2) (18.12.3) completes the proof.

Ad 6. This statement can be proved in the same way as 13.1.4. The counter which has the length $\preceq n$ (Statement 7) can be treated like an additional input head.

Statement 8 is an immediate consequence of Statement 7 and 22.2.

Space results. The statements 2 and 6 are consequences of the space results of 13.7.1 and 13.7.5, resp.

Ad 3 and 7. Evidently, $C = \{0^n 1^n : n \in \mathbb{N}\} \in$ **1-DCA**, and by 8.31 we have $C \notin$ **DSPACE**(s) for $s \prec \log$. ∎

Remark 1. In contrast to 13.6.5 we have **1-C-DTIME**(id) $\subset$ **1-C-DTIME**(Lin). An example for a witness is A^* with $A = \{0^n 1^m : n \geqq m \geqq 1\}$ (see [FipMeRo 68]).

Remark 2. **1-NCA** $\subseteq$ **L** would imply **L** = **NL** since there are 1-NCA languages which are $\leq_m^{\log}$ complete in **NL** (cf. 8.43.3 and Remark 3 after 8.43).

13.7. Theorem. For 2-way counter automata we have the following relationships to time and space complexity classes:

	automata	time	space
1	2:k-DCA	$=$ 2:k-C-DTIME(Lin n^{2k})	$\subseteq$ L
2		$\subseteq$ (k+1)T-DTIME(n^{2k})	$=$ 2:k-C-DSPACE(n^k)
3		$\subseteq$ RAM-DTIME(Lin n^k)	
4	2:multi-DCA	$\subseteq$ P	$=$ L
5	2:k-NCA	$\subseteq$ 2T-DTIME(n^{4k})	$\subseteq$ NL
6		$\subseteq$ RAM-DTIME(Lin $n^{3k}/\log n$)	$=$ 2:k-C-NSPACE(n^{2k})
7		$=$ 2:k-C-NTIME(Lin n^{3k})	
8		$\subseteq$ (k+1)T-NTIME(n^{3k})	
9	2:multi-NCA	$\subseteq$ P	$=$ NL

Proof. Time results.

Ad 1. Because of the space result of Statement 2 the counter of a 2:k-DCA can be assumed to have maximum length n^k. For each fixed length of the counter, a 2:k-DCA can have $\preceq n^k$ different situations. Therefore, any finite computation path is of length $\preceq n^{2k}$.

Ad 2 and 4. These statements are immediate consequences of Statement 1.

Ad 3, 5, and 6. These statements are consequences of 13.20.2, 13.20.5 and 13.20.6.

Ad 7. This statement can be proved in the same way as Statement 1. Actually, finite computation paths of a 2:k-NCA can have cycles, but cutting out all cycles one gets an equivalent (with respect to acceptance) computation path of length $\preceq n^{3k}$.

Ad 8 and 9. These statements are immediate consequences of statements 7 and 5, resp.

Space results.

Ad 1. Because of Statement 2 the contents of the counter of a 2:k-DCA can be stored in binary presentation in exactly the same way as the head positions within $\preceq \log n$ squares of a Turing tape.

Statement 5 can be seen analogously. The statements 2 and 6 can be proved in the same way as the corresponding results for 2-way pushdown automata (see 13.20.1 and 13.20.5, respectively). Statement 4 follows from Statement 1 and the space result of 13.2.3. Statement 9 follows from Statement 5 and the space result of 13.2.8. ∎

13.2.2. Relationships between Different Classes of Counter Automata

We start with some relationships between 2-way counter automata and 2-way finite automata.

13.8. Theorem. For $k \geqq 1$,

1. **2:k-DFA** $\subseteqq$ **2:k-DCA** $\subseteqq$ **2:(2k)-DFA**.
2. **2:k-NFA** $\subseteqq$ **2:k-NCA** $\subseteqq$ **2:(3k)-NFA**.

Proof. The inclusion **2:k-DFA** $\subseteqq$ **2:k-DCA** and **2:k-NFA** $\subseteqq$ **2:k-NCA** are obvious. We show **2:k-DCA** $\subseteqq$ **2:(2k)-DFA**. Because of the space result of 13.7.2 we can choose a 2:k-DCA whose counter has maximum length n^k. Consequently, the counter can be simulated by k additional input heads.

The inclusion **2:k-NCA** $\subseteqq$ **2:(3k)-NFA** can be proved analogously. ∎

Two-way deterministic (nondeterministic) counter automata with $k + 1$ input heads are more powerful than those with k input heads.

13.9. Theorem. For $k \geqq 1$ and $X \in \{D, N\}$, **2:k-XCA** $\subset$ **2:(k+1)-XCA**. ∎

Remark. The witness sets can be chosen to be SLA languages. The proof is similar to the proof of the corresponding result for finite automata (see [MeWa 82]).

As in the case of finite automata, there are some proper inclusions between the different types of counter automata having the same number of heads.

13.10. Theorem. For $k \geqq 1$,

1. **1:k-DCA** $\subset$ **2:k-DCA**.
2. **1:k-DCA** $\subset$ **1:k-NCA**.
3. **1:k-NCA** $\subset$ **2:k-NCA**.
4. [Chb 84] **2-DCA** $\subset$ **2-NCA**.

Proof. Simple counting arguments show $S \notin$ **1:k-NCA** for all $k \geqq 1$. On the other hand, $S \in$ **2-DCA** and $\overline{S} \in$ **1-NCA**. For Statement 4 see [Chb 84]. ∎

Remark. It is not known whether the inclusion **2:k-DCA** $\subseteqq$ **2:k-NCA** is proper for $k > 1$. Theorem 23.18.1.2 shows that **2:k-DCA** $=$ **2:k-NCA** for some $k > 1$ implies **L** $=$ **NL**.

13.3. Pushdown Automata

13.3.1. Relationships to Time and Space Complexity Classes

In order to transfer complexity results from one-head pushdown automata to k-head pushdown automata, a translational method is used. The input words to be recognized by an automaton with k heads will be padded in such a way that they can be recognized by an automaton with one head only. The crucial point is a special structure of the enlarged word which is given by the function f_k defined as follows: Let $w = a_1a_2\dots a_n$, $a_1, \dots, a_n \in \{0, 1\}$, $a_0 = a_{n+1} = \Box$, and $b_1, b_2, \dots, b_k \notin \{0, 1, \Box\}$. We define

$$f_k(w) = \prod_{i_k=0}^{n+1}\left[\left(\dots\prod_{i_2=0}^{n+1}\left[\left(\prod_{i_1=0}^{n+1}[(a_{i_1}, \dots, a_{i_k})\, b_1]\right) b_2^{n+2}\right]\dots\right) b_k^{n+2}\right].$$

13.11. Lemma. [Sud 77a] For $k \geqq 2$ and $X \in \{D, N\}$,

1. $f_k \in$ T-DTIME(Lin n^k) $\cap$ RAM-DTIME(Lin n^k) $\cap$ L.
2. For every 1:k-XPDA A there exists a 1-XPDA A' such that $L(A) \leqq_m^{\log} L(A')$ via f_k.

Proof. The complexity results (Statement 1) are evident. Now we describe how a 1-XPDA A' simulates on the input $f_k(w)$ a 1:k-XDPA A working on the input $w = a_1a_2\dots a_n$. The input tape square of A with the inscription a_j is said to be the input tape square j. The input tape square of A' with the inscription $(a_{i_1}, a_{i_2}, \dots, a_{i_k})$ is said to be the input tape square $(i_1, i_2, \dots, i_k)$. A step of A moving its k heads from the tape squares $i_1, i_2, \dots, i_k$ to the tape squares $i_1', i_2', \dots, i_k'$ will be simulated by A' in a stage of its work beginning (ending) with the same contents of the pushdown store as A and with the head on tape square $(i_1, i_2, \dots, i_k)$ $\left((i_1', i_2', \dots, i_k')\right)$. Since all symbols read by the heads of A are stored in the square $(i_1, i_2, \dots, i_k)$ of its input tape, A' can change the top symbol of its pushdown store in the same way as A. Thus it is sufficient to describe how the input head of A moves from the square $(i_1, i_2, \dots, i_k)$ to the square $(i_1', i_2', \dots, i_k')$. W.l.o.g. we assume that A moves at most one input head in each step. If A moves the jth head, then the input head of A' has to move from the square $(i_1, \dots, i_j, \dots, i_k)$ to the square $(i_1, \dots, i_j + 1, \dots, i_k)$. We observe

a) There is exactly one block of b_j's between the squares $(i_1, \dots, i_j, \dots, i_k)$ and $(i_1, \dots, i_j + 1, \dots, i_k)$.

b) There are exactly $(n + 2) - i_l$ blocks of b_l's between the square $(i_1, \dots, i_j, \dots, i_k)$ and the next block of b_{l+1}'s $(l = 1, \dots, j - 1)$.

c) There are exactly i_l blocks of b_l's between the previous block of b_{l+1} and the square $(i_1, \dots, i_j + 1, \dots, i_k)$.

First A' proceeds successively for $l = 1, 2, \dots, j - 1$ as follows: A' moves its input head up to the next block of b_{l+1}'s and writes one b_{l+1} for every block of b_l's onto its pushdown store. Running through this next block of b_{l+1}'s, A' deletes all $(n + 2) - i_l$ symbols b_{l+1} (see b) from the pushdown store and writes i_l symbols b_{l+1} onto it. Then A' proceeds successively for $l = j - 1, j - 2, \dots, 1$ as follows: A' moves its input head through as many blocks of b_l's as there are b_{l+1}'s in the pushdown store and deletes these b_{l+1}'s from there. After doing so for $l = 1$ the input head has reached the square $(i_1, \dots, i_j + 1, \dots, i_k)$ (see c). ∎

Using this lemma it is not hard to prove the next theorem which is of the same spirit as 6.14.

13.12. Theorem.

1. For $k \geqq 1$, X ∈ {D, N}, Y ∈ {mT: $m \geqq 2$} ∪ {multiT, RAM}, Z ∈ {D, N} and increasing $t \geqq$ id,

$$\textbf{1-XPDA} \subseteqq \textbf{Y-ZTIME}(t) \quad \text{implies} \quad \textbf{1:k-XPDA} \subseteqq \textbf{Y-ZTIME}\big(t(\text{Lin } n^k)\big).$$

2. For $k \geqq 2$, X ∈ {D, N}, Z ∈ {D, N} and increasing $s \geqq \log$

$$\textbf{1-XPDA} \subseteqq \textbf{ZSPACE}(s) \quad \text{implies} \quad \textbf{1:k-XPDA} \subseteqq \textbf{ZSPACE}\big(s(\text{Lin } n^k)\big). \ \blacksquare$$

The proofs of 13.11 and 13.12 can be used almost without modification to prove analogous results for two-way pushdown automata.

13.13. Lemma. Let $k \geqq 2$ and X ∈ {D, N}. For every 2:k-XPDA A there exists a 2-XPDA A' such that $L(A) \leqq_m L(A')$ via f_k. If A works within polynomial time, then A' also works within polynomial time. ∎

13.14. Theorem. 1. For $k \geqq 2$, X ∈ {D, N}, Y ∈ {mT: $m \geqq 2$} ∪ {multiT, RAM}, Z ∈ {D, N} and increasing $t \geqq$ id,

$$\textbf{2-XPDA} \subseteqq \textbf{Y-ZTIME}(t) \quad \text{implies} \quad \textbf{2:k-XPDA} \subseteqq \textbf{Y-ZTIME}\big(t(\text{Lin } n^k)\big).$$

2. For $k \geqq 2$, X ∈ {D, N}, Z ∈ {D, N} and increasing $s \geqq \log$

$$\textbf{2-XPDA} \subseteqq \textbf{ZSPACE}(s) \quad \text{implies} \quad \textbf{2:k-XPDA} \subseteqq \textbf{ZSPACE}\big(s(\text{Lin } n^k)\big). \ \blacksquare$$

13.15. Theorem. For 1-way pushdown automata we have the following relationships to time and space complexity classes:

	automata	time	space
1	**1-DPDA**	$\subseteqq$ **1-T-DTIME**(Lin)	
2	**1:k-DPDA**	= **1:k-PD-DTIME**(Lin)	$\subseteqq$ **DSPACE**$((\log n)^2)$
3		$\subseteqq$ **2-kT-DTIME**(Lin)	$\nsubseteq$ **DSPACE**(s); $s \prec \log$
4	**1-NPDA**	= **1-PD-NTIME**(id)	
5		$\subseteqq$ **1-T-NTIME**(id)	
6		$\subseteqq$ **multiT-DTIME**$(n^{2.496k})$	$\subseteqq$ **DSPACE**$((\log n)^2)$
7	**1:k-NPDA**	$\subseteqq$ **RAM-DTIME**(Lin $n^{2k} \cdot \log n$)	$\nsubseteq$ **DSPACE**(s), $s \prec \log$
8		= **1:k-PD-NTIME**(Lin)	
9		$\subseteqq$ **1-2T-NTIME**(id)	

Proof. Time results.

Ad 2 and 8. Let X ∈ {D, N}. In the proof of 13.33.1 and 13.33.3 (time) it is shown that for every 1:k-XSA A_1 there is an equivalent 1:k-XSA A_4 such that: if w is accepted by A_4, then w is accepted on a computation path having at most $k \cdot n$ writing moves and $k \cdot n$ erasing moves. Since the construction of A_4 causes no additional stack scans, this construction also applies to the case of pushdown automata.

Ad 4. The above argument yields **1-NPDA = 1-PD-NTIME**(Lin). However 12.4 shows even **1-NPDA = 1-PD-NTIME**(id).

The statements 1, 3 and 5 are immediate consequences of the statements 2, 2 and 4, resp. Statement 6 follows from 4.8, 12.11 and 13.12.1. Statement 7 follows from of 4.8, 12.12.1 and 13.12.1.

Ad 9. Statement 8 implies **1:k-NPDA** $\subseteq$ **2-kT-NTIME**(Lin). The equality **multiT-NTIME**(Lin) = **1-2T-NTIME**(id) (see 22.2) completes the proof.

Space results.

The statements 2 and 6 are immediate consequences of 4.8, 12.14 and 13.12.2. The statements 3 and 7 follows from the space result of 13.6.3. ∎

Remark 1. In [Coo 79] the stronger result is proved that **1-DPDA** $\subseteq$ **DSPACE-TIME**$((\log n)^2$, Pol) (see also 25.6).

Remark 2. In contrast to 13.15.4 we have **1-PD-DTIME**(id) $\subset$ **1-PD-DTIME**(Lin). An example for a witness is

$$\{w \# v \# u \# x\colon (x = w^{-1} \vee u = v^{-1}) \wedge w, v, u, x \in \{0, 1\}^*\}.$$

Remark 3. 1-NPDA $\not\subseteq$ **XL** would imply **XL** $\subset$ **P** because of **1-NPDA** $\subseteq$ **P**. But since **2:multi-NPDA** = **P** (cf. 13.20.8) it is even not known whether **2:multi-NPDA** $\not\subseteq$ **XL**, for X $\in$ {D, N}.

Remark 4. 1-NPDA $\subseteq$ **L** would imply **L** = **NL** because there are 1-NPDA languages which are $\leq_m^{\log}$ complete in **NL** (cf. 8.43.3 and Remark 3 after 8.43. See also 23.18 and 23.19).

Remark 5. 1-NPDA $\subseteq$ **XL** would imply **2-auxPD-T-NSPACE-TIME**(log, Pol) $\subseteq$ **XL**, for X $\in$ {D, N} (cf. 14.3).

However, it is not known whether **1-XPDA** $\subseteq$ **YL** (X, Y $\in$ {D, N}). The following theorem, which is a consequence of 14.2, deals with this problem.

13.16. Theorem. [Gre 73] There is an $A \in$ **1-NPDA** such that

$$A \in \mathbf{XL} \leftrightarrow \mathbf{1\text{-}NPDA} \subseteq \mathbf{XL}, \quad \text{for} \quad X \in \{D, N\}.\ ∎$$

Some subclasses of **1-NPDA** which are shown to be included in **L** or **NL** are listed in Remark 3 after 12.14. Now we quote some additional results on the complexity of 1-way PDA languages. For further results see 12.13.

13.17. Theorem. [Col 71] **1-DPDA** $\subseteq$ **IA**^{multi}**-DTIME**(id). ∎

13.18. Theorem. [KaVo 70]

1. **1-DPDA** $\subseteq$ **T-DREV**(id).
2. **1-NPDA** $\subseteq$ **T-DREV**(n^3). ∎

Finally we deal with SLA languages accepted by 1-way PDA.

13.19. Theorem. [GiGr 66], [HaIb 68] For $k \geqq 1$, every SLA language accepted by a 1:k-NPDA is regular. ∎

13.20. Theorem. For 2-way pushdown automata we have the following relationships to time and space complexity classes:

	automata	time	space
1		$\subseteq$ **2T-DTIME**($n^{2k} \cdot \log n$)	= **2:k-PD-DSPACE**(n^k)
2	**2:k-DPDA**	$\subseteq$ **RAM-DTIME**(Lin n^k)	$\subseteq$ **DSPACE**(n^k)
3		$\supseteq$ **multiT-DTIME**($n^{\lfloor k/2 \rfloor}$)	
4	**2:multi-DPDA**	= **P**	$\subseteq$ **PSPACE**
5		$\subseteq$ **2T-DTIME**(n^{4k})	= **2:k-PD-NSPACE**(n^{2k})
6	**2:k-NPDA**	$\subseteq$ **RAM-DTIME**(Lin $n^{3k}/\log n$)*)	$\subseteq$ **DSPACE**(n^{2k})
7		$\supseteq$ **multiT-DTIME**($n^{\lfloor k/2 \rfloor}$)	
8	**2:multi-NPDA**	= **P**	$\subseteq$ **PSPACE**

*) The result **2-NPDA** $\subseteq$ RAM-DTIME(Lin $n^3/\log n$) in [Ryt 84] became known to the authors only during proof reading. By Theorem 13.14.1 this implies **2:k-NPDA** $\subseteq$ **RAM-DTIME**(Lin $n^{3k}/\log n$). However, in the subsequent proof we verify only the upper bound Lin n^{3k}.

Proof. Time results.

Ad 6. We restrict ourselves to the case $k = 1$. The case $k > 1$ can be treated in exactly the same manner or it can be concluded from the case $k = 1$ by Theorem 13.14. Let A be a 2-NPDA which (w.l.o.g.) accepts with an empty pushdown store and the input head on the first input symbol. We consider the work of A on a fixed input word $w = a_1 a_2 \ldots a_n$. In a given situation S of A the (nondeterministic) next step of A depends only on the state $s \in \{s_0, s_1, \ldots, s_{q-1}\}$, the head position $\nu \in \{0, 1, \ldots, n+1\}$ and the top symbol $d \in \Delta$. The tripel (s, ν, d) is said to be the PID (partial instantaneous description) of S. Note that there are $t =_{\mathrm{df}} q \cdot (n+2) \cdot (\operatorname{card} \Delta) \asymp n$ different PID's (in the general case there are $t =_{\mathrm{df}} q \cdot (n+2)^k \cdot (\operatorname{card} \Delta) \asymp n^k$ different PID's). Every computation path beginning with S and having only situations with pushdown words not shorter than that of S, depends only on the PID of S. We define: $(s, \nu, d) \models (s', \nu', d') \leftrightarrow$ there is a computation path $S_1 \vdash_A S_2 \vdash_A \ldots \vdash_A S_r$ $(r \geqq 1)$ of A such that

a) (s, ν, d) is the PID of S_1.

b) (s', ν', d') is the PID of S_r.

c) the pushdown words of S_1 and S_r have the same length, and the pushdown words of $S_2, \ldots, S_{r-1}$ are not shorter than that of S_1.

The relation $\models$ is a reflexive and transitive one. Furthermore, if $s_0(s_1)$ is the initial (the w.l.o.g. only accepting) state of A, we have $w \in L(A) \leftrightarrow (s_0, 1, \square) \models (s_1, 1, \square)$. The simulation idea is, exactly as in [Coo 71b], to generate all (s, v, d, s', v', d') such that $(s, v, d) \models (s', v', d')$ and to see whether $(s_0, 1, \square, s_1, 1, \square)$ will be generated.

Evidently, exactly those (s, v, d, s', v', d') with $(s, v, d) \models (s', v', d')$ will be generated, if first the rules R1 and R2 are applied and then the rules R3 and R4 are applied alternately. We define $a_0 = a_{n+1} = \square$.

R1. $(s, v, d) \models (s, v, d)$ for all PID's (s, v, d).

R2. $(s, v, d) \models (s', v', d')$ for all PID's (s, v, d) and (s', v', d') such that $sa_vd \to s'd'(v' - v)$.

R3. $(s, v, d) \models (s', v', d')$ and $(s', v', d') \models (s'', v'', d'')$ imply $(s, v, d) \models (s'', v'', d'')$.

R4. $(s, v, d) \models (s', v', d')$, $s''a_{v''}d'' \to sd \downarrow (v - v'')$ and $s'a_{v'}d' \to s''' \uparrow (v''' - v')$ imply $(s'', v'', d'') \models (s''', v''', d'')$.

Now we describe an algorithm which is based on the application of the rules R1—R4 and can be easily implemented by a DRAM. For this algorithm we need

— a 0-1-valued function E such that $E(s, v, d, s', v', d') = 1$ iff (s, v, d, s', v', d') is already generated,

— a pushdown list G_3 of such (s, v, d, s', v', d') which are already generated but the application of R3 to $(s, v, d) \models (s', v', d')$ and some other $(s', v', d') \models (s'', v'', d'')$ has not yet been attempted,

— a pushdown list G_4 of such (s, v, d, s', v', d') which are already generated and to which the application of R4 has not yet been attempted,

— for every (s, v, d) a stack list $F_{s,v,d}$ of all (s', v', d') for which (s, v, d, s', v', d') has been deleted from G_3,

— for every (s', v', d') a stack list $H_{s',v',d'}$ of all (s, v, d) for which (s, v, d, s', v', d') has been deleted from G_3.

Algorithm:

1. For all (s, v, d) do
   Place (s, v, d, s, v, d) on list G_4. Set $E(s, v, d, s, v, d) = 1$.
2. For all (s, v, d) do
   For all (s', v', d') such that $sa_vd \to s'd'(v' - v)$ do
   Place (s, v, d, s', v', d') on the lists G_3 and G_4.
   Set $E(s, v, d, s', v', d') = 1$.
3. Let (s, v, d, s', v', d') be the top element of G_3.
   a) Delete (s, v, d, s', v', d') from G_3.
   b) Place (s', v', d') on list $F_{s,v,d}$.
   c) Place (s, v, d) on list $H_{s',v',d'}$.
   d) For all (s'', v'', d'') on $F_{s',v',d'}$ do
   if $E(s, v, d, s'', v'', d'') = 0$ then Place (s, v, d, s'', v'', d'') on the lists G_3 and G_4.
   Set $E(s, v, d, s'', v'', d'') = 1$.

f) **For all** (s'', v'', d'') on $H_{s,v,d}$ **do**
if $E(s'', v'', d'', s', v', d') = 0$ **then** Place $(s'', v'', d'', s', v', d')$ on the lists G_3 and G_4.
Set $E(s'', v'', d'', s', v', d') = 1$.

4. Let (s, v, d, s', v', d') be the top element of G_4.
   a) Delete (s, v, d, s', v', d') from G_4.
   b) **For all** $(s'', v'', d'', s''', v''', d'')$ such that $s''a_{v''}d'' \to sd \downarrow (v - v'')$, $s'a_{v'}d' \to s''' \uparrow (v''' - v')$ and $E(s'', v'', d'', s''', v''', d'') = 0$ **do**
   Place $(s'', v'', d'', s''', v''', d'')$ on the lists G_3 and G_4.
   Set $E(s'', v'', d'', s''', v''', d'') = 1$.
5. **If** the lists G_3 and G_4 are empty **then if** $E(s_0, 1, \boxdot, s_1, 1, \boxdot) = 1$
then accept
else reject
else go to 3.

Finally we discuss the time requirements for a DRAM which implements this algorithm. The steps 1 and 2 require time $\precsim t$ and are executed only once. Each of the steps 3, 4 and 5 is executed at most $\precsim t^2$ times because every element can be placed only once on the lists G_3 and G_4, and after the execution of 3 and 4 at least one element is deleted from one of these lists (except for the final execution of 3, 4 and 5). The execution of step 3 requires time $\precsim r$ and the execution of 4 and 5 requires time $\precsim 1$.

Consequently, the total execution time is $\precsim t^3 \asymp n^3$.

Ad 5 (see [AhHoUl 68]). The algorithm described above can be implemented by a T2-DM M in the following way: To each (s, v, d, s', v', d') we assign a fixed square on the tape of M which is used to store the information whether $E(s, v, d, s', v', d') = 1$, whether (s, v, d, s', v', d') is on the lists G_3 and G_4, whether (s, v, d) is on the list $H_{s',v',d'}$ and whether (s', v', d') is on the list $F_{s,v,d}$. Thus each of the steps 1, 2, 3, 4, and 5 requires time $\precsim n^2$, and the total execution time is $\precsim n^4$. By Theorem 13.14 and Theorem 19.10 we have **2:k-NPDA** $\subseteq$ **2T-DTIME**(n^{4k}).

Ad 2. [Coo 72] We proceed in essentially the same way as in the nondeterministic case (Statement 6). However, since the relation $\models$ is not a function, it may be too expansive to generate all (s, v, d, s', v', d'). Fortunately, it turns out that instead of $\models$ the two relations $\models\!\!\Rightarrow$ and $\overset{\frown}{\models}$ can be used, which are readily seen to be a function in the deterministic case. We define

$$(s, v, d) \models\!\!\Rightarrow (s', v', d') \leftrightarrow (s, v, d) \models (s', v', d') \text{ and } s'a_{v'}d' \to s'' \uparrow v'' \text{ for some } s'', v'' \text{ or } (s', v', d') = (s_1, 1, \boxdot).$$

$$(s, v, d) \overset{\frown}{\models} (s', v', d') \leftrightarrow sa_vd \to s'd'(v' - v) \text{ or } d = d' \text{ and there are } (s'', v'', d''), (s''', v''', d''') \text{ such that } sa_vd \to s''d'' \downarrow (v'' - v), (s'', v'', d'') \models (s''', v''', d''') \text{ and } s'''a_{v'''}d''' \to s' \uparrow (v' - v''').$$

Note that $w \in L(A) \leftrightarrow (s_0, 1, \boxdot) \models\!\!\Rightarrow (s_1, 1, \boxdot)$. The simulation idea is to generate all (s, v, d, s', v', d') such that $(s, v, d) \models\!\!\Rightarrow (s', v', d')$ or $(s, v, d) \frown (s', v', d')$, and then to check whether $(s_0, 1, \boxdot, s_1, 1, \boxdot)$ with $(s_0, 1, \boxdot) \models\!\!\Rightarrow (s_1, 1, \boxdot)$ is generated.

Evidently, exactly those (s, v, d, s', v', d') for which $(s, v, d) \models\!\!\Rightarrow (s', v', d')$ or $(s, v, d) \frown (s', v', d')$ holds will be generated, if first the rules R1′ and R2′ are applied and then the rules R3′ and R4′ are applied alternately.

R1′ $(s, v, d) \models\!\!\Rightarrow (s, v, d)$ for all (s, v, d) such that $sa_vd \to s' \uparrow v$ for some s', v' or $(s, v, d) = (s_1, 1, \boxdot)$.

R2′ $(s, v, d) \frown (s', v', d')$ for all (s, v, d, s', v', d') such that $sa_vd \to s'd'(v' - v)$.

R3′ $(s, v, d) \frown (s', v', d')$ and $(s', v', d') \models\!\!\Rightarrow (s'', v'', d'')$ imply $(s, v, d) \models\!\!\Rightarrow (s'', v'', d'')$.

R4′ $(s, v, d) \models\!\!\Rightarrow (s', v', d')$, $s''a_{v''}d'' \to sd \downarrow (v - v'')$ and $s'a_{v'}d' \to s''' \uparrow (v''' - v')$ imply $(s'', v'', d'') \frown (s''', v''', d'')$.

Now we describe an algorithm which is based on the application of the rules R1′–R4′ and can be easily implemented by a DRAM. For this algorithm we need

- a function E such that $E(s, v, d) = (s', v', d')$ if $(s, v, d) \models\!\!\Rightarrow (s', v', d')$ and (s, v, d, s', v', d') is already generated,
- a pushdown list G_3 of such (s, v, d, s', v', d') with $(s, v, d) \frown (s', v', d')$, which are already generated but the application of R3′ to $(s, v, d) \frown (s', v', d')$ and some $(s', v', d') \models\!\!\Rightarrow (s'', v'', d'')$ has not yet been attempted,
- a pushdown list G_4 of such (s, v, d, s', v', d') with $(s, v, d) \models\!\!\Rightarrow (s', v', d')$, which are already generated and to which the application of R4′ has not yet been attempted,
- for every (s', v', d') a pushdown list $H_{s',v',d'}$ of all (s, v, d) with $(s, v, d) \frown (s', v', d')$ for which the first trial to apply R3′ to $(s, v, d) \frown (s', v', d')$ and some $(s', v', d') \models\!\!\Rightarrow (s'', v'', d'')$ failed.

Algorithm:

1. **For all** (s, v, d) **such that** $sa_vd \to s' \uparrow (v' - v)$ **for some** s', v' **do**
   Place (s, v, d, s, v, d) on list G_4. Set $E(s, v, d) = (s, v, d)$.
   Set $E(s_1, 1, \boxdot) = (s_1, 1, \boxdot)$.
2. **For all** (s, v, d) **do**
   For all (s', v', d') **such that** $sa_vd \to s'd'(v' - v)$ **do**
   Place (s, v, d, s', v', d') on list G_3.
3. Let (s, v, d, s', v', d') be the top element of G_3.
   a) Delete (s, v, d, s', v', d') from G_3.
   b) **If** $E(s', v', d') = (s'', v'', d'')$ **then** α) Place (s, v, d, s'', v'', d'') on list G_4
   Set $E(s, v, d) = (s'', v'', d'')$.
   For all (s''', v''', d''') in $H_{s,v,d}$ **do**
   β) Delete (s''', v''', d''') from $H_{s,v,d}$.
   Place $(s''', v''', d''', s, v, d)$ on G_3
   else (i.e. if $E(s', v', d')$ is still undefined)
   Place (s, v, d) on list $H_{s',v',d'}$.

4. Let (s, v, d, s', v', d') be the top element of G_4.
   a) Delete (s, v, d, s', v', d') from G_4.
   b) **For** all $(s'', v'', d'', s''', v''', d'')$ such that $s''a_{v''}d'' \to sd \downarrow (v - v'')$ and $s'a_{v'}d' \to s''' \uparrow (v''' - v')$ **do**
      Place $(s'', v'', d'', s''', v''', d'')$ on list G_3.
5. **If** the lists G_3 and G_4 are empty **then if** $E(s_0, 1, \boxdot) = (s_1, 1, \boxdot)$
   then accept
   else reject
   else got to 3.

Finally we discuss the time requirements for a DRAM which implements this algorithm. The steps 1 and 2 require time $\preceq t$ and are executed only once. Each of the steps 3a, 3bα, 3bβ, 4 and 5 can be implemented by a DRAM within time $\preceq 1$. Since placing a tuple (s, v, d, s', v', d') on list G_4 is connected with the definition of $E(s, v, d)$, every (s, v, d, s', v', d') can be placed on G_4 only once. Therefore, and since $\vDash$ is a function, every (s, v, d, s', v', d') can be placed on G_3 by 4b only once. Moreover, (s, v, d, s', v', d') can be placed on G_3 by 2 only once, and these two cases exclude each other. If, after the deletion of a tuple (s, v, d, s', v', d') from G_3 by **3a**, the rule R3′ cannot be applied, then (s, v, d) is placed on $H_{s',v',d'}$ by **3b**. Later, (s, v, d) can be brought back from $H_{s',v',d'}$ to G_3 only if R3′ is applicable to (s, v, d) $\vDash (s', v', d')$ and some $(s', v', d') \models\!\!\Rightarrow (s'', v'', d'')$. Thus, every (s, v, d) can be placed on $H_{s',v',d'}$ only once and therefore (s, v, d, s', v', d') can be placed on G_3 by 3bβ only once. Consequently, step 3bβ can be executed only $\preceq t$ times, and every (s, v, d, s', v', d') can be placed on G_3 altogether at most twice. Since the execution of step 3 (4) includes a deletion of some (s, v, d, s', v', d') from G_3 (G_4), the steps 3a, 3bα, 4 and 5 can be executed only $\preceq t$ times ($\vDash$ and $\models\!\!\Rightarrow$ are functions). Consequently, the total execution time is $\preceq t \asymp n$, and by Theorem 13.14 we have **2:k-DPDA** $\subseteq$ **RAM-DTIME**(n^k).

Ad 1 (see also [AhHoUl 68]). We discuss the time requirements of a T2-DM which implements the algorithm described in the proof of Statement 2. Since the function E, the lists G_3, G_4 and $H_{s',v',d'}$ can be stored within total space $\preceq n \cdot \log n$ on a Turing tape, each of the steps 3a, 3bα, 3bβ, 4 and 5 can be implemented by a T2-DM within time $\preceq n \cdot \log n$. As shown in the proof of Statement 2, these steps are executed only $\preceq n$ times. The steps 1 and 2 are executed only once and can be implemented by a T2-DM within time $\preceq n^2 \cdot \log n$. By Theorem 13.14 and Theorem 19.10 we have **2:k-DPDA** $\subseteq$ **2T-DTIME**$(n^{2k} \cdot \log n)$.

Ad 3. [Coo 71b] First, for a given 1T-DM M working within time $t = n^k$, an equivalent DTM M' working within time t^2 and in the normalized manner shown in Figure 13.1 can easily be constructed. Let $w = a_1 a_2 \ldots a_n$ $(a_1, a_2, \ldots, a_n \in \{0, 1\})$ be the input word. Defining $a_j = \square$ for $j > n$, the tape square j has the symbol a_j for $j \geqq 1$ at the beginning of the calculation of M' which starts its work scanning tape square 1. Let s_0, s_1, and s_2 be the initial state, the accepting state, and the rejecting state of M, resp. Now we describe the work of a 2:(2k)-DPDA A simulating M'. The automaton A will compute consecutively for $\tau = 1, 2, \ldots, t^2$ the quadruples $(1^{i_\tau}$,

$1^{j_\tau}, s_\tau, b_\tau$) on the top of its pushdown store, where $i_\tau - 1$ is the number of reversals of M''s head up to the τth step, j_τ is the number of the tape square which is scanned by M''s head at the τth step, s_τ is the new state of M' at the τth step, and b_τ is the symbol written by M''s head at the τth step (for i_τ and j_τ see Figure 13.1). If A has already computed ($1^{i_{\tau-1}}, 1^{j_{\tau-1}}, s_{\tau-1}, b_{\tau-1}$), then for the computation of ($1^{i_\tau}, 1^{j_\tau}, s_\tau,$

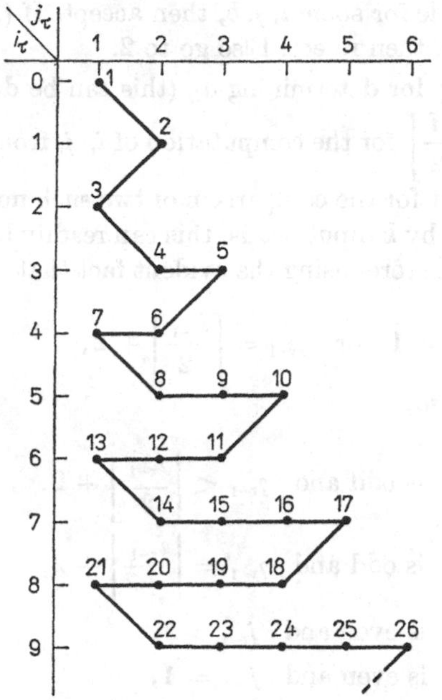

Figure 13.1. Position of the head of M' at the τth step ($\tau = 1, 2, 3, \ldots$)

b_τ) the symbol written in tape square j_τ is needed. If M' has not visited the tape square j_τ before the τth step (this is the case iff $j_\tau = \left\lfloor \frac{i_\tau}{2} \right\rfloor + 2$), then a_{j_τ} is the contents of this tape square. If M''s preceding visit to tape square j_τ has been made at the τ'th step, then $b_{\tau'}$ is the contents of this tape square. In this case A starts the simulation of M' once more where it uses its pushdown store above ($1^{i_{\tau-1}}, 1^{j_{\tau-1}}, s_{\tau-1}, b_{\tau-1}$) only. If ($1^{i_{\tau'}}, 1^{j_{\tau'}}, s_{\tau'}, b_{\tau'}$) is computed, then ($1^{i_{\tau-1}}, 1^{j_{\tau-1}}, s_{\tau-1}, b_{\tau-1}$) and ($1^{i_{\tau'}}, 1^{j_{\tau'}}, s_{\tau'}, b_{\tau'}$) are the top two quadruples of the pushdown store, and ($1^{i_\tau}, 1^{j_\tau}, s_\tau, b_\tau$) can be computed.

Now we give a more detailed algorithm for the 2:(2k)-DPDA A which simulates M'.

1. Write ($1^0, 1^1, s, b$) on the pushdown store, where $s_0 a_1 \to sb +$ is an instruction of M'.
2. If ($1^i, 1^j, s, b$) is the top quadruple of the pushdown store and $j = \left\lfloor \frac{i}{2} \right\rfloor + 1$ and $i = i_{\tau-1}$, $j = j_{\tau-1}$ for some τ, then replace ($1^i, 1^j, s, b$) by ($1^{i_\tau}, 1^{j_\tau}, s', b'$) on the pushdown store where $sa_{j_\tau} \to s'b' -$ is an instruction of M', and go to 5.

3. If $(1^i, 1^j, s, b)\ (1^{i'}, 1^{j'}, s', b')$ are the top two quadruples of the pushdown store and $i = i_{\tau-1}$, $j = j_{\tau-1}$, $i' = i_{\tau'}$, $j' = j_{\tau'}$ for some τ (τ' being the moment of the preceding visit to tape square $j_\tau = j'$) then replace $(1^i, 1^j, s, b)\ (1^{i'}, 1^{j'}, s', b')$ by $(1^{i_\tau}, 1^{j_\tau}, s'', b'')$ on the pushdown store, where $sb' \to s''b''\sigma$ ($\sigma \in \{-, +\}$) is an instruction of M', and go to 5.
4. Go to 1.
5. If $(1^i, 1^j, s_1, b)$ is the top quadruple for some i, j, b, then accept. If $(1^i, 1^j, s_2, b)$ is the top quadruple for some i, j, b, then reject. Else go to 2.

The input heads of A are used only for determining a_{j_τ} (this can be done with one head only), for the computation of $\left\lfloor \frac{i}{2} \right\rfloor$, for the computation of i_τ, j_τ from $i_{\tau-1}, j_{\tau-1}$, for the computation of $i_{\tau'}$ from i_τ, j_τ and for the comparison of two such numbers. Since one such number can be represented by k input heads, this can readily be carried out by $2k$ input heads and the pushdown store, using the evident fact that

$$i_\tau = \begin{cases} i_{\tau-1} + 1 & \text{if} \quad j_{\tau-1} = 1 \quad \text{or} \quad j_{\tau-1} = \left\lfloor \frac{i_{\tau-1}}{2} \right\rfloor + 2, \\ i_{\tau-1} & \text{otherwise}, \end{cases}$$

$$j_\tau = \begin{cases} j_{\tau-1} + 1 & \text{if} \quad i_{\tau-1} \text{ is odd and } j_{\tau-1} < \left\lfloor \frac{i_{\tau-1}}{2} \right\rfloor + 2, \\ j_{\tau-1} - 1 & \text{if} \quad i_{\tau-1} \text{ is odd and } j_{\tau-1} = \left\lfloor \frac{i_{\tau-1}}{2} \right\rfloor + 2, \\ j_{\tau-1} - 1 & \text{if} \quad i_{\tau-1} \text{ is even and } j_{\tau-1} > 1, \\ j_{\tau-1} + 1 & \text{if} \quad i_{\tau-1} \text{ is even and } j_{\tau-1} = 1, \end{cases}$$

$$i_{\tau'} = \begin{cases} i_\tau - 2 & \text{if} \quad j_\tau = 1 \quad \text{or} \quad j_\tau = \left\lfloor \frac{i_\tau}{2} \right\rfloor + 1, \\ i_\tau - 1 & \text{otherwise}. \end{cases}$$

Note that τ' is defined only if $j_\tau < \left\lfloor \frac{i_\tau}{2} \right\rfloor + 2$. However, only in this case M' has to compute $i_{\tau'}$ (by step 3).

Statement 7 follows immediately from Statement 3. The statements 4 and 8 are immediate consequences of the statements 3 and 5.

Space results

Ad 1. [GrHaIb 67] Let A be a 2:k-DPDA having q states and r pushdown symbols. We consider the computation of A on an input word w of length n. Let $l(\tau)$ be the length of the pushdown word and PID(τ) the PID of A (see the proof of the time result of Statement 6) at the τth moment of this computation. Note that there are at most $t = q \cdot r(n + 2)^k$ different PID's of A. Assume that $\max l(\tau) > t$. Then there are moments τ_0 and τ_1 such that $\tau_0 < \tau_1$, $1 \leqq l(\tau_0) = \min\limits_{\tau_0 \leqq \tau \leqq \tau_1} l(\tau) < l(\tau_1)$ and PID(τ_0) = PID(τ_1). Consequently, $l(\tau_0 + i(\tau_1 - \tau_0)) = l(\tau_0) + i(l(\tau_1) - l(\tau_0))$ for all $i \geqq 1$, i.e. A does not halt on input w. Hence, if A decides a language, then A halts on every input and does not use more space than t in its pushdown store. The hint that linear speed-up for the space requirements of PD-stores is possible (cf. Remark 1 after 18.11) completes the proof.

Ad 2. This statement is an immediate consequence of Statement 1.

Ad 5. [GrHaIb 67] Let A be a 2:k-NPDA accepting with an empty pushdown store and having q states and r pushdown symbols. We consider all computation paths β on an input word w of length n. Let $l(\beta, \tau)$ be the length of the pushdown word and PID(β, τ) be the PID of A on the path β at the τth moment (see the proof of the time result of Statement 6). Note that there are at most $t = q \cdot r \cdot (n + 2)^k$ different PID's of A. Assume that $m_0 =_{\text{df}} \min \{\max_\tau l(\beta, \tau) : \beta \text{ is an accepting path}\} > t^2$. Let β_0 be an accepting path such that $\max_\tau l(\beta_0, \tau) = m_0$ and let ϱ be a moment such that $l(\beta_0, \varrho) = m_0$. Now we define $\tau_i =_{\text{df}} \max \{\tau : \tau \leqq \varrho \wedge l(\beta_0, \tau) = i\}$ and $\sigma_i =_{\text{df}} \min \{\tau : \tau \geqq \varrho \wedge l(\beta_0, \tau) = i\}$ for $i = 0, 1, \ldots, m_0$. Hence we have $\tau_0 < \tau_1 < \ldots < \tau_{m_0} = \varrho = \sigma_{m_0} < \ldots < \sigma_1 < \sigma_0$, $l(\beta_0, \tau_i) = l(\beta_0, \sigma_i) = i$ and $\min_{\tau_i < \tau < \sigma_i} l(\beta_0, \tau) > i$. Since there are only t different PID's of A, at most two of the pairs $(\text{PID}(\beta_0, \tau_1), \text{PID}(\beta_0, \sigma_1))$, $(\text{PID}(\beta_0, \tau_2), \text{PID}(\beta_0, \sigma_2)), \ldots, (\text{PID}(\beta_0, \tau_{m_0}), \text{PID}(\beta_0, \sigma_{m_0}))$ coincide; say $\text{PID}(\beta_0, \tau_i) = \text{PID}(\beta_0, \tau_j)$, $\text{PID}(\beta_0, \sigma_i) = \text{PID}(\beta_0, \sigma_j)$ and $1 \leqq i < j \leqq m_0$.

Consequently, the following computation path β_1 is also an accepting one: A works from 0 to τ_i as on path β_0 from 0 to τ_i. Then A works from τ_i to $\tau_i + (\sigma_j - \tau_j)$ as on path β_0 from τ_j to σ_j. This is possible because of $\text{PID}(\beta_0, \tau_i) = \text{PID}(\beta_0, \tau_j)$. Finally, A works from $\tau_i + (\sigma_j - \tau_j)$ as on path β_0 from σ_i. This is possible because of $\text{PID}(\beta_0, \sigma_j) = \text{PID}(\beta_0, \sigma_i)$ (see Figure 13.2).

The effect of this construction is that $\text{card}\, \{\tau : l(\beta_1, \tau) = m_0\} < \text{card}\, \{\tau : l(\beta_0, \tau) = m_0\}$. If $\text{card}\, \{\tau : l(\beta_1, \tau) = m_0\} > 0$, then we proceed with β_1 as with β_0. After finitely many of such steps we get an accepting computation path β_d such that $\max_\tau l(\beta_d, \tau) < m_0$. This contradicts the minimality of m_0. Hence there is an accepting path having only pushdown words of length $\leqq t^2$. The hint that linear speed-up for the space requirements of PD-stores is possible completes the proof.

The statements 4 and 8 follow from the statements 2 and 6, resp.

Ad 6. The proof of Statement 5 (time) yields the space bound $t^2 \asymp n^{2k}$. ∎

Remark 1. Because of **2:multi-DPDA** = **P** it is not known whether **2:multi-DPDA** $\subseteq$ **L**. However, in [Gal 74] it is shown that **2-DPDA** $\neq$ **L**.

Remark 2. In [Mon 83] it is shown that $L \in \mathbf{P}$ implies

$$\{1^{\text{dya}^{-1}(w)} : w \in L\} \in \mathbf{2\text{-}DPDA} \quad \text{for} \quad L \subseteq \{1, 2\}^*.$$

The following result can be considered as a refinement of the equality **2-auxPD-T-XSPACE**(log) = **2-auxPD-C-XSPACE**(Pol) (see 20.13 and 20.15) and can be proved in the same way.

13.21. Theorem. [Kam 72] For X $\in$ {D, N} and $k \geqq 1$,

$$\mathbf{2\text{:}k\text{-}XPDA} = \mathbf{2\text{-}auxPD\text{-}C\text{-}XSPACE}(n^{k-1}). \; ∎$$

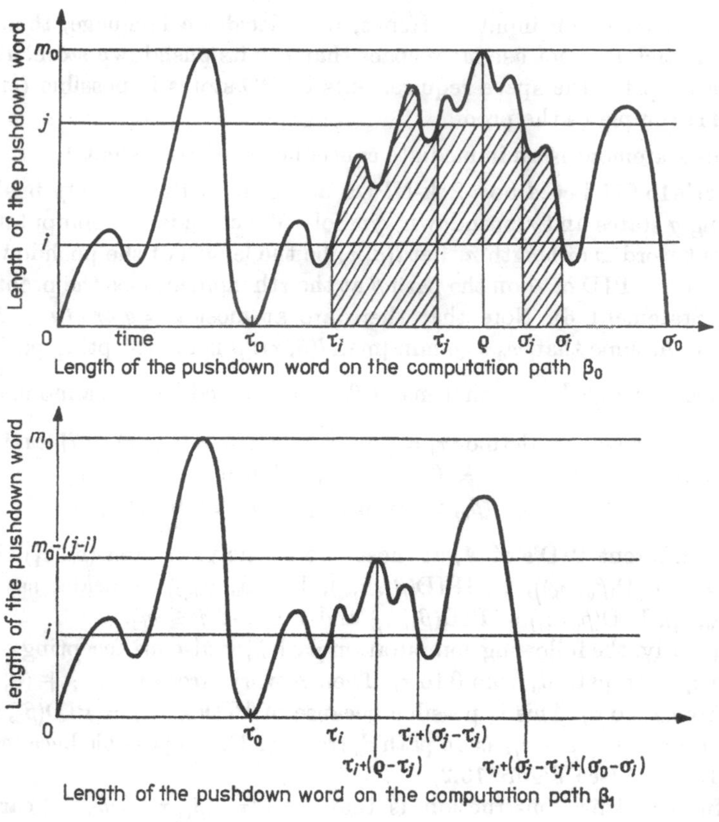

Figure 13.2

Next we state a result on the reversal complexity of 2-way PDA languages.

13.22. Theorem. [KaVo 70]

1. **2-DPDA** $\subseteq$ **T-DREV**$(n^2 \cdot \log n)$.
2. **2-NPDA** $\subseteq$ **T-DREV**(n^4). ∎

Finally we mention a result on 2-way alternating PDA languages.

13.23. Theorem. [LaLiSt 78]

1. **2-APDA** = **DEXPTIME**.
2. **2:multi-APDA** = **DTIME**(2^{Pol}). ∎

13.3.2. Relationships between Different Classes of Pushdown Automata

We start with the hierarchies with respect to the number of input heads.

13.24. Theorem.

1. [Chb 85] For $k \geqq 1$, **1:k-DPDA** $\subset$ **1:(k+1)-DPDA**.
2. [Iba 73a] For $k \geqq 1$ and $\mathbf{X} \in \{\mathrm{D}, \mathrm{N}\}$, **2:k-XPDA** $\subset$ **2:(k+1)-XPDA**. ∎

Remark 1. A witness set for the properness of the second inclusion can be chosen to be a SLA language. The proof is similar to the proof of the corresponding result for finite automata (cf. [MeWa 82]). For an easier proof, using the possibilities of the pushdown store, see [Sud 77d].

Now we mention some relationships between deterministic and nondeterministic PDA.

13.25. Theorem.
1. [Mon 81] For $k \geqq 1$, **2:k-NPDA** $\subseteqq$ **2:(3k+1)-DPDA**.
2. [Kin 81a] For $k \geqq 1$, **1:k-NPDA** $\subseteqq$ **2:3k-DPDA** (for $k = 1$ see also [Sud 77d]).
3. [Chb 85] For $k \geqq 1$, **1:k-DPDA** $\subset$ **1:k-NPDA** (for $k = 1$ see also [Col 71]). ∎

Remark. On 13.25.1: From 13.20.5 and 13.20.3 we can only conclude **2:k-NPDA** $\subseteqq$ **2:(8k)-DPDA**. On 13.25.3: The set S can serve as a witness for **1-DPDA** $\subset$ **1-NPDA**.

Finally we speak about some relationships between PDA and more restricted types of automata.

13.26. Theorem. For $k \geqq 1$,
1. [Sud 77a] **2:(2k)-NFA** $\subseteqq$ **2:k-NPDA.**
2. [Sud 77a] **2:k-NFA** $\subseteqq$ **2:2k-DPDA.**
3. [Kin 81a] **1:k-AFA** $\subseteqq$ **2:k-DPDA.**
4. [Kin 81a] **1:k-NPDA** $\subseteqq$ **1:3k-AFA.**
5. [Kin 81a] **2:k-AFA** $\subseteqq$ **2:k-NPDA** $\subseteqq$ **2:3k-AFA.**

Proof. We only outline the proofs of the statements 1 and 2.

Ad 1. The simulating 2:k-NPDA A' operators in two stages. In the first stage A' simulates the 2:(2k)-NFA A, where A' uses its k input heads to do the work of the first k input heads of A and guesses nondeterministically what the other k input heads of A read. These read symbols and the moves of these heads are recorded in the pushdown store of A'. In the second stage A' checks whether these symbols coincide with the symbols which are read when the stored moves of the other k input heads are executed (in reverse order).

Ad 2. The simulating 2:(2k)-DPDA A' tries systematically all computations of the 2:k-NFA A (but at most the first $q \cdot (n + 2)^k$ steps of each of them where q is the number of states of A). The pushdown store of A' is used to record the computation of A most recent attempted. Thus the next computation to be tried can easily be determined. A' uses the first k heads to perform the simulation and the other k heads to count the number of steps of the computation (just attempted).

For Statements 3, 4 and 5 see [Kin 81a]. ∎

Remark. It is not known whether **2:multi-XFA** = **2:multi-XPDA**, for X ∈ {D, N}. Moreover, it is even not known whether **2:multi-XFA** $\supseteqq$ **2-DPDA**. However, because of 13.2.3 (space), 13.2.8 (space), 13.20.4 (time), 13.20.8 (time) and 13.14.2

we have for X ∈ {D, N},

$$\mathbf{XL} = \mathbf{P} \leftrightarrow \mathbf{2{:}multi\text{-}XFA} = \mathbf{2{:}multi\text{-}XPDA} \leftrightarrow \mathbf{2{:}multi\text{-}XFA} \supseteqq \mathbf{2\text{-}DPDA}.$$

13.27. Theorem. [DuGa 81] **2-DCA** $\subset$ **2-DPDA**. ∎

13.4. Nonerasing Stack Automata

13.4.1. Relationships to Time and Space Complexity Classes

13.28. Theorem. For 1-way nonerasing stack automata we have the following relationships to time and space complexity classes:

	automata	time	space
1	**1:k-DNESA**	= **1:k-NES-DTIME**(Lin n^2)	= **1:k-NES-DSPACE**(id)
2		$\subseteqq$ **2-kT-DTIME**(n^2)	$\subseteqq$ **DSPACE**(id)
3	**1:k-NNESA**	= **1:k-NES-NTIME**(Lin n^2)	= **1:k-NES-NSPACE**(id)
4		$\subseteqq$ **2-kT-NTIME**(n^2)	$\subseteqq$ **DSPACE**(id)

Proof. All results or proofs are special cases of the corresponding results or proofs of 13.33. ∎

13.29. Theorem. [HoUl 67], [Iba 71] For $k \geqq 1$,
1. **2:k-DNESA** = **DSPACE**($n^k \cdot \log n$).
2. **2:k-NNESA** = **NSPACE**(n^{2k}).
3. **2:multi-DNESA** = **2:multi-NNESA** = **PSPACE**.

Proof. a) We show **DSPACE**($n^k \cdot \log n$) $\subseteqq$ **2:k-DNESA**. Let M be a DTM accepting some language within space $n^k \cdot \log n$. Then there is a DTM M' accepting the same language within space $c \cdot n^k \cdot \log n$, using the worktape symbols □, 1, 2 only, and having at each step of its computation on an input word $w \in \{1, 2\}^*$ only worktape inscriptions of the form ...□ □ □w_1□ □ □..., where the leftmost symbol of $w_1 \in \{1, 2\}^*$ is written in the same tape square as the leftmost symbol of w at the beginning of the computation.

For later quotation of this proof we emphasize that we shall use in the following construction of a 2:k-DNESA A, simulating M', only the facts that A has a nonerasing stack, that it can read the input with one input head, and that it has the ability to count up to $t = c \cdot n^k$ without using the stack.

The 2:k-DNESA A will generate successively all ID's of the computation of M' on the input w. To this end we divide an ID $w_1'sw_1''$ of M' (where s is a state and $w_1', w_1'' \in \{1, 2\}^*$) into t blocks of length $\log n$: $w_1'sw_1'' = v_1v_2 \ldots v_{i-1}v_i'sv_i''v_{i+1} \ldots v_t$, in such a way that there is a t_0 such that $i \leqq t_0 \leqq t$, $|v_j| = \log n$ for $j = 1, \ldots, t_0 - 1$,

$|v_{t_0}| \leqq \log n$, and $v_j = \mathrm{e}$ for $j = t_0 + 1, \ldots, t$ (where $v_i =_{\mathrm{df}} v_i' s v_i''$). The ID $w_1' s w_1''$ will be recorded in the stack of A as

$$1^{\mathrm{dya}^{-1}(v_1)} * 1^{\mathrm{dya}^{-1}(v_2)} * \ldots * 1^{\mathrm{dya}^{-1}(v_{i-1})} * 1^{\mathrm{dya}^{-1}(v_i')} s 1^{\mathrm{dya}^{-1}(v_i''^{-1})} * 1^{\mathrm{dya}^{-1}(v_{i+1}^{-1})} * \ldots * 1^{\mathrm{dya}^{-1}(v_t^{-1})}.$$

Different ID's will be seperated by $|$.

First, A generates in its stack the initial ID of M' on w, and then A will generate in its stack each next ID in turn until an accepting or rejecting ID is generated. Then A accepts or rejects w correspondingly.

α) Generation of the initial ID of M' in the stack of A. The 2:k-DNESA A has to generate the word

$$u = | s_0 1^{\mathrm{dya}^{-1}(v_1^{-1})} * 1^{\mathrm{dya}^{-1}(v_2^{-1})} * \ldots * 1^{\mathrm{dya}^{-1}(v_t^{-1})}$$

in its stack where $w = v_1 v_2 \ldots v_t$.

To do so, the symbols of the input word w must be transferred to the stack. For this, at least in the case $k = 1$, the input head cannot stay on the symbol to be transferred (as we shall see). Therefore A generates first not the word u, but the word u mixed with some symbols $+$ in such a way that the number

(number of $+$ on the left of the rightmost $*$) $-$ (number of $+$ on the right of the rightmost $*$)

determines the position of the symbol to be transferred. A starts by writing $+^{\log n} * | s_0$ on the stack. After the transfer of an input symbol, A writes $+$ on the stack, and after the transfer of $\log n$ input symbols, A writes in addition $*$. In this way the words $v_1, v_2, \ldots$ are read each right-to-left and transferred onto the stack in the desired form (apart from the interspersed symbols $+$). To transfer an input symbol $b \in \{1, 2\}$ to the stack means: A copies the number of 1's on the right of the rightmost $*$ and adds 1^b. All this can be done by one input head and the stack head.

If all input symbols are transferred, then further symbols $*$ are added to the stack until it contains $t - 1$ symbols $*$. After adding $|$, the word between the two symbols $|$ is reproduced without the symbols $+$. Thereby the current block of 1's to be copied can be found by counting t symbols $*$ from the top. The length of this block can be stored with one input head while the stack head is running to the top. Now the word u stands on the right of the rightmost symbol $|$.

β) Generation of the next ID of M' in the stack of A. The blocks can be reproduced as described at the end of item α. If the block $1^{\mathrm{dya}^{-1}(v_i')} s 1^{\mathrm{dya}^{-1}(v_i''^{-1})}$ is to be reproduced with the necessary changes, then using one input head A checks whether $\mathrm{dya}^{-1}(v_i''^{-1})$ is even or odd, i.e. whether the first symbol of v_i'' is 2 or 1. Then, corresponding to the next move function of M', the state s is changed and operations of the kind $x \to 2x + 1$, $x \to 2x + 2$ or $x \to \left\lfloor \frac{x-1}{2} \right\rfloor$ are possibly executed for $x = \mathrm{dya}^{-1}(v_i')$, $\mathrm{dya}^{-1}(v_i''^{-1})$. Blocks of the form

$$1^{\mathrm{dya}^{-1}(v_{i-1})} * s 1^{\mathrm{dya}^{-1}(v_i^{-1})} \quad \text{or} \quad 1^{\mathrm{dya}^{-1}(v_i)} s * 1^{\mathrm{dya}^{-1}(v_{i+1}^{-1})}$$

are treated analogously.

b) We show **NSPACE(n^{2k}) $\subseteq$ 2:k-NNESA**. We proceed analogously to item a) with the following differences: An ID of length $\leqq c^2 \cdot n^{2k}$ will be divided into $t = c \cdot n^k$ blocks of length t, which will be written into the stack in uncoded form. The reproduction of a block will be made letterwise. The current block to be copied can be found as in item a) by counting t symbols $*$ from the top. The position of the symbol in the block to be reproduced is guessed nondeterministically. Then running back to the top, A counts the number of letters (except $*$, $+$ and $|$) modulo t. If the stack head reaches the top with 0 modulo t, then the guess was correct.

c) We show **2:k-DNESA $\subseteq$ DSPACE($n^k \cdot \log n$)**. Let A be a 2:k-DNESA that w.l.o.g. finishes its work on every input with the stack head on the top. We consider an input word w of length n. A partial ID (for short: PID) of A is a description of the head positions, the state and the top symbol of A in a certain situation. There are at most $t = c \cdot (n + 2)^k$ different PID's.

It will be found that an mT-DM M simulating A need not store the whole stack word of A (which may be too long), but only the corresponding PID, if it disposes of a table saying with what PID the stack head of A will return to the top if it leaves the top with a given PID. Such a table can be computed recursively by A during its work. At the beginning this table is empty. If in some step of A the stack word is not enlarged, then the table remains unchanged. If the stack is enlarged by pushing one symbol, then the new table can easily be computed from the old one. Such a table consists of t pairs of PID's, each PID being of length $\log t$. Thus A can be simulated by M within space $\preceq t \cdot \log t \preceq n^k \cdot \log n$.

d) We show **2:k-NNESA $\subseteq$ NSPACE(n^{2k})**. We proceed as in item c). However, in the nondeterministic case the stack head can return in more than one, although not more than t different situations. Therefore the tables must be constituted as follows: For every PID there is a row and a column. Where row K and column K' cross, there is a 1 iff the stack head having left the top in a situation with the PID K can return to the top in a situation with the PID K'. Such a table can be stored within space $\preceq t^2 \preceq n^{2k}$.

e) Statement 3 is an immediate consequence of the statements 1 and 2. ∎

Finally we mention some results for 2-way alternating NESA. Note that there is no difference from the results for 2-way alternating SA.

13.30. Theorem. [LaLiSt 78]

1. **2-ANESA = DTIME($2^{2^{\mathrm{Lin}}}$)**.
2. **2:multi-ANESA = DTIME($2^{2^{\mathrm{Pol}}}$)**. ∎

13.4.2. Relationships between Different Classes of Nonerasing Stack Automata

Because of 13.29, 13.28.4 (space) and the space hierarchy results (see 17.23 and 17.32), we immediately get the following results.

13.31. Theorem. For $k \geq 1$ and X $\in$ {D, N}, **2:k-XNESA $\subset$ 2:(k+1)-XNESA**. ∎

13.32. Theorem. For $k \geqq 1$,

1. **2:k-NNESA $\subset$ 2:(4k)-DNESA $\subset$ 2:(2k+1)-NNESA.**
2. **1:multi-NNESA $\subset$ 2-DNESA.** ∎

13.5. Stack Automata

13.5.1. Relationships to Time and Space Complexity Classes

13.33. Theorem. [GiGrHa 67], [HoUl 68b] For 1-way stack automata we have the following relationships to time and space complexity classes:

	automata	time	space
1	**1:k-DSA**	= **1:k-S-DTIME**(Lin n^2)	= **1:k-S-DSPACE**(id)
2		$\subsetneqq$ **2-kT-DTIME**(n^2)	$\subsetneqq$ **DSPACE**(id)
3	**1:k-NSA**	= **1:k-S-NTIME**(Lin n^2)	= **1:k-S-NSPACE**(id)
4		$\subsetneqq$ **2-kT-NTIME**(n^2)	$\subsetneqq$ **DSPACE**(id)

Proof. Time results.

Ad 1. In what follows, we consider w.l.o.g. only 1:k-DSA which move in each step either one of the input heads or the stack head, and which do not alter any stack symbol (unless erasing it).

Let A_1 be a 1:k-DSA. We construct an equivalent 1:k-DSA A_2 having the following property, which is said to be property P_1: if the stack head leaves the top, it does not return until at least one of the input heads has moved. A_2 simulates A_1 as follows: if A_1 writes a symbol on the stack, then A_2 writes this symbol together with a table, which gives for each state s the information whether, and in what state, the stack head will return to the top without a move of any input head, if it leaves the top in state s. Evidently, there is only a bounded number of such tables and, when writing a new symbol, A_2 can determine the new table from this symbol and from the old table. Now, having these tables, the stack head of A_2 need not leave the top unless an input head has to move.

Since A_2 cannot make more than a bounded number of consecutive writing moves without moving any input head and without cycling, A_2 can be simulated by a 1:k-DSA A_3 having property P_1 and which can, after writing a stack symbol, only either erase this symbol again or leave the top with its stack head or move some input head (block encoding of the stack word).

Now we construct an equivalent 1:k-DSA A_4 working as A_3, but avoiding the consecutive writing and erasing of a stack symbol. Consequently, between two writing moves, A_4 has to move an input head.

Therefore, without cycling, A_4 can make at most $k \cdot n$ writing moves and hence at most $k \cdot n$ erasing moves. Now A_4 can work at most $\precsim n$ steps for a fixed stack word and fixed head positions (since the stack word has a length of at most $\lesssim n$). Consequently, A_4 works at most $\lesssim n^2$ steps.

Ad 3. In what follows, we consider w.l.o.g. only 1:k-NSA which move in each step either one of the input heads or the stack head, and which do not alter any stack symbol (unless erasing it).

Let A_1 be a 1:k-NSA. As in the proof of Statement 1, we construct an equivalent 1:k-NSA A_2 whose computation paths all have the property P_1.

Now we construct an equivalent 1:k-NSA A_3 such that, if w is accepted by A_3, then w is accepted by a computation path β having the following property, which is said to be property P_2: β has property P_1 and, between two input head moves, all erasing moves precede all writing moves. Consequently, between two input head moves in β, the 1:k-NSA A_3 can either consecutively:

1. move the stack head to the top (if it is not already there),
2. erase some top symbols,
3. write some new top symbols,
4. move (possibly) the stack head into the stack

or it can move the stack head only inside the stack.

A_3 works as follows: If A_2 writes a stack symbol, then A_3 does nondeterministically both:

— A_3 works as A_2, i.e. first A_3 also writes this symbol.

— A_3 goes directly to those states in which A_2 can be if it erases this symbol after some time (without moving any input head in the meantime, but many symbols way be written and erased on the stack). Since A_2 works in the meantime without reading further input symbols, A_3 can "know" these states.

Now we construct an equivalent 1:k-NSA A_4 such that, if w is accepted by A_4, then w is accepted by a computation path β having the following property, whichi s said to be property P_3: β has property P_2 and has stack words of length $\leqq 2k \cdot n$.

A_4 simulates A_3 as follows: If A_3 writes consecutively the word v on its stack, then A_4 guesses nondeterministically in what portions v will be erased later (i.e. there is no move of any input head and no writing between the erasure of the symbols of such a portion) and writes instead of such a portion p a table which answers, for each k-tuple $(a_1, \ldots, a_k)$ of input symbols and for each state s, the following question: In what states and in what directions the stack head leaves p (with either no erasing or total erasing) if it enters p with the state s from the left(right)-hand side and while the input heads read $a_1, \ldots, a_k$ in fixed positions. Note that there is only a bounded number of such tables.

Using these tables, A_4 can simulate the work of A_3. Since there are at most $k \cdot n$ writing phases and at most $k \cdot n$ erasing phases, A_4 writes at most $2k \cdot n$ tables during the simulation of a path of A_3 having property P_2 and having no cycles, provided A_4 makes all guesses correctly. Consequently, there are also at most $2k \cdot n$ erasing moves. Finally, on computation paths without cycles, between two input head moves there can be at most $\lesssim n$ steps having the stack head inside the stack. Note that for any computation path there is an equivalent one without cycles. Thus, if w is accepted by A_4, then w is accepted by a computation path of at most $\leq n^2$ steps.

The statements 2 and 4 are immediate consequences of the statements 1 and 3, resp.

Space results

For the statements 1 and 3 see the proofs of the corresponding time results. Statement 2 follows immediately from Statement 1.

Ad 4. Following the proof of Statement 3 (time result), we can start with a 1:k-NSA A such that if w is accepted by A, then w is accepted by a computation path having property P_3. As we have seen in this proof, a computation path having property P_3 and having no cycles consists of at most $c \cdot n$ writing moves, at most $c \cdot n$ erasing moves, and at most $c \cdot n$ scans of the stack word (for some $c > 0$). Moreover, all moves of the input heads are made during such a stack scan (let us consider the case that the stack head scans the top and an input head moves as a special case of a stack scan) and the stack head can be inside the stack for at most $c \cdot n$ steps without any input head moving. Now we represent such a computation path β by a sequence x_β encoding

— a writing move by the written symbol,

— an erasing move by —,

— a scan of the stack word including moves of the input heads $i_1, \ldots, i_r$ ($i_1, \ldots, i_r \in \{1, \ldots, k\}$) in this order by $i_1, \ldots, i_r$.

Thus, x_β is a word of length at most $d \cdot n$ for some $d > 0$. Now a suitable DTM M can generate within space $d \cdot n$ all these sequences x_β in a systematic manner. After generating such a sequence, M checks whether this sequence is the representation of some accepting computation path having property P_3. This can be done within space n, because the stack words of such computation paths have maximal length $c \cdot n$ and all possibilities for the work of A inside the stack between two input head moves can be tried successively within space n. ∎

Remark 1. In contrast to 13.33.3 (time), for the case of stacks using one stack symbol only (SC-stack counter) it is shown in [GiRo 74] that **1-NSCA** = **1-CS-NTIME**(Lin).

Remark 2. For X ∈ {D, N}, every 1-XSA language is the image under a nonerasing homomorphism of an **XL** language (see [KiWr 78]).

It will be found that the 2:k-XSA languages for every k and X ∈ {D, N} can be characterized in terms of time complexity classes. In order to get these results we show first a relationship to T-auxPD-XSPACE complexity classes.

13.34. Theorem. [Coo 71 b] For $k \geqq 1$,

1. **2:k-DSA** = **T-auxPD-DSPACE**($n^k \cdot \log n$).
2. **2:k-NSA** = **T-auxPD-NSPACE**(n^{2k}).
3. **2:multi-DSA** = **2:multi-NSA** = **T-auxPD-DSPACE**(Pol).

Proof. a) We show **T-auxPD-DSPACE**($n^k \cdot \log n$) $\subsetneqq$ **2:k-DSA** using some techniques described in the proof of **DSPACE**($n^k \cdot \log n$) $\subsetneqq$ **2:k-DNESA** (see 13.29.1)*). Let M be a T-PD-DM using only $c \cdot n^k \cdot \log n$ squares of the Turing tape, having only the tape symbols $\Box$, 1, 2, and having at each step of the computation on an input word $w \in \{1, 2\}^*$ only Turing tape inscriptions of the form $\ldots \Box\Box\Box w_1 \Box\Box\Box \ldots$, where the leftmost symbol of $w_1 \in \{1, 2\}^*$ is written in the same tape square as the leftmost symbol of w at the beginning of the computation. For later quotation of this proof we emphasize that we shall use in the following construction of a 2:k-DSA A, simulating M, only the facts that A has a stack, that it can read the input with one input head, and that it has the ability to count up to $t = c \cdot n^k$ without using the stack.

If M has at some step of its computation the pushdown word $a_1 a_2 \ldots a_r$, where $a_1, a_2, \ldots, a_r$ are pushdown symbols, then A will have the stack word

$$u_1(w_1, v_1, a_1)\; u_2(w_2, v_2, a_2) \ldots u_r(w_r, v_r, a_r),$$

where

- w_i is the special encoding (used in item a) of the proof of 13.29) of the ID of the Turing tape at that moment at which the pushdown head has previously scanned the cell with the symbol a_i,
- v_i is the special encoding of the guessed ID of the Turing tape at that moment at which the pushdown head will next scan the cell with the symbol a_i, and
- u_i is some garbage which will be ignored during the further work of A.

W.l.o.g. M moves its pushdown head at each step. We describe now how A simulates the next step of M.

If M writes the symbol a on its pushdown store, then A writes (w_{r+1}, v_{r+1}, a), where w_{r+1} can be generated from w_r by the method described in item a) of the proof of 13.29, and v_{r+1} is the first word (with respect to some fixed order) which can be considered as the special encoding of an ID of the Turing tape.

If M erases the top symbol a_r of the pushdown store, then

1. A erases $,v_r, a_r)$.
2. A checks whether v_{r-1} is the ID following w_r in M, using the method of generating the next ID in reverse and erasing w_r.
3. A erases u_r.
4. A marks $(w_{r-1}, v_{r-1}, a_{r-1})$ in order to ignore it after the simulation of the current step of M. Thus $(w_{r-1}, v_{r-1}, a_{r-1})$ belongs then to u_{r-1}.
5. If the test in 2 has been successful then the new $(w'_{r-1}, v'_{r-1}, a_{r-1})$ will be generated on the top of the stack. This can be done by copying v_{r-1} ($w'_{r-1} = v_{r-1}$) and generating the first word which can be considered as the special encoding of an ID of the Turing tape (this will be v'_{r-1}).
6. If the test in 2 has not been successful, then that part of the computation of A that begins with the generation of $(w_{r-1}, v_{r-1}, a_{r-1})$ will be repeated, but using instead of the guessed ID v_{r-1} the ID v'_{r-1} which is the next with respect to the

*) The reader is recommended to read first the proof of 13.29.1 in order to understand the details of this proof.

fixed order. The triple $(w_{r-1}, v'_{r-1}, a_{r-1})$ can be generated on the top of the stack by copying w_{r-1} and a_{r-1} and computing from v_{r-1} the next ID (with respect to the fixed order).

b) The inclusion **T-auxPD-NSPACE**$(n^{2k}) \subseteq$ **2:k-NSA** can be proved in the same way as the inclusion **T-auxPD-DSPACE**$(n^k \cdot \log n) \subseteq$ **2:k-DSA** where the generation of the next ID is not executed as in the proof of **DSPACE**$(n^k \cdot \log n) \subseteq$ **2:k-DNESA** but as in the proof of **NSPACE**$(n^{2k}) \subseteq$ **2:k-NNESA** (see 13.29).

c) We show **2:k-DSA** $\subseteq$ **T-auxPD-DSPACE**$(n^k \cdot \log n)$. A 2:k-DSA A can be simulated by a T-PD-DM M in essentially the same way as a 2:k-DNESA is simulated by a DTM (see item c) of the proof of 13.29). The only difference is that if A erases the top symbol, then M does not know the new table and the new top symbol. Therefore, if A writes a symbol on the top, then, before writing this symbol, M also writes the old table on the pushdown store. If A erases the top symbol, then M also erases this symbol and transfers the new table to the Turing tape.

d) The inclusion **2:k-NSA** $\subseteq$ **T-auxPD-NSPACE**(n^{2k}) can be proved in the same way as the inclusion **2:k-DSA** $\subseteq$ **T-auxPD-DSPACE**$(n^k \cdot \log n)$, but the tables have the form described in item d) of the proof of 13.29.

e) Statement 3 is an immediate consequence of the statements 1 and 2. ∎

As a consequence of 13.34 and 20.13 we have

13.35. Theorem. [Coo 71b], [Iba 71] For $k \geq 1$,

1. **2:k-DSA** = **DTIME**$(2^{\text{Lin}\, n^k \cdot \log n})$.
2. **2:k-NSA** = **DTIME**$(2^{\text{Lin}\, n^{2k}})$.
3. **2:multi-DSA** = **2:multi-NSA** = **DTIME**(2^{Pol}). ∎

Finally we mention some results for 2-way alternating SA.

13.36. Theorem. [LaLiSt 78]

1. **2-ASA** = **DTIME**$(2^{2^{\text{Lin}}})$.
2. **2:multi-ASA** = **DTIME**$(2^{2^{\text{Pol}}})$. ∎

13.5.2. Relationships between Different Classes of Stack Automata

Because of 13.35, 13.33.4 (space) and the time hierarchy results (see 17.35) we immediately get the following results.

13.37. Theorem. For $k \geq 1$ and X $\in$ {D, N}, **2:k-XSA** $\subset$ **2:(k+1)-XSA**. ∎

Remark. For stack-counter automata (see Remark 1 after 13.33) hierarchies with respect to the number of heads have been established in [Miy 80] for the one-way case and in [MeWa 82] for the two-way case.

13.38. Theorem. For $k \geqq 1$,

1. **2:k-NSA** $\subset$ **2:(2k)-DSA** $\subset$ **2:(k+1)-NSA**.
2. **1:multi-NSA** $\subset$ **2-DSA**. ∎

13.6. Checking Stack Automata

13.6.1. Relationships to Time and Space Complexity Classes

13.39. Theorem. For 1-way checking stack automata we have the following relationships to time and space complexity classes:

	automata	time	space
1	**1:k-DCSA**	= **1:k-CS-DTIME**(Lin n^2)	= **1:k-CS-DSPACE(id)**
2		$\subseteqq$ **2-kT-DTIME**(n^2)	$\subseteqq$ **L**
3	**1:k-NCSA**	$\subseteqq$ **1:k-CS-NTIME**(Lin n^2)	= **1:k-CS-NSPACE(id)**
4		$\subseteqq$ **2-kT-NTIME**(n^2)	$\subseteqq$ **DSPACE**(id)

Proof. All results and proofs (except for the space result of Statement 2) are special cases of the corresponding results or proofs of 13.33. The space result of Statement 2 is a consequence of 13.40.1. ∎

13.40. Theorem.

1. [Iba 71] **2:multi-DCSA** = **L**.
2. For $k \geqq 1$, **2:k-DCSA** $\subseteqq$ **2-kT-DTIME**(n^{2k}).
3. [Fim 69], [Iba 71] For $k \geqq 1$, **2:k-NCSA** = **NSPACE**(n^k).
4. **2:multi-NCSA** = **PSPACE**.

Proof. Ad 1. Because of **L** $\subseteqq$ **2:multi-DFA** (see 13.2) it is sufficient to show **2:multi-DCSA** $\subseteqq$ **L**. Let A be a **2:k-DCSA**. Since there can be at most $t = q \cdot n^k$ different situations outside the stack, A can make at most $c \cdot t$ steps in the writing phase (without cycling). Hence, the stack is of maximum length $c \cdot t$, and the position of the stack head in the checking phase can be written down within space $\log t$.

A 2-DTM M can simulate A as follows: M works like A but without writing down any stack symbol. The position of the stack head of A is stored on the Turing tape. If in the checking phase the stack head of A reads a new symbol, then M simulates the whole writing phase of A (without writing anything) up to that moment in which this symbol is written by A.

Ad 2. In the proof of Statement 1 we have seen that a 2:k-DCSA A performs at most $\lesssim n^k$ steps in the writing phase. Therefore the stack is of length $\lesssim n^k$ and A performs at most $\lesssim n^{2k}$ steps in the checking phase.

Ad 3. First we show **NSPACE(n^k) $\subseteq$ 2:k-NCSA**. Let M be an NTM working within space n^k such that M uses only tape squares on the left (inclusive) of the tape square in which the first input symbol is initially written.

For later quotation of the proof we emphasize that we shall use in the following construction of a 2:k-DCSA A, which simulates M, only the facts that A has a checking stack, that it can read the input with one input head, and that it has the ability to count up to $t = n^k$ on inputs of length n without using the stack.

In the writing phase, A writes nondeterministically some word that consists of worktape symbols of M, states of M and a new symbol $+$. In the checking phase A checks whether this word is the sequence of ID's of length $t + 1$ of M (separated by symbol $+$) corresponding to a possible computation path of M on the given input w. To this end A checks whether

1. the stack word is of form $w_1 + w_2 + \dots + w_r$ where $|w_i| = t + 1$ and no $+$ is in w_i,
2. $w_1 = s_0 w \,\square^{t-n}$ where s_0 is the initial state of M,
3. w_{i+1} is a possible next ID of the ID w_i, and
4. s_1 occurs in w_r where s_1 is the accepting state of M.

If this is fulfilled, A accepts w.

Now we show **2:k-NCSA $\subseteq$ NSPACE(n^k)**. Let A be a 2:k-NCSA. Without loss of generality we assume that: 1. every accepting computation of A ends in a situation in which the stack head scans the leftmost square (containing the bottom symbol $\boxdot$) and the input heads scan the first input symbol). 2. In each step A moves either some of its input heads or the stack head. 3. The set of states which A uses during its writing phase (for short: writing states) and the set of states which A uses during its checking phase are disjoint. Roughly speaking, an NTM M can simulate the work of A as follows: it guesses nondeterministically and successively all crossing sequences of the stack (including not only the states and the directions of the crossings, but also the positions of the input heads) and checks whether any two adjacent crossing sequences are compatible. More precisely: A crossing sequence of a boundary between two tape squares of the stack of A is a sequence

$$(\sigma_1, s_1, n_{1,1}, \dots, n_{1,k}),\ (\sigma_2, s_2, n_{2,1}, \dots, n_{2,k}),\ \dots,\ (\sigma_m, s_m, n_{m,1}, \dots, n_{m,k})$$

where $\sigma_i \in \{\rightarrow, \leftarrow\}$ is the direction of the ith crossing of the stack head over this boundary, s_i is the state of this crossing and $n_{i,1}, \dots, n_{i,k}$ are the positions of the input heads during this crossing. Since we can restrict ourselves to computations of A without cycles, such a crossing sequence consists of at most $\lesssim n^k$ tuples. However, the description of such a crossing sequence in the given form can be of length $n^k \cdot \log n$. Therefore M writes down the crossing sequences in such a manner that the tuples are ordered with respect to the positions of the input heads, i.e. first all tuples with head positions $0, 0, \dots, 0$, then all tuples with head positions $0, 0, \dots, 1$, etc. and finally all tuples with head positions $n + 1, n + 1, \dots, n + 1$. The tuples with different head positions are separated by the marker $*$. Thus the head positions in the tuples can be omitted without loss of information. Such a description of a crossing sequence is of length $\lesssim n^k$.

The compatibility of two adjacent (guessed) crossing sequences is checked by M as follows: For every tuple in the left (right) crossing sequence with the stack head direction $\rightarrow$ ($\leftarrow$), the NTM M simulates the work of A, beginning with the situation which corresponds to this tuple, until the point when the stack head leaves the square between the two boundaries in question. M checks whether the tuple belonging to the situation in which the stack head leaves the square occurs in the corresponding crossing sequence and whether it is not already marked. If so, M marks this tuple. If after doing so all tuples in the left (right) crossing sequence with the stack head direction $\leftarrow$ ($\rightarrow$) are marked, then these crossing sequences are compatible. Note that A reads $\square$ in a certain square of the stack if and only if the stack head enters this square with a writing state. Then A replaces this symbol by another symbol, which will remain there during the further computation.

The NTM M now works as follows: First it guesses the crossing sequence ($\rightarrow$, initial state, 1, 1, ..., 1), ($\leftarrow$, accepting state, 1, 1, ..., 1) [written down as $**\ldots*$ ($\rightarrow$, initial state), ($\leftarrow$, accepting state) $**\ldots*$] belonging to the left boundary of the leftmost square of the stack (actually this crossing sequence is empty, but for simplicity we assume that the stack head crosses this boundary left-to-right with the starting state in the first step, and right-to-left with an accepting state in the last step). Then M guesses a further (the next to the right) crossing sequence, checks whether they are compatible, and erases the left one. M continues in such a manner until it decides nondeterministically to guess a last crossing sequence. This crossing sequence has to be empty. If all checks end sucessfully, then M accepts the input word on this computation path. Evidently, in this case and only in this case, the guessed crossing sequences describe an accepting computation of A.

Statement 4 is an immediate consequence of Statement 3. ∎

Finally we quote some results on automata having a certain number of stores of type CS and at most one store of type C, PD, S or NES (see [VoWa 81]).

13.41. Theorem. For $l \geqq 1$,

1. **2:multi-lCS-DA = L.**
2. **2:multi-multiCS-DA = L.** ∎

13.42. Theorem. For $l \geqq 1$,

1. **2:multi-lCS-C-DA = L.**
2. **2:multi-multiCS-C-DA = L.** ∎

13.43. Theorem. For $l \geqq 1$,

1. **2:multi-lCS-PD-DA = DTIME**$\left(2^{2^{\cdot^{\cdot^{2^{\mathrm{Pol}}}}}}\Big\}l\right)$.
2. **2:multi-multiCS-PD-DA** $= \mathscr{E}$. ∎

Remark. Results similar to those of Theorem 13.43 are proved in [Eng 83] using "iterated pushdown automata".

13.44. Theorem. For $k, l \geqq 1$,

1. **2:k-lCS-S-DA** $=$ **DTIME**$\left(2^{2^{\cdot^{\cdot^{\cdot^{2^{\mathrm{Lin}\, n^k \cdot \log n}}}}}}\right\}2l+1\Big)$.
2. **2:multi-lCS-S-DA** $=$ **DTIME**$\left(2^{2^{\cdot^{\cdot^{\cdot^{2^{\mathrm{Pol}}}}}}}\right\}2l+1\Big)$.
3. **2:multi-multiCS-S-DA** $= \mathscr{E}$. ∎

13.45. Theorem. For $k, l \geqq 1$,

1. **2:k-lCS-NES-DA** $=$ **DSPACE**$\left(2^{2^{\cdot^{\cdot^{\cdot^{2^{\mathrm{Lin}\, n^k \cdot \log n}}}}}}\right\}l\Big)$.
2. **2:multi-lCS-NES-DA** $=$ **DSPACE**$\left(2^{2^{\cdot^{\cdot^{\cdot^{2^{\mathrm{Pol}}}}}}}\right\}l\Big)$.
3. **2:multi-multiCS-NES-DA** $= \mathscr{E}$. ∎

Remark. The accepting power of the nondeterministic counterparts of the deterministic automata investigated above leads beyond the subrecursive area. The same can be said for automata having more than one store of type C, PD, S or NES. Actually, we have

2:k-X_1-X_2-DA $=$ **REC** for $k \geqq 1$, $X_1, X_2 \in \{C, PD, S, NES\}$

and

2:k-X_1-X_2-NA $=$ **RE** for $k \geqq 1$, $X_1, X_2 \in \{C, PD, S, NES, CS\}$.

13.6.2. Relationships between Different Classes of Checking Stack Automata

The proof of 13.40.1 readily yields

13.46. Theorem. For $k \geqq 1$, **2:k-DFA** $\subseteq$ **2:k-DCSA** $\subseteq$ **2:(4k)-DFA**. ∎

Two-way deterministic (nondeterministic) checking stack automata with $k + 1$ input heads are more powerful than those with k input heads.

13.47. Theorem. For $k \geqq 1$ and $X \in \{D, N\}$, **2:k-XCSA** $\subset$ **2:(k+1)-XCSA**. ∎

Remark. Witness sets for the proper inclusion in 13.47 (deterministic case) can be chosen to be SLA-languages. The proof is similar to the proof of the corresponding result for finite automata (cf. [MeWa 82]). The nondeterministic case of 13.47 is an immediate consequence of 13.40.3 and the nondeterministic space hierarchy (cf. 17.32).

The following theorem is an immediate consequence of 13.39 and 13.40.

13.48. Theorem.

1. **1:multi-DCSA** $\subseteq$ **2:multi-DCSA** $\subset$ **2-NCSA**.
2. **1:multi-NCSA** $\subseteq$ **2-NCSA**. ∎

13.7. Summary of the Time and Space Results

A. Deterministic 1-way automata

1:k-	multiT-DTIME	RAM-DTIME	DSPACE
DFA	$\subseteq$ Lin ($\subseteq$ id)	$\subseteq$ id	$\subseteq \log n$ (= 0)
DCA	$\subseteq$ Lin	$\subseteq$ id	$\subseteq \log n$
DPDA	$\subseteq$ Lin	$\subseteq$ id	$\subseteq (\log n)^2$
DCSA	$\subseteq n^2$	$\subseteq n^2/\log n$	$\subseteq \log n$
DNESA	$\subseteq n^2$	$\subseteq n^2/\log n$	$\subseteq$ id
DSA	$\subseteq n^2$	$\subseteq n^2/\log n$	$\subseteq$ id

B. Nondeterministic 1-way automata

1:k-	multiT-DTIME	RAM-DTIME	multiT-NTIME	DSPACE	NSPACE
NFA	$\subseteq n^{k+1}$ ($\subseteq$ id)	$\subseteq n^k$	$\subseteq$ id	$\subseteq (\log n)^2$ (= 0)	$\subseteq \log n$ (= 0)
NCA	$\subseteq n^{k+2}$ ($\subseteq n^2$)	$\subseteq n^{k+2}/\log n$ ($\subseteq n^2$)	$\subseteq$ id	$\subseteq (\log n)^2$	$\subseteq \log n$
NPDA	$\subseteq n^{2.496k}$	$\subseteq n^{2k} \cdot \log n$	$\subseteq$ id	$\subseteq (\log n)^2$	$\subseteq (\log n)^2$
NCSA	$\subseteq 2^{\text{Lin}}$	$\subseteq 2^{\text{Lin}}$	$\subseteq n^2$	$\subseteq$ id	$\subseteq$ id
NNESA	$\subseteq 2^{\text{Lin}}$	$\subseteq 2^{\text{Lin}}$	$\subseteq n^2$	$\subseteq$ id	$\subseteq$ id
NSA	$\subseteq 2^{\text{Lin}}$	$\subseteq 2^{\text{Lin}}$	$\subseteq n^2$	$\subseteq$ id	$\subseteq$ id

C. Deterministic 2-way automata

2:k-	multiT-DTIME	RAM-DTIME	DSPACE
DFA	$\subseteq n^k$	$\subseteq n^k/\log n$ ($\subseteq$ id) for $k \geqq 2$	$\subseteq \log n$ (= 0)
DCA	$\subseteq n^{2k}$	$\subseteq n^k$	$\subseteq \log n$
DPDA	$\subseteq n^{2k} \cdot \log n$	$\subseteq n^k$	$\subseteq n^k$
DCSA	$\subseteq n^{2k}$	$\subseteq n^{2k}/\log n$	$\subseteq \log n$
DNESA	$\subseteq 2^{\text{Lin}\, n^k \cdot \log n}$	$\subseteq 2^{\text{Lin}\, n^k \cdot \log n}$	$= n^k \cdot \log n$
DSA	$= 2^{\text{Lin}\, n^k \cdot \log n}$	$= 2^{\text{Lin}\, n^k \cdot \log n}$	$\subseteq 2^{\text{Lin}\, n^k \cdot \log n}$

D. Nondeterministic 2-way automata

2:k-	multiT-DTIME	RAM-DTIME	multiT-NTIME	DSPACE	NSPACE
NFA	$\subseteq n^{2k}$ ($\subseteq$ id)	$\subseteq n^k$	$\subseteq n^k$	$\subseteq (\log n)^2$ $(= 0)$	$\subseteq \log n$ $(= 0)$
NCA	$\subseteq n^{4k}$	$\subseteq n^{3k}/\log n$	$\subseteq n^{3k}$	$\subseteq (\log n)^2$	$\subseteq \log n$
NPDA	$\subseteq n^{4k}$	$\subseteq n^{3k}/\log n$	$\subseteq n^{4k}$	$\subseteq n^{2k}$	$\subseteq n^{2k}$
NCSA	$\subseteq 2^{\mathrm{Lin}\, n^k}$	$\subseteq 2^{\mathrm{Lin}\, n^k}$	$\subseteq 2^{\mathrm{Lin}\, n^k}$	$\subseteq n^{2k}$	$= n^k$
NNESA	$\subseteq 2^{\mathrm{Lin}\, n^{2k}}$	$\subseteq 2^{\mathrm{Lin}\, n^{2k}}$	$\subseteq 2^{\mathrm{Lin}\, n^{2k}}$	$\subseteq n^{4k}$	$= n^{2k}$
NSA	$= 2^{\mathrm{Lin}\, n^{2k}}$	$= 2^{\mathrm{Lin}\, n^{2k}}$	$\subseteq 2^{\mathrm{Lin}\, n^{2k}}$	$\subseteq 2^{\mathrm{Lin}\, n^{2k}}$	$\subseteq 2^{\mathrm{Lin}\, n^{2k}}$

Remark. The results in brackets correspond to the one-head case (i.e. $k = 1$). All RAM-DTIME results are understood to have the factor Lin. All RAM-DTIME results in the tables A, B and C having the form $n^m/\log n$ follow from the corresponding multiT-DTIME results by using Theorem 19.31.3.

§ 14. Complexity Classes and Reducibility

In this section we study such problems as "Which complexity classes are closed under certain given complexity bounded reducibilities?", "What about complete and hard sets for them?", "Which complexity classes can be characterized as the reducibility closure of suitable other families of languages?"

14.1. Closure with Respect to Reducibility

14.1.1. Reducibility Closure

We start with the observation that every reflexive, transitive reducibility notion $\leqq$ yields a closure operator $\mathcal{R}_{\leqq}$ by the definition

$$\mathcal{R}_{\leqq}(M) =_{\mathrm{df}} \left\{X : \bigvee_{Y \in M} (X \leqq Y)\right\}.$$

In what follows the closure operators are denoted by the same symbolism as the corresponding reducibility notions. For example, $\mathcal{R}_{\mathrm{hom}}$, $\mathcal{R}_{\mathrm{m}}^{\log}$ are the closure operators belonging to $\leqq_{\mathrm{hom}}$, $\leqq_{\mathrm{m}}^{\log}$, resp.

We first state a general result about the characterization of reducibility closures in terms of complexity classes.

14.1. Theorem.

1. For X $\in$ {D, N} and $s \geqq \log$,

$$\mathbf{XSPACE}(s(\mathrm{Pol})) = \mathcal{R}_{\mathrm{pad}}(\mathbf{XSPACE}(s)) = \mathcal{R}_{\mathrm{T}}^{\log}(\mathbf{XSPACE}(s)).$$

2. For X $\in$ {D, N} and $t(n) \geqq c \cdot n$ for some $c > 1$,

$$\mathbf{multiT\text{-}XTIME}(t(\mathrm{Pol})) = \mathcal{R}_{\mathrm{pad}}(\mathbf{multiT\text{-}XTIME}(t)) = \mathcal{R}_{\mathrm{T}}^{\mathrm{P}}(\mathbf{multiT\text{-}XTIME}(t)).$$

Proof. Ad 1. As $\mathcal{R}_{\mathrm{pad}}(M) \subseteq \mathcal{R}_{\mathrm{m}}^{\log}(M) \subseteq \mathcal{R}_{\mathrm{T}}^{\log}(M)$ for every M we have only to show
(a) $\mathbf{XSPACE}(s(\mathrm{Pol})) \subseteq \mathcal{R}_{\mathrm{pad}}(\mathbf{XSPACE}(s))$ and
(b) $\mathcal{R}_{\mathrm{T}}^{\log}(\mathbf{XSPACE}(s)) \subseteq \mathbf{XSPACE}(s(\mathrm{Pol}))$.

To prove (a) let $A \in \mathbf{XSPACE}(s(\mathrm{Pol}))$. Then there exists some $k > 0$ such that $A \in \mathbf{XSPACE}(s(n^k))$. The conditions of 6.16 are satisfied, and we get $A_{n^k} \in \mathbf{XSPACE}(s)$. Because of $A \leqq_{\mathrm{pad}} A_{n^k}$, Statement (a) is proved.

To prove (b) let $C \in \mathcal{R}_{\mathrm{T}}^{\log}(\mathbf{XSPACE}(s))$. This means, that there exist some set $B \in \mathbf{XSPACE}(s)$ and some oracle machine $M^{(B)}$ deciding (if X = D)/accepting (if X = N) the set C within logarithmic space. $M^{(B)}$ may generate queries of polynomial length, and thus a direct combination of $M^{(B)}$ and a machine deciding/accepting B would yield a machine which could use polynomial space. This is too much for small s. However, as in the proof of 6.1.1 we can construct a machine using at most $s(\mathrm{Pol})$ workspace.

Ad 2. This statement can be proved as Statement 1, but using for (b) the direct combination of $M^{(B)}$ and a machine deciding/accepting B. ∎

Our next aim is the characterization of complexity classes in terms of reducibility closure. The very nature of the following two theorems is the existence of complete sets in certain complexity classes, and thus they could appear in § 14.2.2. Nevertheless, because of their theoretical consequences they play a special role, and we deal with them here separately.

We start with a characterization of **CF** = **1-PD-NTIME**(id) (cf. 12.4) and **1-PD-DTIME**(id).

For the Dyck language D_2 (i.e. the language defined by the grammar with the productions $S \to SS$, $S \to Z_1 S \bar{Z}_1$, $S \to Z_2 S \bar{Z}_2$, $S \to e$) we define the languages $\mathrm{L_N}(\mathrm{D}_2)$ and $\mathrm{L_D}(\mathrm{D}_2)$:

$$\mathrm{L_N}(\mathrm{D}_2) = \Big\{[w_{1,1} \# \ldots \# w_{1,k_1}] \ldots [w_{s,1} \# \ldots \# w_{s,k_s}] : s, k_1, \ldots, k_s \in \mathbb{N} \wedge w_{i,j} \in \{Z_1, Z_2, \bar{Z}_1, \bar{Z}_2\}^* \wedge \bigvee_{w \in \mathrm{D}_2} \bigvee_{i_1} \ldots \bigvee_{i_s} Z_1 Z_2 Z_1 w_{1,i_1} \ldots w_{s,i_s} \bar{Z}_2 \sqsubseteq w\Big\},$$

$$\mathrm{L_D}(\mathrm{D}_2) = \Big\{\big[w_{1,1}^1 \# \ldots \# w_{1,r_{1,1}}^1 \oplus \ldots \oplus w_{t_1,1}^1 \# \ldots \# w_{t_1,r_{1,t_1}}^1\big] \ldots \big[w_{1,1}^s \# \ldots \# w_{1,r_{s,1}}^s \oplus \ldots \oplus w_{t_s,1}^s \# \ldots \# w_{t_s,r_{s,t_s}}^s\big] : s, t_i, r_{i,j} \in \mathbb{N} \wedge w_{i,j} \in \{Z_1, Z_2\}^* \wedge \bigvee_{w \in \mathrm{D}_2} \bigvee_{i_1} \ldots \bigvee_{i_s} \bigvee_{j_1} \ldots \bigvee_{j_s} Z_1 Z_2 Z_1 \bar{Z}_1^{i_1} \bar{Z}_2 \bar{Z}_1^{j_1} w_{i_1,j_1}^1 \ldots \bar{Z}_1^{i_s} \bar{Z}_2 \bar{Z}_1^{j_s} w_{i_s,j_s}^s \bar{Z}_2 \sqsubseteq w\Big\}.$$

It is easy to see that $\mathrm{L_N}(\mathrm{D}_2) \in \mathbf{CF}$ and $\mathrm{L_D}(\mathrm{D}_2) \in \mathbf{1\text{-}PD\text{-}DTIME}(\mathrm{id})$.

14.2. Theorem.
1. [Gre 73] $\mathcal{R}_{\text{hom}}(\mathrm{L_N(D_2)}) = \mathbf{CF}$.
2. $\mathcal{R}_{\text{hom}}(\mathrm{L_D(D_2)}) = \mathbf{1\text{-}PD\text{-}DTIME}(\mathrm{id})$.

Proof. Ad 1. "$\subseteq$". If $A \in \mathcal{R}_{\text{hom}}(\mathrm{L_N(D_2)})$, then $A = h^{-1}(\mathrm{L_N(D_2)})$ for some homomorphism h. As $\mathrm{L_N(D_2)} \in \mathbf{CF}$ and $\mathbf{CF}$ is closed under inverse homomorphism (cf. 4.12) we obtain $A \in \mathbf{CF}$.

"$\supseteq$". For $A \in \mathbf{CF}$ let M be a 1-NPDA accepting A in realtime (cf. 12.4) and having the states s_0 (accepting state), s_1 (initial state), $s_2, \ldots, s_r$ and the PD-symbols $A_1 = \square, A_2, \ldots, A_t$. For $x \in \Sigma$ we define $h_N(x) = [w_1 \# w_2 \# \ldots \# w_s]$ where

$$\begin{aligned}\{w_1, w_2, \ldots, w_s\} = {} & \{\bar{Z}_1^i \bar{Z}_2 \bar{Z}_1^j Z_1^j Z_2 Z_1^m Z_2 Z_1^k \colon s_i x A_j \to s_k A_m \downarrow + \text{ is an instruction of } M\} \\ & \cup \{\bar{Z}_1^i \bar{Z}_2 \bar{Z}_1^j Z_1^k \colon s_i x A_j \to s_k \uparrow + \text{ is an instruction of } M\} \\ & \cup \{\bar{Z}_1^i \bar{Z}_2 \bar{Z}_1^j Z_1^m Z_2 Z_1^k \colon s_i x A_j \to s_k A_m + \text{ is an instruction of } M\}.\end{aligned}$$

Defining further $h_N(x_1 \ldots x_n) = h_N(x_1) \ldots h_N(x_n)$ we obtain

$$w \in A \leftrightarrow h_N(w) \in \mathrm{L_N(D_2)}.$$

Ad 2. The proof is analogous to that of Statement 1, but instead of h_N it uses the homomorphism h_D defined by

$$h_D(x) = [w_{1,1} \# \ldots \# w_{1,t} \oplus \ldots \oplus w_{r,1} \# \ldots \# w_{r,t}]$$

where for $i = 0, \ldots, r$ and $j = 1, \ldots, t$

$$w_{i,j} = \begin{cases} Z_1^j Z_2 Z_1^m Z_2 Z_1^k & \text{if } s_i x A_j \to s_k A_m \downarrow + \text{ for some } k, m, \\ Z_1^k & \text{if } s_i x A_j \to s_k \uparrow + \text{ for some } k, \\ Z_1^m Z_2 Z_1^k & \text{if } s_i x A_j \to s_k A_m + \text{ for some } k, m. \end{cases}$$ ∎

Remark. By the preceding theorem it is obvious that no context-free language has a greater time or space complexity than $\mathrm{L_N(D_2)}$. Therefore it is justified to call $\mathrm{L_N(D_2)}$ a *hardest context-free language*.

14.3. Theorem. [Sud 78]
1. $\mathcal{R}_m^{\log}(\mathrm{L_N(D_2)}) = \mathcal{R}_m^{\log}(\mathbf{CF}) = \mathbf{2\text{-}auxPD\text{-}T\text{-}NSPACE\text{-}TIME}(\log, \mathrm{Pol})$.
2. $\mathcal{R}_m^{\log}(\mathrm{L_D(D_2)}) = \mathcal{R}_T^{\log}(\mathbf{DCF}) = \mathbf{2\text{-}auxPD\text{-}T\text{-}DSPACE\text{-}TIME}(\log, \mathrm{Pol})$.

Proof. Ad 2. "$\subseteq$". Here we only have to show that $\mathcal{R}_T^{\log}(\mathbf{DCF}) \subseteq \mathbf{2\text{-}auxPD\text{-}T\text{-}DSPACE\text{-}TIME}(\log, \mathrm{Pol})$. Let $A \in \mathcal{R}_T^{\log}(\mathbf{DCF})$. Then there exists a $B \in \mathbf{DCF} = \mathbf{1\text{-}DPDA}$ such that $A \leq_T^{\log} B$. Consequently there exist an oracle 2-DTM $M_1^{(B)}$ deciding A within logarithmic space and a 1-DPDA M_2 deciding B. Now we combine $M_1^{()}$ and M_2 in such a way that the oracle tape of $M_1^{()}$ and the input tape of M_2 are omitted. The symbols which are written by $M_1^{()}$ on its oracle tape are immediately processed by M_2 as input symbols. A question of $M_1^{()}$ to the oracle can be answered by looking for an accepting state of M_2 at this moment. Thus the running time of $M_1^{()}$ is enlarged by at most a constant factor (cf. 13.15.2) and the new machine, which is a 2-PD-T-DM, works (as $M_1^{(B)}$) within logarithmic space and polynomial time.

"$\supseteq$". We have to show that **2-auxPD-T-DSPACE-TIME**(log, Pol) $\subseteq \mathcal{R}_m^{\log}(L_D(D_2))$. Following the proof of 13.2.3 (space) we observe that also for a 2-auxPD-T-DM which works in polynomial time a log-bounded Turing tape can equivalently be replaced by several input heads, i.e. **2-auxPD-T-DSPACE-TIME**(log, Pol) $\subseteq$ **2:multi-PD-DTIME**(Pol). Now, by 13.11 and 13.13, for every language $A \in$ **2:multi-PD-DTIME** (Pol) there exists a $B \in$ **2-PD-DTIME**(Pol) such that $A \leqq_m^{\log} B$.

Thus, we still have to show that **2-PD-DTIME**(Pol) $\subseteq \mathcal{R}_m^{\log}(L_D(D_2))$. Let $A \in$ **2-PD-DTIME**(Pol). Then, w.l.o.g. A can be decided by a 2-DPDA M within polynomial time and whose head advances in each step and reverses its direction only at the end of the input word (and it does so ad infinitum), i.e. when scanning $\square$. If M processes the input w within $p(n)$ steps (p being a polynomial and $n = |w|$) then its input head reads during the computation process the whole infinite sequence $(w \square w^{-1} \square)^\omega$ in a one-way manner although a final state is reached already while reading $\tilde{w} = (w \square w^{-1} \square)^{p(n)}$. Consequently, we can conclude as in the proof of Theorem 14.2.2 that

$$w \in A \leftrightarrow h_D(\tilde{w}) \in L(D_2),$$

where $h_D(\tilde{w})$ can be computed from w within logarithmic space.

Ad 1. The proof for $\leqq_m^{\log}$ can be made as for Statement 2. In the case of $\leqq_T^{\log}$ we do not succeed in combining a 2-DTM $M_1^{()}$ and a 1-NPDA M_2 accepting B because it could happen that the simulation of the oracle returns on some path the answer "no" in a case where the correct answer (on another path) should be "yes", and this wrong answer could lead to a false acceptance of the input word. ∎

Remark 1. So far we do not know whether the properties of the inclusion $\mathcal{R}_m^{\log}(\mathbf{CF}) \subseteq \mathcal{R}_T^{\log}(\mathbf{CF})$ is proper. However, we have: $\mathcal{R}_m^{\log}(\mathbf{CF}) = \mathcal{R}_T^{\log}(\mathbf{CF}) \leftrightarrow \mathcal{R}_m^{\log}(\mathbf{CF})$ is closed under complementation (cf. [Sud 77c]).

Remark 2. Another characterization of $\mathcal{R}_m^{\log}(\mathbf{CF})$ follows from Remark 1 after 20.36.

Remark 3. In [Ern 76], [ArSu 76] and [Asv 80] languages belonging to $\mathcal{R}_m^{\log}(\mathbf{CF})$ are described.

Remark 4. Theorem 14.3 belongs to a broader context of similar results (see also Theorem 12.19):

$\mathcal{R}_m^{\log}(\mathbf{REG}) = \mathcal{R}_m^{\log}(\mathbf{EDOL}) = \mathbf{L}$,		$\mathcal{R}_m^{\mathrm{Nlog}}(\mathbf{REG}) = \mathbf{NL}$,	
$\mathcal{R}_m^{\log}(\mathbf{LIN}) = \mathbf{NL}$	[Sud 75],	$\mathcal{R}_m^{\mathrm{Nlog}}(\mathbf{LIN}) = \mathbf{NL}$	[Lag 84],
$\mathcal{R}_m^{\log}(\mathbf{EDTOL}) = \mathbf{NL}$	[JoSk 77b], [Sud 77b],	$\mathcal{R}_m^{\mathrm{Nlog}}(\mathbf{EDTOL}) = \mathbf{NP} = \mathcal{R}_m^{\mathrm{Nlog}}(\mathbf{NP})$	[Lag 84],
$\mathcal{R}_m^{\log}(\mathbf{EOL}) = \mathcal{R}_m^{\log}(\mathbf{CF})$	[Sud 77b],	$\mathcal{R}_m^{\mathrm{Nlog}}(\mathbf{EOL}) = \mathcal{R}_m^{\mathrm{Nlog}}(\mathbf{CF}) = \mathcal{R}_m^{\mathrm{Nlog}}(\mathbf{DCF})$	[Lag 84],
$\mathcal{R}_m^{\log}(\mathbf{ETOL}) = \mathbf{NP}$	[Lee 75a],	$\mathcal{R}_m^{\mathrm{Nlog}}(\mathbf{ETOL}) = \mathbf{NP}$	[Lag 84].

Theorem 14.3 can be interpreted as follows: The free use of a pushdown store during a log-space and polynomial time bounded computation can be normalized in such a manner that the pushdown is used only at the end of this computation. Furthermore, this result supports the conjecture **CF** $\nsubseteq$ **NL** because **CF** $\subseteq$ **NL** implies $\mathcal{R}_{\mathrm{m}}^{\log}(\mathbf{CF}) = $ **2-auxPD-T-NSPACE-TIME**(log, Pol) $\subseteq$ **NL**, and it seems unlikely that an auxiliary pushdown store does not increase the power of a log-space and polynomial time bounded 2-NTM.

Finally we mention that Theorem 11.14 gives a characterization of **RUD** in terms of $\mathcal{R}_{\mathrm{T}}^{\mathrm{NTIME,Lin}}$.

14.1.2. Closure under Reducibility

The definition of closure under reducibility (cf. p. 47) can be rephrased as follows: A class $\mathscr{C}$ is closed under $\leqq$-reducibility $\leftrightarrow$ $\mathcal{R}_{\leqq}(\mathscr{C}) = \mathscr{C}$.

We start with a theorem saying which time and space complexity classes are known to be closed under various reducibilities.

14.4. Theorem. If $t \geqq t_0$ $(s \geqq s_0)$ stands in the square of row r and column c in Table 14.1, then the complexity class in row r is closed under the reducibility in column c for all $t \geqq t_0$ $(s \geqq s_0)$. ∎

From Theorem 14.1 we conclude that certain complexity classes are not closed under such reducibilities which are too "coarse" in comparison with the "finer" bounds of the classes:

14.5. Theorem.

1. Let $\mathrm{X} \in \{\mathrm{D}, \mathrm{N}\}$ and $s \geqq \log$. If **XSPACE**$\big(s(\mathrm{Lin})\big) \subset$ **XSPACE**$\big(s(\mathrm{Pol})\big)$, then **XSPACE**$\big(s(\mathrm{Lin})\big)$ is not closed under $\leqq_{\mathrm{pad}}$.
2. Let $\mathrm{X} \in \{\mathrm{D}, \mathrm{N}\}$ and $t(n) \geqq c \cdot n$ for some $c > 1$. If **multiT-XTIME**$\big(t(\mathrm{Lin})\big) \subset$ **multiT-XTIME**$\big(t(\mathrm{Pol})\big)$, then **multiT-XTIME**$\big(t(\mathrm{Lin})\big)$ is not closed under $\leqq_{\mathrm{pad}}$. ∎

The following theorem deals with the closure of some important special complexity classes under various reducibilities.

14.6. Theorem. In Table 14.2 the symbol + (−) in the square of row r and column c indicates that the complexity class in row r is (not) closed under the reducibility in column c. If i is in the square of row r and column c, then it is not known whether the complexity class in row r is closed under the reducibility in column c, and Statement i gives consequences of a positive and a negative solution ($\mathrm{X} \in \{\mathrm{D}, \mathrm{N}\}$).

The numbers in Table 14.2 are explained as follows:

1. $+ \leftrightarrow$ **P** $\subseteq$ **DSPACE**$(\log^k n)$
   $k = 1$: $+ \leftrightarrow$ **L** = **P**.
2. $+ \rightarrow$ **NP** $\subseteq$ **DSPACE**$(\log^k n)$
   $- \rightarrow$ **DSPACE**$(\log^k n) \subset$ **NTIME**$(2^{\mathrm{Lin}\log^k n})$ and hence
   L $\subset$ **NP**
   $k = 1$: $+ \leftrightarrow$ **L** = **NP**.

Table 14.1

	$\leqq_{\mathrm{hom}}$	$\leqq_{\mathrm{m}}^{\mathrm{log\text{-}lin}}$	$\leqq_{\mathrm{pad}}$	$\leqq_{\mathrm{m}}^{\mathrm{log}}$	$\leqq_{\mathrm{m}}^{\mathrm{P}}$	$\leqq_{\mathrm{m}}^{\mathrm{NP}}$	$\leqq_{\mathrm{T}}^{\mathrm{log}}$	$\leqq_{\mathrm{T}}^{\mathrm{P}}$	$\leqq_{\mathrm{T}}^{\mathrm{NP}}$
DSPACE(s(Lin))	$s \geqq 0$	$s \geqq \log$							
DSPACE(s(Pol))	$s \geqq 0$	$s \geqq \log$	$s \geqq \log$	$s \geqq \log$	$s \geqq \mathrm{id}$	$s \geqq \mathrm{id}$	$s \geqq \log$	$s \geqq \mathrm{id}$	$s \geqq \mathrm{id}$
NSPACE(s(Lin))	$s \geqq 0$	$s \geqq \log$							
NSPACE(s(Pol))	$s \geqq 0$	$s \geqq \log$	$s \geqq \log$	$s \geqq \log$	$s \geqq \mathrm{id}$	$s \geqq \mathrm{id}$			
multiT-DTIME(t(Lin))	$t \geqq \mathrm{id}$								
DTIME(Pol t(Lin))	$t \geqq \mathrm{id}$	$t \geqq \mathrm{id}$							
DTIME(Pol t(Pol))	$t \geqq \mathrm{id}$	$t \geqq \mathrm{id}$	$t \geqq \mathrm{id}$	$t \geqq \mathrm{id}$	$t \geqq \mathrm{id}$	$t(n) \geqq 2^n$	$t \geqq \mathrm{id}$	$t \geqq \mathrm{id}$	$t(n) \geqq 2^n$
multiT-NTIME(t(Lin))	$t \geqq \mathrm{id}$								
NTIME(Pol t(Lin))	$t \geqq \mathrm{id}$	$t \geqq \mathrm{id}$							
NTIME(Pol t(Pol))	$t \geqq \mathrm{id}$	$t \geqq \mathrm{id}$	$t \geqq \mathrm{id}$	$t \geqq \mathrm{id}$	$t \geqq \mathrm{id}$	$t \geqq \mathrm{id}$			

Table 14.2

	$\le_{\text{hom}}$	$\le_m^{\text{log-lin}}$	$\le_{\text{pad}}$	$\le_m^{\log}$	$\le_m^{P}$	$\le_m^{NP}$	$\le_T^{\log}$	$\le_T^{P}$	$\le_T^{NP}$
DSPACE$(\log^k n)$, $k \ge 1$	+	+	+	+	1	2	+	3	2
NSPACE$(\log^k n)$, $k \ge 1$	+	+	+	+	4	5	6	7	5
DSPACE(Pollog)	+	+	+	+	8	9	+	8	9
XSPACE(n^k), $k \ge 1$	+	+	−	−	−	−	−	−	−
PSPACE	+	+	+	+	+	+	+	+	+
multiT-DTIME(n^k), $k \ge 2$	+	10	−	−	−	−	−	−	−
multiT-NTIME(n^k), $k \ge 1$	+	11	−	−	−	−	−	−	−
P	+	+	+	+	+	12	+	+	12
NP	+	+	+	+	+	+	13	13	13
XTIME$(2^{\text{Lin}\, n^k})$, $k \ge 1$	+	+	−	−	−	−	−	−	−
DTIME(2^{Pol})	+	+	+	+	+	+	+	+	+
NTIME(2^{Pol})	+	+	+	+	+	+	14	14	14

3. $+ \to$ $\mathbf{P} \subseteq \mathbf{DSPACE}(\log^k n)$
   $- \to$ $\mathbf{DSPACE}(\log^k n) \subset \mathbf{DTIME}(2^{\mathrm{Lin}\,\log^k n})$ and hence
   $\mathbf{L} \subset \mathbf{P}$
   $k = 1$: $+ \leftrightarrow \mathbf{L} = \mathbf{P}$.
4. $+ \leftrightarrow$ $\mathbf{P} \subseteq \mathbf{NSPACE}(\log^k n)$
   $k = 1$: $+ \leftrightarrow \mathbf{NL} = \mathbf{P}$.
5. $+ \to$ $\mathbf{NP} \subseteq \mathbf{NSPACE}(\log^k n)$
   $- \to$ $\mathbf{NSPACE}(\log^k n) \subset \mathbf{NTIME}(2^{\mathrm{Lin}\,\log^k n})$ and hence
   $\mathbf{NL} \subset \mathbf{NP}$
   $k = 1$: $+ \leftrightarrow \mathbf{NL} = \mathbf{NP}$.
6. $+ \leftrightarrow$ $\mathbf{coNSPACE}(\log^k n) = \mathbf{NSPACE}(\log^k n)$.
7. $+ \to$ $\mathbf{P} \subseteq \mathbf{NSPACE}(\log^k n)$
   $- \to$ $\mathbf{NSPACE}(\log^k n) \subset \mathbf{DTIME}(2^{\mathrm{Lin}\,\log^k n})$ and hence
   $\mathbf{NL} \subset \mathbf{P}$
   $k = 1$: $+ \leftrightarrow \mathbf{NL} = \mathbf{P}$.
8. $+ \leftrightarrow$ $\mathbf{P} \subseteq \mathbf{DSPACE}(\mathrm{Pol}\,\log)$.
9. $+ \leftrightarrow$ $\mathbf{NP} \subseteq \mathbf{DSPACE}(\mathrm{Pol}\,\log)$.
10. $+ \to$ $\mathbf{L} \subseteq \mathbf{multiT\text{-}DTIME}(n^k)$
    $- \to$ $\mathbf{L} \nsubseteq \mathbf{multiT\text{-}DTIME}(n^{k-1})$.
11. $+ \to$ $\mathbf{L} \subseteq \mathbf{multiT\text{-}NTIME}(n^k)$
    $- \to$ $\mathbf{L} \nsubseteq \mathbf{multiT\text{-}NTIME}(n^{k-1})$.
12. $+ \leftrightarrow$ $\mathbf{P} = \mathbf{NP}$.
13. $+ \leftrightarrow$ $\mathbf{coNP} = \mathbf{NP}$.
14. $+ \leftrightarrow$ $\mathbf{coNTIME}(2^{\mathrm{Pol}}) = \mathbf{NTIME}(2^{\mathrm{Pol}})$.

Proof. All positive results are consequences of 14.4, all negative results are consequences of 14.5. It remains to prove the statements 1—14. We outline the proof of two sample statements.

Ad 1. First assume $\mathbf{P} \subseteq \mathbf{DSPACE}(\log^k n)$. Let $A \leq_m^P B$ via $f \in \mathrm{P}$ and let $B \in \mathbf{DSPACE}(\log^k n)$. Defining $A_f = \{w \# m \# a\colon a$ is the m-th symbol of $f(w)\}$ we obtain $A_f \in \mathbf{P}$ and hence $A_f \in \mathbf{DSPACE}(\log^k n)$. From the latter we easily conclude $f \in \mathrm{DSPACE}(\log^k n)$. As in the proof of the transitivity of $\leq_T^{\log}$ (6.1.1) we can see that $A \in \mathbf{DSPACE}(\log^k n)$.

Now assume that $\mathbf{DSPACE}(\log^k n)$ is closed under $\leq_m^P$. Since $A \leq_m^P \{1\}$ for all $A \in \mathbf{P}$ and $\{1\} \in \mathbf{DSPACE}(\log^k n)$ we obtain $\mathbf{P} \subseteq \mathbf{DSPACE}(\log^k n)$.

Similar ideas are used to prove the statements 2, 3, 4, 5, 7, 8, 9, 10 and 11. Note that the results $\mathbf{XL} \subset \mathbf{YP}$($\mathrm{X}, \mathrm{Y} \in \{\mathrm{D}, \mathrm{N}\}$) are derived from the statements which immediately precede them by translational results (see 26.5).

Ad 13, taken for $\leq_T^P$. First assume $\mathbf{coNP} = \mathbf{NP}$. Let $A \leq_T^P B$ and $B \in \mathbf{NP}$. The set A can be accepted by a nondeterministic TM M within polynomial time as follows: M works like a deterministic oracle TM M' reducing A to B within polynomial time. If M' makes a query to the oracle B, then M starts nondeterministically acceptors for B and $\overline{B}$. Only one of them can have accepting paths on this input. On these paths M continues to simulate M'. Now assume that **NP** is closed under $\leq_T^P$. Since $A \leq_T^P \bar{A}$ and $\bar{A} \in$ **NP** for all $A \in$ **coNP** we obtain $A \in$ **NP**.

The statements 6 and 14 can be proved analogously. ∎

14.2. Hard and Complete Sets

14.2.1. General Results

In this section some statements about the existence and possible types of complete sets are presented.

We start with the observation that the result 11.5 about AFL-generators in certain complexity classes is nothing else than a result about the existence of $\leq_{hom}$-complete sets in these complexity classes:

14.7. Theorem.

1. Let $X \in \{D, N\}$, and let $s \geq id$ be a fully 2-T-DSPACE-constructible increasing function such that $k \cdot s(n) \leq s(k \cdot n)$ for $k, n \in \mathbb{N}$. Then A_s^X is $\leq_{hom}$-complete in **XSPACE**$(s(\text{Lin}))$.
2. Let $t(n) \geq 2^n$ be a fully multiT-DTIME-constructible increasing function such that $k \cdot t(n) \leq t(k \cdot n)$ for $k, n \in \mathbb{N}$ and $t(n)^2 \leq t(l \cdot n)$ for some l and all $n \in \mathbb{N}$. Then B_t^D is $\leq_{hom}$-complete in **multiT-DTIME**$(t(\text{Lin}))$.
3. Let $t \geq id$ be a fully 3T-DTIME-constructible superadditive function. Then B_t^N is $\leq_{hom}$-complete in **multiT-NTIME**$(t(\text{Lin}))$. ∎

Remark. The remark after 11.5 shows how to modify $A_{n^{1/k}}^X$ to get a $\leq_{hom}$-complete set in **XSPACE**$(n^{1/k})$ for $k \geq 1$.

The next result shows that sets which are $\leq_m^{\text{log-lin}}$-complete in $\Phi(t(\text{Lin}))$ can be shown to be $\leq_m^{\log}$-complete in $\Phi(t(\text{Pol}))$ for certain measures Φ and certain bounds t. This result can be applied to the $\leq_{hom}$-complete sets of the preceding theorem.

14.8. Theorem.

1. Let $X \in \{D, N\}$ and $s \geq \log$. If A is $\leq_m^{\text{log-lin}}$-complete in **XSPACE**$(s(\text{Lin}))$, then A is $\leq_m^{\log}$-complete in **XSPACE**$(s(\text{Pol}))$.
2. Let $X \in \{D, N\}$ and $t \geq id$. If A is $\leq_m^{\text{log-lin}}$-complete in **multiT-XTIME**$(t(\text{Lin}))$, then A is $\leq_m^{\log}$-complete in **multiT-XTIME**$(t(\text{Pol}))$.

Proof. We prove statement 1, statement 2 can be proved analogously. Let $B \in$ **X-SPACE**$(s(n^m))$ for some $m \in \mathbb{N}$. By 6.16 the padded set B_{n^m} belongs to **XSPACE**$(s(n))$ From $B \leq_{pad} B_{n^m}$ and $B_{n^m} \leq_m^{\text{log-lin}} A$ we conclude $B \leq_m^{\log} A$. ∎

Again using padding, arbitrary complete sets in complexity classes of the form $\Phi(t(\text{Pol}))$ can be converted to relatively simple complete sets in these complexity classes.

14.9. Theorem.

1. Let $\leqq \in \{\leqq_m^{\log}, \leqq_m^P, \leqq_T^{\log}, \leqq_T^P\}$, $X \in \{D, N\}$ and $s \geqq \log$. If **XSPACE**$(s(\mathrm{Pol}))$ has a $\leqq$-complete set, then there exists such a set in **XSPACE**$(s(\mathrm{Lin}))$.
2. Let $\leqq \in \{\leqq_m^{\log}, \leqq_m^P, \leqq_T^{\log}, \leqq_T^P\}$, $X \in \{D, N\}$ and $s \geqq \mathrm{id}$. If **multiT-XTIME**$(t(\mathrm{Pol}))$ has a $\leqq$-complete set, then there exists such a set in **multiT-XTIME**(t). ∎

Our next theorem deals with the existence of complete sets in some important special complexity classes.

Table 14.3. The existence of complete sets in some important complexity classes

		$\leqq_{hom}$	$\leqq_m^{\log\text{-lin}}$	$\leqq_m^{\log}$	$\leqq_m^P$	$\leqq_T^{\log}$	$\leqq_T^P$
1	**L**	?	+	+	×	+	×
2	**NL**	?	?	+	×	×	×
3	**DSPACE**(Pol log)	—	—	—	×	—	×
4	**XSPACE**(n^k), $k \geqq 1$	+	+	×	×	×	×
5	**PSPACE**	—	—	+	+	+	+
6	**REALTIME**	—	×	×	×	×	×
7	**P**	—	?	+	+	+	+
8	**multiT-NTIME**(n^k), $k \geqq 1$	+	×	×	×	×	×
9	**NP**	—	?	+	+	×	×
10	**XTIME**$(2^{\mathrm{Lin}\, n^k})$, $k \geqq 1$	+	+	×	×	×	×
11	**DTIME**(2^{Pol})	—	—	+	+	+	+
12	**NTIME**(2^{Pol})	—	—	+	+	×	×

14.10. Theorem. In Table 14.3 the symbol + (—, ?) in the square of row r and column c indicates that the complexity class in row r possesses (does not possess, is not known to possess, resp.) a complete set with respect to the reducibility in column c. The symbol × in this square indicates that the complexity class in row r is not closed or not known to be closed under the reducibility in column c.

Proof. Positive results. The positive results in the rows 4—5 and 8—12 are consequences of 14.7 and 14.8. For row 2 (row 7) note that GAP(SP) is $\leqq_m^{\log}$-complete in **NL**(**P**) (cf. 8.43).

Negative results. All these results are shown using Lemma 3.4. For example: From **DSPACE**(Pol log) $= \bigcup_{k \geqq 1}$ **DSPACE**$(\log^k n)$, **DSPACE**$(\log^k n) \subset$ **DSPACE**(Pol log), $k \geqq 1$, and the fact that **DSPACE**$(\log^k n)$ is closed under $\leqq_m^{\log}$ (14.6) we can conclude that **DSPACE**(Pol log) cannot have a $\leqq_m^{\log}$-complete set. In the case of **REALTIME** $= \bigcup_{k \geqq 1}$ **kT-DTIME**(id) one has to use the obvious fact that every **kT-DTIME**(id) is closed under $\leqq_{hom}$ and S. O. Aanderaa's result **kT-DTIME**(id) $\subset$ **(k+1)T-DTIME**(id) for all $k \geqq 1$ (8.37). ∎

For the possibility of the existence of sets complete in **NP** ∩ **coNP** see [Kow 84].

We conclude this subsection by mentioning that certain complete sets cannot be SLA-languages.

14.11. Theorem. [Hart 78], [BeHa 77] A $\leq_m^{\log}$-complete set in **NLINSPACE**, **PSPACE**, **DEXPTIME** and **EXPSPACE** cannot be sparse (and therefore cannot be an SLA-language). ∎

Regarding NP-complete sparse sets see 24.16.

14.2.2. Examples

14.2.2.1. Space Classes

Before giving several lists of complete and hard problems we give some general definitions. Let $\mathfrak{A}$ be a class of automata or grammars. If $\mathfrak{A}$ is a class of automata, a superscript 1 means the class of automata from $\mathfrak{A}$ restricted to a single letter alphabet.

$$\mathrm{NE}(\mathfrak{A}) =_{\mathrm{df}} \{A : A \in \mathfrak{A} \wedge \mathrm{L}(A) \neq \emptyset\} \text{ (non empty)},$$
$$\mathrm{INF}(\mathfrak{A}) =_{\mathrm{df}} \{A : A \in \mathfrak{A} \wedge \mathrm{L}(A) \text{ infinite}\} \text{ (infinite)},$$
$$\mathrm{EQ}(\mathfrak{A}) =_{\mathrm{df}} \{(A, B) : A, B \in \mathfrak{A} \wedge \mathrm{L}(A) = \mathrm{L}(B)\} \text{ (equivalent)},$$
$$\mathrm{EW}(\mathfrak{A}) =_{\mathrm{df}} \{A : A \in \mathfrak{A} \wedge e \in \mathrm{L}(A)\} \text{ (empty word)},$$
$$\mathrm{MEMBER}_x(\mathfrak{A}) =_{\mathrm{df}} \{A : A \in \mathfrak{A} \wedge x \in \mathrm{L}(A)\}, \quad x \in \Sigma^*,$$
$$\mathrm{MEMBER}(\mathfrak{A}) =_{\mathrm{df}} \{(A, x) : A \in \mathfrak{A} \wedge x \in \mathrm{L}(A)\},$$
$$\mathscr{P}(\mathfrak{A}) =_{\mathrm{df}} \{A : A \in \mathfrak{A} \wedge \mathscr{P}(\mathrm{L}(A)) = 1\}, \text{ where } \mathscr{P} \text{ is a predicate on } \mathfrak{P}(\Sigma^*).$$

In all subsequent theorems we assume that $\mathscr{P}$ is nonconstant on $\{\mathrm{L}(A) : A \in \mathfrak{A}\}$, and $\mathscr{P}(\emptyset) = 0$.

The following and further results for problems of these types can be found in [Jon 75], [Gal 76a], [Hun 76], [HuRoSz 76], [HuRo 77] and [JoLa 77].

14.12. Theorem. The following problems are $\leq_m^{1\text{-}\log}$-complete [-hard] in **L**:

1GAP, [UGAP],

NE(1-DFA^1), INF(1-DFA^1), EW(1-DFA^1).

Proof. See [HaImMa 78], [LePa 80], [Jon 75], [Jon 75], [Gal 76a], resp. ∎

14.13. Theorem. The following problems are $\leq_m^{\log}$-complete [-hard] in **NL**:

GAP, $\mathrm{GAP_o}$, $\mathrm{GAP_{ao}}$,

NE(1-DFA), NE(1-NFA), INF(1-NFA), EW(1-NFA), [$\mathscr{P}$(1-NFA)].

Proof. See [Savi 73a], [Sud 73], [Sud 73], [Jon 75], [Jon 75], [Jon 75], [Gal 76a], Hun 76], resp. ∎

Remark 1. Further problems which are $\leq_m^{\log}$-complete in **NL** can be found in [JoLiLa 76], [HaMa 81] and [Waa 81].

Remark 2. For **SSPACE**(log) see Remark 4 after 8.43.

Remark 3. [Hon 80a] contains a natural problem which is complete in **DLINSPACE** with respect to a constant space linear time m-reducibility.

For the next result concerning **PSPACE** we define some further problems.

The game introduced in § 25.3.2.1.1 gives rise to the following decision problem investigated in [GiLeTa 79]:

PEBBLE GAME $=_{df}$ $\{(G, k)$: G is a dag with indegree bounded by 2, and a distinguished node of G can be pebbled with no more than k pebbles$\}$.

14.14. Theorem. The following problems are $\leq_m^{\log}$-complete [-hard] in **PSPACE**. These of the last two rows are even $\leq_m^{\log\text{-lin}}$-complete [-hard] in **NLINSPACE**:

GHEX, NODE KAYLES, GEOGRAPHY, PEBBLE GAME,
EQ(1-NFA), NE(2-NFA), EW(1-NCSA), [$\mathscr{P}$(1-NNESA)],
NE(1-NNESA),
MEMBER(1-NNESA), [$\mathscr{P}$(1-DNESA)], MEMBER$_x$(1-DNESA),
INEQ(0, ∪, ·, ², ¬),
B_ω, EQUALITY, IIC, S4,
INEQ(0, 1, ∪, ·, *), NEC(0, 1, ∪, ·, *), NE(0, 1, ∪, ·, *, ∩),
[SUC$_1$, $\mathscr{T}$(ℕ, <), WELLORD, LEXORD].

Proof. See [EvTa 76], [Sch 76], [Sch 76], [GiLeTa 79], [HuRo 77], [Gal 76a], [Gal 76a], [Hun 76], [Hun 76], [Hun 76], [Hun 76], [Hun 76], [StMe 73], [StMe 73], [Sto 77], [Sta 79], [Lad 77], [StMe 73], [MeSt 72], [Für 80], [Fer 74], [Fer 74], [Fer 74], [Fer 74], resp. ∎

Remark 1. In [LiSi 78] it is proved that even GEOGRAPHY on planar bipartite graphs of maximum degree 3 is $\leq_m^{\log}$-complete in **PSPACE**.

Remark 2. In [Lin 78] a more general pebble game is considered, which is played on dags where the nodes are labelled with ∧ and ∨. The related problem is also $\leq_m^{\log}$-complete in **PSPACE**. If the pebble game is played in such a way that every vertex must be pebbled exactly once, then the related problem is NP-complete (see [Meyh 79]). For further problems related to pebble games see [Sud 81] and [KaAdIw 79].

Remark 3. In [Stor 83] and [FrGaJoScYe 78] problems which are related to chess and checkers are shown to be $\leq_m^{\log}$-complete in **PSPACE**.

Remark 4. Some graph-theoretical problems which are members of **NL** (for example GAP and UGAP) become PSPACE-complete when the graphs are described in a succinct manner by special descriptional languages (see [GaWi 83], [Wag 84b]). Similar results can be found in [Var 82] and [BeOtWi 83].

14.15. Theorem. [Hun 77] If the first order theory $\mathscr{T}$ has a model in which some fixed predicate of $\mathscr{T}$ is interpreted nontrivially, then $\mathscr{T}$ is $\leq_m^{\log}$-hard for **PSPACE**. ∎

Remark. If M is a finite Boolean algebra, then $\mathcal{T}(M)$ is $\leqq_m^{\log}$-complete in **PSPACE** ([Koz 80b]).

14.16. Theorem. The following problems are $\leqq_m^{\log}$-complete in $\mathbf{DSPACE}(2^{\mathrm{Pol}})$, those of the last two rows are even $\leqq_m^{\log\text{-lin}}$-complete in **EXPSPACE**:

ABA,
INEQ(0, 1, ∪, ·, *, ∩), NEC(0, 1, ∪, ·, *, ∩),
INEQ(0, 1, ∪, ·, *, ^{2}), NEC(0, 1, ∪, ·, *, ^{2}).

Proof. See [Berp 79], [Für 80], [Für 80], [MeSt 72], [MeSt 72], resp. ∎

14.17. Theorem. The following problems are $\leqq_m^{P}$-hard for $\mathbf{DSPACE}\left(2^{2^{\cdot^{\cdot^{2}}}}\right\}\mathrm{Lin(log)}\big)$:

INEQ(0, 1, ∪, ·, ⁻),
$\mathrm{SUC}_2(\sqsubseteq)$, FUN, PFT.

Proof. See [Sto 74], [MeSt 72], [Mey 74], [Mey 74], resp. ∎

14.18. Theorem. The following problems are $\leqq_m^{P}$-complete [-hard] in $\mathbf{DSPACE}\left(2^{2^{\cdot^{\cdot^{2}}}}\right\}\mathrm{Lin}\big)$:

WS1S, [LINORD, $\mathcal{T}_M(\mathbb{N}, <)$].

Proof. See [Mey 73], [Mey 74], [Sto 74], resp. ∎

14.2.2.2. Time Classes

14.19. Theorem. The following problems are $\leqq_m^{\log}$-complete [-hard] in **P**:

SP, LP, AGAP,
[$\mathcal{P}$(1-XPDA)], EW(1-XPDA), X ∈ {D, N},
NE(cfg), INF(cfg), MEMBER_z(cfg).

Proof. See [Coo 73a], [DoLiRe 79], [Imm 79], [Hun 76], [Hun 76], [JoLa 77], [JoLa 77], [JoLa 77], resp. ∎

14.20. Theorem. [Schn 78] The following problems are $\leqq_m^{\mathrm{DTIME},n\cdot\mathrm{Pol}(\log n)}$-complete in $\mathbf{NTIME}(n \cdot \mathrm{Pol}(\log n))$:

SAT, 3-SAT, 3-COLOUR, INDEPENDENT SET, EXACT COVER. ∎

Remark. For further such problems see [KoKu 84].

For the next result we need some definitions.

SUBGRAPHISOM = {(G_1, G_2): G_1, G_2 are directed graphs and there exists a subgraph G' of G_2 which is isomorphic to G_1}.

A scheduling problem is given by numbers k, n, m (number of hours, teachers, classes, resp.) a matrix $R = (R_{ij})_{n,m}$ (R_{ij} is the number of hours teacher i is required to teach class j) and sets $T_1, \ldots, T_n \subseteq \{1, \ldots, k\}$ (T_i is the set of hours during which teacher i is available) and sets $C_1, \ldots, C_m \subseteq \{1, \ldots, k\}$ (C_j is the set of hours during which class j is available). A schedule for the problem $(k, R, T_1, \ldots, T_n; C_1, \ldots, C_m)$ is a function

$$f: \{1, \ldots, n\} \times \{1, \ldots, m\} \mapsto \mathfrak{P}(\{1, \ldots, k\})$$

($f(i, j)$ is the set of hours teacher i teaches class j) such that $f(i, j) \subseteq T_i \cap C_j$, card $f(i, j) = R_{ij}$, $f(i, j) \cap f(i, j') = \emptyset$ for $j \neq j'$ and $f(i, j) \cap f(i', j) = \emptyset$ for $i \neq i'$.

Now we can define

$$\text{SCHEDULE} = \{(k, R, T_1, \ldots, T_n; C_1, \ldots, C_m)\colon \text{there exists a schedule for this problem}\}.$$

A further problem is

$$\text{BANDWIDTH} =_{\mathrm{df}} \{(V, E, k)\colon G = (V, E) \text{ is a graph} \wedge k \in \mathbb{N} \wedge B(G) \leqq k\},$$

where $B(G) = \min\limits_{f\colon V \xrightarrow{1\text{-}1} \mathbb{N}} \max\limits_{(x,y)\in E} |f(x) - f(y)|$ is the "bandwidth" of G. The value $\max\limits_{(x,y)\in E} |f(x) - f(y)|$ is called the bandwidth of the representation f of G.

In the following we make use of the notion of NP-completeness defined as follows: A set is called *NP-complete* if it is $\leqq_{\mathrm{m}}^{\mathrm{P}}$-complete in NP.

The following list contains only a small number of NP-complete problems. A much more complete list can be found in [GaJo 79], which is continued in [Joh 81–].

14.21. Theorem. The following problems are $\leqq_{\mathrm{m}}^{\log}$-complete in **NP**:

SAT, KNAPSACK, SOS, PARTITION, BIN PACKING,
SCHEDULE, EXACT COVER, SET COVER,
BANDWIDTH, COLOUR, TRAVELLING SALESMAN,
CLIQUE, NODE COVER, INDEPENDENT SET,
HAMILTONIAN CIRCUIT, SUBGRAPHISOM,
NE(2-NFA^1), EW(1-DCSA),
INEQ(0, ∪, ·, *), INEQ(0, 1, ∪, ·).

Proof. See [Coo 71a], [Kar 72], [Kar 72], [Kar 72], [GaJo 79], [EvItSh 75], [Kar 72], [Kar 72], [GaGrJoKn 78], [Kar 72], [Kar 72], [Kar 72], [Kar 72], [Kar 72], [Kar 72], [Kor 79], [Gal 76a], [Hun 76], [StMe 73], [StMe 73], resp. ∎

Remark 1. The problem 2-SAT (SAT restricted to formulas with at most two literals per clause) belongs to **P** (see [EvItSh 75]), whereas 3-SAT^1 (unary presentation of the indices) is still NP-complete (see [Coo 71a], [ShBe 74], resp.) For another NP-complete restriction of 3-SAT see [Lic 82].

Remark 2. KNAPSACK is a special case of LIQ($\mathbb{G}$). For this and related problems see 7.22.

Remark 3. KNAPSACK restricted to two variables belongs to **P** (see [HiWo 76]).

Remark 4. 2-BIN PACKING (see definition on p. 167) is NP-complete.

Remark 5. For further results on scheduling problems see [EvItSh 75], [Ull 75], [GaJo 75], and [ItRo 77].

Remark 6. 3-COLOUR (see definition on p. 167) restricted to planar graphs is still NP-complete (see [GaJoSt 74]) 2-COLOUR is in **P**. In [LaWi 83] reasons for the different behaviour of 2-COLOUR and 3-COLOUR are studied.

Remark 7. TRAVELLING SALESMAN restricted to rectangular networks in the Euclidian plane is still NP-complete (see [Pap 75]).

Remark 8. The restrictions of NODE COVER to graphs of degree $d \leqq 3$ or to planar graphs of degree $d \leq 6$ are NP-complete (see [GaJoSt 74]).

Remark 9. HAMILTONIAN CIRCUIT remains NP-complete when restricted to planar graphs. This is contained in [Ple 79] and improves a result from [GaJoTa 76].

Remark 10. When graph problems which are complete in certain complexity classes are restricted to graphs with bandwidth constraints, then these restricted problems often become complete (or hard) in space restricted subclasses of the given classes (see [Sud 80], [MoSu 80], [MoSu 81]). For a given graph problem C and $f \in \mathbb{R}$ we define the bandwidth restricted version of C by

$$C(f) = \{G\colon G \in C \text{ and the bandwidth of the representation of } G \text{ is bounded by } f(v) \text{ where } v \text{ is the number of vertices of } G\}.$$

Examples. NODE COVER(f) is $\leqq_m^{\log}$-complete in **NTIME-SPACE**(Pol, f). For space complexity classes we have: if $f(2^{2^n})$ is T-DSPACE-constructible and $\log\log n \leqq f(n) \leqq \log n$, then

$$\{\text{GAP}(2^{kf})\colon k \geqq 1\} \text{ is } \leqq_m^{\log}\text{-complete in } \textbf{NSPACE}(f)$$

and

$$\{\text{AGAP}(2^{kf})\colon k \geqq 1\} \text{ is } \leqq_m^{\log}\text{-complete in } \textbf{ASPACE}(f).$$

($\mathcal{A} \subseteqq \mathcal{C}$ is $\leqq$-complete in $\mathcal{C}$ iff for every $C \in \mathcal{C}$ there exists an $A \in \mathcal{A}$ with $C \leqq A$.)

Remark 11. Problems complete in **NP** with respect to several random polynomial time reducibilities are contained in [AdMa 77], [AdMa 79] and [VaVa 82].

Remark 12. In [PaYa 82] several problems are shown to be $\leqq_m^P$-complete in **NP** $\wedge$ **coNP**, for instance EXACT CLIQUE $=_{df} \{(G, k)\colon G$ is a finite graph $\wedge\, k \in \mathbb{N} \wedge G$ has a maximal clique of size $k\}$. For further problems $\leqq_m^P$-complete in **NP** $\wedge$ **coNP** see [Wag 84a].

14.22. Theorem.

1. B_k is $\leqq_m^{\log}$-complete in Σ_k^P.
2. The inequivalence problems for integer expressions over $\cup$ and $+$, SEMILIN and for context-free grammars with single letter terminal alphabet are $\leqq_m^{\log}$-complete in Σ_2^P.

Proof. See [StMe 73], [StMe 73], [Huy 80], [Huy 84], resp. ∎

Remark 1. In [Koz 77a] an analogue of B_k in finite algebras is defined.

Remark 2. Further problems which are $\leqq_m^{\log}$-complete in Σ_k^P ($k \geqq 1$) can be found in [Wag 84a] and [Wag 84b].

Remark 3. In contrast to recursion theory (Δ_2 has provably no complete sets) Δ_2^P has $\leqq_m^P$-complete sets [Pap 82]. Examples are UNIQUE MIN TRAVELLING SALESMAN $=_{df} \{(V, f)$: V finite set $\wedge f: V^2 \to \mathbb{N} \wedge$ there exists a uniquely determined permutation π of V such that $\sum_{v \in V} f(v, \pi(v))$ is minimal$\}$ and the problems UNIQUE MAX LIQ($\mathfrak{G}$) and UNIQUE MAX(0, 1) KNAPSACK which are defined analogously.

We define

MAJORITY SAT = {H: H is a Boolean expression satisfied by more than half of all the assignments of its variables}.

14.23. Theorem. [Gil 74] MAJORITY SAT is $\leqq_m^P$-complete in **RTIME**(Pol).

Remark. For further problems $\leqq_m^P$-complete in **RTIME**(Pol) see [Simj 75].

14.24. Theorem. The following problems are $\leqq_m^{\log}$-complete [-hard] in **DTIME**(2^{Pol}):

SC,

[$\mathcal{P}$(1-XSA)], MEMBER$_x$(1-XSA), NE(1-XSA), X $\in$ {D, N}.

Proof. See [ChSt 76], [Hun 76], [Hun 76], [Hun 76], resp. ∎

Remark. Many other games which are $\leqq_m^{\log}$-complete in **DTIME**(2^{Pol}) are contained in [ChSt 76] and [StCh 79]. Certain generalizations of checkers, chess and go are $\leqq_m^P$-complete in DTIME(2^{Pol}) ([Robs 84], [FrLi 81] and [Robs 83]).

14.25. Theorem. The following problems are $\leqq_m^{\log}$-complete [-hard] in **NTIME**(2^{Pol}), those of the last two rows are even $\leqq_m^{\text{log-lin}}$-complete [-hard] in **NEXPTIME**:

MON,

INEQ(0, 1, $\cup$, $\cdot$, ^{2}),

[SUC$_2$, INJ, RA].

Proof. See [Mey 74], [StMe 73], [Fer 74], [Fer 74], [FeRa 75], resp. ∎

Remark. Some graph-theoretical problems which are **NP**-complete (for example COLOUR, CLIQUE, NODE COVER and INDEPENDENT SET) become $\leqq_m^{\log}$-complete in **NTIME**(2^{Pol}) when the graphs are described in a succinct manner by special descriptional languages (see [GaWi 83], [Wag 84b]). Similar results can be found in [Var 82] and [BeOtWi 83].

14.26. Theorem. The following problems are $\leqq_m^{\text{log-lin}}$-hard in **NTIME**($2^{2^{\text{Lin}}}$):

INJM, SUC$_1$(M), PA.

Proof. See [Fer 74], [Fer 74], [FimRa 74], resp. ∎

Chapter V
Properties of Complexity Measures

Compared with the original aim of the theory, namely the investigation of the complexity of problems, the investigation of complexity measures and their properties as studies in § 15 and § 16 is of minor importance. It seems that it has been carried out only for the reason of mathematical completeness and has been to some extent a diversion.

On the other hand, there are really weighty topics like hierarchies and speed-up which must be considered as properties of complexity measures. They are dealt with in § 17 and § 18. Moreover, some of the results described in § 15, like union and honesty theorems, turn out to be indispensable, or at least useful auxiliary tools, for other investigations. For example, the honesty theorem allows us to solve the problem of upward hierarchy results in a very satisfactory way. The material of § 16 has mainly been developed in connection with the efforts to find a precise definition of "natural" complexity measures. This goal has not been reached, but the various properties of complexity measures which have been investigated in connection with this problem deserve further attention.

§ 15. General Properties of Complexity Measures

This section is devoted to properties which are shared by all Blum measures. This is expressed by the fact that most of the theorems of this section have the logical structure "For all Blum measures (φ, Φ) it holds that ..." (although we sometimes drop this quantification).

15.1. *Enumeration Properties*

We start with some notions. Let $\mathscr{C} \subseteq \mathbb{P}_1$, $B \subseteq \mathbb{N}$.

Definition. $\mathscr{C}$ is called *presentable* by $B \leftrightarrow \mathscr{C} = \{\varphi_i : i \in B\}$.

$\mathscr{C}$ is called *recursively enumerable* (r.e.) $\leftrightarrow$ $\mathscr{C}$ is presentable by some r.e. set.

$\mathscr{C}$ is called *h-enumerable* with respect to $\Phi \leftrightarrow \bigvee_{B \in \Sigma_1} \left(\mathscr{C} = \{\varphi_i : i \in B\} \wedge \bigwedge_{i \in B} \Phi_i \leq_{\text{ae}} h\right)$.

A special case of presentability of $\mathscr{C}$ is that by the complete index set

$$\Omega\mathscr{C} =_{\mathrm{df}} \{i \colon \varphi_i \in \mathscr{C}\}.$$

Remark. By a simple padding argument it follows that recursively enumerable sets can always be represented even by recursive sets.

15.1.1. Recursive Enumerability of Complexity Classes

Our first question is whether complexity classes of Blum measures are always recursively enumerable. Some time and space measures for example have only recursively enumerable classes (cf. 15.5). A positive answer would be very satisfying because the complexity classes would be easy to handle. Unfortunately, the answer is negative as shown in [Lewf 71] and [LaRo 72]. We prove more generally that every Blum measure has a (pathological) "submeasure" which has not only r.e. complexity classes.

Definition. The measure (ψ, Ψ) is called a *submeasure* of the Blum measure (φ, Φ) iff there is a $g \in \mathbb{R}_1$ such that

(1) $\psi_i = \varphi_{g(i)}$,

(2) $\Psi_i = \Phi_{g(i)}$,

(3) ψ is an acceptable numbering of $\mathbb{P}_1$.

Note that any submeasure of a Blum measure is also a Blum measure.

15.1. Theorem. [Hart 73] Every Blum measure has a submeasure which has not only r.e. complexity classes.

Proof. Let (φ, Φ) be a Blum measure. Consider $f(n, i) \equiv i$ and apply the *s-m-n*-Theorem to obtain an $s \in \mathbb{R}_1$ such that $\varphi_{s(i)}(n) \equiv i$. W.l.o.g. (see 2.6) we can assume s to be strictly increasing. Now a submeasure (ψ, Ψ) and $t_0 \in \mathbb{R}_1$ are determined in such a way that exactly the constant functions $\varphi_{s(i)} \equiv i$ with $i \in \overline{\mathrm{K}}$ belong to $\Psi(t_0)$, where $\mathrm{K} = \{i \colon i \in \mathrm{D}_i\}$ is the halting problem. To this end we define the following function which is obviously partial recursive ($\mathrm{T}(i, j, n)$ means the predicate "M on input j halts within n steps"):

$$\xi(j, k, l, n) =_{\mathrm{df}} \begin{cases} \varphi_k(n) & \text{if } \bigl(n \in \mathrm{D}_l \wedge \Phi_j(n) > \varphi_l(n)\bigr) \\ & \quad \vee \bigvee_i \bigl(k = s(i) \wedge \sim \mathrm{T}(i, i, n)\bigr), \\ 1 + \varphi_j(n) & \text{if } n \in \mathrm{D}_l \wedge \Phi_j(n) \leqq \varphi_l(n) \\ & \quad \wedge \Bigl(k \notin \mathrm{R}_s \vee \bigvee_i \bigl(k = s(i) \wedge \mathrm{T}(i, i, n)\bigr)\Bigr), \\ \text{undefined} & \text{otherwise}. \end{cases}$$

Appealing to the *s-m-n*-Theorem and the recursion theorem we obtain a function $b \in \mathbb{R}_2$ such that

$$\varphi_{b(k,l)}(n) = \xi\bigl(b(k, l), k, l, n\bigr).$$

Now the function

$$\varphi_{c(l)}(n) = \sup \{\Phi_{b(s(i),l)}(n) : s(i) \leqq n \wedge \sim \mathrm{T}(i, i, n)\}$$

is total because $\sim \mathrm{T}(i, i, n)$ implies

$$\varphi_{b(s(i),l)}(n) = \xi\big(b(s(i), l), s(i), l, n\big) = \varphi_{s(i)}(n) = i$$

and thus $\Phi_{b(s(i),l)}(n)$ is defined. Choose l_0 with the property $\varphi_{l_0} = \varphi_{c(l_0)}$ (resursion theorem) and define $t_0 =_{\mathrm{df}} \varphi_{l_0}$ and $g(k) =_{\mathrm{df}} b(k, l_0)$. Then we get

$$\varphi_{g(k)}(n) = \begin{cases} \varphi_k(n) & \text{if} \quad \Phi_{g(k)}(n) > t_0(n) \vee \bigvee_i \big(k = s(i) \wedge \sim \mathrm{T}(i, i, n)\big), \\ 1 + \varphi_{g(k)}(n) & \text{if} \quad \Phi_{g(k)}(n) \leqq t_0(n) \wedge \Big[k \notin \mathrm{R}_s \vee \bigvee_i \big(k = s(i) \wedge \mathrm{T}(i, i, n)\big)\Big]. \end{cases}$$

The condition of the second line is impossible. Hence, for all $k \in \mathbb{N}$

$$\varphi_{g(k)} = \varphi_k,$$

$$\Phi_{g(k)}(n) \leqq t_0(n) \to \bigvee_i \big(k = s(i) \wedge \sim \mathrm{T}(i, i, n)\big).$$

Define $\psi_k =_{\mathrm{df}} \varphi_{g(k)}$. This implies $\varphi_k = \psi_k$ which shows ψ to be an acceptable numbering of $\mathbb{P}_1$. Now we consider the submeasure (ψ, Ψ), where $\Psi_k = \Phi_{g(k)}$. We obtain

$$\begin{aligned} \Psi(t_0) &= \{\psi_k : \Psi_k \leqq_{\mathrm{ae}} t_0\} = \{\varphi_{g(k)} : \Phi_{g(k)} \leqq_{\mathrm{ae}} t_0\} \\ &\subseteqq \{\varphi_k : \bigvee_i (k = s(i) \wedge M_i \text{ never halts on input } i)\} \\ &= \{\varphi_{s(i)} : i \in \overline{K}\}. \end{aligned}$$

On the other hand, if $i \in \overline{K}$, then $\varphi_{s(i)} = \psi_{s(i)} \in \Psi(t_0)$ because $\Psi_{s(i)} = \Phi_{g(s(i))} = \Phi_{b(s(i),l_0)} \leqq_{\mathrm{ae}} \varphi_{c(l_0)} = t_0$. Finally, $\Psi(t_0)$ is not r.e. because otherwise $\overline{K} = \{h(0) : h \in \Psi(t_0)\}$ would be r.e., which is known to be false (see 2.14). ∎

Remark. It is far easier to construct a single example of a measure which has a nonenumerable complexity class ([Hart 73]).

Thus the property of having only r.e. complexity classes is not a general property of Blum measures. It is dealt with together with some similar properties (conformity and finite invariance) in § 16. On the other hand, however, every measure must have infinitely many recursively enumerable complexity classes. To show this we consider $\mathbf{F} =_{\mathrm{df}} \{f : f \text{ total} \wedge f =_{\mathrm{ae}} 0\}$ and prove

15.2. Lemma. For every Blum measure (φ, Φ) there exists a $t_0 \in \mathbb{R}_1$ such that $\mathbf{F} \subseteqq \Phi(t_0)$.

Proof. The lemma remains true if $\mathbf{F}$ is replaced by any r.e. set $\mathbf{C} \subseteqq \mathbb{R}_1$. It is evident that $\mathbf{F}$ is recursively enumerable: $\mathbf{F} = \{\varphi_{s(i)} : i \in \mathbb{N}\}$. Define $t_0(x) =_{\mathrm{df}} \max_{i \leqq x} \Phi_{s(i)}(x)$. Then $\mathbf{F} \subseteqq \Phi(t_0)$. ∎

15.3. Theorem. [Bor 72] For every Blum measure (φ, Φ) there exists a $t_0 \in \mathbb{R}_1$ (t_0 can be chosen according to the previous lemma) such that $\Phi(t)$ is recursively enumerable for all $t \geqq t_0$.

Proof. Let $\varphi_j \geqq t_0$ be a recursive function. The idea is to modify for each $u \in \mathbb{N}$ the function φ_i, namely to change it into a function $\varphi_{g_j(i,u)} \in \mathbf{F} \subseteqq \Phi(\varphi_j)$, if and only if its complexity exceeds φ_j for at least one argument between u and x. This yields a recursive enumeration of $\Phi(\varphi_j)$. More formally, define

$$\varphi_{g_j(i,u)}(x) =_{\mathrm{df}} \begin{cases} \varphi_i(x) & \text{if } \bigwedge_{y \leqq x} \big((y \leqq u \wedge \Phi_i(y) \leqq u) \vee \Phi_i(y) \leqq \varphi_j(y)\big), \\ 0 & \text{otherwise}. \end{cases}$$

Then

$$\Phi(\varphi_j) = \{\varphi_{g_j(i,u)} : i, u \in \mathbb{N}\}. \blacksquare$$

Remark. It is also possible to give a uniform enumeration of all complexity classes $\Phi(\varphi_j)$ for recursive $\varphi_j \geqq t_0$.

Following [LaRo 72] we define

$$A(i, j, u, x) \leftrightarrow \bigwedge_{y \leqq x} \big[(y \leqq u \wedge \Phi_i(y) \leqq u) \vee (\Phi_j(y) \leqq x \to \Phi_i(y) \leqq \varphi_i(y))\big]$$

and

$$\varphi_{g(i,j,u)}(x) = \begin{cases} \varphi_i(x) & \text{if } A(i, j, u, x) \wedge \Phi_i(x) \leqq \max\big(u, \varphi_j(x), \Phi_j(x)\big), \\ 0 & \text{otherwise}. \end{cases}$$

Then $\Phi(\varphi_j) = \{\varphi_{g(i,j,u)} : i, u \in \mathbb{N}\}$.

15.1.2. The Quality of Enumeration

In this subsection we study the quality of recursive enumerations of complexity classes. To gauge it we use the notion of h-enumerability.

The enumeration $\{\varphi_i : i \in \mathrm{B}\}$ of the complexity class $\Phi(t)$ would be "bad" if some Φ_i ($i \in \mathrm{B}$) would be much larger than t. It turns out that in practice relatively good enumerations can almost always be found. More precisely:

15.4. Theorem. [Mcir 69] For every Blum measure Φ there exists an $h \in \mathbb{R}_2$ such that for sufficiently large $t \in \mathbb{R}_1$ the class $\Phi(t)$ is $h \square t$-enumerable.

Proof. We simply continue the second version of the proof of 15.3 indicated in the preceding remark. Define

$$h(x, z) =_{\mathrm{df}} \sup \{\Phi_{g(i,j,u)}(x) : i, j, u \leqq x \wedge [\Phi_i(x) \leqq z \vee \sim A(i, j, u, x)]\}.$$

Now $\varphi_{g(i,j,u)} = \varphi_i$ or $\varphi_{g(i,j,u)} \in \mathbf{F}$. In either case $\Phi_{g(i,j,u)}(x)$ for almost all x is contained in the set under the supremum operation for $h(x, \varphi_j(x))$. Hence

$$\Phi_{g(i,j,u)}(x) \leqq_{\mathrm{ae}} h\big(x, \varphi_j(x)\big). \blacksquare$$

As an illustration we state

15.5. Theorem. [LaRo 72]

1. T-DSPACE(s) is s-enumerable for all recursive $s \geqq \mathrm{id}$.
2. T-DTIME(t) is t^2-enumerable for all recursive $t \geqq \mathrm{id}$.
3. multiT-DTIME(t) is $t \log t$-enumerable for all recursive $t \geqq \mathrm{id}$.

Proof. Ad 1. We show that the enumeration constructed in the remark after 15.3 can be performed in the required way. Let $s = \varphi_j$ be given where w.l.o.g. $\varphi_j \leqq$ T-DSPACE$_j$ can be assumed. We describe a DTM $M_{g(i,j,u)}$ (in the following called M for short) computing $\varphi_{g(i,j,u)}$ within space s. We make use of the fact that there exists an $r \in \mathbb{R}_1$ such that $\varphi_i = \varphi_{r(i)}$, $M_{r(i)}$ works without loops and T-DSPACE$_i$ = T-DSPACE$_{r(i)}$.

On input x the machine M works as follows

1. (Checking $A(i, j, u, x)$)
1.1. $y = 0$.
1.2. **If** $y \leqq u$ and $\varphi_{r(i)}(y)$ can be computed within space u **then** go to 1.5.
1.3. **If** $s(y)$ cannot be computed within space x **then** go to 1.5.
1.4. **If** $\varphi_{r(i)}(y)$ can be computed within space $s(y)$ **then** go to 1.5
else output 0.
Stop.
1.5. **If** $y < x$ **then** $y \leftarrow y + 1$. Go to 1.2.
2. (Checking T-DSPACE$_i(x) \leqq \max\big(u, \varphi_j(x), \text{T-DSPACE}_j(x)\big)$. Because of the assumption $\varphi_j \leqq$ T-DSPACE$_j$ we have only to check T-DSPACE$_i(x) \leqq \max\big(u, \text{T-DSPACE}_j(x)\big)$.)
2.1. $AS = x$ ("available space").
2.2. **If** $\varphi_{r(i)}(x)$ can be computed within space $\max(u, AS)$
then output $\varphi_{r(i)}(x)$. Stop.
2.3. **If** $s(x)$ can be computed within space AS **then** output 0. Stop.
2.4. $AS \leftarrow AS + 1$. Go to 2.2.

If T-DSPACE$_i \leqq_{\mathrm{ae}} s$, then for sufficiently large u the machine $M_{g(i,j,u)}$ computes φ_i within space s almost everywhere. If T-DSPACE$_i(x) > s(x)$ for $x > u$, then $\varphi_{g(i,j,u)}$ becomes a function which is almost everywhere equal to zero, because for sufficiently large arguments it will be detected in 1.4 that $\varphi_i(x)$ cannot be computed within $s(x)$ squares.

Ad 2 and 3. Similar algorithms can be developed for T-DTIME and multiT-DTIME. ∎

Theorem 15.4 and the first statement of 15.5 suggest the question of whether for any measure Φ the classes $\Phi(t)$ are always t-enumerable provided t is sufficiently large. This is not the case, and, consequently, 15.4 cannot be improved, as is shown by the next theorem generalizing a result in [LaRo 72].

15.6. Theorem. Every Blum measure Φ has a submeasure Ψ with the property that there exist arbitrarily large t such that $\Psi(t)$ is not t-enumerable. ∎

Remark. Note that for any measure Φ there are always arbitrarily large functions t such that $\Phi(t)$ is t-enumerable. This is an immediate consequence of 17.4 and 15.4.

Theorems 15.6 and 15.1 concern the existence of suitable submeasures. As a counterpart of both of these theorems we now show that any measure can be embedded into a measure having only r.e. complexity classes which are, moreover, all t-enumerable.

15.7. Theorem. [LaRo 72] Every Blum measure Φ is a submeasure of some Blum measure Ψ with

$$\bigwedge_{t\in\mathbb{R}_1} \Psi(t) = \mathbf{F} \cup \Phi(t) \quad \text{and} \quad \bigwedge_{t\in\mathbb{R}_1} \Psi(t) \text{ is } t\text{-enumerable.}$$

Proof. Using the terminology of the second version of the proof of 15.3 (given in the remark after 15.3) we define

$$\psi_{2i} = \varphi_i,$$

$$\psi_{2i+1} = \varphi_{g(i_1,i_2,i_3)}, \quad \text{where} \quad \langle i_1, i_2, i_3\rangle = i,$$

$$\Psi_{2i}(x) = \Phi_i(x),$$

and

$$\Psi_{2i+1}(x) = \begin{cases} \Phi_{i_1}(x) & \text{if} \quad A(i_1, i_2, i_3, x) \wedge \Phi_{i_1}(x) \leqq \max\left(i_3, \varphi_{i_2}(x), \Phi_{i_2}(x)\right), \\ 0 & \text{otherwise}. \end{cases}$$

ψ is an acceptable enumeration, (ψ, Ψ) is a Blum measure and (φ, Φ) is a submeasure of (ψ, Ψ). This can be easily verified. We show $\Psi(\varphi_j) = \mathbf{F} \cup \Phi(\varphi_j)$ for all $\varphi_j \in \mathbb{R}_1$.

a) Let $f \in \Psi(\varphi_j)$. Then either $f = \psi_{2k} = \varphi_k$ and $\Phi_k = \Psi_{2k} \leqq_{\text{ae}} \varphi_j$ and thus $f \in \Phi(\varphi_j)$, or $f = \psi_{2k+1}$. In the latter case either $f = \psi_{2k+1} = \varphi_{g(k_1,k_2,k_3)} = \varphi_{k_1}$ and $\Phi_{k_1} = \Psi_{2k+1} \leqq_{\text{ae}} \varphi_j$ and hence $f \in \Phi(\varphi_j)$, or $f = \psi_{2k+1} \in \mathbf{F}$.

b) Let $f \in \mathbf{F} \cup \Phi(\varphi_j)$. If $f \in \mathbf{F}$, then there is some i such that $f = \psi_i$ and $\Psi_i =_{\text{ae}} 0$. Hence $f \in \Psi(\varphi_j)$. If $f \in \Phi(\varphi_j)$, then there is some i such that $f = \varphi_i = \psi_{2i}$ and $\Psi_{2i} = \Phi_i \leqq_{\text{ae}} \varphi_j$. Hence $f \in \Psi(\varphi_j)$.

That $\Psi(\varphi_j) = \{\psi_{2\langle i,j,u\rangle+1} : i, u \in \mathbb{N}\}$ is a φ_j-presentation is evident because either $\psi_{2\langle i,j,u\rangle+1} = \varphi_i$ and $\Psi_{2\langle i,j,u\rangle+1} = \Phi_i \leqq_{\text{ae}} \varphi_j$ or $\psi_{2\langle i,j,u\rangle+1} \in \mathbf{F}$ and $\Psi_{2\langle i,j,u\rangle+1} =_{\text{ae}} 0$. ∎

The next result contrasts with 15.4.

15.8. Theorem. [Wer 74] For every Blum measure Φ, every $t \in \mathbb{R}_1$ such that $\Phi(t)$ is infinite and every $h \in \mathbb{R}_2$ there exists an r.e. set $S \subseteqq \Phi(t)$ which is not $h \square t$-enumerable. ∎

Remark. The set S can be constructed by diagonalization.

15.1.3. Presentation of Complexity Classes

Although in general not all complexity classes are r.e. a general presentation statement can be proved:

15.9. Theorem. [LaRo 72]

1. Every complexity class $\Phi(t)$ can be presented by some $C \in \Pi_1$.
2. $\mathbb{P}_1 \setminus \Phi(t)$ is always r.e.

Proof. Ad 1. Applying 2.5 and 2.6 to $\varphi_k(i, x) =_{\text{df}} \varphi_i(x)$ we obtain a strictly increasing function $s \in \mathbb{R}_1$ such that $\varphi_{s(i)} = \varphi_i$ and $s > \text{id}$. Define for a given $t \in \mathbb{R}_1$ the set

$$E =_{\text{df}} \left\{s^{[j]}(i) : \bigvee_n \Phi_i(n) > t(n) + (j \dot{-} n)\right\}.$$

E is obviously r.e., and it is easy to verify the statement

$$\bigvee_r (\varphi_r = \varphi_i \wedge r \in \mathrm{R}_s \setminus E) \leftrightarrow \varphi_i \in \mathbb{R}_1 \wedge \Phi_i \leqq_{\text{ae}} t.$$

Hence the Π_1-set $\mathrm{R}_s \setminus E$ is a presentation for $\Phi(t)$.

Ad 2. For a given $t \in \mathbb{R}_1$ we would like to have an $s \in \mathbb{R}_1$ such that

$$\varphi_{s(i)} = \begin{cases} \varphi_i & \text{if} \quad \varphi_i \notin \Phi(t), \\ \text{some function from } \mathbb{P}_1 \setminus \mathbb{R}_1 & \text{otherwise}. \end{cases}$$

Then $\{\varphi_{s(i)} : i \in \mathbb{N}\} = \mathbb{P}_1 \setminus \Phi(t)$ would be the desired recursive enumeration. A possible definition of $\varphi_{s(i)}$ is the following:

$$\varphi_{s(i)}(n) = \begin{cases} \varphi_i(n) \quad \text{if} \quad \bigwedge_{k \leqq n} \Big[\bigvee_{m < n} \big(\Phi_k(m) \leqq \max\big(n, t(m)\big) \wedge \varphi_k(m) \neq \varphi_i(m)\big) \\ \qquad\qquad \vee \bigvee_{m \geqq n} \big(\Phi_k(m) \leqq t(m) \wedge \varphi_k(m) \neq \varphi_i(m)\big)\Big] \\ \qquad\qquad \text{and } \varphi_i(n) \text{ is defined}, \\ \text{undefined otherwise}. \blacksquare \end{cases}$$

Regarding presentations by complete index sets, it is easy to verify

15.10. Theorem. [Lewf 71] $\Omega\Phi(t) \in \Sigma_3 \cap \Pi_3$ for all Φ and all $t \in \mathbb{R}_1$. ∎

We close this subsection with some remarks concerning complexity classes of partial recursive functions. The following types of classes with partial recursive names $\tau \in \mathbb{P}_1$ have been studied in the literature:

$$\Phi(\tau) =_{\text{df}} \{\varphi_i : \mathrm{D}_\tau \subseteq \mathrm{D}_i \wedge \Phi_i \leqq_{\text{ae}} \tau\},$$

$$\Phi_{\text{p}}(\tau) =_{\text{df}} \{\varphi_i : \Phi_i \leqq_{\text{ae}} \tau\}.$$

Note that for $t \in \mathbb{R}_1$ the class $\Phi(t)$ coincides with $\Phi(t)$ as defined in § 5, whereas $\Phi_{\text{p}}(t)$ may contain partial functions with cofinite domain.

The following results should be compared with 15.10 and 15.9.

15.11. Theorem. [Rob 74a]

1. $\Omega\Phi(\tau) \in \Sigma_3$ for all $\tau \in \mathbb{P}_1$.
2. $\Omega\Phi(\tau)$ is Σ_3-complete for some Φ and some arbitrarily large $\tau \in \mathbb{P}_1$.
3. $\Phi(\tau)$ is always presentable by Π_1-sets, but $\mathbb{P}_1 \setminus \Phi(\tau)$ need not be r.e. ∎

For further results see [Rob 74a].

15.2. *An Embedding Property for Program Classes*

Together with complexity classes of various types, program classes have also been studied (see [Lewf 71], [Rob 74a], [LaRo 72], [Emd 78]). They are defined as $I(\tau) = \{i: \mathrm{D}_\tau \subseteq \mathrm{D}_i \wedge \Phi_i \leqq_{\mathrm{ae}} \tau\}$ and $I_{\mathrm{p}}(\tau) = \{i: \Phi_i \leqq_{\mathrm{ae}} \tau\}$ for $\tau \in \mathbb{P}_1$. Instead of developing the whole known theory of these classes, we restrict ourselves to one single result which is basic for the next two subsections.

We start with an auxiliary result:

15.12. Lemma. $I(\tau), I_{\mathrm{p}}(\tau) \in \Sigma_2$ for all $\tau \in \mathbb{P}_1$.

Proof. Because of

$$i \in I(\varphi_k) \leftrightarrow \mathrm{D}_k \subseteq \mathrm{D}_i \wedge \Phi_i \leqq_{\mathrm{ae}} \varphi_k$$
$$\leftrightarrow \bigvee_u \bigvee_v \bigwedge_x \bigwedge_y \Big(\Phi_k(x) = y \to \big[(x \leqq u \wedge \Phi_i(x) \leqq v) \vee \Phi_i(x) \leqq \varphi_k(x)\big]\Big)$$

we obtain $I(\varphi_k) \in \Sigma_2$, and because of

$$i \in I_{\mathrm{p}}(\varphi_k) \leftrightarrow \Phi_i \leqq_{\mathrm{ae}} \varphi_k$$
$$\leftrightarrow \bigvee_u \bigwedge_x \bigwedge_y \big(\Phi_k(x) = y \to [x \leqq u \vee \Phi_i(x) \leqq \varphi_k(x)]\big)$$

we obtain $I_{\mathrm{p}}(\varphi_k) \in \Sigma_2$. ∎

Theorem 15.13 (15.14) will show that every Σ_2-set X (every Σ_2-set $X \subseteq \Omega\mathbb{R}_1$) can be embedded into a least $I_{\mathrm{p}}(\tau)$, $\tau \in \mathbb{P}_1$ $\big(I(t), t \in \mathbb{R}_1\big)$.

Analysing the proof of the Honesty Theorem (15.21) given in [MoMe 74]. P. van Emde Boas found out that this embedding property is the proper mathematical core of both the Union Theorem (15.16) and the Honesty Theorem (15.21) (see [Emd 78]).

Since we are mainly interested in complexity classes of total functions we concentrate on 15.14.

For stating the theorems we need the closure operators Γ_{p} and Γ defined by

$$\Gamma_{\mathrm{p}}(X) = \bigcap \{I_{\mathrm{p}}(\sigma): X \subseteq I_{\mathrm{p}}(\sigma) \wedge \sigma \in \mathbb{P}_1\}$$

and

$$\Gamma(X) = \bigcap \{I(s): X \subseteq I(s) \wedge s \in \mathbb{R}_1\}.$$

15.13. Theorem. For every family $\Big\{X_j: X_j = \Big\{i: \bigvee_x \bigwedge_y A(j, i, x, y) = 1\Big\} \wedge j \in \mathbb{N}\Big\}$ of Σ_2-sets ($A \in \mathbb{R}_4$) there exists an $s \in \mathbb{R}_1$ such that

1. $\Gamma_{\mathrm{p}}(X_j) \subseteq I_{\mathrm{p}}(\varphi_{s(j)})$,
2. $\{\varphi_{s(j)}: j \in \mathbb{N}\}$ is a measured set. ∎

15.14. Theorem. For every family $\Big\{X_j: X_j = \Big\{i: \bigvee_x \bigwedge_y A(j, i, x, y) = 1\Big\} \wedge j \in \mathbb{N}\Big\}$ of Σ_2-sets ($A \in \mathbb{R}_4$) there exists an $s \in \mathbb{R}_1$ such that

1. $X_j \subseteq \Omega\mathbb{R}_1 \to \varphi_{s(j)} \in \mathbb{R}_1 \wedge \Gamma(X_j) = I(\varphi_{s(j)})$,
2. $\{\varphi_{s(j)} : j \in \mathbb{N}\}$ is a measured set.

Proof. First we neglect the dependency of the given Σ_2-sets on j. The proof consists of three steps. For a given Σ_2-set X the first step describes the computation of a function t. In the second step properties of t will be derived. In Step 3 it will be proved that if $X \subseteq \Omega\mathbb{R}_1$, then $\Gamma(X) = I(t)$. It is clear that if X depends recursively on the parameter j, the function t becomes a $t_j = \varphi_{s(j)}$ depending also recursively on j.

Step 1. As an auxiliary tool we consider a priority queue (PQ) which contains at any moment a finite sequence of elements from $\mathbb{N}^3 \times \{b, w\}$, where b and w are interpreted as "black" and "white" markers, resp. Furthermore, one of the elements of PQ can be labelled by the additional marker m. The queue PQ is initially empty. If ξ and η are elements in PQ and ξ stands before η, then ξ has a higher priority than η. Thus the top element in PQ has the highest priority.

Let X be given in the form $X = \left\{i : \bigvee_x \bigwedge_y A(j, i, x, y)\right\}$ for some $A \in \mathbf{REC}$ and some $j \in \mathbb{N}$. Two different processes, Test and Comp, occur in the computation of t. They both have access to PQ. The flowchart represented in Figure 15.1 shows how they are coordinated.

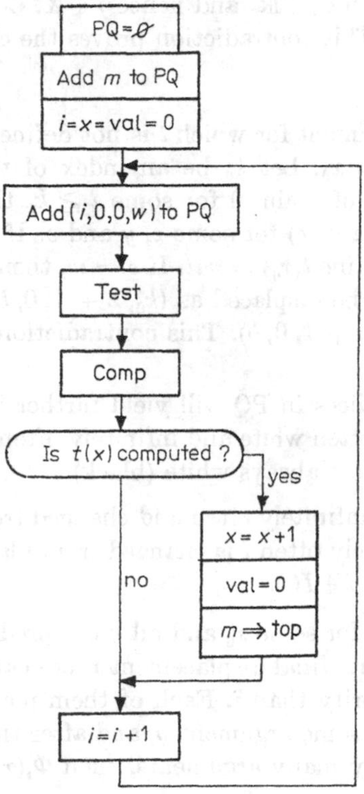

Figure 15.1

A call of Test causes the following actions: For every element of the form (l, z, y, w) in PQ (if for some z, y the element (l, z, y, w) is in PQ, then we will say that l is "white") the value $A(j, l, z, y)$ is computed. If $A(j, l, z, y) = 1$, then (l, z, y, w) is replaced by $(l, z, y + 1, w)$. If not, (l, z, y, w) is cancelled, and $(l, z + 1, 0, b)$ is added as the last element of PQ (in this case we say that l has been displaced).

A call of Comp causes an attempt to compute $t(x)$ according to the following algorithm:

1. **If** no element stands behind m in PQ **then** leave Comp.
2. **If** (k, z, y, w) stands behind m in PQ
   then if $\Phi_k(x) \leq val$ **then** move m behind (k, z, y, w) and go to 1
   else $val = val + 1$ and leave Comp.
3. **If** $(k, z, 0, b)$ stands behind m in PQ
   then if $\Phi_k(x) > val$ **then** $t(x) = val$
   cancel all elements $(j, x, 0, b)$ of PQ such that $t(x) < \Phi_j(x)$ and add $(j, x, 0, w)$ to PQ. Leave Comp.
   else move m behind $(k, z, 0, b)$. Go to 1.

Step 2. Assume $X \subseteq \Omega\mathbb{R}_1$. We prove some auxiliary claims.

Claim 0. It is impossible that m stands infinitely long before some index j.

Proof. Assume that m stands infinitely long before j. This is possible only if j is white and $\Phi_j(x) = \infty$ for some x. But then $\varphi_j \notin \mathbb{R}_1$ and hence $j \notin X$. Consequently, j will fail the test and will be displaced. This contradiction proves the claim. ▯

Claim 1. $t \in \mathbb{R}_1$.

Proof. Assume that x_0 is the smallest argument for which t is not defined. Then the algorithm works infinitely long with $x = x_0$. Let k_0 be an index of the nowhere defined function (hence $k_0 \notin X$). Because of Claim 0 for some $i \geq k_0$ the situation will be reached that m stands before (k_0, z, y, c) for some z, y and c. If $c = b$, then because of $val < \Phi_{k_0}(x_0) = \infty$ we must define $t(x_0) = val$. If $c = w$, then (k_0, z, y, w) will fail the test after some time and will be displaced as $(k_0, z + 1, 0, b)$. But then $t(x_0)$ will be defined when m reaches $(k_0, z + 1, 0, b)$. This contradiction proves the claim. ▯

An analysis of the behaviour of the indices in PQ will yield further insight. The index i is called *unstable* if it is infinitely often white and infinitely often black. It is called *white stable* (*black stable*) if it is almost always white (black).

Case 1. i is unstable. Hence i is displaced infinitely often and changed from white to black (because of failing Test), and infinitely often i is changed from black to white (because of $t < \Phi_i$). This shows $i \notin X$ and $i \notin I(t)$.

Case 2. i is white stable. This means that for some x_0 and all y the predicate $A(j, i, x_0, y)$ is satisfied and, hence, $i \in X$. After its final displacement there can exist only finitely many black indices of higher priority than i. Each of them may cause the termination of the computation of $t(y)$ for some argument y, and after this it will be displaced. If x is larger than these finitely many arguments, then $\Phi_i(x)$ cannot be larger than $t(x)$ (cf. step 2 of Comp), and hence $\Phi_i \leqq_{ae} t$. As $i \in X \subseteq \Omega\mathbb{R}_1$ we conclude $\varphi_i \in \mathbb{R}_1$ and thus $i \in I(t)$.

Case 3. i is black stable. Then $\Phi_i \leqq_{ae} t$ and hence $\varphi_i \in \mathbb{R}_1$ implies $i \in I(t)$. Eventually, only stable elements will stand before i. This means: For almost all x it holds that $\Phi_i(x) = 0$ or, if at least one white stable element stands before i, then $\Phi_i(x)$ is majorized by

$$\max \{\Phi_\nu : \nu \text{ is white stable and has a higher priority than } i\}.$$

Consequently (by Case 2) $\Phi_i(x) =_{ae} 0$ or there exist indices $i_1, \ldots, i_k \in I(t) \cap X$ such that $\Phi_i(x) \leqq_{ae} \max \{\Phi_{i_1}(x), \ldots, \Phi_{i_k}(x)\}$. In both cases we obtain $i \in I(t)$.

Claim 2. $X \subseteq I(t)$.

Proof. Assume $i \in X$. Then $\varphi_i \in \mathbb{R}_1$. Case 1 is impossible. In the Cases 2 and 3 we get $i \in I(t)$. □

Claim 3. If $i \in I(t)$, then there exist finitely many indices $i_1, \ldots, i_k \in X$ such that $\Phi_i \leqq_{ae} \max \{\Phi_{i_1}, \ldots, \Phi_{i_k}\}$ or $\Phi_i =_{ae} 0$.

Proof. Assume $i \in I(t)$. Case 1 is impossible. In Case 2, $i \in X \cap I(t)$. Hence $k = 1$ and $i_1 = i$ suffices. In Case 3 our assertion is already proved. □

Step 3. We prove that $\Gamma(X) = I(t)$ whenever $X \subseteq \Omega\mathbb{R}_1$. $\Gamma(X) \subseteq I(t)$ because, on account of Claim 2, $I(t)$ is a constituent of the intersection $\bigcap_{X \subseteq I(u)} I(u) = \Gamma(X)$. To prove the converse inclusion $I(t) \subseteq \Gamma(X)$ let $i \in I(t)$ be given. By Claim 3, $\Phi_i =_{ae} 0$ and hence $i \in \Gamma(X)$ or $\Phi_i \leqq_{ae} \max (\Phi_{i_1}, \ldots, \Phi_{i_k})$ for suitably chosen $i_1, \ldots, i_k \in X$. If $X \subseteq I(u)$, then $\Phi_i \leqq_{ae} \max (\Phi_{i_1}, \ldots, \Phi_{i_k}) \leqq_{ae} u$. This implies $i \in I(u)$ and, consequently, $I(t) \subseteq \Gamma(X)$.

Now we pay attention to the dependency of X on j. We get $t = t_j = \varphi_{s(j)}$ for some $s \in \mathbb{R}_1$, and we prove: The class $\{\varphi_{s(j)} : j \in \mathbb{N}\}$ is a measured set. To decide "$\varphi_{s(j)}(x) = y$" our procedure is applied to X_j. If in the process of computing $t_j(x)$ the auxiliary variable *val* reaches the value $y + 1$, then we conclude $\varphi_{s(j)}(x) \neq y$. If the process stops earlier, then the question "$\varphi_{s(j)}(x) = y$?" can be answered. ∎

Remark 1. Γ is a closure operator on the Σ_2-subsets of $\Omega\mathbb{R}_1$ whose closed sets are exactly the program classes $I(t)$.

Remark 2. It should be noticed that our proof does not show $X \subseteq I_p(\tau)$ for $X \nsubseteq \Omega\mathbb{R}_1$.

15.3. *Closure Properties*

Is the set of all complexity classes of a Blum measure Φ closed under reasonable operations like union and intersection?

There can be no theorem asserting that the union of two complexity classes is always a complexity class. An example:

$$\text{Let} \quad t_i(n) = \begin{cases} n^2 & \text{if} \quad n \equiv i(2), \\ n & \text{otherwise}, \end{cases} \qquad i = 0, 1.$$

Assume that there is a $t \in \mathbb{R}_1$ such that **T-DTIME**$(t_0) \cup$ **T-DTIME**$(t_1) =$ **T-DTIME**(t). Since $S_0 \in$ **T-DTIME**(t_0) and $S_1 \in$ **T-DTIME**(t_1) we conclude $S_0, S_1 \in$ **T-DTIME**(t). Because of 8.13, $S_0 \in$ **T-DTIME**(t) implies $t \succeq t_0$. Similarly, because of $S_1 \in$ **T-DTIME**(t) we get $t \succeq t_1$. Hence, $t(n) \succeq n^2$. But then $S = S_0 \cup S_1 \in$ **T-DTIME**(t), whereas $S \notin$ **T-DTIME**$(t_0) \cup$ **T-DTIME**(t_1). This contradiction shows that **T-DTIME**$(t_0) \cup$ **T-DTIME**(t_1) cannot be a complexity class with respect to T-DTIME. More generally one proves using diagonalization methods

15.15. Theorem. [HaSt 65], [McMe 69] For every Blum measure Φ there are arbitrarily large t_1, t_2 such that $\Phi(t_1) \cup \Phi(t_2)$ is not a complexity class with respect to Φ. ∎

However, under suitable good additional assumptions even infinite unions of complexity classes may be proved to be themselves complexity classes.

15.16. Union Theorem. [McMe 69] Let $r \in \mathbb{R}_1$ such that $\varphi_{r(i)} \in \mathbb{R}_1$ and $\varphi_{r(i)} \leqq \varphi_{r(i+1)}$ for all $i \in \mathbb{N}$. Then there exists a $t \in \mathbb{R}_1$ such that $\Phi(t) = \bigcup_{i \in \mathbb{N}} \Phi(\varphi_{r(i)})$. Moreover, an index of t can effectively be computed from an index of r.

Proof. The Union Theorem is an immediate consequence of the following union theorem for program classes, which is, in its turn, a consequence of 15.14. ∎

15.17. Theorem. [Emd 78] Let $r \in \mathbb{R}_1$ such that $\varphi_{r(i)} \in \mathbb{R}_1$ and $\varphi_{r(i)} \leqq \varphi_{r(i+1)}$ for all $i \in \mathbb{N}$. Then there exists a $t \in \mathbb{R}_1$ such that $I(t) = \bigcup_{i \in \mathbb{N}} I(\varphi_{r(i)})$. Moreover, an index of t can effectively be computed from an index of r.

Proof. Let $r = \varphi_k$. The proof of 15.12 shows that there exists a seven place recursive predicate A such that

$$I(\varphi_{\varphi_k(i)}) = \left\{j : \bigvee_u \bigvee_v \bigwedge_x \bigwedge_y A(k, i, j, u, v, x, y)\right\}.$$

This yields a Σ_2-description of $\bigcup_{i \in \mathbb{N}} I(\varphi_{\varphi_k(i)})$:

$$\bigcup_{i \in \mathbb{N}} I(\varphi_{\varphi_k(i)}) = \left\{j : \bigvee_i \bigvee_u \bigvee_v \bigwedge_x \bigwedge_y A(k, i, j, u, v, x, y)\right\}.$$

Consequently, according to 15.14, there exists an $s \in \mathbb{R}_1$ such that

$$I(\varphi_{s(k)}) = \Gamma\Big(\bigcup_{i \in \mathbb{N}} I(\varphi_{\varphi_k(i)})\Big), \quad \text{i.e.} \quad \bigcup_{i \in \mathbb{N}} I(\varphi_{\varphi_k(i)}) \subseteqq I(\varphi_{s(k)}).$$

To show the reverse implication, let $j \in I(\varphi_{s(k)})$. By Claim 3 of the proof of 15.14, $\Phi_j =_{\text{ae}} 0$ and hence $j \in \bigcup_{i \in \mathbb{N}} I(\varphi_{\varphi_k(i)})$ or there exist finitely many $i_1, \ldots, i_m \in \bigcup_{i \in \mathbb{N}} I(\varphi_{\varphi_k(i)})$ such that $\Phi_j \leqq_{\text{ae}} \max(\Phi_{i_1}, \ldots, \Phi_{i_m})$. As the sets $I(\varphi_{\varphi_k(i)})$ form a chain with respect to inclusion, $i_1, \ldots, i_m$ must belong to one single $I(\varphi_{\varphi_k(l)})$. Consequently, $\Phi_j \leqq_{\text{ae}} \varphi_{\varphi_k(l)}$, and thus

$$j \in I(\varphi_{\varphi_k(l)}) \subseteqq \bigcup_{i \in \mathbb{N}} I(\varphi_{\varphi_k(i)}). \ ∎$$

A nontrivial application of the Union Theorem shows that $\mathbb{P}r_1$ is a space complexity class. This follows from these facts: $\mathbb{P}r_1 = \bigcup_{i \in \mathbb{N}} \mathscr{E}_1^i$ (2.19.4) and $\mathscr{E}_1^i = \text{DSPACE}(\mathscr{E}_1^i)$ (10.20.3), $f \leqq a_{k+1} \in \mathscr{E}_1^{k+1}$ for all $f \in \mathscr{E}^k$ (2.19.1 and 2.17) and $a_k \leqq a_{k+1}$ for all $k \geqq 3$.

Further results concerning the Union Theorem can be found in [BaYo 73].

A union theorem can also be proved for complexity classes of partial functions of the type $\Phi_p(\tau) = \{\varphi_i : \Phi_i \leqq_{ae} \tau\}$ (whose functions may have a smaller domain than τ has). This follows from 15.13.

In the case where the growing sequence of complexity classes arises from restrictions of a given function φ_i, the union function can be shown to be a restriction of φ_i:

15.18. Theorem. [Bar 77] Let $r \in \mathbb{R}_1$ and $\varphi_i \in \mathbb{P}_1$ with infinite domain such that

$$\bigwedge_{n \in \mathbb{N}} \varphi_{r[n+1](i)} \subset \varphi_{r[n](i)}.$$

Then there exists an $A \in \mathbf{RE}$ such that

$$\Phi_p(\varphi_{i|A}) = \bigcup_{n \in \mathbb{N}} \Phi_p(\varphi_{r[n](i)}). \blacksquare$$

Remark. There exists an $r \in \mathbb{R}_1$ such that for sufficiently large $\varphi_i \in \mathbb{P}_1$ with infinite domain the properties $\bigwedge_{n \in \mathbb{N}} \varphi_{r[n+1](i)} \subset \varphi_{r[n](i)}$ and $\Phi_p(\varphi_{r[n](i)}) \subset \Phi_p(\varphi_{r[n+1](i)})$ are satisfied.

We now consider the intersection operation. Although $I(t_1) \cap I(t_2) = I(\min(t_1, t_2))$ is always true, we have no corresponding result for complexity classes:

15.19. Theorem. [LaRo 72] There exist a Blum measure Φ and arbitrarily large s, $t \in \mathbb{R}_1$ such that the intersection $\Phi(s) \cap \Phi(t)$ is not a complexity class with respect to Φ. $\blacksquare$

And there is also no possibility of proving an intersection result similar to 15.16:

15.20. Theorem. [Bas 72] For every Blum measure Φ there exists an $r \in \mathbb{R}_1$ such that $\varphi_{r(i)} \in \mathbb{R}_1$ and $\varphi_{r(i)} > \varphi_{r(i+1)}$ for all $i \in \mathbb{N}$, and for all $t \in \mathbb{R}_1$

$$\Phi(t) \neq \bigcap_{i \in \mathbb{N}} \Phi(\varphi_{r(i)}). \blacksquare$$

An analogous result for the classes $\Phi_p(\tau)$, $\tau \in \mathbb{P}_1$, has been proved in [Bar 77].

15.4. Naming Properties

Some of the results which can be formulated in terms of complexity classes heavily depend on their names (for example hierarchy results, cf. § 17) or are actually results about their names. For example, the Compression Theorem (17.10) can only be formulated for names from a measured set. The question is how many complexity classes are lost by a restriction to names from a measured set. It is the contents of the next result that this is no restriction at all, i.e. there is always a measured set such that all complexity classes can be named by functions of this set.

15.21. Naming or Honesty Theorem. [McMe 69] There exists an $s \in \mathbb{R}_1$ such that
1. $\{\varphi_{s(i)} : i \in \mathbb{N}\}$ is a measured set and
2. $\bigwedge_i \big(\varphi_i \in \mathbb{R}_1 \to \varphi_{s(i)} \in \mathbb{R}_1 \wedge \Phi(\varphi_i) = \Phi(\varphi_{s(i)})\big)$.

Proof. This theorem is an immediate consequence of the next theorem, which we can call Naming or Honesty Theorem for program classes. ∎

15.22. Theorem. There exists an $s \in \mathbb{R}_1$ such that
1. $\{\varphi_{s(i)} : i \in \mathbb{N}\}$ is a measured set and
2. $\bigwedge_i \big(\varphi_i \in \mathbb{R}_1 \to I(\varphi_i) = I(\varphi_{s(i)})\big)$.

Proof. First of all we notice (cf. proof of 15.12) that there exists a recursive function h such that $I(\varphi_i) = X_{h(i)}$ where $X_0, X_1, \ldots$ is a standard enumeration of Σ_2. By 15.14, there exists a measured set $\{\varphi_{g(i)} : i \in \mathbb{N}\}$ such that $\Gamma\big(I(\varphi_i)\big) = \Gamma(X_{h(i)}) = I(\varphi_{g(h(i))})$. Define $s = g \circ h$. If $\varphi_i \in \mathbb{R}_1$, then $I(\varphi_i) \subseteq \Omega\mathbb{R}_1$, and $I(\varphi_i)$ is closed with respect to Γ (Remark 1 after 15.14). Hence, $I(\varphi_i) = \Gamma\big(I(\varphi_i)\big) = I(\varphi_{s(i)})$. ∎

Definition. A function s with the properties stated in 15.21 is called an ***honesty procedure on*** $\mathbb{R}_1$.

Theorem 15.21 can be generalized to the classes $\Phi_p(\tau)$, $\tau \in \mathbb{P}_1$:

15.23. General Honesty Theorem. [McMe 69] There exists an $s \in \mathbb{R}_1$ such that
1. $\{\varphi_{s(i)} : i \in \mathbb{N}\}$ is a measured set and
2. $\bigwedge_i \Phi_p(\varphi_i) = \Phi_p(\varphi_{s(i)})$.

Proof. This theorem is proved with the help of 15.13 in exactly the same manner as 15.21 is proved with the help of 15.14. ∎

Remark. 15.23 can be proved in such a way that the function s satisfies Theorem 15.21 (cf. [MoMe 74]).

Definition. A function $s \in \mathbb{R}_1$ with the properties stated in 15.23 is called an ***honesty procedure on*** $\mathbb{P}_1$.

In contrast to 15.23, there is no naming theorem for the classes $\Phi(\tau)$.

15.24. Theorem. [Rob 76] If $s \in \mathbb{R}_1$ and $\{\varphi_{s(i)} : i \in \mathbb{N}\}$ is a measured set, then there exist arbitrarily large φ_i such that $\Phi(\varphi_i) \neq \Phi(\varphi_{s(i)})$. ∎

Honesty procedures and their properties have been studied extensively. We close the section with some selected results without proofs.

The first result concerns renamings of classes $\Phi_p(\tau)$ by restricting the domains of the names in a nontrivial way.

15.25. Theorem. [BaRo 76]
1. There exists an $s \in \mathbb{R}_1$ such that for every $\varphi_i \in \mathbb{P}_1$ with infinite domain, $D_i \setminus D_{s(i)}$ is r.e. and infinite and $\Phi_p(\varphi_i) = \Phi_p(\varphi_{s(i)})$.
2. No s satisfying these two conditions can be an honesty procedure on $\mathbb{P}_1$. ∎

15.26. Theorem. [Con 72] No honesty procedure on $\mathbb{R}_1$ can be a total effective operator. ∎

This means that for an honesty procedure s there must exist numbers i, j such that $\varphi_i = \varphi_j$ and $\varphi_{s(i)} \neq \varphi_{s(j)}$.

The next result shows that honesty procedures can be found that produce names which have infinitely often large distance from the old names.

15.27. Theorem. [Wer 71] For any $r \in \mathbb{R}_1$ there exists an honesty procedure s on $\mathbb{R}_1$ such that

$$\bigwedge_{\substack{\varphi_i \in \mathbb{R}_1 \\ \varphi_i \text{ sufficiently large}}} (r \circ \varphi_i \leqq_{\text{io}} \varphi_{s(i)} \wedge \varphi_{s(i)} \leqq_{\text{io}} \varphi_i). \ \blacksquare$$

Furthermore, there exist honesty procedures on $\mathbb{R}_1$ such that the new name is "very seldom" larger than the old one.

15.28. Theorem. [MoMe 74] There exists an honesty procedure s on $\mathbb{R}_1$ such that for every $\varphi_i \in \mathbb{R}_1$

$$\underline{\lim}_{x \to \infty} \frac{1}{x} \operatorname{card} \{y \leqq x : \varphi_i(y) < \varphi_{s(i)}(y)\} = 0. \ \blacksquare$$

There is an honesty procedure on $\mathbb{R}_1$ that yields for every complexity class honest names, which are infinitely often arbitrarily large and any honesty procedure on $\mathbb{P}_1$ has of necessity this property:

15.29. Theorem. [MoMe 74]

1. There exists an honesty procedure s on $\mathbb{R}_1$ such that

$$\bigwedge_{t \in \mathbb{R}_1} \bigwedge_{h \in \mathbb{R}_1} \bigvee_{\varphi_i =^* t} \lim_{x \to \infty} \frac{1}{x} \operatorname{card} \{y \leqq x : \varphi_{s(i)}(y) < h(y)\} = 0.$$

2. For any honesty procedure s on $\mathbb{P}_1$,

$$\bigwedge_{t \in \mathbb{R}_1} \bigwedge_{h \in \mathbb{R}_1} \bigvee_{\varphi_i =^* t} \underline{\lim}_{x \to \infty} \frac{1}{x} \operatorname{card} \{y \leqq x : \varphi_{s(i)}(y) < h(y)\} = 0. \ \blacksquare$$

Finally we mention that there exist honesty procedures preserving monotonicity.

15.30. Theorem. [MoMe 74] There exists an honesty procedure s on $\mathbb{R}_1$ such that

$$\bigwedge_{\varphi_i \in \mathbb{R}_1} (\varphi_i \geqq \mathrm{id} \wedge \varphi_i \text{ monotone} \rightarrow \varphi_{s(i)} \text{ monotone}). \ \blacksquare$$

§ 16. Properties of Specific Measures

In this section we collect measure-theoretic properties which are valid only for specific measures. The first subsection concerns the space measure. The properties discussed in subsections 2—7 are connected with the attempt to define "natural" complexity measures. They all reflect certain natural features of reasonable complexity measures. It turns out, however, that they altogether are not consistent. So, the state of the art is, that no definition convincingly represents the notion of naturalness. There are even suspicions that this may not exist at all. Anyway, selecting a subclass of natural measures remains so far a matter of individual taste.

16.1. Step Counting Functions of T-DSPACE and Optimality

Let us call a function $f \in \mathbb{R}_1$ *simple* iff it satisfies the following honesty condition:

$$\bigvee_{\varphi_i = f} \text{T-DSPACE}_i \leqq f.$$

Following L. A. Levin we define: $M \subseteqq \mathbb{R}_1$ is called a *canonical set* iff

1. M consists only of simple functions,
2. $f, g \in M \wedge h$ simple $\wedge\, h \gtrsim \min(f, g) \to h \in M$,
3. $\{i : \varphi_i \in M \wedge \varphi_i$ is a step counting function$\} \in \Sigma_2$.

16.1. Theorem. [Lev 74] M is a canonical set if and only if there exists a set A such that $M = \{\text{T-DSPACE}_i : \varphi_i = c_A\}$. ∎

Since for every $g \in \mathbb{R}_1$ the set $\{\text{T-DSPACE}_i : \text{T-DSPACE}_i \gtrsim g\}$ is a canonical set, we obtain as a corollary

16.2. Theorem. For every $g \in \mathbb{R}_1$ there exists a set A such that

$$\{\text{T-DSPACE}_i : \varphi_i = c_A\} = \{\text{T-DSPACE}_i : \text{T-DSPACE}_i \gtrsim g\}. \;∎$$

Remark 1. The weaker version of this result where $=$ is replaced by $\subseteqq$ is Rabin's theorem 17.1 for T-DSPACE.

Remark 2. From 16.2 we conclude the following statement which is contained in the Compression Theorem 17.9: If g is h-honest, then there exists a set A such that

$$\bigwedge_{\varphi_i = c_A} \text{T-DSPACE}_i \gtrsim g \wedge \bigvee_{\varphi_j = c_A} \text{T-DSPACE}_j \leqq_{\text{ae}} h \,\square\, g.$$

Remark 3. Since for every $g \in \mathbb{R}_1$ and every $r \in \mathbb{R}_2$ there exist canonical subsets M of $\{\text{T-DSPACE}_i : \text{T-DSPACE}_i \gtrsim g\}$ such that for every $\text{T-DSPACE}_i \in M$ there exists a $\text{T-DSPACE}_j \in M$ with $\text{T-DSPACE}_j \lesssim r \,\square\, \text{T-DSPACE}_i$, Theorem 16.2 yields as a corollary the Speed-Up Theorem 18.1.

The following theorem is the converse of 16.2.

16.3. Theorem. For every $f \in \mathbb{R}_1$ there exists a $g \in \mathbb{R}_1$ such that

$$\{\text{T-DSPACE}_i : \varphi_i = f\} = \{\text{T-DSPACE}_i : \text{T-DSPACE}_i \gtrsim g\}. \;∎$$

These results can be suitably extended to arbitrary Blum measures. Related work has been done in [MeWi 78].

As a corollary of 16.2 we obtain the optimality of T-DSPACE.

Definition. 1. The function $t \in \mathbb{R}_1$ is called an *optimal step counting function* for $f \in \mathbb{R}_1$ with respect to the Blum measure (φ, Φ) iff

$$\bigvee_i (\varphi_i = f \wedge \Phi_i = t) \quad \text{and} \quad \bigwedge_j (\varphi_j = f \to \Phi_j \gtrsim t).$$

2. A Blum measure (φ, Φ) is *optimal* iff every total Φ_i is optimal with respect to (φ, Φ) for some 0-1-valued function.

Applying 16.2 to the total step counting functions g of T-DSPACE we obtain immediately

16.4. Theorem. T-DSPACE is optimal. ∎

Remark 1. This result has been proved in [Tra 65] by diagonalization.

Remark 2. It is easy to construct measures which are not optimal.|

16.2. Flowchart Measures

It has been proposed to consider flowchart measures as typical representatives of natural measures [Hart 73]. In different settings flowchart measures have been studied in [Bak 74] and [Lis 75a]. The function and predicate symbols are interpreted by functions from a fixed set $F \subseteq \mathbb{R}$, and every such set yields a *type of flowchart algorithm* and a corresponding notion of *flowchart computability.* Such interpreted flowcharts compute only partial recursive functions. We consider only such interpretations (we call them *universal*) which render all partial recursive functions computable. Universal interpretations are constructed in the proof of 16.5 and in the proof of 16.14.3.

Given a universal interpretation, all interpreted flowcharts of this type can be effectively enumerated in a standard way as $F_0, F_1, \ldots$ If F_i computes φ_i and we define

$$\Phi_i(x) = \begin{cases} \text{length of the computation described by } F_i \text{ on input } x \text{ if this computation halts,} \\ \text{undefined otherwise,} \end{cases}$$

then (φ, Φ) is called the *flowchart measure* belonging to this notion of flowchart computability. Thus, flowchart measures can be considered as the time measures of the corresponding types of flowchart algorithm. Note that every flowchart measure is a Blum measure. We also consider flowcharts where every node is valuated by 0 or 1 (its "cost") in such a way that nodes with outdegree 1 within a cycle must be valuated by 1. Now the generalized flowchart measure is defined by the number of executions of functions corresponding to nodes with value 1 for halting computations, and it is undefined otherwise. For further generalizations see [Lis 75a].

The investigations of the following subsections will show that not every Blum measure can be a flowchart measure (see for instance Remark 1 after 16.18). However:

16.5. Theorem. [Weih 74] Every Blum measure is a submeasure of a generalized flowchart measure.

Proof. Let (φ, Φ) be a Blum measure. We consider flowcharts over

$$\{\mathrm{e}, \mathrm{s}\} \cup \{M_k : k \in \mathbb{N}\} \cup \{u_k : k \in \mathbb{N}\}$$

where e denotes the constant function whose value is the empty word, $s(x) = x + 1$,

$$M_k(x, y) = \begin{cases} 1 & \text{if} \quad \Phi_k(x) = y, \\ 0 & \text{otherwise}, \end{cases} \qquad u_k(x, y) = \begin{cases} \varphi_k(x) & \text{if} \quad \Phi_k(x) = y, \\ 0 & \text{otherwise}. \end{cases}$$

The flowchart $F_{g(k)}$ (see Figure 16.1) computes $\varphi_k(x)$. If the nodes (0), (1) and (2) are free of cost and (3) has cost 1, the value of the corresponding flowchart measure is exactly $\Phi_k(x)$. ∎

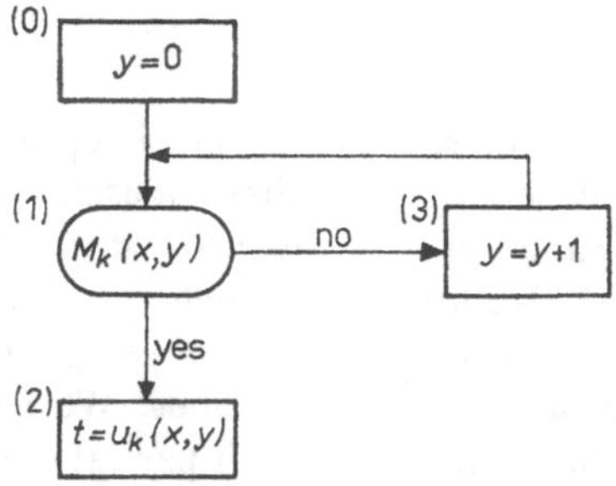

Figure 16.1

Remark 1. This result has been proved independently by [Buc 74] (cf. also [Lis 75a]).

Remark 2. It is not hard to see (cf. [Hec 75]) that Theorem 16.5 holds also true for flowchart measures in the following weaker version: For every Blum measure (φ, Φ) there exist a flowchart measure (ψ, Ψ) and a function $g \in \mathbb{R}_1$ and $c \in \mathbb{N}$ such that

$$\bigwedge_{i \in \mathbb{N}} \Big(\varphi_i = \psi_{g(i)} \wedge \bigwedge_{\Phi_i(x) \geq c} \Phi_i(x) = \Psi_{g(i)}(x)\Big).$$

Remark 3. If we would define a measure to be natural if and only if it has only r.e. complexity classes, then every natural measure would have nonnatural submeasures as follows from 15.1. Although we do not adopt this definition, it seems nevertheless reasonable to look for a notion of naturalness which implies that the class of the natural measures is not closed with respect to forming submeasures. Flowchart measures do have this property.

16.3. Finite Invariance and Conformity

In § 15 we have proved that not every measure has the property that all its complexity classes are r.e. To show this we constructed a rather pathological measure which hardly anybody would acknowledge as reflecting essential features of computing. If we assume that natural measures should have the property of having only r.e. complexity classes, then the question arises of which properties imply the recursive enumerability of all complexity classes of a measure. A first suggestion comes from analyzing the proof of 15.1 which shows that the possibility of getting unbounded changes in complexity by finite changes in the computed functions, is responsible for the result 15.1. This is impossible for measures having the property of being finitely invariant.

Definition. Φ is *finitely invariant* $\leftrightarrow$

$$\bigwedge_{f\in\mathbb{R}_1}\bigwedge_{g\in\mathbb{R}_1}\bigwedge_{t\in\mathbb{R}_1}\big(f =_{\mathrm{ae}} g \wedge f \in \Phi(t) \to g \in \Phi(t)\big).$$

Another property which is related to finite invariance is that of conformity. Both these properties have been introduced in [Lewf 71].

Definition. Φ is *conform* $\leftrightarrow$

$$\bigwedge_{t_1\in\mathbb{R}_1}\bigwedge_{t_2\in\mathbb{R}_1}\big(\Phi(t_1) \neq \emptyset \wedge \Phi(t_2) \neq \emptyset \to \Omega\Phi(t_1) \equiv_1 \Omega\Phi(t_2)\big).$$

16.6. Theorem. [Lewf 71] **If a Blum measure is finitely invariant, then it is conform and all its complexity classes are r.e.**

Proof. 1. Finite invariance implies conformity. Let Bound denote the class of all recursive functions bounded by some constant. We show: $\Omega\Phi(t) \equiv_1 \Omega$ Bound for all $t \in \mathbb{R}_1$ such that $\Phi(t) \neq \emptyset$.

a) $\Omega\Phi(t) \leqq_1 \Omega$ Bound (this is valid without the assumption of finite invariance): Define

$$\varphi_{r(i)}(x) = \begin{cases} \mu\langle j, a, b\rangle \Big[j, a, b \leqq x \wedge \bigwedge\limits_{z\leqq x} \varphi_j(z) = \varphi_i(z) \\ \qquad \wedge \bigwedge\limits_{z\leqq a} \Phi_j(z) \leqq b \wedge \bigwedge\limits_{a<z\leqq x} \Phi_j(z) \leqq t(z)\Big] \\ \qquad\qquad \text{if such a triple } (j, a, b) \text{ exists and } \{0, \ldots, x\} \subseteqq \mathrm{D}_i, \\ x \qquad\quad \text{if such a triple does not exist and } \{0, \ldots, x\} \subseteqq \mathrm{D}_i, \\ \text{undefined} \quad \text{if } \{0, \ldots, x\} \nsubseteq \mathrm{D}_i. \end{cases}$$

It follows that $i \in \Omega\Phi(t) \leftrightarrow r(i) \in \Omega$ Bound and thus $\Omega\Phi(t) \leqq_1 \Omega$ Bound via r.

b) Ω Bound $\leqq_1 \Omega\Phi(t)$: We choose an increasing function $h \in \mathbb{R}_2$ according to 5.24 such that $\varphi_i(x) \leqq_{\mathrm{ae}} h(x, \Phi_i(x))$ and define $g(x) = h(x, t(x))$. Furthermore, let f be some function from $\Phi(t)$. Define

$$\varphi_{s(i)}(x) = \begin{cases} f(x) & \text{if } \varphi_i(x) \leqq \max\{\varphi_i(0), \ldots, \varphi_i(x \dot- 1)\} \text{ and} \\ & \quad \{0, \ldots, x\} \subseteqq \mathrm{D}_i, \\ g(x) + 1 & \text{if this condition is not satisfied but} \\ & \quad \{0, \ldots, x\} \subseteqq \mathrm{D}_i, \\ \text{undefined} & \text{if } \{0, \ldots, x\} \nsubseteq \mathrm{D}_i. \end{cases}$$

From $i \in \Omega$ Bound we conclude $\varphi_{s(i)} =_{\mathrm{ae}} f$. As $f \in \Phi(t)$ and Φ is finitely invariant, $\varphi_{s(i)} \in \Phi(t)$ and hence $s(i) \in \Omega\Phi(t)$. If $i \notin \Omega$ Bound, then $\varphi_{s(i)}$ is either not total or infinitely often larger than g. Thus, $\varphi_{s(i)} \notin \Phi(t)$, and hence Ω Bound $\leqq_1 \Omega\Phi(t)$.

2. Finite invariance implies the recursive enumerability of all complexity classes. Let $t \in \mathbb{R}_1$ be given and choose $f \in \Phi(t)$. Define

$$\varphi_{s(i,a)}(x) = \begin{cases} \varphi_i(x) & \text{if } \bigwedge_{z \leqq x} \left[\left(z < a \wedge \Phi_i(z) \leqq a\right) \vee \Phi_i(z) \leqq t(z)\right], \\ f(x) & \text{otherwise}. \end{cases}$$

Then either $\varphi_{s(i,a)} =_{\mathrm{ae}} f$, in which case $\varphi_{s(i,a)} \in \Phi(t)$ because of the finite invariance of Φ, or $\varphi_{s(i,a)} = \varphi_i$. In this latter case, $\varphi_i \in \Phi(t)$ and hence $\varphi_{s(i,a)} \in \Phi(t)$. This shows that s enumerates a subset of $\Phi(t)$. On the other hand, when $g \in \Phi(t)$, then there exist a j and a b such that $\varphi_j = g$ and $\bigwedge_z \left[\left(z < b \wedge \Phi_j(z) \leqq b\right) \vee \Phi_j(z) \leqq t(z)\right]$. Therefore, $g = \varphi_{s(j,b)}$. This shows $\Phi(t) = \{\varphi_{s(i,a)}: i, a \in \mathbb{N}\}$. ∎

Remark 1. Recursive enumerability of all complexity classes does not imply finite invariance. For example, let Φ be a Blum measure with r.e. complexity classes and let i_0 be an index of an arbitrary recursive function. Define

$$\Phi_i'(x) = \begin{cases} 0 & \text{if } i = i_0, \\ \Phi_i(x) + x & \text{otherwise}. \end{cases}$$

Φ' is also a Blum measure, as can be easily verified. It is not finitely invariant, because a single change of φ_{i_0} causes infinitely many changes in its complexity. But all complexity classes are r.e. because $\Phi'(t) = \{\varphi_{i_0}\}$ for $t \not\geqq \mathrm{id}$ and $\Phi'(t) = \Phi(t \dot- \mathrm{id}) \cup \{\varphi_{i_0}\}$ otherwise.

Remark 2. The measure Φ' constructed in Remark 1 is not conform because $\Omega\Phi'(0)$ is only in Π_2 and hence $\Omega\Phi'(0) \leqq_1 \Omega$ Total (see 2.16). On the other hand, $\Omega\Phi'(t) \leqq_1 \Omega$ Bound for every $t \in \mathbb{R}_1$ according to part 1a of the proof of 16.6. Furthermore, part 1b of this proof shows Ω Bound $\leqq_1 \Omega\Phi'(t)$ for sufficiently large $t \in \mathbb{R}_1$ (take $f = 0$ and use the fact $\mathbf{F} \subseteqq \Phi'(t)$ as stated in Lemma 15.2). Hence $\Omega\Phi'(t) \equiv_1 \Omega$ Bound for sufficiently large $t \in \mathbb{R}_1$. But 2.16 states Ω Bound $\not\equiv_1 \Omega$ Total. This shows that the recursive enumerability of all complexity classes does not imply conformity.

Remark 3. Conformity does not imply finite invariance. For example, let Φ = T-DSPACE and choose $f, t_1, t_2 \in \mathbb{R}_1$ in such a way that $f \in \Phi(t_2) \setminus \Phi(t_1)$. Let $\varphi_{i_0} = f$. Define

$$\Psi_i(x) = \begin{cases} t_1(x) & \text{if } i = i_0, \\ \Phi_i(x) & \text{otherwise}. \end{cases}$$

Ψ is evidently a Blum measure, but it is not finitely invariant because no finite modification of f can have space complexity bounded by t_1. Hence no finite modification of f can belong to $\Psi(t_1)$. The conformity of Ψ is easily verified.

Remark 4. As the measure Ψ of Remark 3 also has thoroughly r.e. complexity classes we conclude that even recursive enumerability of all classes and conformity together do not imply finite invariance.

Remark 5. It seems to be an open problem whether conformity implies the recursive enumerability of all complexity classes.

Remark 6. Finite invariance does not imply the property of being a flowchart measure. This follows from 16.10.2 and 16.12.

The demand that finite changes in functions should cause no change in their complexity is perhaps a little bit too strong. The following slightly weakened version has nearly the same degree of practical justification as the original notion of finite invariance:

Definition. The measure Φ is called *weakly finitely invariant* $\leftrightarrow$

$$\bigwedge_{f \in \mathbb{R}_1} \bigwedge_{g \in \mathbb{R}_1} \bigwedge_{t \in \mathbb{R}_1} \big(f \in \Phi(t) \wedge f =_{\text{ae}} g \rightarrow g \in \Phi(t + \text{Const})\big).$$

This suggests a weakening of conformity:

Definition. Φ is called *weakly conform* $\leftrightarrow$

$$\bigwedge_{t_1 \in \mathbb{R}_1} \bigwedge_{t_2 \in \mathbb{R}_1} \big(\Phi(t_1 + \text{Const}) \neq \emptyset \wedge \Phi(t_2 + \text{Const}) \neq \emptyset \rightarrow \Omega\Phi(t_1 + \text{Const}) \\ \equiv_1 \Omega\Phi(t_2 + \text{Const})\big).$$

Before proving the next theorem we need an auxiliary result:

16.7. Lemma. For any flowchart measure the finite sets can be decided with constant complexity.

Proof. Let F_i be a flowchart deciding the finite set E and let $k =_{\text{df}} \max_{x \in E} \Phi_i(x)$. Now, every flowchart can be brought into the form of a (possibly infinite) tree. We do this for F_i and consider that initial tree $F_i^{(k)}$ which has on every path at most k nodes different from the start node. It is evident that those inputs x for which the computation follows a path which has been pruned cannot belong to E. These paths have therefore to lead to a negative answer. Thus, completing each pruned path of $F_i^{(k)}$ by a node giving this negative answer, yields a flowchart deciding E with constant complexity. ∎

It is easy to construct generalized flowchart measures which are not finitely invariant, and it is unknown whether every flowchart measure is finitely invariant. However,

16.8. Theorem. Every flowchart measure is weakly finitely invariant.

Proof. Let f be computed by the flowchart F and let g be a finite modification of f, i.e. $E =_{\text{df}} \{x: f(x) \neq g(x)\}$ is finite. To compute g, we first choose a flowchart F' deciding E with constant complexity. This is possible by 16.7. Now combine F' with F and a flowchart G for g in such a way that the exits of F' with positive answer are linked with the start node of G and the exits of F' with negative answer are connected with the start node of F. This new flowchart computes g, and its computational complexity is almost everywhere given by $c + \Phi_F(x)$, where Φ_F is the complexity belonging to F and c is a constant. ∎

As immediate consequences we have

16.9. Corollary.

1. Flowchart measures are weakly conform.
2. If Φ is a flowchart measure, then for all $t \in \mathbb{R}_1$ the classes $\Phi(t + \text{Const})$ are recursively enumerable.

Proof. This follows from 16.8 and a "weak" version of 16.6 which can be proved in the same way as 16.6. ∎

We believe that the weak versions of conformity and finite invariance should be considered as more fundamental, more realistic ones than the original notions, and that we should be satisfied with the enumerability of the classes $\Phi(t + \text{Const})$ instead of demanding recursive enumerability of the classes $\Phi(t)$.

16.4. Compositionality

It seems natural that the complexity of a function which is composed by an effective operation from two other ones should be computable from the values and complexities of the operands by a function depending only on the operation. We make this more precise by giving an example. Let ω be an operation on $\mathbb{P}_1$ which is effectively described by some $h \in \mathbb{R}_2$: $\varphi_{h(i,j)} = \omega(\varphi_i, \varphi_j)$. Then for reasonable measures Φ we have an $s \in \mathbb{R}_7$ such that

$$\Phi_{h(i,j)}(x) = s\big(x, \varphi_i(x), \varphi_j(x), \Phi_i(x), \Phi_j(x), \Phi_i(\varphi_j(x)), \Phi_j(\varphi_i(x))\big).$$

We call this a compositionality with respect to ω. Here the idea is that s depends only on ω and on the way ω is carried out.

For example, addition $\varphi_i + \varphi_j$ can be carried out by a DTM in such a way that $\varphi_i(x) + \varphi_j(x) = \varphi_{a(i,j)}(x)$ and

$$\text{T-DSPACE}_{a(i,j)}(x) = \text{T-DSPACE}_i(x) + \text{T-DSPACE}_j(x)$$

is satisfied.

One of the most important operations on $\mathbb{P}_1$, namely substitution, suggests a compositionality property of the following form:

Definition. Let $\varphi_{h(i,j)} = \varphi_i \circ \varphi_j$. The measure Φ is said to have the property of (h, s)-*compositionality* (with respect to substitution) $\leftrightarrow$

$$\Phi_{h(i,j)}(x) = s\big(\Phi_i(\varphi_j(x)), \Phi_j(x)\big).$$

Obviously, substitution can be performed by Turing machines in such a way that we have

$$\text{Tmulti-DTIME}_{h(i,j)}(x) = \text{Tmulti-DTIME}_i(\varphi_j(x)) + \text{Tmulti-DTIME}_j(x)$$

and

$$\text{T-DSPACE}_{h'(i,j)}(x) = \text{T-DSPACE}_i(\varphi_j(x)) + \text{T-DSPACE}_j(x)$$

for suitable h, h' describing the substitution.

Compositionality has been investigated intensively in [Lis 75b], [Lis 76] and [Lis 77] under the name "conservation property". In this subsection we confine ourselves to compositionality with respect to substitution.

16.10. Theorem. [Lis 75b]

1. There exist Blum measures which do not satisfy any (h, s)-compositionality.
2. Flowchart measures satisfy the property of $(h, +)$-compositionality for suitably chosen h.

Proof. Ad 1. We use the measure Φ' of Remark 1 after 16.6 and assume $\varphi_{i_0}(x) = x + 1$. As $\varphi_{h(i_0,i_0)} \neq \varphi_{i_0}$ we conclude $\Phi'_{h(i_0,i_0)}(x) \geqq x$. On the other hand, $s\big(\Phi'_{i_0}(\varphi_{i_0}(x)), \Phi'_{i_0}(x)\big) = s(0, 0)$. This shows $\Phi'_{h(i_0,i_0)}(x) \neq s\big(\Phi'_{i_0}(\varphi_{i_0}(x)), \Phi'_{i_0}(x)\big)$ for every $s \in \mathbb{R}_2$ and for every function h with the property $\varphi_{h(i,j)} = \varphi_i \circ \varphi_j$.

Ad 2. If $F_{h(i,j)}$ is that flowchart which is constructed by an immediate concatenation of F_i and F_j, the claimed compositionality property is obvious. ∎

Let (ψ, Ψ) be a submeasure of (φ, Φ). As ψ is also a Gödel numbering of $\mathbb{P}_1$, there exist two functions $g, f \in \mathbb{R}_1$ such that $\psi_i = \varphi_{g(i)}$ and $\varphi_i = \psi_{f(i)}$. The equations

$$\psi_i \circ \psi_j = \varphi_{g(i)} \circ \varphi_{g(j)} = \varphi_{h(g(i),g(j))} = \psi_{f(h(g(i),g(j)))}$$

suggest the following definition. Let the Blum measure (φ, Φ) satisfy the property of (h, s)-compositionality. This property carries over to the submeasure (ψ, Ψ) of (φ, Φ) if there exist $f, g \in \mathbb{R}_1$ such that $\psi_i = \varphi_{g(i)}$, $\Psi_i = \Phi_{g(i)}$, $\varphi_i = \psi_{f(i)}$ and

$$\Psi_{f(h(g(i),g(i)))}(x) = s\big(\Psi_i(\psi_j(x)), \Psi_j(x)\big).$$

16.11. Theorem. [Lis 76] If the Blum measure (φ, Φ) satisfies the property of (h, s)-compositionality, then this property does not carry over to all submeasures of (φ, Φ). ∎

In the sense of Remark 3 after 16.5 compositionality is a candidate for a naturalness property.

Many further results about compositionality can be found in [Lis 75b]. We restrict ourselves to the relations between compositionality and further properties of measures.

16.12. Theorem. [Lis 77] Finite invariance does not imply compositionality.

Proof. The set $\{f : f =_{\text{ae}} s\}$ is r.e., where $s(x) = x + 1$. Let $A \in \mathbf{REC}$ be chosen in such a way that $\{f : f =_{\text{ae}} s\} = \{\varphi_i : i \in A\}$. Define for a given finitely invariant Blum measure Φ:

$$\Phi'_i(x) =_{\text{df}} \begin{cases} 0 & \text{if } i \in A, \\ \Phi_i(x) + x & \text{otherwise}. \end{cases}$$

Then, obviously, Φ' is also a Blum measure, and it is finitely invariant. But it does not satisfy the property of compositionality, as can be proved in exactly the same way as statement 1 of 16.10. ∎

Remark. In [Lis 77] it is also shown that 16.12 remains true if "finite invariance" is replaced by "properness", "parallelity" or "density". These properties will be discussed in the following subsections.

On the other hand, compositionality with respect to substitution alone is not strong enough to imply other desirable properties:

16.13. Theorem. [Lis 77] Compositionality with respect to substitution implies neither conformity nor the property of having only r.e. complexity classes.

Proof. Take $f \in \mathbb{R}_1$ such that $\varphi_{f(j)}$ is the constant function j and such that $R_f \in \mathbf{REC}$. Let T' be the predicate defined by

$$T'(j, m) = 1 \leftrightarrow M_j \text{ stops on input } j \text{ after } m \text{ steps}.$$

Now, using an arbitrary Blum measure Φ'' we construct Φ' by

$$\Phi'_i(x) = \begin{cases} 0 & \text{if } \bigvee_j [i = f(j) \wedge \sim T'(j, x)], \\ \Phi''_i(x) + x + 1 & \text{otherwise}. \end{cases}$$

Φ' is also a Blum measure.

Choose furthermore a function $h \in \mathbb{R}_2$ with the following three properties

(1) $\varphi_{h(i,j)} = \varphi_i \circ \varphi_j$,

(2) $\langle n, m\rangle < \langle n', m'\rangle \to 0 < h(n, m) < h(n', m')$,

(3) $R_h \cap R_f = \emptyset$.

This can be achieved by the padding technique.

Now a function Φ is defined by induction:

$$\Phi_i(x) = \begin{cases} \Phi_{l(k)}\big(\varphi_{r(k)}(x)\big) + \Phi_{r(k)}(x) + 1 & \text{if } h\big(l(k), r(k)\big) = i, \\ \Phi'_i(x) & \text{otherwise}. \end{cases}$$

Φ is a Blum measure which satisfies the property

$$\Phi_{h(i,j)}(x) = \Phi_i\big(\varphi_j(x)\big) + \Phi_j(x) + 1.$$

We first prove that $\Phi(0)$ is not r.e. By definition

$$\begin{aligned} \Phi(0) &= \{\varphi_i : \Phi_i \leqq_{\text{ae}} 0\} = \{\varphi_i : \Phi'_i \leqq_{\text{ae}} 0\} \\ &= \Big\{\varphi_{f(j)} : \sim\bigvee_x T'(j, x)\Big\} = \{\varphi_{f(j)} : j \in \bar{K}\}. \end{aligned}$$

As $\bar{K} = \{\varphi_{f(j)}(0) : j \in \bar{K}\}$ is not r.e., $\Phi(0)$ cannot be r.e.

That Φ is not conform can be shown as follows. By the Padding Theorem 2.6 we obtain a 1-1 function g such that

$$\varphi_{g(i)}(x) = \begin{cases} \varphi_i(x) & \text{if } \varphi_i(0) = \ldots = \varphi_i(x) \wedge \sim T'\big(\varphi_i(x), x\big) \wedge i \in R_f, \\ \text{not defined otherwise}. \end{cases}$$

Then, evidently, $\Omega\Phi(0) \leqq_1 \Omega\,\text{Total}$ via g.

As $\Omega\,\text{Total} \in \Pi_2$ (see 2.16) the rest of the proof follows in the same way as in Remark 2 after 16.6. ∎

Remark. In [Lis 77] it is shown that the definition of Φ can be modified in such a way that, in addition to 16.13, it does not possess any of the properties parallelity, properness or density (see following subsections). Various further results concerning compositionally can be found there.

16.5. Parallelity

Parallelity is a special case of compositionality which is very welcome because it allows nice constructions.

Definition. Φ is called *parallel* iff there exists an $h \in \mathbb{R}_2$ such that

$$\varphi_{h(i,j)}(x) = \begin{cases} \varphi_i(x) & \text{if} \quad \Phi_i(x) \leqq \Phi_j(x), \\ \varphi_j(x) & \text{otherwise} \end{cases}$$

and

$$\Phi_{h(i,j)}(x) = \min \{\Phi_i(x), \Phi_j(x)\}.$$

In view of the next result, we should not insist on considering parallelity to be a characteristic property of what we would like to call natural measures.

16.14. Theorem.

1. T-DSPACE and multiT-DTIME are parallel.
2. [Bis 78] T-DTIME is not parallel.
3. Flowchart measures need not be parallel.

Proof. The parallelity of T-DSPACE is obvious. For multiT-DTIME one uses first a multitape TM which realizes the parallel combination of both original TM's and which has two heads on the first tape (holding the input). Then by 19.11 one can convert this machine into a multitape TM which has only one head per tape.

Ad 2. We consider a DTM M_i deciding $\mathbb{N}^2$ in realtime and a DTM M_j deciding $\overline{\mathbb{N}^2}$ in such a way that the head returns back to the first symbol of the input word as soon as it reads the symbol $\square$. Obviously, if T-DTIME is parallel, the machine $M_{h(i,j)}$ resulting from M_i and M_j by parallel combination decides $A =_{\text{df}} \{(m, n): m, n \in \mathbb{N} \wedge n \leqq m + 1\}$ in realtime. This contradicts the nonregularity of $A' = \{1^m \square 1^n: m, n \in \mathbb{N} \wedge n \leqq m + 1\}$.

Ad 3. DTM's can be considered as flowcharts processing words from $\Sigma^*\{|\}\Sigma^*$, where $|$ ($\notin \Sigma$) indicates the head position. These flowcharts have the interpretation $\{L_a: a \in \Sigma\} \cup \{R_a: a \in \Sigma\} \cup \{T_a: a \in \Sigma\}$ where $L_a(ux \mid yv) =_{\text{df}} u \mid xav$, $R_a(ux \mid yv) = uxa \mid v$ and

$$T_a(u \mid xv) =_{\text{df}} \begin{cases} 1 & \text{if} \quad x = a, \\ 0 & \text{if} \quad x \neq a. \end{cases}$$

Note that every flowchart of this type can be interpreted as a DTM. Obviously T-DTIME and the corresponding flowchart measure are linearly related.

Now assuming parallelity for this measure we obtain a linear upper bound on T-DTIME for A' (see proof of Statement 2). This contradicts 8.19.2. ∎

16.6. *Properness*

It seems reasonable to expect that natural measures should have the property that determining the computational complexity of an algorithm is no more expensive than running the algorithm itself. This property is called properness and is defined as follows.

Definition. Φ is *proper* $\leftrightarrow \bigwedge_{\Phi_i \in \mathbb{R}_1} \left(\Phi_i \in \Phi(\Phi_i)\right)$.

16.15. Theorem.

1. T-DSPACE is proper.
2. multiT-DTIME is proper.
3. RAM-DTIME is not proper.
4. [Bak 74], [Lis 75a] No flowchart measure is proper.

Proof. Statements 1 and 2 are clear.

Ad 3 (4). The identity function is computable on a DRAM (a flowchart of any type of flowchart algorithm) within time $2x$ (zero time). But to compute the function $2x$ (the zero function) a DRAM (a flowchart) needs more than $2x$ steps (at least one step). ∎

Remark. multiT-DTIME is proper only if functions over a single letter alphabet are considered (thus, the values of the step counting functions must be represented in unary as assumed in this section).

It seems reasonable to slightly weaken the definition of properness, by saying that Φ is *weakly proper* if $\Phi_i \in \Phi(\mathrm{Lin}\ \Phi_i)$ for all $\Phi_i \in \mathbb{R}_1$. This property is still almost as natural as strict properness, but is a property possessed by more measures:

16.16. Theorem.

1. RAM-DTIME is weakly proper.
2. DREV is weakly proper.
3. If in the flowchart measure Φ the successor function is computable with constant complexity, then Φ is weakly proper. ∎

16.17. Theorem.

1. Parallelity does not imply properness.
2. Properness does not imply parallelity.

Proof. Ad 1. Let $\Phi' =$ T-DSPACE and $\varphi_{i_0} = \mathrm{id}$. Define

$$\Phi_i(x) = \begin{cases} 0 & \text{if } i = i_0, \\ \Phi_i'(x) & \text{otherwise}. \end{cases}$$

Then Φ is a measure which is parallel but not proper because $\Phi_{i_0} \notin \Phi(\Phi_{i_0}) = \Phi(0)$.

Ad 2. Let Φ = T-DSPACE. We choose arbitrary numbers i_0, i_1 such that $\varphi_{i_0}, \varphi_{i_1} \in \mathbb{R}_1$ and an infinite recursive $A \subseteqq \Sigma^*$ with infinite complement. We define the new measure (ψ, Ψ) by

$$\psi_0 = \varphi_{i_0}, \qquad \Psi_0(x) = \begin{cases} 0 & \text{if } x \in A, \\ \Phi_{i_0}(x) & \text{otherwise}, \end{cases}$$

$$\psi_1 = \varphi_{i_1}, \qquad \Psi_1(x) = \begin{cases} 0 & \text{if } x \notin A, \\ \Phi_{i_1}(x) & \text{otherwise}, \end{cases}$$

$$\psi_2 = \Psi_0, \qquad \Psi_2 = \Psi_0,$$

$$\psi_3 = \Psi_1, \qquad \Psi_3 = \Psi_1,$$

$$\psi_{i+4} = \varphi_i, \qquad \Psi_{i+4} = \Phi_i \qquad (i = 0, 1, 2, \ldots).$$

The reader easily verifies that (ψ, Ψ) is a proper measure which is, however, not parallel. If Ψ were parallel then the parallelization of ψ_0 and ψ_1 would require the step counting function 0, but Ψ does not contain the step counting function 0 at all. ∎

The next result shows that parallelity and properness together do not imply all properties discussed in the preceding subsections.

16.18. Theorem. [Lis 77] There exists a measure which is parallel and proper but which is neither conform nor does it have the property of compositionality nor does it have only r.e. complexity classes. ∎

Remark 1. Parallelity and properness together are not strong enough to force a measure to be a flowchart measure. This follows from 16.18 and 16.10.2.

Remark 2. In [LaRo 72] the weaker statement is proved that parallelity and properness together do not imply that all complexity classes are r.e.

We close the subsection with a further consequence of properness.

16.19. Theorem. [McMe 69] If Φ is a proper measure whose step counting functions satisfy the condition $\Phi_i \geqq \mathrm{id}$, then not every complexity class with recursive name can be named by a Φ_i.

Proof. By 5.24 there exists an increasing $h \in \mathbb{R}_1$ such that $\varphi_i(x) \leqq_{ae} h(\Phi_i(x))$ whenever $\Phi_i \geqq \mathrm{id}$. Let us assume that every complexity class can be named by a $\Phi_i \in \mathbb{R}_1$. Then by 15.23, with $r = h$, we conclude the existence of Φ_i and Φ_j such that $\Phi(\Phi_i) = \Phi(\Phi_j)$ and $h \circ \Phi_i <_{io} \Phi_j$. Because of the properness, $\Phi_j \in \Phi(\Phi_j) = \Phi(\Phi_i)$. Hence there exists a k such that $\varphi_k = \Phi_j$ and $\Phi_k \leqq_{ae} \Phi_i$. Consequently, $\Phi_j(x) = \varphi_k(x) \leqq_{ae} h(\Phi_k(x)) \leqq_{ae} h(\Phi_i(x))$. Altogether we have obtained the contradiction $h(\Phi_i(x)) <_{io} \Phi_j(x) \leqq_{ae} h(\Phi_i(x))$. ∎

16.7. *Density*

Another property studied in connection with naturalness of measures is density.

Definition. Φ is called *dense* on $F \subseteqq \mathbb{R}_1 \leftrightarrow$

$$\bigwedge_{s \in F} \bigwedge_{t \in F} \bigvee_{u \in F} \big(\Phi(s) \subset \Phi(t) \rightarrow \Phi(s) \subset \Phi(u) \subset \Phi(t)\big).$$

However, we agree with P. R. YOUNG (see [You 71]), who expressed the feeling that the study of density will yield less insight into computational complexity than the study of the other properties investigated here. It might however be of interest to find out whether specific measures are dense.

16.20. Theorem. [BoCoHo 79] T-DSPACE is dense on $\mathbb{R}_1$.

Proof. We prove the density of T-DSPACE only on the set of the (total) step counting functions of T-DSPACE. This restricted version of the result was already known from [Tra 65]. Let $s, t \in \mathbb{R}_1$ be step counting functions for T-DSPACE such that T-DSPACE$(s) \subset$ T-DSPACE(t). Since $t \lessdot_{io} s$ would imply T-DSPACE$(s) \setminus$ T-DSPACE$(t) \neq \emptyset$ (see Remark 3 after 17.23) we have $s \lesssim t$. As the linear speed-up also holds for T-DSPACE when functions are computed we obtain $s \lessdot_{io} t$. Hence there exists an unbounded function $f \in \mathbb{R}_1$ such that $s \cdot f \lessdot_{io} t$. We choose an i such that $\varphi_i = f$ and define

$$g(n) = \max\left(1, \sup\{f(x) : x \leqq n \wedge \text{T-DSPACE}_i(x) \leqq n\}\right).$$

Obviously g is increasing, unbounded and computable within linear space, and we have $s \cdot g \lessdot_{io} t$. Now we define $u(n) = \min\{s(n) \cdot g(n), t(n)\}$. Since $s \cdot g$ is a step counting function of T-DSPACE and this measure is parallel, u is also a step counting function of T-DSPACE. Furthermore, $s \lesssim u \lesssim t$ and $s \lessdot_{io} u \lessdot_{io} t$. By Remark 3 after 17.23 we obtain T-DSPACE$(s) \subset$ T-DSPACE$(u) \subset$ T-DSPACE(t). ∎

Remark. In [Lis 75c] density has been defined for the classes $\Phi(\tau)$, $\tau \in \mathbb{P}_1$, and it has been proved that T-DSPACE is dense on $\mathbb{P}_1$.

We mention two further results from [BoCoHo 69] which are also proved in [You 71] by an easier construction.

16.21. Theorem.
1. There exists a Blum measure which is dense on $\mathbb{R}_1$, but which is not dense on the set of its recursive step counting functions.
2. There exists a Blum measure which is dense on the set of its recursive step counting functions, but which is not dense on $\mathbb{R}_1$. ∎

§ 17. Hierarchies

In this section we investigate the structure of complexity hierarchies, i.e. the structure of the partial order $(\{\Phi(t) : t \in \mathbb{R}_1\}, \subseteqq)$ of all complexity classes (with recursive bounding functions) for any given Blum measure Φ.

There are two main directions in the study of complexity hierarchies, namely to establish general hierarchy results for all Blum measures, and special results for Blum measures of particular importance, such as space and time measures.

In both cases the most important question is: what changes of the bounding function are sufficient to cause a change of the complexity class. It is an interesting fact that this does not depend so much on the complexity class as on the chosen bounding function (name) for this class.

Hierarchy results for specific Blum measures have many applications in all parts of the theory of computational complexity. This emphasizes the importance of this section, which is one of the classical parts of this theory.

Hierarchy results for honesty classes are not mentioned here. A survey on these results can be found in [Emd 74].

17.1. Hierarchies for Arbitrary Blum Measures

17.1.1. Upward Hierarchy Results

In this subsection we investigate the question of how to enlarge the name (i.e. the bounding function) of a complexity class in order to get a larger complexity class. This kind of hierarchy results will be called *upward hierarchy results*. First we show that it is always possible to find a larger complexity class for any given complexity class of a Blum measure (φ, Φ)*).

17.1. Theorem. [Rab 60] For every $\psi \in \mathbb{P}_1$ there is a partial recursive predicate $\chi \in \mathbb{P}_1$ such that $D_\chi = D_\psi$ and

$$\bigwedge_{\varphi_i = \chi} \bigwedge^{\text{ae}}_{x \in D_\psi} \Phi_i(x) > \psi(x).$$

Proof. In a diagonalization process we define the function χ, together with two auxiliary sequences $\{N_n : n \in \mathbb{N}\}$ and $\{K_n : n \in \mathbb{N}\}$ of sets of natural numbers, in such a way that χ will differ from all functions φ_j such that $\bigvee^{\text{io}}_{x \in D_\psi} \Phi_j(x) \leqq \psi(x)$. N_n will be the set of all indices j such that in the first n stages of the process $\chi(x) =_{\text{df}} 1 \dot{-} \varphi_j(x)$ has been defined for some x, and K_n will be the set of all arguments x for which $\chi(x)$ has already been defined in the first n stages.

Now we choose an index i_0 such that $\varphi_{i_0} = \psi$. The function χ can be computed by the following algorithm.

Stage 0. Let $\langle x, y \rangle = 0$.

If $\Phi_{i_0}(x) \leqq y$ **then** $\chi(x) = \begin{cases} 1 \dot{-} \varphi_0(x) & \text{if} \quad \Phi_0(x) \leqq \psi(x), \\ 0 & \text{otherwise}, \end{cases}$

$$N_0 = \begin{cases} \{0\} & \text{if} \quad \Phi_0(x) \leqq \psi(x), \\ \emptyset & \text{otherwise}, \end{cases}$$

$$K_0 = \{x\},$$

else $N_0 = \emptyset, \quad K_0 = \emptyset.$

Stage $n + 1$. Let $\langle x, y \rangle = n + 1$.

*) Throughout § 17.1 (φ, Φ) is an arbitrary Blum measure.

If $x \notin K_n$ **and** $\Phi_{i_0}(x) \leqq y$ **then** $M_{n+1} = \{j : j \leqq n + 1 \wedge j \notin N_n \wedge \Phi_j(x) \leqq \psi(x)\}$

$$\chi(x) = \begin{cases} 1 \dot{-} \varphi_{j_0}(x) & \text{if } M_{n+1} \neq \emptyset \wedge j_0 = \min M_{n+1}, \\ 0 & \text{if } M_{n+1} = \emptyset, \end{cases}$$

$$N_{n+1} = \begin{cases} N_n \cup \{j_0\} & \text{if } M_{n+1} \neq \emptyset \wedge j_0 = \min M_{n+1}, \\ N_n & \text{if } M_{n+1} = \emptyset, \end{cases}$$

$$K_{n+1} = K_n \cup \{x\},$$

else $N_{n+1} = N_n, \quad K_{n+1} =_{\text{df}} K_n$.

Obviously we have $D_\chi = D_\psi$. Since there is nothing to prove for a finite D_ψ we can assume that D_ψ is infinite.

Assume furthermore that there is an index i such that $\varphi_i = \chi$ and $\Phi_i(x) \leqq \psi(x)$ for infinitely many $x \in D_\psi$. Let $x_1, x_2, \ldots$ be an infinite sequence with $\Phi_i(x_t) \leqq \psi(x_t)$ for $t = 1, 2, \ldots$, and let $n_t = \langle x_t, \Phi_{i_0}(x_t)\rangle$ for $t = 1, 2, \ldots$ It is evident that the sequence $n_1, n_2, \ldots$ cannot be bounded by a constant. Let $n_{t_1} \geqq i$. Then $i \in M_{n_{t_1}}$. If $i = \min M_{n_{t_1}}$, then $\chi(x_t) = 1 \dot{-} \varphi_i(x_{t_1})$ contrary to the assumption. Otherwise $= \min M_{n_{t_1}} < i$, but this j will never appear in the further M_n's. Hence, for $n_{t_2} > n_{t_1}$ we have $i \in M_{n_{t_2}}$ but $j \notin M_{n_{t_2}}$. Now the same argument is repeated for n_{t_2}. After a finite number of steps it will appear that $i = \min M_{n_{t_r}}$ for some t_r and then $\chi(x_{t_r}) = 1 \dot{-} \varphi(x_{t_r})$, contradicting the assumption. ∎

The proof of Theorem 17.1 shows that the following effective version of this theorem also holds true.

17.2. Theorem. There is an $s \in \mathbb{R}_1$ such that for every $\varphi_j \in \mathbb{P}_1$

1. $\varphi_{s(j)}$ is a partial recursive predicate,
2. $D_{s(j)} = D_j$,
3. $\bigwedge_{\varphi_i = \varphi_{s(j)}} \bigwedge^{\text{ae}}_{x \in D_j} \Phi_i(x) > \varphi_j(x)$.

Proof. The procedure for computing $\chi(x)$ in the proof of Theorem 17.1 is recursive in x as well as in i_0 (the given index for the function ψ). Therefore there is an $\alpha \in \mathbb{P}_2$ such that $\alpha(i_0, x) = \chi(x)$. By the *s-m-n*-Theorem (2.5) there is an $s \in \mathbb{R}_1$ such that $\alpha(i_0, x) = \varphi_{s(i_0)}(x)$. ∎

17.3. Corollary. There is an $r \in \mathbb{R}_1$ such that card $D_i = \aleph_0$ implies $\boldsymbol{\Phi}(\varphi_i) \subset \boldsymbol{\Phi}(\varphi_{r(i)})$.

Proof. By the preceding theorem there is an $s \in \mathbb{R}_1$ with the following property: For every $i \in \mathbb{N}$ such that card $D_i = \aleph_0$ there is a set A such that $\varphi_{s(i)} = c_A$ and $A \notin \boldsymbol{\Phi}(\varphi_i)$. However, $A \in \boldsymbol{\Phi}(\Phi_{s(i)})$ and by the *s-m-n*-theorem there is an $s' \in \mathbb{R}_1$ such that $\Phi_i(x) = \Phi(i, x) = \varphi_{s'(i)}(x)$. Therefore we can choose $r =_{\text{df}} s' \circ s$. ∎

Remark. With $\boldsymbol{\Phi}(\varphi_i) \subset \boldsymbol{\Phi}(\varphi_{r(i)})$ we also have $\Phi(\varphi_i) \subset \Phi(\varphi_{r(i)})$. Therefore, we shall generally prove proper inclusions for the $\boldsymbol{\Phi}$-classes and equalities for the Φ-classes.

Corollary 17.3 ensures that for every $t \in \mathbb{R}_1$ there is a $t' \in \mathbb{R}_1$ such that $\boldsymbol{\Phi}(t) \subset \boldsymbol{\Phi}(t')$. The question arises of how large the difference between t and t' has to be, and whether this difference can be uniformly bounded for all t by a recursive function. The following theorem shows that the latter is not possible.

17.4. Theorem. For every $g \in \mathbb{R}_1$ such that $g \geqq \mathrm{id}$ there are arbitrarily large increasing functions $t \in \mathbb{R}_1$ such that $\Phi(t) = \Phi(g \circ t)$. ∎

Theorem 17.4 is a consequence of the following stronger theorem which states that there is no i such that Φ_i is infinitely often between t and $g \circ t$, i.e. between t and $g \circ t$ there is a gap of computations.

17.5. Gap Theorem. [Tra 67], [Bor 72]. For every $g \in \mathbb{R}_2$ such that $g(x, y) \geqq y$ there are arbitrarily large increasing functions $t \in \mathbb{R}_1$ such that

$$\bigwedge_i \bigwedge_x^{\mathrm{ae}} \left(\Phi_i(x) \leqq t(x) \vee \Phi_i(x) > g\big(x, t(x)\big)\right).$$

Proof. Let $s \in \mathbb{R}_1$ be an arbitrarily large function. We define

$$t(0) = s(0),$$

$$t(x+1) = \min \Big\{y \colon y \geqq s(x+1) \wedge y \geqq t(x) \wedge \bigwedge_{i \leqq x} \big(\Phi_i(x+1) \leqq y \vee \Phi_i(x+1) > g(x+1, y)\big)\Big\}.$$

Since $\Phi_i(x+1) > g(x+1, y)$ means $\Phi_i(x+1) \nleqq g(x+1, y)$ the set on which the minimum is taken is recursive. It remains to show that this set is not empty. Set

$$y_1 = \max \big\{s(x+1), t(x), \sup \{\Phi_i(x+1) \colon i \leqq x \wedge x+1 \in \mathrm{D}_i\}\big\}.$$

If $\Phi_i(x+1)$ is defined for $i \leqq x$, then $\Phi_i(x+1) \leqq y_1$, else $\Phi_i(x+1) > g(x+1, y_1)$. Consequently, y_1 is in the set and t is recursive.

Furthermore, the inequality $t(x) < \Phi_i(x) \leqq g\big(x, t(x)\big)$ can hold for $x < i$ only. Therefore $\bigwedge_{x \geqq i} \left(\Phi_i(x) \leqq t(x) \vee \Phi_i(x) > g\big(x, t(x)\big)\right)$. ∎

It is interesting to see how "small" (relative to g) an increasing function t can be found satisfying Theorem 17.5. A slight modification of the definition of the function t in the proof of this theorem yields:

17.6. Theorem. Let $g, s, r \in \mathbb{R}_1$ such that $g \geqq \mathrm{id}$, g, s increasing and $r(x+1) - r(x)$ increasing and unbounded. Then there is an increasing $t \in \mathbb{R}_1$ such that $s(x) \leqq t(x) \leqq g^{[r(x)]}\big(s(x)\big)$ for all x and

$$\bigwedge_i \bigwedge_x^{\mathrm{ae}} \left(\Phi_i(x) \leqq t(x) \vee \Phi_i(x) > g\big(t(x)\big)\right),$$

and consequently $\Phi(t) = \Phi(g \circ t)$. ∎

Examples. For every $\varepsilon > 0$ there are increasing $t_1, t_2, t_3 \in \mathbb{R}_1$ such that

$$\bigwedge_x x \cdot \log x \leqq t_1(x) \leqq (1+\varepsilon) \cdot x \cdot \log x \quad \text{and} \quad \Phi(t_1) = \Phi(t_1 + 1),$$

$$\bigwedge_x x^x \leqq t_2(x) \leqq x^{(1+\varepsilon)\cdot x} \quad \text{and} \quad \Phi(t_2) = \Phi(2t_2),$$

$$\bigwedge_x 2^{x^x} \leqq t_3(x) \leqq 2^{x^{(1+\varepsilon)\cdot x}} \quad \text{and} \quad \Phi(t_3) = \Phi(t_3^2)$$

The Gap Theorem states that for a given $t \in \mathbb{R}_1$ a function $t' \in \mathbb{R}_1$ with $\Phi(t) \subset \Phi(t')$ cannot be bounded uniformly (for all $t \in \mathbb{R}_1$) by $g \circ t$, where g is a recursive function. However, it might be possible for t' to be bounded uniformly by $F(t)$ where F was a total effective operator. (Note that Corollary 17.3 does not answer this question in the affirmative, since the function r in this corollary can give $\varphi_{r(i)} \neq \varphi_{r(j)}$ though $\varphi_i = \varphi_j$. Consequently, r does not define a total effective operator.) The following theorem shows that this is not possible.

17.7. Theorem. For every total effective operator F such that $F(t) \geqq t$ for all $t \in \mathbb{R}_1$ there are arbitrarily large increasing functions $t \in \mathbb{R}_1$ such that $\Phi(t) = \Phi(F(t))$. ∎

This theorem is a consequence of the following stronger theorem which states that there is no i such that $t <_{\text{io}} \Phi_i \leqq_{\text{ae}} F(t)$.

17.8. Operator Gap Theorem. [Con 72] For every effective operator F such that $F(t) \geqq t$ for all $t \in \mathbb{R}_1$ there are arbitrarily large increasing functions $t \in \mathbb{R}_1$ such that

$$\bigwedge_i \left(\Phi_i \leqq_{\text{ae}} t \vee \Phi_i >_{\text{io}} F(t)\right). \ ∎$$

We omit the rather complicated proof of this theorem.

Remark. The Operator Gap Theorem cannot be strengthened in such a way that $\bigwedge_i \bigwedge_x^{\text{ae}} \left(\Phi_i(x) \leqq t(x) \vee \Phi_i(x) > F(t)(x)\right)$ holds (analogously to the Gap Theorem). For examples see [Blu 67a] and [Con 72].

Since the functions $t' \in \mathbb{R}_1$ with $\Phi(t) \subset \Phi(t')$ cannot be uniformly bounded in t for all $t \in \mathbb{R}_1$ by a recursive function, the question arises whether this can be done for all t which are in a suitable subclass G of $\mathbb{R}_1$ large enough to contain a name for every complexity class.

This is really the case. First we show that such a uniform recursive bound can be found for a class $G \subseteqq \mathbb{R}_1$ if and only if G is a measured set.

17.9. Compression Theorem. [Blu 67a] The set $G = \{\gamma_i : i \in \mathbb{N}\}$ is measured if and only if there exists an $h \in \mathbb{R}_3$ and a $g \in \mathbb{R}_1$ such that for all i

a) $\varphi_{g(i)}$ is a partial recursive predicate such that $\mathrm{D}_{g(i)} = \mathrm{D}_{\gamma_i}$,

b) $\bigwedge\limits_{\varphi_j = \varphi_{g(i)}} \bigwedge\limits_{x \in \mathrm{D}_{g(i)}}^{\text{ae}} \Phi_j(x) > \gamma_i(x)$,

c) $\bigwedge\limits_{x \in \mathrm{D}_{g(i)}} \Phi_{g(i)}(x) \leqq h(i, x, \gamma_i(x))$.

Proof. 1. Let $\{\varphi_{r(i)} : i \in \mathbb{N}\}$ be a measured set and define $\gamma_i = \varphi_{r(i)}$. By Theorem 17.2 there exists an $s \in \mathbb{R}_1$ such that $\varphi_{s(r(i))}$ is a partial recursive predicate, $\mathrm{D}_{r(i)} = \mathrm{D}_{s(r(i))}$ and

$$\bigwedge_{\varphi_j = \varphi_{s(r(i))}} \bigwedge_{x \in \mathrm{D}_{r(i)}}^{\text{ae}} \Phi_j(x) > \varphi_{r(i)}(x).$$

Hence the function $g =_{\text{df}} s \circ r$ satisfies the conditions a) and b). Furthermore we define

$$h(i, x, y) = \begin{cases} \Phi_{g(i)}(x) & \text{if} \quad y \geqq \gamma_i(x), \\ 0 & \text{otherwise}, \end{cases}$$

and we get $\Phi_{g(i)}(x) \leqq h(i, x, \gamma_i(x))$, and $h \in \mathbb{R}_1$ because of $D_{\gamma_i} = D_{g(i)}$ and the decidability of "$y \geqq \gamma_i(x)$".

2. Let $\{\gamma_i : i \in \mathbb{N}\} \subseteq \mathbb{P}_1$, $h \in \mathbb{R}_3$ and $g \in \mathbb{R}_1$ such that a) and c) hold true. The predicate "$\gamma_i(x) = y$" can be decided as follows: Compute $h(i, x, y)$. If $\Phi_{g(i)}(x) > h(i, x, y)$, then $\gamma_i(x) \neq y$. If $\Phi_{g(i)}(x) \leqq h(i, x, y)$, then $x \in D_{g(i)} = D_{\gamma_i}$. Compute $\gamma_i(x)$ and check whether $\gamma_i(x) = y$. ∎

From the preceding theorem we easily obtain the following:

17.10. Theorem. Let $G \subseteq \mathbb{P}_1$ be a measured set.

1. There is an $h' \in \mathbb{R}_2$ which is increasing in the second argument such that for every $t \in G \cap \mathbb{R}_1$ there exists a set A such that

$$\bigwedge_{\varphi_j = c_A} \Phi_j >_{\text{ae}} t \quad \text{and} \quad \bigvee_{\varphi_j = c_A} \Phi_j \leqq_{\text{ae}} h' \,\square\, t,$$

and consequently $\Phi(t) \subset \Phi(h' \,\square\, t)$.

2. For every increasing unbounded $t' \in \mathbb{R}_1$ there is an increasing $h'' \in \mathbb{R}_1$ such that for every $t \in G \cap \mathbb{R}_1$, $t \geqq t'$, there exists a set A such that

$$\bigwedge_{\varphi_j = c_A} \Phi_j >_{\text{ae}} t \quad \text{and} \quad \bigvee_{\varphi_j = c_A} \Phi_j \leqq_{\text{ae}} h'' \circ t,$$

and consequently $\Phi(t) \subset \Phi(h'' \circ t)$.

Proof. Ad 1. Define $h'(x, y) = \max_{i \leqq x} h(i, x, y)$, where h is the function defined in the proof of Theorem 17.9. Since h is increasing in the third argument, the function h' is increasing in the second argument.

Ad 2. Define $h''(y) = \max_{t'(x) \leqq y} h'(x, y)$, where h' is the function defined in the proof of statement 1. ∎

The Naming Theorem (15.21) shows that there is a measured set G such that

$$\{\Phi(t) : t \in \mathbb{R}_1\} = \{\Phi(t) : t \in G \cap \mathbb{R}_1\}.$$

This and Theorem 17.10 yield the following theorem:

17.11. Theorem. There is an $h \in \mathbb{R}_2$ and an $s \in \mathbb{R}_1$ such that for every $\varphi_i \in \mathbb{R}_1$

$$\Phi(\varphi_i) = \Phi(\varphi_{s(i)}) \subset \Phi(h \,\square\, \varphi_{s(i)}),$$

and

$$\Phi(\varphi_i) = \Phi(\varphi_{s(i)}) \subset \Phi(h \,\square\, \varphi_{s(i)}). ∎$$

Finally, we investigate a question which arises from Theorem 17.10. This theorem shows that for a measured set G there is an $h \in \mathbb{R}_1$ such that for sufficiently large $t \in G \cap \mathbb{R}_1$ the condition $t' \geqq_{\text{ae}} h \circ t$ implies $\Phi(t) \subset \Phi(t')$. The question of whether the condition $t' \geqq_{\text{ae}} h \circ t$ can be replaced by the condition $t' \geqq_{\text{io}} h \circ t \wedge t' \geqq t$ must be answered in the negative because of the following theorem. For this theorem we need the notion of s-immunity and a lemma concerning this notion.

Let $s \in \mathbb{R}_1$. A set A is called strongly s-immune iff

$$\bigwedge_{A'} (A' \setminus A \text{ finite} \wedge A' \text{ infinite} \rightarrow \Phi\text{-Comp } A' >_{\text{io}} s).$$

17.12. Lemma. [Con 71a], [FlSt 74b]. For every sufficiently large $s \in \mathbb{R}_1$ there exists a strongly s-immune set. ∎

Now we are ready to prove the theorem:

17.13. Theorem. [Con 71a], [McMe 69] Let $h \in \mathbb{R}_1$ such that $h \geqq \text{id}$. Then

$$\bigwedge_{t \in \mathbb{R}_1} \bigvee_{t' \in \mathbb{R}_1} \big(t' \geqq_{\text{io}} h \circ t \wedge t' \geqq t \wedge \Phi(t) = \Phi(t')\big).$$

Proof. Let A be an arbitrary infinite recursive set. Define

$$t'(x) = \begin{cases} t(x) & \text{if } x \notin A, \\ h\big(t(x)\big) & \text{if } x \in A. \end{cases}$$

The idea of the proof is the following: If $t <_{\text{io}} \Phi_j \leqq_{\text{ae}} t'$, then the complexity of $A' =_{\text{df}} \{x : t(x) < \Phi_j(x)\}$ (note that $A' \setminus A$ is finite and A' is infinite) can be bounded by a function depending only on t. But A can be chosen to be a w-immune set where w can be chosen arbitrarily large. Therefore it can be achieved that the complexity of A' is arbitrarily high. This contradicts the boundedness of the complexity of A'.

More precisely, we conclude as follows: Because of the definition of t' we have $t \leqq t' \leqq h \circ t$ and $t' \geqq_{\text{io}} h \circ t$. Assume that $\Phi(t) \subset \Phi(t')$. Then there is a $j \in \mathbb{N}$ such that $D_j = \mathbb{N}$ and $t <_{\text{io}} \Phi_j \leqq_{\text{ae}} t'$. Since $\{\Phi_i : i \in \mathbb{N}\}$ is a measured set there exists an $r \in \mathbb{R}_1$ and an $h' \in \mathbb{R}_2$, which is increasing in the second argument, such that $\Phi_i = \varphi_{r(i)}$ and $\Phi_{r(i)} \leqq_{\text{ae}} h' \,\square\, \varphi_{r(i)} = h' \,\square\, \Phi_i$ (see 5.26).

In particular, we have $\Phi_{r(j)} \leqq_{\text{ae}} h' \,\square\, \Phi_j \leqq_{\text{ae}} h' \,\square\, t' \leqq h' \,\square\, (h \circ t)$. By the s-m-n-Theorem there is an $s \in \mathbb{R}_1$ such that

$$\text{sgn}\,\big(\varphi_i(x) \dot{-} \varphi_l(x)\big) = \varphi_{s(i,l)}(x).$$

Now define $A' = \{x : t(x) < \Phi_j(x)\}$. Obviously, $A' \setminus A$ is finite and A' is infinite. Furthermore, choose $k \in \mathbb{N}$ such that $\varphi_k = t$. We have

$$c_{A'}(x) = \text{sgn}\,\big(\Phi_j(x) \dot{-} t(x)\big) = \text{sgn}\,\big(\varphi_{r(j)}(x) \dot{-} \varphi_k(x)\big) = \varphi_{s(r(j),k)}(x)$$

and 5.27 yields a recursive v which is increasing in its second and third argument such that

$$\begin{aligned} \Phi_{s(r(j),k)}(x) &\leqq_{\text{ae}} v\big(x, \Phi_{r(j)}(x), \Phi_k(x)\big) \\ &\leqq_{\text{ae}} v\Big(x, h'\big(x, h(t(x))\big), \Phi_k(x)\Big). \end{aligned} \qquad (*)$$

The function $w(x) =_{\text{df}} v\Big(x, h'\big(x, h(t(x))\big), \Phi_k(x)\Big)$ depends only on the function t and is independent of A, j and A' (j and A' are derived from A). Therefore we can choose A to be a w-immune set. Since A' is an infinite subset of A and $s\big(r(j), k\big)$ is an index of $c_{A'}$ we get $\Phi_{s(r(j),k)} >_{\text{io}} w$, contradicting $(*)$. ∎

17.1.2. Downward Hierarchy Results

In this subsection we investigate the question of how to diminish the name of a complexity class in order to get a smaller complexity class. This kind of hierarchy results will be called *downward hierarchy results.* At first glance this question seems to be more or less identical to the question for upward hierarchy results. However, we find that there are essential differences between these two questions. Generally it can be said that the knowledge we possess at present about downward hierarchy results is not as satisfactory as our knowledge about upward hierarchy results.

Similar to the upward hierarchy result **17.11**, there is a set $G \subseteqq \mathbb{R}_1$ containing arbitrarily large functions, and a recursive function $h \leqq \mathrm{id}$ such that $\Phi(h \circ t) \subset \Phi(t)$ for all $t \in G$ (this is a consequence of **17.15**). However, the following theorem shows that there is no such set G which contains a name for every recursively named complexity class. This means that a downward analogue of the (upward) Compression Theorem (17.9) does not hold.

17.14. Theorem. [Bor 72], [Con 71a] Let $h \in \mathbb{R}_1$ be an unbounded increasing function such that $h \leqq \mathrm{id}$. Then there are arbitrarily large increasing functions $t \in \mathbb{R}_1$ such that

$$\bigwedge_{t' \in \mathbb{R}_1} \big(\Phi(t) = \Phi(t') \rightarrow \Phi(h \circ t') = \Phi(t')\big).$$

Proof. Let h_1 be an increasing function such that

$$h_1\big(h(x)\big) > x \tag{1}$$

(for example choose $h_1(x) =_{\mathrm{df}} \min \{y : h(y) > x\}$). Since $\{\Phi_i : i \in \mathbb{N}\}$ is a measured set and h_1 is increasing, the set $\{h_1 \circ \Phi_i : i \in \mathbb{N}\}$ is measured too. By Theorem 17.10.1 there is an $h_2 \in \mathbb{R}_2$ with the following properties: h_2 is increasing in the second argument and for every $\Phi_i \in \mathbb{R}_1$ there is an $f_i \in \mathbb{R}$ such that

$$\bigwedge_{\varphi_j = f_i} \Phi_j \geqq_{\mathrm{ae}} h_1 \circ \Phi_i \quad \text{and} \quad \bigvee_{\varphi_j = f_i} \Phi_j \leqq_{\mathrm{ae}} h_2 \,\square\, (h_1 \circ \Phi_i). \tag{2}$$

The function $h_3(x, y) =_{\mathrm{df}} \max \{h_2(x, y), y\}$ is also increasing in the second argument, and we have $h_3(x, y) \geqq y$, $h_3(x, y) \geqq h_2(x, y)$ and therefore

$$\bigvee_{\varphi_j = f_i} \Phi_j \leqq_{\mathrm{ae}} h_3 \,\square\, (h_1 \circ \Phi_i). \tag{3}$$

Because of (1) we get

$$g(x, y) =_{\mathrm{df}} h_3\big(x, h_1(y)\big) \geqq h_3\Big(x, h_1\big(h(y)\big)\Big) \geqq h_3(x, y) \geqq y.$$

Now by the Gap Theorem (17.5) there are arbitrarily large increasing $t \in \mathbb{R}_1$ such that

$$\Phi_i \leqq_{\mathrm{ae}} g \,\square\, t = h_3 \,\square\, (h_1 \circ t) \quad \text{implies} \quad \Phi_i \leqq_{\mathrm{ae}} t. \tag{4}$$

Assume that there is a $t \in \mathbb{R}_1$ satisfying (4) and a $t' \in \mathbb{R}_1$ such that

$$\Phi(t) = \Phi(t') \quad \text{and} \quad \Phi(h \circ t') \subset \Phi(t'). \tag{5}$$

Hence there is a $\Phi_k \in \mathbb{R}_1$ such that

$$h \circ t' <_{\text{io}} \Phi_k \leqq_{\text{ae}} t. \tag{6}$$

Because of (2), (6) and (1) we have for every $\varphi_j = f_k$

$$\Phi_j \geqq_{\text{ae}} h_1 \circ \Phi_k \geqq_{\text{io}} h_1 \circ h \circ t' > t'$$

and therefore

$$f_k \notin \Phi(t'). \tag{7}$$

Because of (3) there is a $\varphi_j = f_k$ such that $\Phi_j \leqq_{\text{ae}} h_2 \circ (h_1 \circ \Phi_k)$. By (6) and (4) we conclude $\Phi_j \leqq_{\text{ae}} t$ and $f_k \in \Phi(t)$. By (5) we get $f_k \in \Phi(t')$ which contradicts (7). ∎

There is still another difference between the upward and downward hierarchy results. Theorem 17.13 says that there is no $t \in \mathbb{R}_1$ such that $t' \geqq_{\text{io}} h \circ t \wedge t' \geqq t$ implies $\Phi(t) \subset \Phi(t')$ for a suitable $h \in \mathbb{R}_1$. Contrary to this result we have the following downward hierarchy result.

17.15. Theorem. [Con 71a] There is a recursive $h \leqq \text{id}$ and there are arbitrarily large functions t such that

$$\bigwedge_{t' \in \mathbb{R}_1} \big(t' \leqq_{\text{io}} h \circ t \wedge t' \leqq t \to \Phi(t') \subset \Phi(t)\big).$$

Proof. By Theorem 17.10.2 and the Naming Theorem (15.21) there is a measured set G and a recursive increasing $h' \geqq \text{id}$ such that

$$\bigwedge_{A \in \mathbf{REC}} \ \bigvee_{\substack{g \in G \cap \mathbb{R}_1 \\ g \geqq \text{id}}} A \in \Phi(g)$$

and

$$\bigwedge_{\substack{g \in G \cap \mathbb{R}_1 \\ g \geqq \text{id}}} \ \bigvee_{B \in \mathbf{REC}} \Big(\bigwedge_{\varphi_j = c_B} \Phi_j >_{\text{ae}} g \wedge \bigvee_{\varphi_j = c_B} \Phi_j \leqq_{\text{ae}} h' \circ g\Big). \tag{1}$$

Defining $h(x) = \min\{y : h'(y) \geqq x\}$ we get $h(x) \leqq x$ and $h\big(h'(x)\big) \leqq x$. Now we show that

1. $G' =_{\text{df}} \{h' \circ g : g \in G \cap \mathbb{R}_1 \wedge g \geqq \text{id}\}$ contains arbitrarily large functions and
2. for all $t \in G'$

$$\bigwedge_{t' \in \mathbb{R}_1} \big(t' \leqq_{\text{io}} h \circ t \wedge t' \leqq t \to \Phi(t') \subset \Phi(t)\big).$$

Ad 1. Let $s \in \mathbb{R}_1$ be an arbitrary function. Define $r(x) = \min\{y : h'(y) \geqq s(x)\}$. By Theorem 17.1 there is a recursive A such that $\bigwedge_{\varphi_i = c_A} \Phi_i >_{\text{ae}} r$. Because of (1) there is a $g \in G \cap \mathbb{R}_1$, $g \geqq \text{id}$, such that $A \in \Phi(g)$. Consequently, there is a $\varphi_k = c_A$ such that $\Phi_k \leqq_{\text{ae}} g$, and we get

$$h'\big(g(x)\big) \geqq_{\text{ae}} h'\big(\Phi_k(x)\big) \geqq_{\text{ae}} h'\big(r(x)\big) \geqq s(x).$$

Ad 2. Let $h' \circ g \in G'$ and $t' \in \mathbb{R}_1$ such that $t' \leqq_{\text{io}} h \circ h' \circ g$ and $t' \leqq h' \circ g$. By (1) there is a recursive set B such that

$$\bigwedge_{\varphi_j = c_B} \Phi_j >_{\text{ae}} g \quad \text{and} \quad \bigvee_{\varphi_j = c_B} \Phi_j \leqq_{\text{ae}} h' \circ g.$$

Consequently, $\varphi_j = c_B$ implies $\Phi_j >_{\text{ae}} g \geqq h \circ h' \circ g \geqq_{\text{io}} t'$ and therefore $B \in \Phi(h' \circ g) - \Phi(t')$. The hint that $t' \leqq h' \circ g$ completes the proof. ▮

17.1.3. Minimal Growth Rate

In this subsection we state some results on gaps in the "low end" of the complexity hierarchies. A Blum measure (φ, Φ) is said to have a *minimal growth rate* if there is a function $t \in \mathbb{R}_1$ such that $\underline{\lim}_{n \to \infty} t(n) = \infty$ and $\Phi_i \geqq_{\text{io}} t$ for all unbounded $\Phi_i \in \mathbb{R}_1$. Our first theorem shows that every Blum measure has a minimal growth rate.

17.16. Theorem. [McMe 69] There is a $t \in \mathbb{R}_1$ such that $\underline{\lim}_{n \to \infty} t(n) = \infty$ and

$$\bigwedge_i (\Phi_i \text{ unbounded on } D_i \to \Phi_i \geqq_{\text{io}} t). \ ▮$$

17.17. Corollary. There is a $t \in \mathbb{R}_1$ such that $\underline{\lim}_{n \to \infty} t(n) = \infty$ and

$$\Phi(t) = \Phi(\text{Const}). \ ▮$$

Given certain suppositions the function t can be chosen to be increasing.

17.18. Theorem. [Bor 72] If the predicate "$\Phi_i < k$" is decidable, then there is an unbounded increasing $t \in \mathbb{R}_1$ such that

1. $\bigwedge_{\Phi_i \in \mathbb{R}_1} (\Phi_i \text{ unbounded and increasing} \to \Phi_i \geqq t)$,
2. $\bigwedge_{\Phi_i \in \mathbb{R}_1} (\Phi_i \text{ unbounded} \to \Phi_i \geqq_{\text{io}} t)$. ▮

17.19. Corollary. If the predicate "$\Phi_i < k$" is decidable, then there is an unbounded increasing $t \in \mathbb{R}_1$ such that

$$\Phi(t) = \Phi(\text{Const}). \ ▮$$

Remark. It is not hard to see that the predicate "$\Phi_i < k$" is decidable for Φ = 2-T-DSPACE, 1-T-DSPACE. And actually, for these measures we already know increasing minimal growth rates with $t = \log\log$ and $t = \log$, resp. (cf. 8.28 and 8.3).

17.1.4. The Structure of Complexity Hierarchies

In this subsection we state some results on the order-theoretical structure of $\mathcal{P}(\Phi) =_{\text{df}} (\{\Phi(t) : t \in \mathbb{R}_1\}, \subseteqq)$. It is a simple consequence of Theorem 17.1 that there are chains of type ω in $\mathcal{P}(\Phi)$. But the structure of $\mathcal{P}(\Phi)$ is still more complex.

17.20. Theorem. [McCr 69] Every countable partial order can be order-isomorphically embedded in $\mathscr{P}(\Phi)$. ∎

This result has been strengthened in the sense that the complexity classes which are images of the order-isomorphism are seperated by an arbitrarily large total effective operator.

17.21. Theorem. [Aln 73a], [Mol 76] Let $[\mathbb{N}, \prec]$ be a partial order, $r \in \mathbb{R}_1$ and let F be a total effective operator. Then there is an $s \in \mathbb{R}_1$ such that

(1) $\varphi_{s(i)} \in \mathbb{R}_1$ and $\mathrm{R}_{s(i)} \subseteqq \{0, 1\}$,

(2) $\Phi_{s(i)}$ is strictly increasing,

(3) $\varphi_{s(i)} \notin \Phi(r)$,

(4) $j \prec i \to \Phi(F(\Phi_{s(j)})) \subset \Phi(\Phi_{s(i)})$,

(5) $j \not\preceq i \to \Phi(\Phi_{s(j)}) \not\subseteqq \Phi(F(\Phi_{s(i)}))$. ∎

A stronger result can be found in [Aln 73a].

17.2. Hierarchies for Specific Measures

In § 17.1.1 we have seen that for every Blum measure Φ and every measured set G there is a function $h \in \mathbb{R}_1$ such that $\Phi(t) \subset \Phi(h \circ t)$ for all sufficiently large $t \in G \cap \mathbb{R}_1$. The main question in this subsection is to find, for specific measures Φ (especially space and time measures) and suitable measured sets G, small functions $h \in \mathbb{R}_1$ such that $h \geqq \mathrm{id}$ and $\Phi(t) \subset \Phi(h \circ t)$ or $h \leqq \mathrm{id}$ and $\Phi(h \circ t) \subset \Phi(t)$. Since a set $G \subseteqq \mathbb{P}_1$ is measured if and only if the complexity of its functions can be uniformly bounded by the functions themselves (cf. 5.26), the sets G we want to use have to contain only functions which are in some sense honest. Indeed, we shall use sets G with a property like $G \cap \mathbb{R}_1 = \{t : t$ is Φ-constructible$\}$, $\{t : t$ is fully Φ-constructible$\}$, $\{t : t(|w|) \in \Phi(t)\}$ or other similar ones.

17.2.1. Deterministic Space Measures

We start with the space measure for 2-way TM's. The proof of the following theorem is based on diagonalization arguments and will serve as a model for many proofs in this and the following subsections. The following lemma is necessary for this theorem.

17.22. Lemma. [Sip 78] Let $M_0, M_1, M_2, \ldots$ be a Gödel numbering of all 2-DTM's and let 2-T-DSPACE be the corresponding space measure. There is an $r \in \mathbb{R}_1$ such that

1. $\varphi_{r(i)} = \varphi_i$,
2. $\text{2-T-DSPACE}_{r(i)} = \text{2-T-DSPACE}_i$,
3. the computation of $M_{r(i)}$ do not have nontrivial cycles, i.e. if $K_0, K_1, K_2, \ldots$ is the computation of $M_{r(i)}$ on an input w, then there are no n_1, n_2, n_3 such that $n_1 < n_2 < n_3$, $K_{n_1} = K_{n_3}$ and $K_{n_1} \neq K_{n_2}$. ∎

17.23. Theorem. [StHaLe 65] If s is 2-T-DSPACE-constructible, then

$$\text{2-T-DSPACE}(s) \setminus \bigcup_{s' \prec_{\text{io}} s} \text{2-T-DSPACE}(s') \neq \emptyset.$$

Proof. We first prove the theorem using the stronger supposition of full 2-T-DSPACE constructibility, since this proof is more transparent.

Let $\{M_i\}$ be a Gödel numbering of all 2-DTM's and 2-T-DSPACE the corresponding space measure. By Lemma 17.22 there is an r such that $\varphi_i = \varphi_{r(i)}$, $\text{2-T-DSPACE}_i = \text{2-T-DSPACE}_{r(i)}$ and $M_{r(i)}$ does not cycle. Now we construct a 2-DTM M which works in 4 stages.

Stage 1. We choose $k \in \mathbb{N}$ such that $\text{2-T-DSPACE}_k(w) = s(|w|)$ for all w. During the first stage M works like M_k, but it marks all tape squares scanned by the worktape head. Thus, at the end of this stage a piece of length $s(|w|)$ is marked on the worktape of M if w is the input.

Stage 2. If 0^m1^i is the final part of length $\min(|w|, s(|w|))$ of the input word w (for some $m, i \in \mathbb{N}$), then M writes the binary presentation of i on its worktape. Otherwise M rejects w.

Stage 3. Let D_i be the space used for the computation of the program code of $M_{r(i)}$. If $D_i \leqq s(|w|)$, then M computes this code and writes it on the first track of the worktape. Otherwise M rejects w.

Stage 4. On the second track of its worktape M simulates the work of $M_{r(i)}$ on w, where the worktape symbols and states of $M_{r(i)}$ are encoded by words from $\{0, 1\}^{l_i}$ (for a suitable l_i). Therefore, if $M_{r(i)}$ needs S tape squares for its work on w, then M needs $l_i.S$ tape squares for the complete simulation of this work. However, M simulates the work of $M_{r(i)}$ only as long as no more than $s(|w|)$ tape squares of its tape are needed, i.e. as long as $M_{r(i)}$ does not need more than $\lfloor s(|w|)/l_i \rfloor$ tape squares. Since $M_{r(i)}$ does not cycle, it either stops within this number of tape squares or leaves this zone. If $M_{r(i)}$ stops, then M can simulate the work of $M_{r(i)}$ on w completely and accepts w iff $M_{r(i)}$ rejects it. If $M_{r(i)}$ leaves the given amount of tape squares, then M rejects w.

Evidently, M accepts a certain language L within space s. We assume that L can also be accepted by a 2-DTM M_j within space s' such that $s' \prec_{\text{io}} s$. We choose m in such a way that $n =_{\text{df}} m + j$ fulfils the inequalities

1. $j \leqq s(n)$,
2. $D_j \leqq s(n)$ and
3. $l_j \cdot s'(n) \leqq s(n)$.

Now consider the work of M on the input word 0^m1^j. Since $j \leqq s(n)$ the final part of length $\min(n, s(n))$ is $0^{m'}1^j$ for some $m' \geqq 0$. Consequently, the binary presentation of j will be written on the tape during stage 2. Since $D_j \leqq s(n)$ the program code of $M_{r(j)}$ can be computed in stage 3. Since $l_j \cdot s'(n) \leqq s(n)$ the computation of $M_{r(j)}$ can be completely simulated by M. Consequently we have

$$0^m1^j \in L \leftrightarrow 0^m1^j \in \mathrm{L}(M_j) \leftrightarrow 0^m1^j \in \mathrm{L}(M_{r(j)}) \leftrightarrow 0^m1^j \notin \mathrm{L}(M).$$

This contradicts the fact that M accepts L.

The proof using the weaker supposition of 2-T-DSPACE-constructibility has basically the same structure as that above. The main difference can be outlined as follows: the number i is determined as the smallest number not greater than $|w|$ and such that

1. $D_i \leqq$ 2-T-DSPACE$_k(w)$ and $l_i \cdot$ 2-T-DSPACE$_i(w) \leqq$ 2-T-DSPACE$_k(w)$,
2. this number i has not been used during the work of M on a smaller input word (with respect to some given order of Σ^* which is compatible with the word length) and this can be checked within space 2-T-DSPACE$_k(w)$. ∎

Remark 1. Because of the linear speed-up for the measure 2-T-DSPACE (cf. 18.11) the condition $s' \prec_{io} s$ in Theorem 17.23 cannot be weakened.

Remark 2. In the "low end" of the 2-T-DSPACE hierarchy we have **2-T-DSPACE**(0) = **2-T-DSPACE**(s) for $s \prec \log\log$ by 8.28. The proof of 8.28 shows that there is no i such that $0 \prec_{io}$ 2-T-DSPACE$_i \prec \log\log$. Consequently, the supposition of Theorem 17.23 is not fulfilled for functions s such that $0 \prec_{io} s \prec \log\log$. However, there are fully 2-T-DSPACE-constructible s such that $0 \prec_{io} s \preceq \log\log$ (cf. 5.6.3). Consequently, **2-T-DSPACE**(0) $\subset$ **2-T-DSPACE**(log log) (see Remark 1 after 8.29).

Remark 3. Under stronger suppositions than those made in Theorem 17.23 it can be shown that even subsets of 1^* are in **2-T-DSPACE**(s) $\setminus$ **2-T-DSPACE**(s'). Such suppositions are

a) s is fully 2-T-DSPACE-constructible and $s' \prec s$ (see [Sei 77b]) or
b) s, s' are fully 2-T-DSPACE-constructible and $s' \prec_{io} s$ (see [Sei 77b]), or
c) s is fully 2-T-DSPACE-constructible and there is an $a \in \mathbb{R}_1$ such that $s \circ a \prec s' \circ a$ (see [HaBe 75]).

17.24. Corollary. Let s be 2-T-DSPACE-constructible.

1. $s' \prec_{io} s \leftrightarrow$ **2-T-DSPACE**(s) $\setminus$ **2-T-DSPACE**(s') $\neq \emptyset$,
2. $s' \preceq s$ implies

$$s' \prec_{io} s \leftrightarrow \textbf{2-T-DSPACE}(s') \subset \textbf{2-T-DSPACE}(s). \ \blacksquare$$

Theorem 17.23 is a downward hierarchy result. The corresponding upward hierarchy result is somewhat weaker.

17.25. Theorem. If s is 2-T-DSPACE-constructible and unbounded, and if $g \in \mathbb{R}_1$ is increasing and unbounded, then

$$\textbf{2-T-DSPACE}(s) \subset \textbf{2-T-DSPACE}(s \cdot g).$$

Proof. Define $g'(n) = \max(1, g(\sup\{m : m \leqq n \wedge \log m \leqq s(n) \wedge g(m)$ can be computed within space $s(n)\}))$. Since s is 2-T-DSPACE-constructible, the function $s \cdot g'$ is also 2-T-DSPACE-constructible. Since s and g are unbounded, g' is also unbounded. Consequently, $s \prec_{io} s \cdot g'$ and by Theorem 17.23 we have **2-T-DSPACE**(s) $\subset$ **2-T-DSPACE**($s \cdot g'$). Since g is increasing, we have $g' \leqq g$ and hence **2-T-DSPACE**(s) $\subset$ **2-T-DSPACE**($s \cdot g$). ∎

Remark. The supposition "s is unbounded" actually means $s \succeq_{io} \log\log$ (see 8.28 or Remark 2 after 17.23).

Now we investigate the space measure for 1-T-DTM's.

17.26. Lemma. Let $M_0, M_1, M_2, \ldots$ be a Gödel numbering of all 1-DTM's and let 1-T-DSPACE be the corresponding space measure. There is an $r \in \mathbb{R}_1$ such that

1. $\varphi_{r(i)} = \varphi_i$,
2. $\text{1-T-DSPACE}_{r(i)} = \max\{\text{1-T-DSPACE}_i, \log\}$,
3. the computations of $M_{r(i)}$ do not have nontrivial cycles. ∎

17.27. Theorem. [StHaLe 65] If $s \succeq \log$ is fully 2-T-DSPACE-constructible, then

$$\textbf{1-T-DSPACE}(s) \setminus \bigcup_{s' \prec_{io} s} \textbf{1-T-DSPACE}(s') \neq \emptyset .$$

Proof. We proceed as in the proof of 17.23 (using Lemma 17.26 instead of Lemma 17.22) where stage 2, which is executed first, is modified as follows: M reads the input w. If $w = 0^m 1^i$ for some $m, i \in \mathbb{N}$, then the binary notation of m and i is written on the worktape. Otherwise M rejects w.

Now the stages 1, 3 and 4 described in the proof of 17.23 are executed, M using the binary notations of m and i on the worktape instead of $0^m 1^i$ on the input tape. ∎

Remark. Corollary 8.4 states that $\textbf{1-T-DSPACE}(0) = \textbf{1-T-DSPACE}(s)$ for $s \prec \log$. The set $\{0^n 1^n : n \in \mathbb{N}\}$ is a witness for $\textbf{1-T-DSPACE}(0) \subset \textbf{1-T-DSPACE}(\log)$.

Because of the linear speed-up for the measure 1-T-DSPACE (cf. 18.11), the condition $s' \prec_{io} s$ in Theorem 17.27 cannot be weakend.

17.28. Corollary. Let s be fully 2-T-DSPACE-constructible.

1. $s' \prec_{io} s \leftrightarrow \textbf{1-T-DSPACE}(s) \setminus \textbf{1-T-DSPACE}(s') \neq \emptyset$,
2. $s' \preceq s$ implies

$$s' \prec_{io} s \leftrightarrow \textbf{1-T-DSPACE}(s') \subset \textbf{1-T-DSPACE}(s). \; ∎$$

Analogously to Theorem 17.25 one can show the following theorem:

17.29. Theorem. If s is fully 2-T-DSPACE-constructible and unbounded, and if $g \in \mathbb{R}_1$ is increasing and unbounded, then

$$\textbf{1-T-DSPACE}(s) \subset \textbf{1-T-DSPACE}(s \cdot g). \; ∎$$

Remark. Since the measures 1-T-DSPACE and 2-T-DSPACE coincide for bounding functions $s \geqq \text{id}$, the importance of 17.27, 17.28 and 17.29 lies in the range $s \prec_{io} \text{id}$. Furthermore, the measure T-DSPACE need not be treated separately, since it is defined for bounding functions $s \geqq \text{id}$ only, and in this range it coincides with the measure 2-T-DSPACE.

Now we state some hierarchy results for space measures of Turing machines with restricted cardinality of the worktape alphabet. The proofs are essentially the same as in the case of Turing machines with unrestricted cardinality of the worktape alphabet. The difference to the proof of 17.23 and 17.27 is that the simulating machine M can simulate the other machines without encoding their worktape symbols. Only their states must be encoded. Therefore there is a $c > 1$ such that $l_i < c$ for all $i \in \mathbb{N}$ (see stage 4 of the proofs). Consequently we have the following theorem:

17.30. Theorem. Let $k \geqq 1$ and $l \geqq 3$. There is a $c > 1$ such that for every 2-kT$_l$-DSPACE-constructible unbounded $s \in \mathbb{R}_1$,

1. $\textbf{2-kT}_l\textbf{-DSPACE}\left(\frac{1}{c} \cdot s\right) \subset \textbf{2-kT}_l\textbf{-DSPACE}(s) \subset \textbf{2-kT}_l\textbf{-DSPACE}(c \cdot s)$.

2. $\textbf{1-kT}_l\textbf{-DSPACE}\left(\frac{1}{c} \cdot s\right) \subset \textbf{1-kT}_l\textbf{-DSPACE}(s) \subset \textbf{1-kT}_l\textbf{-DSPACE}(c \cdot s)$,

   for $s \succeq \log$. ∎

Remark 1. For the case $s(n) = a \cdot n^r$ (where a is a rational number and $r \in \mathbb{N}$) it is shown in [IbSa 75] that the proper inclusions of Theorem 17.30.2 are valid for all $c > 1$.

Remark 2. A separate treatment of i-multiT$_l$-DSPACE is not necessary because of **i-multiT$_l$-DSPACE**(s) = **i-T-DSPACE**(s) for all $s \geqq 0$, $l \in \mathbb{N}$ and $i = 1, 2$.

Remark 3. In [Žák 79] a hierarchy result for a space measure for 2-kT$_l$-DMs which also takes the number of states into account is investigated.

We conclude this subsection with a result on the 2-auxPD-T-DSPACE hierarchy.

17.31. Theorem. If $s \geqq \log$ is fully 2-T-DSPACE-constructible, then

$$\textbf{2-auxPD-T-DSPACE}(s) \setminus \bigcup_{s' \prec_{io} s} \textbf{2-auxPD-T-DSPACE}(s') \neq \emptyset.$$

Proof. First, **2-auxPD-T-DSPACE**(s) = **multiT-DTIME**$(2^{\mathrm{Lin}\, s})$ by 20.13. Furthermore, if s is fully 2-T-DSPACE-constructible, then $2^{k \cdot s(|w|)} \in$ multiT-DTIME$(2^{k \cdot s} \cdot k \cdot s)$ for some $k \geqq 2$, and $s' \prec_{io} s$ implies $2^{l \cdot s'(n)} \prec_{io} 2^{k \cdot s(n)}$ for all $l \geqq 0$. Consequently, by 17.40 there is a language in **multiT-DTIME**$(2^{k \cdot s}) \setminus \bigcup_{s' \prec_{io} s}$ **multiT-DTIME**$(2^{\mathrm{Lin}\, s'})$. ∎

Remark. By similar arguments it can be shown that Theorem 17.31 remains valid if PD is replaced by S, NES or C.

17.2.2. Nondeterministic Space Measures

It is not hard to see that for the proof of hierarchy results for nondeterministic measures the diagonalization arguments fail in general. Most of the hierarchy results for nondeterministic measures are proved using translational techniques as, for instance, the following theorem:

17.32. Theorem. [Sei 77b] If s is fully 2-T-DSPACE-constructible and
1. $s'(n + 1) \prec s(n)$ or 2. $s'(n + 1) \preceq s(n)$ and $s' \prec s$, then

$$\textbf{2-T-NSPACE}(s) \setminus \textbf{2-T-NSPACE}(s') \neq \emptyset.$$

Proof. First we suppose $s'(n + 1) \prec s(n)$. From this we conclude $s \succ 0$ and by 5.6.2 we get $s \succeq \log$. If $s' \prec_{io} \log$, then we have **2-T-NSPACE**$(s) \setminus$ **2-T-NSPACE**$(s') \neq \emptyset$ by 8.30. Hence we can suppose $s' \succeq \log$.

Let the programs of 2-NTM's be encoded in such a way that the set PC of all program codes is a regular subset of 01 {00, 11}* 10. By M_e we denote the 2-NTM whose program has the code e. Obviously, there is a "universal" 2-NTM U such that

$$\mathrm{L}(U) = \{ex : e \in PC \wedge x \in L(M_e)\}$$

and

$$\bigwedge_{e \in PC} \bigvee_{c_e \in \mathbb{N}} \bigwedge_x^{\mathrm{ae}} \text{2-T-NSPACE}_U(ex) \leqq c_e \cdot \text{2-T-NSPACE}_{M_e}(x).$$

We set $L_1 =_{\mathrm{df}} \{x : x \in \mathrm{L}(U) \wedge \text{2-T-NSPACE}_U(x) \leqq s(|x|)\}$. Since s is fully 2-T-DSPACE-constructible we have $L_1 \in \textbf{2-T-NSPACE}(s)$. We assume that $L_1 \in \textbf{2-T-NSPACE}(s')$.

Let $L_2 \subseteq 1^*$ be an arbitrary recursive set. Then there is a fully 2-T-DSPACE-constructible $s_1 \in \mathbb{R}_1$ such that $L_2 \in \textbf{2-T-DSPACE}(s_1)$. We define

$$L_3 \in \{e1^l0^k : e \in PC \wedge (k < 2^{s_1(l)} \to e1^l0^{k+1} \in L_1) \wedge (k \geqq 2^{s_1(l)} \to 1^l \in L_2)\}.$$

Since s_1 is fully 2-T-DSPACE-constructible and $s' \geq \log$ we have $L_3 \in \textbf{2-T-NSPACE}(s'(n+1))$.

Lemma 1. There exists an $e_0 \in PC$ and a $c > 0$ such that

$$\mathrm{L}(M_{e_0}) = \{x : e_0x \in L_3\}$$

and

$$\text{2-T-NSPACE}_{M_{e_0}}(n) \leqq_{\mathrm{ae}} c \cdot s'(|e_0| + n + 1).$$

Proof of Lemma 1. Evidently there is a recursive function f such that $f(PC) \subseteq PC$ and $M_{f(e)}$ works as follows: $M_{f(e)}$ writes e on its worktape and then works according to the program of M_e.

Since $L_3 \in \textbf{2-T-NSPACE}(s'(n+1))$, there is a 2-NTM M_{e_1} such that $\mathrm{L}(M_{e_1}) = L_3$ and $\text{2-T-NSPACE}_{M_{e_1}}(n) \leqq s'(n+1)$.

Let e_2 be the program code of a 2-NTM which works as follows: if x is on the input tape and e is on the worktape, then M_{e_2} writes $f(e)$ on its worktape, erases e, works like M_{e_1} on the input $f(e)x$ and, if M_{e_1} stops, it erases $f(e)$. Put $e_0 =_{\mathrm{df}} f(e_2)$.

The 2-NTM M_{e_0} works on the input x as follows:

a) M_{e_0} writes e_2 on the worktape,

b) M_{e_0} writes e_0 on its worktape and erases e_2,

c) M_{e_0} works like M_{e_1} on e_0x and, if M_{e_1} stops, it erases e_0.

Therefore, $x \in \mathrm{L}(M_{e_0}) \leftrightarrow e_0x \in \mathrm{L}(M_{e_1}) = L_3$ and

$$\begin{aligned}\text{2-T-NSPACE}_{M_{e_0}}(n) &\leqq c' + \text{2-T-NSPACE}_{M_{e_1}}(|e_0| + n) \quad \text{for some } c' > 0,\\ &\leqq_{\mathrm{ae}} c \cdot s'(|e_0| + n + 1) \quad \text{for some } c > 0. \;\square\end{aligned}$$

Lemma 2. There exists an $l_0 \in \mathbb{N}$ such that for all $l \geqq l_0$ and all $k \geqq 0$

$$1^l0^k \in \mathrm{L}(M_{e_0}) \leftrightarrow 1^l \in L_2.$$

Proof of Lemma 2. Case 1. If $k \geqq 2^{s_1(l)}$, then

$$1^l0^k \in \mathrm{L}(M_{e_0}) \leftrightarrow e_0 1^l 0^k \in L_3 \leftrightarrow 1^l \in L_2 .$$

Case 2. For $k < 2^{s_1(l)}$ we prove the assertion by induction on k running down from $k = 2^{s_1(l)}$ to $k = 0$. For the induction step, assume $1^l0^{k+1} \in L(M_{e_0}) \leftrightarrow 1^l \in L_2$ for some $k < 2^{s_1(l)}$.

Let l_0 be so large that the following inequalities are valid for $l \geqq l_0$:

$$\begin{aligned} \text{2-T-NSPACE}_U(e_0 1^l 0^{k+1}) &\leqq c_{e_0} \cdot \text{2-T-NSPACE}_{M_{e_0}}(1^l0^{k+1}) \\ &\leqq c_{e_0} \cdot c \cdot s'(|e_0 1^l 0^{k+1}| + 1) \quad \text{(by Lemma 1)} \\ &\leqq s(|e_0 1^l 0^{k+1}|). \end{aligned} \tag{1}$$

Now we conclude

$$\begin{aligned} 1^l0^k \in \mathrm{L}(M_{e_0}) &\leftrightarrow e_0 1^l 0^k \in L_3 && \text{(by Lemma 1)} \\ &\leftrightarrow e_0 1^l 0^{k+1} \in L_1 && \text{(by the definition of } L_3) \\ &\leftrightarrow e_0 1^l 0^{k+1} \in \mathrm{L}(U) && \text{(by the definition of } L_1 \text{ and because of (1))} \\ &\leftrightarrow 1^l 0^{k+1} \in \mathrm{L}(M_{e_0}) && \text{(by the universality of } U) \\ &\leftrightarrow 1^l \in L_2 && \text{(by the induction hypothesis).} \ \square \end{aligned}$$

Let M be a 2-NTM which works on the input x as follows: If $x \notin 1^*$, then M rejects. If $x = 1^l$ such that $l < l_0$, then M accepts 1^l iff $1^l \in L_2$, using its finite memory. If $x = 1^l$ such that $l \geqq l_0$, then M works like M_{e_0}. Consequently, $\mathrm{L}(M) = L_2$ and

$$\begin{aligned} \text{2-T-NSPACE}_M(n) &\leqq \text{2-T-NSPACE}_{M_{e_0}}(n) \\ &\leqq c \cdot s'(|e_0| + n + 1) \quad \text{(by Lemma 1)} \\ &\leqq_{\mathrm{ae}} s(|e_0| + n)_{\mathrm{df}} = \hat{s}(n). \end{aligned}$$

Hence we have $\mathbf{REC} \cap \mathfrak{P}(1^*) \subseteqq \mathbf{2\text{-}T\text{-}NSPACE}(\hat{s}) \subseteqq \mathbf{2\text{-}T\text{-}DSPACE}(\hat{s}^2)$ (see 23.1.2). Then $\hat{s}$ is not recursive by Corollary 17.3. This contradicts the supposition that s is fully 2-T-DSPACE-constructible.

Now we suppose $s'(n+1) \lesssim s(n)$ and $s'(n) \prec s(n)$. Defining $s''(n) =_{\mathrm{df}} s(n+1)$ we get $s'(n+1) \prec s''(n)$. By the part of the theorem proved above there is a language $L \in \mathbf{2\text{-}T\text{-}NSPACE}(s'') \setminus \mathbf{2\text{-}T\text{-}NSPACE}(s') = \mathbf{2\text{-}T\text{-}NSPACE}\big(s(n+1)\big) \setminus \mathbf{2\text{-}T\text{-}NSPACE}(s')$. Consequently, $L_0 =_{\mathrm{df}} \{0x : x \in L\} \in \mathbf{2\text{-}T\text{-}NSPACE}(s) \setminus \mathbf{2\text{-}T\text{-}NSPACE}\big(s'(n-1)\big)$.

Case 1: $L \in \mathbf{2\text{-}T\text{-}NSPACE}(s)$. Then $L \in \mathbf{2\text{-}T\text{-}NSPACE}(s) \setminus \mathbf{2\text{-}T\text{-}NSPACE}(s')$.

Case 2: $L \notin \mathbf{2\text{-}T\text{-}NSPACE}(s)$. By the linear speed-up for 2-T-NSPACE and by $s'(n+1) \lesssim s(n)$ we get $L \notin \mathbf{2\text{-}T\text{-}NSPACE}\big(s'(n+1)\big)$. Consequently, $L_0 \notin \mathbf{2\text{-}T\text{-}NSPACE}(s')$, i.e. $L_0 \in \mathbf{2\text{-}T\text{-}NSPACE}(s) \setminus \mathbf{2\text{-}T\text{-}NSPACE}(s')$. ∎

Remark 1. A weaker result for polynomial bounding functions can be found in [Iba 72].

Remark 2. Under stronger suppositions than those made in Theorem 17.32 it can be shown that even subsets of 1^* are in **2-T-NSPACE**$(s) \setminus$ **2-T-NSPACE**(s'). Such suppositions are for instance: there is an increasing function f such that $0 \leqq_{io} f \leqq$ id, $\{1^k 0^{f(k)}: k \in \mathbb{N}\} \in \mathbf{L}$ and $s'(n + f(n)) < s(n)$ (see [Sei 77b]).

In the sublogarithmic range a stronger result can be shown by counting arguments.

17.33. Theorem. [HoUl 69b] If $s \leqq \log$ is fully 2-T-DSPACE-constructible, then $s' <_{io} s$ implies

$$\textbf{2-T-DSPACE}(s) \setminus \textbf{2-T-NSPACE}(s') \neq \emptyset.$$

Proof. By 8.30 we have $L_s =_{df} \{u2^k u^{-1}: |u| = 2^{s(|u2^k u^{-1}|)} \wedge u \in \{0,1\}^*\} \notin$ **2-T-NSPACE**(s') for $s' <_{io} s$ and $s(n) \leqq \log \frac{n}{2}$. If s is fully 2-T-DSPACE-constructible, then $L_s \in$ **2-T-DSPACE**(s). ∎

Remark. The "low end" of the 2-T-NSPACE hierarchy is similar to that of the 2-T-DSPACE hierarchy (cf. Remark 2 after 17.23). By 8.29 we have **2-T-NSPACE**(0) = **2-T-NSPACE**(s) for $s < \log\log$, and by 5.6.3 and 17.33 we have **2-T-NSPACE**(0) $\subset$ **2-T-NSPACE**(log log) (see Remark 1 after 8.29).

Next we state a result on the 2-T-ASPACE hierarchy which follows immediately from 20.36 and 17.40.

17.34. Theorem. [ChSt 76] If $s \geqq \log$ is fully 2-T-DSPACE-constructible, then $s' <_{io} s$ implies

$$\textbf{2-T-ASPACE}(s) \setminus \textbf{2-T-ASPACE}(s') \neq \emptyset. \blacksquare$$

Finally we mention some results on the measure 1-auxPD-T-NSPACE. Because of **1-auxPD-T-NSPACE**(s) = **2-auxPD-T-NSPACE**(s) = **2-auxPD-T-DSPACE**(s) for $s \geqq$ id, Theorem 17.31 holds true also for **1-auxPD-T-NSPACE** and $s \geqq$ id. In [ChWa 84] it is shown that for every increasing unbounded recursive function s there exists an increasing unbounded recursive function $s' < s$ such that **1-auxPD-T-NSPACE**$(s') \subset$ **1-auxPD-T-NSPACE**(s). Thus, 1-auxPD-T-NSPACE does not have a minimal growth rate (see § 17.1.3.). The same is true for a probabilistic space measure investigated in [KaVe 84].

17.2.3. Deterministic Time Measures

Most of the hierarchy results for deterministic time measures are shown, as was the case for deterministic space measures, by diagonalization arguments. But, at least for the Turing machine time measures, these arguments do not yield such good results as they do for space measures. This phenomenon is caused by the fact that the time measures of deterministic Turing machines do not have both the property of parallelity (see § 16.5) and the ability of cheap simulation. For a detailed investigation of this matter see [HaHo 71a]. We start with the T-DTIME hierarchy.

17.35. Theorem. [Hart 68a] If $t(|w|) \in$ T-DTIME$(t \cdot \log t)$ and $t(n) \succeq n^2$, then

$$\textbf{T-DTIME}(t \cdot \log t) \setminus \bigcup_{t' \prec_{io} t} \textbf{T-DTIME}(t') \neq \emptyset.$$

Proof. The proof is similar to that of Theorem 17.23. The following changes are necessary:

First, the function r is not needed in this proof, i.e. $r(i)$ must be replaced by i.

In stage 1 the machine M computes $t(|w|)$ within time $\preceq t(|w|) \cdot \log t(|w|)$. In the other stages, instead of the space restriction $s(|w|)$, the time restriction $t(|w|)$ is observed. This is done in such a way that the binary presentation of $t(|w|)$ (the "clock") is moved with the head. After every step of the actual computation of M as described in the stages 2, 3 and 4, this binary presentation is diminished by 1. Thus, for one step of the actual computation the machine M needs $\preceq \log t(|w|)$ steps. Therefore M needs altogether $\preceq t(|w|) \cdot \log t(|w|)$ steps. In stage 3 an additional check is made: that the length of the program code of M_i does not exceed $\log t(|w|)$. During the simulation in stage 4 the program code of M_i is moved simultaneously with the binary presentation of $\log t(|w|)$.

The number n must be chosen in such a way that it in addition fulfils the inequality "length of the program code of $M_i \leq \log t(n)$". ∎

Remark 1. Theorem 17.35 remains valid if "$t(|w|) \in$ T-DTIME$(t \cdot \log t)$" is replaced by "t is fully T-DTIME-constructible" or by "t is fully 2-T-DTIME-constructible" because each of these conditions implies the original condition (see 5.4.3).

Remark 2. The following proposition can be shown in a similar manner: If $t(|w|) \in$ T-DTIME(t) and $t(n) \succeq n^2$, then

$$\textbf{T-DTIME}(t) \setminus \bigcup_{t' \cdot \log t' \prec_{io} t} \textbf{T-DTIME}(t') \neq \emptyset.$$

Remark 3. The analogues of Theorem 17.35 for 2-T-DTIME and 1-T-DTIME can be shown by the same method, where in the latter case $\log t(n)$ must be replaced by $\max\{n, \log t(n)\}$.

Remark 4. The analogues of Theorem 17.35 for PD-T-DTIME, 1-PD-T-DTIME, and 2-PD-T-DTIME can also be shown by the same method. In [IgHo 74] it is proved that if $1 \leq p < q \leq 2$, then **2-PD-T-DTIME**$(n^p) \subset$ **2-PD-T-DTIME**(n^q).

17.36. Corollary. If $t(|w|) \in$ T-DTIME$(t \cdot \log t)$, then $t' \prec_{io} t$ and $t' \preceq t \cdot \log t$ imply

$$\textbf{T-DTIME}(t') \subset \textbf{T-DTIME}(t \cdot \log t). ∎$$

By translational methods similar to those in [RuFi 65] Corollary 17.36 can be strengthened in the range from n^2 to 2^n (see [Vog 78], for many single results of this kind see [HoUl 79]).

17.37. Theorem. If t is an increasing function such that $t(n) \geq n^2$, $t(n+1) \leq c \cdot t(n)$ for some $c > 0$ and $t(|w|) \in$ T-DTIME$\left(t\left(\frac{n}{\lfloor 2 \rfloor}\right) \Big/ (\log n)^2 \cdot \log t(n)\right)$, then for every $\varepsilon > 0$

$$\textbf{T-DTIME}(t/(\log t)^{\varepsilon}) \subset \textbf{T-DTIME}(t) \subset \textbf{T-DTIME}(t \cdot (\log t)^{\varepsilon}).$$

Proof. (Sketch). By a translational lemma for T-DTIME (analogous to 6.18.1) it is not hard to show

Lemma 1. If s and t are increasing functions such that $t(n) \geqq n^2$, $s(n) \leqq t(n)$, $t(n+1) \leqq c \cdot t(n)$ for some $c > 0$ and $s(|w|)$, $t(|w|) \in \text{T-DTIME}\left(t\left(\frac{n}{\lfloor 2 \rfloor}\right) \Big/ (\log n)^2 \cdot \log t(n)\right)$, then $\textbf{T-DTIME}(t \cdot s) \subset \textbf{T-DTIME}(t \cdot s^2)$ implies $\textbf{T-DTIME}(t) \subset \textbf{T-DTIME}(t \cdot s)$. □

From Lemma 1 one can conclude

Lemma 2. If s and t are as in Lemma 1, then $\textbf{T-DTIME}(t) \subset \textbf{T-DTIME}(t \cdot s^k)$ for some $k \geqq 1$ implies $\textbf{T-DTIME}(t) \subset \textbf{T-DTIME}(t \cdot s)$. □

Now we choose $m \in \mathbb{N}$ such that $\varepsilon > \frac{1}{m}$ and $s(n) = (\log t(n))^{1/m}$. By Corollary 17.36 we have $\textbf{T-DTIME}(t) \subset \textbf{T-DTIME}(t \cdot s^{m+1})$ and Lemma 2 yields $\textbf{T-DTIME}(t) \subset \textbf{T-DTIME}(t \cdot s) \subseteq \textbf{T-DTIME}(t \cdot (\log t)^\varepsilon)$. The other proper inclusion can be shown in the same manner. ∎

Remark. The suppositions of Theorem 17.37 are fulfilled by a large class of functions, for instance by n^c for $c > 2$ and d^n for $d > 1$.

Counting arguments yield very good results in the range from $n \cdot \log n$ to n^2.

17.38. Theorem. [Hen 65] If $n \cdot \log n \lesssim t(n) \leqq \left\lfloor \frac{n^2}{2} \right\rfloor$ and $t(|w|) \in \text{T-DTIME}(t)$, then

$$\textbf{T-DTIME}(t) \setminus \bigcup_{t' \prec_{io} t} \textbf{T-DTIME}(t') \neq \emptyset.$$

Proof. Since $t(|w|) \in \text{T-DTIME}(t)$, we have also $t''(|w|) \in \text{T-DTIME}(t)$ where $t''(n) =_{\text{df}} \left\lfloor \frac{t(n)}{n} \right\rfloor$.

In [Hen 65] it has been proved using the method of the proof of 8.13 that for $t' \prec_{io} t$ the set $D_{t''} = \{w \#^k w \colon |w| = t''(k + 2 \cdot |w|) \wedge w \in \{0, 1\}^*\}$ does not belong to $\textbf{T-DTIME}(t')$. On the other hand, because of the constructibility of t'' we obviously have $D_{t''} \in \textbf{T-DTIME}(t)$. ∎

Remark. Corollary 8.19 states that $\textbf{T-DTIME}(\text{id}) = \textbf{T-DTIME}(t)$ if $n \lesssim t(n) \prec n \cdot \log n$. The above theorem shows that $\textbf{T-DTIME}(\text{id}) \subset \textbf{T-DTIME}(n \cdot \log n)$ (see Remark 2 after 8.19).

For further results on the T-DTIME hierarchy see [Hart 77]. Now we deal with Turing machines with at least two tapes. By somewhat complicated diagonalization arguments the following theorem on the kT-DTIME-hierarchy ($k \geqq 2$) can be shown, which improves an earlier result in [Pau 77].

17.39. Theorem. [Für 82] Let $k \geqq 2$ and $t(|w|) \in \text{kT-DTIME}(t)$, then

$$\textbf{kT-DTIME}(t) \setminus \bigcup_{t' \prec_{io} t} \textbf{kT-DTIME}(t') \neq \emptyset. \ ∎$$

The following hierarchy result for multiT-DTIME can also be proved according to the model of the proof of Theorem 17.23.

17.40. Theorem. [HeSt 66] If $t \geqq \text{id}$ and $t(|w|) \in \text{multiT-DTIME}(t \cdot \log t)$, then

$$\textbf{multiT-DTIME}(t \cdot \log t) \setminus \bigcup_{t' \prec_{io} t} \textbf{multiT-DTIME}(t') \neq \emptyset.$$

Proof. To show this theorem the proof of 17.23 must be altered as follows: The simulating machine M is a kT-DTM where $k \geqq 4$ is chosen in such a way that

$t(|w|) \in$ kT-DTIME$(t \cdot \log t)$. In stage 1, M computes $t(|w|)$ and marks $t(|w|) \cdot \log t(|w|)$ tape squares on the first tape. This can be done within $\lesssim t(|w|) \cdot \log t(|w|)$ steps. In stage 3 the program code of the 2T-DM M_i is possibly computed and stored on tape 2. In stage 4 the machine M simulates the work of M_i on w on the tapes 3 and 4 within $t(|w|) \cdot \log t(|w|)$ steps. Thus we get

$$\mathbf{kT\text{-}DTIME}(t \cdot \log t) \setminus \bigcup_{t' \prec_{io} t \cdot \log t} \mathbf{2T\text{-}DTIME}(t') \neq \emptyset.$$

If $t' \prec_{io} t$, then $t' \cdot \log t' \prec_{io} t \cdot \log t$, and by Theorem 19.16 we get

$$\bigcup_{t' \prec_{io} t} \mathbf{multiT\text{-}DTIME}(t') \subseteq \bigcup_{t' \cdot \log t' \prec_{io} t \cdot \log t} \mathbf{2T\text{-}DTIME}(t' \cdot \log t')$$
$$\subseteq \bigcup_{t' \prec_{io} t \cdot \log t} \mathbf{2T\text{-}DTIME}(t'). \blacksquare$$

Remark 1. By 5.4.3 the condition "$t(|w|) \in$ multiT-DTIME$(t \cdot \log t)$" in Theorem 17.40 can be replaced by "t is fully multiT-DTIME-constructible".

Remark 2. In the same way as for Theorem 17.40 it can be shown that if $t \geqq$ id and t is multiT-DTIME-constructible, then $\mathbf{multiT\text{-}DTIME}(t) \setminus \bigcup_{t' \cdot \log t' \prec_{io} t} \mathbf{multiT\text{-}DTIME}(t') \neq \emptyset$.

Remark 3. In [SeFiMe 77] it is shown that under stronger suppositions even subsets of 1^* are contained in $\mathbf{multiT\text{-}DTIME}(t) \setminus \mathbf{multiT\text{-}DTIME}(t')$. Such suppositions are
1. t is fully multiT-DTIME-constructible and $t' \cdot \log t' \prec t$, or
2. t, t' are fully multiT-DTIME-constructible and $t' \cdot \log t' \prec_{io} t$.

Remark 4. By 22.5.2 and 18.12.4 we have $\mathbf{multiT\text{-}DTIME}(\mathrm{id}) \subset \mathbf{multiT\text{-}DTIME}((1 + \varepsilon) \cdot n)$ for all $\varepsilon > 0$.

Remark 5. In [Strn 68] a very strong hierarchy result for the measure 1^*-multiT-DTIME has been shown for the range from n to $\frac{n^2}{\log^2 n}$ (see p. 124).

In the same manner as for the measure T-DTIME (cf. 17.37) the hierarchy result for multiT-DTIME can be strengthened in the range from n to 2^n.

17.41. Theorem. If t is an increasing function such that $t(n) \geqq c_1 \cdot n$ for some $c_1 > 1$, $t(n+1) \leqq c_2 \cdot t(n)$ for some $c_2 > 1$ and $t(|w|) \in$ multiT-DTIME$\left(t\left(\left\lfloor \frac{n}{2} \right\rfloor\right) / (\log n)^2\right)$, then, for every $\varepsilon > 0$,

$$\mathbf{multiT\text{-}DTIME}(t/(\log t)^\varepsilon) \subset \mathbf{multiT\text{-}DTIME}(t) \subset \mathbf{multiT\text{-}DTIME}(t \cdot (\log t)^\varepsilon). \blacksquare$$

Finally we note a result on random access machine time hierarchies which improves an earlier result of [CoRe 72]. Here we get better results than for Turing machines because random access machines have the property of parallelity and admit cheap simulation.

17.42. Theorem. [SuZa 73] There is a $c > 1$ with the following property: if $t \geqq$ id and $t(|w|) \in$ RAM-DTIME(t), then $\mathbf{RAM\text{-}DTIME}(t) \subset \mathbf{RAM\text{-}DTIME}(c \cdot t)$. $\blacksquare$

Remark. Hierarchy results for random access stored program machines can be found in [SuZa 73] and [Hart 71].

17.2.4. Nondeterministic Time Measures

By translational methods similar to those used in the proof of Theorem 17.32 and by the use of the equality **multiT-NTIME**(t) = **2T-NTIME**(t) (cf. 19.19) the following theorem can be proved:

17.43. Theorem. [SeFiMe 77] If $t \geqq \mathrm{id}$ is fully multiT-DTIME-constructible, then

$$\mathbf{multiT\text{-}NTIME}(t) \setminus \bigcup_{t'(n+1) \prec t(n)} \mathbf{multiT\text{-}NTIME}(t') \neq \emptyset. \blacksquare$$

Remark 1. In [SeFiMe 77] it is shown that under stronger suppositions even subsets of 1^* are in **multiT-NTIME**$(t) \setminus$ **multiT-NTIME**(t'). Such suppositions are, for example: $t \geqq \mathrm{id}$ is fully multiT-DTIME-constructible, and f is a strictly increasing function such that $\{1^{f(n)}: n \in \mathbb{N}\} \in$ **multiT-DTIME**(id) and $t'(n + f^{-1}(n)) \prec t(n)$ where $f^{-1}(n) = \min\{k: f(k) \geqq n\}$.

Remark 2. A weaker result for polynomial bounding functions had already been shown in [Coo 73b].

Remark 3. T-NM's with fixed but sufficiently large number of states can accept more sets within time bound $a_2t(n)$ than within $a_1t(n)$ where $0 < a_1 < a_2$ and t is of the form

$$t(n) = n^{\alpha}(\log n)^{\beta_1}(\log\log n)^{\beta_2}\ldots$$

such that $n \log n \preceq t(n) \preceq n^2$ ([Kob 83]).

Finally we state a time hierarchy result for alternating Turing machines which can be shown by standard diagonalization arguments.

17.44. Theorem. [ChSt 76] If t and t' are fully multiT-DTIME-constructible and $t' \prec_{\mathrm{io}} t$, then

$$\mathbf{multiT\text{-}ATIME}(t) \setminus \mathbf{multiT\text{-}ATIME}(t') \neq \emptyset. \blacksquare$$

Hierarchies for relativized time measures are investigated in [Mor 81].

17.2.5. Further Measures

In this subsection we state some hierarchy results for crossing, reversal, return, and dual return measures. We start with the crossing measures.

According to the model of the proof of Theorem 17.23 it is possible to show the following theorem:

17.45. Theorem. If $t \geqq \mathrm{id}$ and $t(|w|) \in$ T-DCROS(Lin t), then

$$\mathbf{T\text{-}DCROS}(\mathrm{Lin}\, t) \setminus \bigcup_{t' \prec_{\mathrm{io}} t} \mathbf{T\text{-}DCROS}(t') \neq \emptyset. \blacksquare$$

Remark 1. The "factor" Lin appears because linear speed-up for T-DCROS is not known.

Remark 2. By 8.16 we have **T-DCROS**(0) = **T-DCROS**(t) for $t \prec \log$.

Remark 3. By 21.1.2 we have **1-T-DCROS**(Lin t) = **T-DCROS**(Lin t) for $t \geqq \mathrm{id}$. Therefore a separate treatment of the 1-T-DCROS hierarchy is not necessary. For 2-T-DCROS one can get, by the same method, a similar result even for $t \geqq 0$.

From the equality **T-NCROS**(t) = **NSPACE**(t) for $t \geqq \mathrm{id}$ (cf. 21.3.1) and the hierarchy result for NSPACE (cf. 17.32) we conclude the following theorem:

17.46. Theorem. If t is a fully 2-T-DSPACE-constructible function such that $t \geqq \mathrm{id}$ and $t'(n+1) \prec t(n)$ (or $t'(n+1) \preceq t(n)$ and $t' \prec t$), then

$$\textbf{T-NCROS}(t) \setminus \textbf{T-NCROS}(t') \neq \emptyset. \blacksquare$$

Remark. By 21.1.2 we have **1-T-NCROS**(Lin t) = **T-NCROS**(Lin t) for $t \geqq \mathrm{id}$. Therefore a separate treatment of the 1-T-NCROS hierarchy is not necessary. Because of **2-T-NCROS**(t) = **NSPACE**$(t(n) \cdot \log n)$ for $t \geqq 0$ (cf. 21.3.2) one can get a similar result for 2-T-NCROS even for $t \geqq 0$.

In [Wec 79] a crossing measure for 2T-DM's is introduced, and a hierarchy for constant bounding functions is shown.

We continue with some results on reversal measures. For the measure T-DREV it is known that $t \prec \log$ implies **T-DREV**(0) = **T-DREV**(t) because of **T-DREV**(t) $\subseteqq$ **T-DCROS**(t) and Remark 2 after 17.45.

For the measure 1-T-DREV we state the following theorem whose first two statements follow from 8.27.

17.47. Theorem. [Hart 68b]

1. **1-T-DREV**(k) $\subset$ **1-T-DREV**($k+1$) for $k \geqq 0$.
2. There are t such that $0 \prec t \prec \mathrm{id}$ and **1-T-DREV**(t) $\subset$ **1-T-DREV**($t+1$).
3. There are arbitrarily large functions t such that

$$\textbf{1-T-DREV}(t) \setminus \bigcup_{t'^2 \prec t} \textbf{1-T-DREV}(t') \neq \emptyset. \blacksquare$$

For hierarchy results for the measures 1-multiC-DREV and 1-multiC-NREV see [Hro 84].

Finally we state some results on return and dual return measures. Because of **i-T-DCROS**(Lin t) = **i-T-DRET**(Lin t) for sufficiently large t and $i = 0, 1, 2$ (cf. 21.8) the hierarchy results for i-T-DCROS carry over to i-T-DRET.

For constant bounding functions we have the following theorem.

17.48. Theorem.

1. [Pec 77] **T-DRET**(k) $\subset$ **T-DRET**($k+1$), for $k \geqq 1$.
2. [BrSa 77] **1-T-DRET**(k) $\subset$ **1-T-DRET**($k+1$), for $k \geqq 1$.
3. [Hib 67] **T-DDUR**(k) $\subset$ **T-DDUR**($k+1$), for $k \geqq 1$. $\blacksquare$

In the nondeterministic case the hierarchy results for 2-auxPD-T-DSPACE (cf. 17.31) carry over to T-NRET and 1-T-NRET because of **T-NRET**(t) = **1-T-NRET**(t) = **2-auxPD-T-DSPACE**(t) **for sufficiently large** t **(cf. 21.9, 21.11, and 20.13).**

17.49. Theorem. If t is fully 2-T-DSPACE-constructible, $t \geqq \text{id}$ and $i = 0$ ($t(n) \geqq n \cdot \log n$ and $i = 1$), then

$$\textbf{i-T-NRET}(t) \setminus \bigcup_{t' \prec_{\text{io}} t} \textbf{i-T-NRET}(t') \neq \emptyset. \ \blacksquare$$

Remark 1. Because of **2-T-NRET**(Lin t) = **2-auxPD-T-NSPACE**$(t(n) \cdot \log n)$ for $t \geqq \text{id}$ (cf. 21.11.2) a similar hierarchy result is true for 2-T-NRET.

Remark 2. Theorem 17.49 remains valid if RET is replaced by DUR (cf. the remark after 21.12).

Remark 3. For the "low ends" of the T-NRET and T-NDUR hierarchies see 12.1 and 12.5.

17.2.6. Double Complexity Measures

Here we restrict ourselves to the space-time double measure. For special classes of the time-reversal double measure see § 22.1 and § 22.2. Combining the proofs of the corresponding space and time hierarchies (cf. 17.23 and 17.35) we get the following theorem.

17.50. Theorem. [Gli 71] If $s \geqq \text{id}$ and $t \geqq \text{id}$ are fully T-DSPACE-TIME-constructible, then

$$\bigcup_{\substack{s' \prec s \\ t' \prec t}} \textbf{T-DSPACE-TIME}(s', t') \subset \textbf{T-DSPACE-TIME}(s, t \cdot \log t). \ \blacksquare$$

Since hierarchy results with a fixed s or a fixed t are connected with the space-time problem it is very hard to get such results. Only results are known in which s and t do not differ very much.

17.51. Theorem. [Für 82] For $k \geqq 2$, $s > \text{id}$ and $s(|w|) \in \text{kT-DTIME}(s)$ and for all rationals p and q such that $0 < p < q < 1$,

$$\textbf{kT-DSPACE-TIME}(s, t_p) \subset \textbf{kT-DSPACE-TIME}(s, t_q),$$

where $t_p(n) =_{\text{df}} 2^{2^{\cdot^{\cdot^{2}}}} \Big\} \lfloor p \cdot \log^* s(n) \rfloor$. ∎

In [Nek 73] a hierarchy result for the double measure multiT-DSPACE-TIME is proved for fixed time bound $t = \text{id}$ and space bounds $s \leqq \text{id}$. As a special case of this result we have

$$\textbf{multiT-DSPACE-TIME}(n^{r_1}, \text{id}) \subset \textbf{multiT-DSPACE-TIME}(n^{r_2}, \text{id})$$

for $0 < r_1 < r_2 \leqq 1$.

In [Hro 84] hierarchy results for the measures 1-multiC-DTIME-REV and 1-multiC-NTIME-REV are proved.

Finally we note that in [Berl 77] a gap result for the space-time double hierarchy of alternating Turing machines with restricted alternation depth is shown.

§ 18. Speed-Up

The naive belief that every function must have a program with smallest complexity (with respect to a given measure) cannot be justified. There exist functions f such that for every program computing f there is another one which has almost everywhere a smaller complexity. Furthermore, for any given amount of complexity a function f can be chosen in such a way that improvements of this amount can always be achieved.

For specific measures, in particular for space and time measures, general speed-up is of great importance, i.e. the fact that for every program with a sufficiently large step counting function there exists an equivalent program whose step counting function is by a constant factor smaller.

This and some related questions will be treated in this section. Note that a survey of the most important speed-up results can also be found in [Emd 75].

18.1. Speed-Up

8.1.1. The Speed-Up Theorem

We say that a function $f \in \mathbb{R}_1$ has *h-speed-up* if and only if

$$\bigwedge_{\varphi_i = f} \bigvee_{\varphi_j = f} h \square \Phi_j <_{ae} \Phi_i .$$

The first theorem shows that for every $h \in \mathbb{R}_2$ there are arbitrarily complex functions having h-speed-up.*)

18.1. Speed-Up Theorem. [Blu 67a] For every $h \in \mathbb{R}_2$ and every $g \in \mathbb{R}_1$ there is a recursive set A such that

$$\bigwedge_{\varphi_i = c_A} \Phi_i >_{ae} g \wedge \bigwedge_{\varphi_i = c_A} \bigvee_{\varphi_j = c_A} h \square \Phi_j <_{ae} \Phi_i .$$

Proof. Since it is sufficient to prove the theorem for those $h \in \mathbb{R}_2$ and $g \in \mathbb{R}_1$ which are strictly increasing (in all arguments), we can restrict ourselves to the case where h does not really depend on its first argument, i.e. to the case $h \in \mathbb{R}_1$.

Following [HaHo 71a] we prove the theorem first for the Blum measure T-DSPACE and then we use the fact that all Blum measures are recursively related to obtain the general result. Since for every $f \in \mathbb{R}_1$ there is an $f' \in \mathbb{R}_1$ such that $f \leqq f'$

*) In this section (φ, Φ) is an arbitrary Blum measure.

and $f' \in$ T-DSPACE(f'), we can assume that $h \in$ T-DSPACE(h) and $g \in$ T-DSPACE(g). Furthermore, assume $h(n) \geqq n^2$. Define

$$r(0) = g(0) + 2,$$
$$r(n+1) = \max\{h(r(n)) + 1, g(2n+2)\}, \quad \text{for} \quad n \geqq 0.$$

Consequently, $r \in$ T-DSPACE(r) and for $n \geqq 0$

$$r(n+1) > h(r(n)), \qquad r(n) \geqq g(2n) \quad \text{and} \quad r(n) \geqq 2^n. \tag{1}$$

We shall define a set $A \subseteq \mathbb{N}$ such that

$$\bigwedge_{\varphi_i = c_A} \text{T-DSPACE}_i(n) \geqq_{\text{ae}} r(n \dot{-} i) \tag{2}$$

and

$$\bigwedge_k \bigvee_{\varphi_j = c_A} \text{T-DSPACE}_j(n) \leqq_{\text{ae}} r(n \dot{-} k). \tag{3}$$

Then, by (2) and (1) we get for every $\varphi_i = c_A$

$$\text{T-DSPACE}_i(n) \geqq_{\text{ae}} r(n \dot{-} i) \geqq g(2n \dot{-} 2i) >_{\text{ae}} g(n),$$

and by (2), (1) and (3) we find for every $\varphi_i = c_A$ a $\varphi_j = c_A$ such that

$$\text{T-DSPACE}_i(n) \geqq r(n \dot{-} i) >_{\text{ae}} h\big(r((n \dot{-} i) \dot{-} 1)\big) \geqq_{\text{ae}} h\big(\text{T-DSPACE}_j(n)\big).$$

Thus, it remains to define a set $A \subseteq \mathbb{N}$ which fulfils (2) and (3). We define the function c_A in such a way that c_A differs from all functions φ_i with T-DSPACE$_i(n)$ $<_{\text{io}} r(n \dot{-} i)$. Simultaneously with $c_A(n)$, we define the sets K_n and N_n, where N_n is the set of those indices i for which we defined $c_A(m)$ to be not equal to $\varphi_i(m)$ for some $m \leqq n$:

$$K_0 =_{\text{df}} \emptyset, \qquad c_A(0) =_{\text{df}} 0, \qquad N_0 =_{\text{df}} \emptyset,$$

and for $n \geqq 0$

$$K_{n+1} =_{\text{df}} \{i : i \leqq n+1 \wedge i \notin N_n \wedge \text{T-DSPACE}_i(n+1) < r((n+1) \dot{-} i)\},$$

$$c_A(n+1) =_{\text{df}} \begin{cases} 1 \dot{-} \varphi_{\min K_{n+1}}(n+1) & \text{if} \quad K_{n+1} \neq \emptyset, \\ 0 & \text{if} \quad K_{n+1} = \emptyset, \end{cases}$$

$$N_{n+1} =_{\text{df}} \begin{cases} N_n \cup \{\min K_{n+1}\} & \text{if} \quad K_{n+1} \neq \emptyset, \\ N_n & \text{if} \quad K_{n+1} = \emptyset. \end{cases}$$

Now consider an index i such that T-DSPACE$_i(n) <_{\text{io}} r(n \dot{-} i)$. Then i must be the minimum of some K_m and hence $\varphi_i \neq c_A$. Consequently, A fulfils condition (2).

Now we show that A fulfils condition (3) too. Let $k \in \mathbb{N}$. We construct a DTM M_j such that $\varphi_j = c_A$ and T-DSPACE$_j(n) \leqq_{\text{ae}} r(n \dot{-} k)$. Choose $n_0 \in \mathbb{N}$ such that $\{0, 1, \ldots, k\} \cap \bigcup_{n \in \mathbb{N}} N_n \subseteq N_{n_0}$. For each $n \leqq n_0$, M_j has stored in its finite control the value of $c_A(n)$. In order to compute the value of $c_A(n)$ for $n > n_0$, M_j computes first K_{n_0+1}, N_{n_0+1}, K_{n_0+2}, N_{n_0+2}, ..., K_n as follows: Assume the set N_m $(n_0 \leqq m < n)$ is already computed (N_{n_0} is stored in the finite control of M_j). In order to compute K_{m+1},

M_j checks for $i = k + 1, k + 2, \ldots, m + 1$ whether T-DSPACE$_i(m + 1) < r\big((m + 1) \dot{-} i\big)$. This is done as follows:

a) M_j computes $r\big((m + 1) \dot{-} i\big)$ within space $r\big((m + 1) \dot{-} i\big) \leqq r(n \dot{-} i) \leqq r(n \dot{-} k)$.

b) M_j computes the program code of M_i. This can be done by (1) within space $i \leqq m + 1 \leqq n \leqq_{\text{ae}} 2^{n \dot{-} k} \leqq r(n \dot{-} k)$.

c) M_j simulates the work of M_i as long as M_i does not use more than $r\big((m + 1) \dot{-} i\big)$ tape squares. Since M_j must encode the worktape symbols and the states of M_i, the simulation can be performed within space $i \cdot r\big((m + 1) \dot{-} i\big) \leqq n \cdot r(n \dot{-} i)$ $\leqq n \cdot r\big(n \dot{-} (k + 1)\big) = n \cdot r\big((n \dot{-} k) \dot{-} 1\big) \leqq_{\text{ae}} 2^{(n \dot{-} k) \dot{-} 1} \cdot r\big((n \dot{-} k) \dot{-} 1\big)$ $\leqq \big(r\big((n \dot{-} k) \dot{-} 1\big)\big)^2 \leqq h\big(r\big((n \dot{-} k) \dot{-} 1\big)\big) < r(n \dot{-} k)$ because of $k + 1 \leqq i \leqq m + 1 \leqq n$, $n^2 \leqq h(n)$ and (1).

Hence, M_j computes K_{m+1} within space $r(n \dot{-} k)$. Consequently, the set N_{m+1} can be computed in the same amount of space. After computing K_n in this manner, M_j computes $c_A(n)$ as follows: If $K_n = \emptyset$, then $c_A(n) = 0$. If $i_0 = \min K_n$, then we have T-DSPACE$_{i_0}(n) < r(n \dot{-} i_0) \leqq r(n \dot{-} k)$, and $c_A(n) = 1 \dot{-} \varphi_{i_0}(n)$ can be computed within space $r(n \dot{-} k)$.

Now we treat the general case of an arbitrary Blum measure (ψ, Ψ). Because of 5.21.1 there are a strictly increasing $r_1 \in \mathbb{R}_1$ and an $s_1 \in \mathbb{R}_1$ such that for all i

$$\psi_{s_1(i)} = \varphi_i \quad \text{and} \quad \Psi_{s_1(i)} \leqq_{\text{ae}} r_1 \circ \text{T-DSPACE}_i .$$

Likewise by 5.21.2 there are an $r_2 \in \mathbb{R}_2$ which is strictly increasing in the second argument, and an $s_2 \in \mathbb{R}_1$, such that for all i

$$\varphi_{s_2(i)} = \psi_i \quad \text{and} \quad \text{T-DSPACE}_{s_2(i)} \leqq_{\text{ae}} r_2 \,\square\, \Psi_i .$$

Set $h' = r_2 \,\square\, (h \circ r_1)$ and $g' =_{\text{df}} r_2 \,\square\, g$. Let A be a recursive set such that

$$\bigwedge_{\varphi_i = c_A} \text{T-DSPACE}_i >_{\text{ae}} g' \quad \text{and} \quad \bigwedge_{\varphi_i = c_A} \bigvee_{\varphi_j = c_A} h' \circ \text{T-DSPACE}_j <_{\text{ae}} \text{T-DSPACE}_i .$$

Now $\psi_i = c_A$ implies $\varphi_{s_2(i)} = c_A$ and there is a $\varphi_j = c_A$ such that $h' \circ$ T-DSPACE$_j$ $<_{\text{ae}}$ T-DSPACE$_{s_2(i)}$. Consequently,

$$r_2 \,\square\, \Psi_i \geqq_{\text{ae}} \text{T-DSPACE}_{s_2(i)} >_{\text{ae}} g' = r_2 \,\square\, g$$

and

$$r_2 \,\square\, \Psi_i \geqq_{\text{ae}} \text{T-DSPACE}_{s_2(i)} >_{\text{ae}} h' \circ \text{T-DSPACE}_j$$
$$= r_2 \,\square\, (h \circ r_1 \circ \text{T-DSPACE}_j) \geqq r_2 \,\square\, (h \circ \Psi_{s_1(j)}) .$$

Since r_2 is strictly increasing in the second argument, we have

$$\Psi_i >_{\text{ae}} r \quad \text{and} \quad \Psi_i >_{\text{ae}} h \circ \Psi_{s_1(j)} . \;\blacksquare$$

Remark 1. Defining $r_i(n) =_{\text{df}} r(n \dot{-} i)$ for the function r defined in the proof of 18.1 we get for $\Phi =$ T-DSPACE

$$\bigwedge_{\varphi_i = c_A} \bigvee_{j} r_j \leqq_{\text{ae}} \Phi_i \quad \text{and} \quad \bigwedge_{i} \bigvee_{\varphi_j = c_A} \Phi_j \leqq_{\text{ae}} r_i .$$

Such a sequence $\{r_i : i \in \mathbb{N}\}$ is said to be a *complexity sequence for the function c_A with respect to the measure* (φ, Φ) (cf. [MeFip 72]). More generally, this notion can be defined for arbitrary $\varphi \in \mathbb{P}_1$ instead of c_A and arbitrary Blum measures (φ, Φ). The notion of complexity sequence has been characterized in [MeFip 72], [Schn 74a] and [ScSt 72].

Remark 2. Theorem 18.1 shows that the complexity of recursive functions cannot always be characterized by single functions but always by sequences of functions (complexity sequences, see Remark 1). Conversely, if a few weak and simple properties are satisfied, then such a sequence characterizes the complexity of some recursive function (see [MeWi 78], for related ideas see § 16.1).

The following theorem asserts that speed-up by any total recursive operator is possible.

18.2. Theorem. [MeFip 72] For every total effective operator F and every function $g \in \mathbb{R}_1$ there is a recursive set A such that

$$\bigwedge_{\varphi_i = c_A} \Phi_i >_{\text{ae}} g \quad \text{and} \quad \bigwedge_{\varphi_i = c_A} \bigvee_{\varphi_j = c_A} F(\Phi_j) <_{\text{ae}} \Phi_i . \blacksquare$$

Remark 1. In [Blu 67a] the special case $F(f) = f \circ f$ of Theorem 18.2 is already proved.

Remark 2. There are also speed-up results for subrecursive classes of functions. In [CoBo 72] it is shown that a theorem analogous to 18.1 holds true for the class of primitive recursive functions and for the time measure of the *loop programs* introduced in [MeRi 67] (see p. 45). In [Aln 76] and [Aln 77] this is proved for a wide variety of subrecursive languages.

18.1.2. Effective Speed-Up

It is of interest to know whether, for a given function having speed-up, there is an effective procedure for finding faster programs for this function. In other words: Assume that f has h-speed-up. Does there exist a $\sigma \in \mathbb{P}_1$ such that $\{i : \varphi_i = f\} \subseteq D_\sigma$ and

$$\bigwedge_{\varphi_i = f} (\varphi_{\sigma(i)} = f \wedge h \mathbin{\square} \Phi_{\sigma(i)} <_{\text{ae}} \Phi_i)?$$

For small functions h this may be possible (for example take the measure T-DSPACE or multiT-DTIME: from a given program of a function f another program of this function can be effectively constructed which is faster than the given program by a linear factor (cf. 18.11 and 18.12). However, the following theorem shows that for larger functions h this is impossible for every function f.

18.3. Theorem. [Blu 71] Let $h \in \mathbb{R}_2$ be sufficiently large and $f \in \mathbb{R}_1$. Then there is no $\sigma \in \mathbb{P}_1$ such that $\{i : \varphi_i = f\} \subseteq D_\sigma$ and

$$\bigwedge_{\varphi_i = f} (\varphi_{\sigma(i)} = f \wedge h \mathbin{\square} \Phi_{\sigma(i)} <_{\text{ae}} \Phi_i) . \blacksquare$$

Remark 1. For the measure multiT-DTIME the condition "h sufficiently large" can be replaced by "$h(x, y) \geqq (x^4 + y)^2$" (cf. [Blu 71]).

Remark 2. At first glance one could think that the construction of the faster machine M_j in the proof of 18.1 is effective. This is not true since the number n_0 cannot be effectively computed.

Remark 3. For effective speed-up in classes of provably equivalent programs see [You 77].

Furthermore, for sufficiently large h there is no f having h-speed-up such that a faster program for f can be chosen from a recursively enumerable set of programs for f.

18.4. Theorem. [HaHo 71a] Let $h \in \mathbb{R}_2$ be sufficiently large and $f \in \mathbb{R}_1$. Then there is no recursively enumerable set B such that

$$\bigwedge_{j \in B} \varphi_j = f \wedge \bigwedge_{\varphi_i = f} \bigvee_{j \in B} h \,\square\, \Phi_j \leqq_{\text{ae}} \Phi_i . \ \blacksquare$$

The following theorem shows that for every total effective operator F a function f having *F-speed-up* can be found for which the number of a faster program can be effectively bounded by the number of the given program for f.

18.5. Theorem. [MeFip 72] Let F be a total effective operator. There is a recursive set A and a $\sigma \in \mathbb{P}_1$ such that $\{i : \varphi_i = c_A\} \subseteqq D_\sigma$ and

$$\bigwedge_{\varphi_i = c_A} \bigvee_{j \leqq \sigma(i)} \big(\varphi_j = c_A \wedge F(\Phi_j) <_{\text{ae}} \Phi_i\big) . \ \blacksquare$$

Remark. Theorem 18.5 does not hold for all recursive functions having F-speed-up. In [HeYo 71] it is shown that for every $h \in \mathbb{R}_2$ such that $h(x, y) \geqq y$ there is a recursive set A such that c_A has h-speed-up but the number of a faster program for c_A cannot be effectively bounded by the number of the given program for c_A. For related results see [MeFip 72] and [CoHa 71].

The following theorem shows that it is not possible to bound effectively both the number of the faster program and the number of arguments x for which the inequality $(h \,\square\, \Phi_j)(x) < \Phi_i(x)$ does not hold.

18.6. Theorem. [Schn 73a] Let $h \in \mathbb{R}_2$ be sufficiently large, and let A be a recursive set. Then there is no $\sigma \in \mathbb{P}_1$ such that $\{i : \varphi_i = c_A\} \subseteqq D_\sigma$ and

$$\bigwedge_{\varphi_i = c_A} \bigvee_{j \leqq \sigma(i)} \Big(\varphi_j = c_A \wedge h \,\square\, \Phi_j <_{\text{ae}} \Phi_i \wedge \operatorname{card} \big\{x : h\big(x, \Phi_j(x)\big) \geqq \Phi_i(x)\big\} \leqq \sigma(i)\Big) . \ \blacksquare$$

Unlike the speed-up, the *pseudo speed-up* is effective. A function $f \in \mathbb{R}_1$ has *pseudo h-speed-up* if and only if

$$\bigwedge_{\varphi_i = f} \bigvee_{\varphi_j \in \mathbb{R}_1} (\varphi_j =_{\text{ae}} f \wedge h \,\square\, \Phi_j <_{\text{ae}} \Phi_i) .$$

18.7. Theorem. [Blu 71] There are $g, s \in \mathbb{R}_2$ such that for every $\varphi_k \in \mathbb{R}_2$ which fulfils $\varphi_k \geqq g$ and which is increasing in the second argument, and for every $f \in \mathbb{R}_1$ which has pseudo φ_k-speed-up,

$$\bigwedge_{\varphi_i = f} (\varphi_{s(k,i)} \in \mathbb{R}_1 \wedge \varphi_{s(k,i)} =_{\text{ae}} f \wedge \varphi_k \,\square\, \Phi_{s(k,i)} <_{\text{ae}} \Phi_i) . \ \blacksquare$$

18.1.3. General Speed-Up

A Blum measure (φ, Φ) has *general h-speed-up* if and only if

$$\bigwedge_{\varphi_i \in \mathbb{R}_1} \bigvee_{\varphi_j = \varphi_i} h \square \Phi_j \leqq_{\text{ae}} \Phi_i .$$

This condition can be reformulated in terms of complexity classes.

18.8. Theorem. Let $h \in \mathbb{R}_2$ be strictly increasing in its second argument and $h(x, y) \geqq y$. Then (φ, Φ) has general h-speed-up if and only if

$$\bigwedge_{\Phi_i \in \mathbb{R}_1} \bigvee_{t \in \mathbb{R}_1} \big(h \square t \leqq \Phi_i \wedge \Phi(t) = \Phi(h \square t) = \Phi(\Phi_i)\big).$$

Proof. 1. Let (φ, Φ) have general h-speed-up, and let $\varphi_i \in \mathbb{R}_1$. Then the function t defined by $t(x) = \max\{y : h(x, y) \leqq \Phi_i(x)\}$ is recursive and it satisfies the inequality $h \square t \leqq \Phi_i$. We prove $\Phi(\Phi_i) = \Phi(t)$. For this end it is sufficient to prove $\Phi(\Phi_i) \subseteqq \Phi(t)$ since $\Phi(t) \subseteqq \Phi(h \square t) \subseteqq \Phi(\Phi_i)$ follows from the assumption and from the inequality $h \square t \leqq \Phi_i$. Let $f \in \Phi(\Phi_i)$. Then there exists a k such that $f = \varphi_k$ and $\Phi_k \leqq_{\text{ae}} \Phi_i$ and because of the h-speed-up there exists a $\varphi_{k'} = \varphi_k$ with $h \square \Phi_{k'} \leqq_{\text{ae}} \Phi_i$ By the definition of t we get $\Phi_{k'} \leqq_{\text{ae}} t$. Consequently, $f \in \Phi(t)$.
2. Assume that (φ, Φ) is a measure with the property

$$\bigwedge_{\Phi_i \in \mathbb{R}_1} \bigvee_{t \in \mathbb{R}_1} \big(h \square t \leqq \Phi_i \wedge \Phi(t) = \Phi(\Phi_i)\big).$$

Let $\varphi_i \in \mathbb{R}_1$. Because of $\varphi_i \in \Phi(\Phi_i) = \Phi(t)$ there exists a j such that $\varphi_i = \varphi_j$ and $\Phi_j \leqq_{\text{ae}} t$ and hence $h \square \Phi_j \leqq_{\text{ae}} h \square t \leqq_{\text{ae}} \Phi_i$. ∎

For every $h \in \mathbb{R}_2$ there does exist a Blum measure having general h-speed-up.

18.9. Theorem. Let $h \in \mathbb{R}_2$ such that $h(x, y) \geqq y$. Then there is a Blum measure (φ, Φ) having general h-speed-up.

Proof. Let (ψ, Ψ) be an arbitrary Blum measure. We define

$$\varphi_i = \psi_{r(i)} \quad \text{and} \quad \Phi_i(x) = \underbrace{h\Big(x, h\Big(x, \dots h\big(x,}_{x \dot{-} l(i) \text{ times}} \Psi_{r(i)}(x)\big) \dots\Big)\Big).$$

It can easily be verified that (φ, Φ) is also a Blum measure. Let $\varphi_i \in \mathbb{R}_1$. Defining $j = \langle \mathrm{l}(i) + 1, \mathrm{r}(i) \rangle$ we get

$$\varphi_j = \psi_{r(j)} = \psi_{r(i)} = \varphi_i$$

and

$$\begin{aligned} h\big(x, \Phi_j(x)\big) &= \underbrace{h\Big(x, h\Big(x, h\Big(x, \dots h\big(x,}_{1 + x \dot{-} (\mathrm{l}(i)+1) \text{ times}} \Psi_{r(i)}(x)\big) \dots\Big)\Big)\Big) \\ &= \underbrace{h\Big(x, h\Big(x, \dots h\big(x,}_{x \dot{-} \mathrm{l}(i) \text{ times}} \Psi_{r(i)}(x)\big) \dots\Big)\Big) \quad \text{for} \quad x \geqq \mathrm{l}(i) + 1 \\ &= \Phi_i(x). \ ∎ \end{aligned}$$

However, not only such pathological measures, but also more natural measures, have general h-speed-up, although with relatively small h. Our first example is a measure having general polynomial speed-up.

18.10. Theorem. [FipMeRo 68] The measure i-multiC-XSPACE ($i = 1, 2$: X $\in$ {D, N}) has general n^2-speed up, i.e. for $i = 1, 2$, X $\in$ {D, N}, and every $s \in \mathbb{R}_1$,

$$\textbf{i-multiC-XSPACE}(s) = \textbf{i-multiC-XSPACE}(\mathrm{Pol}\ s).$$

Proof. Let M be an i-kC-XM working with maximum counter length $s(n)$ on inputs of length n. For every $x \in \mathbb{N}$ there are integers x_1, x_2 and σ such that $x = x_1^2 + 2x_2 + \sigma$, where $x_2 \leqq x_1 \leqq \sqrt{x}$ and $\sigma \in \{0, 1\}$. Therefore the work of M can be simulated by an i-$(2k)$C-XM M' having for one counter C of M with length x two counters C_1 and C_2 with length $x_1 - x_2$ and x_2, resp., and storing σ in its finite control. The operations $x \leftarrow x + 1$ in C is replaced in M' by the following acitivities which can be performed within one step:

if $\sigma = 0$ **then** $\sigma = 1$
  else if $C_1 = 0$ **then** C_2 and C_1 change their roles,
    $C_1 \leftarrow C_1 + 1$,
   else $C_2 \leftarrow C_2 + 1, \sigma = 0$.

The operation $x = x \dot{-} 1$ in C is replaced in M' by the following activities which can be performed within one step:

if $\sigma = 1$ **then** $\sigma = 0$
  else if $C_2 = 0$ **then** C_1 and C_2 change their roles,
    $C_2 \leftarrow C_2 \dot{-} 1$,
   else $C_2 \leftarrow C_2 \dot{-} 1, C_1 \leftarrow C_1 + 1, \sigma = 1$.

The test $C = 0$ can be replaced in M' by the tests $C_1 = 0$, $C_2 = 0$, $\sigma = 0$, which can be performed within one step. Evidently, M' works with maximum counter length $\sqrt{s(n)}$ for inputs of length n. ∎

Remark. The general polynomial speed-up for counter machines is caused by the unary presentation of an integer in a counter. If the length of the binary presentation is measured, then the polynomial speed-up becomes a linear speed-up.

Now we will consider general *linear speed-up*, a property held by some important measures. We start with space measures.

18.11. Theorem. [HaSt 65], [StHaLe 65] For $i = 0, 1, 2$, $k \geqq 1$, X $\in$ {D, N} and $s \geqq 0$ ($s \geqq \mathrm{id}$ for $i = 0$),

$$\textbf{i-kT-XSPACE}(s) = \textbf{i-kT-XSPACE}(\mathrm{Lin}\ s),$$

$$\textbf{i-multiT-XSPACE}(s) = \textbf{i-multiT-XSPACE}(\mathrm{Lin}\ s).$$

Proof. It is sufficient to prove $\textbf{i-kT-XSPACE}(m \cdot s) \subseteq \textbf{i-kT-XSPACE}(s)$ for arbitrary $m \in \mathbb{N}$. Let M be an i-kT-XM working within space $m \cdot s$. The worktape is subdivided into m-tuples of adjacent squares (blocks), and a new i-kT-XM M' is designed which holds the contents of a whole block of M in each square, and which simulates M step by step. Note that for TM's without input tape the work of M' starts with a phase where the input word is contracted (thus M' needs at least space id) whereas for TM's with input the worktape can be used from the very beginning in this contracted form. ∎

Remark 1. Linear speed-up also holds for the measure i-Z_1-Z_2- ... -Z_k-XSPACE, where $i \in \{0, 1, 2\}$, $Z_1, \ldots, Z_k \in \{\text{T, PD, C, S, NES, CS}\}$ and for $X \in \{\text{D, N}\}$ (in the case $i = 0$ for bounding functions $s \geqq \text{id}$). The proof for stores of type PD, S, NES or CS is the same as for Turing tapes. For the counters, instead of x, the integer $\left\lfloor \frac{x}{m} \right\rfloor$ for sufficiently large m is stored, the remainder being stored in the finite control. This remark remains true if some of the stores Z_i are replaced by auxZ_i.

Remark 2. Theorem 18.11 also holds for complexity classes of functions.

Remark 3. The linear speed-up in Theorem 18.11 is not possible without an enlargement of the worktape alphabet as follows from 17.30.

Now we will deal with time measures. Note that we have here the linear speed-up only for sufficiently large bounding functions, therefore it is not a general linear speed-up.

18.12. Theorem. [HaSt 65]

1. For $X \in \{\text{D, N}\}$ and every t such that $t(n) \succ n^2$,

$$\text{T-XTIME}(t) = \text{T-XTIME}(\text{Lin } t).$$

2. For $X \in \{\text{D, N}\}$, $k \geqq 2$ and every t such that $t(n) \geqq_{\text{ae}} c \cdot n$ for some real number $c > 1$,

$$k\text{T-XTIME}(t) = k\text{T-XTIME}(\text{Lin } t).$$

3. For $X \in \{\text{D, N}\}$, $k \geqq 1$ and every t such that $t(n) \geqq_{\text{ae}} c \cdot n$ for some real number $c > 1$,

$$1\text{-}k\text{T-XTIME}(t) = 1\text{-}k\text{T-XTIME}(\text{Lin } t).$$

4. For $i = 0, 1, 2$, $X \in \{\text{D, N}\}$ and every t such that $t(n) \geqq_{\text{ae}} c \cdot n$ for some real number $c > 1$,

$$i\text{-multiT-XTIME}(t) = i\text{-multiT-XTIME}(\text{Lin } t).$$

Proof. The possibility of linear time speed-up rests upon a refinement of the method of tape contraction described in the preceding proof. A refinement of the method is necessary to prevent the following phenomenon. The head of M might oscillate between adjacent squares of different blocks, and then the head of M' would oscillate in the same way, and there would be no guarantee for a speed-up.

Ad 1. Let M be an XTM working within time $m \cdot t(n)$ for some $m \in \mathbb{N}$. For given $\varepsilon > 0$ we describe an XTM M' which is equivalent to M and works within time $t(n) + \varepsilon \cdot n^2$.

The simulation will be done step by step in such a way that the inscription

$$\ldots \square \square \square\, b_1 b_2 b_3 \ldots b_{i-2} b_{i-1} b_i b_{i+1} b_{i+2} \ldots b_{l-1} b_l \,\square \square \square \ldots$$

of the tape of M (where every b_i consists of $3m$ tape symbols of M and the head of M scans a symbol belonging to $b_{i-1}b_ib_{i+1}$) is represented on the tape of M' in the form

$$\ldots \square \square \square\, b_1b_2b_3 \ldots b_{i-2}b_{i+2} \ldots b_{l-1}b_l \,\square \square \square \ldots$$

where every b_i is written in only one tape square of M', the head of M' scans b_{i+2} and $b_{i-1}b_ib_{i+1}$ is stored in the finite control of M'.

The machine M works within two phases.

Phase 1 (Preconditioning of M'). Within $\varepsilon \cdot n^2$ steps, M' generates the tape inscription which corresponds to the initial situation of M.

Phase 2 (step by step simulation). Assume that the head of M scans a symbol belonging to b_i. Since $b_{i-1}b_ib_{i+1}$ is stored in the finite control M' can simulate within one step at least $3m$ steps of the work of M. When the head of M leaves $b_{i-1}b_ib_{i+1}$ to the right, then M' writes b_{i-1} and b_i into the tape squares holding b_{i+2} and b_{i+3} and stores $b_{i+1}b_{i+2}b_{i+3}$ in its finite control. In such a way the desired form of the tape inscription of M' is restored within two steps. The case that the head of M leaves $b_{i-1}b_ib_{i+1}$ to the left is treated analogously, where three steps are required instead of two steps.

Ad 2, 3 and 4. The proof of Statement 1 is used with the following changes. For kT-XM's with $k \geqq 2$ the preconditioning phase can be carried out in time $(1+\varepsilon) \cdot n$ by copying the input in contracted form on another worktape. This idea works also for proving Statement 4. For 1-kT-DM's a preconditioning phase is not necessary, but since the n input steps cannot be sped up we obtain a speed-up only for $t(n) \geqq c \cdot n$ for some $c > 1$. ∎

Remark 1. For the measure 2-kT-DTIME the general linear speed-up is not known. The method of condensing the tape contents does not apply because the steps including a move of the input head cannot be sped up (the latter is true also for the measure 1-kT-DTIME, but in that case there are at most n such moves, i.e. at most n steps, which cannot be sped up).

Remark 2. The general linear speed-up also holds for the measure i-Z_1-Z_2- ... -Z_k-XTIME, where $i = 0, 1$; $Z_1, \ldots, Z_k \in$ {T, PD, C, S, NES, CS} and X $\in$ {D, N}, for bounding functions $t(n) \geqq_{ae} c \cdot n$ for some real number $c > 1$ (for bounding functions $t(n) \geq n^2$ if $i = 0$ and at most one Z_j is in {T, PD, S, NES}). The method is the same as that for the corresponding space measure (cf. Remark 1 after 18.11).

Remark 3. Since the construction in the proof of 18.12 also yields the general linear speed up for the space measures, general linear speed-up also holds for the space-time double measure. For example, for $s \geqq 0$ and $t(n) \geqq_{ae} c \cdot n$ for some real number $c > 0$,

$$\textbf{1-multiT-DSPACE-TIME}(s, t) = \textbf{1-multiT-DSPACE-TIME}(\text{Lin } s, \text{Lin } t).$$

Remark 4. Since the condensing of the worktape contents does not cause additional reversals of the worktape heads (see the proof of 18.12) we have

$$\textbf{1-kT-XTIME-REV}(t, s) = \textbf{1-kT-XTIME-REV}(\text{Lin } t, s),$$

$$\textbf{1-multiT-XTIME-REV}(t, s) = \textbf{1-multiT-XTIME-REV}(\text{Lin } t, s)$$

for $k \geqq 1$, X $\in$ {D, N}, $t(n) \geqq_{ae} c \cdot n$ for some $c > 1$ and $s \geqq 0$.

Remark 5. In [Scha 77] it is claimed that linear speed-up for T-DTIME in the range between $n \cdot \log n$ and n^2 does not hold.

Remark 6. The statements 2, 3 and 4 of Theorem 18.12 do not hold for X = D and $t = \text{id}$. This follows from Theorem 22.5 and the evident relationships **1-kT-DTIME**(id) = **(k+1)T-DTIME**(id) and **1-kT-DTIME**(Lin) $\subsetneqq$ **(k+1)T-DTIME**(Lin). Furthermore, Statement 3 (Statement 2) is not known to be valid for X = N, $k = 1$ ($k = 2$) and $t = \text{id}$. However, in all other cases the statements 2, 3 and 4 of Theorem 18.12 do hold true for $t = \text{id}$. This is shown by the following theorem.

18.13. Theorem. [BoGr 70]

1. For k $\geqq$ 3 and $t \geqq \text{id}$, **kT-NTIME**(t) = **kT-NTIME**(Lin t).
2. For k $\geqq$ 2 and $t \geqq \text{id}$, **1-kT-NTIME**(t) = **1-kT-NTIME**(Lin t).
2. For i = 0, 1, 2 and $t \geqq \text{id}$, **i-multiT-NTIME**(t) = **i-multiT-NTIME**(Lin t).

Proof. Let M be a Turing machine working within time $m \cdot t(n)$. The simulating Turing machine M' does in a nondeterministic manner both: it works like a Turing machine which, in the case $t(n) \geqq 2n$, speeds the computation time from $m \cdot t(n)$ to $t(n)$ by the method of Theorem 18.12, and it works like the Turing machine which is constructed in the proof of **multiT-NTIME**($2mn$) = **1-PD-CS-NTIME**(n) (cf. Theorem 22.2). Now, for a given input of length n which is accepted by M, if $t(n) \geqq 2n$, then the first method gives an accepting computation path of maximum length $t(n)$. If $n \leqq t(n) \leqq 2n$, then the second method gives an accepting computation path of length $n \leqq t(n)$. ∎

Remark. The general linear speed-up for $t \geqq \text{id}$ also holds for the time measures of those types of nondeterministic machines which have, in addition to an input tape, at least one PD store and one CS store or at least three PD stores (cf. 22.2). The general linear speed-up also holds for IA^d-XTIME (for $d \geqq 1$, X $\in$ {D,N}, and bounding functions $t(n) \geqq_{ae} c \cdot n$ for some $c > 1$) and for the time measure for Markov algorithms (for bounding functions $t \succeq \text{id}$, cf. [Zej 71 a]). General linear speed-up does not hold for i-RAM-DTIME (not even for large bounding functions, cf. 17.42).

At the end of this section we state some speed-up results for further measures.

18.14. Theorem. [Fip 65]

1. For X $\in$ {D, N} and $t \geqq 1$, **T-XREV**(t) = **T-XREV**(Lin t).
2. For X $\in$ {D, N} and $t \succeq \text{id}$, **1-T-XREV**(t) = **1-T-XREV**(Lin t).
3. For $t \geqq 0$, **2-T-NREV**(t) = **2-T-NREV**(Lin t). ∎

Remark. Theorem 17.47 shows that the linear speed-up for 1-T-DREV does not in general hold for bounding functions below id. Statement 3 of Theorem 18.14 follows from the relationships between the nondeterministic space measure and the measure 2-T-NREV (cf. 21.6).

In a similar manner we get the following speed-up results (cf. 21.1, 21.3, 21.9, and 21.11).

18.15. Theorem.

1. **T-NCROS**(t) = **T-NCROS**(Lin t) for $t \geqq \mathrm{id}$.
2. **1-T-NCROS**(t) = **1-T-NCROS**(Lin t) for $t \geqq \mathrm{id}$.
3. **2-T-NCROS**(t) = **2-T-NCROS**(Lin t) for $t \geqq 0$.
4. **T-NRET**(t) = **T-NRET**(Lin t) for $t \geqq \mathrm{id}$.
5. **1-T-NRET**(t) = **1-T-NRET**(Lin t) for $t \geqq n \cdot \log n$. ∎

18.2. Partial Speed-Up

Let $h \in \mathbb{R}_2$ and $\varphi \in \mathbb{P}_1$. The function φ has *partial h-speed-up* if and only if

$$\bigwedge_{\varphi_i=\varphi} \bigvee_{\varphi_j=\varphi} h \circ \Phi_j <_{\text{io}} \Phi_i .$$

For fixed functions h this notion has not been investigated as intensively as the following notion: A function $\varphi \in \mathbb{P}_1$ has *partial speed-up* if and only if it has partial h-speed-up for all $h \in \mathbb{R}_2$. Obviously, a recursive function cannot have partial speed-up. The notion of partial speed up can be extended to recursively enumerable subsets of $\mathbb{N}$. We define

$$\mathbf{PS} = \{A : A \text{ recursively enumerable} \wedge \chi_A \text{ has partial speed-up}\}.$$

Furthermore, one can ask whether the partial speed-up can be effective. A function $\varphi \in \mathbb{P}_1$ has *effective partial speed-up* if and only if there exists an $s \in \mathbb{R}_2$ such that

$$\bigwedge_{\varphi_k \in \mathbb{R}_2} \bigwedge_{\varphi_i=\varphi} (\varphi_{s(k,i)} = \varphi \wedge \varphi_k \circ \Phi_{s(k,i)} <_{\text{io}} \Phi_i).$$

For recursively enumerable subsets of $\mathbb{N}$ we define

$$\mathbf{PS}_{\text{eff}} = \{A : A \text{ recursively enumerable} \wedge \chi_A \text{ has effective partial speed-up}\}.$$

From the fact that any two Blum measures can be recursively bounded by each other (cf. 5.21), it follows immediately that the classes **PS** and $\mathbf{PS}_{\text{eff}}$ do not depend on the chosen Blum measure. This indicates that in spite of their complexity theoretic definitions the notions of partial speed-up are concepts of the theory of recursive functions rather than of complexity theory. This observation is confirmed by some of the following theorems.

For the next theorem we define $A^{\langle 1\rangle} = \{i : \mathrm{D}_i \cap \bar{A} \neq \emptyset\}$ (the "weak jump" of A) and $A^{\langle i+1\rangle} = (A^{\langle i\rangle})^{\langle 1\rangle}$.

Furthermore, we need the following notions.

1. Let $A = \{a_0, a_1, \ldots\}$ be an infinite subset of $\mathbb{N}$ and $a_0 < a_1 < a_2 < \ldots$ A is called *hyperimmune* iff $\sim\bigvee_{f \in \mathbb{R}_1} \bigwedge_{n \in \mathbb{N}} a_n \leqq f(n)$.
2. A is called *hypersimple* iff A is r.e. and $\bar{A}$ is hyperimmune.
3. A is called *hyperhyperimmune* iff A is infinite and

$$\sim\bigvee_{f \in \mathbb{R}_1} \left[\bigwedge_u (\mathrm{D}_{f(u)} \text{ finite} \wedge \mathrm{D}_{f(u)} \cap A \neq \emptyset) \wedge \bigwedge_u \bigwedge_v (u \neq v \to \mathrm{D}_{f(u)} \cap \mathrm{D}_{f(v)} = \emptyset)\right].$$

4. A is called *hyperhypersimple* iff A is r.e. and $\bar{A}$ is hyperhyperimmune.
5. A is called *productive* iff there exists an $f \in \mathbb{R}_1$ such that $D_x \subseteq A \to f(x) \in A \setminus D_x$.
6. A is called *creative* iff A is r.e. and $\bar{A}$ is productive.
7. A is called *subcreative* iff A is r.e. and

$$\bigvee_{f \in \mathbb{R}_1} \bigwedge_{j \in \mathbb{N}} \bigvee_{x \in \mathbb{N}} [(x \in D_j \cap A \vee x \in \overline{D_j \cup A}) \wedge D_{f(j)} = A \cup \{x\}].$$

The following relationships between these notions are known.

18.16. Theorem. [Rog 67], [BlMa 73]

1. Every hyperhypersimple set is hypersimple but not vice versa.
2. Every creative set is subcreative (note that the notions of creativity and $\leqq_m$-completeness in Σ_1 coincide).
3. The classes of the subcreative sets and the hyperhypersimple sets are disjoint.
4. There exist subcreative sets which are hypersimple. ∎

18.17. Theorem.

1. [BlMa 73] A r.e. set A is in **PS** iff there is no $s \in \mathbb{R}_1$ such that
$$\bigwedge_i (D_i \cap \bar{A} = D_{s(i)} \cap \bar{A} \wedge (D_i \subset A \to D_{s(i)} \text{ is finite})).$$
2. [Soa 77] A r.e. set A is in **PS** iff $A^{\langle 1 \rangle} \in \Sigma_2 \cap \Pi_2$.
3. [Ben 79] A r.e. set $A \neq \mathbb{N}$ is in **PS** iff $A^{\langle 2 \rangle}$ is not $\leqq_1$-complete in Σ_2.
4. [Ben 79] A r.e. set A is in $\mathbf{PS}_{\mathrm{eff}}$ iff $A^{\langle 1 \rangle}$ is $\leqq_1$-complete in Σ_2.
5. [BlMa 73] A r.e. set A is in $\mathbf{PS}_{\mathrm{eff}}$ iff it is subcreative. ∎

From 18.16 we obtain

18.18. Corollary.

1. Every creative set is in $\mathbf{PS}_{\mathrm{eff}}$.
2. No hyperhypersimple set is in $\mathbf{PS}_{\mathrm{eff}}$.
3. There are hypersimple sets which are in $\mathbf{PS}_{\mathrm{eff}}$. ∎

Between the sets from **PS** on the one hand and the hypersimple and hyperhypersimple sets on the other hand, resp., the following relationships are known.

18.19. Theorem. [Marq 75]

1. Every hyperhypersimple set is in **PS**.
2. There are hypersimple sets which are not in **PS**. ∎

18.20. Corollary. $\mathbf{PS}_{\mathrm{eff}} \subset \mathbf{PS}$. ∎

For connection between speedable sets and complexity sequences see [BeSo 78].

By definition, a function $\varphi \in \mathbb{P}_1$ does not have partial speed-up if there is a function $h \in \mathbb{R}_2$ and a $\varphi_i = \varphi$ such that

$$\bigwedge_{\varphi_j = \varphi} \Phi_i \leqq_{\mathrm{ae}} h \circ \Phi_j.$$

In this case we say that i is an h-optimal program for φ, or simply an optimal program for φ. I. Filotti has observed that a function $\varphi \in \mathbb{P}_1$ has only optimal programs if it has one optimal program (cf. [Marq 75]). Consequently, a recursive function has only optimal programs. In [Aln 73b] it is shown that it is not possible to find effectively an h-optimal program for a function with h-optimal programs.

Still another question in this connection must be answered in the negative, namely the question of whether an arbitrary partial recursive function has an optimal sequence of programs. The sequence $\{\varphi_{s(i)} : i \in \mathbb{N}\}$, $s \in \mathbb{R}_1$, is said to be an *optimal sequence of programs* of $\varphi \in \mathbb{P}_1$ if and only if

$$\bigwedge_i \varphi_{s(i)} = \varphi \wedge \bigwedge_{\varphi_j = \varphi} \bigvee_i \Phi_{s(i)} \leqq_{\text{ae}} \Phi_j .$$

In [Tra 67] it is shown that for every Blum measure there are recursive functions which do not have an optimal sequence of programs.

A special kind of partial speed-up is "levelling". A function $\varphi \in \mathbb{P}_1$ has *r-g-levelling* ($r, g \in \mathbb{R}_1$, $g > r$) if and only if

$$\bigwedge_{\varphi_i = \varphi} \bigvee_{\varphi_j = \varphi} \bigvee_x^{\text{io}} \big(\Phi_j(x) \leqq r(x) \wedge \Phi_i(x) > g(x)\big).$$

A function $\varphi \in \mathbb{P}_1$ is said to be *levelable* if and only if there is an $r \in \mathbb{R}_1$ such that φ has r-g-levelling for all $g \in \mathbb{R}_1$ such that $g > r$.

A function $\varphi \in \mathbb{P}_1$ has effective r-g-levelling if and only if there is an $s \in \mathbb{R}_1$ such that

$$\bigwedge_{\varphi_i = \varphi} \Big(\varphi_{s(i)} = \varphi \wedge \bigvee_x^{\text{io}} \big(\Phi_{s(i)}(x) \leqq r(x) \wedge \Phi_i(x) > g(x)\big)\Big).$$

A function $\varphi \in \mathbb{R}_1$ is said to be effectively levelable if and only if there is an $r \in \mathbb{R}_1$ and an $s \in \mathbb{R}_2$ such that

$$\bigwedge_{\substack{\varphi_l \in \mathbb{R}_1 \\ \varphi_l > r}} \bigwedge_{\varphi_i = \varphi} \Big(\varphi_{s(l,i)} = \varphi \wedge \bigvee_x^{\text{io}} \big(\Phi_{s(l,i)}(x) \leqq r(x) \wedge \Phi_i(x) > \varphi_l(x)\big)\Big).$$

For recursively enumerable subsets of $\mathbb{N}$ we define

$$\mathbf{LE} = \{A : A \text{ recursively enumerable} \wedge \chi_A \text{ levelable}\},$$

$$\mathbf{LE}_{\text{eff}} = \{A : A \text{ recursively enumerable} \wedge \chi_A \text{ effectively levelable}\}.$$

Again, these classes do not depend on the chosen Blum measure because any two Blum measures can be recursively bounded by each other (cf. 5.21). As an immediate consequence of the definition of $\mathbf{LE}$ and $\mathbf{LE}_{\text{eff}}$ we have

18.21. Proposition.

1. Every (effectively) levelable function has (effective) partial speed-up.
2. $\mathbf{LE} \subseteqq \mathbf{PS}$ and $\mathbf{LE}_{\text{eff}} \subseteqq \mathbf{PS}_{\text{eff}}$. ∎

In 18.18.1 we have stated that every creative set is in $\mathbf{PS}_{\text{eff}}$. We even have

18.22. Theorem. [Blu 71] Every creative set is in $\mathbf{LE}_{\text{eff}}$. ∎

The next theorem shows that LE_{eff} is properly included in PS_{eff}.

18.23. Theorem. [BlMa 73] $\mathbf{PS}_{\text{eff}} \setminus \mathbf{LE} \neq \emptyset$. ∎

Levelability is actually a recursion-theoretic notion:

18.24. Theorem. [BlMa 73] A recursively enumerable set A is in **LE** if and only if there is an $s \in \mathbb{R}_1$ and a sequence $(M_j)_{j \in \mathbb{N}}$ of recursive sets such that

$$\bigwedge_i \varphi_{s(i)} = c_{M_i} \wedge \bigwedge_{B \in \mathbf{REC}} \left(B \subseteq A \to \bigvee_i (M_i \cap \overline{B} \text{ is infinite} \wedge M_i \subset A)\right). \blacksquare$$

Similar characterizations of **LE** and $\mathbf{LE}_{\text{eff}}$ as given in 18.17 for **PS** and $\mathbf{PS}_{\text{eff}}$ can be found in [Fil 72], [Soa 77] and [Ben 79].

Though a recursive set cannot be levelable, there are recursive sets which have r-g-levelling for functions r and g which are small compared with the complexity of the set. This is shown by the next theorem, which is due to A. R. Meyer.

18.25. Theorem. [Sto 74] Let $\log \leqq h < s$ and let s be fully 2-T-DSPACE constructible. Then there is an $A \in \mathbf{DSPACE}(s)$ which has effective 0-h-levelling. ∎

The property of effective log-h-levelling is translated by a certain kind of $\leqq_m^{\text{log-lin}}$ reducibility.

18.26. Theorem. [Sto 74] Let $A \leqq_m^{\text{log-lin}} B$ via f, where f is a one-one function such that $|f(x)| \geqq |x|$ and

$$f^{-1}(y) = \begin{cases} x & \text{if } f(x) = y, \\ e & \text{if } y \notin R_f \end{cases}$$

is in **L**. Furthermore, let $h \geqq \log$ be an increasing function. If A has effective log-h-levelling, then B has effective log-$h(\lceil c \cdot n \rceil)$-levelling for some $c > 0$. ∎

Since the $\leqq_m^{\text{log-lin}}$-completeness is in many cases shown by reductions which fulfil the suppositions of Theorem 18.26, the corresponding complete sets have log-h-levelling for suitable functions h.

18.27. Corollary. [Sto 74] NEC(0, 1, ∪, ·, *, ²) is effectively log-c^n-levelable for some $c > 0$ (with respect to the measure 2-T-DSPACE). ∎

Related results can be found in [Berl 76].

Part 3

Relationships between Different Measures

Chapter VI
Simulation

Simulation results have been developed in order to compare the computational power or different types of machine. They allow the translation of lower and upper bounds from one measure to another.

We first investigate realistic space measures and realistic time measures (§ 19). Measures which are not realistic or which are not known to be realistic are considered in § 20, where we deal with measures of machines with weak or powerful instructions, such as SRAM's or MRAM's, machines with auxiliary stores, recursive and parallel machines and machines with modified acceptance such as alternating and probabilistic machines.

Further measures for Turing machines are dealt with in § 21, and § 22 focuses on relationships between various realtime and linear time classes.

§ 19. Realistic Space and Time Measures

This section deals with the comparison of different realistic space measures and of different realistic time measures (see p. 66 and p. 70, resp.). Some results, especially those on the relationships between the time measures of different types of Turing machines date from the beginning of the theory of computational complexity. New and interesting methods have been developed for their proof and this methods have later been fruitfully applied to many other problems. The importance of the results themselves lies in the possibility of more easily finding efficient algorithms for certain problems. For example, in most cases it is much easier to construct a multihead Turing machine which solves a given problem than a time equivalent multi-tape Turing machine with only one head per tape. The existence of the latter then follows by the theorem on the real-time simulation of multi-head Turing machines by multi-tape Turing machines with only one head per tape.

19.1. Comparison of Space Measures

It is easy to see that for $s \geqq 0$ and X $\in$ {D, N}, all classes **2-kTh-XSPACE**(s) coincide, i.e. **2-kTh-XSPACE**(s) = **2-T-XSPACE**(s) for all $k, h \geqq 1$. In order to prove the above equality one constructs for a given 2-kTh-XM M_1 an equivalent 2-T-XM M_2

using no more space than M_1 and having a larger worktape alphabet. Thus if we fix the number of worktape symbols to be not greater than m, we would expect that it is not in general possible to prove

$$\textbf{2-kT}_{\mathbf{m}}\textbf{h-XSPACE}(s) \subseteq \textbf{2-T}_{\mathbf{m}}\textbf{-XSPACE}(s)$$

(although $\textbf{2-kT}_{\mathbf{m}}\textbf{h-XSPACE}(\text{Lin } s) = \textbf{2-T}_{\mathbf{m}}\textbf{-XSPACE}(\text{Lin } s) = \textbf{XSPACE}(s)$, i.e. $\textbf{2-kT}_{\mathbf{m}}\textbf{h-XSPACE}$ is a realistic space measure). This is actually the case, and we shall now see under what conditions an increase of the number of worktape symbols, the number of worktapes or the number of worktape heads, resp., can enlarge the space complexity class for a fixed bounding function s. First we fix the number k of worktapes. For $k = 1$ we have

19.1. Theorem. [Sei 77a], [Sei 77b]

1. For $m \geqq 2$ and unbounded fully 2-T-DSPACE constructible s,
   1. $\textbf{2-T}_{\mathbf{m}}\textbf{-DSPACE}(s) \subset \textbf{2-T-DSPACE}(s)$,
   2. $\textbf{2-T}_{\mathbf{m}}\textbf{-NSPACE}(s) \subset \textbf{2-T-NSPACE}(s)$ if $1 < s(n+1) \lesssim s(n)$.
2. For sufficiently large m and unbounded fully $\textbf{2-T}_{\mathbf{m}}\textbf{-DSPACE}$ constructible s,
   1. $\textbf{2-T}_{\mathbf{m}}\textbf{-DSPACE}(s) \subset \textbf{2-T}_{\mathbf{m+1}}\textbf{-DSPACE}(s)$ if $s(n+1) - s(n) < s(n)$,
   2. $\textbf{2-T}_{\mathbf{m}}\textbf{-NSPACE}(s) \subset \textbf{2-T}_{\mathbf{m+1}}\textbf{-NSPACE}(s)$ if $s(n+1) - s(n) < s(n)$. ∎

Remark 1. Note that the condition $s(n+1) \lesssim s(n)$ $\big(s(n+1) - s(n) < s(n)\big)$ excludes all functions such that $s(n) >_{\text{io}} 2^{c \cdot n}$ for all $c > 0$ $(s(n) \geqq_{\text{io}} 2^{c \cdot n}$ for some $c > 0)$.

Remark 2. The proof of the above results is similar to the proof of the corresponding hierarchy results for 2-T-DSPACE and 2-T-NSPACE in § 17 (i.e. hierarchy with respect to the bounding function s). Special cases of 19.1.2 can be found in [Iba 74].

The above results also apply to the case of more worktapes. This follows from Corollary 19.3, which is a consequence of the following evident lemma (stated first in [IbSa 75]).

19.2. Lemma. For $k, m \geqq 2$, $X \in \{D, N\}$ and arbitrary s,

1. $\textbf{2-T}_{\mathbf{m}^k}\textbf{-XSPACE}(s) \subseteq \textbf{2-kT}_{\mathbf{m}}\textbf{-XSPACE}(s)$,
2. $\textbf{2-kT}_{\mathbf{m}}\textbf{-XSPACE}(s) \subseteq \textbf{2-T}_{\mathbf{m}^k+1}\textbf{-XSPACE}(s)$. ∎

19.3. Corollary. For $k, m_0 \geqq 2$, $X \in \{D, N\}$ and arbitrary s,

$$\textbf{2-T}_{\mathbf{m}}\textbf{-XSPACE}(s) \subset \textbf{2-T}_{\mathbf{m+1}}\textbf{-XSPACE}(s) \quad (\subset \textbf{2-T-XSPACE}(s), \text{ resp.})$$

for all $m \geqq m_0$ implies

$$\textbf{2-kT}_{\mathbf{m}}\textbf{-XSPACE}(s) \subset \textbf{2-kT}_{\mathbf{m+1}}\textbf{-XSPACE}(s) \quad (\subset \textbf{2-kT-XSPACE}(s), \text{ resp.})$$

for all $m \geqq m_0$. ∎

Next we fix the number m of worktape symbols. As a further consequence of Lemma 19.2 we have

19.4. Corollary. For $k \geqq 1$, $m_0 \geqq 2$, X $\in$ {D, N} and arbitrary s,

$$\mathbf{2\text{-}T_m\text{-}XSPACE}(s) \subset \mathbf{2\text{-}T_{m+1}\text{-}XSPACE}(s) \quad (\subset \mathbf{2\text{-}T\text{-}XSPACE}(s), \text{ resp.})$$

for all $m \geqq m_0$ implies

$$\mathbf{2\text{-}kT_m\text{-}XSPACE}(s) \subset \mathbf{2\text{-}(k+1)T_m\text{-}XSPACE}(s)$$
$$(\subset \mathbf{2\text{-}multiT_m\text{-}XSPACE}\ (s), \text{ resp.})$$

for all $m \geqq m_0$. ∎

Remark. From 19.2 we conclude the equality

$$\mathbf{2\text{-}multiT_m\text{-}XSPACE}(s) = \mathbf{2\text{-}T\text{-}XSPACE}(s)$$

for $m \geqq 2$, $X \in$ {D, N} and arbitrary s.

Except for 19.5.2.2 which can be shown by the usual diagonalization argument (cf. the hierarchy result for 2-T-DSPACE in § 17), all statements in the next theorem are immediate consequences of 19.1 and 19.4.

19.5. Theorem.

1. For $k \geqq 1$, $m \geqq 2$ and unbounded fully 2-T-DSPACE constructible s,
   1. $\mathbf{2\text{-}kT_m\text{-}DSPACE}(s) \subset \mathbf{2\text{-}multiT_m\text{-}DSPACE}(s)$,
   2. $\mathbf{2\text{-}kT_m\text{-}NSPACE}(s) \subset \mathbf{2\text{-}multiT_m\text{-}NSPACE}(s)$ if $1 < s(n+1) \leq s(n)$.
2. For $k \geqq 1$, sufficiently large m and unbounded fully 2-T_m-DSPACE constructible s,
   1. $\mathbf{2\text{-}kT_m\text{-}DSPACE}(s) \subset \mathbf{2\text{-}(k+1)T_m\text{-}DSPACE}(s)$ if $s(n+1) - s(n) < s(n)$,
   2. $\mathbf{2\text{-}kT_m\text{-}DSPACE}(s) \subset \mathbf{2\text{-}(k+2)T_m\text{-}DSPACE}(s)$ if $\log n < s(n)$,
   3. $\mathbf{2\text{-}kT_m\text{-}NSPACE}(s) \subset \mathbf{2\text{-}(k+1)T_m\text{-}NSPACE}(s)$ if $s(n+1) - s(n) < s(n)$. ∎

Next we fix both the number m of worktape symbols as well as the number k of worktapes. The following theorem shows that worktape symbols cannot be replaced by additional worktape heads (contrary to the case of worktapes which can replace worktape symbols and vice versa, cf. Lemma 19.2).

19.6. Theorem. [Sei 77a], [Sei 77b] For $k \geqq 1$, sufficiently large m, X $\in$ {D, N} and fully 2-T_{m+1}-DSPACE constructible s such that $s(n+1) - s(n) < s(n)$,

$$\mathbf{2\text{-}kT_m multi\text{-}XSPACE}(s) \subset \mathbf{2\text{-}kT_{m+1}\text{-}XSPACE}(s). \ ∎$$

However, an increase in the number of worktape heads can enlarge the space complexity class in some cases.

19.7. Theorem. [Sei 77a], [Sei 77b] For $h \geqq 1$, $m \geqq 2$, X $\in$ {D, N} and unbounded fully 2-T_mh-DSPACE constructible s,

$$\mathbf{2\text{-}T_m h\text{-}XSPACE}(s) \subset \mathbf{2\text{-}T_m multi\text{-}XSPACE}(s)$$
$$\text{if } s(n+1) - s(n) \leq \log s(n) \text{ and } 1 < s(n). \ ∎$$

For related results, especially for SLA witnesses for the proper inclusions stated in this subsection, see [Sei 77a], [Sei 77b].

19.2. Comparison of Time Measures

19.2.1. Turing Machines, Tapes, Heads and Dimension

19.2.1.1. Number of Tapes versus Number of Heads

In this subsection we shall see how a number of Turing tapes can time-equivalently be replaced by heads working on one tape, and vice versa.

First it is easy to see that the work on k tapes can be simulated in the same time on one tape, where the same number of heads work on k tracks in the same way as they worked originally on the k different tapes.

19.8. Theorem. For $i = 0, 1, 2$, $k \geqq 1$, $h \geqq 1$, $d \geqq 1$, $\mathrm{X} \in \{\mathrm{D}, \mathrm{N}\}$ and $t \geqq \mathrm{id}$,

$$\textbf{i-kT}^{\textbf{d}}\textbf{h-XTIME}(t) \subseteq \textbf{i-T}^{\textbf{d}}(\textbf{k} \cdot \textbf{h})\textbf{-XTIME}(t). \ \blacksquare$$

The proof of the converse is much more complicated. First we consider the case of Turing machines with one-dimensional tapes.

19.9. Lemma. [Stoß 70] For $i = 0, 1, 2$, $k \geqq 1$, $h \geqq 2$, $\mathrm{X} \in \{\mathrm{D}, \mathrm{N}\}$ and $t \geqq \mathrm{id}$,

$$\textbf{i-kTh-XTIME}(t) \subseteq \textbf{i-(k} \cdot \textbf{h)T-XTIME}(\mathrm{Lin}\, t).$$

Proof. Since the several tapes of an i-kTh-XM will be simulated separately and simutaneously it will be sufficient to show **i-Th-XTIME**(t) $\subseteq$ **i-hT-XTIME**$(\mathrm{Lin}\, t)$.

Let M_1 be an i-Th-XM working within time $t(n)$. It is easy to construct an equivalent i-Th-XM M_2 working within time $c_2 \cdot t(n)$, for some $c_2 > 1$, for which in each step only one head is active, i.e. only one head can write and move. All other i-Th-XM's constructed in this proof (i.e. M_3 and M_4) will also have this property.

Next we construct an equivalent (to M_2) i-Th-XM M_3 working within time $c_3 \cdot t(n)$, for some $c_3 > 1$, whose heads $H_1, H_2, \ldots, H_h$ fulfil in every step of the computation the condition $|H_1| \leqq |H_2| \leqq \ldots \leqq |H_h|$ where $|H_j|$ is the number of the tape square scanned by H_j at this moment ($j = 1, \ldots, h$). The machine M_3 proceeds as follows: At the beginning of the computation the condition is fulfilled automatically with $|H_1| = |H_2| = \ldots = |H_h| = 1$. Further, if M_2 performs a step in which H_j is active, then the head H_j of M_3 checks by writing a test symbol whether there are still other heads on the same tape square. If this is the case and the simulation of the present step of M_2 would yield $|H_j| < |H_k|$ for some $k < j$ (or $|H_j| > |H_k|$ for some $k > j$), then H_j interchanges its role with that of H_k. If there are several such k, this will be done with the smallest (greatest) of them.

Now we construct an equivalent (to M_3) i-Th-XM M_4 working within time $c_4 \cdot t(n)$, for some $c_4 > 1$, whose heads $H_1', H_2', \ldots, H_h'$ fulfil at the beginning of the computation the condition $|H_j'| = j$ ($j = 1, \ldots, h$) and in every step of the computation the condition $|H_1'| < |H_2'| < \ldots < |H_h'|$. The machine M_4 will work in such a way that, before the simulation of each step of M_3, the head H_j' will be situated $j - 1$ tape squares to the right of that square on which the head H_j is situated. At the beginning of the computation this condition is fulfilled by $|H_j'| = j$. In order to simulate a step of M_3 in which H_j is active, M_4 first moves its heads $H_1', H_2', \ldots, H_j'$ consecutively

$j - 1$ tape squares to the left, then H'_j simulates the activity of H_j, and finally M_4 moves $H'_j, H'_{j-1}, \ldots, H'_1$ consecutively $j - 1$ tape squares to the right. Thus the order $|H_1| \leqq |H_2| \leqq \ldots \leqq |H_h|$ of the heads of M_3 carry over to the heads of M_4 and moreover there will never be two heads of M_4 simultaneously on the same tape square.

Finally we construct an equivalent (to M_4) i-hT-XM M_5 working within time $c_5 \cdot t(n)$, for some $c_5 > 0$. The machine M_5 (having the heads $H''_1, H''_2, \ldots, H''_h$ on the tapes $1, 2, \ldots, h$, resp.) will work in such a way that before the simulation of each step of M_4 we have the following situation (*):

There is a decomposition of the Turing tapes into regions $R_1, R_2, \ldots, R_h$ such that

a) the region R_1 is infinite to the left, the region R_j is neighbouring to R_{j+1} ($j = 1, \ldots, j - 1$), and R_h is infinite to the right,

b) the head H''_j is situated in the region R_j, and $|H'_j| = |H''_j|$ ($j = 1, \ldots, h$),

c) the contents of the region R_j of the tape j of M_5 will be identical with the contents of the region R_j of the tape of M_4 except for an additional marker L in the leftmost tape square of R_j (if ($\neq 1$) and an additional marker R in the rightmost tape square of R_j (if $j \neq h$). Outside R_j the tape j of M_5 will have only symbols $\square$ ($j = 1, \ldots, h$).

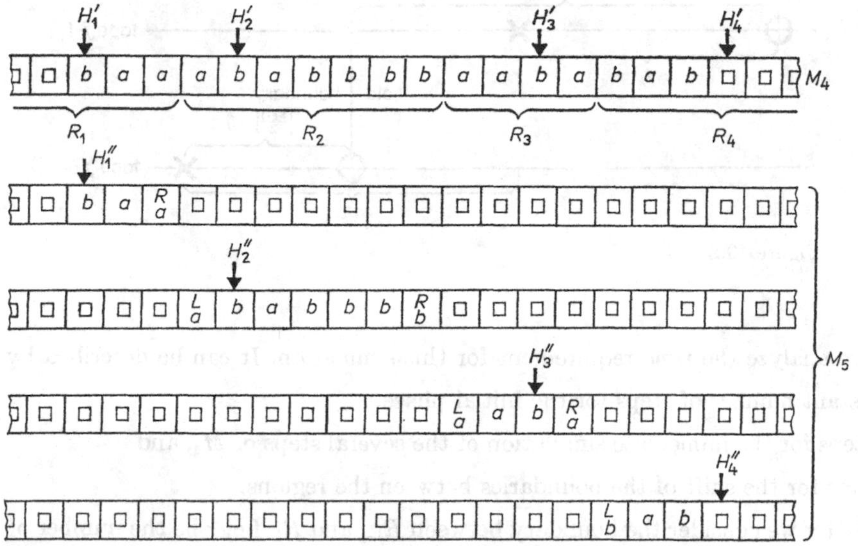

Figure 19.1

Figure 19.1 shows an example for the case $h = 4$. In an initial phase M_5 moves H''_j to the tape square j ($j = 2, \ldots, h$). If the tape of M_4 is the input tape (i.e. in the case $i = 0$), then the tape h will be the input tape of M_5 and the contents of the tape square j of the tape h will be replaced by $\square$ and transferred to the tape square j of the tape j ($j = 1, \ldots, h - 1$). Furthermore, in the tape square j of the tape j the markers L (if $j \neq 1$) and R (if $j \neq h$) will be added ($j = 1, \ldots, h$). Thus the situation (*) for the simulation of the first step of M_4 is reached.

We describe now how M_5 simulates a step of M_4 and restores the situation (*). Assume that in this step of M_4 the head H'_j is active.

Case 1. H'_j does not cross the boundaries of R_j in this step. Then this step of M_4 is simulated directly by M_5 without changing the region R_j.

Case 2. H'_j crosses a boundary of R_j (say, the left one) in this step. Then R_{j-1} and R_j will be redefined in such a way that the boundary between them is moved to the left. More exactly the new boundary between R_{j-1} and R_j will be between the tape squares $\left\lfloor \frac{|H'_{j-1}| + |H'_j|}{2} \right\rfloor$ and $\left\lfloor \frac{|H'_{j-1}| + |H'_j|}{2} \right\rfloor + 1$ (i.e. approximately in the middle between the head positions $|H'_{j-1}|$ and $|H'_j|$, where H'_j is the head position after the execution of the present step). This can be done by moving forth and back the heads H''_{j-1} and H''_j using the marker R on tape $j-1$. After the rearrangement of the regions M_5 can simulate the current step of M_4 without crossing any boundary by H''_j. Moreover, the situation (*) is restored.

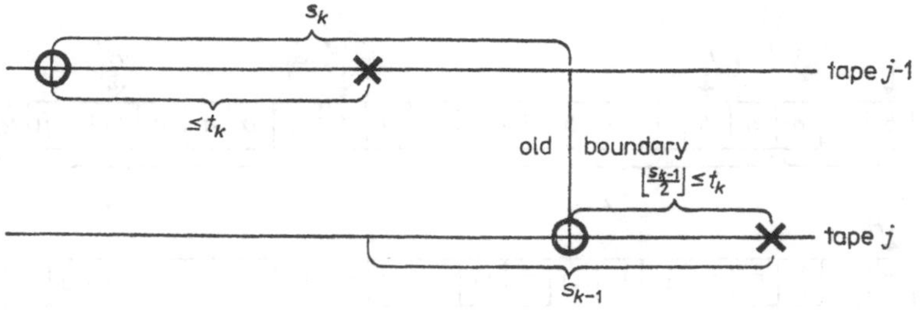

Figure 19.2

Now we analyze the time requirement for this simulation. It can be described by

a) a constant number of steps for the initial phase,

b) $t(n)$ steps for the immediate simulation of the several steps of M_4, and

c) the time for the shift of the boundaries between the regions.

For the latter we consider the boundary between R_{j-1} and R_j. Let r be the number of shifts of this boundary, and let t_k be the number of steps of M_4 between the $(k-1)$th and the kth shift ($k = 1, \ldots, r$). Consequently, $\sum_{k=1}^{r} t_k \leqq c_4 \cdot t(n)$. Futher let s_k be the distance between H'_{j-1} and H'_j before the kth shift ($k = 1, \ldots, r$). Obviously, M_5 needs at most $2 \cdot s_k$ steps for the kth shift. The worst possible case is that H'_{j-1} moves between the $(k-1)$th and the kth shift of the boundary in each step to the left (see Figure 19.2 where $\times$ denotes the head positions before the $(k-1)$th shift and $\circ$ denotes the head positions before the kth shift).

Hence, $s_k \leqq t_k + \left\lceil \frac{s_{k-1}}{2} \right\rceil \leqq t_k + (t_k + 1)$, and M_5 needs at most $2 \cdot (2t_k + 1)$ steps to perform the kth shift. For all shift of this boundary

$$\sum_{k=1}^{r} 2 \cdot (2t_k + 1) \leqq 5 \cdot \sum_{k=1}^{r} t_k \leqq 5 \cdot c_4 \cdot t(n) \quad \text{steps}$$

are needed. Consequently, for all $h - 1$ boundaries M_5 needs $5 \cdot c_4 \cdot (h - 1) \cdot t(n)$ steps. ∎

As an immediate consequence of the preceding lemma and the linear speed-up for i-kT-XTIME (cf. 18.12) we have

19.10. Theorem.

1. For $i = 0, 1$, $k \geqq 1$, $h \geqq 2$, $X \in \{D, N\}$ and $t(n) \geqq (1 + \varepsilon) \cdot n$ where $\varepsilon > 0$,

$$\textbf{i-kTh-XTIME}(t) \subseteqq \textbf{i-(k} \cdot \textbf{h)T-XTIME}(t).$$

2. For $k \geqq 1$, $h \geqq 2$, $X \in \{D, N\}$ and $t(n) \geqq (1 + \varepsilon) \cdot n$ where $\varepsilon > 0$,

$$\textbf{2-kTh-XTIME}(t) \subseteqq \textbf{2-(k} \cdot \textbf{h + 1)T-XTIME}(t). \ ∎$$

Remark. Since in the proof of Lemma 19.9 the tapes are simulated seperately and simultaneously, Theorem 19.10 can be generalized as follows.

For $i = 0, 1$, $r \geqq 1$, $k_1, \ldots, k_r$, $h_1, \ldots, h_r \geqq 1$, $X \in \{D, N\}$ and $t(n) \geqq (1 + \varepsilon) \cdot n$, for some $\varepsilon > 0$,

$$\textbf{i-k}_1\textbf{Th}_1\textbf{-k}_2\textbf{Th}_2\textbf{-}\ldots\textbf{-k}_r\textbf{Th}_r\textbf{-XTIME}(t) \subseteqq \textbf{i-}\left(\sum_{j=1}^{r} \textbf{k}_j \cdot \textbf{h}_j\right)\textbf{T-XTIME}(t).$$

Because of 19.8 the converse is also true. The resulting equality can be reformulated in the following way: The power of a Turing machine without input tape or with a one-way input tape with the time restriction $t(n) \geqq (1 + \varepsilon) \cdot n$, for some $\varepsilon > 0$, depends only on the number of its heads and does not depend on the number of its tapes and the distribution of these heads on the tapes.

Theorem 19.10 excludes only very small time bounds of which the most important is $t = \text{id}$. In these cases one can trade one head only for more than one tape.

19.11. Theorem. [LeSe 77] For $i = 0, 1, 2$, $k \geqq 1$, $h > 1$, $X \in \{D, N\}$ and $t \geqq \text{id}$,

$$\textbf{i-kTh-XTIME}(t) \subseteqq \textbf{i-k(4h-4)T-XTIME}(t). \ ∎$$

Remark 1. Theorem 19.11 improves an earlier result of P. C. Fischer, A. R. Meyer, and A. L. Rosenberg (cf. [FipMeRo 68]) in which $8h - 7$ stands for the factor $4h - 4$.

Remark 2. In the nondeterministic case most of the results of 19.10 and 19.11 can be improved by the equations **multiT-NTIME**(t) = **1-2T-NTIME**(t) (cf. 19.20).

Now we state some results for Turing machines with multidimensional worktapes.

19.12. Lemma. [LeSe 77] For $i = 0, 1, 2,\ k \geq 1,\ d \geq 2,\ h \geq 2,\ \mathrm{X} \in \{\mathrm{D}, \mathrm{N}\}$ and $t \geq \mathrm{id}$,

1. $\mathbf{i\text{-}kT^dh\text{-}XTIME}(t) \subseteq \mathbf{i\text{-}3kT^d(h\text{-}1)\text{-}multiT\text{-}XTIME}(t)$.
2. $\mathbf{i\text{-}kT^dh\text{-}XTIME}(t) \subseteq \mathbf{1\text{-}k\binom{h}{3}T^d3\text{-}multiT\text{-}XTIME}(t)$. ∎

As an immediate consequence we have

19.13. Theorem. For $i = 0, 1, 2, k \geq 1, d \geq 2, h \geq 2, \mathrm{X} \in \{\mathrm{D}, \mathrm{N}\}$ and $t \geq \mathrm{id}$,

1. $\mathbf{i\text{-}kT^dh\text{-}XTIME}(t) \subseteq \mathbf{i\text{-}(3^{h-1} \cdot k)T^d\text{-}multiT\text{-}XTIME}(t)$.
2. $\mathbf{i\text{-}kT^dh\text{-}XTIME}(t) \subseteq \mathbf{i\text{-}9k\binom{h}{3}T^d\text{-}multiT\text{-}XTIME}(t)$. ∎

Remark 1. For $h \geq 6$ statement 19.13.2 is better than statement 19.13.1.

Remark 2. Theorem 19.13.1 yields as a special case $\mathbf{2\text{-}T^22\text{-}DTIME}(t) \subseteq \mathbf{2\text{-}3T^2\text{-}multiT\text{-}DTIME}(t)$. In [Pau 81] it has been shown that this cannot be improved to $\mathbf{2\text{-}T^22\text{-}DTIME}(t) \subseteq \mathbf{2\text{-}2T^2\text{-}DTIME}(t)$.

19.14. Corollary. For $i = 0, 1, 2, d \geq 1, \mathrm{X} \in \{\mathrm{D}, \mathrm{N}\}$ and $t \geq \mathrm{id}$,

$$\mathbf{i\text{-}multiT^dmulti\text{-}XTIME}(t) = \mathbf{multiT^d\text{-}XTIME}(t) = \mathbf{T^dmulti\text{-}XTIME}(t). \ ∎$$

19.2.1.2. Reductions of the Number of Tapes and Heads

The question we shall deal with in this subsection is what amount of time has to be paid for simultaneously reducing the number of tapes and heads of a Turing machine. Because of the results of the preceding subsection this question is equivalent to the following question concerning multitape Turing machines with only one head per tape: How can multitape TM's be simulated by multitape TM's with a smaller number of tapes. Contrary to the preceding subsection in which the loss of tapes could be equalized by adding heads to the remaining tapes we have to expect that the reduction of tapes with only one head each must be paid for by larger time bounds.

We start with the reduction to one tape.

19.15. Theorem. For $d \geq 1, \mathrm{X} \in \{\mathrm{D}, \mathrm{N}\}$ and $t \geq \mathrm{id}$,

$$\mathbf{multiT^d\text{-}XTIME}(t) \subseteq \mathbf{T^d\text{-}XTIME}(t^2).$$

Proof. Let M be a $\mathrm{kT^d}$-XM for some $k > 1$. We describe the work of a $\mathrm{T^d}$-XM M simulating M. The tape of M' has k tracks, and after the simulation of a step of M the ith track of every tape square of M' has the same content as the corresponding tape square of the ith tape of M. A step of M will be simulated by M' in such a way that the head of M' visits all k (marked) tape squares which correspond to tape squares scanned by the heads of M.

For $d > 1$ the head of M' can find the way between these k head positions by marked directed paths. These paths are updated during the simulation. If such a path between two head positions intersects itself, the superflous cycle can be deleted. Since a path between two head positions has at most length $2 \cdot t(n)$, one step of M can be simulated by M' within $\preceq t(n)$ steps. ∎

Remark. It is not hard to see that a "compact" encoding of the contents of the one-dimensional tapes in a d-dimensional cube of a d-dimensional tape can reduce the length of the paths between the head positions (see the above proof) to be not longer than $2 \cdot t(n)^{1/d}$. Thus one step of a kT-XM can be simulated by a suitable T^d-XM within $\lesssim t(n)^{1/d}$ steps. Hence we have for $d \geqq 1$, $X \in \{D, N\}$ and $t(n) \gtrsim n^2$,

$$\mathbf{multiT\text{-}XTIME}(t) \subseteqq \mathbf{T^d\text{-}XTIME}(t^{1+1/d}).$$

Now we deal with the reduction to two tapes. First the deterministic case.

19.16. Theorem. [HeSt 66] For $t \geqq \mathrm{id}$,

$$\mathbf{multiT\text{-}DTIME}(t) \subseteqq \mathbf{T\text{-}PD\text{-}DTIME}(t \cdot \log t).$$

Proof. Let M be a kT-DM working within time $t(n)$. We describe the work of a T-PD-DM M' simulating M. The Turing tape of M' will have two tracks for each tape of M. We focus our attention on the two tracks of M' corresponding to the jth tape of M. The pushdown store is used only to transport information on the Turing tape. The symbol written in a square of the jth tape will be recorded in exactly one tape square of M' and there either in the lower track or in the upper track. The lower (upper) track of a tape square of M' contains either exactly one such symbol or it is empty.

The two tracks corresponding to the jth tape of M will be subdivided into the blocks $\ldots B_{-2}, B_{-1}, B_0, B_1, B_2, \ldots$ where B_0 consists of one tape square and B_i consists of $2^{|i|-1}$ tape squares for $i > 0$. A block B_i is said to be

- empty iff not only the upper track but also the lower track of each tape square of B_i is empty,
- half-full iff the upper track of each tape square of B_i is empty, and the lower track contains the contents of some square of the jth tape,
- full iff both the lower and the upper track of each tape square of B_i contains the contents of some square of the jth tape.

Every block B_i will be empty, half-full or full. The block B_0 will always be half-full. Its lower track will contain the contents of the square which is scanned by the head of the jth tape.

Hence the contents of all tape squares scanned by the k heads of M at some moment will be contained at the corresponding moment in B_0 and the next step of M can easily be simulated within time $\lesssim 1$. However, after this the contents of the two tracks (corresponding to the jth tape) may have to be shifted in order to bring the contents of the new square scanned by the head of the jth tape into the block B_0.

When the simulation starts the contents of the jth tape are in the lower track of the corresponding tape squares, the upper track will be empty. Later, after every shift, the contents of the jth tape at the corresponding moment can be obtained by writing the contents of the blocks $\ldots B_{-2}, B_{-1}, B_0, B_1, B_2, \ldots$ one after another where for $i < 0$ ($i > 0$) the content of the upper track of B_i is written to the right (left) of that of the lower track.

A shift to the right corresponding to a move of the head of the jth tape to the left will be done according to the following algorithm. A shift to the left will be performed analogously.

1. Let B_i be the first block to the right of B_0 which is not full. If B_i is half-full (empty), then bring the contents of B_{i-1} into the upper (lower) track of B_i where the contents of the lower track of B_{i-1} is brought to the right of the contents of the upper track of B_{i-1}. If $i > 1$, then go to 1.
2. Let B_i be the first block to the left of B_0 which is not empty. If B_i is half-full (full), then bring the contents of the lower (upper) track of B_i into the lower track of the blocks $B_{i+1}, B_{i+2}, \ldots, B_{-1}, B_0$. Stop.

Now the contents of the tape square scanned by the head of the jth track is recorded in the lower track of B_0. Figure 19.3 shows an example in which four shiftes to the right and then four shiftes to the left are performed.

Finally it is obvious that

— the contents of B_i can be changed at most once per $2^{|i|-1}$ steps of M,

— any change of the contents of B_i according to the above algorithm can be performed using the pushdown store within $c \cdot 2^{|i|}$ steps of M' for some $c > 1$,

— only the contents of the blocks B_i such that $|i| \leqq \log t(n) + 1$ can be changed.

Therefore M' needs

$$k \cdot \sum_{i=1}^{\log t(n)} \cdot \frac{t(n)}{2^{|i|-1}} \cdot c \cdot 2^{|i|} \lesssim t(n) \cdot \log t(n) \quad \text{steps}$$

for the simulation of the kT-DM M. ∎

19.17. Corollary. For $t \geqq \mathrm{id}$,

$$\textbf{multiT-DTIME}(t) \subseteqq \textbf{2T-DTIME}(t \cdot \log t) . \ ∎$$

Remark 1. In [IgHo 74] it is shown that in the above theorem the Turing tape and the pushdown store can be replaced by two stacks.*)

Remark 2. In [PiFi 76] Corollary 19.17 has been proved for oblivious 2T-DM's working within time Lin $t \log t$.

19.18. Theorem.

1. [PaSeSi 80] For $d \geqq 2$ and $\mathrm{X} \in \{\mathrm{D}, \mathrm{N}\}$,

$$\textbf{multiT}^{\mathbf{d}}\textbf{-XTIME}(t) \subseteqq \textbf{T-T}^{\mathbf{d}}\textbf{-XTIME}(t^{1+1/d-1/(d^3-d^2+d)}) .$$

2. [Lou 81] For $d \geqq 2$,

$$\textbf{multiT}^{\mathbf{d}}\textbf{-DTIME}(t) \subseteqq \textbf{2T}^{\mathbf{d}}\textbf{-DTIME}(t^{1+1/d-1/d^2} \cdot \log t) . \ ∎$$

Remark 1. Theorem 19.18.1 is valid even for $\mathrm{T^d}$multi-XM's with head-to-head jumps.

---

*) If a stack holds the input w then the initial stack word is w^{-1} where the head scans the top symbol. However, in the results above, 2S can be replaced by 1-2S, and PD-CS can be replaced by 1-PD-CS.

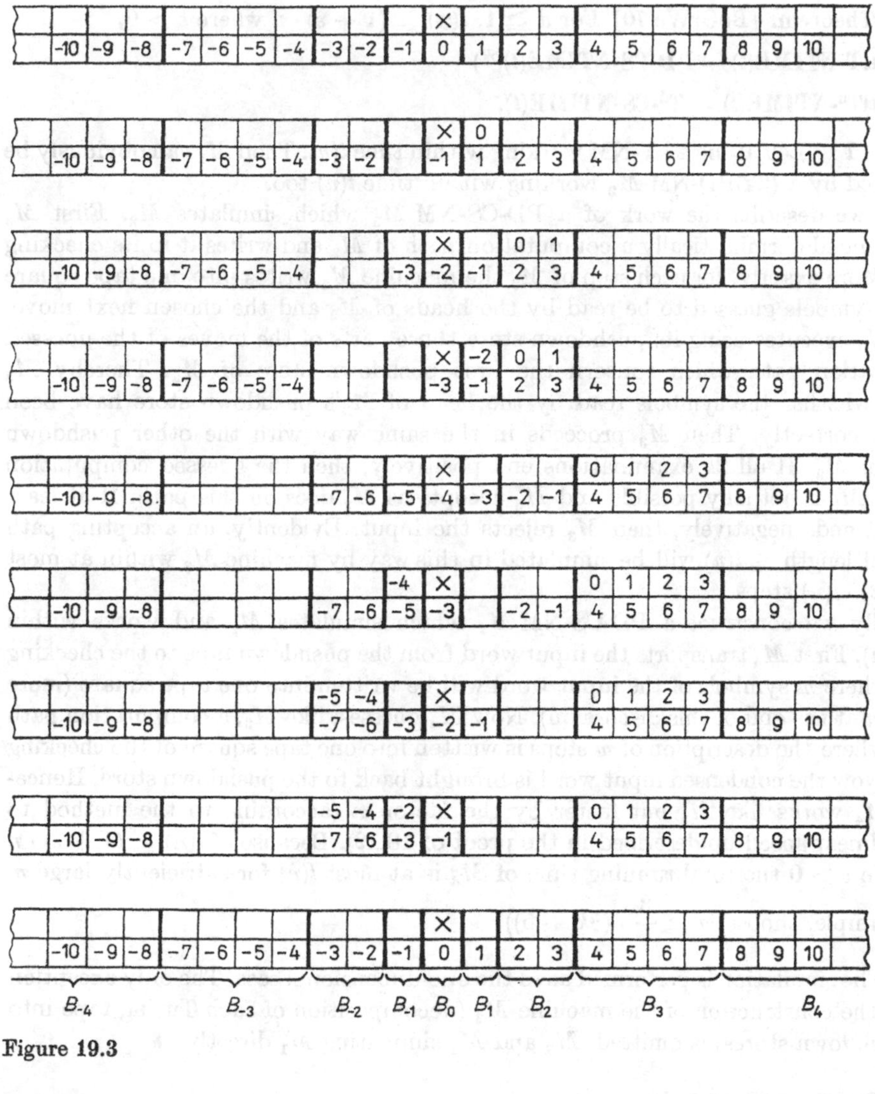

Figure 19.3

Remark 2. By information-theoretic methods the following result has been shown in [PaSeSi 80]. For $h > h' \geqq 1$, $d, d' \geqq 1$, $t \geqq \mathrm{id}$, $g \geqq \mathrm{id}$, if every $h\mathrm{T}^d$-DM working in time t can be simulated step by step by a $\mathrm{T}^{d'}h'$-DM within time $g \circ t$, then $g(n) \geq \frac{n^{1+\alpha}}{\log^\beta n}$, where

$$\alpha = \frac{\left(\frac{d}{d'} - \frac{1}{d}\right)\left(1 - \frac{h'}{h}\right)}{(d+1)\left(1 - \frac{h'}{h}\right) + \frac{d \cdot h'}{d' \cdot h}} \quad \text{and} \quad \beta = \frac{\frac{d \cdot h'}{d' \cdot h}}{(d+1)\left(1 - \frac{h'}{h}\right) + \frac{d \cdot h'}{d' \cdot h}}.$$

In the nondeterministic case the two-tape simulation of multitape Turing machines can be performed without loss of time.

19.19. Theorem. [BoGrWe 70] For $d \geqq 1$, $t(n) \geqq (1 + \varepsilon) \cdot n$ where $\varepsilon > 0$,

1. **multiT-NTIME**(t) = **PD-CS-NTIME**(t),*)
2. **multiT**$^{\mathbf{d}}$**-NTIME**(t) = **T**$^{\mathbf{d}}$**-CS-NTIME**(t).

Proof. Ad 1. Let M_1 be a kT-NM working within time $t(n)$. Then M_1 can obviously be simulated by a (2k)PD-NM M_2 working within time $t(n)$ too.

Now we describe the work of a PD-CS-NM M_3 which simulates M_2. First M_3 guesses nondeterministically a computation path of M_2 and writes it in its checking stack. More exactly, for each step of M_2 the machine M_3 writes into one tape square the $2k$ symbols guessed to be read by the heads of M_2 and the chosen next move. Next M_3 executes using its pushdown store those parts of the moves of the guessed computation path which concern the first pushdown store of M_2. Thereby M_3 checks whether the symbols read by the head of M_3's pushdown store have been guessed correctly. Then M_3 proceeds in the same way with the other pushdown stores of M_2. If all $2k$ examinations end positively, then the guessed computation path of M_2 is actually possible and M_3 accepts as M_2 does on this path. If at least one test ends negatively, then M_3 rejects the input. Evidently, an accepting path of M_2 of length $\leqq t(n)$ will be simulated in this way by machine M_3 within at most $(4k + 2) \cdot t(n)$ steps.

Finally we construct a PD-CS-NM M_4 which simulates M_2 and works within time $t(n)$. First M_4 transports the input word from the pushdown tape to the checking stack where m symbols of the input word will be written into one tape square (more about m at the end of this discussion). Now M_4 guesses, like M_3, a computation path of M_2 where the description of m steps is written into one tape square of the checking stack. Now the condensed input word is brought back to the pushdown store. Henceforth M_4 works like M_3 but faster by the factor m (according to the method to obtain linear speed-up described in the proof of 18.12). Because of $t(n) \geqq (1 + \varepsilon) \cdot n$ for some $\varepsilon > 0$ the total running time of M_4 is at most $t(n)$ for sufficiently large m (for example, choose $m \geqq \frac{3}{\varepsilon} \cdot (8k + 5)$).

Ad 2. The simulation is performed as in the one-dimensional case. The only exception is that the construction of the machine M_2 (decomposition of each Turing tape into two pushdown stores) is omitted, M_3 and M_4 simulating M_1 directly. ∎

19.20. Corollary. For $t \geqq \mathrm{id}$,

$$\textbf{multiT-NTIME}(t) = \textbf{1-PD-CS-NTIME}(t).$$

Proof. Let M be a kT-NM working within time t. A 1-PD-CS-NM M' can simulate M as follows. For an input w it does nondeterministically both: it simulates M as described in the proof of 19.19 and it simulates M as described in the proof of 22.2.

---

*) If a stack holds the input w then the initial stack word is w^{-1} where the head scans the top symbol. However, in the results above, 2S can be replaced by 1-2S, and PD-CS can be replaced by 1-PD-CS.

If M accepts w and $t(|w|) \geqq 2 \cdot |w|$ ($t(|w|) < 2 \cdot |w|$), then the first (second) possibility yields an accepting computation of length $t(|w|)$. ∎

19.21. Corollary. For $d \geqq 1$, $t(n) \geqq (1 + \varepsilon) \cdot n$ where $\varepsilon > 0$,

$$\mathbf{multiT^d\text{-}NTIME}(t) = \mathbf{2T^d\text{-}NTIME}(t). \ ∎$$

Remark. The method of the proof of 19.19 also yields for $t(n) \geqq (1 + \varepsilon) \cdot n$, $\varepsilon > 0$,

$$\mathbf{multiT\text{-}NTIME\text{-}REV}(t, \mathrm{Const}) = \mathbf{1\text{-}PD\text{-}CS\text{-}NTIME\text{-}REV}(t, \mathrm{Const}).$$

19.2.1.3. Reduction of the Dimension

In this subsection we deal with the simulation of Turing machines with multi-dimensional tapes by Turing machines having tapes of less dimension, where dimension one is of special interest. We start with this case. The results depend on the number of tapes allowed for the simulating machine.

19.22. Theorem. For $X \in \{D, N\}$ and $t \geqq \mathrm{id}$,

$$\mathbf{multiT^{multi}\text{-}XTIME}(t) \subseteq \mathbf{T\text{-}XTIME}(t^2 \cdot \log t).$$

Proof. Let M be a $\mathrm{kT^d}$-DM working within time $t(n)$. We describe the work of a DTM M' simulating M. The nondeterministic case can be treated in the same manner. M' will have 2 tracks for every tape of M as well as d additional tracks. During the simulation of M the head of M' will write the sequence of symbols written by the jth head of M on the $(2j-1)$th track and the sequence of head moves made by the jth head of M on the $(2j)$th track. At the beginning of the simulation of the τth step of M the head of M' is situated on the right-hand end of these sequences. First M' checks whether the tape square reached by the previous move of the jth head of M has already been visited before. To this end the moves of that head in each of the d directions (stored on the second track) will be added up separately on the d additional tracks in binary notation. Thus M' can find the symbol read by the jth head at the beginning of the τth step of M within time $\precsim t(n) \cdot \log t(n)$. The machine M' will do so for $j = 1, \ldots, k$ in succession. Now M' can record the symbols printed by the heads and the directions of the moves made by the heads of M in the τth step into the next square. Consequently, the total computation time of M' is $\precsim t(n)^2 \cdot \log t(n)$. ∎

Remark. In [Für 78] an improvement of Theorem 19.22 (t^2 instead of $t^2 \cdot \log t$) has been announced.

The following theorem can be regarded as a generalization of Theorem 19.16 where the idea of the proof of 19.16 can be applied in this general case too.

19.23. Theorem. [Stoß 71] For $d \geqq 1$, $X \in \{D, N\}$ and $t \geqq \mathrm{id}$,

$$\mathbf{multiT^d\text{-}XTIME}(t) \subseteq \mathbf{T\text{-}PD\text{-}XTIME}(t^{2-1/d} \cdot \log t). \ ∎$$

19.24. Corollary. For $d \geqq 1$, $X \in \{D, N\}$ and $t \geqq id$,

1. $\mathbf{multiT^d\text{-}XTIME}(t) \subseteq \mathbf{2T\text{-}XTIME}(t^{2-1/d} \cdot \log t)$,
2. $\mathbf{multiT^{multi}\text{-}XTIME}(t) \subseteq \mathbf{2T\text{-}XTIME}(t^2)$. ∎

The next theorem shows that using a simulating machine with more one-dimensional tapes the factor $\log t$ in the above result can be omitted. We do not include the rather complicated proof.

19.25. Theorem. [PiFi 76] For $d \geqq 1$, $X \in \{D, N\}$ and $t \geqq id$,

$$\mathbf{multiT^d\text{-}XTIME}(t) \subseteq \mathbf{multiT\text{-}XTIME}(t^{2-1/d}). \ ∎$$

Remark 1. The first result of this type can be found in [HaSt 65] where $\mathbf{multiT^d\text{-}XTIME}(t) \subseteq \mathbf{multiT\text{-}XTIME}(t^2)$ has been shown. In [Für 78] the result $\mathbf{multiT^d\text{-}XTIME}(t) \subseteq \mathbf{2T\text{-}XTIME}(t^{2-1/d})$ for $d \geqq 2$ and $X \in \{D, N\}$ has been announced. This would improve 19.23 too.

Remark 2. At least for $t = \mathrm{Lin}$ this result cannot be improved in the deterministic case, since in [Hen 66] sets B_d for $d \geqq 2$ are given such that $B_d \in \mathbf{T^d\text{-}DTIME}(\mathrm{Lin})$ and $B_d \notin \mathbf{multiT\text{-}DTIME}(t)$ for $t \prec n^{2-1/d}$. However, for nondeterministic machines a better result can be established in the case $t(n) = 2n$.

19.26. Theorem. [Mon 77c] $\mathbf{T^{multi}\text{-}NTIME}(2n) \subseteq \mathbf{multiT\text{-}NTIME}\big(n(\log n)^2\big)$. ∎

Finally we state some results on the reduction of multi-dimensional ($(d+1)$-dimensional, resp.) tapes to d'-dimensional tapes for $d' \geqq 2$.

19.27. Theorem. For $d \geqq d' \geqq 2$, $t \geqq id$ and $\varepsilon > 0$,

1. [Lou 81] $\mathbf{multiT^d\text{-}DTIME}(t) \subseteq \mathbf{2T^{d'}\text{-}DTIME}(t^{1+1/d-1/d \cdot d'} \cdot \log t)$.
2. [Lou 81] $\mathbf{multiT^d\text{-}DTIME}(t) \subseteq \mathbf{multiT^{d'}\text{-}DTIME}\big(t^{1+1/d'-1/d}\,\mathrm{Pol}(\log t)\big)$.
3. $\mathbf{multiT^{multi}\text{-}DTIME}(t) \subseteq \mathbf{2T^{d'}\text{-}DTIME}(t^{1+1/d'})$.
4. [Gri 79b] $\mathbf{multiT^d\text{-}NTIME}(t) \subseteq \mathbf{multiT^{d'}\text{-}NTIME}(t^{1+1/d'-1/d+\varepsilon})$.
5. $\mathbf{multiT^{multi}\text{-}NTIME}(t) \subseteq \mathbf{multiT^{d'}\text{-}NTIME}(t^{1+1/d'})$. ∎

Remark 1. (see [Lou 81]) Statement 1 is also true for $d' \geqq d$.

Remark 2. Statement 4 can be found in [Gri 79b] in a more general form: For $d > d'$, $t \geqq id$, $s \geqq id$, and $\varepsilon > 0$,

$$\mathbf{multiT^d\text{-}NTIME\text{-}SPACE}(t, s) \subseteq \mathbf{multiT^{d'}\text{-}NTIME\text{-}SPACE}(t \cdot s^{1/d'-1/d+\varepsilon}, s^{1+\varepsilon}).$$

Remark 3. For a lower time bound on the step-by-step simulation by machines with smaller dimension and smaller number of heads see Remark 2 after 19.18.

Remark 4. Results analogous to 19.27.2 and Remark 2 after 19.18 for simulations of deterministic by probabilistic machines are contained in [Pip 82].

19.2.2. Other Types of Machines

19.2.2.1. Random Access Machines

In this subsection we compare the time measures of SRAM and RAM with each other and with the time measures of multi-dimensional Turing machines.

First we deal with the relationships between the time measures of SRAM's and RAM's.

19.28. Theorem. For $X \in \{D, N\}$ and $t \geqq id$,

1. $\mathbf{SRAM\text{-}XTIME}(t) \subseteq \mathbf{RAM\text{-}XTIME}(\mathrm{Lin}\, t)$,
2. $\mathbf{RAM\text{-}XTIME}(t) \subseteq \mathbf{SRAM\text{-}XTIME}(\mathrm{Lin}\, t^2)$.

Proof. Statement 1 is evident, the proof of statement 2 is roughly outlined as follows: Let M be a RAM working within time t. The register i of M is represented in the simulating SRAM M' by the registers $r(i, 0)$, $r(i, 0) + 1$, $r(i, 1)$, $r(i, 1) + 1, \ldots$, $r(i, |\mathbf{R}_i| + 1)$, $r(i, |\mathbf{R}_i| + 1) + 1$. The contents $\mathbf{R}_i$ of the register i of M is stored bitwise in the registers $r(i, 2), r(i, 3), \ldots, r(i, |\mathbf{R}_i| + 1)$. The register $r(i, j) + 1$ is a pointer to the register $r(i, j + 1)$. The register $r(i, 0)$ $\big(r(i, 1)\big)$ is a pointer to the register $r(2i, 0)$ $\big(r(2i + 1, 0)\big)$. All registers of the form $r(i, 0)$ and $r(i, 1)$ build a pointer tree which also allows the simulation of instructions with indirect addressing within time $\lesssim t(n)$. ∎

Remark. Theorem 19.28 remains valid if RAM is replaced by BRAM. Now we compare the time measures of SRAM and multi-dimensional Turing machines. A weaker version of 19.29.1 can be found in [Gri 77].

19.29. Theorem. For $X \in \{D, N\}$ and $t \geqq id$,

1. $\mathbf{SRAM\text{-}XTIME}(t) \subseteq \mathbf{T^d\text{-}PD\text{-}XTIME}(t^{1+1/d} \cdot \log t)$.
2. [Schö 78] $\mathbf{multiT^{multi}\text{-}XTIME}(t) \subseteq \mathbf{SRAM\text{-}XTIME}(\mathrm{Lin}\, t)$.

Proof. Ad 1. We restrict ourselves to the deterministic case. The nondeterministic case can be treated analogously.

Let M be a DSRAM working within time $t(n)$. As a first step we construct a DSRAM M' simulating M as follows: first M' transfers the symbols $a_1, a_2, \ldots, a_n \in \{0, 1\}$ of the input $a_1 a_2 \ldots a_n$ to the registers $1, 2, \ldots, n$, resp. During the simulation the register $n + 2 + i$ of M' will have the same contents as the register i of M for $i \geqq 0$ while the registers 0 and $n + 1$ of M' will hold the position of the input head of M and the number $n + 2$, resp. Thus M' can easily simulate M within time $\lesssim t(n)$.

Now we describe the work of a $\mathrm{T^d}$-PD-DM M'' simulating the work of M' after the transfer of the input symbols $a_1, a_2, \ldots, a_n$ to the registers $1, 2, \ldots, n$. The pushdown store of M'' will serve only to transport information on the Turing tape. To store the contents of the registers of M' the machine M'' will use on its Turing tape only the squares $(i_1, \ldots, i_d)$ with $i_1, \ldots, i_d \geqq 0$. The squares $(i_1, \ldots, i_{d-1}, 0)$, $(i_1, \ldots, i_{d-1}, 1)$, $(i_1, \ldots, i_{d-1}, 2), \ldots$ are said to be the row $(i_1, \ldots, i_{d-1})$. Now we choose a mapping $\hat{c}\colon \mathbb{N}^d \xrightarrow{1\text{-}1} \mathbb{N}$ and mappings $\hat{c}_1, \ldots, \hat{c}_d\colon \mathbb{N} \mapsto \mathbb{N}$ such that

a) $\hat{c}\big(\hat{c}_1(x), \ldots, \hat{c}_d(x)\big) = x$,

b) $\hat{c}_j\big(\hat{c}(x_1, \ldots, x_d)\big) = x_j$, for $j = 1, \ldots, d$,

c) $\hat{c}_j(z) \lesssim z^{1/d}$, for $j = 1, \ldots, d$,

d) $\hat{c}, \hat{c}_1, \ldots, \hat{c}_d \in \mathrm{P}$.

It is not hard to see that such mappings really exist. After the simulation of a step of M' the row $(i_1, \ldots, i_{d-1})$ of the Turing tape of M'' will contain (beginning with the square $(i_1, \ldots, i_{d-1}, 0)$) the word

$$\mathbf{R}_{\hat{c}(i_1,\ldots,i_{d-1},0)} \# \mathbf{R}_{\hat{c}(i_1,\ldots,i_{d-1},1)} \# \ldots \# \mathbf{R}_{\hat{c}(i_1,\ldots,i_{d-1},j)} \#$$

where j is the largest number such that the register $\hat{c}(i_1, \ldots, i_{d-1}, j)$ has already been used.

Evidently, the contents of any register of M' and the address of any register of M' used during the computation on an input of length n can be bounded by $t(n) + n + a$ where a is a suitable constant. For the initialization the input symbols $a_1, a_2, \ldots, a_n$ must be brought into the corresponding rows. This can be done within time $\precsim n^{1+1/d} \cdot \log n$.

For the simulation of one step of M' in which the register i is involved M'' performs the following activities.

a) M'' computes in the squares $(0, 0, \ldots, -1)$, $(1, 0, \ldots, -1)$, $(2, 0, \ldots, -1), \ldots$ the values $\hat{c}_1(i), \ldots, \hat{c}_d(i)$ and stores $1^{\hat{c}_d(i)} \# \ldots \# 1^{\hat{c}_1(i)}$ on the pushdown store. This takes $\precsim (\log t(n))^m + t(n)^{1/d}$ for some $m > 0$.

b) Using the contents of its pushdown store M'' moves the head to the row $(\hat{c}_1(i), \ldots, \hat{c}_{d-1}(i))$ within time $\precsim t(n)^{1/d}$.

c) Using the contents of the pushdown store M'' moves the head within the row $(\hat{c}_1(i), \ldots, \hat{c}_{d-1}(i))$ to the word R_i which is situated before the $(\hat{c}_d(i) + 1)$th $\#$. If the number of symbols $\#$ in this row is smaller than $\hat{c}_d(i) + 1$, then the missing symbols $\#$ are introduced. This can be done within time $\precsim t(n)^{1/d} \cdot \log t(n)$.

d) M'' transfers the word R_i into the pushdown store (within time $\precsim \log t(n)$) or modifies it. If necessary, the contents of the row $(\hat{c}_1(i), \ldots, \hat{c}_{d-1}(i))$ on the right of $\mathbf{R}_i$ must be shifted to the right by the help of the pushdown store. This can be done within time $\precsim t(n)^{1/d} \cdot \log t(n)$.

Hence, the total computation time of M'' is $\precsim t(n)^{1+1/d} \cdot \log t(n)$.

Ad 2. We show only **multiT-XTIME**$(t) \subseteq$ **SRAM-XTIME**(Lin t). The case of multidimensional tapes, which is much more complicated, can be found in [Schö 78].

Let M be a kT-DM working within time $t(n)$. We construct a DSRAM M'' simulating M within time $\precsim t(n)$. The nondeterministic case can be treated analogously.

First, there exists obviously a kT-DM M' simulating M within time $\precsim t(n)$, using only the tape squares $2, 3, 4, \ldots$ on each tape and having the worktape alphabet $\{0, 1\}$. Our DSRAM M'' will actually simulate M'. The contents of the square i on the jth tape will be stored in the register $k \cdot i + j$. The registers $1, \ldots, k$ will hold the position of the k heads and the registers 0 and $k + 1$ are used for organization.

For the initialization M'' transfers the input symbols $a_1, a_2, \ldots, a_n$ to the registers $2k + 1, 3k + 1, \ldots, (n + 1)\, k + 1$.

Now it is clear how M'' can simulate one step of M' within $\precsim 1$ steps. ∎

Remark 1. In [Mon 77c] it is shown that

$$\textbf{SRAM-NTIME}(\mathrm{Lin}) \subseteq \textbf{multiT-NTIME}\big(n \cdot (\log n)^2\big).$$

Remark 2. Many models similar to the SRAM's have been studied in the literature. The *storage modification machines* investigated in [Schö 73], [Schö 78] can be simulated by SRAM's with a linear increase of the worktime and vice versa. The machines investigated in [KoUs 58] can be simulated by SRAM's with a linear increase of the worktime. The converse is not known (cf. [Schö 78]).

The next theorem compares the time measures of RAM's and multidimensional Turing machines. For the proof of this theorem we need a new notion. For $k \geq 2$, a kT-DM M *decides the set A within time $t(n)$ in a block-respecting manner* if for any $w \in \Sigma^*$ the following holds: The machine M starting its work with w on the first tape and a word $v \in \Sigma^*$ of length $\leq t(|w|)$ on the kth tape

a) halts within $t(|w|)$ steps,
b) reaches an accepting (rejecting) situation if $w \in A$ (if $w \notin A$),
c) crosses with its heads the boundaries between the tape squares $i \cdot |v|$ and $i \cdot |v| + 1$ $(i = \ldots, -2, -1, 0, 1, 2, \ldots)$ only at times which are integer multiples of $|v|$.

For $i = \ldots, -2, -1, 0, 1, 2, \ldots$ the region between the tape squares $(i - 1) \cdot |v| + 1$ and $i \cdot |v|$ (inclusively) is said to be a block.

19.30. Lemma. [HoPaVa 75] For every kT-DM M deciding a set A within time $t(n)$ there is a (k + 1)T-DM M' deciding the set A within time $\preceq t(n)$ in a block-respecting manner.

Proof. Let M work on an input of length n. The actual simulation of M will be done on the tapes $1, \ldots, k$ of M'. The head on the tape $k + 1$ simply moves back and forth over the word v and serves as a clock to indicate multiples of $|v|$ time units. The block boundaries on the tapes $1, \ldots, k$ are marked during the simulation as follows: whenever the rightmost (leftmost) marked boundary is crossed to the right (left), the simulation is interrupted and the next boundary to the right (left) is marked with the help of the clock. If either head on one of these tapes attempts to cross a (marked) block boundary at a time other than a multiple of $|v|$, the heads on the tapes $1, \ldots, k$ temporarily halt until the head on tape $k + 1$ indicates the next multiple of $|v|$.

A difficulty arises when a tape head of M returns too fast to the boundary crossed the previous time, for instance, when a tape head simply moves back and forth between two adjacent tape cells which are in different blocks. In this case the simulation by M' would be slower by a factor of $|v|$.

This difficulty is overcome as follows: If the head on tape j of M moves into a block, then M' writes on two new tracks of this block u_r^{-1} and u_l^{-1} where u_r and u_l are the inscriptions of the adjacent blocks. This can easily be done in a block-respecting way within time $\preceq |v|$. Now at least $|v|$ steps of M can be simulated by M' without crossing any block boundary on tape j. If the jth head of M leaves the three blocks whose inscriptions are stored in the actual block on the jth tape of M', then M' updates these three blocks and erases the two additional tracks. This can also be done in a block-respecting manner within time $\preceq |v|$. ∎

19.31. Theorem.

1. For $d \geq 1$, X $\in$ {D, N} and $t \geq \mathrm{id}$,

$$\mathbf{RAM\text{-}XTIME}(t) \subseteq \mathbf{T^d\text{-}PD\text{-}XTIME}(t^{2+1/d}).$$

2. For X ∈ {D, N} and $t \geqq \mathrm{id}$,

$$\textbf{multiT}^{\textbf{multi}}\textbf{-XTIME}(t) \subseteqq \textbf{RAM-XTIME}(\mathrm{Lin}\, t).$$

3. [HoPaVa 75] For $t(n) \geqq n \cdot \log n$ such that $\log t(|w|) \in$ RAM-DTIME(Lin $t/\log t$),

$$\textbf{multiT-DTIME}(t) \subseteqq \textbf{RAM-DTIME}(\mathrm{Lin}\, t/\log t).$$

Proof. Statement 1 can be proved by suitable modifications and refinements of the proof of 19.29.1. Statement 2 is an immediate consequence of 19.29.2.
Ad 3. Let A be in **multiT-DTIME**(t). By Lemma 19.30 there is a kT-DM M accepting A within time $t(n)$ in a block-respecting manner. Without loss of generality the work-tape alphabet of M consists only of two symbols and the marker for the block boundaries. M will use block length $\frac{1}{4k+2} \cdot \log t(n)$.

The $(2k+1)$-tuple $K = (s, i_1, \ldots, i_k, w_1, \ldots, w_k)$, where s is the state, $i_1, \ldots, i_k$ are integers denoting the head positions and $w_1, \ldots, w_k \in \{0, 1\}^*$ are the contents of the actual blocks of a given situation of M, is said to be the *PID* (partial instantaneous description) of that situation of M. Let K' denote the PID of that situation which M, started in a situation with PID K, will reach as soon as one of the k heads leaves its block but at most after $\frac{1}{4k+2} \cdot \log t(n)$ steps.

A DRAM M' simulating M will work in two phases: a precomputing phase and a simulation phase. During the precomputing phase a search tree is generated which allows M' to find for any given PID K the PID K' within $\precsim 1$ steps. This will be crucial for the time efficient simulation stated in the theorem.

The machine M' starts its precomputing phase by computing the block length $m = \frac{1}{4k+2} \cdot \log t(n)$ which is possible within time $\precsim \frac{t(n)}{\log t(n)}$. Since the values of the components of the PID's are bounded by $\left\lceil t(n)^{\frac{1}{4k+2}} \right\rceil$ there exist at most $m^{2k+1} \precsim \sqrt{t(n)}$ PID's. They correspond in a canonical way to the leaves of a tree of height $2k+1$ and degree m. This tree is implemented as follows: Every vertex of the tree is represented by a register, and whenever a vertex (which is not a leaf) is represented by register i, then the registers $i+1, \ldots, i+m$ contain the addresses of the registers representing the m descendants of this vertex. If a leaf corresponds to the PID K, then the register representing this leaf will contain the address of the first component of K' (whose components are stored in $2k+1$ consecutive registers).

First M' implements this tree (apart from the contents of the registers representing the leaves) which can be done within $\precsim \sqrt{t(n)}$ steps. Then M' systematically generates all PID's, and to every PID K it computes K' which is stored in some $2k+1$ consecutive registers $j, j+1, \ldots$ For every K this can be done within $\precsim \log t(n)$ steps (using the method of the proof of 19.29.2). The address j is stored into the register representing the leaf corresponding to K. This register can be found within a number of steps which is proportional to the height of the tree, i.e. $\precsim 1$ steps. Thus, the precomputing time is $\precsim \sqrt{t(n)} \cdot \log t(n)$.

For the simulation of M by M' the tape contents of M are stored in a doubly linked list of registers of M' in the following way: Per block M' uses one register to contain the block contents and two link registers containing the addresses of the registers for the left and right neighbour block. The simulation starts with the initialization of this list.

Now note that the block-respecting machine M makes at least $\frac{1}{4k+2} \cdot \log t(n)$ steps between the moments when tape heads cross boundaries. Such a phase is simulated by M' as follows: Knowing the PID K of the corresponding situation the machine M' uses its search tree to find K'. Knowing the addresses of the registers containing the actual blocks M' updates the contents of these blocks with the help of K', and the links enable M' to determine the new PID. This takes altogether $\precsim 1$ steps. Thus M' works within $\precsim \frac{t(n)}{\log t(n)}$ steps. ∎

Remark 1. Theorem 19.31.1 remains valid if RAM is replaced by BRAM.

Remark 2. The proof of 19.29.1 yields for RAM's with logarithmic cost criterion: For $d \geqq 1$, $X \in \{D, N\}$ and $t \geqq id$,

$$\mathbf{RAM\text{-}XTIME}^{\log}(t) \subseteq \mathbf{T^d\text{-}PD\text{-}XTIME}(t^{1+1/d}).$$

The case $d = 1$ for this result and for 19.31.1 can be found in [CoRe 72].

In [Wie 83] it is shown that

$$\mathbf{RAM\text{-}DTIME}^{\log}(t) \subseteq \mathbf{multiT\text{-}DTIME}(t^2/\log\log t)$$

and

$$\mathbf{RAM\text{-}DTIME}^{\log}(t) \subseteq \mathbf{multiT\text{-}NTIME}(t \cdot \log t).$$

Remark 3. For multi-dimensional Turing tapes the method of 19.31.3 yields: For $t(n) \geqq n \cdot \log n$ such that $\log t(|w|) \in$ RAM-DTIME$(\mathrm{Lin}\ t/(\log t)^{1/d})$,

$$\mathbf{multiT^d\text{-}DTIME}(t) \subseteq \mathbf{RAM\text{-}DTIME}(\mathrm{Lin}\ t/(\log t)^{1/d})$$

(see [Gri 79b]).

Remark 4. In [CoRe 72] the *random access stored program machines* (*RASP's*) are compared with RAM's. Contrary to RAM's the program of a RASP is stored in some of its registers. Thus the program of a RASP can be changed during the computation. Nevertheless, with respect to the acceptance of sets this ability of the RASP's is not really an advantage: for $X \in \{D, N\}$ and $t \geqq id$,

$$\mathbf{RAM\text{-}XTIME}(\mathrm{Lin}\ t) = \mathbf{RASP\text{-}XTIME}(\mathrm{Lin}\ t).$$

However, in [Hart 71] it is shown that this does not remain true in the case of the computation of functions.

19.2.2.2. Iterative Arrays

In this subsection we state some relationships between the time measure of iterative arrays and other time measures which we have already investigated.

19.32. Theorem.

1. For $d \geqq 1$, $X \in \{D, N\}$ and $t \geqq \mathrm{id}$, $\mathbf{IA^d\text{-}XTIME}(t) \subseteq \mathbf{T\text{-}XTIME}(t^{d+1})$.
2. [CoAa 69] For $t \geqq \mathrm{id}$, $\mathbf{multiT\text{-}DTIME}(t) \subseteq \mathbf{IA^1\text{-}DTIME}(\mathrm{Lin}\ t)$.
3. [Sei 77c] For $d \geqq 1$ and $t(n) \geqq n^d$, $\mathbf{multiT\text{-}NTIME}(t) \subseteq \mathbf{IA^d\text{-}NTIME}(\mathrm{Lin}\ t^{1/d})$. ∎

Remark. Because of 19.32.3 and the hierarchy results for multiT-NTIME (cf. 17.43) Statement 19.32.1 cannot be improved in such a way that there is a $d' < d$ such that $\mathbf{IA^d\text{-}NTIME}(t) \subseteq \mathbf{multiT\text{-}NTIME}(t^{d'})$.

19.33. Corollary. $\mathbf{NP} = \mathbf{IA^{multi}\text{-}NTIME}(\mathrm{Lin})$. ∎

19.34. Theorem. [SeWe 76] For $X \in \{D, N\}$ and $t \geqq \mathrm{id}$, $\mathbf{IA^1\text{-}XTIME}(t) \subseteq \mathbf{BRAM\text{-}XTIME}(\mathrm{Lin}\ t)$. ∎

A lower time bound for on-line simulations of IA^d by probabilistic $\mathrm{IA}^{d'}$ $(d' < d)$ can be found in [PaSi 83].

19.2.2.3. Further Results

In this subsection we state some results on some modifications of the usual model of Turing machines, which have a limited random access, by the ability of the heads to jump.

The first modification allows the head to jump to markers. This ability actually increases the power of the model.

A *tabulator tape* (*TAB*) is a Turing tape with one head which can not only move one step to the left or right, but can also jump to the nearest tape square with some marker on it. More precisely, there can be different kinds of markers, and the head can jump to the nearest tape square having a marker of the desired kind. A tabulator tape using only one marker of each kind is abbreviated by TAB^1.

19.35. Theorem. [Weic 74] For $X \in \{D, N\}$ and $t \geqq \mathrm{id}$,

1. $\mathbf{multiT\text{-}XTIME}(t) \subseteq \mathbf{1\text{-}TAB^1\text{-}XTIME}(\mathrm{Lin}\ t)$.
2. $\mathbf{1\text{-}TAB\text{-}XTIME}(t) \subseteq \mathbf{T\text{-}XTIME}(t^2)$. ∎

Remark. $1\text{-}\mathrm{TAB}^1$-DM's are, at least for linear time bounds, more powerful than multiT-DM's, i.e. $\mathbf{multiT\text{-}DTIME}(\mathrm{Lin}) \subset \mathbf{1\text{-}TAB^1\text{-}DTIME}(\mathrm{Lin})$. Statement 19.36 which is due to B. Monien can also be found in [Weic 74].

19.36. Theorem.

1. For $X \in \{D, N\}$, $\varepsilon > 0$ and $t \geqq \mathrm{id}$ such that $t(|w|) \in \mathbf{1\text{-}TAB\text{-}DTIME}(\mathrm{Lin}\ t)$,

$$\mathbf{SRAM\text{-}XTIME}(t) \subseteq \mathbf{1\text{-}TAB\text{-}XTIME}(\mathrm{Lin}\ t^{1+\varepsilon}).$$

2. [Weic 74] For $X \in \{D, N\}$ and $t \geqq \mathrm{id}$,

$$\mathbf{TAB\text{-}XTIME}(t) \subseteq \mathbf{SRAM\text{-}XTIME}(\mathrm{Lin}\ t \cdot \log t).$$

$$\mathbf{TAB^1\text{-}XTIME}(t) \subseteq \mathbf{SRAM\text{-}XTIME}(\mathrm{Lin}\ t). \quad ∎$$

For further results see [Weic 71], [Weic 74].

The second modification is based on multi-head Turing machines. Every head of the Turing tape has the ability to jump to the position of another head. As it is shown in [SaVi 77] this ability does not increase the power of one-dimensional Turing machines.

§ 20. Further Time and Space Measures

In this section we investigate those time and space measures which are not realistic or for which it is not yet known whether they are realistic. Most of these measures are based on machines whose computational power is enlarged by powerful new instructions, auxiliary stores or modified notions of acceptance. Thus in some cases the inclusion **DTIME**(Pol t) $\subseteq$ **DSPACE**(Pol t) or the inclusion **DSPACE**(s) $\subseteq$ **DTIME** ($2^{\mathrm{Lin} s}$) becomes an equality if the corresponding new measure is used on the left-hand side. In this manner we get some interesting material concerning the general space-time problem.

20.1. Weak and Powerful Instructions

20.1.1. Counter Machines

A counter machine can change the contents of each store by at most 1 in each step. This is relatively weak compared with the ability of machines having a realistic time measure (for example Turing machines). Thus the inclusion

$$\mathbf{XSPACE}(\log t) \subseteq \mathbf{XTIME}(\mathrm{Pol}\, t)$$

(cf. 5.10) becomes the equality

$$\mathbf{XSPACE}(\log t) = \mathbf{2\text{-}multiC\text{-}XTIME}(\mathrm{Pol}\, t)\,.$$

The equation

$$\mathbf{XSPACE}(\log t) = \mathbf{2\text{-}multiC\text{-}XSPACE}(\mathrm{Pol}\, t)$$

comes from the unary presentation of numbers in the counters. Note that at least the ideas of the proofs in this subsection are due to [FipMeRo 68]; some generalizations and special results can be found in [Gre 76].

20.1. Theorem. For $i = 1, 2$, $\mathrm{X} \in \{\mathrm{D}, \mathrm{N}\}$ and $t \geq \mathrm{id}$,

$$\begin{aligned}\mathbf{i\text{-}T\text{-}XSPACE}(\log t) &= \mathbf{i\text{-}3C\text{-}XTIME}(\mathrm{Pol}\, t) = \mathbf{i\text{-}multiC\text{-}XTIME}(\mathrm{Pol}\, t)\\ &= \mathbf{i\text{-}3C\text{-}XSPACE}(\mathrm{Pol}\, t) = \mathbf{i\text{-}multiC\text{-}XSPACE}(\mathrm{Pol}\, t)\,.\end{aligned}$$

Proof. The inclusions

$$\mathbf{i\text{-}3C\text{-}XTIME}(\mathrm{Pol}\, t) \subseteq \mathbf{i\text{-}3C\text{-}XSPACE}(\mathrm{Pol}\, t)\,,$$

$$\mathbf{i\text{-}3C\text{-}XTIME}(\mathrm{Pol}\, t) \subseteq \mathbf{i\text{-}multiC\text{-}XTIME}(\mathrm{Pol}\, t)\,,$$

$$\mathbf{i\text{-}3C\text{-}XSPACE}(\mathrm{Pol}\, t) \subseteq \mathbf{i\text{-}multiC\text{-}XSPACE}(\mathrm{Pol}\, t)$$

and

$$\textbf{i-multiC-XTIME}(\mathrm{Pol}\, t) \subseteq \textbf{i-multiC-XSPACE}(\mathrm{Pol}\, t)$$

are evident.

Furthermore, a natural number not greater than $t(n)^k$ ($k \in \mathbb{N}$) can be stored in binary notation within space $\log t(n)$. Thus, **i-multiC-XSPACE**(Pol t) $\subseteq$ **i-T-XSPACE**($\log t$).

It remains to show that **i-T-XSPACE**($\log t$) $\subseteq$ **i-3C-XTIME**(Pol t). Let M be an i-XTM which works within space $\log t$. As in the proof of Theorem 5.1 (ad 7) we construct an i-XTM M' such that

a) $L(M') = L(M)$,

b) M' has only the worktape symbols 0 and 1, and

c) for every worktape inscription $\ldots 000 w_1 \overset{\downarrow}{a} w_2 000 \ldots$ ($w_1, w_2^{-1} \in 1\{0, 1\}^* \cup \{e\}$, $a \in \{0, 1\}$, and $\downarrow$ marks the head position) during the work of M' on an input of length n we have $|w_1|, |w_2| \leq m \cdot \log t(n) = \log t(n)^m$, for some $m > 0$ (for X = N this holds only for a suitable computation).

Now we construct an i-3C-XM M'' which simulates M' in such a way that this worktape inscription is represented as follows: counter 1 of M'' contains $\mathrm{bin}^{-1} w_1$, counter 2 of M'' contains $\mathrm{bin}^{-1} w_2^{-1}$, and a is stored in the finite memory of M''. In order to simulate a step of M', operations of the form $k \leftarrow 2k$, $k \leftarrow 2k + 1$ and $k \leftarrow \left\lfloor \frac{k}{2} \right\rfloor$ must be executed where k is the contents of counter 1 or counter 2 of M''. This can be done with the help of counter 3 within time $\lesssim k$. Since $\log \mathrm{bin}^{-1} w_1 \leq |\mathrm{bin}\,\mathrm{bin}^{-1} w_1| = |w_1| \leq \log t(n)^m$ we get $\mathrm{bin}^{-1} w_1 \leq t(n)^m$. Analogously, we get $\mathrm{bin}^{-1} w_2^{-1} \leq t(n)^m$. Consequently, M'' can simulate one step of M' within $\lesssim t(n)^m$ steps. Since M' can have at most $\lesssim t(n)^m$ different worktape inscriptions, it works within time $\lesssim n \cdot t(n)^m \lesssim t(n)^{m+1}$. Hence M'' works within time $\lesssim t(n)^{2m+1}$. ∎

Remark. For i-2C-XM's we have only the result **i-3C-XTIME**(t) $\subseteq$ **i-2C-XTIME**($2^{\mathrm{Lin}\, t}$) and consequently **i-T-XSPACE**($\log t$) $\subseteq$ **i-2C-XTIME**($2^{\mathrm{Pol}\, t}$) for $i = 1, 2$, X $\in$ {D, N} and $t \geq \mathrm{id}$. The trick is to represent the contents k, l and m of three counters as $2^k 3^l 5^m$ in one counter. With the help of a further counter, a change of an exponent by 1 can be executed within $\lesssim 2^k 3^l 5^m$ steps.

The following theorem relates the time measures of Turing machines and counter machines. The second statement is a consequence of Theorem 20.1, the first statement is proved in [Vit 82] where it is shown that the simulating TM's can even be constructed to be oblivious (i.e. their head positions are functions of time and input length only).

20.2. Theorem. For $i = 1, 2$, X $\in$ {D, N} and $t \geq \mathrm{id}$,

1. **i-multiC-XTIME**(t) $\subseteq$ **i-T-XTIME**(t).
2. **i-T-XTIME**(t) $\subseteq$ **i-3C-XTIME**($2^{\mathrm{Lin}\, t}$). ∎

Now we compare analogous measures for counter machines with different input modes.

20.3. Theorem.

1. For $k \in \{3, 4, 5, \ldots, \text{multi}\}$ and $t(n) \geqq 2^n$,

$$\textbf{1-kC-XSPACE}(\text{Pol } t) = \textbf{2-kC-XSPACE}(\text{Pol } t) = \textbf{1-kC-XTIME}(\text{Pol } t) = \textbf{2-kC-XTIME}(\text{Pol } t).$$

2. For $k \in \{3, 4, 5, \ldots, \text{multi}\}$, $X \in \{D, N\}$ and $\text{id} \leqq s$, $\log s <_{\text{io}} \text{id}$,

$$\textbf{1-kC-XSPACE}(\text{Pol } s) \subset \textbf{2-kC-XSPACE}(\text{Pol } s),$$

$$\textbf{1-kC-XSPACE}(s) \subset \textbf{2-kC-XSPACE}(s).$$

3. For $k \in \{3, 4, 5, \ldots, \text{multi}\}$, $X \in \{D, N\}$ and $(n^2 \leqq t(n), \log t(n) <_{\text{io}} n)$ or $(2n \leqq t(n), \log t(n) <_{\text{io}} \sqrt{n})$,

$$\textbf{1-kC-XTIME}(\text{Pol } t) \subset \textbf{2-kC-XTIME}(\text{Pol } t),$$

$$\textbf{1-kC-XTIME}(t) \subset \textbf{2-kC-XTIME}(t).$$

Proof. Statement 1 is a consequence of 20.1 and 5.2.

Ad 2 and 3. It is obvious that $S \in \textbf{2-C-DSPACE-TIME}(n, n^2)$. Assume that $S \in$ **1-multiC-NSPACE**(Pol s) or **1-multiC-NTIME**(Pol s). By 20.1 we obtain $S \in$ **1-T-NSPACE**($\log s$), and by 8.2 we obtain $\log s \geq \text{id}$. Thus it remains to prove statement 3 for $2n \leqq t(n)$, $\log t(n) <_{\text{io}} \sqrt{n}$. The set $\{(0^{|w|}1)^{|w|}\, ww^{-1} : w \in \{0, 1\}^*\}$ is in **2-3C-DTIME**($2n$). The same argument as above shows that it cannot be in **1-multiC-NTIME**(Pol t) for $\log t(n) <_{\text{io}} \sqrt{n}$. ∎

Comparisons between deterministic and nondeterministic counter machines can be found in 23.7 and 24.6.

Now we deal with some trade-offs between time and space on the one hand and the number of counters on the other hand. A careful analysis of the proof of 20.1 yields

20.4. Theorem. For $i = 1, 2$, $k \geqq 4$, $X \in \{D, N\}$ and $s, t \geqq \text{id}$,

$$\textbf{i-kC-XSPACE-TIME}(s, t) \subseteqq \textbf{i-3C-XSPACE-TIME}(s^k, t \cdot s^k \log s). \ ∎$$

20.5. Corollary. For $i = 1, 2$, $k \geqq 4$, $X \in \{D, N\}$ and $s, t \geqq \text{id}$,

1. **i-kC-XSPACE**(s) $\subseteqq$ **i-3C-XSPACE**(s^k).
2. **i-kC-XTIME**(t) $\subseteqq$ **i-3C-XTIME**($t^{k+1} \cdot \log t$). ∎

We state the following supplementary result:

20.6. Theorem. [Gre 76] For $i = 1, 2$, $k \geqq 1$, $r \geqq 1$, $m \geqq 1$, $X \in \{D, N\}$ and $s, t \geqq \text{id}$,

$$\textbf{i-(m}^{\textbf{r}}\textbf{k)C-XSPACE-TIME}(s, t) \subseteqq \textbf{i-(r + m + k − 1)C-XSPACE-TIME}(s^{m^r}, t \cdot s^{m^r}). \ ∎$$

On the other hand, the space requirements of a counter machine can essentially be reduced by adding new counters.

20.7. Theorem. For $i = 1, 2$, $k \geqq 1$, $r \geqq 1$, $X \in \{D, N\}$ and arbitrary s, t,

$$\mathbf{i\text{-}kC\text{-}XSPACE\text{-}TIME}(s^r, t) \subseteq \mathbf{i\text{-}(r \cdot k)C\text{-}XSPACE\text{-}TIME}(s, t).$$

Proof. The proof for $r = 2$ is carried out in the proof of Theorem 18.10. The proof for $r > 2$ is more complicated, but it is based on the same principle. ∎

20.8. Corollary. For $i = 1, 2$, $k \geqq 1$, $r \geqq 1$, $X \in \{D, N\}$ and $s \geqq 0$,

1. $\mathbf{i\text{-}kC\text{-}XSPACE}(s^r) \subseteq \mathbf{i\text{-}(k \cdot r)C\text{-}XSPACE}(s)$.
2. $\mathbf{i\text{-}multiC\text{-}XSPACE}(\mathrm{Pol}\, s) = \mathbf{i\text{-}multiC\text{-}XSPACE}(s)$. ∎

Some results about the relationships between special time and space complexity classes of counter machines can be found in [FipMeRo 68], [Gre 76] and [Mon 78b]. Especially for realtime and linear time complexity classes, see § 22.2.

Counter machines without input tape are sometimes called *Minsky machines*. The input $w \in \{1, 2\}^*$ is given in one of the counters as $\mathrm{dya}^{-1}(w)$. It is not hard to see that

$$\begin{aligned}&\mathbf{multiC\text{-}XSPACE\text{-}TIME}(s(\mathrm{Lin}), t(\mathrm{Lin}))\\ &= \mathbf{2\text{-}multiC\text{-}XSPACE\text{-}TIME}(s(2^{n+\mathrm{Const}}), t(2^{n+\mathrm{Const}}))\end{aligned}$$

for $X \in \{D, N\}$ and increasing $s, t \geqq \mathrm{id}$. By Theorem 20.1 we obtain

$$\begin{aligned}\mathbf{multiC\text{-}XSPACE}(\mathrm{Pol}\, s(\mathrm{Lin})) &= \mathbf{multiC\text{-}XTIME}(\mathrm{Pol}\, s(\mathrm{Lin}))\\ &= \mathbf{XSPACE}(\log s(2^{n+\mathrm{Const}}))\end{aligned}$$

for $X \in \{D, N\}$ and increasing $s \geqq \mathrm{id}$. In particular, for $X \in \{D, N\}$,

$$\mathbf{multiC\text{-}XSPACE}(\mathrm{Pol}) = \mathbf{multiC\text{-}XTIME}(\mathrm{Pol}) = \mathbf{XLINSPACE}.$$

20.1.2. Successor RAM's

In § 5 we have shown that $\mathbf{SRAM\text{-}XSPACE}(\mathrm{Lin}\, s) = \mathbf{SRAM'\text{-}XSPACE}(\mathrm{Lin}\, s) = \mathbf{XSPACE}(s)$ and $\mathbf{SRAM\text{-}XTIME}(\mathrm{Pol}\, t) = \mathbf{XTIME}(\mathrm{Pol}\, t)$, $X \in \{D, N\}$ and $s, t \geqq \mathrm{id}$. The latter equation shows that SRAM-XTIME is a realistic time measure. However, this result could only be established by extensive use of indirect addressing. Without indirect addressing we obtain another result, since the operations $+1$ and -1 are weak compared with full addition and full subtraction.

First we state a result which can easily be verified.

20.9. Theorem. For $X \in \{D, N\}$ and $t \geqq \mathrm{id}$,

$$\mathbf{SRAM'\text{-}XTIME}(\mathrm{Pol}\, t) = \mathbf{1\text{-}multiC\text{-}XTIME}(\mathrm{Pol}\, t). \ ∎$$

Consequently, we obtain the same relationships to the space measure as we have stated in § 20.1.1 for counter machines.

20.10. Theorem. For $X \in \{D, N\}$ and $t \geqq \mathrm{id}$,

$$\mathbf{SRAM'\text{-}XTIME}(\mathrm{Pol}\, t) = \mathbf{1\text{-}T\text{-}XSPACE}(\log t). \ ∎$$

The next theorem compares the measure SRAM′-XTIME with the time measure for 1-XTM's.

20.11. Theorem. For X ∈ {D, N} and $t \geq$ id,
1. **SRAM′-XTIME**(t) ⊆ **1-T-XTIME**(Lin $t \cdot \log t$).
2. **1-T-XTIME**(t) ⊆ **SRAM′-XTIME**($2^{\text{Lin}\, t}$).

Proof. The first inclusion can be obtained straightforewardly by storing all register contents of a XSRAM′ on the worktape of a 1-XTM. The second inclusion is a consequence of 20.2 and 20.9. ∎

20.1.3. Multiplication RAM's and Concatenation RAM's

In 5.2 we have shown that for X ∈ {D, N} and $s \geq$ id,

$$\textbf{XSPACE}(s) = \textbf{MRAM-XSPACE}(\text{Lin}\, s) = \textbf{MRAM}'\textbf{-XSPACE}(\text{Lin}\, s)$$
$$= \textbf{CRAM-XSPACE}(\text{Lin}\, s) = \textbf{CRAM}'\textbf{-XSPACE}(\text{Lin}\, s).$$

However, it is not known whether the time measures for multiplication and concatenation RAM's are polynomially related to the general time measure. Since multiplication and concatenation in connection with Boolean operations are relatively powerful, we are able to show that these time measures are polynomially related to the general space measure.

20.12. Theorem. [HaSi 76], [PrSt 76] Let X = D and $t \geq$ id or X = N, $t \geq$ id and $t(|w|) \in$ DSPACE(Pol t). Then

$$\textbf{DSPACE}(\text{Pol}\, t) = \textbf{MRAM-XTIME}(\text{Pol}\, t) = \textbf{MRAM}'\textbf{-XTIME}(\text{Pol}\, t)$$
$$= \textbf{CRAM-XTIME}(\text{Pol}\, t) = \textbf{CRAM}'\textbf{-XTIME}(\text{Pol}\, t).$$

Proof. 1. We show **MRAM-XTIME**(Pol t) ⊆ **XSPACE**(Pol t) for X ∈ {D, N}. It is easy to see that MRAM-XTIME and the time measure for XMRAM's without input tape (the input beeing given in one register) are polynomially related. Thus we can start with an XMRAM M without input tape working within time t^k. First we construct an NMRAM M' such that a) L(M') = L(M), b) M' works within time $\lesssim t^{2k}$, and c) M' uses only the first $2t(n)^k + d$ registers (for suitable $d \geq 0$). We assume without loss of generality that M does not use register 0. Since in a single step M can change the contents of at most one register, it can change at most $t(n)^k$ registers during its work on an input of length n. The contents of these registers will be stored by M' in the registers $d + 1, d + 3, d + 5, \ldots$; their addresses will be stored in the registers $d, d + 2, d + 4, \ldots$ If an instruction of M refers to the register i, then M' searches in the registers $d, d + 2, d + 4, \ldots$ for this address. If it is found in register $d + 2j$, then the contents $\mathbf{R}_i$ of register i of M can be found in register $d + 2j + 1$ of M'. If it is not found, then $\mathbf{R}_i$ = e. When M changes the contents of a register which has not previously been changed, M' stores address and new contents of this register in two new neighbouring registers. Thus M' can simulate one step of M within $\lesssim t(n)^k$ steps and uses only the first $d + 2t(n)^k$ registers. Now we construct an XTM M'' which simulates M' within space Pol t. For X = N the NTM M'' first guesses nondeterministically the sequence of instructions corresponding to an accepting computation of M' on an input of length n. Then M'' checks deterministically whether this sequence of instructions is possible on this input. For $\mathbf{X} = \mathbf{D}$ the DTM M'' performs successively for $i = 1, 2, 3, \ldots$ first all sequences of instructions corresponding to an accepting computation of length i, and then all sequences of instructions corresponding to a rejecting computation of length i.

Since only in the case of the conditional jump the order in which the instructions are executed depends on the contents of the register, only in this case does M'' have anything to check. We assume without loss of generality that M' only uses conditional jumps of the kind

$$\text{IF } (\mathbf{R}_i = 0) \text{ THEN GOTO } k .$$

In this case the machine M'' has to check whether $\mathbf{R}_i = \mathrm{e}$. This can be done by studying the "history" of the information stored in register i. Since words of length $r = \max\{n, a\} \cdot 2^{t(n)^{2k}}$ can be generated by M' within $t(n)^{2k}$ steps (a is the length of the largest constant in the program of M'), there can appear registers the contents of which cannot be stored completely within space $\mathrm{Pol}\, t(n)$. Thus the history of the contents of register i is studied bitwise. For $i \geqq 0$, $\sigma \geqq 1$ and $\tau \geqq 0$ define

$$\mathbf{R}_i(\sigma, \tau) = \begin{cases} \text{the } \sigma\text{th symbol of } \mathbf{R}_i(\tau) \text{ if } 1 \leqq \sigma \leqq |\mathbf{R}_i(\tau)|, \\ \mathrm{e} \quad \text{otherwise}, \end{cases}$$

where $\mathbf{R}_i(\tau)$ is the contents of register i after step τ of M'. Thus, if in the τth step of M' the instruction IF ($\mathbf{R}_i = 0$) THEN GOTO k is executed, then M'' has to check whether $\mathbf{R}_i(1, \tau) = \mathrm{e}$. We show that this can be done recursively within space $\mathrm{Pol}\, t(n)$.

The computation of a bit $\mathbf{R}_i(\sigma, \tau)$, for $i \leqq d + 2t(n)^k$, $\sigma \leqq r$ and $\tau \leqq t(n)^{2k}$, is performed according to the recursion formulas given below, where only bits $\mathbf{R}_{i'}(\sigma', \tau')$ are used for which $i' \leqq d + 2t(n)^k$, $\sigma' \leqq r$ and $\tau' < \tau$. Hence every such triple (i', σ', τ') can be written down within space $\lesssim t(n)^{2k}$. Furthermore, the recursion is carried out in a depth-first manner (with respect to τ) such that, for every $\tau' < \tau$, at most a bounded number of triples (i', σ', τ') need to be stored at the same time. Thus M'' works within space $\lesssim t(n)^{4k}$.

As an example for all other cases we deal with the change of the register i in the τth step by the execution of the instructions $\mathbf{R}_i \leftarrow \mathbf{R}_{\mathbf{R}_j}$, $\mathbf{R}_i \rightarrow \mathbf{R}_j \wedge \mathbf{R}_l$, and $\mathbf{R}_i \leftarrow \mathbf{R}_j + \mathbf{R}_l$.

a) $\mathbf{R}_i \leftarrow \mathbf{R}_{\mathbf{R}_j}$. We have $\mathbf{R}_i(\sigma, \tau) = \mathbf{R}_{\mathrm{bin}^{-1}\mathbf{R}_j}(\sigma, \tau - 1)$ where

$$\mathbf{R}_j = \mathbf{R}_j(1, \tau - 1)\, \mathbf{R}_j(2, \tau - 1) \ldots \mathbf{R}_j(r, \tau - 1).$$

It is true $\mathbf{R}_j$ can be of length r and this would be too long. However, we do not need $\mathbf{R}_j$ but $\mathrm{bin}\,\mathrm{bin}^{-1}\mathbf{R}_j$ (i.e. $\mathbf{R}_j$ without the leading 0's), and by the supposition about M' we have $\mathrm{bin}^{-1}\mathbf{R}_j \leqq d + 2t(n)^k$.

b) $\mathbf{R}_i \leftarrow \mathbf{R}_j \wedge \mathbf{R}_l$. We have

$$\mathbf{R}_i(\sigma, \tau) = \begin{cases} 1 \quad \text{if} \quad \mathbf{R}_j(\sigma, \tau - 1) = \mathbf{R}_l(\sigma, \tau - 1) = 1, \\ \mathrm{e} \quad \text{if} \quad \mathbf{R}_j(\sigma, \tau - 1) = \mathbf{R}_l(\sigma, \tau - 1) = \mathrm{e}, \\ 0 \quad \text{otherwise}. \end{cases}$$

c) $\mathbf{R}_i \leftarrow \mathbf{R}_j + \mathbf{R}_l$. First $\sigma_j = \mu\sigma(\mathbf{R}_j(\sigma, \tau - 1) = \mathrm{e}) \dot{-} 1$ and $\sigma_l = \mu\sigma(\mathbf{R}_l(\sigma, \tau - 1) = \mathrm{e}) \dot{-} 1$ are computed. Because of $\sigma_j, \sigma_l \leqq r$, these numbers can be written down within space $\lesssim t(n)^{2k}$. We define

$$\mathrm{CARRY}(0) = 0$$

and, for $0 < \sigma \leqq \max(\sigma_j, \sigma_l)$,

$$\mathrm{CARRY}(\sigma) = \begin{cases} 1 & \text{if } \mathbf{R}_j(\sigma_j - \sigma + 1, \tau - 1) + \mathbf{R}_l(\sigma_l - \sigma + 1, \tau - 1) \\ & \qquad + \mathrm{CARRY}(\sigma - 1) \geqq 2, \\ 0 & \text{otherwise} \end{cases}$$

and

$$\sigma' = \max(\sigma_j, \sigma_l) + \mathrm{CARRY}(\max(\sigma_j, \sigma_l)).$$

Then, for $\sigma = 0, \ldots, \sigma' - 1$,

$$\mathbf{R}_i(\sigma' - \sigma, \tau) = \mathbf{R}_j(\sigma_j - \sigma, \tau - 1) \oplus \mathbf{R}_l(\sigma_l - \sigma, \tau - 1) \oplus \mathrm{CARRY}(\sigma),$$

where $\mathbf{R}_m(\sigma, \tau - 1) =_{\mathrm{df}} 0$ for $m = j, l$ and $\sigma \leqq 0$.

2. We show DSPACE(Pol t) $\subseteq$ CRAM′-DTIME(Pol t).

A. We estimate the CRAM′-DTIME-complexity of some functions, which will be used for what follows.

a) Let $\mathbf{R}_1 = v$ and $\mathbf{R}_2 = w$. The following program computes $\mathbf{R}_0 = w^{|v|}$ within $\lesssim |v|$ steps:

1 $\mathbf{R}_0 \leftarrow \mathrm{e}$
2 $\mathbf{R}_1 \leftarrow 1\mathbf{R}_1$
3 $\mathbf{R}_3 \leftarrow 10$
4 IF ($\mathbf{R}_3 > \mathbf{R}_1$) THEN STOP
5 $\mathbf{R}_0 \leftarrow \mathbf{R}_0\mathbf{R}_2$
6 $\mathbf{R}_3 \leftarrow \mathbf{R}_3 0$
7 GOTO 4

b) Let $\mathbf{R}_1 = w$ and $\mathbf{R}_2 = \mathrm{bin}\, k$. The following program computes $\mathbf{R}_0 = w^k$ within $\lesssim \log k$ steps:

1 $\mathbf{R}_0 \leftarrow \mathrm{e}$
2 $\mathbf{R}_3 \leftarrow 0^{|\mathbf{R}_2|}1$
3 $\mathbf{R}_2 \leftarrow 0\mathbf{R}_2$
4 IF ($\mathbf{R}_3 > \mathbf{R}_2$) THEN STOP
5 $\mathbf{R}_4 \leftarrow \mathbf{R}_2 \wedge \mathbf{R}_3$
6 IF ($R_4 = 0$) THEN GOTO 8
7 $\mathbf{R}_0 \leftarrow \mathbf{R}_0\mathbf{R}_1$
8 $\mathbf{R}_1 \leftarrow \mathbf{R}_1\mathbf{R}_1$
9 $\mathbf{R}_3 \leftarrow \mathbf{R}_3 0$
10 $\mathbf{R}_2 \leftarrow 0\mathbf{R}_2$
11 GOTO 4

Note that line 2 can be carried out by a) within $\lesssim |\mathbf{R}_2| = |\mathrm{bin}\, k| \lesssim \log k$ steps.

c) The words $w^{|v_1|+|v_2|} = w^{|v_1|}w^{|v_2|}$ and $w^{|v_1|\cdot|v_2|} = (w^{|v_1|})^{|v_2|}$ can be computed from w, v_1 and v_2 by a DCRAM′ within $\lesssim |v_1| + |v_2|$ steps.

d) The words $w^{k_1+k_2} = w^{k_1}w^{k_2}$ and $w^{k_1 \cdot k_2} = (w^{k_1})^{k_2}$ can be computed from w, k_1 and k_2 by a DCRAM′ within $\lesssim \log k_1 + \log k_2$ steps.

e) The words $0^{|w|}$ and $1^{|w|}$ can be computed from w by a DCRAM′ within a constant number of steps because of $0^{|w|} = w \wedge \sim w$ and $1^{|w|} = \sim 0^{|w|}$.

f) Let $\mathrm{ERASE}(w_1, w_2)$ $\big(\mathrm{ERASE}'(w_1, w_2)\big)$ be that word of length $|w_2| \dot{-} |w_1|$ which originates from w_2 by erasing its first (last) $|w_1|$ symbols. In what follows we would like to be able to compute $\mathrm{ERASE}(w_1, w_2)$ $\big(\mathrm{ERASE}'(w_1, w_2)\big)$ within $\lesssim |w_1|$ steps by a DCRAM′. But it is not clear how this can be done.

However, we are able to convert a DCRAM′ M having the instructions

$$\mathbf{R}_0 \leftarrow \mathrm{ERASE}(\mathbf{R}_1, \mathbf{R}_2) \quad \text{and} \quad \mathbf{R}_0 \leftarrow \mathrm{ERASE}'(\mathbf{R}_1, \mathbf{R}_2)$$

and working within time t' into an equivalent DCRAM′ M' without these instructions which works within time $c \cdot t'$ for some $c > 0$. This machine works as follows: Every register k of M corresponds to two registers k and k' of M' such that $\mathbf{R}'_k = 0^m\mathbf{R}_k0^m$ and $\mathbf{R}'_{k'} = 1^m$ for some $m \geqq 0$ after each step of the simulation. At the beginning of the simulation we have $m = 0$ for all registers. Table 20.1 describes how an instruction of M is replaced by a small program of M'. Because of e) the new DCRAM′ runs by at most a constant factor more slowly than the original DCRAM′.

g) The number $l \cdot 2^k$ can be computed from k and l within $\lesssim \log k$ steps because of $\mathrm{bin}\, l \cdot 2^k = (\mathrm{bin}\, l)0^k$. The number $2^k \dot{-} 2^l$ can be computed from k and l within $\lesssim \log k + \log l$ steps because of $\mathrm{bin}\,(2^k \dot{-} 2^l) = 1^{k \dot{-} l}0^l = \mathrm{ERASE}(1^l, 1^k)0^l$.

Table 20.1.

The instruction of M	must be replaced in M' by
$\mathbf{R}_0 \leftarrow \mathrm{ERASE}(\mathbf{R}_1, \mathbf{R}_2)$	$\mathbf{R}'_0 \leftarrow 0^{2\cdot\lvert\mathbf{R}'_{1'}\rvert+\lvert\mathbf{R}'_{2'}\rvert}\,\mathbf{R}_2$ $\wedge\, 0^{\lvert\mathbf{R}'_1\rvert+2\cdot\lvert\mathbf{R}'_{2'}\rvert}\,(1^{\lvert\mathbf{R}'_1\rvert+2\cdot\lvert\mathbf{R}'_{2'}\rvert} \vee 1^{2\cdot\lvert\mathbf{R}'_{1'}\rvert+\lvert\mathbf{R}'_2\rvert})$ $\mathbf{R}'_{0'} \leftarrow 1^{\lvert\mathbf{R}'_1\rvert}\,\mathbf{R}'_{2'}\,\mathbf{R}'_{2'}$
$\mathbf{R}_0 \leftarrow \mathrm{ERASE}'(\mathbf{R}_1, \mathbf{R}_2)$	$\mathbf{R}'_0 \leftarrow 0^{\lvert\mathbf{R}'_1\rvert+\lvert\mathbf{R}'_{2'}\rvert}\,\mathbf{R}_2$ $\wedge\, (1^{\lvert\mathbf{R}'_1\rvert+2\cdot\lvert\mathbf{R}'_{2'}\rvert} \vee 1^{2\cdot\lvert\mathbf{R}'_{1'}\rvert+\lvert\mathbf{R}'_2\rvert})\,0^{\lvert\mathbf{R}'_1\rvert+2\cdot\lvert\mathbf{R}'_{2'}\rvert}$ $\mathbf{R}'_{0'} \leftarrow 1^{\lvert\mathbf{R}'_1\rvert}\,\mathbf{R}'_{2'}\,\mathbf{R}'_{2'}$
$\mathbf{R}_0 \leftarrow \mathbf{R}_1\mathbf{R}_2$	$\mathbf{R}'_0 \leftarrow 0^{\lvert\mathbf{R}'_1\rvert}\,\mathbf{R}'_2\,0^{2\cdot\lvert\mathbf{R}'_{1'}\rvert} \vee 0^{\lvert\mathbf{R}'_{1'}\rvert+\lvert\mathbf{R}'_{2'}\rvert}\,\mathbf{R}'_1$ $\mathbf{R}'_{0'} \leftarrow \mathbf{R}'_{1'}\mathbf{R}'_{1'}\mathbf{R}'_{2'}$
$\mathbf{R}_0 \leftarrow \mathbf{R}_1 \wedge \mathbf{R}_2$	$\mathbf{R}'_0 \leftarrow 0^{\lvert\mathbf{R}'_{2'}\rvert}\,\mathbf{R}'_1\,0^{\lvert\mathbf{R}'_{2'}\rvert} \wedge 0^{\lvert\mathbf{R}'_{1'}\rvert}\,\mathbf{R}'_2\,0^{\lvert\mathbf{R}'_{1'}\rvert}$ $\mathbf{R}'_{0'} \leftarrow \mathbf{R}'_{1'}\mathbf{R}'_{2'}$
$\mathbf{R}_0 \leftarrow \sim\mathbf{R}_1$	$\mathbf{R}'_0 \leftarrow (\sim 0^{\lvert\mathbf{R}'_{1'}\rvert}\,\mathbf{R}'_1\,0^{\lvert\mathbf{R}'_{1'}\rvert}) \oplus (1^{2\cdot\lvert\mathbf{R}'_{1'}\rvert}\,0^{\lvert\mathbf{R}'_1\rvert} \oplus 0^{\lvert\mathbf{R}'_1\rvert}\,1^{2\cdot\lvert\mathbf{R}'_{1'}\rvert})$ $\mathbf{R}'_{0'} \leftarrow \mathbf{R}'_{1'}\mathbf{R}'_{1'}$
$\mathbf{R}_0 \leftarrow \lvert w\rvert$	$\mathbf{R}'_0 \leftarrow w$ $\mathbf{R}'_{0'} \leftarrow \mathrm{e}$
IF ($\mathbf{R}_1 = 0$) THEN GOTO ...	IF ($\mathbf{R}'_1 = 0$) THEN GOTO ...
IF ($\mathbf{R}_1 > \mathbf{R}_2$) THEN GOTO ...	IF($0^{\lvert\mathbf{R}'_{2'}\rvert}\,\mathbf{R}'_1\,0^{\lvert\mathbf{R}'_{2'}\rvert} > 0^{\lvert\mathbf{R}'_{1'}\rvert}\,\mathbf{R}'_2\,0^{\lvert\mathbf{R}'_{1'}\rvert}$) THEN GOTO ...

h) Let $\mathrm{EXT}(k_1, k_2, k_3, w) =_{\mathrm{df}} v_1 0^{k_3} v_2 0^{k_3} \ldots v_r 0^{k_3}$ where $r = 2^{k_1}$, $v_1 v_2 \ldots v_r \sqsubseteq w0^\omega$ and $|v_1| = |v_2| = \ldots = |v_r| = k_2$. The function EXT can be computed within $\preceq (k_1 + \log k_2 + \log k_3) \cdot k_1$ steps by a DCRAM' which in the stage l of its work brings the words v_j with $j \equiv 1(2^{k_1-l})$ into their final positions with the help of "masks" $m_l =_{\mathrm{df}} 1^{k_2 \cdot 2^{k_1-l}} 0^{(k_2+2k_3) \cdot 2^{k_1-l}}$. More precisely, the DCRAM' works according to the recursion formula

$$w_0 = w \wedge 1^{k_2 \cdot 2^{k_1}},$$

$$w_l = \left(w_{l-1} \wedge m_l^{2^{l-1}}\right) \vee 0^{k_3 \cdot 2^{k_1-l}} \left(w_{l-1} \wedge 0^{k_2 \cdot 2^{k_1-l}} m_l^{2^{l-1}}\right), \qquad l \geqq 1,$$

$$\mathrm{EXT}(k_1, k_2, k_3, w) = \mathrm{ERASE}'\left(\mathrm{ERASE}\left(1^{(k_2+k_3) \cdot 2^{k_1}}, w_{k_1}\right), w_{k_1}\right),$$

where the values 2^{k_1-l} and 2^{l-1} are computed recursively by the formulas

$$\mathrm{bin}\, 2^{k_1-0} = 10^{k_1}, \qquad \mathrm{bin}\, 2^{k_1-(l+1)} = \mathrm{ERASE}'(1, \mathrm{bin}\, 2^{k_1-l}), \quad \text{for} \quad l \geqq 0,$$

and

$$\mathrm{bin}\, 2^{1-1} = 1, \qquad \mathrm{bin}\, 2^{(l+1)-1} = (\mathrm{bin}\, 2^{l-1})0, \quad \text{for} \quad l \geqq 1.$$

i) Let $\mathrm{MULT}(k_1, k_2, k_3, w) = v_1^{2^{k_3}} v_2^{2^{k_3}} \ldots v_r^{2^{k_3}}$ where $r = 2^{k_1}$, $v_1 v_2 \ldots v_r \sqsubseteq w0^\omega$ and $|v_1| = |v_2| = \ldots = |v_r| = k_2$. The function MULT can be computed within $(k_1 + \log k_2 + k_3) \cdot k_1$ steps by a DCRAM' which works according to the recursion formula

$$w_0 = \mathrm{EXT}\left(k_1, k_2, (2^{k_3} - 1) \cdot k_2, w\right),$$

$$w_{l+1} = w_l \vee 0^{k_2 \cdot 2^l} w_l, \quad l \geqq 0,$$

$$\mathrm{MULT}(k_1, k_2, k_3, w) = \mathrm{ERASE}'(1^{(2^{k_3}-1) \cdot k_2}, w_{k_3}).$$

Note that the multiplication $(2^{k_3} - 1) \cdot k_2$ need not be carried out, since it appears in $\mathrm{EXT}\left(k_1, k_2, (2^{k_3} - 1) \cdot k_2, w\right)$ only as an exponent and can consequently be treated as under d).

j) The function $\mathrm{DISJ}(k_1, k_2, w) =_{\mathrm{df}} \bigvee_{l=0}^{2^{k_2}-1} \mathrm{ERASE}(1^{k_1 \cdot l}, w)$ can be computed within $\preceq (\log k_1 + k_2) \cdot k_2$ steps by a DCRAM' which works according to the recursion formula

$$w_1 = w,$$

$$w_{l+1} = w_l \vee \mathrm{ERASE}(1^{k_1 \cdot 2^{k_2-l}}, w_l), \qquad l \geqq 1,$$

$$\mathrm{DISJ}(k_1, k_2, w) = w_{k_2+1}$$

where the value 2^{k_2-l} is treated as the value 2^{k_1-l} under h).

k) The function $\mathrm{DISJ}'(k_1, k_2, w) =_{\mathrm{df}} \bigvee_{l=0}^{k_2-1} \mathrm{ERASE}(1^{k_1 \cdot l}, w)$ can be computed within $\preceq \log k_1 + k_2$ steps by a DCRAM' which works according to the recursion formula

$$w_1 = w,$$

$$w_{l+1} = w_l \vee \mathrm{ERASE}(1^{l \cdot k_1}, w), \qquad l \geqq 1,$$

$$\mathrm{DISJ}'(k_1, k_2, w) = w_{k_2}.$$

l) An $n \times n$ matrix with elements 0 and 1 is given in the form

$$A = a_{11}a_{12}\ldots a_{1n}a_{21}a_{22}\ldots a_{2n}\ldots a_{n1}a_{n2}\ldots a_{nn}.$$

Let $a_i = a_{i1}a_{i2}\ldots a_{in}$ and $\bar{a}_i = a_{1i}a_{2i}\ldots a_{ni}$. By $C = A \circ B$ we denote the $(\vee, \wedge)$-product of the $n \times n$ matrices A and B, i.e. $c_{ij} = \bigvee_{l=1}^{n} a_{il} \wedge b_{lj}$. If $n = 2^m$, then $C = A \circ B$ can be computed within $\preceq m^2$ steps by a DCRAM′ according to the following formulas

$$A_1 = \mathrm{MULT}(m, 2^m, m, A) = a_1^n a_2^n \ldots a_n^n,$$

$$B_1 = \mathrm{EXT}(2m, 1, 2^m - 1, B) = b_{11}0^{n-1}b_{12}0^{n-1}\ldots b_{nn}0^{n-1},$$

$$B_2 = \mathrm{ERASE}'\left(1^{2^{3m}-2^{2m}}, \mathrm{DISJ}(2^{2m} - 1, m, B_1)\right) = \bar{b}_1\bar{b}_2\ldots\bar{b}_n,$$

$$B_3 = \mathrm{MULT}(0, 2^{2m}, m, B_2) = (\bar{b}_1\bar{b}_2\ldots\bar{b}_n)^n,$$

$$C_1 = A_1 \wedge B_3 = d_{11}d_{12}\ldots d_{1n}\ldots d_{n1}d_{n2}\ldots d_{nn}$$

where $d_{ij} = (a_{i1} \wedge b_{1j}) \ldots (a_{in} \wedge b_{nj})$,

$$C_2 = \mathrm{DISJ}(1, m, C_1) \wedge (10^{2^m-1})^{2^{2m}} = c_{11}0^{n-1}c_{12}0^{n-1}\ldots c_{nn}0^{n-1},$$

$$C_3 = \mathrm{DISJ}(2^m - 1, m, C_2) \wedge (1^{2^m}0^{2^{2m}-2^m})^{2^m} = c_1 0^{(n-1)\cdot n} c_2 0^{(n-1)\cdot n} \ldots c_n 0^{(n-1)\cdot n},$$

$$C = \mathrm{ERASE}'\left(1^{2^{3m}-2^{2m}}, \mathrm{DISJ}(2^{2m} - 2^m, m, C_3)\right).$$

B. Now we actually show that **DSPACE**(Pol t) $\subseteq$ **CRAM′-DTIME**(Pol t). For similar reasons to those mentioned at the beginning of the proof we can use CRAM′'s without input tape. Let M be a DTM working within space $t(n)^k$ and, without loss of generality, in such a way that a) it starts its work in tape square 1 (which holds the first symbol of the input) and with state s_0, b) it does not move its head to the left to tape square 1, and c) if it accepts (rejects) the input, then it finishs its work in tape square 1 with state $s_1(s_2)$ and with an empty tape. Let "code" be a block encoding of length 2^r of all tape symbols and states of M over the alphabet $\{0, 1\}^*$, where code(0) $= 0^{2^r}$ and code(1) $= 10^{2^r-1}$. We shall use only ID's of length $u' =_{\mathrm{df}} 2^{k\cdot|\mathrm{bin}\, t(n)|} \geqq t(n)^k + 1$; for shorter tape inscriptions we add □'s to the right. Then $u =_{\mathrm{df}} 2^r \cdot u'$ is the length of the codes of these ID's.

The work of M can be described by the Boolean function f_M where $f_M(y_{1,1}, \ldots, y_{1,2^r}, y_{2,1}, \ldots, y_{2,2^r}, y_{3,1}, \ldots, y_{3,2^r}, y'_{1,1}, \ldots, y'_{1,2^r}, y'_{2,1}, \ldots, y'_{2,2^r}, y'_{3,1}, \ldots, y'_{3,2^r}) = 1$ if there are $y_1, y_2, y_3, y'_1, y'_2, y'_3$ such that code $y_i = y_{i,1}\ldots y_{i,2^r}$ and code $y'_i = y'_{i,1}\ldots y'_{i,2^r}$ for $i = 1, 2, 3$, and the part $y_1y_2y_3$ of an ID of M can be replaced by $y'_1y'_2y'_3$ in order to get the next ID.

a) Computation of u' and u. For the moment we assume that the DCRAM′ which we are constructing has access to the value $t(n)$. In this case the DCRAM′ can compute u' and u within $\preceq \log t(n)$ steps since bin $u' = 1(0^{|\mathrm{bin}\, t(n)|})^k$ and bin $u =$ (bin u')0^r.

b) Generation of the $2^u \times 2^u$ matrix $A_M = a_{1,1}a_{1,2}, \ldots, a_{1,2^u}, \ldots a_{2^u,1}a_{2^u,2}, \ldots, a_{2^u,2^u}$,

where

$$a_{i,j} = \begin{cases} 1 & \text{if there are ID's } K, K' \text{ such that } \mathrm{code}(K) = \mathrm{bin}_u(i-1), \\ & \mathrm{code}(K') = \mathrm{bin}_u(j-1) \text{ and } K \underset{M}{\vdash} K', \\ 0 & \text{otherwise.} \end{cases}$$

First, the sequence $b_u = v_1 v_2 \ldots v_{2^u}$, where $v_j = \mathrm{bin}_u(j-1)$, can be generated within $\lesssim u^3$ steps by a DCRAM' which works according to the recursion formula

$$b_1 = 01,$$
$$a_{l+1} = \mathrm{ERASE}'\big(1, 0\mathrm{EXT}(l, l, 1, b_l)\big),$$
$$b_{l+1} = a_{l+1}\big((10^l)^{2^l} \vee a_{l+1}\big).$$

Now the matrix A_M can be generated within $\lesssim u^2$ steps by a DCRAM' which works according to the following formulas:

$$A_1 = \mathrm{MULT}(u, u, u, b_u) = v_1^{2^u} v_2^{2^u} \ldots v_{2^u}^{2^u},$$
$$A_2 = \mathrm{MULT}(0, u \cdot 2^u, u, b_u) = (v_1 v_2 \ldots v_{2^u})^{2^u},$$
$$A_3 = \mathrm{f}_M\big(A_1, \mathrm{ERASE}(1, A_1), \mathrm{ERASE}(1^2, A_1), \ldots, \mathrm{ERASE}(1^{3 \cdot 2^r - 1}, A_1), A_2,$$
$$\mathrm{ERASE}(1, A_2), \ldots, \mathrm{ERASE}(1^{3 \cdot 2^r - 1}, A_2)\big),$$
$$A_4 = \sim \mathrm{DISJ}'(2^r, u' - 2, \sim A_3) \wedge (10^{u-1})^{2^{2u}}$$
$$= a_{1,1} 0^{u-1} a_{1,2} 0^{u-1} \ldots a_{2^u, 2^u} 0^{u-1},$$
$$A_5 = \mathrm{EXT}(2u, u, 2^{2u} - u, A_4) = a_{1,1} 0^{2^{2u}-1} a_{1,2} 0^{2^{2u}-1} \ldots a_{2^u, 2^u} 0^{2^{2u}-1},$$
$$A_M = \mathrm{ERASE}'\big(1^{2^{4u} - 2^{2u}}, \mathrm{DISJ}(2^{2u} - 1, 2 \cdot u, A_5)\big).$$

For the computation of the values $u' - 2$, $u - 1$, and $2^{2u} - u$, we use $u' = 2^{k \cdot |\mathrm{bin}\, t(n)|}$, $u = 2^{k \cdot |\mathrm{bin}\, t(n)| + r}$ and Ag).

c) Generation of the $2^u \times 2^u$ matrix $A_M^* = a_{1,1}^*, a_{1,2}^*, \ldots a_{1,2^u}^*, \ldots, a_{2^u,1}^*, a_{2^u,2}^*, \ldots, a_{2^u,2^u}^*$, where

$$a_{i,j}^* = \begin{cases} 1 & \text{if there are ID's } K, K' \text{ such that } \mathrm{code}(K) = \mathrm{bin}_u(i-1), \\ & \mathrm{code}(K') = \mathrm{bin}_u(j-1) \text{ and } K \overset{*}{\underset{M}{\vdash}} K', \\ 0 & \text{otherwise.} \end{cases}$$

Obviously we have $A_M^* = \bigvee_{l=1}^{2^u} A_M^l = (A_M \vee I_{2^u})^{2^u}$ where $I_{2^u} = 1(0^{2^u}1)^{2^u - 1}$. Therefore A_M^* can be generated by a DCRAM' within $\lesssim u^3$ steps by u $(\vee, \wedge)$-multiplications of $2^u \times 2^u$ matrices.

d) Decision of whether $w = x_1 \ldots x_n \in \mathrm{L}(M)$. The code of the initial ID of M is $\mathrm{bin}_u b =_{\mathrm{df}} \mathrm{code}(s_0)\mathrm{code}(x_1) \ldots \mathrm{code}(x_n)\mathrm{code}(\square)^{u'-n-1}$, the code of the accepting ID of M is $\mathrm{bin}_u c =_{\mathrm{df}} \mathrm{code}(s_1)\mathrm{code}(\square)^{u'-1}$. Hence the DCRAM' has to check whether $a_{b+1,c+1}^* = 1$. Consequently, $w \in \mathrm{L}(M) \leftrightarrow A_M^* \wedge 0^{b \cdot 2^u + c_1} \neq 0^{2^{2u}}$. This can be tested by a DCRAM' within $\lesssim u$ steps if it already has the words $\mathrm{bin}_u b$ and $\mathrm{bin}_u c$. Because

$$\mathrm{bin}_u c = \mathrm{code}(s_1)\mathrm{code}(\square)^{2^{k \cdot \lceil \mathrm{bin}\, t(n) \rceil} - 2^0}$$

and

$$\mathrm{bin}_u b = \mathrm{code}(s_0)\mathrm{code}(x_1)\ldots\mathrm{code}(x_n)\mathrm{code}(\square)^{2^m-n-1}\mathrm{code}(\square)^{2^{k\cdot|\mathrm{bin}\,t(n)|}-2^m},$$

where $m = \mu m'$ ($2^{m'} \geqq n + 1$), these words can be computed by a DCRAM′ within $\lesssim n + \log t(n)$ steps (cf. Ag) and Ab)).

Thus we have obtained a CRAM′ M' which decides $L(M)$ within $\lesssim n + \log t(n) + u^3 \lesssim t(n)^{3k}$ steps, if it has access to the value $t(n)$. Since the latter is not in general the case, M' uses successively the values $t' = 1, 2, 3, \ldots$ instead of $t(n)$, and M' checks not only whether $a^*_{b+1,c+1} = 1$, but also whether $a^*_{b+1,d+1} = 1$, $\mathrm{bin}_u d = \mathrm{code}(s_2)$ $\mathrm{code}(\square)^{u'-1}$ being the code of the rejecting ID of M. At the latest for $t' = t(n)$, one of the checks ends positively, and M' can correspondingly accept or reject. In such a manner M' works within a total of $\lesssim \sum_{t'=1}^{t(n)} t'^{3k} \lesssim t(n)^{3k+1}$ steps.

3. We show **CRAM-XTIME**(Pol t) $\subseteqq$ **MRAM-XTIME**(Pol t) and **CRAM′-XTIME** (Pol t) $\subseteqq$ **MRAM-XTIME**(Pol t) for X ∈ {D, N}. It is sufficient to show that the instruction $\mathbf{R}_0 \leftarrow \mathbf{R}_1\mathbf{R}_2$ can be replaced by a constant number of instructions of an MRAM′. Since $2^{|\mathbf{R}_2|} \cdot \mathbf{R}_1 + \mathbf{R}_2$ is equal to $\mathbf{R}_1\mathbf{R}_2$ without its leading 0's and since $\mathbf{R}_1 \neq \mathbf{R}_1 \vee 1$ if and only if $\mathbf{R}_1$ has leading 0's, we can replace the instruction $\mathbf{R}_0 \leftarrow \mathbf{R}_1\mathbf{R}_2$ by

1 IF ($\mathbf{R}_1 \neq \mathbf{R}_1 \vee 1$) THEN GOTO 4

2 $\mathbf{R}_0 \leftarrow 2^{|\mathbf{R}_2|} \cdot \mathbf{R}_1 + \mathbf{R}_2$

3 GOTO 8

4 IF ($\mathbf{R}_1 = \mathrm{e}$) THEN GOTO 7

5 $\mathbf{R}_0 \leftarrow \left(2^{|\mathbf{R}_2|} \cdot (\mathbf{R}_1 \vee 1) + \mathbf{R}_2\right) \oplus 1$

6 GOTO 8

7 $\mathbf{R}_0 \leftarrow \mathbf{R}_2$

where the instruction $\mathbf{R}_0 \leftarrow 2^{|\mathbf{R}_1|}$ can be replaced by

1 IF ($\mathbf{R}_1 = \mathrm{e}$) THEN GOTO 4

2 $\mathbf{R}_0 \leftarrow \left((\mathbf{R}_1 \oplus \mathbf{R}_1) \vee 1\right) \cdot 2$

3 GOTO 5

4 $\mathbf{R}_0 \leftarrow 1$ ∎

Remark 1. In [HaSi 76] the MRAM part of the preceding theorem is proved, in [PrSt 76] the result for "vector RAM's" is proved which differ from the concatenation RAM's used here.

Remark 2. Similar results for various other types of RAM's can be found in [Simj 79].

20.2. *Turing Machines with Auxiliary Stores*

The starting point for the investigation of Turing machines with auxiliary stores was the paper [Coo 71 b] in which S. A. Cook showed that the addition of an auxiliary pushdown store enlarges the ability of space-bounded Turing machines in such a way that **2-auxPD-T-DSPACE**(s) = **DTIME**$(2^{\mathrm{Lin}\,s})$ can be shown for $s \geqq \log$. S. A. Cook's result has stimulated investigations of Turing machines having other auxiliary stores.

20.2.1. Auxiliary Pushdown Stores

20.13. Theorem. [Coo 71 b]

1. For $s \geqq \log$,

$$\textbf{2-auxPD-T-DSPACE}(s) = \textbf{DTIME}(2^{\mathrm{Lin}\,s}).$$

2. For $s \geqq \log$ such that $s(|w|) \in \mathbf{DTIME}(2^{\mathrm{Lin}\,s})$,

$$\textbf{2-auxPD-T-NSPACE}(s) = \textbf{DTIME}(2^{\mathrm{Lin}\,s}).$$

Proof. We show **DTIME**$(2^{\mathrm{Lin}\,s}) \subseteqq$ **2-auxPD-T-DSPACE**(s) for $s \geqq \log$. The proof of 13.20.3 (time) has shown that a DTM which works within time $t = n^k$ can be simulated by a 2:(2k)-DPDA. For this proof it was sufficient that this 2:(2k)-DPDA could read the input in a 2-way manner, that it had a pushdown store and that it could store, compare, and divide by 2 a bounded number of natural numbers not greater than t. If we take $t = 2^{c \cdot s(n)}$, then this can be done by a 2-PD-T-DM which uses at most $\log t \asymp s(n)$ squares on its worktape, where quadruples (i, j, s, b)instead of $(1^i, 1^j, s, b)$ are stored in the pushdown store.

Now we show **2-auxPD-T-NSPACE**$(s) \subseteqq$ **DTIME**$(2^{\mathrm{Lin}\,s})$ for $s \geqq \log$ such that $s(|w|) \in \mathbf{DTIME}(2^{\mathrm{Lin}\,s})$. The proof of 13.20.5 (time) has shown that a 2:k-NPDA M can be simulated by a 2T-DM M' which works within time n^{4k}. The machine M' writes down successively all $(s', v', d', s'', v'', d'')$ such that $(s', v', d') \models (s'', v'', d'')$ (s', s'' being states, d', d'' being pushdown symbols and v', v'' being input head positions). Since there are at most $\lesssim n^k$ input head positions, there are at most $\lesssim n^{2k}$ such sextuples. To find a new sextuple takes M' at most $\lesssim n^{2k}$ steps.

If we start from a 2-PD-T-NM which works within space s then M' has to write down all $(s', v', d', s'', v'', d'')$ such that $(s', v', d') \models (s'', v'', d'')$, where here each v includes an input head position, a worktape head position and a worktape inscription. Since there are at most $\lesssim 2^{c_1 \cdot s(n)}$ such v (n being the input length), there are at most $\lesssim 2^{2c_1 s(n)}$ such sextuples. To find a new sextuple takes M' at most $\lesssim 2^{c_2 s(n)}$ steps. Thus M' works within time $\lesssim 2^{c_3 \cdot s(n)}$. Note that M' must have the value $s(n)$ in order to generate only those sextuples whose worktape inscriptions have a length not greater than $s(n)$.

Finally, we show **2-auxPD-T-DSPACE**$(s) \subseteqq$ **DTIME**$(2^{\mathrm{Lin}\,s})$ for $s \geqq \log$. The 2T-DM M' works in the same way as when simulating a 2-PD-T-NM, but instead of the value $s(n)$ it uses successively the values $\hat{s} = 1, 2, \ldots$, and M' checks not only whether $(s_0, \nu_0, \boxdot, s_1, \nu_1, \boxdot)$ is generated but also whether $(s_0, \nu_0, \boxdot, s_2, \nu_1, \boxdot)$ is generated for some ν_1 (s_0, s_1, s_2 being the initial, accepting and rejecting state, resp., ν_0 describing the initial situation on input tape and worktape). At the latest for $\hat{s} = s(n)$, one of these two sextuples will be generated, and M' can accept or reject correspondingly. In such a manner M' works

$$\sum_{\hat{s}=1}^{s(n)} 2^{c_3 \cdot \hat{s}} \leq 2^{c_3(s(n)+1)} \leq 2^{c_4 \cdot s(n)} \text{ steps for some } c_4 > 0. \blacksquare$$

The proof of 20.13 shows (see also [Harj 79])

20.14. Theorem. For X ∈ {D, N} and $s \geqq \log$ such that, for X = N, $s(|w|) \in$ **DTIME** $(2^{\text{Lin}\, s})$,

$$\mathbf{DTIME}(2^{\text{Lin}\, s}) = \mathbf{2\text{-}auxPD\text{-}T\text{-}XSPACE\text{-}TIME}(s, 2^{2^{\text{Lin}\, s}}). \blacksquare$$

20.15. Theorem. Now we deal with other types of Turing machines having an auxiliary pushdown store.

1. For $l \geqq 1$ and $s \geqq \log$,

$$\begin{aligned}\mathbf{DTIME}(2^{\text{Lin}\, s}) &= \mathbf{2\text{-}auxPD\text{-}PD\text{-}DSPACE}(s)\\ &= \mathbf{2\text{-}auxPD\text{-}S\text{-}DSPACE}(s)\\ &= \mathbf{2\text{-}auxPD\text{-}lC\text{-}DSPACE}(2^{\text{Lin}\, s}).\end{aligned}$$

2. For $l \geqq 1$ and $s \geqq \log$ such that $s(|w|) \in \mathbf{DTIME}(2^{\text{Lin}\, s})$,

$$\begin{aligned}\mathbf{DTIME}(2^{\text{Lin}\, s}) &= \mathbf{2\text{-}auxPD\text{-}PD\text{-}NSPACE}(s)\\ &= \mathbf{2\text{-}auxPD\text{-}S\text{-}NSPACE}(s)\\ &= \mathbf{2\text{-}auxPD\text{-}lC\text{-}NSPACE}(2^{\text{Lin}\, s}).\end{aligned}$$

Proof. Let $s \geqq \log$, and let $s(|w|) \in \mathbf{DTIME}(2^{\text{Lin}\, s})$ in the case X = N. From 20.13 we obtain $\mathbf{DTIME}(2^{\text{Lin}\, s}) = \mathbf{2\text{-}auxPD\text{-}T\text{-}XSPACE}(s)$.

a) We show $\mathbf{2\text{-}auxPD\text{-}T\text{-}XSPACE}(s) \subseteq \mathbf{2\text{-}auxPD\text{-}PD\text{-}XSPACE}(s)$. The inscription of the Turing tape of a 2-PD-T-XM M can be stored, with a marker for the head position, on the second pushdown store of a 2-PD-PD-XM M'. If the head position or the inscription of the Turing tape of M is changed, then M' can simulate this activity by bringing the right part (i.e. the part to the right of the marker) of the second pushdown store to the first (the auxiliary) pushdown store and then back to the second one.

b) The inclusions $\mathbf{2\text{-}auxPD\text{-}PD\text{-}XSPACE}(s) \subseteq \mathbf{2\text{-}auxPD\text{-}S\text{-}XSPACE}(s) \subseteq \mathbf{2\text{-}auxPD\text{-}T\text{-}XSPACE}(s)$ are evident.

c) We show $\mathbf{2\text{-}auxPD\text{-}PD\text{-}XSPACE}(s) \subseteq \mathbf{2\text{-}auxPD\text{-}C\text{-}XSPACE}(2^{\text{Lin}\, s})$. Let M be a 2-PD-PD-XM working within space $c \cdot s$ and having ,without loss of generality, only the stack symbols $\boxdot$, 1, 2, $\square$. Consequently, a stack word $w \in \boxdot\{1, 2\}^*$ can be considered as the dyadic presentation of the number $\text{dya}^{-1}w$, which can be stored in the counter of a 2-PD-C-XM M'. A change of the top symbol of the second pushdown store of M corresponds to one of the operations $x \leftarrow 2x + 1$, $x \leftarrow 2x + 2$, $x \leftarrow x + 1$, $x \leftarrow x \dot{-} 1$ and $x \leftarrow \left\lfloor \frac{x}{2} \right\rfloor$ of the contents x of the counter of M'. These operations can actually be performed by M' using the pushdown store.

d) The inclusion $\mathbf{2\text{-}auxPD\text{-}lC\text{-}XSPACE}(2^{\text{Lin}\, s}) \subseteq \mathbf{2\text{-}auxPD\text{-}T\text{-}XSPACE}(s)$ is evident. ∎

The equation $\mathbf{2\text{-}auxPD\text{-}C\text{-}DSPACE}(\text{Pol}\, s) = \mathbf{DTIME}(\text{Pol}\, s)$ for $s \geqq \text{id}$ (cf. 20.15) suggests the question of whether more exact relationships between 2-auxPD-C-DSPACE and realistic time measures can be shown. From 13.20.2 (time) and 13.21 we get the first statement of the following theorem.

20.16. Theorem.

1. For $k \geqq 0$, **2-auxPD-C-DSPACE**$(n^k) \subseteq$ **RAM-DTIME**(Lin n^{k+1}).
2. [Mon 74c] For $k \geqq 1$, **multiT-DTIME**$(n^k) \subseteq$ **2-auxPD-C-DSPACE**$(n^{k-1}(\log n)^4)$. ∎

Further related results can be found in [Kam 72], [Mon 74b] and [Mon 75b].

20.2.2. Auxiliary Counters

In this subsection we show that the addition of an auxiliary counter to a space-bounded 2-T-XM does not increase the computational power of this machine.

20.17. Theorem. For X ∈ {D, N} and $s \geqq \log$, **2-auxC-T-XSPACE**(s) = **XSPACE**(s).

Proof. We show **2-auxC-T-DSPACE**$(s) \subseteq$ **DSPACE**(s). Let M be a 2-C-T-DM which on inputs of length n uses at most $s(n)$ squares on its Turing tape. For given contents of the counter, M can be in at most $\lesssim 2^{c \cdot s(n)}$ different situations. The proof of 13.20.1 (space) shows that then the counter is of maximal length $\lesssim 2^{c \cdot s(n)}$. Hence the contents of the counter of M can be stored within space $\lesssim s(n)$.

Now we show **2-auxC-T-NSPACE**$(s) \subseteq$ **NSPACE**(s). We proceed essentially in the same way as above. The proof of 13.20.5 (space) shows: if there is an accepting computation of M, then there is an accepting computation with maximum counter length $\lesssim 2^{2c \cdot s(n)}$. Hence, the contents of the counter on such an accepting path can be stored within space $\lesssim s(n)$.

Finally, the inclusion **XSPACE**$(s) \subseteq$ **2-auxC-T-XSPACE**(s) is evident. ∎

The next theorem can be shown in the same way as Theorem 20.17.

20.18. Theorem. For $l \geqq 2$, X ∈ {D, N} and $s \geqq \log$,

$$\textbf{2-auxC-lC-XSPACE}(2^{\text{Lin}\, s}) = \textbf{XSPACE}(s).\ ∎$$

20.2.3. Auxiliary Nonerasing Stacks

20.19. Theorem. [Iba 71] For $k \geqq 1$ and arbitrary s,

1. **2:k-auxNES-T-DSPACE**(s) = **DSPACE**$(n^k \cdot \log n \cdot 2^{\text{Lin}\, s})$.
2. **2:k-auxNES-T-NSPACE**(s) = **NSPACE**$(n^{2k} \cdot 2^{\text{Lin}\, s})$.

Proof. We show **DSPACE**$(n^k \cdot \log n \cdot 2^{\text{Lin}\, s}) \subseteq$ **2:k-auxNES-T-DSPACE**(s). For the simulation of a DTM M by a 2:k-DNESA M' in the proof of 13.29, we have only used the fact that M' can read the input in a 2-way manner, that it has a nonerasing stack and that it can count up to $t = c \cdot n^k$ without the help of the nonerasing stack. Thus M' is able to simulate a DTM M which works within space $t \cdot \log t \lesssim n^k \cdot \log n$. A 2:k-NES-T-DM M' which uses at most $s(n)$ squares on its Turing tape is able to count up to $t = n^k \cdot 2^{c \cdot s(n)}$ for some $c > 0$ without the use of the nonerasing stack. Consequently, M' is able to simulate a DTM which works within space $t \cdot \log t \lesssim n^k \cdot \log n \cdot 2^{2c \cdot s(n)}$.

Now we show that **2:k-auxNES-T-DSPACE**$(s) \subseteq$ **DSPACE**$(n^k \cdot \log n \cdot 2^{\mathrm{Lin}\, s})$. For the simulation of a 2:k-DNESA M by a DTM M' in the proof of 13.29 it was important that M can be in at most $t = c \cdot n^k$ different situations which have the same stack word, the same position of the stack head and the same input. Thus M' was able to simulate M within space $t \cdot \log t \preceq n^k \cdot \log n$. A 2:k-NES-T-DM M which works within space s can be in at most $t = n^k \cdot 2^{c \cdot s(n)}$ different situations which have the same stack word, the same position of the stack head and the same input. Consequently a DTM M' is able to simulate M within space $t \cdot \log t \preceq n^k \cdot \log n \cdot 2^{2 \cdot c \cdot s(n)}$.

Finally we note that Statement 2 can be shown analogously. ∎

Now we deal with other types of Turing machines having auxiliary nonerasing stacks.

20.20 Theorem. For $k \geqq 1$, $l \geqq 1$ and $s \geqq 1$,

1. **2:k-auxNES-lC-DSPACE**$(s) =$ **DSPACE**$\left(n^k \cdot s(n)^l (\log n + \log s(n))\right)$,
   2:k-auxNES-lC-NSPACE$(s) =$ **NSPACE**$(n^{2k} \cdot s(n)^{2l})$.
2. **2:k-auxNES-PD-DSPACE**$(s) =$ **DSPACE**$(n^k \cdot \log n \cdot 2^{\mathrm{Lin}\, s})$,
   2:k-auxNES-PD-NSPACE$(s) =$ **NSPACE**$(n^{2k} \cdot 2^{\mathrm{Lin}\, s})$.
3. **2:k-auxNES-S-DSPACE**$(s) =$ **DSPACE**$(n^k \cdot \log n \cdot 2^{\mathrm{Lin}\, s})$,
   2:k-auxNES-S-NSPACE$(s) =$ **NSPACE**$(n^{2k} \cdot 2^{\mathrm{Lin}\, s})$.

Proof. The proof of statement 1 is analogous to the proof of 20.19, but uses $t = c \cdot n^k \cdot s(n)^l$.
Ad 2. By 20.19, it is sufficient to show

$$\textbf{2:k-auxNES-PD-XSPACE}(s) = \textbf{2:k-auxNES-T-XSPACE}(s)$$

for X ∈ {D, N}. The inclusion "$\subseteq$" is obvious; the inclusion "$\supseteq$" can be shown in the same way as the inclusion **2-auxPD-T-XSPACE**$(s) \subseteq$ **2-auxPD-PD-XSPACE**(s) (cf. 20.15), but here a word brought from the pushdown to the stack cannot be erased from there again. Therefore it is marked and can thus be ignored in the sequel.

Ad 3. This statement is a consequence of statement 2 and Theorem 20.19. ∎

20.21. Corollary. For $k \geqq 1$, $l \geqq 1$, X ∈ {D, N} and $s \geqq \log$,

$$\begin{aligned}\textbf{DSPACE}(2^{\mathrm{Lin}\, s}) &= \textbf{2:k-auxNES-T-XSPACE}(s)\\ &= \textbf{2:k-auxNES-S-XSPACE}(s)\\ &= \textbf{2:k-auxNES-PD-XSPACE}(s)\\ &= \textbf{2:k-auxNES-lC-XSPACE}(2^{\mathrm{Lin}\, s}).\ ∎\end{aligned}$$

20.2.4. Auxiliary Stacks

20.22. Theorem. [Iba 71] For $k \geqq 1$ and arbitrary s,

1. **2:k-auxS-T-DSPACE**$(s) =$ **T-auxPD-DSPACE**$(n^k \cdot \log n \cdot 2^{\mathrm{Lin}\, s})$.
2. **2:k-auxS-T-NSPACE**$(s) =$ **T-auxPD-NSPACE**$(n^{2k} \cdot 2^{\mathrm{Lin}\, s})$.

The proof is analogous to that of 20.19, but uses 13.34. ∎

20.23. Theorem. For $k \geqq 1$, $l \geqq 1$ and $s \geqq 1$,

1. **2:k-auxS-lC-DSPACE**$(s) =$ **T-auxPD-DSPACE**$\left(n^k \cdot s(n)^l \cdot (\log n + \log s(n))\right)$,
   2:k-auxS-lC-NSPACE$(s) =$ **T-auxPD-NSPACE**$(n^{2k} s(n)^{2l})$.

2. $\mathbf{2{:}k\text{-}auxS\text{-}PD\text{-}DSPACE}(s) = \mathbf{T\text{-}auxPD\text{-}DSPACE}(n^k \cdot \log n \cdot 2^{\mathrm{Lin}\, s})$,
   $\mathbf{2{:}k\text{-}auxS\text{-}PD\text{-}NSPACE}(s) = \mathbf{T\text{-}auxPD\text{-}NSPACE}(n^{2k} \cdot 2^{\mathrm{Lin}\, s})$.
3. $\mathbf{2{:}k\text{-}auxS\text{-}S\text{-}DSPACE}(s) = \mathbf{T\text{-}auxPD\text{-}DSPACE}(n^k \cdot \log n \cdot 2^{\mathrm{Lin}\, s})$,
   $\mathbf{2{:}k\text{-}auxS\text{-}S\text{-}NSPACE}(s) = \mathbf{T\text{-}auxPD\text{-}NSPACE}(n^{2k} \cdot 2^{\mathrm{Lin}\, s})$.

The proof is analogous to that of 20.20. ∎

Since it is obvious that $\mathbf{T\text{-}auxPD\text{-}XSPACE}(s) = \mathbf{2\text{-}auxPD\text{-}T\text{-}XSPACE}(s)$, for $s \geqq \mathrm{id}$ and $\mathrm{X} \in \{\mathrm{D}, \mathrm{N}\}$, by using Theorem 20.13 we get the following corollary.

20.24. Corollary.

1. For $k \geqq 1$, $l \geqq 1$ and arbitrary s,

$$\begin{aligned}\mathbf{DTIME}(2^{\mathrm{Lin}\, n^k \cdot \log n \cdot 2^{\mathrm{Lin}\, s}}) &= \mathbf{2{:}k\text{-}auxS\text{-}T\text{-}DSPACE}(s)\\ &= \mathbf{2{:}k\text{-}auxS\text{-}S\text{-}DSPACE}(s)\\ &= \mathbf{2{:}k\text{-}auxS\text{-}PD\text{-}DSPACE}(s)\\ &= \mathbf{2{:}k\text{-}auxS\text{-}lC\text{-}DSPACE}(2^{\mathrm{Lin}\, s}).\end{aligned}$$

2. For $k, l \geqq 1$ and $s(|w|) \in \mathrm{DTIME}(2^{\mathrm{Lin}\, s})$,

$$\begin{aligned}\mathbf{DTIME}(2^{\mathrm{Lin}\, n^{2k} \cdot 2^{\mathrm{Lin}\, s}}) &= \mathbf{2{:}k\text{-}auxS\text{-}T\text{-}NSPACE}(s)\\ &= \mathbf{2{:}k\text{-}auxS\text{-}S\text{-}NSPACE}(s)\\ &= \mathbf{2{:}k\text{-}auxS\text{-}PD\text{-}NSPACE}(s)\\ &= \mathbf{2{:}k\text{-}auxS\text{-}lS\text{-}NSPACE}(2^{\mathrm{Lin}\, s}). \ ∎\end{aligned}$$

20.25. Corollary. Let $\mathrm{X} = \mathrm{D}$, $k, l \geqq 1$, and $s \geqq \log$ or $\mathrm{X} = \mathrm{N}$, $k, l \geqq 1$, $s \geqq \log$ and $s(|w|) \in \mathrm{DTIME}(2^{\mathrm{Lin}\, s})$,

$$\begin{aligned}\mathbf{DTIME}(2^{2^{\mathrm{Lin}\, s}}) &= \mathbf{2{:}k\text{-}auxS\text{-}T\text{-}XSPACE}(s)\\ &= \mathbf{2{:}k\text{-}auxS\text{-}S\text{-}XSPACE}(s)\\ &= \mathbf{2{:}k\text{-}auxS\text{-}PD\text{-}XSPACE}(s)\\ &= \mathbf{2{:}k\text{-}auxS\text{-}lC\text{-}XSPACE}(2^{\mathrm{Lin}\, s}). \ ∎\end{aligned}$$

20.2.5. Auxiliary Checking Stacks

20.26. Theorem. [Iba 71]

1. For $k \geqq 1$ and $s \geqq \log$,

$$\mathbf{2{:}k\text{-}auxCS\text{-}T\text{-}DSPACE}(s) = \mathbf{DSPACE}(s).$$

2. For $k \geqq 1$ and arbitrary s,

$$\mathbf{2{:}k\text{-}auxCS\text{-}T\text{-}NSPACE}(s) = \mathbf{NSPACE}(n^k \cdot 2^{\mathrm{Lin}\, s(n)}).$$

Proof. Ad 1. We show $\mathbf{2{:}k\text{-}auxCS\text{-}T\text{-}DSPACE}(s) \subseteqq \mathbf{DSPACE}(s)$, the other inclusion is obvious. For the simulation of a 2:k-DCSA M by a 2-DTM M' in the proof of 13.40.1, it was important that the writing phase of a halting computation of M consisted of at most $t = c \cdot n^k$ steps. Thus M' was able to simulate M within space

$\log t \lesssim \log n$. The writing phase of a halting computation of a 2:k-CS-T-DM M which uses at most $s(n)$ squares of the Turing tape can consist of at most $t = n^k \cdot 2^{c_1 \cdot s(n)} \lesssim 2^{c_2 \cdot s(n)}$ steps. Thus a 2-DTM is able to simulate M within space $\log t \lesssim s(n)$.

Ad 2. We show **NSPACE**$(n^k \cdot 2^{\mathrm{Lin}\, s(n)}) \subseteq$ **2:k-auxCS-T-NSPACE**(s). For the simulation of an NTM M which works within space n^k by a 2:k-NCSA M' in the proof of 13.40.3, we have only used the fact that M' can read the input in a 2-way manner, that it has a checking stack and that it can count up to $t = c \cdot n^k$ without the use of the checking stack. A 2:k-CS-T-NM which uses at most $s(n)$ squares of the Turing tape can count up to $t = n^k \cdot 2^{c \cdot s(n)}$ without the use of the checking stack. Thus M' is able to simulate an NTM which works within space $n^k \cdot 2^{c \cdot s(n)}$.

Now we show **2:k-auxCS-T-NSPACE**$(s) \subseteq$ **NSPACE**$(n^k \cdot 2^{\mathrm{Lin}\, s(n)})$. This can be done in the same way as in the proof of **2:k-NCSA** $\subseteq$ **NSPACE**(n^k) (cf. 13.40.3). Here the crossing sequences also include the contents of the Turing tape. Since such contents are of length $\lesssim s(n)$, each of the crossing sequences consists of at most $\lesssim n^k \cdot 2^{c_1 \cdot s(n)}$ tuples and can be written down (using the special description) within space $\lesssim n^k \cdot 2^{c_1 \cdot s(n)} \cdot s(n) \leqq n^k \cdot 2^{c_2 \cdot s(n)}$. ∎

Analogously to 20.26.2 we show

20.27. Theorem. For $k \geqq 1$, $l \geqq 1$ and $s \geqq 2$,

$$\mathbf{2{:}k\text{-}auxCS\text{-}lC\text{-}NSPACE}(s \cdot \mathrm{Pol}\log s) = \mathbf{NSPACE}(n^k \cdot s(n)^l \cdot \mathrm{Pol}\log s(n)).$$

$$\mathbf{2{:}k\text{-}auxCS\text{-}lC\text{-}NSPACE}(\mathrm{Pol}\, s) = \mathbf{NSPACE}(n^k \cdot \mathrm{Pol}\, s). \ ∎$$

20.28. Corollary. For $k \geqq 1$, $l \geqq 1$ and $s \geqq \log$,

$$\begin{aligned}\mathbf{NSPACE}(2^{\mathrm{Lin}\, s}) &= \mathbf{2{:}k\text{-}auxCS\text{-}T\text{-}NSPACE}(s)\\ &= \mathbf{2{:}k\text{-}auxCS\text{-}lC\text{-}NSPACE}(2^{\mathrm{Lin}\, s}). \ ∎\end{aligned}$$

For the deterministic case it can be shown that the addition of some further auxiliary checking stacks and one auxiliary counter to a 2:k-auxCS-T-DM does not essentially increase the power of this type of machine. Note that 2:k-2auxCS-NM's can already accept all recursively enumerable sets.

20.29. Theorem. [VoWa 81] For $k \geqq 1$, $l \geqq 1$ and $s \geqq \log$,

$$\mathbf{2{:}k\text{-}lauxCS\text{-}auxC\text{-}T\text{-}DSPACE}(s) = \mathbf{DSPACE}(s). \ ∎$$

The situation changes if other types of auxiliary store are added.

20.30. Theorem. [VoWa 81] For $k \geqq 1$, $l \geqq 0$ and $s \geqq 1$ such that s is fully T-DSPACE-constructible,

1. $\mathbf{2{:}k\text{-}lauxCS\text{-}auxPD\text{-}T\text{-}DSPACE}(s) = \mathbf{DTIME}\left(2^{2^{\cdot^{\cdot^{2^{\mathrm{Lin}\, s}}}}}\Big\}\, l+1\right)$, for $s \geqq \log$.

2. $\mathbf{2{:}k\text{-}lauxCS\text{-}auxNES\text{-}T\text{-}DSPACE}(s) = \mathbf{DSPACE}\left(2^{2^{\cdot^{\cdot^{2^{\mathrm{Lin}\, n^k \cdot \log n \cdot 2^{\mathrm{Lin}\, s(n)}}}}}}\Big\}\, l\right)$.

3. $\mathbf{2{:}k\text{-}lauxCS\text{-}auxS\text{-}T\text{-}DSPACE}(s) = \mathbf{DTIME}\left(2^{2^{\cdot^{\cdot^{2^{\mathrm{Lin}\, n^k \cdot \log n \cdot 2^{\mathrm{Lin}\, s(n)}}}}}}\Big\}\, 2l+1\right)$. ∎

20.31. Corollary. For $l \geqq 1$ and $s \geqq \log$ such that s is fully T-DSPACE-constructible,

1. $\mathbf{2{:}multi\text{-}lauxCS\text{-}auxPD\text{-}T\text{-}DSPACE}(s) = \mathbf{DTIME}\left(2^{2^{\cdot^{\cdot^{2^{\mathrm{Lin}\,s}}}}}\right\}{\scriptstyle l+1}\Big)$.

2. $\mathbf{2{:}multi\text{-}lauxCS\text{-}auxNES\text{-}T\text{-}DSPACE}(s) = \mathbf{DSPACE}\left(2^{2^{\cdot^{\cdot^{2^{\mathrm{Lin}\,s}}}}}\right\}{\scriptstyle l+1}\Big)$.

3. $\mathbf{2{:}multi\text{-}lauxCS\text{-}auxS\text{-}T\text{-}DSPACE}(s) = \mathbf{DTIME}\left(2^{2^{\cdot^{\cdot^{2^{\mathrm{Lin}\,s}}}}}\right\}{\scriptstyle 2l+2}\Big)$. ∎

20.3. Recursive and Parallel Machines

In this subsection we deal with machines which have the ability to call themselves as a subroutine, i.e. such a machine is able to start by *recursive calls* identical copies of itself, possibly with other inputs. The original machine then continues its work with the results of the work of the copies. Note that a copy started by a recursive call can on its part make recursive calls to identical copies of itself etc.

Such a machine is said to be *recursive* if it can make a new recursive call to a copy only after the copy which has been previously called has finished its work. Recursive machines have been introduced in order to describe algorithms which include calls to themselves as subroutines, for example algorithms using the principle "divide and conquer".

If a machine can make recursive calls to several copies which work in parallel, then this machine is called a *parallel* machine.

20.3.1. Recursive Machines

Recursive machines were introduced for the first time in [Savi 77], in terms of recursive Turing machines. A *recursive Turing machine* (for short: *recursive TM*) consists of a finite control, two two-way read-only input tapes, a worktape, and a write-only output tape; in addition it has the ability to call itself as a subroutine. By such a *recursive call* the work of an identical recursive TM is started, which has the same first input tape as the original recursive TM and whose second input tape is the output tape of the original recursive TM. The second tape of the original machine is always empty. A recursive TM can finish its work by going into an accepting or a rejecting state. A recursive call interrupts the work of the original recursive TM up to the stop of the copy which has been called. Then the original recursive TM continues its work with a new state, which is induced by the final state of the copy. In this way the further work of the original recursive TM depends on the result of the work of the copy. The original recursive TM continues its work with an empty output tape.

A recursive TM which has been called recursively as a copy of an original recursive TM can naturally call on its part a further identical copy of itself etc. The ith machine of such a sequence of identical copies of a recursive TM is called the *ith level of the computation* in question, i.e. the recursive TM called by a recursive TM which is the ith level is called the $(i+1)$th level ($i = 1, 2, 3, \ldots$). Note that the first input tape which holds the input of the whole computation is common to all levels of this computation. Figure 20.1 shows three levels of a computation.

An input is accepted (rejected) by a deterministic recursive TM M if M reaches during its work on this input an accepting (rejecting) state. An input is accepted by a nondeterministic recursive TM M if M reaches on at least one of its computations

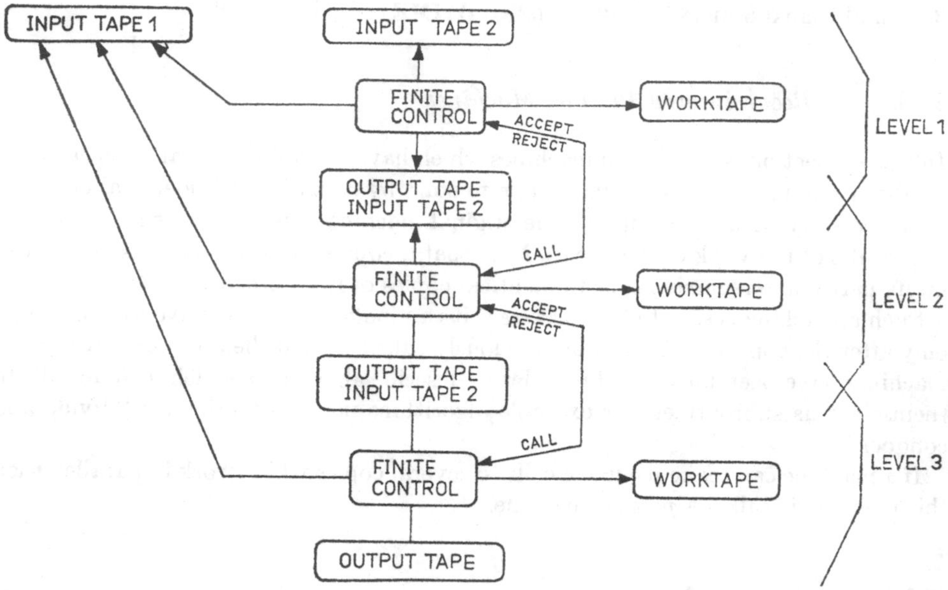

Figure 20.1

on this input an accepting state. The language $L(M)$, the complexity measures **DRECTIME**, **DRECSPACE**, **NRECTIME**, **NRECSPACE** and the corresponding complexity classes are defined in the same way as for "classical" Turing machines. Note that for the space measures the maximum of the space used on the various levels of a computation is taken, where also the output tapes are included.

First, it is easy to see that the time measure for recursive TM's is a realistic time measure.

20.32. Theorem. For $X \in \{D, N\}$ and $t \geqq \mathrm{id}$,

$$\mathbf{XRECTIME}(\mathrm{Pol}\, t) = \mathbf{XTIME}(\mathrm{Pol}\, t). \blacksquare$$

In the same way, one can readily verify that a space measure for recursive TM's which is defined as the sum of the space used in the various levels of a computation would be a realistic space measure. However, we have already defined the space measure for recursive TM's as the maximum of the space used in the various levels of a computation. For this measure we obtain

20.33. Theorem. [Savi 77]
1. For $s \geqq \log$,

$$\mathbf{DRECSPACE}(s) = \mathbf{DTIME}(2^{\mathrm{Lin}\, s}).$$

2. For $s \geqq \log$ such that $s(|w|) \in \mathrm{DTIME}(2^{\mathrm{Lin}\, s})$,

$$\mathbf{NRECSPACE}(s) = \mathbf{DTIME}(2^{\mathrm{Lin}\, s}).$$

Proof. Taking into account Theorem 20.13, it is sufficient to prove

$$\mathbf{XRECSPACE}(s) = \mathbf{2\text{-}auxPD\text{-}T\text{-}XSPACE}(s)$$

for X ∈ {D, N} and $s \geqq \log$.

a) We prove **2-auxPD-T-XSPACE**(s) ⊆ **XRECSPACE**(s). This is especially easy for X = N: Let M be a 2-PD-T-NM working within space s. A recursive NTM M' simulating M works in the same way as M as long as M does not push a symbol onto its pushdown store. When M writes a symbol onto the pushdown store, M' stores the top symbol on its tape, guesses nondeterministically an ID K' and calls a copy, which gets as an input the new top symbol a, the ID K of the current situation and the guessed ID K'. Now the copy, knowing the current top symbol a and the ID K of the current situation of M, is able to simulate M (K' being so far ignored). If M stores a further symbol, the copy in its turn activates a new copy. If M pops a, then the copy compares the ID of the reached situation with K'. If they coincide, the calling copy continues its work with K' otherwise this computation path is unsuccessful. For X = D, the guessing of ID's is replaced by systematic trial.

b) We prove **XRECSPACE**(s) ⊆ **2-auxPD-T-XSPACE**(s) for X ∈ {D, N}. Let M be a recursive TM working within space s. A 2-PD-T-XM M' can simulate M as follows: M' simulates the work of M without using the pushdown store as long as M' does not make a recursive call. If M' makes a recursive call, then it stores the ID of the current situation of M in the pushdown store; only the contents of the output tape of M remain on the Turing tape of M'. Now M' simulates the work of the identical copy of M started by this recursive call. Recursive calls made by this copy are handled by M' in the same way. When the copy finishes its work, then M' continues the simulation of the original machine. This is possible since M' can restore the situation of M of the time of the recursive call using the topmost ID in the pushdown store. ∎

Related work has been done in [StSa 79] and [GuIb 79a].

20.3.2. Parallel Machines

Parallel machines (in the sense of this subsection) were first introduced in [SaSt 76] in terms of *parallel RAM's*. We make precise this concept in terms of *parallel Turing machines* (for short: *parallel TM's*). A parallel Turing machine is defined in the same way as a recursive TM was defined in the preceding subsection, with the only difference that it has two output tapes and can make simultaneous recursive calls to two identical copies, which then work in parallel. The original parallel TM continues its work after the stop of both copies. The language L(M) for a parallel TM M, the complexity measures DPARTIME, NPARTIME, DPARSPACE, NPARSPACE, and the corresponding complexity classes are defined in the same way as for recursive TM's.

Since there is no difference between a recursive TM and a parallel TM with regard to the space used we obtain

20.34. Theorem.

1. For $s \geqq \log$,

$$\mathbf{DPARSPACE}(s) = \mathbf{DTIME}(2^{\mathrm{Lin}\, s}).$$

2. For $s \geqq \log$ such that $s(|w|) \in \mathrm{DTIME}(2^{\mathrm{Lin} s})$,

$$\mathbf{NPARSPACE}(s) = \mathbf{DTIME}(2^{\mathrm{Lin}\, s}). \blacksquare$$

Since an unbounded number of processes can simultaneously run in a parallel TM, the corresponding time measure are polynomially related to the general space measure.

20.35. Theorem. [SaSt 76] For X $\in$ {D, N} and $t \geqq \mathrm{id}$,

$$\mathbf{XPARTIME}(\mathrm{Pol}\, t) = \mathbf{XSPACE}(\mathrm{Pol}\, t).$$

Proof. a) We show $\mathbf{XPARTIME}(t) \subseteqq \mathbf{XSPACE}(t)$. A computation of a parallel TM working on inputs of length n within time $t(n)$ has at most $t(n)$ levels. An XTM M' can simulate the work of M as follows: it simulates M step by step as long as M does not make a recursive call. If M makes a recursive call to two copies of itself, then M' first simulates completely the work of the first copy, while storing the input of the second copy, and then simulates completely the work of the second copy. Recursive calls made by the copies are handled by M' in the same manner. Thus M' has to store in addition to the current situation of M, the sequence of instructions executed so far on smaller levels together with markers which mark the recursive calls not already simulated. Consequently, the XTM M' can work within space $t(n)$.

b) We show $\mathbf{XSPACE}(t) \subseteqq \mathbf{XPARTIME}(\mathrm{Lin}\, t^3)$. Let M be an XTM working on inputs of length n within space $t(n)$. This machine can be simulated by a parallel TM M' as follows. First we assume that M' knows $t(n)$. The question of "$w \in \mathrm{L}(M)$?" can be converted into the question of "$K_0 \overset{2^{c \cdot t(n)}}{\underset{M}{\vdash}} K_1$?" for some $c > 0$, where K_0 is the corresponding initial ID and K_1 is the (without loss of generality) only accepting ID of M. The machine M' subdivides the latter question into the two similar questions "$K_0 \overset{2^{c \cdot t(n)-1}}{\underset{M}{\vdash}} K_2$?" and "$K_2 \overset{2^{c \cdot t(n)-1}}{\underset{M}{\vdash}} K_1$?" Since M' does not know the intermediate ID K_2, all possible ID's of length $t(n)$ are tried in parallel. This process of subdividing is continued by M' until questions of the kind "$K \overset{1}{\underset{M}{\vdash}} K'$?" occur. These questions can be answered by M' immediately. If M' does not know $t(n)$, then it tries this simulation for all $t' = 1, 2, \ldots$ instead of $t(n)$ in parallel. For X = D it asks "$K_0 \overset{2^{t'}}{\underset{M}{\vdash}} K$?" not only for the accepting ID $K = K_1$, but also for the (without loss of generality only) rejecting ID $K = K_1'$. A careful estimation of the time used by M' gives an upper bound of $\precsim t(n)^3$. $\blacksquare$

Remark 1. The paper [Hon 80b] deals with the duality between the time measure for parallel machines and the space measure. This duality can also be maintained for nondeterministic and alternating machines under a modification of the definition of nondeterminism and alternation.

Remark 2. In [Savi 78], parallel RAM's are considered which in addition to arithmetical instructions have instructions which are similar to those of a CRAM. This two-fold increase in the ability of the RAM's (namely parallelity and powerful instructions) increases the computational power considerably. For $t \geqq \log$, nondeterministic machines of this type can do within time Pol t the same work as can be done by ordinary nondeterministic machines within time $2^{\text{Pol}\,t}$.

Remark 3. In [FoWy 78] an essentially different type of parallel machines has been investigated. The instructions of the RAM's used there are basically those of ordinary RAM's. If such a machine initiates the work of an identical copy, it works in parallel with the copy (i.e. it does not halt until the stop of the copy) and it does not receive an answer from the copy. All machines working in parallel have access to a global memory which consists of a infinite series of registers. For $t \geqq \log$, deterministic (nondeterministic) machines of this type can do within time Pol t the same work as can be done by ordinary deterministic (nondeterministic) machines within space Pol t (time $2^{\text{Pol}\,t}$).

Remark 4. An interesting kind of parallelity has been investigated in [Bre 77] and [ČuYu 84] in terms of iterative arrays whose structure is that of infinite binary trees. The results for the time and space measures of these machines are analogous to those of MRAM's (cf. 5.2 and 20.12). In [ČuYu 84] it is shown that the **multiT-NTIME**(t) languages can be accepted by deterministic iterative tree arrays within time $c \cdot t(n)$.

Remark 5. That parallel machines are generally more efficient than sequential ones has been shown in [DyTo 83] where it has been proved that every deterministic Turing machine working within time t can be simulated by a suitable deterministic parallel RAM within time $\sqrt{t}$, for $t \geqq \text{id}$.

Parallelity in the sense that an exponentially growing number of computations running in parallel is set up can be found in a hidden form in some sequentially working machine types. As has been pointed out in [PrSt 76], the shift operations connected with the Boolean operations of a vector RAM or, amounting to essentially the same thing, the concatenation (multiplication) in cooperation with the Boolean operations of a CRAM (MRAM) represent full parallelity. Thus parallelity can appear in sequential machines with powerful instructions. Another possibility is illustrated by the alternating Turing machines, which are dealt with in the next subsection. For a general discussion of the notion of parallelity see also [Gols 78], [Ruz 79a], [Hon 80b] and [DyCo 80].

Theorem 20.35 allows us to transform space to parallel time, but the amount of hardware which is necessary to perform this transformation is not taken into account. In [DyCo 80] this problem is studied for two parallel computation models, for which a measure HARDWARE is considered which reflects the number of *processors* or the amount of *circuitry*. We report some relevant results of this paper without giving detailed definitions. We use the prefix "PAR" to refer to the parallel machines defined in [DyCo 80] and to suggest a wide model invariance of the results, as claimed in [DyCo 80].

The question concerning a bound on hardware when space is transformed to parallel time can be formulated as follows: For what t we have:

$$\mathbf{DSPACE}(s) \subseteq \mathbf{DPARTIME\text{-}HARDWARE}(\mathrm{Pol}\, s, t(\mathrm{Pol}\, s))?$$

Obviously, the inclusion

$$\mathbf{DSPACE}(s) \subseteq \mathbf{DPARTIME\text{-}HARDWARE}(\mathrm{Pol}\, s, 2^{\mathrm{Pol}\, s})$$

is true, but it can be conjectured that

$$\mathbf{DSPACE}(s) \nsubseteq \mathbf{DPARTIME\text{-}HARDWARE}(\mathrm{Pol}\, s, \mathrm{Pol}\, s)$$

because the right-hand side coincides with **DTIME**(Pol s). This conjecture is supported by the following fact shown in [Yaoa 81]: If KNAPSACK is decided by a certain type of parallel algorithm within time $\frac{1}{2}\sqrt{k}$ (k is the number of items), then at least $2^{\frac{1}{2}\sqrt{k}}$ processors are activated.

Some relationships between these and other measures, which allow further insight into the fast parallel simulation of Turing machine computations, can be found in § 21 (21.7.3, 21.14.4 and 21.16).

20.4. *Machines with Modified Acceptance*

In this subsection we deal with nondeterministic machines which accepts inputs in an unconventional manner: the results of the several computations on an input can be in a sense connected. The essential difference between such kinds of acceptance and the acceptance by parallel machines is the following: The parallel machines can make these connections themselves during their work, while in the case of the machines dealt with in this subsection this must be done by an external observer.

20.4.1. Alternating Machines

It is not surprising that the results for the time and space measures for alternating machines are similar to those for parallel machines. Moreover, not only the results but also the proofs are analogous.

The concept of alternating machines has been developed independently in [ChSt 76] and [Koz 76]; the paper [ChKoSt 78] combines the results of these papers.

20.36. Theorem. [ChKoSt 78] For $s \geqq \log$ such that $s(|w|) \in \mathrm{DTIME}(2^{\mathrm{Lin}\, s})$,

$$\mathbf{ASPACE}(s) = \mathbf{DTIME}(2^{\mathrm{Lin}\, s}).$$

The proof can be made as for 20.34.2, i.e. as for 20.33.2. ∎

Remark 1. A comparison of 20.36 with 20.13 gives **ASPACE**(s)=**2-auxPD-T-NSPACE**(s) for all $s \geqq \log$ such that $s(|w|) \in$ **DTIME**($2^{\mathrm{Lin}\, s}$). In [Ruz 78] and [Ruz 79b] it is proved that **ASPACE-TREESIZE**(s, Pol t) = **2-auxPD-T-NSPACE-TIME**(s, Pol t) for $s \geqq \log$ and $t \geqq \mathrm{id}$, where the measure TREESIZE measures the number of nodes in an accepting tree. Strong relationships between ASPACE-TREESIZE and a double measure based on first order logic expressibility can be found in [Imm 80].

Remark 2. Several measures similar to TREESIZE (for example, number of leaves of accepting trees of ATM's) have been considered in [Kin 81b], and complexity classes defined by simultaneous bounds on space and one of these measures are examined. The measure "number of leaves" has been further examined in [InTaTa 82].

20.37. Theorem. [ChKoSt 78]

1. For $s \geqq \mathrm{id}$, $\mathbf{NSPACE}(s) \subseteqq \mathbf{multiT\text{-}ATIME}(\mathrm{Lin}\ s^2)$.
2. For $t \geqq \mathrm{id}$, $\mathbf{multiT\text{-}ATIME}(t) \subseteqq \mathbf{NSPACE}(t)$.
3. For $t \geqq \mathrm{id}$ such that $t(|w|) \in \mathrm{DSPACE}(t)$, $\mathbf{multiT\text{-}ATIME}(t) \subseteqq \mathbf{DSPACE}(t)$.

Proof. Essentially, we can use the proof of the corresponding results for parallel Turing machines (cf. 20.35). The following alterations or remarks are necessary:

Ad 1. The process of guessing an intermediate ID can be made within time $\lesssim t(n)$, since the portion of the intermediate ID which has already been guessed can be enlarged within one step while the parallel machine has to transfer this portion to the output tapes.

Ad 2. For every universal situation all next situations are treated as recursive calls in the proof of 20.35. For existential situations of M the NTM M' guesses nondeterministically which next situation leads to an accepting $\beta \in \mathcal{S}_M$.

Ad 3. A DTM M' works in the same way as the NTM M' above, but precomputes the allowed time $t(n)$ and simulates each computation of M up to an accepting situation but at most $t(n)$ steps. ∎

20.38. Corollary.

1. For $t \geqq \mathrm{id}$, $\mathbf{ATIME}(\mathrm{Pol}\ t) = \mathbf{NSPACE}(\mathrm{Pol}\ t)$.
2. For $t \geqq \mathrm{id}$ such that $t(|w|) \in \mathrm{DSPACE}(\mathrm{Pol}\ t)$,

$$\mathbf{ATIME}(\mathrm{Pol}\ t) = \mathbf{DSPACE}(\mathrm{Pol}\ t). \ ∎$$

Remark 1. The statements 20.37.1 and 20.37.3 give a new proof of the result: $\mathbf{NSPACE}(s) \subseteqq \mathbf{DSPACE}(s^2)$ for $s \geqq \mathrm{id}$ such that $s(|w|) \in \mathrm{DSPACE}(s^2)$ (cf. 23.1.2).

Remark 2. Let $\mathbf{ASPACE\text{-}ALT}(s, t)$ be the class of all languages which can be accepted by alternating TM's working both within space s and with at most t alternations, i.e. making at most t changes from universal states to existential states or vice versa on each path of the computation tree. In [ChKoSt 78] it is shown that

$$\mathbf{ASPACE\text{-}ALT}(\mathrm{Pol}\ s, \mathrm{Pol}\ s) = \mathbf{SPACE}(\mathrm{Pol}\ s)$$

for functions s fulfilling certain constructibility assumptions.

The next theorem shows that the addition of an auxiliary store can enlarge the computational power of alternating machines (cf. 20.36, 20.13, 20.19 and 20.25). It is suprising that nonerasing stacks and unrestricted stacks have the same power in this respect.

20.39. Theorem. [LaLiSt 78]

1. For $s \geqq 1$ such that $s(|w|) \in \mathrm{DTIME}(2^{n \cdot 2^{\mathrm{Lin}\, s}})$,
$$\textbf{2-auxPD-T-ASPACE}(s) = \textbf{DTIME}(2^{n \cdot 2^{\mathrm{Lin}\, s}}).$$
2. For $s \geqq \log$ such that $s(|w|) \in \mathrm{DTIME}\left(2^{2^{2^{\mathrm{Lin}\, s}}}\right)$,
$$\textbf{2-auxNES-T-ASPACE}(s) = \textbf{2-auxS-T-ASPACE}(s) = \textbf{DTIME}\left(2^{2^{2^{\mathrm{Lin}\, s}}}\right). \blacksquare$$

Similar statements concerning the measures ASPACE-TREESIZE and ASPACE-REV with additional auxiliary pushdown or stack tape can be found in [GuIb 82].

In [LaLiSt 78] an interesting restriction of alternating acceptance has been investigated, which can be informally described as follows: For the Σ_k-acceptance (Π_k-acceptance) of an input w by an alternating TM, only such w-accepting $\beta \in \mathcal{S}_M$ are admissible for which in every chain $K_1 \vdash_{\overline{M}} K_2 \vdash_{\overline{M}} \dots \vdash_{\overline{M}} K_r$, $K_1, K_2, \dots, K_r \in \beta$, there are at most $k - 1$ alternations between existential and universal situations, and if there are exactly $k - 1$ alterations, then K_1 is an existential (universal) situation. Obviously, the Σ_1-acceptance is the nondeterministic acceptance. The complexity classes for these kinds of acceptance are denoted in such a way that Σ_r (Π_k) stands for the "A" in the corresponding complexity classes for alternating acceptance.

20.40. Theorem. [LaLiSt 78]

1. For $s \geqq \log$ such that $s(|w|) \in \mathrm{DTIME}(2^{\mathrm{Lin}\, s})$,
$$\textbf{2-auxPD-T-}\Sigma_1\textbf{SPACE}(s) = \textbf{DTIME}(2^{\mathrm{Lin}\, s}),$$
$$\textbf{2-auxPD-T-}\Pi_1\textbf{SPACE}(s) = \textbf{DTIME}(2^{\mathrm{Lin}\, s}),$$
$$\textbf{2-auxPD-T-}\Pi_2\textbf{SPACE}(s) = \textbf{co-NTIME}(2^{\mathrm{Lin}\, s}).$$
2. For $s \geqq \log$ such that $s(|w|) \in \mathrm{DSPACE}(2^{\mathrm{Lin}\, s})$,
$$\textbf{2-auxPD-T-}\Sigma_2\textbf{SPACE}(s) = \textbf{DSPACE}(2^{\mathrm{Lin}\, s}),$$
$$\textbf{2-auxPD-T-}\Sigma_3\textbf{SPACE}(s) = \textbf{DSPACE}(2^{\mathrm{Lin}\, s}),$$
$$\textbf{2-auxPD-T-}\Pi_3\textbf{SPACE}(s) = \textbf{DSPACE}(2^{\mathrm{Lin}\, s}). \blacksquare$$

On the other hand, more sophisticated kinds of alternating acceptance have been developed which give the machines still more power than conventional alternating acceptance. For related investigations the reader is referred to [Ref 79], [PeRe 79], [LaNo 83] and [Pet 80].

Finally we mention some results relating alternating time measures to each other, or to other time measures. The next theorem is the "alternating analogue" of 19.20.

20.41. Theorem. [PaPrRe 79] For $t \geqq \mathrm{id}$, $\textbf{multiT-ATIME}(t) = \textbf{T-ATIME}(t)$. $\blacksquare$

The next theorem is an analogue of 25.3 and 25.1.

20.42. Theorem.

1. [PaPrRe 79] For all $t \geqq \mathrm{id}$,
$$\textbf{T-NTIME}(t) \subseteqq \textbf{T-ATIME}\left(n + \sqrt{t(n)}\right).$$

2. [DyTo 83] For all $t \geqq \text{id}$,

$$\textbf{multiT-DTIME}(t) \subseteqq \textbf{T-ATIME}(t/\log t). \blacksquare$$

Strong relationships between **ASPACE-TIME** and a double measure based on first order logic expressibility is given in [Imm 80].

20.4.2. Probabilistic Machines

The probabilistic mode of acceptance allows an external observer to connect the results of different computations of a machine on a given input. However, the total result is something like the average of the results of the single computations, i.e. they are finally connected only in a numerical manner. That might be the reason for the fact that it has not yet been possible to show that probabilistic machines have the same computational power as alternating machines. The following two theorems show that they have at least the same computational power as nondeterministic machines.

20.43. Theorem. [Gil 74] For fully multiT-DTIME-constructible $t(n) \geqq 4n$,

$$\textbf{multiT-NTIME}(t) \subseteqq \textbf{multiT-RTIME}(t),$$

$$\textbf{comultiT-NTIME}(t) \subseteqq \textbf{multiT-RTIME}(t).$$

Proof. We show that $\textbf{multiT-NTIME}(t) \subseteqq \textbf{multiT-RTIME}(t(n) + 2n + 3)$ for multiT-DTIME-constructible $t \geqq \text{id}$. From this, the result for multiT-NTIME can be concluded using the linear speed-up for multiT-DTIME and multiT-NTIME. The result for comultiT-NTIME can be shown in the same way. Let M_1 be a k_1T-NM working on inputs of length n within $t(n)$ steps and halting (without loss of generality) only in accepting situations, and let M_2 be a k_2T-DM which, on inputs of length n, works exactly $t(n)$ steps. The $(k_1 + k_2)$T-NM M_3 works as follows: First M_3 copies the input onto another tape and moves the two heads back within $2n$ steps. Then M_3 splits its work into two next situations K_1 and K_2. Starting with K_1, the machine M_3 works with k_1 tapes in the same way as M_1, and (simultaneously) with the k_2 remaining tapes as M_2. Now consider an arbitrary computation of M_1. If M_1 does not halt later than M_2, then M_3 halts in the same moment as M_1 with an accepting state. If M_2 halts earlier than M_1, then M_3 works another step, splitting its work into an accepting and a rejecting situation. Starting with K_2, the machine M_3 simulates M_2, but in each step splitting the computation in order to reach in the next step an accepting and a rejecting situation. If M_2 halts, then M_3 splits into two rejecting and one accepting situation, which takes two further steps, as shown in Figure 20.2. This figure represents the scheme of the computations of M_3 on an input of length n where 1 (0) stands for an accepting (a rejecting) situation.

Now it is obvious that a) if M_1 accepts the input w, then p (M_3 accepts w within time $t(n) + 2n + 3) > 1/2$ and b) if M_1 does not accept the input w, then p (M_3 rejects w within time $t(n) + 2n + 3) > 1/2$. ∎

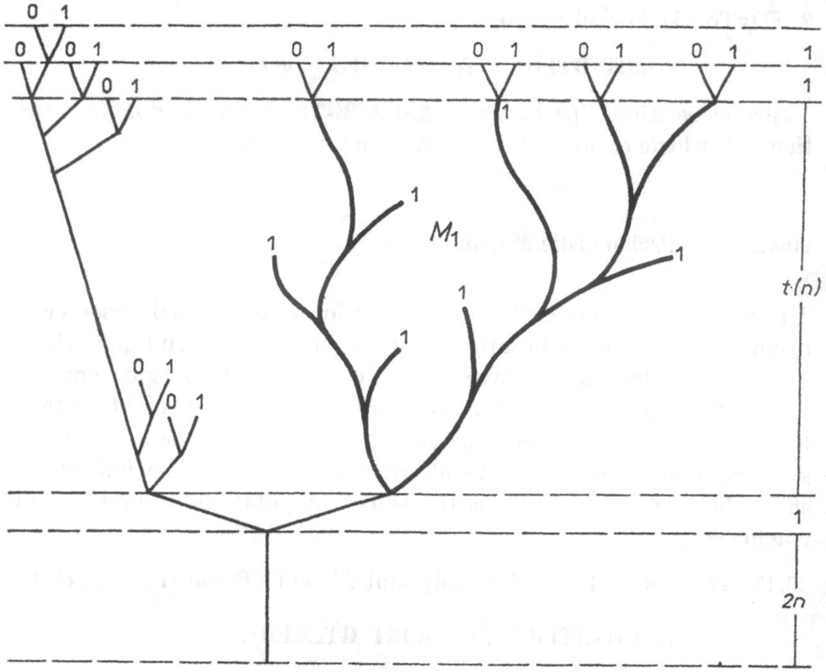

Figure 20.2

Now we state some results on the relationships between probabilistic space measures and deterministic and nondeterministic space measures.

20.44. Theorem. For fully 2-multiT-DSPACE-constructible $s \geqq \log$,

1. [Gil 74] **NSPACE**(s) $\subseteqq$ **RSPACE**(s),
2. [Jun 81] **RSPACE**$'$(s) $\subseteqq$ **DSPACE**(s^2),
3. [Jun 81] **RSPACE**$'$(s) $\subseteqq$ **ASPACE**(s).
4. [Simj 81] **coRSPACE**$'$(s) = **RSPACE**$'$(s). ∎

Remark. The statements 20.44.1 and 20.44.2 give a new proof for the result: **NSPACE**(s) $\subseteqq$ **DSPACE**(s^2) for $s \geqq \log$ (cf. 23.1.2), where the proof here is completely different from that usually given for 23.1.2. Statement 20.44.2 improves a weaker result in [Simj 77]. For a similar result for complexity classes of functions see [SiGiHu 78]. For further results see [RuSiTo 82].

20.45. Corollary. For fully 2-multiT-DSPACE-constructible $s \geqq \log$,

$$\mathbf{DSPACE}(\mathrm{Pol}\, s) = \mathbf{NSPACE}(\mathrm{Pol}\, s) = \mathbf{RSPACE}(\mathrm{Pol}\, s). \ \blacksquare$$

The next theorem shows that the inclusion **DSPACE**(s) $\subseteqq$ **DTIME**($2^{\mathrm{Lin}\, s}$) (cf. 5.10.3) can be strengthened to hold also for RSPACE.

20.46. Theorem. [Gil 74] For $s \geqq \log$,

$$\mathbf{RSPACE}(s) \subseteqq \mathbf{DTIME}(2^{\mathrm{Lin}\, s}).$$

Proof. Let M be a 2-NTM accepting the set L probabilistically within space s. Thus for the work of M on an input w of length n only situations with a worktape inscription of length $\leqq s(n)$ are of interest. Let $K_1, K_2, \ldots, K_m$ be the ID's of these situations, where $m = 2^{c \cdot s(n)}$ for a suitable $c > 0$, K_1 the initial ID, K_{m-1} the (without loss of generality) only rejecting ID and K_m is the (without loss of generality) only accepting ID. Let further, for $i, j \in \{1, \ldots, m\}$,

$$p_{i,j} = \begin{cases} 1/d & \text{if} \quad K_i \vdash_{\overline{M}} K_j \quad \text{and} \quad \operatorname{card}\{K\colon K_i \vdash_{\overline{M}} K\} = d, \\ 0 & \text{if} \quad \sim K_i \vdash_{\overline{M}} K_j. \end{cases}$$

Now we define the matrices $P = (p_{i,j})$ and $P^* = \sum_{i=0}^{\infty} P^i$. Hence we have $w \in L \leftrightarrow p^*_{1,m} > 1/2$ and $w \notin L \leftrightarrow p^*_{1,m-1} > 1/2$. A DTM M' can first compute P and then $P^* = (I - P)^{-1}$ by the usual methods for the inversion of matrices within time $2^{c_1 \cdot s(n)}$, where we assume that $s(n)$ is given. If $s(n)$ is not given, then M' tries successively $s' = 1, 2, \ldots$ instead of $s(n)$ until either $p^*_{1,m} > 1/2$ or $p^*_{1,m-1} > 1/2$.

This will occur at the latest for $s' = s(n)$. Thus M' can work within time $2^{c_2 \cdot s(n)}$. ∎

Finally we note that related concepts have been investigated in [Simj 77] in terms of threshold Turing machines (see p. 449) and in [Val 77a] in terms of counting Turing machines. The time measures of these types of machines are polynomially related to RTIME.

§ 21. Further Measures

In this section we compare the crossing, reversal, return, and dual return measures with each other as well as with time and space measures. We consider them for Turing machines with one worktape and several input modes. (Note that all recursive sets have constant complexity with respect to the crossing measure and the return measure, resp., for computations on Turing machines with more than one worktape, as can easily be verified.)

It is an interesting observation that in most cases the complexity classes of these measures either coincide with certain time or space complexity classes or are between such time and space complexity classes whose coincidence is still an open problem.

Furthermore we deal with measures based on logical notworks. They are strongly related to time, space and reversal.

21.1. Measures for Turing Machine Computations

21.1.1. The Crossing Measures

First we restate 5.15.2 which is an immediate consequence of the definition of the crossing measures: For $X \in \{D, N\}$ and $t \geqq 0$,

$$\text{T-XCROS}(t) \subseteqq \text{1-T-XCROS}(t) \subseteqq \text{2-T-XCROS}(t).$$

The first inclusion can obviously be conversed with a linear factor for bounds $t \geqq \text{id}$.

21.1. Theorem.

1. For $X \in \{D, N\}$, $t \geqq 0$ and $\varepsilon > 0$,

$$\textbf{1-T-XCROS}(t) \subseteq \textbf{T-XCROS}(t(n) + \varepsilon \cdot n).$$

2. For $X \in \{D, N\}$ and $t \geqq \mathrm{id}$,

$$\textbf{T-XCROS}(\mathrm{Lin}\ t) = \textbf{1-T-XCROS}(\mathrm{Lin}\ t). \blacksquare$$

Therefore we shall focus on the measures T-DCROS, T-NCROS, 2-T-DCROS and 2-T-NCROS. As an immediate consequence of the definition of the crossing measures we stated in 5.14.1 and 5.14.2 that $\textbf{i-T-XTIME}(t) \subseteq \textbf{i-T-XCROS}(t)$ for $i = 0, 1, 2$, $X \in \{D, N\}$ and $t \geqq \mathrm{id}$. However we can show that even the corresponding space complexity classes are included in the crossing complexity classes.

21.2. Theorem.

1. [Chy 75] For $X \in \{D, N\}$ and $t \geqq \mathrm{id}$,

$$\textbf{XSPACE}(t) \subseteq \textbf{T-XCROS}(t).$$

2. [Gre 78b] For $X \in \{D, N\}$ and $t \geqq 0$,

$$\textbf{XSPACE}(t(n) \cdot \log n) \subseteq \textbf{2-T-XCROS}(\mathrm{Lin}\ t).$$

Proof. Ad 1. Let M be an XTM accepting a language L within space $t(n)$. We describe an XTM M' which accepts L in the following manner: M' writes down the sequence of ID's (for $X = D$; in the case $X = N$ the machine M' writes down a possible sequence of ID's) corresponding to the given input word w of length n. Since such an ID has the maximum length $t(n)$, the next (a next) ID can be generated by moving the head $\left\lceil \frac{t(n)}{3} \right\rceil$ times forewards and $\left\lceil \frac{t(n)}{3} \right\rceil - 1$ times back, shifting three symbols at once. Therefore every boundary between two tape squares is crossed at most $2 \left\lceil \frac{t(n)}{3} \right\rceil + 1 \leqq_{ae} t(n)$ times.

Ad 2. Let M be a 2-XTM accepting a language L within space $t(n) \cdot \log n$. We outline the work of a 2-XTM M' accepting L such that $\text{2-T-DCROS}_{M'}(n) \preceq t(n)$.

The machine M' works essentially like the machine M' in the above proof but shifting $\left\lceil \frac{\log n}{k} \right\rceil$ symbols at once (for sufficiently large k). These $\left\lceil \frac{\log n}{k} \right\rceil$ symbols can be recorded during the head move to the right as the position of the input head. However, there are two difficulties:

a) For $t(n) \prec_{io} n$ the ID's which are of the form $w_1 s w_2 \# v_1 \mid v_2$ can be too long because of $|w_1 w_2| = n$. To overcome this difficulty we use $\#^{|w_1|} s \# v_1 \mid v_2$ instead of the complete ID. This can be of the same length, but now M' can shift $\#^{|w_1|}$ all at once using the input head.

b) In order to convert a piece of length $\left\lceil \frac{\log n}{k} \right\rceil$ of the word $s \# v_1 \mid v_2$ into its unary presentation too many additional crossings could be necessary. To overcome this difficulty M' writes a new symbol "+" n times between any two adjacent symbols of $s \# v_1 \mid v_2$. The operations $2 \cdot m$, $\left\lfloor \frac{m}{2} \right\rfloor$, $m + c$ and $m \dot{-} c$ (where c is a constant) needed to convert a binary number (word) into its unary presentation and vice versa can now be performed using the +-parts and the input tape with only a constant number of crossings per square boundary. ∎

In the nondeterministic case the inclusions in the above theorem are actually equations.

21.3. Theorem.

1. [Chy 75] For $t \geqq \text{id}$, **NSPACE**(t) = **T-NCROS**(t).
2. [Gre 78b] For $t \geqq 0$, **NSPACE**$\big(t(n) \cdot \log n\big)$ = **2-T-NCROS**(t).

Proof. Ad 1. Because of 21.2.1 we have to show only T-NCROS$(t) \subseteq$ NSPACE(t) for $t \geqq \text{id}$. Let M be an NTM accepting a language L with at most $t(n)$ crossings per tape square boundary. By CS_i we denote the crossing sequence to the right of the tape square i. First M' guesses nondeterministically the leftmost nonempty crossing sequence, say CS_{i_0}, writes it into a first track of the tape squares 1, 2, ... and checks whether it is compatible with the initial content of the tape square i_0. Then M' guesses nondeterministically CS_{i_0+1}, writes it into a second track of the tape squares 1, 2, ..., checks whether CS_{i_0} and CS_{i_0+1} are compatible with each other and with the initial contents of the tape square $i_0 + 1$, erases CS_{i_0} and shifts CS_{i_0+1} to the first track. Then M' guesses nondeterministically CS_{i_0+2} etc. The machine M' continues in this manner until it guesses the empty sequence. M' has also to guess when the first input symbol is to be taken as the initial contents of the tape square whose crossing sequences are stored at this time on the tape. M' accepts if all checks have been finished successfully and an accepting state has occurred anywhere in the guessed sequences. Since M has an accepting computation path with crossing sequences of maximum length $t(n)$, the NTM M' has an accepting computation path using no more than $t(n)$ tape squares.

Ad 2. The proof is almost the same as for Statement 1. The only difference is that we use here, instead of the crossing sequences (having only states), sequences of states and input head positions. Thus such a sequence is of maximum length $t(n) \cdot \log n$ if the corresponding crossing sequence is of maximum length $t(n)$. ∎

Remark 1. Theorem 21.3.2 is proved in [Gre 78b] as a special case of a more general theorem on the crossing measure for computations on two-way Turing machines with one worktape which is initially not empty (the *preset Turing machines*). Theorem 21.3.2 corresponds to the case when only numbers of a given regular set are initially "preset" on the tape.

Remark 2. Let **T-NSPACE-T-CROS**(s, t) be the class of all languages accepted by 2T-NM's working on the first tape within space s and on the second tape with at most t crossings. In [Mei 84b] it is shown that **T-NSPACE-T-CROS**(s, t) = **NSPACE**$(s \cdot t)$, for $s \cdot t \geqq id$ and under certain constructibility assumptions for s and t.

21.1.2. The Reversal Measures

First we restate 5.15.1 which is an immediate consequence of the definition of the reversal measures: For X ∈ {D, N} and $t \geqq 0$,

$$\textbf{T-XREV}(t) \subseteqq \textbf{1-T-XREV}(t) \subseteqq \textbf{2-T-XREV}(t).$$

Furthermore, remember the fact that the linear speed-up does hold for the measures T-DREV and T-NREV for bounds $t \geqq 1$ (cf. 18.14.1). Therefore the inclusion $\textbf{T-XREV}(t) \subseteqq \textbf{1-T-XREV}(t)$ can obviously be conversed for bounds $t \geqq \text{id}$.

21.4. Theorem.

1. For X ∈ {D, N}, $t \geqq 0$ and $\varepsilon > 0$, $\textbf{1-T-XREV}(t) \subseteqq \textbf{T-XREV}\big(t(n) + \varepsilon \cdot n\big)$.
2. For X ∈ {D, N} and $t \geqq \text{id}$, $\textbf{T-XREV}(t) = \textbf{1-T-XREV}(t)$. ∎

Remark. In [Fip 68] it is shown that the bound $t(n) + \varepsilon \cdot n$ in 21.4.1 cannot be improved.

Because of Theorem 21.4 we shall focus on the measures T-DREV, T-NREV, 2-T-DREV and 2-T-NREV. As an immediate consequence of the definition of the reversal measures, we stated in 5.14.1 that $\textbf{i-T-XTIME}(t) \subseteqq \textbf{i-T-XREV}(t)$ for $i = 0, 1, 2$, X ∈ {D, N} and $t \geqq \text{id}$. However, we even have the following relationships to time and space measures. First the deterministic case.

21.5. Theorem.

1. [KaVo 70] For $t \geqq \text{id}$, $\textbf{multiT-DTIME}(t) \subseteqq \textbf{T-DREV}(t)$.
2. [Hart 68b] For $t \geqq \text{id}$, $\textbf{T-DREV}(t) \subseteqq \textbf{T-DSPACE-TIME}(t, t^2)$.

Proof. Ad 1. Using the linear speed-up for T-DREV this statement becomes evident.

Ad 2. Let M be a DTM working with at most $t(n)$ reversals and having q states. We show that $\text{T-DSPACE}_M(n) \preceq t(n)$ and $\text{T-DTIME}_M(n) \preceq t(n)^2$. At the beginning of a computation which does not cycle, the head of M can make at most $q \cdot n + q$ steps without a reversal, namely it can rest $q - 1$ steps on any input symbol before moving further to the right and it can move q steps to the right on the right hand side of the input word. Then, analogously, it can make at most $q \cdot (n + q) + q$ steps without reversal to the left since there can now be $n + q$ nonempty tape squares. In the same way we can continue and we get altogether

$$\text{T-DSPACE}_M(n) \leqq n + q \cdot \big(t(n) + 1\big) \asymp t(n)$$

and

$$\text{T-DTIME}_M(n) \leqq \sum_{i=0}^{t(n)} \big(q \cdot (n + i \cdot q)\big) \asymp t(n)^2. \ \blacksquare$$

Now we treat the nondeterministic case. Because of 5.14.2 and 21.3 we have $\textbf{T-NREV}(t) \subseteqq \textbf{T-NCROS}(t) = \textbf{NSPACE}(t)$ for $t \geqq \text{id}$ and $\textbf{2-T-NREV}(t) \subseteqq \textbf{2-T-NCROS}(t) = \textbf{NSPACE}\big(t(n) \cdot \log n\big)$. Actually these inclusions are equations.

21.6. Theorem.

1. [Chy 75] For $t \geqq \text{id}$, $\textbf{NSPACE}(t) = \textbf{T-NREV}(t)$.
2. [Gre 78b] For $t \geqq 0$, $\textbf{NSPACE}(t(n) \cdot \log n) = \textbf{2-T-NREV}(t)$.

Proof. Ad 1. It remains to show that $\mathbf{NSPACE}(t) \subseteq \mathbf{T\text{-}NREV}(t)$. Let M be a NTM accepting a language L within space $t(n)$ and time $t'(n)$. Without loss of generality M uses only the tape squares 1, 2, 3, ... By $a_{i,j}$ we denote the symbol of the tape square i in the jth step of M. Furthermore, let

$$b_{ij} = \begin{cases} (a_{i,j}, s) & \text{if } M \text{ has in the } j\text{th step the head position } i \text{ and the state } s, \\ a_{i,j} & \text{otherwise.} \end{cases}$$

Now we describe the work of an NTM M' accepting L with $t(n)$ reversals. First M' guesses nondeterministically $b_{1,0}, b_{1,1}, \ldots, b_{1,t'(n)}$ (thereby guessing $t'(n)$ also) and writes them successively into the tape squares $1, 2, 3, \ldots, t'(n) + 1$. Now M' makes the first reversal, guesses nondeterministically $b_{2,t'(n)}, \ldots, b_{2,1}, b_{2,0}$, writes them successively into the tape squares $t'(n) + 2, \ldots, 4, 3, 2$, checks simultaneously whether this sequence is compatible with the sequence $b_{1,t'(n)}, \ldots, b_{1,2}, b_{1,1}$ and erases the latter. Now M' makes the second reversal, guesses nondeterministically $b_{3,0}$, $b_{3,1}, \ldots, b_{3,t'(n)}$ etc. Note that M' has not really to guess the symbols $a_{1,0}, a_{2,0}, \ldots$, $a_{t(n),0}$ because they are already written in the tape squares $1, 2, \ldots, t(n)$ where they are needed.

The machine M' continues in this manner until a sequence without states is guessed. If all checks have been finished successfully, and if an accepting state has occurred anywhere in a guessed sequence, then M' accepts on this computation path. Since there is an accepting computation path of M using only the tape squares $1, 2, \ldots, t(n)$, there is an accepting computation path of M' with only $t(n)$ reversals of the head.

Ad 2. The inclusion $\mathbf{2\text{-}NSPACE}\big(t(n) \cdot \log n\big) \subseteq \mathbf{2\text{-}T\text{-}NREV}(t)$ can be shown using the basic idea of the above proof and the trick of the proof 21.2.2. For details see [Gre 78b]. ∎

Remark 1. Theorem 21.6.2 is a special case of a more general theorem on the reversal measure for computations on *preset Turing machines*, which is proved in [Gre 78b] (cf. the Remark 1 after 21.3).

Remark 2. In [KaVo 70] the reversal measure for computations on Turing machines whose worktape heads cannot rest is investigated. The following theorem is proved: For $t \geqq 0$, $\mathbf{multiT\text{-}DREV}'(t) \subseteq \mathbf{2T\text{-}DREV}'(6t)$. Thereby the prime indicates the restriction to this class of Turing machines.

Remark 3. Let $\mathbf{T\text{-}NSPACE\text{-}T\text{-}REV}(s, t)$ be the class of all languages accepted by 2T-NM's working on the first tape within space s and on the second tape with at most t reversals. In [Mei 84b] it is shown that $\mathbf{T\text{-}NSPACE\text{-}T\text{-}REV}(s, t) = \mathbf{NSPACE}(s \cdot t)$ for $s \cdot t \geqq \mathrm{id}$ and under certain constructibility assumptions for s and t.

Finally we state a theorem about the reversal measure for multitape Turing machines.

21.7. Theorem. For $t \geqq \mathrm{id}$, $s \geqq 0$,

1. $\mathbf{DSPACE}(t) \subseteq \mathbf{multiT\text{-}DREV}(\mathrm{Lin}\, t) \subseteq \mathbf{DTIME}(2^{\mathrm{Lin}\, t})$.
2. [Hon 80b] $\mathbf{multiT\text{-}DREV}(\mathrm{Pol}\, t) = \mathbf{DSPACE}(\mathrm{Pol}\, t)$.

3. [Tat 81] **i-multiC-XREV**(Pol t) = **XTIME**(Pol t) $(i = 1, 2,\ \mathrm{X} \in \{\mathrm{D}, \mathrm{N}\})$.
4. [DyCo 80] **multiT-DREV-SPACE**(Pol t, Pol s) = **DPARTIME-HARDWARE**(Pol t, Pol s). ∎

Remark 1. For nondeterministic TM's with more than one worktape one reversal per tape is sufficient to do all what can be done by nondeterministic TM's. This is reflected by

$$\mathbf{2T\text{-}NREV}(1) = \mathbf{RE}$$

and

$$\mathbf{2T\text{-}NTIME\text{-}REV}(\mathrm{Pol}\, t, 1) = \mathbf{NTIME}(\mathrm{Pol}\, t).$$

The idea is to guess nondeterministically the sequence of ID's on both worktapes and, going back, to check the correctness of the guess by comparing the contents of the two tapes (cf. [BaBo 74]).

Remark 2. Statement 4 is a refinement of Statement 2 because of

$$\mathbf{DSPACE}(\mathrm{Pol}\, s) = \mathbf{DPARTIME}(\mathrm{Pol}\, s) \quad (\text{cf. } 20.35).$$

Remark 3. In [Mei 84a] and [Mei 84b] it has been shown that, under certain constructibility assumptions for s,

$$\mathbf{2\text{-}auxC\text{-}T\text{-}DREV}(\mathrm{Pol}\, s) = \mathbf{DSPACE}(\mathrm{Pol}\, s), \quad \text{for} \quad s \geqq \mathrm{id},$$

$$\mathbf{2\text{-}auxPD\text{-}T\text{-}DREV}(\mathrm{Pol}\, s) = \mathbf{PSPACE}, \quad \text{for} \quad s \geqq \mathrm{id},$$

$$\mathbf{2\text{-}auxNES\text{-}T\text{-}DREV}(s) = \mathbf{DTIME}(2^{\mathrm{Lin}\, s}), \quad \text{for} \quad s \geqq \log,$$

$$\mathbf{2\text{-}auxS\text{-}T\text{-}DREV}(s) = \mathbf{DTIME}(2^{2^{\mathrm{Lin}\, s}}), \quad \text{for} \quad s \geqq \log.$$

These results should be compared with the corresponding space results in § 20.2.

21.1.3. The Return Measures

First we restate 5.15.3 which is an immediate consequence of the definition of the return measures: For $\mathrm{X} \in \{\mathrm{D}, \mathrm{N}\}$ and $t \geqq 0$,

$$\mathbf{1\text{-}T\text{-}XRET}(t) \subseteqq \mathbf{2\text{-}T\text{-}XRET}(t).$$

No obvious relationship between the classes **T-XRET**(t) and **1-T-XRET**(t) is known.

Furthermore we stated in 5.14.2 that **i-T-XCROS**(t) $\subseteqq$ **i-T-XRET**(t) for $i = 0, 1, 2$, $\mathrm{X} \in \{\mathrm{D}, \mathrm{N}\}$ and $t \geqq 0$. At least in the deterministic case this inclusion can be reversed with a linear factor for $t \geqq \mathrm{id}$.

21.8. Theorem.

1. [Chy 77] For $t \geqq \mathrm{id}$, **T-DCROS**(Lin t) = **T-DRET**(Lin t).
2. For $t \geqq \mathrm{id}$, **1-T-DCROS**(Lin t) = **1-T-DRET**(Lin t).
3. For $t(n) \geqq n^2$, **2-T-DCROS**(Lin t) = **2-T-DRET**(Lin t). ∎

Remark 1. We omit the proof, which can be found for Statement 1 in [Chy 77]. The other results can be shown by a slight modification of the method used there.

Remark 2. For functions t such that $2 \lesssim t <_{io}$ id we have **T-XCROS**(t) $\subset$ **T-XRET**(t) and also **T-XCROS**(Lin t) $\subset$ **T-XRET**(Lin t). For a witness take the set S which is obviously in **T-DRET**(2) but not in **NCROS**(Lin t) for $t <_{io}$ id (see the proof of 8.13.1).

Now we take up the nondeterministic case. Here the return measures correspond to the auxiliary pushdown space measure in the same way as the crossing measures correspond to the usual space measure (cf. 21.3). First we consider the measure T-NRET.

21.9. Theorem. [WeBr 79]
1. For $t \geqq 0$, **T-NRET**(t) $\subseteqq$ **1-auxPD-T-NSPACE**(t).
2. For $t \geqq$ id, **T-NRET**(t) $=$ **2-auxPD-T-NSPACE**(t).

Proof. Ad 1. Let M_1 be an NTM accepting a language L within return $t(n)$. It is not so hard to see that there is a NTM M_2 accepting L within return $t(n) + 2$ and having the following properties:

a) the head makes reversals only immediately after it has changed a symbol,

b) M_2 has two disjoint sets S_λ and S_ϱ of states. If the head moves to the left (right), M_2 is in a state of S_λ (S_ϱ). The head moves in every step.

For the exact construction of M_2 see [Wec 76].

Next we construct an NTM M_3 which works as follows: M_3 starts in a tape square $i_0 < 0$ which is not visited by the head of M_2 on the given input, marks this tape square and goes with the state r_1 to the right to the tape square 1. Now M_3 works exactly like M_2. After the stop of M_2 the head of M_3 goes in state r_2 to the left to the tape square i_0 and stops. Thus M_3 makes at most two returns more than M_2. Finally we describe the work of a 1-PD-T-NM M_4 simulating M_3, accepting L and using no more than $t(n)$ squares of its Turing tape.

Similarly to the NTM M' in the proof of 21.3.1 (there M' simulates within space $t(n)$ an NTM M with crossing sequences of maximum length $t(n)$) M_4 guesses nondeterministically the crossing sequences of M_3. Since they can be longer than $t(n)$ they cannot be written down completely. Therefore M' guesses and writes down only that final part of a crossing sequence which begins with the first alteration of the contents of the corresponding tape square. This part is said to be the essential part of the crossing sequence, for short EPCS. Since the head of M_3 moves over the tape squares with the initial contents without reversals, the remaining initial parts of the crossing sequences can be computed by the help of the tables T^ϱ_w and T^λ_w which are defined as follows: for $a_1, a_2, \ldots, a_m \in \Sigma \cup \{\square\}$ (Σ is the input alphabet),

$$T^\varrho_{a_1a_2\ldots a_m} =_{df} \{(s, s')\colon \text{there are } r_0, r_1, \ldots, r_m \in S_\varrho \text{ such that } r_0 = s,\ r_m = s', \text{ and } r_{j-1}a_j \to r_ja_j + 1 \text{ is an instruction of } M_3 \text{ for } j = 1, \ldots, m\},$$

$$T^{\lambda}_{a_1a_2\ldots a_m} = \{(s, s')\colon \text{ there are } r_0, r_1, \ldots, r_m \in S_\lambda \text{ such that } r_0 = s,\ r_m = s', \text{ and } r_j a_j \to r_{j-1} a_j - 1 \text{ is an instruction of } M_3 \text{ for } j = 1, \ldots, m\}.$$

Evidently, $T^{\varrho}_{e} = \{(s, s)\colon s \in S_\varrho\}$, $T^{\lambda}_{e} = \{(s, s)\colon s \in S_\lambda\}$, $T^{\varrho}_{wv} = T^{\varrho}_{w} \circ T^{\varrho}_{v}$ and $T^{\lambda}_{wv} = T^{\lambda}_{w} \circ T^{\lambda}_{v}$ (where "$\circ$" denotes the product of relations). We define $T_w = (T^{\varrho}_{w}, T^{\lambda}_{w})$.

The pushdown store records those parts of the guessed EPCS's which have no connection with another EPCS to the right up to that moment. Moreover it records those tables T_w which will allow these connection to be made later.

More exactly, the machine M_4 works on the input word $a_1a_2\ldots a_n$ according to the following algorithm:

1. Store $T_e r_1 r_2 T_e$ into the pushdown store. (Remember that r_1r_2 is the leftmost nonempty crossing sequence. Furthermore note that the rightmost symbol of a word to be stored in the pushdown store will be the topmost symbol there.) Set $i = 0$.
2. If $i = 0$, then choose nondeterministically $i = 0$ or $i = 1$ ($i = 1$ indicates the beginning of the input word). If $1 \leqq i \leqq n$, then set $i = i + 1$.
3. Guess nondeterministically the sequence of contents of a tape square after the first alteration of its initial contents a_i (where $a_0 = a_{n+1} = \Box$) and the corresponding left (right) EPCS $s_1, s_2, \ldots, s_k$ ($t_1, t_2, \ldots, t_m$). If these sequences are not compatible with each other, then reject. Otherwise, for $j = k, k-1, \ldots, 2, 1$ do
   a) If T_vT_w is on the top, then replace T_vT_w by T_{vw} and go to a).
   b) If rT_w is on the top, then replace rT_w by T_w. If $(r, s_j) \notin T^{\varrho}_{w} \cup T^{\lambda}_{w}$, then reject.
   c) If $\Box T_w$ is on the top of the pushdown store, then reject.
4. If T_vT_w is on the top, then replace T_vT_w by T_{vw} and go to 4.
5. If $i > 0$, $m = 0$, and $\Box T_w$ is on the top, then accept. Otherwise store $Ta_it_1t_2\ldots t_mT_e$ in the pushdown store, erase the sequences guessed in 3 from the Turing tape and go to 2.

Ad 2. Let M be a 2-PD-T-NM accepting a language L and using no more than $t(n)$ squares on its Turing tape. The word $w_1sw_2 \# v_1 \mid v_2 \# a$ is said to be the *partial instantaneous description* (for short PID) of a situation of M in which M has the state s, $w_1w_2 = \Box w \Box$ (w is the input word), the input head scans the first symbol of w_2, the Turing tape has the contents $\ldots\Box\Box\Box v_1v_2\Box\Box\Box\ldots$ where the first symbol of v_1 and the last symbol of v_2 are not equal to $\Box$, the head of the Turing tape scans the first symbol of $v_2\Box$ and a is the top symbol of the pushdown store. For a fixed input word w we define the relation $\frown$ between PID's as follows: $K \frown K'$ iff there is a computation path $S \vdash_M S_1 \vdash_M S_2 \vdash_M \ldots \vdash_M S_r \vdash_M S'$ ($r \geqq 1$) such that $K(K')$ is the PID of S (S'), the pushdown words of S and S' have the same length, and the pushdown words of $S_1, S_2, \ldots, S_r$ are longer. Consequently, if $K \frown K'$, then K and K' have the same top symbols. Now we describe the work of an NTM M' simulating M within return $t(n)$. The NTM M' works in two stages. In the first stage M' guesses nondeterministically a sequence $K_0, K_1, \ldots, K_r$ of PID's which will be the tape contents at the end of the first stage. In the second stage M checks whether K_0 is the

PID of the initial situation, whether there are situations S, S' such that $S \vdash_{\overline{M}} S'$ and K_i (K_{i+1}) is the PID of S (S') for $i = 0, 1, \ldots, \tau - 1$, and whether K_τ is the PID of an accepting situation. However, this does not ensure that $K_0, K_1, \ldots, K_\tau$ is actually the sequence of PID's of a computation path $S_0 \vdash_{\overline{M}} S_1 \vdash_{\overline{M}} \ldots \vdash_{\overline{M}} S_\tau$, because in the case that K_i causes the erasure of the top symbol it is not certain whether K_{i+1} has the correct top symbol (namely the same top symbol as the PID K_j with $K_j \frown K_{j+1}$). Therefore M' guesses and writes down the sequence $K_0, K_1, \ldots, K_\tau$ in such a temporal order that immediately after writing down a guessed PID K_i which causes an enlargement of the pushdown store, M' guesses and writes down a PID K_j which is guessed to satisfy $K_i \frown K_j$ and which has the same top symbol as K_i. Thereby the place for K_j on the Turing tape is also guessed.

This can be done according to an algorithm which is basically the same as the algorithm in the proof of **CF** $\subsetneqq$ **T-NRET**(2) (cf. 12.5). Consequently, during the first stage of the work of M' every tape square is visited at most twice by the head after the first alteration of its contents. The checks in the second stage can be carried out in such a way that every tape square is visited by the head at most $t(n)/2$ times. If all checks have been finished successfully, then M' accepts, otherwise M' rejects. ∎

21.10. Corollary. For all $t \geqq \mathrm{id}$ such that $t(|w|) \in \mathbf{DTIME}(2^{\mathrm{Lin}\, t})$,

$$\mathbf{T\text{-}NRET}(t) = \mathbf{DTIME}(2^{\mathrm{Lin}\, t}).$$

Proof. Theorem 21.9 states **T-NRET**(t) = **2-auxPD-T-NSPACE**(t). Then 20.13 yields the assertion. ∎

Remark. Theorem 21.9 exhibits a difference between **T-DRET**(t) and **T-NRET**(t) in the range $2 \leqq t \prec \mathrm{id}$. The set $\overline{\mathrm{D}^{\#}}$ is obviously in **T-NRET**(2), but $\overline{\mathrm{D}^{\#}} \in$ **T-DRET**(t) for $t \prec \mathrm{id}$ would imply $\mathrm{D}^{\#} \in$ **T-DRET**(t) $\subsetneqq$ **T-NRET**(t) $\subsetneqq$ **1-auxPD-T-NSPACE**(t) which contradicts 8.22.

21.11. Theorem.

1. For $t(n) \geqq n \cdot \log n$, **1-T-NRET**($t$) = **2-auxPD-T-NSPACE**(t).
2. For $t \geqq \mathrm{id}$, **2-T-NRET**(Lin t) = **2-auxPD-T-NSPACE**$\big(t(n) \cdot \log n\big)$.

Proof. All inclusions are shown essentially in the same way as in the proof of 21.9.

Ad 1. In the proof of the inclusion "$\subseteqq$" we use crossing sequences which have not only states but also the information on how many moves to the right the input head has made since the last crossing of the boundary. This enlarges the crossing sequence and thus its essential part by at most n symbols. Since we do not know whether the construction of the machine M_2 (see the proof of 21.9.1) is possible for 1-NTM we omit this construction and we construct the 1-NTM M_3 directly from M_1. This complicates the work with the tables in the construction of the 1-PD-T-NM M_4. Furthermore these tables include not only states, but also head positions of the input head of M_3. Since the description of such a table is of size $n \cdot \log n$, this leads to a space requirement of at least $n \cdot \log n$ on the Turing tape of M_4.

The inclusion "$\supseteqq$" is implicitly shown in the proof of 21.9.2 because the simulating NTM can obviously be replaced by a 1-NTM.

Ad 2. In the pro f the inclusion "$\subseteq$" we use crossing sequences consisting not only of states but also of the position of the input head at the moment of the crossing of the boundary. This enlarges the crossing sequence and thus its essential part by a factor of at most $\log n$. Analogously to the case of 1-NTM we omit the construction of a 2-NTM M_2 and we construct M_3 directly from M_1. The tables in the construction of M_4 include not only states but also head positions of the input head of M_3.

The inclusion "$\supseteq$" can be shown as in the proof of 21.9.2 but the trick of 21.2.2 is used. ∎

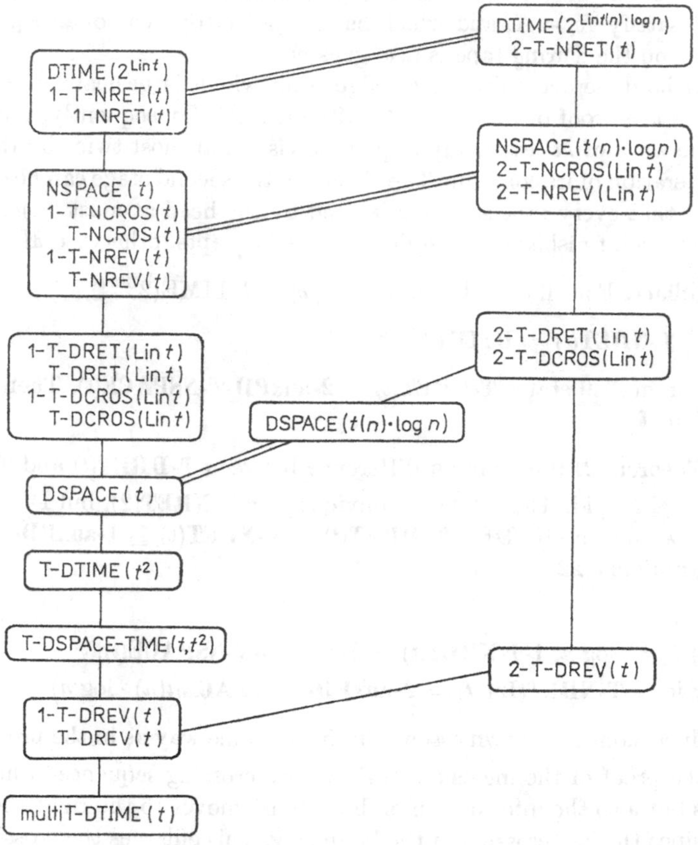

Figure 21.1. / means inclusion, // means proper inclusion, and classes in the same box coincide. All relationships are valid for $t \geqq \mathrm{id}$ with the exception of 1-T-NRET(t) = DTIME$(2^{\mathrm{Lin}\,t})$ which is known only for $t(n) \geqq n \log n$. For some relationships computability assumptions for t are necessary

21.12. Corollary.

1. For all $t(n) \geqq n \cdot \log n$ such that $t(|w|) \in \mathrm{DTIME}(2^{\mathrm{Lin}\,t})$,

$$\textbf{1-T-NRET}(t) = \textbf{DTIME}(2^{\mathrm{Lin}\,t}).$$

2. For all $t \geqq \mathrm{id}$ such that $t(|w|) \in \mathbf{DTIME}(2^{\mathrm{Lin}\, t})$,

$$\mathbf{2\text{-}T\text{-}NRET}(t) = \mathbf{DTIME}(2^{\mathrm{Lin}\, t(n)\cdot \log n}).$$

Remark. The theorems 21.9, 21.11 and their corollaries remain valid if RET is replaced by DUR, i.e. for the dual return measure. The proofs are similar to those for the return measure. For the case without input tape it is even proved in [WeBr 79] and in [Hau 80] that **T-NRET**$(t + \mathrm{Const}) =$ **T-NDUR**$(t + \mathrm{Const})$ for all t.

Finally, Figure 21.1 gives a summary of the known relationships between the measures dealt with in this section.

21.2. Measures Based on Logical Networks

We consider straightline programs to compute Boolean functions having instructions of the form $x_i = \sim x_j$, $x_i = x_j \wedge x_k$ and $x_j = x_j \vee x_k$. If every variable occurs at most once on the left-hand side of an instruction, then such a straightline program is interpreted as a *logical network* (*logical circuit*) as follows. Associate to each of the above instructions a vertex of an (acyclic) digraph labelled with the operation symbol $\sim$ ($\wedge$, $\vee$, resp.) and having exactly one (two, two, resp.) incoming edges labelled with x_j (x_j and x_k, x_j and x_k, resp.) and at least one outcoming edge labelled with x_i. Every edge labelled with x_i must be an outcoming edge of this vertex. If some variable x_m does not occur on the left-hand side of an instruction, then all edges labelled with x_m must be outcoming edges of one vertex, which has indegree zero, and which is called an *input vertex*. A vertex with outdegree zero is called an *output vertex*.

Example. The straightline program

$$x_3 = \sim x_1; \qquad x_4 = x_1 \wedge x_2; \qquad x_5 = x_2 \vee x_3; \qquad x_6 = x_3 \wedge x_4;$$

$$x_7 = x_5 \vee x_6$$

corresponds to the labelled graph shown in Figure 21.2.

For the sake of simplicity we identify a straightline program with the corresponding logical network. The complexity measures SIZE and DEPTH for logical networks C are defined by

$\mathrm{SIZE}(C) =_{\mathrm{df}}$ number of vertices of C which are not input vertices
$=$ number of the instructions in the corresponding straightline program.

$\mathrm{DEPTH}(C) =_{\mathrm{df}}$ length of longest paths in C leading from an input vertex to an output vertex.

Let level(v) be the length of a longest path leading from an input to v. The thickness of C at level i is the number of vertices of level greater than i connected by an edge with a vertex of level $\leq i$.

$$\mathrm{WIDTH}(C) =_{\mathrm{df}} \text{maximal thickness of } C \text{ at all its levels.}$$

To establish a relation between recursive predicates and Boolean functions we consider families $\mathscr{C} = (C_1, C_2, \ldots)$ of networks such that C_n has exactly n input vertices and one output vertex. We define

$$\mathscr{C} \text{ accepts } A \subseteq \{0, 1\}^* \leftrightarrow \bigwedge_n (C_n \text{ computes the Boolean function } \alpha_n(x_1, \ldots, x_n) =_{\mathrm{df}} c_A(x_1 \ldots x_n)).$$

Furthermore, $\mathrm{SIZE}_{\mathscr{C}}(n) =_{\mathrm{df}} \mathrm{SIZE}(C_n)$ and $\mathrm{DEPTH}_{\mathscr{C}}(n) =_{\mathrm{df}} \mathrm{DEPTH}(C_n)$ and $\mathrm{WIDTH}_{\mathscr{C}}(n) =_{\mathrm{df}} \mathrm{WIDTH}(C_n)$.

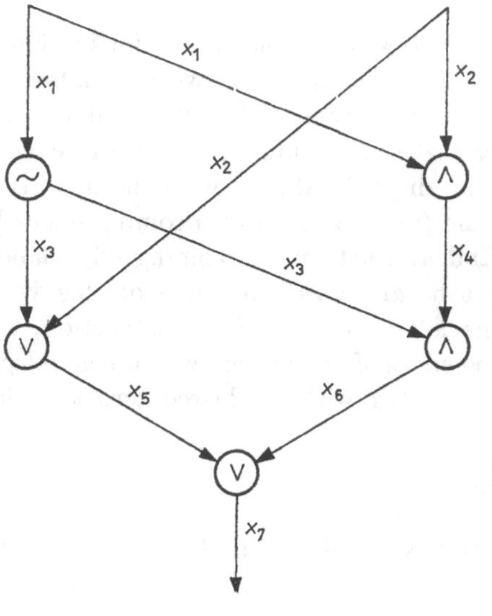

Figure 21.2

If a family $\mathscr{C}$ of logical networks is given, the network C_n is only responsible for inputs of length n, and nothing ensures a uniform treatment of all input lengths. This has the consequence that there are arbitrarily complex sets $A \subseteq \{0, 1\}^*$, even such beyond **RE**, which are accepted by network families of linear depth and size. Take for instance A such that for every n either $A \cap \{0, 1\}^n = \emptyset$ or $A \cap \{0, 1\}^n = \{0, 1\}^n$.

Furthermore, it is known (O. B. Lupanov) that every $A \subseteq \{0, 1\}^*$ can be accepted by a (nonuniform) sequence $\mathscr{C}$ of logical networks such that $\mathrm{SIZE}_{\mathscr{C}}(n) \leq \frac{2^n}{n}$. Although this bound is tight as has been proved by C. E. Shannon, so far no "natural" sequences $\{\alpha_n : n \in \mathbb{N}\}$ are known which require size complexity greater than $4n$. For a detailed survey and further hints to literature see [Lup 63].

To overcome these disparities between TM's and families of networks we require that the latter should admit relatively simple uniform descriptions. This idea has been introduced in [Bor 77] as opposed to the possibility to make the TM's non-uniform as in [Schn 76]. Different uniformity conditions and their influence on the complexity of network families are discussed in [Ruz 79a].

We define the network family $\mathscr{C}$ to be a uniform family of logical networks if $\text{Code}(C_{|w|}) \in \text{DSPACE}(\log \text{SIZE}_{\mathscr{C}})$ where Code is some reasonably chosen encoding of networks by strings.

We introduce the following complexity classes.

$$\mathbf{SIZE}(t) =_{\text{df}} \{A : A \text{ is accepted by some uniform family } \mathscr{C} \text{ of logical networks such that } \text{SIZE}_{\mathscr{C}} \leqq t\},$$

$$\mathbf{DEPTH}(t) =_{\text{df}} \{A : A \text{ is accepted by some uniform family } \mathscr{C} \text{ of logical networks such that } \text{DEPTH}_{\mathscr{C}} \leqq t\},$$

$$\mathbf{WIDTH}(t) =_{\text{df}} \{A : A \text{ is accepted by some uniform family } \mathscr{C} \text{ of logical networks such that } \text{WIDTH}_{\mathscr{C}} \leqq t\},$$

$$\mathbf{SIZE\text{-}DEPTH}(t, s) =_{\text{df}} \{A : A \text{ is accepted by some uniform family } \mathscr{C} \text{ of logical networks such that } \text{SIZE}_{\mathscr{C}} \leqq t \text{ and } \text{DEPTH}_{\mathscr{C}} \leqq s\}.$$

Similarly the complexity classes SIZE-WIDTH, DEPTH-WIDTH and SIZE-DEPTH-WIDTH can be defined.

The following results are known.

21.13. Theorem.

1. [PiFi 76] $\mathbf{multiT\text{-}DTIME}(t) \subseteqq \mathbf{SIZE}(\text{Lin } t \cdot \log t)$, for $t \geqq \text{id}$ such that $t(|w|) \in \text{DSPACE}(\log t)$.
2. [Pip 79] $\mathbf{SIZE}(t) \subseteqq \mathbf{multiT\text{-}DTIME}(\text{Lin } t \cdot \log^2 t)$, for $t \geqq \text{id}$ such that $t \in \text{multiT-DTIME}(\text{Lin } t \cdot \log t)$.
3. [Bor 77] $\mathbf{NSPACE}(s) \subseteqq \mathbf{DEPTH}(\text{Lin } s^2)$, for $s \geqq \log$ such that $s(|w|) \in \text{DSPACE}\big(\max(\log s, \log)\big)$.
4. [Bor 77] $\mathbf{DEPTH}(s) \subseteqq \mathbf{DSPACE}(s)$, for $s \geqq \log$. ∎

The fact that SIZE and multiT-DTIME as well as DEPTH and DSPACE are polynomially related is not known to be extendable to the double measures SIZE-DEPTH and multiT-DTIME-SPACE. However, the following relationships between double measures are known.

21.14. Theorem. For $t \geqq \text{id}$ and $r, s \geqq \log$ such that $t(|w|) \in \text{DSPACE}(\log t)$, $r(|w|) \in \text{DSPACE}\big(\max(\log s, \log)\big)$, $s(|w|) \in \text{DSPACE}\big(\max(\log s, \log)\big)$,

1. [Pip 79] $\mathbf{multiT\text{-}DTIME\text{-}SPACE}(t, s) \subseteqq \mathbf{SIZE\text{-}WIDTH}(t^3, s)$,

$$\mathbf{SIZE\text{-}WIDTH}(t, s) \subseteqq \mathbf{multiT\text{-}DTIME\text{-}SPACE}(\text{Pol } t, s \cdot \log t).$$

2. [Ruz 79a] **multiT-ASPACE-TIME**$(s, t) \subseteq$ **SIZE-DEPTH**$(2^{\text{Lin}\, s}, t)$,

$$\textbf{SIZE-DEPTH}(2^{\text{Lin}\, s}, t) \subseteq \textbf{multiT-ASPACE-TIME}(s, t) \text{ provided } t \geqq \log^2 s.$$

3. [DyCo 80] **multiT-DREV-SPACE**(Pol r, Pol s) = **DEPTH-WIDTH**(Pol r, Pol s).
4. [Pip 79] **multiT-DTIME-REV**$(t, r) \subseteq$ **SIZE-DEPTH**(Pol t, $r \cdot \log^4 t$),

$$\textbf{SIZE-DEPTH}(t, r) \subseteq \textbf{multiT-DTIME-REV}(\text{Pol}\, t, r \cdot \log^2 t). \blacksquare$$

Remark. Compare Statement 3 with 21.7.4.

21.15. Corollary. For $t \geqq \text{id}$, such that $t(|w|) \in \text{DSPACE}(\log t)$,

1. **multiT-DTIME-REV**(Pol t, Pol log t) = **SIZE-DEPTH**(Pol t, Pol log t).
2. **DTIME-SPACE**(Pol t, Pol log t) = **SIZE-WIDTH**(Pol t, Pol log t). ∎

The statements 21.14.2, 21.15.1 and 21.15.2 provide representations for the classes

$$\textbf{NC} =_{\text{df}} \textbf{SIZE-DEPTH}(\text{Pol}, \text{Pol}\log)$$

and

$$\textbf{SC} =_{\text{df}} \textbf{DTIME-SPACE}(\text{Pol}, \text{Pol}\log),$$

which are characterized in [DyCo 80] in terms of parallel machines and in [Ruz 79a] in terms of alternating machines.

Many efforts have been made to separate **SC** from **NC** which is still an open problem and to get deeper insight in the structure of these classes (see for instance [Coo 81], [Coo 83], [McCo 83]).

Concerning the triple measure **SIZE-DEPTH-WIDTH** we have

21.16. Theorem. [DyCo 80] For t, r, s as in Theorem 21.14,

$$\textbf{multiT-DTIME-REV-SPACE}(\text{Pol}\, t, \text{Pol}\, r, \text{Pol}\, s)$$
$$= \textbf{SIZE-DEPTH-WIDTH}(\text{Pol}\, t, \text{Pol}\, r, \text{Pol}\, s). \blacksquare$$

§ 22. Relationships between Specific Complexity Classes

Among the relationships which are in general valid for the complexity classes of one or several complexity measures (especially inclusions and equalities, see §§ 19, 20 21) there are many special results for complexity classes with a small bounding function. Especially the linear time and realtime classes of Turing machines and counter machines have been of interest to researchers. We shall deal with the corresponding results in the subsections 1 and 2. In subsection 3 we shall collect some inclusional relationships between the most important complexity classes, these relationships being proved in other sections.

22.1. Linear Time and Realtime Classes of Turing Machines

In this subsection we deal with the complexity class $\mathbf{Q} = \mathbf{multiT\text{-}NTIME}(\mathrm{id})$ and some subclasses of **Q** defined by the restriction to a fixed number of tapes and to a bounded number of reversals. The class **Q** is of special interest because the sets of **Q** seem to be relatively simple. However, on the other hand they represent in a sense the full power of polynomial time nondeterminism (remember that there are sets in **Q** which are $\leqq_m^{\log}$-complete in **NP**, see 14.9.2).

We start with the study of **Q**. Our aim is to prove the equality **multiT-NTIME**(Lin) = **Q** = **1-3PD-NTIME**(id). To this end we remember the notion of quasirealtime computability. A 1-way k-tape Turing machine works in *quasi-realtime with delay d* iff it moves its input head in each dth step, i.e. in the steps d, $2d$, $3d$, ... of its work. Thus in each of the steps $d + 1$, $2d + 1$, $3d + 1$ it reads a new input symbol. The machine stops after the $(n \cdot d + 1)$th step. A quasi-realtime computation can easily be converted into a realtime computation.

22.1. Lemma. Let $d \geqq 1$, $k \geqq 1$, $Y_1, Y_2, \ldots, Y_k \in \{T, S, PD, C, NES, CS\}$ and $X \in \{D, N\}$. For every 1-Y_1-Y_2- ... -Y_k-XM M_1 accepting a language L in quasi-realtime there exists a 1-Y_1-Y_2- ... -Y_k-XM M_2 accepting L in realtime where for $i = 1, 2, \ldots, k$ the number of reversals of the ith head of M_2 does not exceed the number of reversals of the ith head of M_1.

Proof. The proof is the same as in the general case of linear speed-up by the factor d (see 18.12). While in the general case the input word must be first condensed (because possibly d input symbols must be read within d steps) this is not necessary in the quasi-realtime case with delay d where exactly one input symbol must be read within d steps. ∎

This lemma will be used twice in the proof of the following theorem.

22.2. Theorem. [BoGr 70] **multiT-NTIME**(Lin) = **Q** = **1-3PD-NTIME**(id)
= **1-PD-CS-NTIME**(id).

Proof. We show only the inclusion **multiT-NTIME**(Lin) $\subseteqq$ **1-3PD-NTIME**(id) using the techniques of [BoNiPa 74]. The inclusion **multiT-NTIME**(Lin) $\subseteqq$ **1-PD-CS-NTIME**(id) can be shown by similar techniques (see [BoGr 70]).

a) **1-multiT-NTIME**(Lin) $\subseteqq$ **Q**. First, because of 18.12.4 we have **1-multiT-NTIME** (Lin) = **1-multiT-NTIME**($2n$). Hence for a given $L \in$ **1-multiT-NTIME**(Lin), there is a 1-kT-NM M_1 accepting L within time $2n$. We construct a 1-(k+5)T-NM M_2 accepting L in quasi-realtime with delay 4. For simplicity we assume that 4 is a divisor of the length n of the input word. The other three cases can be treated analogously, and M_2 decides nondeterministically on one of these four cases. The work of M_2 will be subdivided into three phases. The input head of M_2 will move in each 4th step.

Phase 1. M_2 guesses nondeterministically the input word and writes it left-to-right onto the tapes 2 and 4. Let u be the guessed input word. During the $|u|$ steps of this

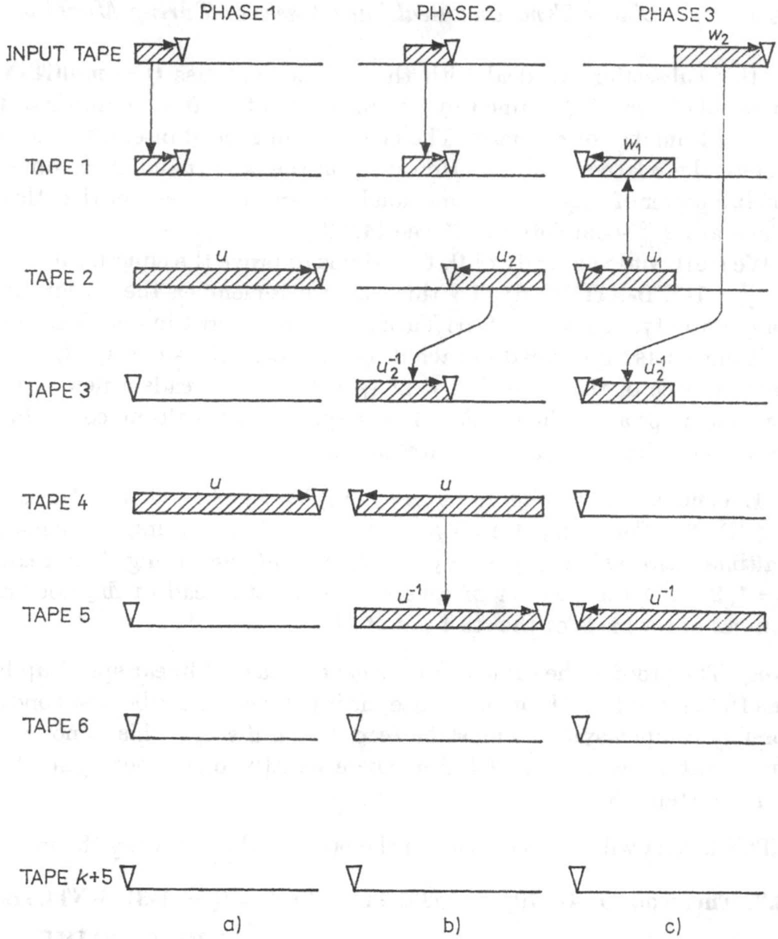

Figure 22.1. By ∇ the head positions at the end of a phase are marked. A horizontal arrow marks the motion of a head during a phase. The other arrows with one (two) vertices stand for the transfer of a word (comparison of two words)

phase M_2 writes the first $\frac{|u|}{4}$ symbols of the real input word w left-to-right onto tape 1. Phase 1 is represented in Figure 22.1a.

Phase 2. The head of tape 4 moves to the left in each step (reading u^{-1}), and simultaneously u^{-1} is written left-to-right onto the fifth tape. The head of tape 2 moves to the left in each second step and simultaneously u_2^{-1} is written left-to-right onto the third tape (by v_1 and v_2 we denote the first and second half, resp., of a word v). During the $|u|$ steps of this phase M_2 writes the next $\frac{|u|}{4}$ symbols left-to-right onto tape 1. Phase 2 is represented in Figure 22.1b.

Phase 3. The head of tape 3 moves to the left in each fourth step where the read word u_2 is compared with w_2. The heads of tapes 1 and 2 also move to the left in each fourth step, where the read words u_1 and w_1 are compared. On the tapes $1, 2, \ldots, k$ the work of M_1 on the input word u is simulated where tape 5 whose head moves right-to-left is used instead of the input tape. If u is accepted by M_1 and the result of the comparisons of u_1 and w_1 and of u_2 and w_2 is positive, then w is accepted by M_2, otherwise M_2 rejects w. Phase 3 is represented in Figure 22.1c.

Now, by Lemma 22.1 there exists a 1-(k+5)T-NM accepting L in realtime.

b) $\mathbf{Q} \subseteq \mathbf{eh}(\Gamma_\cap(\mathbf{1\text{-}NPDA}))$. Since a Turing tape can be replaced time equivalently by two pushdown stores we have $\mathbf{Q} \subseteq \mathbf{1\text{-}multiPD\text{-}NTIME}(\mathrm{id})$. Now Lemma 11.8 yields $\mathbf{1\text{-}multiPD\text{-}NTIME}(\mathrm{id}) \subseteq \mathbf{eh}(\Gamma_\cap(\mathbf{1\text{-}NPDA}))$.

c) $\mathbf{eh}(\Gamma_\cap(\mathbf{1\text{-}NPDA})) \subseteq \mathbf{1\text{-}3PD\text{-}NTIME}(\mathrm{id})$. Since $\mathbf{1\text{-}3PD\text{-}NTIME}(\mathrm{id})$ is evidently closed under e-free homomorphisms it is sufficient to show $\Gamma_\cap(\mathbf{1\text{-}NPDA}) \subseteq \mathbf{1\text{-}3PD\text{-}NTIME}(\mathrm{id})$. Let $A_1, A_2, \ldots, A_k$ be 1-NPDA accepting the languages $L_1, L_2, \ldots, L_k$, resp. The proof of 12.4 shows that we can assume without loss of generality that $A_1, A_2, \ldots, A_k$ work in realtime and accept with empty store. Now we construct a 1-3PD-NM M_1 accepting $L_1 \cap L_2 \cap \ldots \cap L_k$ in quasi-realtime with delay $2k + 4$. The first (second, third) pushdown store is subdivided into three (two, two, resp.) tracks.

For simplicity we assume that $2k + 4$ is a divisor of the length n of the input word. The other $2k + 3$ cases can be treated analogously, and M_1 decides nondeterministically on one of these $2k + 4$ cases. The work of M_1 will be subdivided into three phases. The input head of M_1 will move in each $(2k+4)$th step.

Phase 1. M_1 guesses nondeterministically the input word w and also w^{-1} and writes them (left-to-right) on the first and second track, resp., of all three pushdown stores. Let u_0 and v_0^{-1} be these guessed words. Than M_1 proceeds for $j = 1, 2, \ldots, k$ as follows: M_1 guesses nondeterministically the input word w and also w^{-1} and writes them (left-to-right) on the first and second track, resp., of the pushdown store 1 and 2. Let u_j and v_j^{-1} be these guessed words.

Simultaneously M_1 simulates the work of A_j on the input word u_j using the first track of pushdown store 3. If A_j rejects u_j, then M_1 rejects w. In the case that all guesses are correct the first phase consists of $(k + 1) \cdot n$ steps. The input symbols read during this phase are stored (left-to-right) onto the third track of pushdown store 1, i.e. M_1 writes an input symbol into each $(2k+4)$th square of the third track.

The situation at the end of phase 1 is represented in Figure 22.2a.

Phase 2. M_1 guesses nondeterministically the first half of w and also of w^{-1} and writes them (left-to-right) on the first and second track, resp., of the pushdown stores 1 and 2. Let $(u_{k+1})_1$ and $(v_{k+1}^{-1})_1$ be these guessed words (for a given word x we denote the first and second half of x by x_1 and x_2 resp.). Then M_1 erases $(u_{k+1})_1$ and $(v_{k+1}^{-1})_1$ on pushdown tape 2 and simultaneously writes $(u_{k+1})_1^{-1} = (u_{k+1}^{-1})_2$ and $(v_{k+1}^{-1})_1^{-1} = (v_{k+1})_2$ (left-to-right) onto the second and first track, resp., of pushdown store 1. Simultaneously the head of pushdown tape 3 erases $(u_0)_2$ and $(v_0^{-1})_2$. In the case that all guesses are correct, the second phase consists of n steps. As in phase 1, the input symbols read during this phase are stored in the third track of pushdown store 1. The situation at the end of phase 2 is represented in Figure 22.2b.

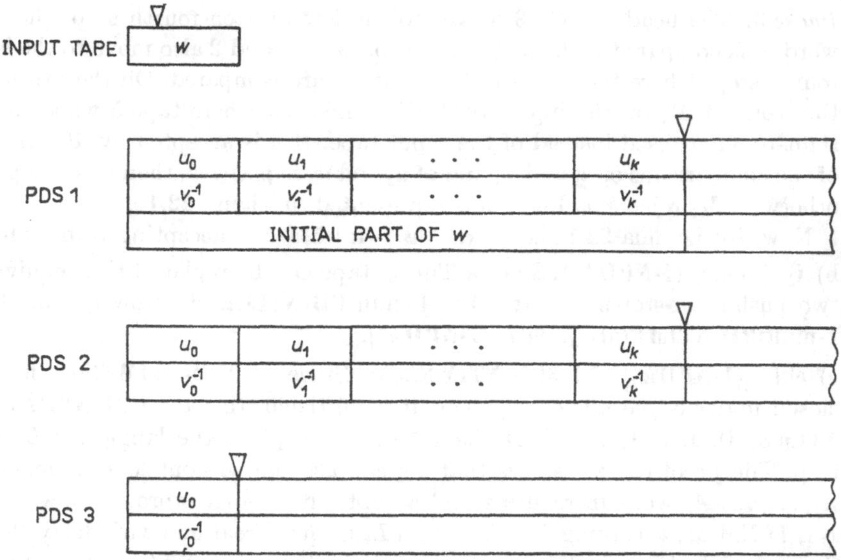

Figure 22.2 a

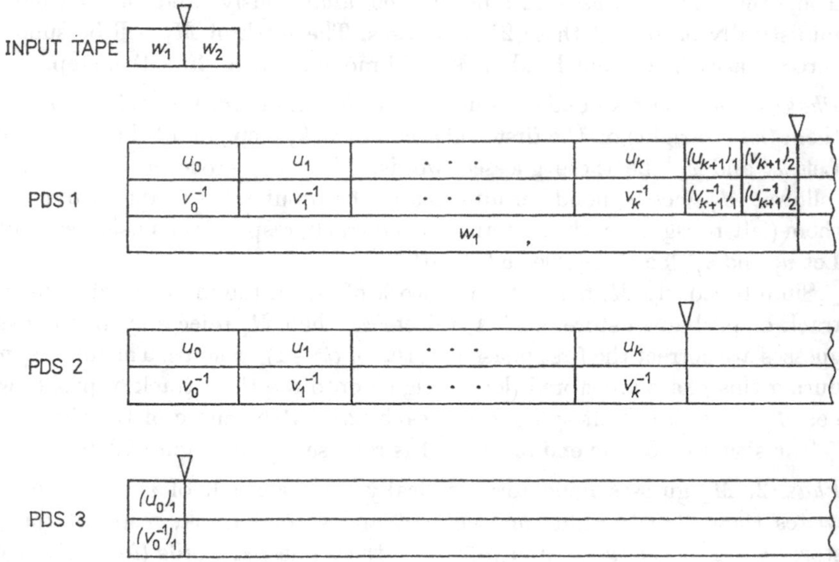

Figure 22.2 b

Phase 3. The heads of the pushdown stores 1 and 2 move to the left in each step where they check whether $u_1 \dots u_k\ (u_{k+1})_1(v_{k+1})_2 = u_0u_1\dots u_k$ and $v_1^{-1}\dots v_k^{-1}\ (v_{k+1}^{-1})_1$ $(u_{k+1}^{-1})_2 = v_0^{-1}v_1^{-1}\dots v_k^{-1}$. Simultaneously the head of pushdown store 3 moves to the left in each $(2k+4)$th step where it checks whether the contents of the third track of tape 1 are equal to the contents of the first track of pushdown store 3 (i.e. whether

$w_1 = (u_0)_1$) and whether the second half of the input word is equal to the contents of the second track of pushdown store 3 (i.e. whether $w_2 = (v_0)_2$). If the result of all these comparisons is positive, then M_1 accepts w, otherwise M_2 rejects w.

Now let us see that M_1 actually accepts $L_1 \cap L_2 \cap \ldots \cap L_k$. In phase 1 the machine M_1 checks whether $u_1 \in L_1, u_2 \in L_2, \ldots, u_k \in L_k$. Since the end of $u_0, u_1, \ldots, u_k, v_0^{-1}, v_1^{-1}, \ldots, v_k^{-1}$ can be considered to be marked, M_1 checks in phase 3 whether $u_0 = u_1 = \ldots = u_k$, $(u_k)_1 = (u_{k+1})_1$, $(u_k)_2 = (v_{k+1})_2$, $v_0 = v_1 = \ldots = v_k$, $(v_k)_2 = (v_{k+1})_2$ and $(v_k)_1 = (u_{k+1})_1$ (and consequently $u_k = v_k$). Furthermore, M_1 checks in phase 3 whether $w = w_1w_2 = (u_0)_1(v_0)_2 = u_0$. Consequently M_1 accepts w iff $w \in L_1 \cap L_2 \cap \ldots \cap L_k$.

Now, by Lemma 22.1 there is a **1-3PD-NM** accepting $L_1 \cap L_2 \cap \ldots \cap L_k$ in realtime. ▮

Remark. Item c of the proof shows that every language of **Q** can be accepted by a **1-3PD-NM** in realtime in such a way that the heads of two of the pushdown stores make at most one reversal each.

By similar methods one can prove

22.3. Theorem. [BrVo 82] **CSQ** $=_{\text{df}}$ **1-multiCS-NTIME**(Lin) = **1-2CS-NTIME**(id). ▮

By Theorem 22.8 we obtain

22.4. Corollary. **RBQ** $\subseteqq$ **CSQ** $\subseteqq$ **Q**. ▮

Now we deal with the relationships between the classes **1-kT-XTIME**(id), **1-kT-XTIME**(Lin), **1-multiT-XTIME**(id) and **1-multiT-XTIME**(Lin) for $k \geqq 1$ and X $\in$ {D, N}. (We can restrict ourselves to the 1-way case because for the realtime classes there is no difference between the 1-way and 2-way input mode.) Theorem 22.2 shows the following relationships for the nondeterministic classes:

$$\textbf{1-2T-NTIME}(\text{id}) = \textbf{1-3T-NTIME}(\text{id}) = \ldots = \textbf{1-multiT-NTIME}(\text{id})$$
$$= \textbf{1-2T-NTIME}(\text{Lin}) = \textbf{1-3T-NTIME}(\text{Lin}) = \ldots = \textbf{1-multiT-NTIME}(\text{Lin}),$$

i.e. there is no hierarchy with respect to the number of tapes. Only the inclusion **1-T-NTIME**(id) $\subseteqq$ **1-2T-NTIME**(id) is proper (see [DuGa 82] where **1-T-NTIME**(id) $\subset$ **1-2PD-NTIME**(id) and even **1-2PD-DTIME**(id) $\setminus$ **1-T-NTIME**(id) $\neq \emptyset$ has been shown).

Furthermore, it is known (see [Maa 84]) that

$$\textbf{1-2T-DTIME}(\text{id}) \nsubseteq \textbf{1-T-DTIME}(t), \quad \text{for} \quad t(n) \prec n^2,$$

and

$$\textbf{1-2T-DTIME}(\text{id}) \nsubseteq \textbf{1-T-NTIME}(n^2/\log^5 n).$$

This means in particular that **Q** $\nsubseteq$ **1-T-NTIME**$(n^2/\log^5 n)$.

At least in the realtime case the deterministic classes have different properties. By theorem 8.37 there is a proper hierarchy with respect to the number of tapes:

$$\textbf{1-T-DTIME}(\text{id}) \subset \textbf{1-2T-DTIME}(\text{id}) \subset \textbf{1-3T-DTIME}(\text{id}) \subset \ldots$$

(Note that, by 8.38, we also have **1-T-DTIME**(id) $\subset$ **1-T2-DTIME**(id) $\subset \ldots$)

Nothing is known about a corresponding hierarchy for the linear time case. However, there are some further proper inclusions.

22.5. Theorem.

1. **1-kT-DTIME**(id) $\subset$ **1-kT-DTIME**(Lin), $k \geqq 1$.
2. **REALTIME** = **1-multiT-DTIME**(id) $\subset$ **1-multiT-DTIME**(Lin) = **LINTIME**.
3. **1-kT-DTIME**(id) $\subset$ **1-kT-NTIME**(id).
4. **1-multiT-DTIME**(id) $\subset$ **1-multiT-NTIME**(id).
5. [PaPiSzTr 83] **LINTIME** = **1-multiT-DTIME**(Lin) $\subset$ **1-multiT-NTIME**(Lin) = **Q**.

Proof. The sets R, R, T and R can serve as witnesses for the statements 1, 2, 3 for the case $k = 1$, and 4, resp. (cf. 8.8.1, 8.8.1, 8.24 and 8.8.1, resp.). Statement 3 for $k \geqq 2$ follows from **1-kT-DTIME**(id) $\subset$ **1-(k+1)T-DTIME**(id) $\subseteqq$ **1—2T-NTIME**(id) (cf. 8.37 and 22.2). For Statement 5 see the proof of Theorem 24.6.10. ∎

For the realtime classes of other types of Turing machines the following results are known.

22.6. Theorem. [Han 66] **T-DTIME**(id) $\subset$ **T²-DTIME**(id). ∎

22.7. Theorem. [Vav 77] For $m \geqq 2$, **1-T$_m$-DTIME**(id) $\subset$ **1-T$_{m+1}$DTIME**(id). ∎

Now we consider the class **RBQ** $=_{df}$ **multiT-NTIME-REV**(id, Const) of all sets which can be accepted by multitape Turing machines working in linear time and with a constant number of reversals. Here we can obtain, by the same methods, analogous results to those in the case of **Q**.

22.8. Theorem. [BoNiPa 74], [BrWa 82]

$$\begin{aligned} \textbf{multiT-NTIME-REV}(\text{Lin, Const}) &= \textbf{RBQ} = \textbf{multiT-NTIME-REV}(\text{id}, 1) \\ &= \textbf{1-3PD-NTIME-REV}(\text{id}, 1). \\ &= \textbf{1-2CS-NTIME-REV}(\text{id}, 2) \end{aligned}$$

Proof. a) **1-multiT-NTIME-REV**(Lin, Const) $\subseteqq$ **RBQ**. This inclusion can be shown by the same construction as that made in item a) of the proof of 22.2. Note that the five new tapes in this construction have a constant number of reversals (namely one reversal per tape).

b) **1-multiT-NTIME-REV**(id, Const) $\subseteqq$ **eh**$(\Gamma_\cap$(**1-PD-NTIME-REV**(id, 1)$)$. Since a Turing tape can be replaced by two pushdown stores without an increase of time and reversal, we have **1-multiT-NTIME-REV**(id, Const) $\subseteqq$ **1-multiPD-NTIME-REV**(id, Const). Let M_1 be a 1-kPD-NM accepting a language L in realtime and with at most $2m - 1$ reversals per store. We construct a 1-(k·m)PD-NM M_2 accepting L in realtime and with at most 1 reversal per store. A pushdown store P of M_1 will be replaced by the pushdown stores $P_1, P_2, \ldots, P_m$ of M_2. At any time exactly one of the PD-stores $P_1, \ldots, P_m$ is active, and then it acts like P. For $i = 1, 2, \ldots, m$ the pushdown store P_i is first active from the $(2i - 2)$th reversal to the $(2i)$th reversal or until it becomes empty again. If P_i becomes empty again, then P_j becomes active again where j is the greatest number such that $j \leqq i$ and P_j is nonempty. Obviously M_2

accepts, in the same time and with only one reversal per pushdown store, the same language as M_1. Hence,

$$\textbf{1-multiPD-NTIME-REV}(\text{id}, \text{Const}) \subseteq \textbf{1-multiPD-NTIME-REV}(\text{id}, 1).$$

Finally, by Lemma 11.8 we have

$$\textbf{1-multiPD-NTIME-REV}(\text{id}, 1) \subseteq \textbf{eh}\big(\Gamma_\cap(\textbf{1-PD-NTIME-REV}(\text{id}, 1))\big).$$

c) $\textbf{eh}\big(\Gamma_\cap(\textbf{1-PD-NTIME-REV}(\text{id}, 1))\big) \subseteq \textbf{1-3PD-NTIME-REV}(\text{id}, 1)$. This inclusion can be shown in the same way as the inclusion $\textbf{eh}(\Gamma_\cap(\textbf{1-PD-NTIME}(\text{id})) \subseteq \textbf{1-3PD-NTIME}(\text{id})$ in the proof of 22.2 (item c) with the following modifications.

1. The hint to 12.4 can be omitted because every $L \in \textbf{1-PD-NTIME-REV}(\text{id}, 1)$ can obviously be accepted by a 1-PD-NM in realtime, with at most one reversal and with an empty store.

2. The 1-3PD-NM M_1 to be constructed accepts $L_1 \cap L_2 \cap \ldots \cap L_k$ in quasi-realtime with delay $4k + 4$ (instead of $2k + 4$). In phase 1 the input word w is guessed $2k + 1$ times (instead of $k + 1$ times). The resulting words $u_0, u_1, \ldots, u_{2k+1}, v_0, v_1, \ldots, v_{2k+1}$ have, in addition, k different markers indicating the places of the input word on which the input head is located while one of the pushdown heads makes a reversal. On the word u_i ($1 \leq i \leq k$) the computation of A_i is simulated only until the reversal of the head of A_i. On the word u_{2k+1-i} ($1 \leq i \leq k$) the remaining part of the computation of A_i is simulated. Hence the head of the third pushdown store makes only one reversal.

d) The proof of the equation $\textbf{RBQ} = \textbf{1-2CS-NTIME-REV}(\text{id}, 2)$ can be found in [BrWa 82]. ∎

Remark 1. An analogous result can be shown for the class RSBQ (see Remark 3 after 11.10). In [BoGrWr 78] the following result is proved:

$$\textbf{RSBQ} = \textbf{1-multiRT-NTIME-RESET}(\text{Lin}, \text{Const}) = \textbf{1-3RT-NTIME-RESET}(\text{id}, 1).$$

Remark 2. An analogous result can be shown for the class $\textbf{CBQ} =_{\text{df}} \textbf{multiT-NTIME-CROS}(\text{id}, \text{Const})$ (see also Remark 1 after 11.10).

In [BrWa 82] the following result is proved:

$$\textbf{multiT-NTIME-CROS}(\text{Lin}, \text{Const}) = \textbf{CBQ} = \textbf{1-3PD-NTIME-CROS}(\text{id}, 3) = \textbf{1-2CS-NTIME-CROS}(\text{id}, 4).$$

Moreover, every language from **CBQ** can be accepted in realtime by a **1-3PD-NM** such that two of the pushdown stores make at most one reversal and the third pushdown store is crossing bounded by 3. Note that $\textbf{RBQ} = \textbf{1-2CS-NTIME-CROS}(\text{id}, 3)$.

Theorem 22.8 shows that for $k_1, k_2 \geq 3$ and $m_1, m_2 \geq 1$,

$$\begin{aligned}\textbf{RBQ} &= \textbf{1-k}_1\textbf{T-NTIME-REV}(\text{id}, m_1) = \textbf{1-k}_2\textbf{T-NTIME-REV}(\text{id}, m_2),\\ &= \textbf{1-k}_1\textbf{T-NTIME-REV}(\text{Lin}, m_1) = \textbf{1-k}_2\textbf{T-NTIME-REV}(\text{Lin}, m_2),\end{aligned}$$

i.e. there is no hierarchy with respect to the number of tapes or with respect to the number of reversals for nondeterministic Turing machines. Only the inclusions $\textbf{1-T-NTIME-REV}(\text{id}, m) \subseteq \textbf{1-2T-NTIME-REV}(\text{id}, m)$ for $m \geq 1$ are proper. (See

[DuGa 82] where **1-2PD-NTIME-REV**(id, 1) $\setminus$ **1-T-NTIME-REV**(id, Const) $\neq \emptyset$ has been shown. From this and 22.8 even **1-T-NTIME-REV**(id, Const) $\subset$ **1-2T-NTIME-REV**(id, 2) follows.)

At least in the realtime case the deterministic classes have different properties.

22.9. Theorem. [BuVa 71] For $k \geqq 1$ and $m \geqq 0$,

1. **1-kT-DTIME-REV**(id, m) $\subset$ **1-(k+1)T-DTIME-REV**(id, m).
2. **1-kT-DTIME-REV**(id, m) $\subset$ **1-kT-DTIME-REV**(id, $m + 1$). ∎

Remark. The special case $k = 1$ and $m \geqq 1$ of 22.9.1 is already proved in [Rab 63], the special case $k = 1$ of 22.9.2 is already proved in [Hart 68b].

Nothing is known about the corresponding relationships between the linear time classes. Further relationships are collected in the next theorem.

22.10. Theorem.

1. **1-multiT-DTIME-REV**(id, Const) = **1-multiT-DTIME-REV**(id, 1).
2. **1-multiT-DTIME-REV**(Lin, Const) = **1-multiT-DTIME-REV**(Lin, 1).
3. **1-multiT-DTIME-REV**(id, Const) $\subset$ **1-multiT-DTIME-REV**(Lin, Const).
4. **1-multiT-DTIME-REV**(Lin, Const) $\subset$ **1-multiT-NTIME-REV**(id, Const).
5. **1-multiT-DTIME-REV**(id, Const) $\subset$ **1-multiT-DTIME**(id).
6. **1-multiT-DTIME-REV**(Lin, Const) $\subset$ **1-multiT-DTIME**(Lin).
7. **1-multiT-DTIME-REV**(Lin, Const) = **1-multiT-DREV**(Const).
8. **1-multiT-DTIME-REV**(Lin, Const) $\cap \mathfrak{P}(1^*) \subseteqq$ **REG**.

Proof. The statements 1 and 2 can be shown as in part b in the proof of 22.8.

Ad 3. $R \in$ **1-T-DTIME-REV**(2n, 1) $\setminus$ **1-multiT-DTIME**(id) (cf. 8.8.1).

Ad 4, 5 and 6. $\{0^{n^2}: n \geqq 1\} \in$ **1-2T-NTIME-REV**(2n, 1) $\cap$ **1-T-DTIME**(id). Because of Statement 8 this language cannot be in **1-multiT-DTIME-REV**(Lin, Const).

We omit the proof of statements 7 and 8 (see [BoYa 77]). ∎

Remark 1. It is not known whether the inclusion **RBQ** $\subseteqq$ **Q** is proper. However, in [DuGa 82] **1-T-NTIME-REV**(id, Const) $\subset$ **1-T-NTIME**(id) has been shown.

Remark 2. A grammatical characterization of constant reversal bounded checking stack automata can be found in [Sir 71].

22.2. Linear Time and Realtime Classes of Counter Machines

We start with a theorem about the relationships between the corresponding classes of multitape Turing machines and multicounter machines.

22.11. Theorem.

1. **1-multiC-DTIME**(id) $\subset$ **1-T-DTIME**(id).
2. **2-multiC-DTIME**(Lin) $\subset$ **2-T-DTIME**(Lin).
3. **2-multiC-NTIME**(Lin) $\subset$ **2-T-NTIME**(Lin).

Proof. Simple counting arguments show that

$$\{u \# v \# w : u, v, w \in \{0, 1\}^* \wedge |u| = |v| = |w| \wedge u = w\} \notin \textbf{2-multiC-NTIME}(\mathrm{Lin}).$$

But obviously this language is in **1-T-DTIME**(id). Theorem 20.2.1 completes the proof. ∎

Now we deal with the relationships between linear time and realtime classes of counter machines.

22.12. Theorem.
1. **1-C-DTIME**(id) ⊂ **1-C-DTIME**(Lin).
2. **1-C-NTIME**(id) = **1-C-NTIME**(Lin).
3. **1-kC-XTIME**(id) ⊂ **1-kC-XTIME**(Lin), $k \geqq 2$, X ∈ {D, N}.
4. **1-multiC-XTIME**(id) ⊂ **1-multiC-XTIME**(Lin), X ∈ {D, N}.
5. **2-kC-XTIME**(id) ⊂ **2-kC-XTIME**(Lin), $k \geqq 1$, X ∈ {D, N}.
6. **2-multiC-XTIME**(id) ⊂ **2-multiC-XTIME**(Lin), X ∈ {D, N}.

Proof. Ad 1. Counting arguments show (cf. [FipMeRo 68]) that $\{0^m 1^n : m \geqq n > 1\}^*$ is not in **1-multiC-DTIME**(id), but it is obviously in **1-C-DTIME**($2n$).
Ad 2. By 13.6.5 we have **1-C-NTIME**(id) = **1-NCA.**
Ad 3 and 4. Counting arguments like that used in § 8.1.1.3 show that

$$\{\#^{\mathrm{dya}^{-1}(w)} w^{-1} : w \in \{1, 2\}^*\} \notin \textbf{1-multiC-NTIME}(\mathrm{id}).$$

But this language is obviously in **1-2C-DTIME**(Lin).
Ad 5 and 6. Since the realtime classes for 1-way and 2-way machines of the same type coincide, it remains to prove the case $k = 1$ for X = N. Let M be a 1-kC-NM accepting $A = \{w \# w^{-1}(\# 1^{|w|})^{|w|} : w \in \{0, 1\}^*\}$. The remark after 8.2 yields 1-kC-NSPACE$_M(n^2 + 3n + 1) \geq \sqrt[k]{A'_{n^2+3n+1}} = 2^{n/k}$ for $A'_{n^2+3n+1} =_{\mathrm{df}} \{0, 1\}^n$. Consequently, $A \notin$ **1-multiC-NTIME**(Lin) ⊇ **2-multiC-NTIME**(id). But A is obviously in **2-C-DTIME**(Lin). ∎

Furthermore, there are hierarchies with respect to the number of counters for all types of 1-way classes.

22.13. Theorem. For $k \geqq 1$ and X ∈ {D, N},
1. [FipMeRo 68] **1-kC-XTIME**(id) ⊂ **1-(k+1)C-XTIME**(id).
2. [Gre 76] **1-kC-XTIME**(Lin) ⊂ **1-(k+1)C-XTIME**(Lin).

Proof. Simple counting arguments show that the language

$$\{0^{n_1} 1\, 0^{n_2} 1 \ldots 1\, 0^{n_{k+1}} \# 0^{n_1} 1\, 0^{n_2} 1 \ldots 1\, 0^{n_{k+1}} : n_1, n_2, \ldots, n_{k+1} \geqq 1\}$$

is not in **1-kC-NTIME**(Lin), but it is obviously in **1-(k+1)C-DTIME**(id). ∎

The next theorem deals with the difference between determinism and nondeterminism.

22.14. Theorem. For $i = 1, 2$, $k \geqq 1$ and $X \in \{D, N\}$,

1. **i-kC-DTIME**(id) $\subset$ **i-kC-NTIME**(id).
2. **i-multiC-DTIME**(id) $\subset$ **i-multiC-NTIME**(id).
3. **i-kC-DTIME**(Lin) $\subset$ **i-kC-NTIME**(Lin).
4. **i-multiC-DTIME**(Lin) $\subset$ **i-multiC-NTIME**(Lin).

Proof. Simple counting arguments show that the set $\{u \# v \# w : u, v, w \in \{0, 1\}^* \wedge u \neq w^{-1}\}$ is even not in **2-multiC-DTIME**(Lin), but it is obviously in **1-C-NTIME**(id) (for $i = 1$ see also [Gre 76]). ∎

Next we deal with the difference between 1-way and 2-way machines.

22.15. Theorem. [Gre 76] For $k \geqq 1$ and $X \in \{D, N\}$,

1. **1-kC-XTIME**(Lin) $\subset$ **2-kC-XTIME**(Lin).
2. **1-multiC-XTIME**(Lin) $\subset$ **2-multiC-XTIME**(Lin).

Proof. See the proof of 22.12.5 and 22.12.6. ∎

Remark. In [FipMeRo 68], [Gre 76] and [Mon 78b] many similar results for linear space and for polynomial time complexity classes for counter machines are proved.

Finally we note a result on nondeterministic counter machines working in linear time and with a constant number of reversals.

22.16. Theorem. [BaBo 74], [Gre 78c]

$$\begin{aligned}\textbf{1-multiC-NTIME-REV}(\text{Lin}, \text{Const}) &= \textbf{1-multiC-NTIME-REV}(\text{id}, 1)\\ &= \textbf{1-multiC-NREV}(\text{Const})\\ &= \textbf{1-multiC-NREV}(1). \;\blacksquare\end{aligned}$$

Remark 1. From 22.10.7 and 22.10.8 we obtain **1-multiT-DREV**(Const) $\cap\, \mathfrak{P}(1^*) \subseteqq$ **REG**. For counter machines we even have **2-multiC-DREV**(Const) $\cap\, \mathfrak{P}(1^*) \subseteqq$ **REG** (see [GuIb 79b]).

Remark 2. A characterization of constant reversal bounded counter machines by Presburger formulas can be found in [GuIb 81].

Remark 3. Subclasses of **1-C-NTIME-REV**(id, 1) with restricted use of nondeterminism are investigated in [IbRo 81].

Remark 4. In [Hro 84] it is shown that **1-multiC-XTIME-REV**(Lin, Const) $\subset$ **1-multiC-XTIME**(Lin) for X = D, N.

22.3. Relationships between the Most Important Complexity Classes

In this subsection we collect some results on the relationships between the most important complexity classes. We start with Figure 22.3 which shows the inclusions between these complexity classes.

We now give some comments on the inclusional relationships. All proper inclusions between space classes and also between time classes follow from the hierarchy results for the corresponding measures (see § 17.2). Now we give some comments on the relationships between the classes represented in Figure 22.3.

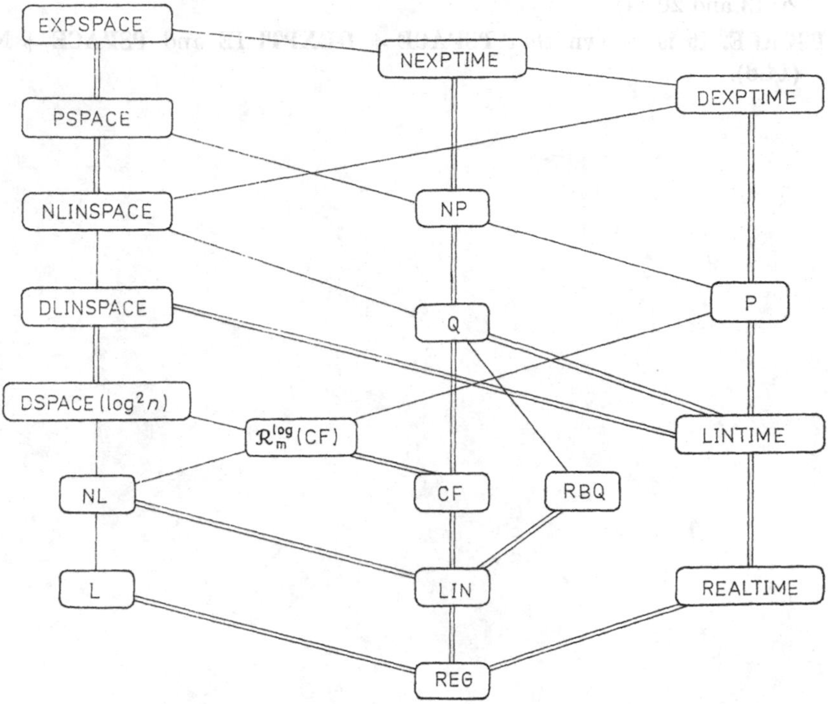

Figure 22.3

REG. The proper inclusions in **L**, **LIN** and **REALTIME** can be verified by $C = \{0^n1^n: n \geqq 1\}$.

LIN. The proper inclusions in **NL**, **CF** and **RBQ** can be verified by the language $\{0^n1^n0^m1^m: n, m \geqq 1\}$. $\mathbf{LIN} \subseteq \mathbf{L}$ implies $\mathbf{L} = \mathbf{NL}$ (23.19).

CF. The proper inclusions in $\mathcal{R}_m^{\log}(\mathbf{CF})$ and **Q** can be verified by the language $\{0^n1^n0^n: n \geqq 1\}$. $\mathbf{CF} \subseteq \mathbf{L}$ implies $\mathcal{R}_m^{\log}(\mathbf{CF}) = \mathbf{L}$ and $\mathbf{CF} \subseteq \mathbf{NL}$ implies $\mathcal{R}_m^{\log}(\mathbf{CF}) = \mathbf{NL}$ (14.6, and 8.43). $\mathbf{CF} \subseteq \mathbf{RBQ}$ implies $\mathbf{RBQ} = \mathbf{Q}$ (11.9 and 11.10).

LINTIME. Note that $\mathbf{LINTIME} \subset \mathbf{DLINSPACE}$ (25.1).

Q. $\mathbf{Q} \subseteq \mathbf{P}$ implies $\mathbf{P} = \mathbf{NP}$ (24.10) and this implies $\mathbf{DEXPTIME} = \mathbf{NEXPTIME}$ (24.3). $\mathbf{Q} = \mathbf{DLINSPACE}$ implies $\mathbf{NP} = \mathcal{R}_m^{\log}(\mathbf{Q}) = \mathcal{R}_m^{\log}(\mathbf{DLINSPACE}) = \mathbf{PSPACE}$ (A 14.1) and this implies $\mathbf{NEXPTIME} = \mathbf{EXPSPACE}$ (26.3.2).

P. $\mathbf{P} = \mathbf{NP}$ implies $\mathbf{DEXPTIME} = \mathbf{NEXPTIME}$ (24.3). $\mathbf{P} = \mathbf{PSPACE}$ implies $\mathbf{DEXPTIME} = \mathbf{EXPSPACE}$ (26.3.2). Furthermore we know $\mathbf{P} \neq \mathbf{DLINSPACE}$ and $\mathbf{P} \neq \mathbf{NLINSPACE}$ (14.6).

NP. **NP** = **PSPACE** implies **NEXPTIME** = **EXPSPACE** (26.3.2). Furthermore we know **NP** ≠ **DLINSPACE** and **NP** ≠ **NLINSPACE** (14.6). $\mathcal{R}_{\mathrm{m}}^{\log}(\mathbf{CF})$ = **2-auxPD-T-NSPACE-TIME**(log, Pol) (14.3). Consequently, **NL** = $\mathcal{R}_{\mathrm{m}}^{\log}(\mathbf{CF})$ means **2-T-NSPACE-TIME**(log, Pol) = **2-auxPD-T-NSPACE-TIME**(log, Pol), and $\mathcal{R}_{\mathrm{m}}^{\log}(\mathbf{CF})$ = **P** means **2-auxPD-T-NSPACE-TIME**(log, Pol) = **2-auxPD-T-NSPACE-TIME**(log, 2^{Pol}) (see 20.13 and 20.14).

PSPACE. It is known that **PSPACE** ≠ **DEXPTIME** and **PSPACE** ≠ **NEXPTIME** (14.6).

Chapter VII
Determinism versus Nondeterminism

From a purely formal point of view nondeterministic algorithms appear as a simple and obvious generalization of the deterministic algorithms. This generalization is suggested by innermathematical reasons, such as the fact that some important classes of languages can only be accepted by nondeterministic machines or are defined in a natural way by existential quantifiers.

There is also, however, an important practical reason which justifies the study of nondeterminism. Nondeterminism captures an essential feature of every problem solving process, namely the possibility of making guesses which can be verified in the further course of the process. This is the practical value of nondeterminism for complexity theory. For practically important problems (see § 7.3, 14.21) it is possible to design relatively fast nondeterministic algorithms, whereas it seems to be impossible to construct equivalent deterministic algorithms without a considerable increase in complexity. This leads us to the question of whether this increase is inevitable or whether nondeterministic and deterministic algorithms operating within equal complexity bounds have the same computational power.

In this chapter we restrict ourselves to the D-ND-problem for space (§ 23) and time (§ 24). The famous special cases of these unsolved problems: the logspace problem, the LBA-problem and the **P-NP** problem, are dealt with in a detailed way. So far, only a few less important special cases of the general problems have been solved (see 23.7, 23.8, 23.10 and 24.6). Several D-ND comparisons for automata can be found in § 13.

§ 23. NSPACE versus DSPACE

23.1. The Problem

The main problem of this section is the question of whether **NSPACE**(s) = **DSPACE**(s) for all or at least for certain $s \in \mathbb{R}$.

Two special cases are of particular interest for reasons which will become clear below, namely $s = log$ and $s = id$. The questions "**NL** = **L**?" and "**NSPACE**(id) = **DSPACE**(id)?" are called the *first logspace problem* and the *first LBA problem*, resp. The latter name comes from the fact that TM's working within linear space are called *linear bounded automata* (LBA). Because of the speed-up result 18.11, LBA's are exactly as powerful as TM's working within space n. Therefore the question

"**NSPACE**(id) $\doteq$ **DSPACE**(id)?" can be reformulated as "Are nondeterministic and deterministic LBA's of equal computational power?".

Before discussing in detail the importance of these two problems for our theory we list some results concerning the general question formulated initially. To begin with, this problem is still open in general. Therefore it is reasonable to slightly modify our question in the following way. "What is the least s' for given s such that **NSPACE**(s) $\subseteq$ **DSPACE**(s')"? The best answer so far is contained in the following theorem.

23.1. Theorem.

1. [Savi 70] Let $s \geqq \log$ and $s(|w|)$, $t(|w|) \in \text{DSPACE}(s \cdot \log t)$. Then

$$\mathbf{NSPACE\text{-}TIME}(s, t) \subseteq \mathbf{DSPACE}(s \cdot \log t).$$

2. If $s(|w|) \in \text{DSPACE}(s^2)$ and $s \geqq \log$, then

$$\mathbf{NSPACE}(s) \subseteq \mathbf{DSPACE}(s^2).$$

3. [MoSu 79] Let $s(|w|) \in \text{L}$ and $\log\log \leqq s \leqq \log$. Then

$$\mathbf{NSPACE}(s) \subseteq \mathbf{DSPACE}\big(s(n) \log n\big).$$

Remark 1. If the honesty condition in 23.1.1 is not satisfied, then the statement can be maintained with the following modification. For every 2-NTM accepting A within space s and simultaneously within time t there exists a 2-DTM working in space $s \cdot \log t$ and accepting (not deciding!) A. Remember that a machine that accepts A yields correct answers for all $w \in A$, but need not answer for $w \notin A$.

Remark 2. The second statement of 23.1 is not true for 1-NTM's working in space below $\sqrt{n}$. This follows from the fact that $\overline{D^{\#}} \in$ **1-T-NSPACE**(log) and $\overline{D^{\#}} \notin$ **1-T-DSPACE**(s) for all $s <_{\text{io}} id$ (cf. 8.2). The failure of W. J. Savitch's proof to work for $s < \sqrt{n}$ is connected with the fact that after having pursued a certain computation path without positive result it might be necessary to reposition the input head in order to try another path. A 1-NTM could do this only by copying the input on the worktape. But this would require space n which is larger than s^2.

Remark 3. For functions s between log log and log Statement 1 of 23.1 only allows us to conclude **NSPACE**(s) $\subseteq$ **NL** $\subseteq$ **DSPACE**($\log^2$). For $s < \log$ this is worse than 23.1.3.

Remark 4. Statement 3 of theorem 23.1 remains true for RSPACE′ (see [Jun 84]).

Proof of Theorem 23.1. Ad 1. Let $M = (S, \Sigma, \delta, s_0, \{s_1\}, \emptyset)$ be a 2-NTM working within space s and time t. Let w be an input of length n. In this proof we write an ID $w_1 s w_1' \# u_1 \mid u_1'$ of M on w in the form bin $|w_1| s \# u_1 \mid u_1'$. Hence there is some $c > 0$ such that $|I| \leqq_{\text{ae}} r =_{\text{df}} c \cdot s(n)$ for all ID's I of M on w.

Let $m =_{\text{df}} \text{card}(S \cup \Sigma \cup \{0, 1, \#, \mid\})$. It is convenient for what follows to interpret the elements of $S \cup \Sigma \cup \{0, 1, \#, \mid\}$ as digits and the ID's as numbers in an m-adic system. Let N be the maximum number which can be written in m-adic presentation using at most r digits.

Let $K_0 = 0s_0 \# \mid$ be the initial ID of M on w and let $K_1 = 0s_1 \# \mid$ be an accepting ID of M on w (without loss of generality we assume that M accepts in a situation corresponding to this ID). The question of whether $w \in L(M)$ is equivalent to the question of whether $K_0 \overset{2^d}{\underset{M}{\vdash}} K_1$ where $d =_{\text{df}} \lceil \log t(n) \rceil$, and this is equivalent to the question of whether $K_0 \overset{2^{d-1}}{\underset{M}{\vdash}} K_2$ and $K_2 \overset{2^{d-1}}{\underset{M}{\vdash}} K_1$ for some intermediate ID K_2. These questions can be subdivided in the same way (divide and conquer!). In such a manner we proceed up to the point where we have to answer questions of the form $K \overset{2^0}{\underset{M}{\vdash}} K'$, which can be answered immediately.

Now we describe the work of a 2-DTM M' which decides $L(M)$ and whose work includes the execution of a recursive procedure for deciding $K \overset{2^k}{\underset{M}{\vdash}} K'$ as described informally above. First M' computes $r = c \cdot s(n)$ and $d = \lceil \log t(n) \rceil$. Then M' marks on its worktape $d + 2$ blocks of length r, and writes K_0 into the first block and K_1 into the $(d+2)$th block. Now M' works according to the following (recursive) algorithm which, given $K \underbrace{\#\ldots\#\ldots\ldots\#\ldots\#}_{k \text{ blocks of length } r} K'$ on the worktape, decides whether $K \overset{2^k}{\underset{M}{\vdash}} K'$. The machine M' needs no more space than these $k + 2$ blocks of length r (as follows by induction on k).

1. **If** $k = 0$ **then** check immediately whether $K \overset{1}{\underset{M}{\vdash}} K'$
   if "yes" **then** return "yes" **else** return "no".
2. $I = 0$.
3. Write I into the $(k+1)$th block.
4. **If** $K \overset{2^{k-1}}{\underset{M}{\vdash}} I$ **then** go to 5
   else if $I < N$ **then** $I \leftarrow I + 1$ and go to 3
   else return "no".
5. Shift I into the second block.
6. **If** $I \overset{2^{k-1}}{\underset{M}{\vdash}} K'$ **then** return "yes"
   else if $I < N$ **then** $I \leftarrow I + 1$ and go to 3
   else return "no".

If M' (applying this algorithm on the worktape inscription $K_0 \underbrace{\#\ldots\#\ldots\ldots\#\ldots\#}_{d \text{ blocks of length } r} K'$) obtains "yes", then w is accepted, otherwise w is rejected.

Statement 2 follows immediately from statement 1 using **NSPACE**$(s) \subseteq$ **NSPACE-TIME**$(s, 2^{\text{Lin}\, t})$ (5.10.2) and linear speed-up (18.11).

Ad 3. We combine some results concerning the graph accessibility problem $\text{GAP}(2^{ds(n)})$ for graphs with bandwidth $2^{ds(n)}$, $d \in \mathbb{N}$. Because of 8.43.6 for every $A \in$ **NSPACE**(s) there exists a $d \in \mathbb{N}$ such that $A \leqq_m^{\log} \text{GAP}(2^{ds(n)})$ provided $s(|w|) \in L$. On the other hand, according to [MoSu 79] all these problems belong to **DSPACE**$(s(n) \cdot \log n)$. As this complexity class is closed under log space reducibility, 23.1.3 follows. ∎

Remark. Another proof of 23.1.3 has been given in [Tom 81].

Statement 1 of 23.1 is valid in a more general way.

23.2. Theorem. [Mon 72] If $s(|w|)$, $t(|w|) \in \text{DSPACE}(s \cdot \log t)$ and $s \geqq \log$, then **2-auxPD-T-NSPACE-TIME**$(s, t) \subseteq$ **DSPACE**$(s \cdot \log t)$. ∎

The *proof* is similar to that for 23.1.1. The only difference is that in order to avoid too long ID's caused by the contents of the pushdown tape the technique of simulating 2-NPDA's by 2T-DM's is applied which we described in the proof of 13.20.5.

We mention some corollaries of 23.1. The first relates our problem to the space-time-problem of § 26.

23.3. Corollary. If $s(|w|) \in \mathrm{DSPACE}(s \log s)$ and $s \geqq \log$, then

$$\mathbf{NSPACE}(s) \subseteqq \mathbf{NSPACE\text{-}TIME}(s, \mathrm{Pol}\, s) \to \mathbf{NSPACE}(s) \subseteqq \mathbf{DSPACE}(s \log s).$$ ∎

A further consequence is the positive solution of the **DSPACE-NSPACE**-problem for certain classes of the form **NSPACE**(Pol s).

23.4. Corollary. If $s(|w|) \in \mathrm{DSPACE}(\mathrm{Pol}\, s)$ and $s \geqq \log$, then

$$\mathbf{DSPACE}(\mathrm{Pol}\, s) = \mathbf{NSPACE}(\mathrm{Pol}\, s).$$

In particular **PSPACE** = **NSPACE**(Pol). ∎

Using the translational lemmas from § 6 we can shift positive results upwards and negative ones downwards.

23.5. Theorem. Let $\log s \leqq s'$, $s(|w|) \in \mathrm{DSPACE}(s)$, $h > id$, and $h(|w|) \in \mathrm{DSPACE}(s \circ h)$. Then

$$\mathbf{NSPACE}(s) \subseteqq \mathbf{DSPACE}(s') \to \mathbf{NSPACE}(s \circ h) \subseteqq \mathbf{DSPACE}(s' \circ h)$$

and

$$\mathbf{DSPACE}(s \circ h) \subset \mathbf{NSPACE}(s \circ h) \to \mathbf{DSPACE}(s) \subset \mathbf{NSPACE}(s).$$

Proof. The second statement is the contraposition of the first statement for $s = s'$, and the first follows from 6.17. ∎

Thus, in order to give a negative solution to the LBA problem one needs "only" prove **DSPACE**$(h) \subset$ **NSPACE**(h) for some $h > id$.

Using 6.21 we obtain the following result:

23.6. Theorem. NLINSPACE = DLINSPACE ↔ NL $\cap\, \mathfrak{P}(1^*) =$ **L** $\cap\, \mathfrak{P}(1^*)$. ∎

Unlike the case of $s \geqq id$ there is a difference between the behaviour of s-space-bounded one-way Turing machines and two-way Turing machines for $s <_{io} \mathrm{id}$ as is pointed out in remark 2 after 23.1. The argument used there leads immediately to a complete solution of the NSPACE-DSPACE-problem for sublinearly space bounded one-way TM's and counter machines. The same problem for two-way TM's is partially solved in 23.9.

23.7. Theorem. For $\log \leqq s <_{io} \mathrm{id}$, $\log \leqq \log s' <_{io} \mathrm{id}$ and $k \in \{3, 4, 5, \ldots, \mathrm{multi}\}$,

1. **1-T-DSPACE**$(s) \subset$ **1-T-NSPACE**(s).
2. **1-kC-DSPACE**(Pol s') $\subset$ **1-kC-NSPACE**(Pol s').
3. **1-kC-DSPACE**$(s') \subset$ **1-kC-NSPACE**(s'). ∎

There are some other known cases with partially or completely solved DSPACE-NSPACE problem:

23.8. Theorem. [Bra 77] For $s <_{io}$ id,

$$\textbf{1-auxPD-T-DSPACE}(s) \subset \textbf{1-auxPD-T-NSPACE}(s).$$

Proof. From 8.5 we know: $B, \bar{B} \notin$ **1-auxPD-T-DSPACE**(s) for $s <_{io}$ id. The set B admits the representation $B = B_1 \cap B_2 \cap B_3$, where

$$B_1 = \{u \# v \# p \# q: |u| = |v| \wedge u, v, p, q \in \{0, 1\}^*\},$$
$$B_2 = \{u \# v \# u^{-1} \# p: u, v, p \in \{0, 1\}^*\},$$
$$B_3 = \{u \# v \# p \# v^{-1}: u, v, p \in \{0, 1\}^*\}.$$

Evidently, $B_1, B_2, B_3 \in$ **DCF**. Hence $\bar{B} = \bar{B}_1 \cup \bar{B}_2 \cup \bar{B}_3 \in$ **CF**, and consequently

$$\bar{B} \in \textbf{1-auxPD-T-NSPACE}(s) \quad \text{for every} \quad s \geqq 0. \blacksquare$$

23.9. Theorem [Kan 83b]. For log log $\lesssim s <$ log, **DSPACE**$(s) \subset$ **NSPACE**(s).

Moreover, there exists a language in **NSPACE**(s) which cannot be *accepted* by an s-space-bounded deterministic TM. ▮

For two-way Turing machines with auxiliary tapes of various types we obtain the following results as corollaries of 20.13, 20.22, 20.21, and 20.26.

23.10. Theorem. For $s \geqq$ log,

1. **2-auxPD-T-DSPACE**$(s) =$ **2-auxPD-T-NSPACE**(s) if $s(|w|) \in$ DTIME$(2^{\text{Lin}\, s})$,
2. **2-auxS-T-DSPACE**$(s) =$ **2-auxS-T-NSPACE**(s),
3. **2-auxNES-T-DSPACE**$(s) =$ **2-auxNES-T-NSPACE**(s),
4. **2-auxCS-T-DSPACE**$(s) \subset$ **2-auxCS-T-NSPACE**(s). ▮

23.2. *The Complement Problem*

A positive solution of the problem "**DSPACE**$(s) =$ **NSPACE**(s)?" would imply **NSPACE**$(s) =$ **coNSPACE**(s). The converse implication need not be true. Rather, we should take into consideration the possibility **NSPACE**$(s) =$ **coNSPACE**(s) and **NSPACE**$(s) \neq$ **DSPACE**(s). In this subsection we deal with the problem "**NSPACE**(s) $=$ **coNSPACE**(s)?" whose special case $s =$ id is known as the *second LBA problem*. This problem as well as the general problem is as yet unsolved. However, in all cases, the classes **NSPACE**(s) and **coNSPACE**(s) have the same deterministic space complexity, i.e.

$$\textbf{NSPACE}(s) \subseteqq \textbf{DSPACE}(s') \leftrightarrow \textbf{coNSPACE}(s) \subseteqq \textbf{DSPACE}(s')$$

for all $s, s' \in \mathbb{R}_1$.

As in the case of theorem 23.5 a translation of results is possible which follows from 6.18.

23.11. Theorem. Let $s \geqq \log$, $s(|w|) \in \mathbf{DSPACE}(s)$, $h > \mathrm{id}$ and $h(|w|) \in \mathbf{DSPACE}(s \circ h)$. Then

$$\mathbf{NSPACE}(s) = \mathbf{coNSPACE}(s) \rightarrow \mathbf{NSPACE}(s \circ h) = \mathbf{coNSPACE}(s \circ h). \blacksquare$$

The complement problem is solved for languages over a single letter alphabet.

23.12. Theorem. [HaBe 75] For every s,

$$\mathbf{NSPACE}(s) \cap \mathfrak{P}(1^*) = \mathbf{coNSPACE}(s) \cap \mathfrak{P}(1^*). \blacksquare$$

A. Hemmerling [Hem 79b] noticed that the complement problem allows an equivalent formulation in terms of different notions of acceptance for nondeterministic TM's.

Definition. The set A is decided by the 2-NTM M within space $s \leftrightarrow_{\mathrm{df}}$

$$\bigwedge_x \bigvee_p \text{2-T-NSPACE}(x, p) \leqq s(|x|) \wedge \bigwedge_x \bigwedge_p \big(\text{2-T-NSPACE}(x, p) < \infty \rightarrow \varphi_M(x, p) = c_A(x)\big).$$

Roughly speaking, deciding machines have the same value on all halting computations on the same input. Furthermore,

$$\mathbf{NSPACE}'(s) =_{\mathrm{df}} \{A : A \text{ is decidable by some 2-NTM within space } s\}.$$

23.13. Theorem. For all $s \in \mathbf{R}_1$,

$$\mathbf{NSPACE}(s) = \mathbf{coNSPACE}(s) \leftrightarrow \mathbf{NSPACE}(s) = \mathbf{NSPACE}'(s).$$

Proof. "$\leftarrow$". Let $A \in \mathbf{NSPACE}(s)$. Because of the supposition there exists a 2-NTM M that decides A within space s. We modify M by exchanging positive and negative answers such that the new machine decides $\bar{A}$. Thus $\bar{A} \in \mathbf{NSPACE}'(s)$. As evidently $\mathbf{NSPACE}'(s) \subseteqq \mathbf{NSPACE}(s)$ we get $\bar{A} \in \mathbf{NSPACE}(s)$. Thus, $\mathbf{coNSPACE}(s) \subseteqq \mathbf{NSPACE}(s)$. This is equivalent to $\mathbf{NSPACE}(s) = \mathbf{coNSPACE}(s)$.

"$\rightarrow$". It is sufficient to prove the nontrivial inclusion $\mathbf{NSPACE}(s) \subseteqq \mathbf{NSPACE}'(s)$. Let $A \in \mathbf{NSPACE}(s)$. According to the supposition there exist 2-NTM's M and M' accepting A and $\bar{A}$, resp., having w.l.o.g. no rejecting states and working within space s.

A 2-NTM M'' to decide A does nondeterministically both the work of M and the work of M'. If M (M') reaches an accepting state, then M'' accepts (rejects). Obviously, M'' decides A within space s. $\blacksquare$

We conclude the subsection with the observation that by theorem 23.13 the problem "$\mathbf{NSPACE}(s) = \mathbf{DSPACE}(s)$?" can be subdivided into the problems "$\mathbf{NSPACE}(s) = \mathbf{coNSPACE}(s)$?" and "$\mathbf{NSPACE}'(s) = \mathbf{DSPACE}(s)$?" each of them being as yet unsolved.

23.3. *The Logspace Problem*

The logspace problem — one of the most complicated problems of complexity theory — is still open. All efforts to solve it have so far led only to even more equivalent reformulations of the problem within quite different frameworks.

On the one hand, this shows the great importance of the problem in the theory of computational complexity. On the other hand, problems turn out to be very hard (in fact as hard as the logspace problem) as soon as they are proved to be equivalent to the logspace problem. (Cf. 23.15 and 23.18 as examples.) Most of the reformulations to be stated now rest on the following theorem and the fact that problems of very different kinds turn out to be $\leqq_m^{\log}$-complete in **NL**.

23.14. Theorem. Let A be $\leqq_m^{\log}$-complete in **NL**. Then

1. $A \in L \leftrightarrow \mathbf{NL} = \mathbf{L}$.
2. $A \in \mathbf{DSPACE}(s) \leftrightarrow \mathbf{NL} \subseteq \mathbf{DSPACE}(s)$ for all $s \geqq \log$ such that $s(\mathrm{Pol}) \precsim s$.
3. $A \in \mathbf{coNL} \leftrightarrow \mathbf{NL} = \mathbf{coNL}$.

Proof. By 3.2 and 3.3. ∎

Remark 1. The relation $\leqq_m^{\log}$ can be replaced in statements 1 and 2 by $\leqq_T^{\log}$ and in Statement 3 by $\leqq_m^{\mathrm{Nlog}}$.

Remark 2. Remember that because of 23.1.2, statement 2 is interesting only if $\log \leqq s <_{io} \log^2$.

The main importance of 23.14 is that it successfully reduces questions concerning whole complexity classes to ones about single specific problems. Although, unfortunately, this reduction has not solved the problems so far, it affords insights into how strongly the logspace problems are entangled with various other problems in the theory of computing.

Each of the problems of 14.13 can play the role of A in 23.14. As an example we mention the graph accessibility problem GAP (8.43).

23.15. Theorem. [Savi 73a] $\mathbf{NL} = \mathbf{L} \leftrightarrow \mathrm{GAP} \in \mathbf{L}$. ∎

As an attempt to show that GAP belongs to **L**, W. J. Savitch interpreted graphs as mazes and defined "maze recognizing automata" which work on graphs, dropping and picking up pebbles on the nodes of the graphs. In [Savi 73a] it is shown that GAP $\in$ **L** if and only if GAP can be accepted by a maze recognizing automaton. With respect to this approach in [Bud 81] it is shown that maze recognizing automata with only two pebbles cannot accept all mazes.

Furthermore, finite automata with pebbles working on mazes have been investigated in [Bud 75], [BlSa 77], [Ass 77], [BlKo 78], [Koz 79], [Rol 79], [Graw 81], [Sze 83] and [HeKr 84]. In [CoRa 78] "jumping automata" for graphs are defined, which are less powerful than deterministic Turing machines, and it is proved that the recognition of GAP by these machines needs $c \cdot \dfrac{\log^2 n}{\log\log n}$ storage. This could give some evidence that GAP $\notin$ **L**. A completely analogous result concerning "random jumping automata" is contained in [BeSi 83]. For another attempt to prove GAP $\notin$ **L** see [AlKaLiLoRa 79]. Because of 8.43.1 the subproblem 1GAP of GAP is $\leqq_m^{1-\log}$-complete in **L**. Therefore 23.15 implies

23.16. Corollary. GAP $\leqq_{\mathrm{m}}^{1-\log}$ 1GAP $\leftrightarrow$ **NL** = **L**. ∎

This in itself trivial consequence of 23.15 reveals however a new aspect: It reduces the logspace problem to 1-DTM's in that for its negative solution it is sufficient to prove the nonexistence of a 1-DTM-computable function that reduces GAP to 1GAP. As in logarithmic space 1-DTM's are strictly less powerful than 2-DTM's one could hope to successfully tackle the problem this way. A first step into this direction is the following result ([Jhn 83]): Let J be the set of all graphs with indegree bounded by 1. Then GAP $\not\leqq_{m}^{1-\log}$ 1GAP $\cap J$.

So far, all known $\leqq_{\mathrm{m}}^{\log}$-complete problems in **NL** are pairwise log-space isomorphic [Hart 78].

23.17. Proposition. If all $\leqq_{\mathrm{m}}^{\log}$-complete problems in **NL** are pairwise log-space isomorphic, then **L** $\neq$ **NL**.

Proof. If **L** = **NL**, then every finite nonempty set is already $\leqq_{\mathrm{m}}^{\log}$-complete in **NL**. But finite sets cannot be isomorphic to infinite ones such as GAP. ∎

Among the equivalent formulations of the logspace problem there can be found some that are formulated completely without using the notion of complexity. This is due to various characterizations of **NL** and **L** by means of automata. By 13.2.3 and 13.2.8 we have **L** = **2:multi-DFA** and **NL** = **2:multi-NFA**. Therefore, the log-ɹpace problem is equivalent to the question of whether every nondeterministic multi-sead finite automaton can be simulated by a deterministic one. The next theorem hhows, among other things, that there are equivalent formulations with even a sestricted number of heads.

23.18. Theorem. [Sud 73]

1. **NL** = **L** is equivalent to each of the following statements:
   (1) **1:2-NFA** $\subseteq$ **2:multi-DFA**,
   (2) **1-NCA** $\subseteq$ **2:multi-DCA**,
   (3) **1-T-NSPACE**(log) $\subseteq$ **L**.
2. **coNL** = **NL** is equivalent to each of the following statements:
   (1) **co1:2-NFA** $\subseteq$ **2:multi-DFA**,
   (2) **co1-NCA** $\subseteq$ **2:multi-DCA**,
   (3) **co1-T-NSPACE**(log) $\subseteq$ **L**.

Proof. Ad 1. The equivalence to (1) follows from **23.14.1** by the facts that GAP_{ao} is $\leqq_{\mathrm{m}}^{\log}$-complete in **NL** (8.43.4) and $\mathrm{GAP}_{ao} \in$ **1:2-NFA** (as can easily be verified). The equivalence to (2) follows from **23.14.1** by the facts that GAP_{o} is $\leqq_{\mathrm{m}}^{\log}$-complete in **NL** (8.43.3), $\mathrm{GAP}_{o} \in$ **1-NCA** and **L** = **2:multi-DCA** (13.7.4). The equivalence of (2) and (3) follows from 1-NCA $\subseteq$ 1-T-NSPACE(log).

Ad 2. These equivalences can be shown in the same way as those of 1. ∎

Remark. In [Mon 78a] a variant of the knapsack problem is considered that is $\leqq_{\mathrm{m}}^{\log}$-complete in **NL**, and is accepted by a 1-NCA using only one zero test at the end of the acceptance process.

23.19. Theorem. [Sud 73] **NL** = **L** $\leftrightarrow$ **LIN** $\subseteq$ **L**.

Proof. The linear language L(S) is $\leq_m^{\log}$-complete in **NL** (8.43.5), and **LIN** $\subseteq$ **NL** (12.3.2). Note that L(S) is only a slight modification of GAP$_0$. ∎

Remark. The weaker statement **CF** $\subseteq$ **L** $\rightarrow$ **NL** $=$ **L** has been proved independently in [Nep 75].

Note that via 23.14.2 the previous theorem could have been stated in the following more general form:

23.20. Theorem.

$$\mathbf{LIN} \subseteq \mathbf{DSPACE}(s) \leftrightarrow \mathbf{NL} \subseteq \mathbf{DSPACE}(s)$$

for all $s \geq \log$, such that $s(\text{Pol}) \lesssim s$. ∎

This formulation even has the advantage that it is more likely that a proof for **LIN** $\subseteq$ **DSPACE**(s) with $s > \log$ will be found than one for **LIN** $\subseteq$ **L**. In fact, **LIN** $\subseteq$ **DSPACE**($\log^2$) (cf. 12.14). Therefore, we obtain W. J. Savitch's result **NL** $\subseteq$ **DSPACE**($\log^2$) as a corollary of 23.20. This is particularly interesting for it shows that W. J. Savitch's result is not only proved by the same method as 12.14, but actually follows from it.

It is also possible to find equivalent formulations of the logspace problem in terms of algebraic closure properties (11.3).

A further consequence of 23.14 is

23.21. Theorem. If **NL** $=$ **L**, then there exists an $f \in \mathbb{R}_1$ such that for every 2-NTM M_i working within logarithmic space, $M_{f(i)}$ is an equivalent 2-DTM working within logarithmic space.

Proof. We choose some complete set A. A direct encoding of all nondeterministic log-space bounded machines into A provides a uniform reducibility method, i.e. there exists some $g \in \mathbb{R}_1$ such that for all M_i with $\text{L}(M_i) \in$ **NL**

$$\text{L}(M_i) \leq_m^{\log} A \text{ via } \varphi_{g(i)}$$

(cf. the proof of 8.43.2 for a typical example of a statement of this kind). Now, by supposition, $A \in$ **L** and therefore there exists a 2-DTM M_m accepting A. As in the proof of 6.1.1, a deterministic logspace bounded machine $M_{f(i)}$ deciding L(M_i) can effectively be constructed from the logspace bounded machines $M_{g(i)}$ and M_m. ∎

23.4. The LBA Problem

The same general remarks on the importance of this problem are valid as for the logspace problem. However, these problems are not independent. As a consequence of 23.5 we have

23.22. Corollary. **L** $=$ **NL** $\rightarrow$ **DLINSPACE** $=$ **NLINSPACE.** ∎

As in the case of the logspace problem, complete sets play an important role.

23.23. Theorem. Let A be $\leqq_m^{\text{log-lin}}$-complete in **NLINSPACE**. Then

1. $A \in$ **DLINSPACE** $\leftrightarrow$ **NLINSPACE** = **DLINSPACE.**
2. $A \in$ **DSPACE**(s) $\leftrightarrow$ **NLINSPACE** $\subseteqq$ **DSPACE**(s) for all $s \geqq id$ such that $s(\text{Lin}) \preceq s$.
3. $A \in$ **coNLINSPACE** $\leftrightarrow$ **NLINSPACE** = **coNLINSPACE.**

Proof. By 3.2 and 3.3. ∎

Remark. Remember that because of 23.1.2, statement 2 is interesting only if id $\leqq s \prec_{io} n^2$.

Now we formulate the LBA problem in terms of automata. From 23.6 and the equalities **XL** = **2:multi-XFA** for X ∈ {D, N} (13.2.3 and 13.2.8) we get

$$\mathbf{NLINSPACE} = \mathbf{DLINSPACE} \leftrightarrow \mathbf{2{:}multi\text{-}NFA} \cap \mathfrak{P}(1^*) = \mathbf{2{:}multi\text{-}DFA} \cap \mathfrak{P}(1^*).$$

The next theorem shows that the number of heads of the nondeterministic automata can be restricted.

23.24. Theorem. [Mon 77b]

1. **NLINSPACE** = **DLINSPACE** $\leftrightarrow$ **2:2-NFA** $\cap\, \mathfrak{P}(1^*) \subseteqq$ **2:multi-DFA.**
2. **coNLINSPACE** = **NLINSPACE** $\leftrightarrow$ **co2:2-NFA** $\cap\, \mathfrak{P}(1^*) \subseteqq$ **2:multi-NFA.**

Proof. By translational methods similar to those used in the proof of 13.4.2 (cf. Remark 1 after 13.4). ∎

Remark. In [HaHu 74] statement 1 is formulated for 2:7-NFA. Note that because of **DLINSPACE** = **EPD2L** (12.19.1.2) and **NLINSPACE** = **CS** = **EP2L** = **EPDT2L** = **EPT2L** (12.15 and 12.19.1.1), the LBA problem can also be formulated in terms of grammars and L-systems.

§ 24. NTIME versus DTIME

24.1. *The Problem*

The main problem of this section is the question of whether **NTIME**(Pol t) = **DTIME** (Pol t) for all or at least for certain $t \in \mathbb{R}_1$. For t = id and $t(n) = 2^n$ we get the open problems "**NP** = **P**?" and "**NEXPTIME** = **DEXPTIME**?", resp. These problems, in particular the first one, are dealt with in the next subsection. In a slightly modified form the general problem can be formulated as follows: For given t find t' as small as possible such that

$$\mathbf{NTIME}(\text{Pol}\, t) \subseteqq \mathbf{DTIME}(\text{Pol}\, t').$$

Corollary 5.12.3 gives the following answer: If $t \geqq$ id and $t(|w|) \in$ DTIME$(2^{\text{Pol}\, t})$, then

$$\mathbf{NTIME}(\text{Pol}\, t) \subseteqq \mathbf{DTIME}(2^{\text{Pol}\, t}).$$

We now consider the complement problem for NTIME classes. We first state that **NTIME**(Pol t) = **DTIME**(Pol t) implies **NTIME**(Pol t) to be closed under complementation. However, on the contrary, **coNTIME**(Pol t) = **NTIME**(Pol t) seems to be consistent with **DTIME**(Pol t) $\subset$ **NTIME**(Pol t). Hence, the following question deserves attention as an independent problem: Does **coNTIME**(Pol t) = **NTIME**(Pol t) hold? For t = id we get the particularly interesting problem of whether **coNP** = **NP** holds.

Positive results for both these problems can always be lifted upwards by the translational methods of § 6.4 (we use 6.17).

24.1. Theorem. If id $\leqq t \leqq t'$ and $t(|w|) \in$ DTIME(Pol t), then for all h such that $h(|w|) \in$ DTIME(Pol($t \circ h$)),

1. **NTIME**(Pol t) $\subseteqq$ **DTIME**(Pol t') $\to$ **NTIME**(Pol $t \circ h$) $\subseteqq$ **DTIME**(Pol $t' \circ h$).
2. **coNTIME**(Pol t) $\subseteqq$ **NTIME**(Pol t) $\to$ **coNTIME**(Pol $t \circ h$) $\subseteqq$ **NTIME**(Pol $t \circ h$). ∎

For the special case $t = t'$ = id we obtain

24.2. Corollary.

1. $\mathbf{P} = \mathbf{NP} \leftrightarrow \bigwedge_{t(|w|) \in \mathrm{DTIME(Pol}\, t)} \mathbf{NTIME}(\mathrm{Pol}\, t) = \mathbf{DTIME}(\mathrm{Pol}\, t)$.
2. $\mathbf{coNP} = \mathbf{NP} \leftrightarrow \bigwedge_{t(|w|) \in \mathrm{DTIME(Pol}\, t)} \mathbf{coNTIME}(\mathrm{Pol}\, t) = \mathbf{NTIME}(\mathrm{Pol}\, t)$. ∎

Remark. One can get rid of the condition $t(|w|) \in$ DTIME(Pol t) if one is satisfied with deterministic acceptance instead of decision. (See also Remark 1 after 23.1.)

24.3. Corollary. **P** = **NP** $\to$ **DEXPTIME** = **NEXPTIME.** ∎

A downward translation of equalities onto the class of SLA-languages is possible as is shown by the next result which can be understood as a converse of 24.3. It could be extended to arbitrary constructible time bounds.

24.4. Theorem. [Boo 74a]

1. The following statements are equivalent:
   (1) **NEXPTIME** = **DEXPTIME.**
   (2) **NP** $\cap\, \mathfrak{P}(1^*)$ = **P** $\cap\, \mathfrak{P}(1^*)$.
   (3) **Q** $\cap\, \mathfrak{P}(1^*) \subseteqq$ **P**.
2. The following statements are equivalent:
   (1) **NEXPTIME** = **coNEXPTIME.**
   (2) **NP** $\cap\, \mathfrak{P}(1^*)$ = **coNP** $\cap\, \mathfrak{P}(1^*)$.
   (3) **Q** $\cap\, \mathfrak{P}(1^*) \subseteqq$ **coNP**.

Proof. The equivalences (1) $\leftrightarrow$ (2) follow immediately from 6.22, (2) $\to$ (3) is trivial. We prove (3) $\to$ (2) by the following padding argument: Let $A \subseteqq 1^*$ be accepted be the NTM M working within time $n^k - n$ for some $k \geqq 2$. Define

$$B = \{1^{m^k} : 1^m \in A\}.$$

Note that $A \leqq_{\mathrm{m}}^{\log} B$. The language B can be accepted by the following 1-3T-NTM M' in time n, and hence $B \in \mathbf{Q}$. For input 1^n the machine M' reads m input symbols and writes them simultaneously on all worktapes, where m is nondeterministically guessed. Using the first two tapes and the input tape it is checked in realtime whether the remaining number of input symbols is exactly $m^k - m$. In the same time on the third tape M is simulated. The input 1^n is accepted by M' iff the test on the first two tapes is positive and M accepts 1^m on tape 3. If now $\mathbf{Q} \cap \mathfrak{P}(1^*) \subseteq \mathbf{P}(\mathbf{coNP})$, then B can be accepted in deterministic polynomial time. But then, because of $A \leq_m^{\log} B$, the language A also belongs to $\mathbf{P}(\mathbf{coNP})$. ∎

Remark. Another necessary and sufficient condition for **NEXPTIME** = **DEXPTIME** can be found in [Boo 79e].

Denoting by **NTIME**′(Pol t) the set of languages decidable by NTM in time Pol t (cf. the definition on p. 420) we get the analogue of 23.13, which also can be proved in the same way as 23.13.

24.5. Theorem.

$$\mathbf{NTIME}(\mathrm{Pol}\, t) = \mathbf{coNTIME}(\mathrm{Pol}\, t) \leftrightarrow \mathbf{NTIME}(\mathrm{Pol}\, t) = \mathbf{NTIME}'(\mathrm{Pol}\, t). \ ∎$$

For specific (i.e. machine-dependent) time measures Φ we also deal with the stronger question "$\mathbf{N}\Phi(t) = \mathbf{D}\Phi(t)$?". Of particular importance is the problem "$\mathbf{Q}$ = **LINTIME**?" concerning the linear time languages.

We do know some results for machine-dependent measures, which will be stated in the next theorem.

24.6. Theorem.

1. **T-DTIME**(id) = **T-NTIME**(id).
2. $\mathbf{T\text{-}DTIME}(t) \subset \mathbf{T\text{-}NTIME}(t)$ for $n \log\log n \leq t < n \log n$ or $n \log n \leq t(n) < n^{\frac{3}{2}} \sqrt{\log n}$.
3. $\mathbf{kT\text{-}DTIME}(\mathrm{id}) \subset \mathbf{kT\text{-}NTIME}(\mathrm{id})$ for all $k \geq 2$.
4. $\mathbf{1\text{-}kT\text{-}DTIME}(\mathrm{id}) \subset \mathbf{1\text{-}kT\text{-}NTIME}(\mathrm{id})$ for all $k \geq 1$.
5. $\mathbf{multiT\text{-}DTIME}(\mathrm{id}) \subset \mathbf{multiT\text{-}NTIME}(\mathrm{id})$.
6. $\mathbf{1\text{-}multiC\text{-}DTIME}(\mathrm{Pol}\, t) \subset \mathbf{1\text{-}multiC\text{-}NTIME}(\mathrm{Pol}\, t)$ provided $\mathrm{id} \leq t <_{\mathrm{io}} 2^{cn}$ for all $c > 0$.
7. $\mathbf{1\text{-}multiC\text{-}DTIME}(t) \subset \mathbf{1\text{-}multiC\text{-}NTIME}(t)$ for $\mathrm{id} \leq t <_{\mathrm{io}} 2^{cn}$ for all $c > 0$.
8. $\mathbf{2\text{-}multiC\text{-}DTIME}(\mathrm{id}) \subset \mathbf{2\text{-}multiC\text{-}NTIME}(\mathrm{id})$.
9. $\mathbf{2\text{-}multiC\text{-}DTIME}(\mathrm{Lin}) \subset \mathbf{2\text{-}multiC\text{-}NTIME}(\mathrm{Lin})$.
10. [PaPiSzTr 83] $\mathbf{LINTIME} \subset \mathbf{Q}$.

Proof. The statements 1, 4, 5, 8, and 9 are already stated in 12.1.2, 22.5.3, 22.5.4, 22.14.2, and 22.14.4, resp.

Ad 2. From 7.10 we know $\bar{\mathrm{C}} \in \mathrm{T\text{-}NTIME}(n \log\log n)$, and by 8.19 we have $\bar{\mathrm{C}} \notin \mathbf{T\text{-}DTIME}(t)$ for $t < n \log n$ because $\bar{\mathrm{C}} \notin \mathbf{REG}$. In [Frv 79e] a set is exhibited which belongs to $\mathbf{T\text{-}NTIME}(n \log n)$ but not to $\mathbf{T\text{-}DTIME}(t)$ for $t < n^{\frac{3}{2}} \cdot \sqrt{\log n}$.

Statement 3 follows from statement 4 by the obvious fact

$$\mathbf{kT\text{-}XTIME}(\mathrm{id}) = \mathbf{1\text{-}(k-1)T\text{-}XTIME}(\mathrm{id}) \text{ for } k \geq 2 \text{ and } \mathrm{X} \in \{\mathrm{D}, \mathrm{N}\}.$$

The statements 6 and 7 follow from $\bar{\mathrm{S}} \in$ **1-C-NTIME**(id) $\setminus$ **1-multiC-DTIME**(t) provided id $\leqq t <_{\mathrm{io}} 2^{cn}$ for all $c > 0$ (see the remark after 8.2).

Ad 10 (very rough sketch)*). The proof is based on the fact that the graphs $G = (V, E)$ defined in the proof of Theorem 25.1 for block respecting machines have the following property: There exists a subset $S \subseteq V$ (a "segregator" for G) such that

1. $\operatorname{card} S \lesssim \dfrac{N}{\log^* N}$
2. every $x \in V \setminus S$ has at most $\dfrac{6N}{\log^* N}$ predecessors in $V \setminus S$.

This property allows simulating a given 1T-DM that works within time $t(n) \geqq n \log^* n$ (t fully 1T-DTIME-constructible) by an alternating machine that works within time $t(n)/\log^* t(n)$ and makes 4 alternations.

Exploiting time hierarchy results for alternation bounded machines one finally concludes the result.

For a general discussion of the method used in this proof see [Kan 83c]. ∎

The following is a step towards separating deterministic from nondeterministic time classes. Let

$$k = \begin{cases} 1 \text{ if } \bigwedge_j \textbf{multiT-DTIME}(n^j) \subset \textbf{multiT-NTIME}(n^j), \\ \min\{j : \textbf{multiT-DTIME}(n^j) = \textbf{multiT-NTIME}(n^j)\} \text{ otherwise.} \end{cases}$$

24.7. Theorem. [Kan 81a] If t and s satisfy the conditions

1. $t(n) \prec n^m$ for some $m \in \mathbb{N}$,
2. $t(n)^{1/k} \succ s(n) \geqq n$,
3. $t(2n) \precsim t(n)$,
4. $t(|w|) \in \text{multiT-DTIME}\left(\dfrac{t(n)}{\log^2 n}\right)$,
5. $s(|w|) \in \text{multiT-DTIME}(s)$,

then

$$\textbf{multiT-DSPACE-TIME}(s, t) \subset \textbf{multiT-NTIME}(t). \ ∎$$

24.8. Corollary. For all $j > k$,

$$\textbf{multiT-DSPACE-TIME}(\mathrm{Pol}(\log), n^j) \subset \textbf{multiT-NTIME}(n^j). \ ∎$$

*24.2. The **P-NP** Problem*

The importance of the general determinism versus nondeterminism problem has been dealt with in the introduction of Chapter VII. This is valid in particular for the special case of the **P-NP** problem which, however, gains additional interest from the

*) The paper became known to the authors only during the last revision of this manuscript.

fact that **P** comes close to what is acknowledged by general consensus to be the class of "feasible" problems. A huge number of problems from various fields of mathematics and computer science are proved to be NP-complete. This means, roughly speaking, that the nondeterministic algorithms running in polynomial time which are known for their solution, can be reconstructed as equivalent deterministic algorithms running also in polynomial time if and only if **P** = **NP**. As many of those problems are of extreme practical importance, the particular significance of the **P-NP** problem for feasibility is evident.

The central role of the **P-NP** problem which concerns machine-independent complexity classes is emphasized by the fact that it can be reformulated in nearly all areas of theoretical computer science. Some examples are listed in this subsection and many others are provided by 24.9 in combination with 14.21.

A further interpretation of the essence of the **P-NP** problem is discussed in § 24.4.

24.2.1. Complete Sets

*24.2.1.1. The **P-NP** Problem in Terms of Complete Sets*

As a special case of 3.2 and 3.3 we get the following theorem which allows to reduce the **P-NP** problem to a question concerning only single languages.

24.9. Theorem.

1. If A is complete in **NP** with respect to $\leqq_{\mathrm{T}}^{\mathrm{P}}$, then

$$\mathbf{P} = \mathbf{NP} \leftrightarrow A \in \mathbf{P}.$$

2. If A is complete in **NP** with respect to $\leqq_{\mathrm{sT}}^{\mathrm{NP}}$, then

$$\mathbf{coNP} = \mathbf{NP} \leftrightarrow A \in \mathbf{coNP}.$$

Note that $\leqq_{\mathrm{T}}^{\mathrm{P}} \subseteq \leqq_{\mathrm{sT}}^{\mathrm{NP}}$ (see 6.9.3).

Proof. To apply 3.2 and 3.3 we need the fact that **P** and **NP** are closed with respect to the used reducibilities which can be found in 14.6. ∎

Remark 1. In statement 2, $\leqq_{\mathrm{sT}}^{\mathrm{NP}}$ cannot be replaced by the other extension $\leqq_{\mathrm{tt}}^{\mathrm{P}}$ of $\leqq_{\mathrm{m}}^{\mathrm{P}}$ because **NP** is not known to be closed with respect to $\leqq_{\mathrm{tt}}^{\mathrm{P}}$.

Remark 2. The use of $\leqq_{\mathrm{sT}}^{\mathrm{NP}}$ instead of $\leqq_{\mathrm{m}}^{\mathrm{P}}$ in statement 2 is justified by the fact that there exist $\leqq_{\mathrm{sm}}^{\mathrm{NP}}$-complete problems in **NP** which are not known to be **NP**-complete (i.e. $\leqq_{\mathrm{m}}^{\mathrm{P}}$-complete in **NP**), for example

$$\left\{(a, b) \colon a, b \in \mathbb{N} \wedge \bigvee_{x,y \in \mathbb{N}} \left((ax+1)y = b\right)\right\}$$

(see [AdMa 77]).

Remark 3. Any problem mentioned in 14.21 can be used in theorem 24.9 instead of A. This shows the great importance of the **P-NP** problem for various fields of mathematics and computer science. The most complete list of **NP**-complete problems published so far is contained in [GaJo 79], which is continued in [Joh 81—].

Remark 4. [HaBe 76] Many of the known **NP**-complete problems are pairwise p-isomorphic. (A and B are p-*isomorphic* iff there exists an $f \in \mathbf{P}$ such that $f: A \xrightarrow{1\text{-}1} B$.) It is not known whether all **NP**-complete problems form a single p-isomorphism class. If yes, $\mathbf{P} \neq \mathbf{NP}$. (See [HaBe 76], for the proof see the similar result 23.17.) If not, one can embed every countable partial ordering in the class C of all p-isomorphism classes of the $\leqq_m^P$-degree of the NP-complete sets where C is ordered by one-one polynomially invertible and size increasing polynomially computable reductions (S. Mahaney and P. Young, see [You 83]; for a precourser see [Mah 81]). This is valid for every $\leqq_m^P$-degree. In particular, **P** does have this rich structure.

Further very detailed investigations about p-isomorphisms between NP-complete sets can be found in [Aus 77], [AuD'APr 76], [AuD'APr 77], [AuD'APr 78] and [AuD'AGaPr 77]. If the Boolean formulas of "low generalized Kolmogorov complexity" are contained in some set $S_0 \subseteq$ SAT and $S_0 \in$ P, then SAT $\setminus S_0$ is an NP-complete set which is not p-isomorphic to SAT ([Hart 84]).

Remark 5. SAT, CLIQUE and other problems remain complete in NP even with respect to a strong reducibility defined by interpretability in logical theories (see [Dah 84]).

The next theorem is a consequence of the existence of suitable **NP**-complete sets.

24.10. Theorem. The following statements are mutually equivalent:

1. $\mathbf{P} = \mathbf{NP}$.
2. $\mathbf{Q} \subseteq \mathbf{P}$.
3. $\mathbf{1\text{-}T\text{-}NTIME}(\mathrm{id}) \subseteq \mathbf{P}$.
4. $\mathbf{T\text{-}NTIME}(n \log n) \subseteq \mathbf{P}$.
5. $\mathbf{1\text{-}NCSA} \subseteq \mathbf{P}$.

Proof. If $\mathbf{P} = \mathbf{NP}$, then the statements 2 to 5 are true because the classes on the left-hand sides are subclasses of **NP**. On the other hand, each of these statements implies 1 because the class on the left-hand side contains an **NP**-complete set.

Ad 2. For **Q** this follows from 14.9.2.

Ad 3. Let 3-$\mathrm{SAT}_\mathrm{d}^1 =_{\mathrm{df}}$ set of all Boolean formulas which can be constructed from formulas from 3-SAT^1 by replacing every variable $x|^k$ by $x|^k\, x|^k$. It is clear that $\mathrm{SAT}^1 \leqq_m^P \mathrm{SAT}_\mathrm{d}^1$, and that $\mathrm{SAT}_\mathrm{d}^1$ is **NP**-complete.

$\mathrm{SAT}_\mathrm{d}^1$ is accepted by a 1-NTM in linear time as follows. When the input head reads the first half of $x|^k\, x|^k$ the worktape head moves k steps to the right and, if the kth square is empty, guesses a truth value and writes it down. When the input head reads the second half the worktape head returns. The truth value that has been found or guessed in square k is used to further evaluate the Boolean formula which is given as input.

Ad 4. By padding one can show as in 14.9.2 that there exist **NP**-complete sets in $\mathbf{T\text{-}NTIME}(n \log n)$.

Ad 5. 3-SAT^1 is **NP**-complete (see Remark 1 after 14.21). We show 3-$\mathrm{SAT}^1 \in \mathbf{1\text{-}NCSA}$. A 1-NCSA to accept 3-$\mathrm{SAT}^1$ guesses first a string of truth values and then using this guess as an assignment to the variables it verifies whether the given Boolean formula is satisfied. ∎

24.11. Theorem. If $\mathbf{P} = \mathbf{NP}$, then there exist an $f \in \mathbb{R}_1$ and a $c > 0$ such that for every NTM M_i working within polynomial time, $M_{f(i)}$ is an equivalent DTM working within polynomial time and size $(M_{f(i)}) \leqq c \cdot \text{size}(M_i)$, where size($M$) denotes the product of the number of symbols and the number of states of M.

Proof. The proof is the same as that of 23.21, but instead of A the set B_{id}^{N} (11.5) is used. ∎

24.2.1.2. The Structure of Complete Languages

24.2.1.2.1. Sparse Languages

We start with the quotation of the following results from [BoWrSeDo 77] concerning a more general notion of completeness than those based on reducibilities.

Definition. Let C_1, C_2 be classes of languages. A is called (C_1, C_2)-*complete* (*inclusion complete* with respect to (C_1, C_2)) iff

$$A \in C_2 \wedge (C_2 \subseteqq C_1 \leftrightarrow A \in C_1).$$

Obviously, every $\leqq_{\text{T}}^{\text{P}}$-complete set of **NP** is (**P**, **NP**)-complete.

From 24.4 we immediately obtain the following two theorems.

24.12. Theorem.

1. ($\mathbf{P} = \mathbf{NP} \leftrightarrow \mathbf{DEXPTIME} = \mathbf{NEXPTIME}$) if and only if there exists a (**P**, **NP**)-complete SLA-language.
2. If there is no (**P**, **NP**)-complete SLA-language, then $\mathbf{P} \neq \mathbf{NP}$ and
$$\mathbf{DEXPTIME} = \mathbf{NEXPTIME}. \;∎$$

24.13. Theorem.

1. ($\mathbf{NP} = \mathbf{coNP} \leftrightarrow \mathbf{NEXPTIME} = \mathbf{coNEXPTIME}$) if and only if there exists an (**NP**, **coNP**)-complete SLA-language.
2. If there is no (**NP**, **coNP**)-complete SLA-language, then $\mathbf{NP} \neq \mathbf{coNP}$ and
$$\mathbf{NEXPTIME} = \mathbf{coNEXPTIME}. \;∎$$

In the remainder of this subsection we consider consequences of the existence of sparse languages which are complete in **NP** with respect to several polynomial time reducibilities.

Definition. $A \subseteqq \Sigma^*$ is called (polynomially) *sparse* if and only if there exists a polynomial p such that

$$\bigwedge_{n \in \mathbb{N}} \left(\text{card}(A \cap \Sigma^n) \leqq p(n)\right).$$

SLA-languages are always sparse.

For our main result we need some auxiliary definitions and results.

Definition. A partial order $\leqq$ on Σ^* is *polynomially well-founded and length related* iff there exists a polynomial p such that

1. if C is a strictly descending chain with maximum element x, then it contains at most $p(|x|)$ elements,
2. $x < y \to |x| < p(|y|)$.

Definition. A is called *self-reducible* (in polynomial time) iff there exists an oracle TM with oracle A accepting A in polynomial time in such a way that on input x only questions are asked which are strictly less than x with respect to some polynomially well-founded and length-related partial order on Σ^*.

24.14. Theorem. [MePa 79] If A is self-reducible and there exist sparse sets B_1, B_2 such that $A \leqq_m^P B_1$ and $\bar{A} \leqq_m^P B_2$, then $A \in \mathbf{P}$.

Proof. Let M be a polynomial time bounded oracle machine accepting A with oracle A by self-reduction, and let $f, g \in \mathbf{P}$ be functions reducing A and $\bar{A}$ to the sparse sets B_1 and B_2, resp. Every input x determines a tree T in the following way:

(1) x becomes the root of T.

(2) To every node y of T all $y_1, \ldots, y_k$ which are asked by M on input y are added as sons of y.

As a selfreducing machine asks only questions about words strictly less than the input, the degrees of the nodes, the depth of T and the length of the labels of the nodes are bounded by a polynomial in $|x|$. This follows from the properties of the polynomially well-founded and length-related partial order. Note that the number of nodes may be exponential in $|x|$. However, using the supposition that A and $\bar{A}$ are polynomially reducible to sparse sets we are able to convert M to a polynomial time algorithm. This can be done as follows: Compute the first question y_1. Then compute in a depth-first search in T the first son of y_1 and proceed in the same way as with y_1. When a leaf y of T is reached, M may compute without further questions and in polynomial time the value $c_A(y)$. Then write the quadruple $(y, f(y), g(y), c_A(y))$ on a list.

When the questions concerning all sons of node z are answered, compute the answer $c_A(z)$ (in polynomial time) and make up a quadruple $(z, f(z), g(z), c_A(z))$ on the list.

When during this procedure a node y' is found such that there exists a y with $(y, f(y), g(y), c_A(y))$ on the list and $f(y') = f(y)$ or $g(y') = g(y)$, then $c_A(y') = c_A(y)$. Therefore no further work within the subtree with root y' is necessary and the tree T may be pruned at node y'. Furthermore, no quadruple for y' need be put on the list. We observe that if $(y, f(y), g(y), c_A(y))$ and $(y', f(y'), g(y'), c_A(y'))$ are on the list, then $f(y) \neq f(y')$ and $g(y) \neq g(y')$. As

$$\text{card}\left(\{f(y) : y \in A \wedge y \text{ node of } T\} \cup \{g(y) : y \in A \wedge y \text{ node of } T\}\right)$$

is polynomially bounded in $|x|$ the list contains only a number of quadruples which is polynomially boundes in $|x|$.

At most a polynomial number (in $|x|$) of steps are needed from the start of the algorithm up to the generation of the first quadruple because the simulated machine

M works in polynomial time and the depth of T is polynomially bounded in $|x|$. The same argument applies to the time between the generation of two successive quadruples. This shows that the list can be generated within polynomial time.

On completion of the list the original question "$x \in A$?" can be answered immediately by table look-up. Otherwise, the simulation of M on x would yield a new quadruple for x not yet on the list. This contradicts the fact that the list is already finished. ∎

24.15. Lemma. If $A \leqq_m^P B$ and $B \subseteqq 1^*$, then there exist sparse sets B_1 and B_2 such that $A \leqq_m^P B_1$ and $\bar{A} \leqq_m^P B_2$.

Proof. Let $A \leqq_m^P B$ via f and $B \subseteqq 1^*$, and we define

$$f'(w) = \begin{cases} f(w) & \text{if } f(w) \in 1^*, \\ 0 & \text{otherwise}. \end{cases}$$

Evidently $B_1 = B$ and $B_2 = (1^* \setminus B) \cup \{0\}$ are sparse and $A \leqq_m^P B_1$ and $\bar{A} \leqq_m^P B_2$ via f'. This proves the lemma. ∎

24.16. Theorem.

1. [Mah 80], [HaMa 80] $\mathbf{P} = \mathbf{NP} \leftrightarrow \bigvee_B (B \text{ sparse} \wedge B \text{ is } \mathbf{NP}\text{-complete})$.
2. [For 79] $\mathbf{P} = \mathbf{NP} \leftrightarrow \bigvee_B (B \text{ sparse} \wedge \bar{B} \text{ is } \mathbf{NP}\text{-complete})$.

Proof. Ad 1. We confine ourselves to the case that B is an SLA-language (see [Berp 78]). Only the nontrivial implication "$\leftarrow$" needs a proof. Let $B \subseteqq 1^*$ be **NP**-complete. Then SAT $\leqq_m^P B$. According to Lemma 24.15 we have SAT $\leqq_m^P B_1$ and $\overline{\text{SAT}} \leqq_m^P B_2$ for some sparse sets B_1, B_2. Furthermore, SAT is self-reducible. To see this let H be a Boolean formula depending on n variables $x_1, \ldots, x_n$, and let H_0 and H_1 denote the formulas $H(0, x_2, \ldots, x_n)$ and $H(1, x_2, \ldots, x_n)$, respectively. Applying $1 \wedge x = x$, $1 \vee x = 1$, $0 \wedge x = 0$, $0 \vee x = x$ we obtain shorter formulas H_0' and H_1', and we have

$$H \in \text{SAT} \leftrightarrow H_0' \in \text{SAT} \vee H_1' \in \text{SAT}.$$

This shows the self-reducibility of SAT. An application of 24.14 yields SAT $\in \mathbf{P}$ and thus $\mathbf{P} = \mathbf{NP}$.

Ad 2. The SLA case can be proved in the same way as the SLA case of statement 1. ∎

It is not known whether in theorem 24.16 NP-completeness can be replaced by completeness with respect to $\leqq_T^P$ or $\leqq_{tt}^P$. Partial answers are contained in [Ukk 83a] (for $\leqq_T^P$-reducibility with a constant number of possible queries and for conjunctive polynomial time tt-reducibility), [Yes 83] (for bounded positive tt-reducibility) and [Yap 83] (for conjunctive and disjunctive polynomial time tt-reducibility). Furthermore, the following results are known.

24.17. Theorem.

1. [KaLi 80] $\bigvee_B (B \text{ sparse} \wedge B\ \leqq_T^P\text{-hard for } \mathbf{NP}) \rightarrow \bigcup_{i \in \mathbf{N}} \Sigma_i^P = \Sigma_2^P$.

2. [KoSc 84] $\bigvee_B$ (B sparse $\wedge$ $B \leqq_{sT}^{NP}$-complete for **NP**) $\to \bigcup_{i\in\mathbb{N}} \Sigma_i^P = \Delta_2^P$.

3. [Lon 82b] $\bigvee_B$ ($\bar{B}$ sparse $\wedge$ $B \leqq_T^{NP}$-complete for **NP**) $\to \bigcup_{i\in\mathbb{N}} \Sigma_i^P = \Delta_2^P$.

4. [KoSc 84] $\bigvee_B$ ($\bar{B}$ sparse $\wedge$ $B \leqq_{sT}^{NP}$-complete for **NP**) $\to \bigcup_{i\in\mathbb{N}} \Sigma_i^P = \Delta_3^P$.

5. [Hart 83a] **EXPSPACE** = **DEXPTIME** $\to \bigvee_B$ (B sparse $\wedge$ $B \leqq_T^P$-hard for **NP**) $\leftrightarrow$ **P** = **NP**.

6. [Yap 83], [Wec 85] $\bigvee_B$ (B sparse $\wedge$ $B \leqq_{sm}^{NP}$-complete in **coNP**) $\leftrightarrow$ **NP** = **coNP**.

7. [Wec 85] $\bigvee_B$ (B sparse $\wedge$ $B \leqq_{sm}^{NP}$-complete in **NP**) $\leftrightarrow$ **NP** = **coNP**. ∎

Remark 1. The weaker version of statement 2 where $\leqq_{sT}^{NP}$ is replaced by $\leqq_T^P$ has been proved in [Mah 80].

Remark 2. Further results in the spirit of statements 6 and 7 for the restriction $\leqq_R$ (random polynomial time reducibility, see [AdMa 77]) of $\leqq_{sm}^{NP}$ can be found in [Wec 85].

Remark 3. In [HaSeIm 83] it is shown that **NP** ∖ **P** (**PSPACE** ∖ **NP**, **PSPACE** ∖ **P**) contains sparse sets if and only if **DEXPTIME** ⊂ **NEXPTIME** (**NEXPTIME** ⊂ **EXPSPACE**, **DEXPTIME** ⊂ **EXPSPACE**, respectively). Comparing this with theorem 24.4.1 we conclude that **NP** ∖ **P** contains sparse sets. This fact relativizes to every oracle as is shown in [HaSeIm 83]. The latter paper also includes the very remarkable result that there exists an oracle B such that $\mathbf{coNP}^B \setminus \mathbf{P}^B$ contains sparse sets but no SLA-languages. Various further results concerning sparse sets and sets of different densities in **NP** and **PSPACE** can be found in [HaYe 83], [Hart 83b] and [HaSeIm 83]. For further results see [Dek 79] and [14.11].

24.2.1.2.2. Almost Polynomial Time Languages

To derive further consequences from 24.14 we need the following definitions.

Definition. 1. For $s \in \mathbb{R}_1$ the set $A \subseteq \Sigma^*$ is called *s-almost polynomial time acceptable* ($A \in s\mathbf{P}$) if there exists a DTM deciding A and working in polynomial time for all but at most $s(n)$ strings of length less than or equal to n.

2. Pol **P** $= \bigcup_{p\in\mathrm{Pol}} p\mathbf{P}$.

It is not hard to see that $A \leqq_m^P B \wedge B \in$ Pol **P** implies that both A and $\bar{A}$ are reducible to sparse sets. Therefore the following theorem can be proved in the same way as 24.16.

24.18. Theorem. [MePa 79]

$$\bigvee_{A} \bigvee_{B \in \mathrm{Pol}\,\mathbf{P}} (A \leqq^{\mathrm{P}}_{\mathrm{m}} B \wedge A \text{ is } \mathbf{NP}\text{-complete}) \leftrightarrow \mathbf{P} = \mathbf{NP}. \blacksquare$$

For a weaker result see [Lyn 75].

A notion similar to sparseness has been introduced in [LaLiRo 78]. This notion is important for the results of this and the next subsection. New proofs of these results are given in [ChMa 80].

Definition. A set A is called *r-interval-easy* if there is a DTM deciding A in time t, and there is a polynomial p such that

$$\bigvee_{n}^{\mathrm{io}} \bigwedge_{y} \big(y \in [n, r(n)] \rightarrow t(y) \leqq p(y)\big).$$

We need the following intuitively clear observation:

24.19. Lemma. If $A \notin \mathbf{P}$, then A cannot be r-interval-easy for arbitrarily large r.

Proof. Let $\psi_1, \psi_2, \ldots$ be an effective numbering of $\{c_B : B \in \mathbf{P}\}$ via the DTM's $M_1, M_2, \ldots$ which work within polynomial time. For $A \notin \mathbf{P}$ we define

$$d_i(n) =_{\mathrm{df}} \mu m\Big(\bigvee_{z} |z| = m > n \wedge \psi_i(z) \neq c_A(z)\Big)$$

and

$$d(n) =_{\mathrm{df}} \max_{i \leqq n} d_i(n).$$

By definition, in $[n, d(n)]$ not one of the machines $M_1, \ldots, M_n$ computes A correctly. Assume now, A is r-interval-easy for $r > d$. Then let M be a machine deciding A in such a way that the time complexity of M on infinitely many intervals of the form $[n, r(n)]$ is bounded by some polynomial p. Then construct M' in the following way: M' accepts x if M accepts x within $p(|x|)$ steps. Otherwise, x is rejected after $p(|x|)$ steps. M' works within polynomial time and hence there exists some M_j equivalent to M'. Consequently, for infinitely many $n > j$ we have

$$d(n) \geqq d_j(n) \geqq r(n).$$

This contradicts the assumption $r > d$. $\blacksquare$

The next result states that interval-easiness is in some sense hereditary downwards with respect to $\leqq^{\mathrm{P}}_{\mathrm{T}}$-reducibility.

24.20. Theorem. [LaLiRo 78] For every $A \in \mathbf{REC}$ and every $s \in \mathbb{R}_1$ there exists an $r \in \mathbb{R}_1$ such that

$$\bigwedge_{B} (A \leqq^{\mathrm{P}}_{\mathrm{T}} B \wedge B \text{ is } r\text{-interval-easy} \rightarrow A \text{ is } s\text{-interval-easy}).$$

Proof. 1. *Preparatory steps.*

a) Let N_2 be a multitape TM deciding A with increasing time bound $t(n) \geqq 2^{n+1}$ where $t(|w|) \in \text{multiT-DTIME}(t)$. We define

$$r(n) = 2^{(t\circ s)^{[m+1]}(m)},, \quad \text{where} \quad m = 2^{2^{2^{n+1}}}.$$

b) Let $A \leqq_T^P B$ via the multitape TM M which is time bounded by the polynomial p, and let N_1 be a multitape TM which decides B in such a way that there exist infinitely many intervals of the form $[n, r(n)]$ such that N_1 is time bounded by the polynomial p_1 on inputs whose lengths are in those intervals. For such an interval $I = [n, r(n)]$ we define $m_k = (t \circ s)^{[k]}(m)$ for $k = 0, 1, \ldots, m$ and obtain the $m + 1$ pairwise disjoint subintervals $J_0 = [m_0, s(m_0)], \ldots, J_m = [m_m, s(m_m)]$ of $I' = [m, \log r(n)]$ which is in its turn a subinterval of I (see Figure 24.1).

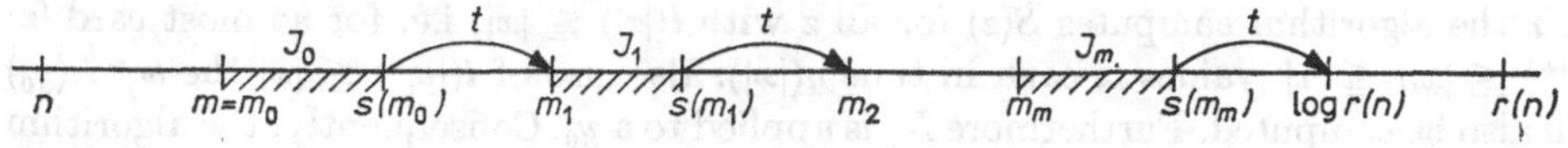

Figure 24.1

c) When M is applied to inputs x with $|x| \in I'$, the lengths of the occurring queries are bounded by $p(|x|) \leqq p(\log r(n)) <_{ae} r(n)$. Hence the length of such a query either belongs to I (and then it can be evaluated in polynomial time by an application of N_1) or it is bounded by n. These latter queries might be difficult to evaluate. However, at least in one interval J_k we will be able to avoid them.

To realize the main idea of the proof we introduce for every input x of M a certain set $S(x)$ which completely describes all possible behaviours of M on x. That means: M is simulated where easy queries are answerd by N_1, and for hard queries u both possibilities $u \in B$ and $u \notin B$ are taken into consideration. Thus we obtain several paths of M on input x. Every path p defines two sets $Q_+(p)$ and $Q_-(p)$ and a value $a(p)$:

$$Q_+(p) = \{u : u \text{ is a query on path } p \wedge p \text{ requires } u \in B \wedge N_1 \text{ runs on } u \text{ more than } p_1(|u|) \text{ steps}\},$$

$$Q_-(p) = \{u : u \text{ is a query on path } p \wedge p \text{ requires } u \notin B \wedge N_1 \text{ runs on } u \text{ more than } p_1(|u|) \text{ steps}\},$$

$$a(p) = \begin{cases} \text{yes} & \text{if } x \text{ would be accepted on } p, \\ \text{no} & \text{if } x \text{ would be rejected on } p. \end{cases}$$

Now we are able to define

$$S(x) = \{(Q_+(p), Q_-(p), a(p)) : p \text{ is a path of the computation of } M \text{ on } x\}.$$

It is important to notice that $S(x) = S(x')$ implies $x \in A \leftrightarrow x' \in A$.

d) If $|x| \in I'$, then for all y such that $|y| \leqq |x|$: the set $S(y)$ can be computed within a number of steps which is polynomially bounded in $|x|$. This can be seen as follows: Since N_1 runs in time p_1 for inputs whose lengths are in I, only queries of length $< n$ can be in $Q_+(p) \cup Q_-(p)$. Therefore $S(y)$ may have at most $2^{2^{n+1}} \leqq \log |x|$ triples. As every triple can be computed within time $p(|y|) \cdot p_1(p(|y|))$, the set $S(y)$ can be computed within time $q(|x|)$ where q is a polynomial.

2. *An algorithm to accept A*

Given input x. Compute $S(x)$ and $S(y)$ for all y such that $t(|y|) \leqq |x|$. If there is such a y such that $S(x) = S(y)$, then apply N_2 to y, otherwise apply N_2 to x.

3. *A is s-interval-easy.*

The number of different sets $S(x)$ with $|x| \in I'$ is bounded by $m = 2^{2^{2^{n+1}}}$, but there exist $m + 1$ intervals $J_0, \ldots, J_m$. Consequently, it is not possible that in each of these intervals a new set $S(x)$ can occur, i.e. there exists an interval, say J_k, such that for every x with $|x| \in J_k$ there exists a y_0 with $t(|y_0|) \leqq |x|$ and $S(x) = S(y_0)$. For such an x the algorithm computes $S(z)$ for all z with $t(|z|) \leqq |x|$, i.e. for at most card $\{z: 2^{|z|+1} \leqq |x|\} \leqq |x|$ values z, each in time $q(|x|)$. Because of $t(|y_0|) \leqq |x|$ the set $S(y_0)$ will also be computed. Furthermore N_2 is applied to a y_0. Consequently, the algorithm runs on x in polynomial time. ∎

As an easily proved consequence of 24.19 and 24.20 we have

24.21. Theorem. $\bigwedge_r \bigvee_B (A \leqq_T^P B \wedge B \text{ is } r\text{-interval-easy}) \rightarrow A \in \mathbf{P}$. ∎

The following quite obvious consequence of 24.21 should be compared with 24.18:

24.22. Theorem.

$\bigvee_A (A \text{ is } \leqq_T^P\text{-complete in } \mathbf{NP} \wedge \bigwedge_r \bigvee_B (A \leqq_T^P B \wedge B \text{ is } r\text{-interval-easy})) \leftrightarrow \mathbf{P} = \mathbf{NP}$. ∎

24.2.1.2.3. The Structure of the $\leqq_T^P$-Degrees

The main results of this section go back to R. Ladner. We prove them according to [LaLiRo 78] whose proofs are easier and lead to partially stronger results.

24.23. Theorem. [Lad 75] For every $B \notin \mathbf{P}$ there exists an $A \in \mathbf{REC}$ such that $A \notin \mathbf{P}$, $A \leqq_m^P B$ and $B \nleqq_T^P A$.

A reformulation of 24.23 in terms of $\leqq_T^P$-degrees emphasizes that there does not exist a minimal nontrivial degree:

24.23′. Theorem. For every $\leqq_T^P$-degree $\mathfrak{b}$ such that $\mathbf{P} <_T^P \mathfrak{b}$ there exists a $\leqq_T^P$-degree $\mathfrak{a}$ such that $\mathbf{P} <_T^P \mathfrak{a} <_T^P \mathfrak{b}$ (note that $\mathbf{P}$ is the least $\leqq_T^P$-degree).

Proof. Theorem 24.21 guarantees the existence of an r such that $B \leqq_T^P C$ implies that C is not r-interval-easy. In particular, B is not r-interval-easy. We further assume that r is strictly increasing and $r(|x|) \in \mathrm{DTIME}(\mathrm{Pol}\, r)$. These properties ensure that

$$D =_{\mathrm{df}} \left\{x: \bigvee_{n \in \mathbf{N}} r^{[2n]}(0) < |x| < r^{[2n+1]}(0)\right\}$$

is infinite and belongs to $\mathbf{P}$.

We put $A = B \cap D$ and verify that A fulfils the requirements of our theorem.

(1) $B \cap D \leqq_m^P B$ is trivial.

(2) $B \not\leq^{\mathrm{P}}_{\mathrm{T}} B \cap D$ because $B \cap D$ is r-interval-easy.
(3) $B \cap D \notin \mathbf{P}$ because $B \cap D \in \mathbf{P}$ would imply that B is r-interval-easy. ∎

From 24.23 we immediately obtain

24.24. Theorem. $\mathbf{P} \neq \mathbf{NP} \leftrightarrow \bigvee_{A \in \mathbf{NP}} (A \notin \mathbf{P} \wedge A$ is not $\leq^{\mathrm{P}}_{\mathrm{T}}$-complete in $\mathbf{NP}$). ∎

Remark 1. The proof of 24.23 shows that it is possible to obtain sets A in the form $B \cap D$ where B is $\leq^{\mathrm{P}}_{\mathrm{T}}$-complete in $\mathbf{NP}$ and $D \in \mathbf{P}$, provided $\mathbf{P} \neq \mathbf{NP}$.

Remark 2. Assuming $\mathbf{P} \neq \mathbf{NP}$ R. Ladner's result invites us to search for natural problems which are in $\mathbf{NP} \setminus \mathbf{P}$ and which are not NP-complete. In [Kar 72] three problems were mentioned which are possible candidates for such sets, namely LIQ(Q), NONPRIME $=_{\mathrm{df}} \{n: n \notin \mathrm{PRIME}\}$ and GRAPHISOM (see § 7.3.4); further candidates are described in [AdMa 77], [AdMa 79], [Hfm 82b] and [VaVa 82].

In the meantime L. G. Khachian has proved LIQ(Q) $\in \mathbf{P}$ (see 7.22). NONPRIME is very unlikely to be $\mathbf{NP}$-complete because it is known that NONPRIME $\in \mathbf{NP} \cap \mathbf{coNP}$. Furthermore, G. L. Miller has given a polynomial time algorithm for NONPRIME using the assumption of the extended Riemann hypothesis (cf. 7.29). This could be taken for some evidence that NONPRIME also belongs to $\mathbf{P}$. (See also 9.24 and the remark after 7.29.) Problems which are equivalent to GRAPHISOM under polynomial time many-one reductions are called isomorphism complete. Such problems are studied in [Boot 78], [Mil 79], [BoCb 79], [Cobn 80], and [KuTr 81]. Most of them are also isomorphism problems like the isomorphism problem for cfg's [HuRo 77] or the isomorphism problem for finitely presented algebras [Koz 77b]. However, in [Koz 78] and [FöMa 79] it is proved that there exist restrictions C_1, C_2 of CLIQUE such that C_1 and C_2 are $\mathbf{NP}$-complete and $C_1 \cap C_2 \equiv^{\mathrm{P}}_{\mathrm{m}}$ GRAPHISOM. For further results on GRAPHISOM see also § 7.3.4.

A stronger version of 24.23 is given by

24.25. Theorem. Let $B \notin \mathbf{P}$ and let $\{C_i: i \in \mathbb{N}\}$ be a recursively enumerable class of recursive sets. Then there exists an $A \in \mathbf{REC}$ such that

$$A \notin \mathbf{P} \wedge A \leq^{\mathrm{P}}_{\mathrm{m}} B \wedge B \not\leq^{\mathrm{P}}_{\mathrm{T}} A \wedge \bigwedge_{i \in \mathbb{N}} (C_i \text{ infinite} \to C_i \nsubseteq A).$$

Proof. In the proof of 24.23, r can be chosen by diagonalization in such a way that for every infinite C_i there exists some $x \in C_i \setminus A$. ∎

As an immediate consequence of this result we obtain

24.26. Theorem. $\mathbf{P} \neq \mathbf{NP} \leftrightarrow \mathbf{NP}$ contains an infinite P-immune set (i.e. an infinite set that does not contain any infinite set from $\mathbf{P}$). ∎

Another application of the method used in the proof of 24.23 yields

24.27. Theorem. [Lad 75] Let $C \neq \emptyset, \Sigma^*$ and $C <^{\mathrm{P}}_{\mathrm{m}} A$. Then there exist B_1, B_2 such that $C <^{\mathrm{P}}_{\mathrm{m}} B_1 <^{\mathrm{P}}_{\mathrm{m}} A$, $C <^{\mathrm{P}}_{\mathrm{m}} B_2 <^{\mathrm{P}}_{\mathrm{m}} A$ and $A = B_1 \cup B_2$ and B_1, B_2 are incomparable. The same is true for $\leq^{\mathrm{P}}_{\mathrm{T}}$.

Proof. Let $C \neq \emptyset, \Sigma^*$ and $C <^{\mathrm{P}}_{\mathrm{m}} A$. Define, for $D \in \mathbf{P}$ which is chosen as in the proof of 24.23, $B_1 = A \cap D$ and $B_2 = A \cap \overline{D}$. Thus $B_1, B_2 \notin \mathbf{P}$, $B_1, B_2 \leq^{\mathrm{P}}_{\mathrm{m}} A$ and $A \not\leq^{\mathrm{P}}_{\mathrm{T}} B_1, B_2$.

Similarly, one shows $C \leqq^{\mathrm{P}}_{\mathrm{m}} B_1, B_2$. B_1 and B_2 are incomparable with respect to $\leqq^{\mathrm{P}}_{\mathrm{T}}$. Otherwise we could easily conclude $A \leqq^{\mathrm{P}}_{\mathrm{T}} B_1$ or $A \leqq^{\mathrm{P}}_{\mathrm{T}} B_2$ which is impossible. ∎

The second statement of the following theorem is a corollary of 24.27.

24.28. Theorem.

1. The set of $\leqq^{\mathrm{P}}_{\mathrm{m}}$-degrees is an upper semilattice.
   The same is true for the $\leqq^{\mathrm{P}}_{\mathrm{T}}$-degrees.
2. Between any two degrees $[A]^{\mathrm{P}}_{\mathrm{m}} < [B]^{\mathrm{P}}_{\mathrm{m}}$ or $[A]^{\mathrm{P}}_{\mathrm{T}} < [B]^{\mathrm{P}}_{\mathrm{T}}$ there exists a chain of the order type of the rationals. ∎

The following theorem is a further consequence of 24.27.

24.29. Theorem. $\mathbf{P} \neq \mathbf{NP} \leftrightarrow \bigvee\limits_{A,B \in \mathbf{NP}} (A \nleqq^{\mathrm{P}}_{\mathrm{T}} B \wedge B \nleqq^{\mathrm{P}}_{\mathrm{T}} A)$. ∎

Without proof we mention that Statement 1 of 24.28 cannot be strengthened:

24.30. Theorem. [Lad 75] The set of all $\leqq^{\mathrm{P}}_{\mathrm{m}}$-degrees (and likewise the set of all $\leqq^{\mathrm{P}}_{\mathrm{T}}$-degrees) do not form a lattice. ∎

Insight into the structure of the $\leqq^{\mathrm{P}}_{\mathrm{T}}$-degrees can be deepened by considering "minimal pairs".

Definition. (A_0, A_1) is called a *minimal pair* $\leftrightarrow$

$$A_0, A_1 \notin \mathbf{P} \wedge \bigwedge_D (D \leqq^{\mathrm{P}}_{\mathrm{T}} A_0 \wedge D \leqq^{\mathrm{P}}_{\mathrm{T}} A_1 \to D \in \mathbf{P}).$$

In [Lad 75] the existence of minimal pairs has been proved. Then in [Mac 76] it has been ensured that a minimal pair can be found in **DEXPTIME**, and in [LaLiRo 78] it has even been shown that below any given $\leqq^{\mathrm{P}}_{\mathrm{T}}$-degree different from **P** there is always a minimal pair.

A very strong minimal pair theorem with many interesting consequences (see Corollary 24.32) has been proved in [Schg 84a].

24.31. Theorem. [Schg 84a] Let B_1, B_2 be recursive sets and let $\mathscr{C} \supseteqq \mathbf{P}$ be a class of recursive sets such that

(1) $B_1, B_2 \notin \mathscr{C}$,

(2) $\mathscr{C}$ is recursively enumerable,

(3) $\mathscr{C}$ is closed under finite variations (i.e. $A \in \mathscr{C} \wedge A \oplus B$ finite $\to B \in \mathscr{C}$).

Then there exist sets A_1, A_2 such that

(4) $A_1, A_2 \notin \mathscr{C}$,

(5) $A_1 \leqq^{\mathrm{P}}_{\mathrm{m}} B_1 \wedge A_2 \leqq^{\mathrm{P}}_{\mathrm{m}} B_2$,

(6) $\{A_1, A_2\}$ is a minimal pair.

Proof. Let $\mathscr{C} = \{C_1, C_2, \ldots\}$ be a recursive enumeration of $\mathscr{C}$. For $j = 1, 2$ define

$$r_j(n) =_{\mathrm{df}} 1 + \max_{i \leqq n} \mu m \Big(\bigvee_z |z| = m \wedge z \in B_j \oplus C_i\Big).$$

The functions r_1 and r_2 are total recursive because of (1) and (3). An immediate consequence of this definition is

(7) For every $i \leqq n$ there exists a word of some length l with $n \leqq l < r_j(n)$ discriminating B_j from C_i.

For $i = 1, 2$ let M_i be a DTM accepting B_i and halting on every input, and let t be an increasing bound on the running times of M_1 and M_2. Choose a fully T-DTIME-constructible increasing function s such that

$$s(n) \geqq 1 + \max\{r_1(n), r_2(n), t(n), 2^n\}.$$

We define for $i = 0, \ldots, 5$

$$S_i =_{\mathrm{df}} \left\{y: \bigvee_{k \in \mathbb{N}} (s^{[6k+i]}(0) \leqq |y| < s^{[6k+i+1]}(0)\right\}.$$

$S_0, \ldots, S_5 \in \mathbf{P}$. We prove $S_0 \in \mathbf{P}$: Let M be a DTM such that for every input x, $s(|x|) = \text{T-DTIME}_M(x)$.

To determine whether $y \in S_0$ compute on M the words $1^{a_1} = 1^{s(0)}$, $1^{a_2} = 1^{s(a_1)}$, $1^{a_3} = 1^{s(a_2)}, \ldots$ until l is found such that $a_l \leqq |y| < a_{l+1}$. Obviously $y \in S_0 \leftrightarrow l \equiv 0\,(6)$. This takes

$$|a_1| + |a_2| + \cdots + |a_l| \leqq \sum_{i=1}^{|y|} i \leqq |y|^2 \text{ steps}.$$

Define

$$A_1 =_{\mathrm{df}} B_1 \cap S_0 \quad \text{and} \quad A_2 =_{\mathrm{df}} B_2 \cap S_3.$$

We prove that A_1, A_2 have the properties (4), (5) and (6).

Ad (4). Assume $A_1 \in \mathscr{C}$. Then $A_1 = C_i$ for some $i \in \mathbb{N}$. Choose k such that $n =_{\mathrm{df}} s^{[6k]}(0) > i$. From (7) we conclude the existence of some $z \in B_1 \oplus C_i$ such that $n = s^{[6k]}(0) \leqq |z| < r_1(n) \leqq s(n) = s^{[6k+1]}(0)$. Hence, $z \in S_0$ and therefore, $z \in A_1 \leftrightarrow z \in B_1$. From $A_1 = C_i$ it follows that $z \in C_i \leftrightarrow z \in B_1$, but this contradicts $z \in B_1 \oplus C_i$. Consequently, $A_1 \notin \mathscr{C}$, and similarly $A_2 \notin \mathscr{C}$.

Ad (5). Let x_0 be a fixed element in $\overline{B}_1$ ($\overline{B}_1$ is not empty because $B_1 \notin \mathbf{P}$). Defining the polynomial time computable function

$$f(x) = \begin{cases} x & \text{if } x \in S_0 \\ x_0 & \text{otherwise} \end{cases}$$

we get $A_1 \leqq_{\mathrm{m}}^{\mathrm{P}} B_1$ via f. In the same way $A_2 \leqq_{\mathrm{m}}^{\mathrm{P}} B_2$ is proved.

Ad (6). Assume $D \leqq_{\mathrm{T}}^{\mathrm{P}} A_1$ and $D \leqq_{\mathrm{T}}^{\mathrm{P}} A_2$, and let Q_1, Q_2 be deterministic polynomial time oracle machines performing these reductions within time bounded by the polynomial $p(n) \leqq 2^n$. The following algorithm accepts D:

> Inputs belonging to $S_2 \cup S_3 \cup S_4$ are given as inputs to Q_1, and for every query $y \in S_0$ the machine Q_1 calls M_1 as a subroutine. Likewise, inputs belonging to $S_5 \cup S_0 \cup S_1$ are given as inputs to Q_2, and for every query $y \in S_3$ the machine Q_2 calls M_2 as a subroutine.

We show that this algorithm works in polynomial time. The decisions "$x \in S_2 \cup S_3 \cup S_4$?", "$y \in S_0$?" and "$y \in S_3$?" can be made in polynomial time.

Case 1. $x \in S_2 \cup S_3 \cup S_4$. Then, for some k, $s^{[6k+2]}(0) \leqq |x| < s^{[6k+5]}(0)$. For every query y occurring during the computation on input x it holds

$$|y| \leqq p(|x|) \leqq 2^{|x|} < 2s^{[6k+5]}(0) \leqq s\big(s^{[6k+5]}(0)\big) = s^{[6k+6]}(0).$$

A query which requires simulating M_1 belongs to S_0. If its length is bounded by $s^{[6k+6]}(0)$ it, must in fact be bounded by $s^{[6k+1]}(0)$. From this we conclude that simulating M_1 takes no more than

$$t(|y|) \leqq t\big(s^{[6k+1]}(0)\big) < s\big(s^{[6k+1]}(0)\big) = s^{[6k+2]}(0) \leqq |x|$$

steps. This proves that in Case 1 the algorithm works in polynomial time.

Case 2. $x \in S_5 \cup S_0 \cup S_1$. In this case the arguments are completely analogous. ∎

24.32. Corollary.

1. [LaLiRo 78] $\mathbf{P} \neq \mathbf{NP} \leftrightarrow \bigwedge_{B \in \mathbf{NP} \setminus \mathbf{P}} \bigvee_{A_0, A_1 \leqq_m^P B} \{A_0, A_1\}$ is a minimal pair.
2. [Schg 84a] If $\mathbf{NP} \neq \mathbf{coNP}$, then there exists a minimal pair $\{A_1, A_2\}$ such that $A_1 \in \mathbf{NP} \setminus \mathbf{coNP}$ and $A_2 \in \mathbf{coNP} \setminus \mathbf{NP}$.
3. [Schg 84a] Let t_1 and t_2 be functions which eventually majorize each polynomial such that $\mathbf{T\text{-}DTIME}(t_1) \setminus \mathbf{T\text{-}DTIME}(t_2) \neq \emptyset$. Then there exists a minimal pair in $\mathbf{T\text{-}DTIME}(t_1) \setminus \mathbf{T\text{-}DTIME}(t_2)$.

Note that statement 3 implies that there exist arbitrarily complex minimal pairs. Further consequences are contained in [Schg 84].

Very detailed examinations of structural properties of polynomial time degrees can be found in [Amb 84], [Amb 85] and [AmFlHu 84].

24.2.2. Further Reformulations of the P-NP-Problem

24.2.2.1. Relations to Algebraic Closure Properties and Automata

From theorem 24.10 we know that $\mathbf{P} = \mathbf{NP} \leftrightarrow \mathbf{Q} \subseteq \mathbf{P}$. Because of $\mathbf{Q} = \mathbf{eh}(\mathbf{CF} \wedge \mathbf{CF} \wedge \mathbf{CF})$ (see 11.9) the following question arises: Can the class $\mathbf{eh}(\mathbf{CF} \wedge \mathbf{CF} \wedge \mathbf{CF})$ in the above equivalence be replaced by $\mathbf{RBQ} = \mathbf{eh}(\mathbf{LIN} \wedge \mathbf{LIN} \wedge \mathbf{LIN})$ (see 11.10) or even by $\mathbf{eh}(\mathbf{LIN} \wedge \mathbf{LIN}) \subset \mathbf{RBQ}$ (see Remark 2 after 11.10)? This question is answered in the affirmative by

24.33. Theorem. $\mathbf{P} = \mathbf{NP} \leftrightarrow \mathbf{eh}(\mathbf{LIN} \wedge \mathbf{LIN}) \subseteq \mathbf{P}$.

Proof. "$\rightarrow$" This implication is an immediate consequence of the above remark.

"$\leftarrow$". To show this implication it is sufficient to show that for every $A \in \mathbf{NP}$ there exists a $B \in \mathbf{eh}(\mathbf{LIN} \wedge \mathbf{LIN})$ such that $A \leqq_{\text{pad}} B$. We use an idea of [BaBo 74]. Let M be an NTM accepting A within time $n^k + k - 2$ for some $k \geqq 2$. For an ID K of M we set $\bar{K} = \bar{h}(K)$ and $\hat{K} = \hat{h}(K)$ where $\bar{h}, \hat{h}$ are homomorphisms defined by

$$\bar{h}(a) = \begin{cases} a & \text{if } a \text{ is a tape symbol of } M, \\ \mathrm{e} & \text{if } a \text{ is a state of } M \end{cases} \quad \text{and} \quad \hat{h}(a) = \hat{a}.$$

We define

$$B_1 = \{\tilde{K}_1 \# \# \hat{K}_3 \# \ldots \# \hat{K}_4^{-1} \# \hat{K}_2^{-1} \#: K_1, K_2, K_3, \ldots \text{ are ID's of } M \wedge K_1 \text{ is an initial ID of } M \wedge K_{2i-1} \vdash_M K_{2i} \text{ for all } i\},$$

$$B_2 = \{\tilde{K}_1 \# \# \hat{K}_3 \# \ldots \# \hat{K}_4^{-1} \# \hat{K}_2^{-1} \#: K_1, K_2, K_3, \ldots \text{ are ID's of } M \wedge \text{ one of these ID's is accepting } \wedge K_{2i} \vdash_M K_{2i+1} \text{ for all } i\}.$$

Obviously, $B_1, B_2 \in$ **LIN** and $K_1, K_2, K_3, K_4, \ldots$ is an accepting computation of M if and only if $\tilde{K}_1 \# \# \hat{K}_3 \# \ldots \# \hat{K}_4^{-1} \# \hat{K}_2^{-1} \#$ is in $B_1 \cap B_2$. Now we define the nonerasing homomorphism h by $h(a) = a$ for $a \neq \square$, $h(\square) = +$ ($+$ being a new symbol) and $h(\hat{a}) = +$. It is easy to see that $w \in A \leftrightarrow w +^{|w|^{2k}+(2k-1)|w|+k^2} \in h(B_1 \cap B_2)$. ∎

The reader should notice that 11.3.4 is a theorem formulating necessary and sufficient conditions for **P** = **NP** in terms of algebraic closure properties. 11.3.2 and 11.3.3 yield at least sufficient conditions.

The next result relates the **P-NP** problem to **2-DPDA**. Remember that **P** = **2:multi-DPDA** (13.20.4).

24.34. Theorem. [Gal 74] **P** = **NP** $\leftrightarrow$ here exists a $\leq_m^{\log}$-complete set in **NP** belonging to **2-DPDA**.

Proof. "$\rightarrow$". If **P** = **NP**, then SAT $\in$ **P** = **2:multi-DPDA**. Hence SAT $\in$ **2:k-DPDA** for some $k \in \mathbb{N}$, and by 13.13 and 13.11.1 we obtain SAT $\leq_m^{\log} B$ for some $B \in$ **2-DPDA**. "$\leftarrow$". This implication is obvious. ∎

24.2.2.2. Miscellaneous

24.2.2.2.1. Logical Networks

Theorem 21.13 shows that the measure SIZE for logical networks is polynomially related to DTIME. This yields

24.35. Theorem. **P** = **NP** iff every set in **NP** has a polynomially bounded uniform size complexity. ∎

In particular, 24.35 shows that **P** = **NP** implies that every set in **NP** has polynomially bounded *non*uniform size complexity. For sets from **R** ⊆ **NP** this is already true without the supposition **P** = **NP** (see [Adl 78] and for the relativized case [Wil 83]), where **R** = $\{A$: A is accepted by a polynomially bounded probabilistic TM with zero error probability for inputs not in $A\}$. As an example (see 9.42), NONPRIME has polynomial nonuniform size complexity. What can be concluded from polynomial nonuniform size complexity of the sets of **NP** and **PSPACE** is stated in the following theorem.

24.36. Theorem. [KaLi 80]

1. If every set of **NP** has polynomially bounded nonuniform size complexity, then $\bigcup_{i \geq 0} \Sigma_i^{\mathrm{P}} = \Sigma_2^{\mathrm{P}}$.

2. If every set of **PSPACE** has polynomially bounded nonuniform size complexity, then $\mathbf{PSPACE} = \Sigma_2^{\mathrm{P}} \cap \Pi_2^{\mathrm{P}}$. ∎

Remark. With respect to the sufficient condition in statement 1 we note that $(\Sigma_2^{\mathrm{P}} \cap \Sigma_2^{\mathrm{P}}) \setminus \mathbf{SIZE}(\mathrm{Lin}(n^k)) \neq \emptyset$ for every $k \in \mathbb{N}$ (see [Kan 81b]). If an appropriate notion of relativized circuits is defined as in [Wil 83], this result relativizes and contrasts with the fact that a relativized version of Δ_2^{P} has linearly bounded circuits (see [Wil 83]).

24.2.2.2.2. Spectra

Characterizations of **NEXPTIME** by spectra of formulas of the first order predicate calculus (see [JoSe 74]) and of **NP** by generalized spectra of certain formulas of the second order predicate calculus (see [Fag 73]) make it possible to reformulate the questions "**NEXPTIME** = **coNEXPTIME**?" and "**NP** = **coNP**?" by asking whether the complements of spectra and generalized spectra are always again spectra or generalized spectra, resp. The first of these questions about spectra has already been asked in 1956 by G. Asser (see [Ass 56]).

24.2.2.2.3. Provability

One might also expect that considering the restricted classes provable-**P** and provable-**NP** (of sets being provably in **P** or provably in **NP**, resp.) could bring some advantage in solving the **P**-**NP** problem. However, it turns out that provable-**P** = **P** and provable-**NP** = **NP** for all reasonable approaches to these provability restricted classes (see [Bak 79]).

24.2.2.2.4. A Measure for Nondeterminism

The interesting and obvious idea of studying directly the phenomenon of nondeterminism has been pursued in [KiFi 77].

A step in a computation of an NTM is called nondeterministic if in this step a strict choice between several next ID's is possible. Let us define $A \in$ **T-NTIME-NONDET**(t, g) if there exists an NTM accepting A in such a way that for any $x \in A$ there exists a successful path of length bounded by $t(|x|)$ with no more than $g(|x|)$ nondeterministic steps (guesses). For short we write

$$\mathbf{Q}_g = \mathbf{multiT\text{-}NTIME\text{-}NONDET}(\mathrm{id}, g),$$

$$\mathbf{NP}_g = \bigcup_{k \in \mathbb{N}} \mathbf{T\text{-}NTIME\text{-}NONDET}(n^k, g).$$

So for example, $\mathbf{Q}_0 = \mathbf{REALTIME}$, $\mathbf{Q}_n = \mathbf{Q}$, $\mathbf{NP}_{\mathrm{Pol}} = \mathbf{NP}$, $\mathbf{NP}_0 = \mathbf{NP}_{\log} = \mathbf{P}$ (because there is only a polynomial number of paths each of polynomial length so that a direct deterministic simulation in polynomial time is possible).

Many **NP**-complete sets are readily seen to be in $\mathbf{NP}_n$. On the other hand, V. R. Pratt's algorithm for PRIME (7.24) is in $\mathbf{NP}_{n^2}$ although PRIME is unlikely to be **NP**-complete (see Remark 2 after 24.24).

In [KiFi 77] a refinement of the inclusion **REALTIME** $\subset$ **Q** with respect to growing NONDET has been proved. For the sake of simplicity we quote only a special case of their result.

24.37. Theorem. [KiFi 77]

$$\mathbf{REALTIME} = \mathbf{Q}_0 \subset \ldots \subset \mathbf{Q}_{\log^{[k]} n} \subset \ldots \subset \mathbf{Q}_{\log\log n} \subset \mathbf{Q}_{\log n} \subseteqq \mathbf{Q}_n = \mathbf{Q}\,. \blacksquare$$

The last of these inclusions is not known to be strict.

Likewise it is unknown whether at least one of the inclusions in the chain

$$\mathbf{NP}_{\log n} \subseteqq \mathbf{NP}_n \subseteqq \mathbf{NP}_{n^2} \subseteqq \ldots \subseteqq \mathbf{NP}_{n^k} \subseteqq \ldots \subseteqq \mathbf{NP}_{\mathrm{Pol}} = \mathbf{NP}$$

is strict ($\mathbf{P} \neq \mathbf{NP}$ would follow) or is an equation. That this latter problem is very hard is indicated by 28.5. What is known about these open questions is summarized in

24.38. Theorem. [KiFi 77]
1. $\mathbf{Q}_{\log n} = \mathbf{Q}_n \rightarrow \mathbf{P} = \mathbf{NP}$.
2. $\mathbf{P} = \mathbf{NP} \leftrightarrow \mathbf{NP}_{\log n} = \mathbf{NP}_n$.

Proof. Statement 1 can be obtained by usual padding. The nontrivial implication "$\leftarrow$" of statement 2 is valid because $\mathbf{NP}_n$ contains **NP**-complete sets. $\blacksquare$

24.2.2.2.5. Relative Succinctness of Representations of Sets from P

For language classes in general very many different representations by machines are possible. For example, P can be represented by polynomially clocked deterministic TM's, i.e. by deterministic TM's running exactly $k + n^k$ steps for some $k \in \mathbb{N}$ (using a "clock"). It can also be represented by polynomially clocked nondeterministic TM's. If one goes from one representation R_1 to some other, perhaps more advantageous representation R_2, this could entail a considerable increase of the size of the machines (size is defined in 24.11). Let us say that the succinctness of representing languages belonging to some class F by machines of R_2 relative to R_1 is bounded by $f \in \mathbb{R}_1$ if

$$\bigwedge_{\substack{A \in R_1 \\ L(A) \in F}} \bigvee_{A' \in R_2} \big(\mathrm{L}(A) = \mathrm{L}(A') \wedge \mathrm{size}(A') \leqq f(\mathrm{size}(A))\big).$$

24.39. Theorem. [HaBa 79]
1. **If $\mathbf{P} \neq \mathbf{NP}$, then the succinctness of representing languages from P by polynomially clocked deterministic TM's relative to polynomially clocked nondeterministic TM's is not recursively bounded.**
2. **If $\mathbf{P} = \mathbf{NP}$, then the succinctness of representing languages from P by deterministic TM's with polynomial clocks relative to nondeterministic TM's with polynomial clocks is linearly bounded.**

***Proof.* The proof of statement 1 is by diagonalization whereas statement 2 is proved in the same way as 24.11. It is easy to see that the resulting machine can be attached with a suitable clock.** $\blacksquare$

Further succinctness results can be found in [HaBa 79] and [Hart 80].

24.2.2.2.6. An Arithmetical Representation of NP

24.40. Theorem. 1. [KeHo 80] $A \in \mathbf{NP}$ **iff there exist** $k, m \in \mathbb{N}$ **and polynomials** $p, p_1, \ldots, p_k, q_1, \ldots, q_m$ **such that**

$$w \in A \leftrightarrow \bigvee_{y_1 \leqq 2^{p_1(|w|)}} \cdots \bigvee_{y_k \leqq 2^{p_k(|w|)}} \bigwedge_{z_1 \leqq q_1(|w|)} \cdots \bigwedge_{z_m \leqq q_m(|w|)} p(\mathrm{dya}^{-1}(w), y_1, \ldots, y_k, z_1, \ldots, z_m) = 0.$$

2. [HoKe 83], [Juk 82] $A \in \mathbf{NP}$ iff there exist $n \in \mathbb{N}$ and polynomials p_1, p_2, p_3, p such that

$$w \in A \leftrightarrow \bigvee_{y_1 < 2^{p_1(|w|)}} \bigwedge_{z < p_2(|w|)} \bigvee_{y_2 < 2^{p_3(|w|)}} \cdots \bigvee_{y_n < 2^{p_3(|w|)}} p(\mathrm{dya}^{-1}(w), y_1, \ldots, y_n, z) = 0. \blacksquare$$

For related results see [AdMa 76]. For a similar characterization of RUD see 4.19.

24.3. The Polynomial Time Hierarchy

Theorem 14.6 shows that NP is not known to be closed under $\leqq_{\mathrm{T}}^{\mathrm{P}}$ nor under $\leqq_{\mathrm{T}}^{\mathrm{NP}}$. This suggests the investigation of the classes $\mathcal{R}_{\mathrm{T}}^{\mathrm{P}}(\mathbf{NP}) = \mathbf{P}^{\mathbf{NP}}$ and $\mathcal{R}_{\mathrm{T}}^{\mathrm{NP}}(\mathbf{NP}) = \mathbf{NP}^{\mathbf{NP}}$ (cf. definition of the reducibility closure in § 14.1.1). Iterating this step the following classes have been defined in [MeSt 72]:

$$\Delta_0^{\mathrm{P}} =_{\mathrm{df}} \Sigma_0^{\mathrm{P}} =_{\mathrm{df}} \Pi_0^{\mathrm{P}} =_{\mathrm{df}} \mathbf{P},$$
$$\Delta_{k+1}^{\mathrm{P}} =_{\mathrm{df}} \mathbf{P}^{\Sigma_k^{\mathrm{P}}} = \mathcal{R}_{\mathrm{T}}^{\mathrm{P}}(\Sigma_k^{\mathrm{P}}),$$
$$\Sigma_{k+1}^{\mathrm{P}} =_{\mathrm{df}} \mathbf{NP}^{\Sigma_k^{\mathrm{P}}} = \mathcal{R}_{\mathrm{T}}^{\mathrm{NP}}(\Sigma_k^{\mathrm{P}}),$$
$$\Pi_{k+1}^{\mathrm{P}} =_{\mathrm{df}} \mathbf{co}\Sigma_{k+1}^{\mathrm{P}}.$$

These sets form the *polynomial time hierarchy*. The following elementary properties of the classes of this hierarchy immediately follow from the definitions.

24.41. Theorem.

1. $\Delta_1^{\mathrm{P}} = \mathbf{P}$.
2. $\Sigma_1^{\mathrm{P}} = \mathbf{NP}$, $\Pi_1^{\mathrm{P}} = \mathbf{coNP}$.
3. $\Sigma_k^{\mathrm{P}} \cup \Pi_k^{\mathrm{P}} \subseteq \Delta_{k+1}^{\mathrm{P}} \subseteq \Sigma_{k+1}^{\mathrm{P}} \cap \Pi_{k+1}^{\mathrm{P}}$ for $k \geqq 0$.
4. $\Sigma_k^{\mathrm{P}}, \Pi_k^{\mathrm{P}}$ are closed under $\leqq_{\mathrm{m}}^{\mathrm{P}}$ for $k \geqq 0$.
5. Δ_k^{P} is closed under $\leqq_{\mathrm{T}}^{\mathrm{P}}$ for $k \geqq 0$. ∎

Note that $\bigcup_{k \geqq 0} \Sigma_k^{\mathrm{P}} = \mathbf{EXRUD}$ (see 11.20.2).

Since **PSPACE** is closed under $\leqq_{\mathrm{T}}^{\mathrm{NP}}$ (Theorem 14.6) we have

24.42. Theorem. $\bigcup_{k \geqq 0} \Sigma_k^{\mathrm{P}} \subseteq \mathbf{PSPACE}$. ∎

The relationships of statement 3 are illustrated in Figure 24.2. The main fault of this hierarchy is that none of the inclusions represented in the figure is known to be proper.

24.43. Theorem. [Sto 77] If $\Sigma_k^{\mathrm{P}} = \Pi_k^{\mathrm{P}}$ or $\Sigma_k^{\mathrm{P}} = \Sigma_{k+1}^{\mathrm{P}}$ for some $k \geqq 1$, then $\Pi_j^{\mathrm{P}} = \Sigma_j^{\mathrm{P}} = \Delta_{j+1}^{\mathrm{P}} = \Sigma_k^{\mathrm{P}}$ for all $j \geqq k$.

Remark. For the special case $k = 1$ this is a consequence of the fact that $\mathbf{NP} = \mathbf{coNP}$ implies that $\mathbf{NP}$ is closed under $\leqq_T^{NP}$ (see 14.6).

Proof. 1. Assume first $\Sigma_k^P = \Sigma_{k+1}^P$. This means that Σ_k^P is closed under $\leqq_T^{NP}$. Then $\Sigma_k^P = \Sigma_{k+1}^P = \dots$ and thus, by definition, $\Pi_k^P = \Pi_{k+1}^P = \dots$ But then by 24.41.3 we obtain $\Sigma_k^P = \Pi_k^P = \Delta_{k+1}^P = \Sigma_{k+1}^P = \Pi_{k+1}^P = \dots$

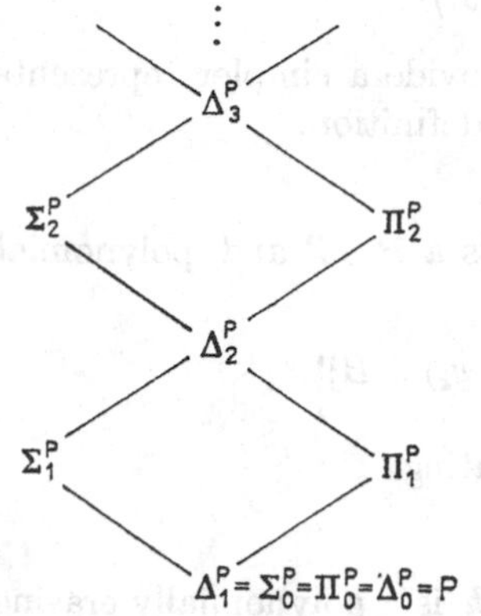

Figure 24.2. The polynomial time hierarchy

2. If $\Sigma_k^P = \Pi_k^P$, then $\Sigma_k^P = \Sigma_{k+1}^P$ can be shown as follows: Let $A \in \Sigma_{k+1}^P$. Then $A \leqq_T^{NP} B$ for some $B \in \Sigma_k^P$ and, likewise, $B \leqq_T^{NP} C$ for some $C \in \Sigma_{k-1}^P$. In general, from these two reducibility statements it does not follow that $A \leqq_T^{NP} C$ is true, because a composition of a machine M_A accepting A with oracle B and a machine M_B accepting B with oracle C requires also definitive answers of M_B to inputs $x \notin B$ which, however, cannot be ensured by M_B. The situation changes completely if also $\bar{B} \leqq_T^{NP} D$ for some $D \in \Sigma_{k-1}^P$ (which is true if $\Sigma_k^P = \mathbf{co}\Sigma_k^P = \Pi_k^P$.) In this case a nondeterministic oracle machine M can easily be constructed deciding (cf. definition before 23.13) B with oracle $E = \{2x: x \in C\} \cup \{2x + 1: x \in D\}$ which obviously also belongs to Σ_{k-1}^P. Now the composition of M_A and M shows $A \leqq_T^{NP} E$, i.e. $A \in \mathcal{R}_T^{NP}(\Sigma_{k-1}^P) = \Sigma_k^P$. ∎

The following statement is both: a corollary to 24.43 and to 11.20.

24.44. Corollary. $\mathbf{P} = \mathbf{NP} \to \mathbf{P} = \bigcup_{k \geqq 0} \Sigma_k^P$. ∎

It could be possible that the hierarchy is finite notwithstanding $\mathbf{P} \neq \mathbf{NP}$.

The question of whether the inclusion $\bigcup_{k \geqq 0} \Sigma_k^P \subseteq \mathbf{PSPACE}$ is strict is connected with the "properness" of the hierarchy.

24.45. Theorem. [Wra 77a]

$$\bigcup_{k \geqq 0} \Sigma_k^P = \mathbf{PSPACE} \to \bigvee_k (\Sigma_k^P = \Sigma_{k+1}^P = \dots = \mathbf{PSPACE}).$$

Proof. Let A be $\leqq_m^{log}$-complete in **PSPACE** (14.14). $\bigcup_k \Sigma_k^P = \text{PSPACE}$ implies $A \in \Sigma_{k_0}^P$ for some k_0. Therefore $\bigcup_k \Sigma_k^P = \mathbf{PSPACE} \subseteq \Sigma_{k_0}^P$ because every Σ_j^P is closed under $\leqq_m^{log}$-reducibility (24.41.4). ∎

There are various connections between the properties of the polynomial time hierarchy (e.g. its properness) and a high and a low hierarchy within **NP**. These hierarchies are defined by making use of a polynomially bounded version of the recursion theoretic jump operator [Schg 83a]. See also [BaBoSc 84a].

The next result shows that the polynomial time hierarchy is an analogue of the arithmetic hierarchy. In what follows we will need quantifiers $\bigvee^p$, $\bigwedge^p$ or, if unspecified, Q^p bounded by the polynomial p. They are defined as follows

$$\mathsf{Q}^p_y B(x, y) \leftrightarrow \mathsf{Q}_y\big(|y| \leqq p(|x|) \wedge B(x, y)\big).$$

Both statements of the following theorem provide a simpler representation of the sets of the hierarchy than is possible by their definition.

24.46. Theorem.

1. [StMe 73] For $k \geqq 1$, $A \in \Sigma_k^{\mathrm{P}} \leftrightarrow$ there exists a $B \in \mathbf{P}$ and polynomials $p_1, \ldots, p_k$ such that

$$A = \Big\{x\colon \bigvee_{y_1}^{p_1} \bigwedge_{y_2}^{p_2} \ldots \mathsf{Q}_{y_k}^{p_k} [(x, y_1, \ldots, y_k) \in B]\Big\}$$

   where the sequence of quantifiers is alternating.

2. [Wra 77a] For $k \geqq 1$,

$$A \in \Sigma_{k+1}^{\mathrm{P}} \leftrightarrow A \in \{h(B)\colon B \in \Pi_k^{\mathrm{P}} \wedge h \text{ is a polynomially erasing homomorphism}\}.$$

Proof. Ad 1. "$\rightarrow$" (Induction on k). For $k = 0$ the assertion is trivial. Assume that it is valid for $k - 1$. Let $A \in \Sigma_k^{\mathrm{P}} = \mathbf{NP}^{\Sigma_{k-1}^{\mathrm{P}}}$. Then there is some $B \in \Sigma_{k-1}^{\mathrm{P}}$ and an oracle NTM M accepting A with oracle B within polynomial time p. In what follows we say "u encodes the finite set $\{u_1, \ldots, u_n\}$" if $u = u_1 \# u_2 \# \ldots \# u_n \#$, where $\#$ is some new symbol. And we say "w in u" if $w = u_i$ for some $i \leqq n$ and $u = u_1 \# \ldots \# u_n \#$. Then we have

$$\begin{aligned} x \in A \leftrightarrow \bigvee_y^{p^2} \bigvee_u^{p^2} \bigvee_v^{p^2} & [u, v \text{ encode finite disjoint sets (of queries)} \\ & \wedge\ y \text{ describes an accepting computation of } M \text{ on } x \\ & \wedge \bigwedge_w^p (y \text{ requires a positive answer to a query } w \rightarrow w \text{ in } u) \\ & \wedge \bigwedge_w^p (y \text{ requires a negative answer to a query } w \rightarrow w \text{ in } v) \\ & \wedge\ u \in (B \#)^* \wedge v \in (\overline{B} \#)^*]. \end{aligned}$$

It is easy to see that $B \in \Sigma_{k-1}^{\mathrm{P}} \leftrightarrow (B \#)^* \in \Sigma_{k-1}^{\mathrm{P}}$ and $B \in \Pi_{k-1}^{\mathrm{P}} \leftrightarrow (B \#)^* \in \Pi_{k-1}^{\mathrm{P}}$. Consequently we can replace "$u \in (B \#)^*$ $\big(v \in (\overline{B} \#)^*\big)$ by a formula of the form

$$\bigvee_{x_1}^{p_1} \bigwedge_{x_2}^{p_2} \ldots \mathsf{Q}_{x_{k-1}}^{p_{k-1}} A(x_1, \ldots, x_{k-1}) \Big(\bigwedge_{x_1}^{p_1} \bigvee_{x_2}^{p_2} \ldots \mathsf{Q}_{x_{k-1}}^{p_{k-1}} A(x_1, \ldots, x_{k-1})\Big)$$

where $A \in \mathbf{P}$. The remaining predicates used in the inner of the above formula are obviously in $\mathbf{P}$. Thus, using standard methods we can easily convert this formula into a formula having the desired form.

"$\leftarrow$". (Induction on k). For $k = 0$ the assertion is trivial. Assume that it is valid for $k-1$. Consider

$$A = \left\{x: \bigvee_{y_1}^{p_1} \bigwedge_{y_2}^{p_2} \dots \mathsf{Q}_{y_k}^{p_k}[(x, y_1, \dots, y_k) \in B]\right\} \quad \text{and} \quad B \in \mathbf{P}.$$

Define

$$C = \left\{(x, y_1): \bigwedge_{y_2}^{p_2} \bigvee_{y_3}^{p_3} \dots \mathsf{Q}_{y_k}^{p_k}[(x, y_1, \dots, y_k) \in B]\right\}$$

Then $A = \left\{x: \bigvee_{y_1}^{p_1} (x, y_1) \in C\right\}$ and, by hypothesis, $C \in \Pi_{k-1}^{\mathrm{P}}$. An oracle NTM can accept A with oracle C in polynomial time as follows: For input x some y_1 with $|y_1| \leqq p_1(|x|)$ is guessed. Then, using the oracle C, it is checked whether $(x, y_1) \in C$. This shows

$$A \in \mathbf{NP}^{\Pi_{k-1}^{\mathrm{P}}} = \mathbf{NP}^{\Sigma_{k-1}^{\mathrm{P}}} = \Sigma_k^{\mathrm{P}}.$$

Ad 2. "$\rightarrow$". Let $A \in \Sigma_k^{\mathrm{P}}$. Then, by statement 1, A admits a representation $A = \left\{x: \bigvee_y^p (x, y) \in C\right\}$ where $C \in \Pi_{k-1}^{\mathrm{P}}$. Define $C' = \{x \# \hat{y}: (x, y) \in C\}$ where $\hat{y} = \hat{a}_1 \dots \hat{a}_n$ for $y = a_1 \dots a_n$. Evidently $C \in \Pi_{k-1}^{\mathrm{P}}$ too, and the homomorphism h defined by $h(a) = a$, $h(\hat{a}) = e$, $h(\#) = e$ is polynomially erasing on C' and maps C' onto A.

"$\leftarrow$". Let $A = h(B)$, $B \in \Pi_{k-1}^{\mathrm{P}}$ and h be polynomially erasing on B. This means $x \in A \leftrightarrow \bigvee_y^q \big(x = h(y) \wedge y \in B\big)$ where q is a polynomial satisfying $|y| \leqq q\big(|h(y)|\big)$ for all $y \in B$. By statement 1 this predicate is obviously in Π_{k-1}^{P}. ∎

A further characterization of the Σ_k^{P} by generalized spectra can be found in [Sto 77] (cf. R. Fagin's result described in § 24.2.2.2.2). A further algebraic characterization of $\bigcup_{k \geqq 0} \Sigma_k^{\mathrm{P}}$ is formulated as 11.20.2.

The polynomial time hierarchy provides a tool for a complexity classification of problems beyond **NP**. To this end the problems of 14.22 are helpful. Note that there are not too many natural problems known to be complete in some Σ_k^{P} for $k > 1$.

A polynomial-time hierarchy based not only on the existential and universal quantifier but also on the "counting quantifier" C defined by $\overset{k}{\underset{y}{\mathrm{C}}}\, H(x, y) \leftrightarrow \operatorname{card} \{y: H(x, y)\} \geqq k$ (and thus including the class **mP**) has been investigated in [Wag 84a]. The classes of this hierarchy allow the complexity classification of problems in which counting is involved. For related results see [Zac 83].

An interesting analogue of the analytical hierarchy has been studied in [Simj 76] which leads to a characterization of the classes $\mathrm{NTIME}\left(2^{2^{\cdot^{\cdot^{\cdot^{2}}}}}\Big\}_i^{\mathrm{Pol}}\right)$.

24.4. *Existence Problems, Search Problems and Counting Problems*

Theorem 24.46.1 yields as a special case that the sets $A \in \mathbf{NP}$ are characterized as $A = \left\{x: \bigvee_y^q (x, y) \in B\right\}$ where $B \in \mathbf{P}$ and q is a polynomial. In the terminology of § 7.1, A is the existence problem belonging to

$$\{(x, y): |y| \leqq q(|x|) \wedge (x, y) \in B\}.$$

We call A the existence problem for (B, q). The corresponding search problem is called the search problem for (B, q).

Theorem 24.46.1 admits the interpretation that the difference between **P** and **NP**, if any, is reflected as the difference between verifying a proposed solution and checking the existence of a solution. So far, the only known method for the latter is exhaustive search which evidently requires exponential time. If $\mathbf{P} = \mathbf{NP}$, this inefficient method could be avoided.

Note that a "good" algorithm for solving the existence problem does not necessarily provide us with a "good" algorithm for solving the corresponding search problem.

But a "good" solution of the search problem for (B, q) yields a "good" solution of the corresponding existence problem A for (B, q). Obviously, if there is a search function f solving the search problem for (B, q) such that $f \in \mathrm{DTIME}(\mathrm{Pol}\, t)$, then $A \in \mathbf{DTIME}(\mathrm{Pol}\, t)$. Thus, the search problems are at least as hard as the corresponding existence problems.

24.47. Theorem. [Schn 79] Let A be the existence problem for (B, q). If A is self-reducible, then the complexities of the existence problem A and the search problem for (B, q) are polynomially related. ∎

Similar results can be found in [BoDe 76].

A further class of problems which are of highly practical importance are the counting problems (see p. 100). Let A be the existence problem for some (B, q). The counting function $n_{B,q}$ for (B, q) is defined by

$$n_{B,q}(x) =_{\mathrm{df}} \mathrm{card}\,\{y : |y| \leqq q(|x|) \wedge (x, y) \in B\}.$$

The obvious equivalence $x \in A \leftrightarrow n_{B,q}(x) \neq 0$ shows that the counting problem for (B, q) is at least as difficult as the existence problem for (B, q). Now let $\#\mathrm{P} =_{\mathrm{df}} \{n_{B,q} : B \in \mathbf{P} \wedge q \text{ is a polynomial}\}$ (in [Val 77a] #P is defined as a class of sets acceptable by polynomial time "counting Turing machines"). The inclusions $\mathrm{P} \subseteq \#\mathrm{P} \subseteq \mathrm{PSPACE}$ are evident.

Defining the notion of #P-completeness for a suitable notion of polynomial time reducibility for functions it turns out that $n_{B,q}$ is #P-complete for many (B, q) with **NP**-complete existence problem.

That many **NP**-complete problems have equally hard counting problems follows from the fact that they admit *parsimonious* reductions to each other (i.e. such polynomial time computable reductions preserving the number of solutions). There are even (B, q) whose existence problems are in **P** having a #P-complete counting problem, for example perfect matchings in bipartite graphs. For these results see [Val 77a] and [Val 77b]. Since there is a great interest in GRAPHISOM (see § 24.2.1.2.3) we mention the following result. Let graphisom $= \{(G, G', f) : G, G'$ are finite graphs and f is an isomorphism from G onto $G'\}$. Then GRAPHISOM and $n_{\mathrm{graphisom}}$ are polynomially related (cf. [Math 79]).

Another possibility of studying the difficulty of counting is to consider the sets $m(B, q) = \{(x, y) : n_{B,q}(x) \geqq y\}$. We define

$$\mathbf{mP} = \{m(B, q) : B \in \mathbf{P} \wedge q \text{ is a polynomial}\}$$

which is defined in [Simj 77] as the polynomial time complexity class of "threshold machines". Such a machine is a nondeterministic TM which accepts an input (x, k) in time $t(|x|)$ if and only if there exist at least k accepting computations of at most $t(|x|)$ steps. Since **mP** = **RTIME**(Pol) (see [Simj 77]) we obtain from 20.43 and the evident inclusion **RTIME**(Pol) $\subseteq$ **PSPACE**

24.48. Theorem. **NP** $\cup$ **coNP** $\subseteq$ **mP** $\subseteq$ **PSPACE**. ∎

This theorem gives some evidence that counting problems can be more difficult than the corresponding existence problems.

There have been made several attempts to compare the class mP and the related class #P with the polynomial-time hierarchy. Up to now it is not known whether e.g. mP $\subseteq \bigcup_{k\geq 0} \Sigma_k^{\mathrm{P}}$ or $\Delta_2^{\mathrm{P}} \subseteq$ mP. However, the following results have been shown:

1. [PaZa 82] The Δ_2^{P} languages which are accepted by deterministic polynomial-time-oracle machines with only a logarithmically bounded number of queries to the oracle belong to a certain extension of mP.
2. [Sip 83] The languages from mP = RTIME(Pol) which are accepted by probabilistic polynomial-time machines with bounded error probability $\varepsilon < \frac{1}{2}$ (for suitable $\varepsilon \geq 0$), usually called BPP, belong to $\Sigma_2^{\mathrm{P}} \cap \Pi_2^{\mathrm{P}}$.
3. [Sto 83] The #P functions have Δ_3^{P}-approximations, i.e. for every $f \in$ #P and every $d \geq 1$ there exists a $c \geq 1$ and an f' which can be computed by a deterministic polynomial time Turing machine with an oracle from Σ_2^{P} and for which $|f(w) - f'(w)| \leq \frac{c}{|w|^d} f(w)$ holds true.

For related results see [Zac 83].

24.5. Independence Results

The immense difficulty of the **P**-**NP** problem has led to the suspicion that in the same way as there exist different geometries (with and without the axiom of parallelity), or set theories (with and without the generalized continuum hypothesis), there might exist both a computing theory with **P** = **NP** and a computing theory with **P** $\neq$ **NP**. The first result in this direction is

24.49. Theorem. [HaHo 76] There exists an oracle A such that $\mathbf{P}^A = \mathbf{NP}^A$ is neither provable nor disprovable in set theory.

Proof. Let $\boldsymbol{F}$ be an axiomatizable formal system for set theory. We use the shorthand $(\mathbf{N})\mathbf{P}^{\varphi_i} =_{\mathrm{df}} (\mathbf{N})\mathbf{P}^A$ if $c_A = \varphi_i$. Let B, C be recursive sets such that $\mathbf{P}^B = \mathbf{NP}^B$ and

$\mathbf{P}^C \neq \mathbf{NP}^C$ (28.4) and define

$$f(x, j) =_{\mathrm{df}} \begin{cases} 1 & \text{if } x \in C \text{ and among the first } x \text{ proofs in } F \\ & \qquad \text{there is one for } \mathbf{P}^{\varphi_j} = \mathbf{NP}^{\varphi_j}, \\ & \text{or } x \in B \text{ and among the first } x \text{ proofs in } F \\ & \qquad \text{there is one for } \mathbf{P}^{\varphi_j} \neq \mathbf{NP}^{\varphi_j}, \\ 0 & \text{otherwise.} \end{cases}$$

The function f is obviously recursive. An application of the s-m-n-Theorem yields an $s \in \mathbb{R}_1$ such that $\varphi_{s(j)}(x) = f(x, j)$, and the recursion theorem guarantees the existence of an i_0 such that $\varphi_{i_0}(x) = f(x, i_0)$. The set defined by φ_{i_0} will be denoted by A, i.e. $\varphi_{i_0} = c_A$.

If $\mathbf{P}^A = \mathbf{NP}^A$ (i.e. $\mathbf{P}^{\varphi_{i_0}} = \mathbf{NP}^{\varphi_{i_0}}$) is provable in F, let x_0 be the number of the shortest proof. Then for all $x \geqq x_0$ we obtain $\varphi_{i_0}(x) = 1 \leftrightarrow x \in C$, i.e. A and C differ at most by a finite set. But then $\mathbf{P}^A = \mathbf{NP}^A$ contradicts $\mathbf{P}^C \neq \mathbf{NP}^C$. In the same way one shows that $\mathbf{P}^A \neq \mathbf{NP}^A$ cannot be provable in F. ∎

How to interpret this result? First of all it does not say anything about a possible independence of $\mathbf{P} = \mathbf{NP}$ from some powerful theory like set theory. Second, the result is perhaps not too surprising because the diagonalization argument involved in the construction of the oracle makes use of provability, and this could of course account for the unprovability of the statement $\mathbf{P}^A = \mathbf{NP}^A$. See for instance the approach in [Háj 77] where just the very difference between provable and unprovable statements is used to prove also **24.49**. (See also [Gra 80].)

Since it is hard to obtain the independence of $\mathbf{P} = \mathbf{NP}$ or other interesting questions of set theory, one naturally restricts attention to less powerful theories such as Peano Arithmetic. Of course the theories used for independence results should be relevant for computer science. However, what theories can be acknowledged to serve as an adequate logical basis for computer science? In [Lip 78] it has been suggested that even a suitable fragment of Peano Arithmetic is sufficient for reasoning in computer science. But in this fragment finite sets become undecidable, whereas undecidable sets can become decidable within linear time (see [JoYo 80]) which shows that this theory is not powerful enough for computer science. That even Peano Arithmetic may not be sufficient is suggested by a result in [Odo 79]: the question of termination of relatively simple programs in strongly typed programming languages is independent of (even an extension of) Peano Arithmetic.

In [JoYo 81] it has been pointed out that the independence of $\mathbf{P} = \mathbf{NP}$ of weak fragments of arithmetic need not be connected with the **P-NP** problem at all, because in some of these theories the only computable functions are those of **P**, or because it is not clear whether in these theories the equivalence of all standard machine models can be proved and if so, whether the polynomial time computations are model invariant. For a further discussion of this topic see [JoYo 81] and [Reg 83].

Notwithstanding, the independence of $\mathbf{P} = \mathbf{NP}$ of fragments of PA or other theories has been investigated (see [Lip 79], [MiLi 79], [Saz 79], [MiLi 80], [Saz 80], [Kano 84] and [Kow 84]).

The results of these papers could be precursors for stronger results.

Chapter VIII
Space and Time

One of the major open problems in complexity theory is to determine the exact relationship between space and time complexity. What is known when space and time are understood as machine-oriented measures is presented in § 25.

We also include the space-time trade-offs in this section because they shed some light on the relationships between space and time, although, strictly speaking, the trade-offs should be considered as statements about upper and lower bounds on the space-time double measure.

In particular we have included results concerning the space-time trade-offs for pebble games because many space-time trade-offs for TM's and RAM's can be derived from the space-time trade-offs for pebble games on graphs from suitably chosen classes.

If the space-time problem is considered for the general machine-independent measures (§ 26), then it resembles to some extent the problems dealt with in § 23 and § 24. The methods and the kind of results are strongly parallel to those of these sections.

§ 25. The Space-Time Problem for Machine-Oriented Measures

The relations between space and time have two different aspects depending on whether these measures are understood as specific, machine-oriented or as general measures which are model invariant within the class of realistic types of machines. This section is devoted to the first aspect whereas the next section deals with the second one.

25.1. Results for Realistic Machine Types

The following results for machine-oriented measures cannot be maintained for the general time and space measures because the XTIME-classes are always of the form **XTIME**(Pol t), which cannot sense logarithmic factors, square roots or similar changes of t.

Our first result concerns multiT-DM's.

25.1. Theorem. [HoPaVa 75] For $t \geqq \mathrm{id}$, **multiT-DTIME**(t) $\subseteq$ **DSPACE** $\left(\frac{t}{\log t}\right)$.

Proof. Let M be a kT-DM working in time t and assume first $t(|w|) \in \mathrm{DSPACE}\left(\frac{t}{\log t}\right)$. Because of 19.30 we can assume w.l.o.g. that M is block respecting, and M will use block length $t(n)^{2/3}$. Hence the whole computation time of M is divided into $t(n)^{1/3}$ intervals of length $t(n)^{2/3}$ each, and there are $t(n)^{1/3}$ steps where crossings over block boundaries are possible.

To prove the theorem we construct an mT-DM M' which generates in a lexicographical way pairs (σ, η) of sequences, where σ and η consist of $t(n)^{1/3}$ states of M and of $t(n)^{1/3}$ k-tuples of head positions of M, respectively. The lengths of σ and η are bounded by $c_1 \cdot t(n)^{1/3} \log t(n)$ for some constant c_1. For every generated pair (σ, η) the machine M' behaves as follows:

The states and head positions in σ and η are interpreted as the states and head positions of M at the ends of the intervals, and, using this information, M' tries to simulate the work of M interval by interval. Whenever an interval is simulated the machine compares the reached states and head positions with the guessed ones in σ and η. If they do not coincide, M' generates the next pair (σ, η) and tries again to simulate M. Otherwise the simulation of the next interval follows. The input is accepted by M' if and only if, after the simulation of the final interval, the accepting state of M appears.

Knowing state and head positions at the beginning of an interval, M' is able to repeat exactly the work of M in this interval if, in addition, it knows the inscriptions of the actual blocks at the beginning of this interval, which are the results of at most k earlier intervals. The main idea of the proof is not to store all possible block inscriptions but to recompute those which are needed. Thus storage can be saved, at the cost of time.

Now we give the details of the proof. From the sequence η of k-tuples of head positions we construct a graph G with nodes $0, 1, 2, \ldots, t(n)^{1/3}$ (node i corresponds to the ith interval of the computation of M). For every $i < t(n)^{1/3}$ the graph G has an edge leading from i to $i + 1$. Furthermore, for every head there is an edge from $j > 0$ to i in G if and only if that block which is scanned during interval i has been scanned by this head during interval j $(< i)$ but not during some interval between j and i. Finally, there is an edge from 0 to i if and only if at least one of the heads scans during interval i an interval which has not been scanned so far by this head. Thus, every node of G has indegree at most $k + 1$ (cf. figures 25.1 and 25.2).

Note that G can easily be constructed from η and that space $t(n)^{1/3} \log t(n)$ suffices to write down an encoding of G.

The graph is constructed in such a way that the antecedents of i are exactly those time intervals from which information for interval i is needed. This corresponds to the rule of the pebble game "a node can be pebbled if all of its antecedents are pebbled" (cf. § 25.3.2.1.1).

Thus, to compute interval $t(n)^{1/3}$ (the final aim of the simulation) we need to store at most $c \cdot \frac{t(n)^{1/3}}{\log t(n)}$ block inscriptions (each of length $t(n)^{2/3}$) because $c \cdot \frac{t(n)^{1/3}}{\log t(n)}$ pebbles are sufficient to pebble the vertex $t(n)^{1/3}$ of G (25.7). Altogether, the storage needed for the simulation is bounded by $c \cdot \frac{t(n)}{\log t(n)}$.

It remains to describe how the pebble game controls the simulation of M by M' and to show that this can be done in the indicated space. First of all it is necessary to compute in a deterministic way a successful sequence of pebble configurations. It is easy to describe an NTM which pebbles the vertex $t(n)^{1/3}$ if G is given as input. It works as follows: In a nondeterministic way it makes moves according to the rules of the game, taking care that the number of pebbles is always bounded by $c \cdot \frac{t^{1/3}(n)}{\log t(n)}$. This NTM does not need more space than the length of the input graph, i.e. it works within space $t(n)^{1/3} \log t(n)$.

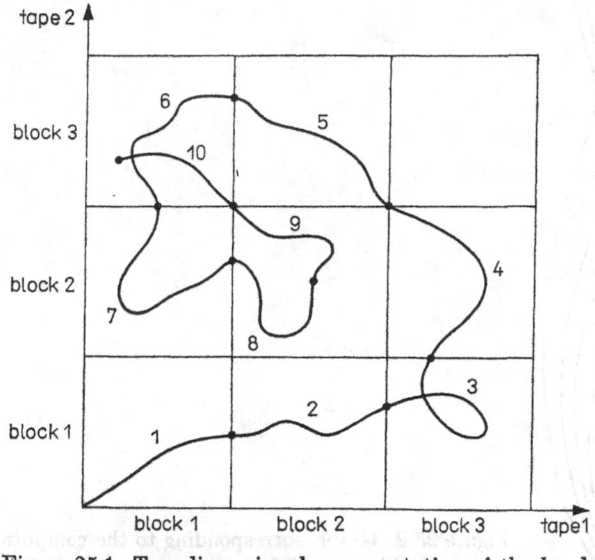

Figure 25.1. Two-dimensional representation of the head motions of a 2T-DM. The numbers indicate the time intervals

The machine M' works on the graph G as a DTM which simulates this NTM within space $t(n)^{2/3} \cdot \log^2 t(n) \leq \frac{t(n)}{\log t(n)}$. This is possible by theorem 23.1.2.

Whenever a new pebble configuration is computed, the simulation of the work of M is continued as follows: When node i has been pebbled, M' works like M in interval i. At the end of this interval the contents of all actual blocks are stored, and it is checked whether the ith components in σ and η had been guessed correctly. When a pebble is removed from node i, the block contents belonging to interval i are erased from the tape.

So far the proof needs the constructibility assumption of t. However, by the usual reasoning we get rid of this additional assumption: We do not start using the block

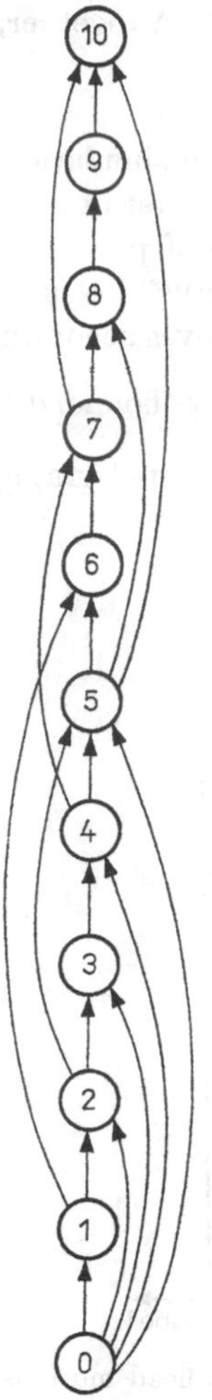

Figure 25.2. Graph corresponding to the computation represented in Fig. 25.1

respecting machine M with block length $t(n)^{2/3}$ but with length $\lambda = 1$. Now M' tries to simulate M (assuming M to work $\sqrt{\lambda}$ intervals per λ steps). If this is impossible, λ is increased by 1 and the simulation of M is tried again etc. ∎

Remark 1. The proof does not work for NTM's because one cannot be sure that the recomputations always yield the same results.

Remark 2. In [AdLo 80] there is another proof of 25.1 using overlap arguments (cf. § 8.1.4).

Remark 3. Theorem 25.1 can also be obtained as a consequence of Theorems 20.42.2 and 20.37.3.

The proof method of 25.1 can be applied to several other machine models:

25.2. Theorem.

1. [PaRe 79] For all $t \geqq \mathrm{id}$,

$$\mathbf{RAM\text{-}TIME}^{\log}(t) \subseteq \mathbf{DSPACE}\left(\frac{t}{\log t}\right).$$

2. [Pip 82] For all $t \geqq \mathrm{id}$,

$$\mathbf{multiT^{multi}DTIME}(t) \subseteq \mathbf{DSPACE}\left(\frac{t}{\log t}\right). \blacksquare$$

For one-tape TM's better results are known, which are furthermore valid also for the nondeterministic model. A weaker version (NSPACE instead of DSPACE) of the following result has been already proved in [HoUl 68a].

25.3. Theorem. [Pat 72]

1. For $t(n) \geqq n^2$ such that $t(|w|) \in \mathrm{DSPACE}(\sqrt{t})$,

$$\mathbf{T\text{-}NTIME}(t) \subseteq \mathbf{T\text{-}DSPACE}(\sqrt{t}).$$

2. For $t \geqq \mathrm{id}$ such that $t(|w|) \in \mathrm{DSPACE}(\sqrt{t(n) \log n})$,

$$\mathbf{2\text{-}T\text{-}NTIME}(t) \subseteq \mathbf{2\text{-}T\text{-}DSPACE}(\sqrt{t(n) \log n}). \blacksquare$$

By the same technique we can prove

25.4. Theorem.

1. For $t \geqq n^2$,

$$\mathbf{T\text{-}DTIME}(t) \subseteq \mathbf{T\text{-}DSPACE}(\sqrt{t}).$$

2. For $t \geqq \mathrm{id}$,

$$\mathbf{2\text{-}T\text{-}DTIME}(t) \subseteq \mathbf{2\text{-}T\text{-}DSPACE}(\sqrt{t(n) \log n}). \blacksquare$$

Remark. In [KuMa 83] it is shown that even

$$\mathbf{T\text{-}DTIME}(t) \subseteq \mathbf{T\text{-}DSPACE\text{-}TIME}(\sqrt{t}, t^2),$$

$$\mathbf{T\text{-}NTIME}(t) \subseteq \mathbf{T\text{-}NSPACE\text{-}TIME}(\sqrt{t}, t^{3/2}), \quad \text{for all} \quad t(n) \geqq n^2,$$

and

$$\mathbf{2\text{-}T\text{-}DSPACE\text{-}TIME}(\sqrt{t(n)} \cdot \log n, t^2), \quad \text{for all} \quad t(n) \geqq n.$$

For related results see [IbMo 83].

25.2. Results for Other Machine Types

Although the exact relations between time and space for realistic machines are not known, and also for the general measures nothing is known (cf. § 26), complete and satisfying results can be presented for machines which are not realistic or not known to be realistic.

The problem can be formulated this way: What has to be changed in realistic machines in order to influence the relation of time and space in such a way that the trivial inclusions **TIME**(Pol t) $\subseteq$ **SPACE**(Pol t) and **SPACE**(s) $\subseteq$ **TIME**($2^{\mathrm{Lin}(s)}$) can be improved to equalities?

Let us first consider the inclusion **TIME**(Pol t) $\subseteq$ **SPACE**(Pol t). A first possibility is to try to increase the value of time by allowing much stronger instructions like multiplication in RAM's or by unrestricted use of parallelity. Taking into account 5.2, the theorems 20.12, 20.35, 20.38 provide examples of this kind: For X $\in$ {D, N} and $t \geqq \mathrm{id}$ with suitable constructibility properties,

$$\mathbf{MRAM\text{-}XTIME}(\mathrm{Pol}\, t) = \mathbf{MRAM\text{-}DSPACE}(\mathrm{Pol}\, t) = \mathbf{DSPACE}(\mathrm{Pol}\, t),$$

$$\mathbf{CRAM\text{-}XTIME}(\mathrm{Pol}\, t) = \mathbf{CRAM\text{-}DSPACE}(\mathrm{Pol}\, t) = \mathbf{DSPACE}(\mathrm{Pol}\, t),$$

$$\mathbf{XPARTIME}(\mathrm{Pol}\, t) = \mathbf{DSPACE}(\mathrm{Pol}\, t),$$

$$\mathbf{ATIME}(\mathrm{Pol}\, t) = \mathbf{DSPACE}(\mathrm{Pol}\, t).$$

A second possibility is to try to weaken the value of space of multiTM's by using the tapes only as counters. Note that this weakens also the value of time. In this way we can interprete 20.1: For $i = 1, 2$ and $k \in \{3, 4, \ldots, \mathrm{multi}\}$, X $\in$ {D, N} and $t \geqq \mathrm{id}$,

$$\mathbf{i\text{-}kC\text{-}XTIME}(\mathrm{Pol}\, t) = \mathbf{i\text{-}kC\text{-}XSPACE}(\mathrm{Pol}\, t).$$

As to the second inclusion, again both possibilities can be illustrated by examples.

Increasing the value of space can be attempted by using an auxiliary PD-tape or by letting work the machine in a recursive or parallel regime (cf. 20.13, 20.32, 20.33 and 20.36): For X $\in$ {D, N} and $s \geqq \log$ with suitable constructibility properties,

$$\mathbf{2\text{-}auxPD\text{-}T\text{-}XSPACE}(s) = \mathbf{2\text{-}PD\text{-}T\text{-}DTIME}(2^{\mathrm{Lin}\, s}) = \mathbf{DTIME}(2^{\mathrm{Lin}\, s}),$$

$$\mathbf{XRECSPACE}(s) = \mathbf{XRECTIME}(2^{\mathrm{Lin}\, s}) = \mathbf{DTIME}(2^{\mathrm{Lin}\, s}),$$

$$\mathbf{ASPACE}(s) = \mathbf{DTIME}(2^{\mathrm{Lin}\, s}).$$

A real increase of the value of space is achieved by adding an auxiliary S-tape (20.25). For X $\in$ {D, N} and $s \geqq \log$ with suitable constructibility properties,

$$\mathbf{XSPACE}(s) \subset \mathbf{2\text{-}auxS\text{-}T\text{-}XSPACE}(s) = \mathbf{2\text{-}S\text{-}T\text{-}DTIME}(2^{2^{\mathrm{Lin}\, s}}) = \mathbf{DTIME}(2^{2^{\mathrm{Lin}\, s}}).$$

The intuitive feeling that the weaker instructions of an SRAM′ lead to a diminished value of time is reflected by 20.10 (use also 5.2): For X $\in$ {D, N} and $s \geqq \mathrm{id}$,

$$\mathbf{SRAM'\text{-}SPACE}(\mathrm{Lin}\, s) = \mathbf{SRAM'\text{-}TIME}(2^{\mathrm{Lin}\, s}).$$

25.3. *Space-Time Trade-Offs*

In this subsection we deal with the space-time double complexity of specific problems. Consequently, lower (upper) bounds for a given problem are elements of $\mathbb{R}_1 \times \mathbb{R}_1$, and, as a rule, we have the effect of incomparability of different lower (upper) bounds of this problem.

If the simultaneous lower space and time bounds of a problem A are given in the form

$$A \in \tau\text{-}\mathbf{DSPACE\text{-}TIME}(s, t) \to f\big(s(n), t(n), n\big) \geqq_{\mathrm{ae}} r(n)$$

where f is a function which is strictly increasing in the first two arguments and $r \in \mathbb{R}_1$, then this can also be interpreted as describing a space-time trade-off:

If $A \in \tau\text{-}\mathbf{DSPACE\text{-}TIME}(s_1, t_1)$ and the space (time) bound $s_1(t_1)$ is replaced by the smaller bound $s_2(t_3)$ such that $f\big(s_2(n), t_1(n), n\big) <_{\mathrm{io}} r(n)$ $\Big(f\big(s_1(n), t_3(n), n\big) <_{\mathrm{io}} r(n)\Big)$, then $t_1(s_1)$ has to be increased at least up to a t_2 (s_3) satisfying $f\big(s_2(n), t_2(n), n\big) \geqq_{\mathrm{ae}} r(n)$ $\Big(f\big(s_3(n), t_3(n), n\big) \geqq_{\mathrm{ae}} r(n)\Big)$ (see Figure 25.3).

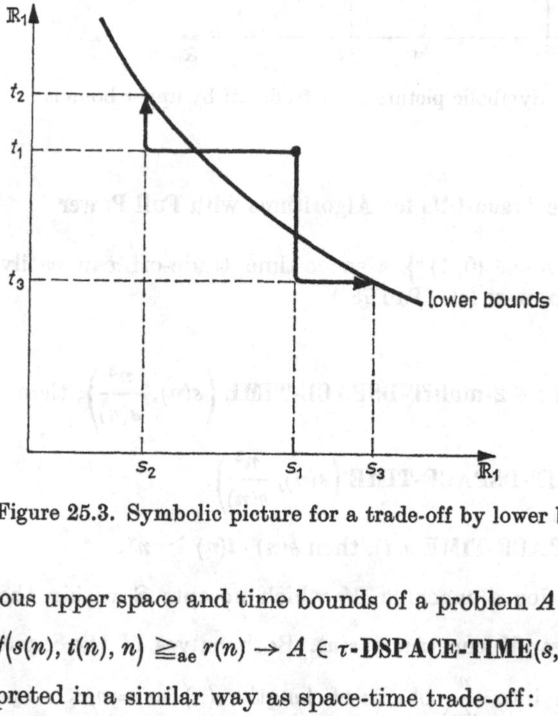

Figure 25.3. Symbolic picture for a trade-off by lower bounds

Simultaneous upper space and time bounds of a problem A given in the form

$$f\big(s(n), t(n), n\big) \geqq_{\mathrm{ae}} r(n) \to A \in \tau\text{-}\mathbf{DSPACE\text{-}TIME}(s, t)$$

can be interpreted in a similar way as space-time trade-off:

If $f\big(s_1(n), t_1(n), n\big) \geqq_{\mathrm{ae}} r(n)$ and more space (time) is available, say $s_2(t_3)$, then A can be solved within every smaller time (space) bound $t_2(s_3)$ satisfying $f\big(s_2(n), t_2(n), n\big) \leqq_{\mathrm{ae}} r(n)$ $\Big(f\big(s_3(n), t_3(n), n\big) \leqq_{\mathrm{ae}} r(n)\Big)$ (see Figure 25.4).

The trade-off is the more useful the better the corresponding bounds are, i.e. the closer the lower bounds come to the upper bounds of A.

Typical examples are presented by the next theorems.

Although this topic belongs to § 8 about lower bounds we treat trade-offs here, because the very fact of a certain exchangeability of time and space resources in solving specific problems seems to be one further important aspect of the relationship between time and space.

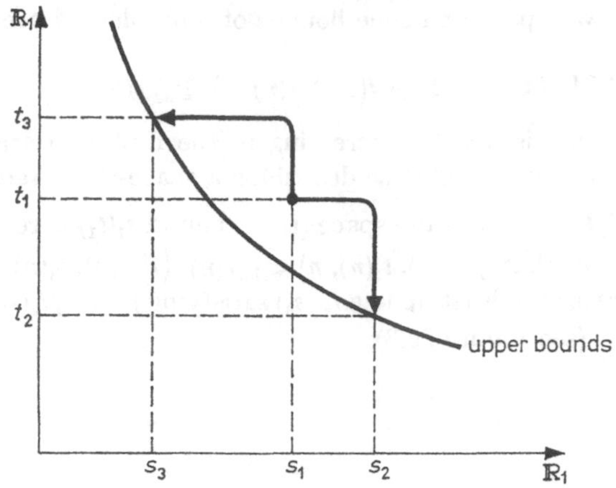

Figure 25.4. Symbolic picture for a trade-off by upper bounds

25.3.1. Space-Time Trade-Offs for Algorithms with Full Power

For $S = \{w : w = w^{-1} \wedge w \in \{0, 1\}^*\}$ a space-time trade-off can easily be demonstrated following A. Cobham (cf. [Pau 78]).

25.5. Theorem.

1. If $\log \leqq s \leqq \text{id}$ and $s \in$ **2-multiT-DSPACE-TIME** $\left(s(n), \dfrac{n^2}{s(n)}\right)$, then

$$S \in \textbf{2-multiT-DSPACE-TIME}\left(s(n), \frac{n^2}{s(n)}\right).$$

2. If $S \in$ **2-multiT-DSPACE-TIME**(s, t), then $s(n) \cdot t(n) \geq n^2$.

Proof. Ad 1. We describe a machine M which accepts S within the complexity bounds $s(n), \dfrac{n^2}{s(n)}$. First M constructs $s(n)$. Both halves of the input word w are thought to be divided into $\dfrac{n}{s(n)}$ blocks of length $s(n)$: $w = v_1 v_2 \ldots u_2 u_1$. Then, for $i = 1, \ldots, \dfrac{n}{s(n)}$, M copies the block v_i from the first half of the input word and the word u_i from the second half of the input word onto different tapes and checks symmetry. The blocks are found with the help of a binary counter of length $\log \dfrac{n}{s(n)} \leqq s(n)$. Evidently M works within space $s(n)$ and within time $\dfrac{n^2}{s(n)}$.

Ad 2. When the proof of 8.13 is repeated, with the notion of crossing sequence from § 8.1.2.4, a lower bound $c \cdot \dfrac{n^2}{s(n)}$ for the time complexity is obtained. ∎

As 25.5 can be proved in the same way for

$$\hat{S} =_{\mathrm{df}} \{w2w^{-1}: w \in \{0, 1\}^*\} \cup \{w22w^{-1}: w \in \{0, 1\}^*\}$$

and since $\hat{S} \in$ **DCF**, the second statement of 25.5 is valid also for **DCF**:

$$\mathbf{DCF} \subseteqq \mathbf{2\text{-}multiT\text{-}DSPACE\text{-}TIME}(s, t) \to s(n)\, t(n) \geq n^2.$$

That the first statement of 25.5 can be maintained also for **DCF** is shown in [BrVe 80a], [BrVe 80b], and [Meh 80]. These three papers improve the inclusion **DCF** $\subseteqq$ **2-multiT-DSPACE-TIME**$(\log^2, \mathrm{Pol})$ which is stated in Remark 1 after 13.15.

25.6. Theorem. [BrVe 80], [BrCoMeVe 83] **DCF** $\subseteqq$ **2-multiT-DSPACE-TIME**$\left(s(n), \dfrac{n^2}{s(n)}\right)$, provided $\log^2 \leqq s \leqq \mathrm{id}$ and $s \in$ **2-multiT-DSPACE-TIME**$\left(s(n), \dfrac{n^2}{s(n)}\right)$. ∎

For related results see [Ver 81].

In [BoCo 80] a space-time trade-off of the form $s(n) \cdot t(n) \geq \dfrac{n^2}{\log n}$, $\log \leqq s \leqq \mathrm{id}$, for sorting n numbers from $\{1, \ldots, n^2\}$ is proved. This is valid for any realistic space measure of a machine type which avoids parallel work.

25.3.2. Space-Time Trade-Offs for Algorithms of Restricted Power

For restricted types of algorithms, space-time trade-offs can usually be found more easily. This concerns in particular combinatorial problems where the algorithms under consideration are nothing else than sequences of appropriate, problem-oriented, primitive operations of a few different types. A typical example is the "pebble game". Speaking about such problems deviates from our principle to exclusively present material concerning algorithms with full computational power. This is, however, justified by the important applications of pebble game results to the space-time problem for machines with full power.

25.3.2.1. The Pebble Game

25.3.2.1.1. Definition and First Results

The *pebble game* introduced in [PaHe 70] is played on directed acyclic graphs (dag's). The rules of the game are:

(R1) A vertex v can be pebbled with a free pebble whenever all predecessors of v are pebbled. (In particular, every node with indegree 0 can always be pebbled.)

(R2) A pebble can always be removed.

A move of the game is a single application of one of these rules. The aim of the game is to pebble any given node of the graph. The problem is to do this with a minimal number of pebbles or within a minimal number of moves.

Our first results concern the number of pebbles necessary and (or) sufficient to pebble certain graphs. We fix $d \geqq 2$ and restrict ourselves to graphs of indegree bounded by d.

For a graph G we denote by $V(G)$ and $E(G)$ the set of vertices and the set of edges of G, resp. Furthermore,

$$\mathcal{G}_d =_{\text{df}} \{G: G \text{ is a dag} \wedge \text{indeg } G \leqq d\},$$

$$\mathrm{S}(G) =_{\text{df}} \min \{k: \text{each node of } G \text{ can be pebbled with } k \text{ pebbles}\},$$

$$\mathrm{T}(G) =_{\text{df}} \min \{k: \text{each node of } G \text{ can be pebbled within } k \text{ moves}\}.$$

We use S and T as the space and the time measures, resp., for pebbling families of graphs and define the following complexity classes:

$$\mathbf{S}(s) =_{\text{df}} \left\{\mathcal{G}: \mathcal{G} \subseteq \mathcal{G}_d \wedge \bigwedge_{G \in \mathcal{G}} \big(G \text{ has } n \text{ nodes} \to \mathrm{S}(G) \leqq s(n)\big)\right\},$$

$$\mathbf{T}(t) =_{\text{df}} \left\{\mathcal{G}: \mathcal{G} \subseteq \mathcal{G}_d \wedge \bigwedge_{G \in \mathcal{G}} \big(G \text{ has } n \text{ nodes} \to \mathrm{T}(G) \leqq t(n)\big)\right\},$$

$$\mathbf{ST}(s, t) =_{\text{df}} \{\mathcal{G}: \mathcal{G} \subseteq \mathcal{G}_d \wedge \text{each node of every } G \in \mathcal{G} \text{ with } n \text{ nodes can be pebbled within } t(n) \text{ moves using no more than } s(n) \text{ pebbles}\}.$$

Sometimes we also make use of the Comp-notation introduced in § 5.

25.7. Theorem. [HoPaVa 75] $\mathcal{G}_d \in \mathbf{S}\left(\mathrm{Lin}\, \dfrac{n}{\log n}\right)$.

Proof. Let $S_d(n) =_{\text{df}} \max \{S(G): G \in \mathcal{G}_d \wedge \text{card } V(G) = n\}$. Then we have to show $S_d(n) \leqq c \dfrac{n}{\log n}$ for some $c > 0$.

We define $R_d(n) = \min \{\text{card } E(G): S(G) = n \wedge \text{indeg } G \leqq d\}$ and show $R_d(n) \geqq c'n \log n$ for some $c' > 0$. Because of the bounded indegree this amounts to

$$\min \{\text{card } V(G): S(G) = n \wedge \text{indeg } G \leqq d\} \geqq \frac{c'}{d} n \log n .$$

This implies that, for graphs with $\dfrac{c'}{dn} n \log n$ vertices, n pebbles are sufficient, because more pebbles are necessary only for graphs with more vertices. But this means $S_d\left(\dfrac{c'}{d} n \log n\right) \leqq n$ and consequently

$$S_d(n) \leqq c \cdot \frac{n}{\log n}$$

for some $c > 0$.

To prove $R_d(n) \geqq c'n \log n$ it is sufficient to show the recurrence

$$R_d(n) \geqq 2R_d\left(\frac{n}{2} - d\right) + \frac{n}{4d} \tag{1}$$

which has a solution $R_d(n) \geqq cn \log n$ as is easily verified.

Let $G = (V, E, v_0) \in \mathcal{G}_d$ be a dag requiring n pebbles and having minimal number $R_d(n)$ of edges. We define two subgraphs $G_1 = (V_1, E_1)$ and $G_2 = (V_2, E_2)$ of G as follows:

$V_1 =_{\text{df}}$ set of those vertices of G which can be pebbled using no more than $\frac{n}{2}$ pebbles,

$V_2 =_{\text{df}} V \setminus V_1, \quad E_1 =_{\text{df}} E \cap (V_1 \times V_1), \quad E_2 =_{\text{df}} E \cap (V_2 \times V_2),$

$A =_{\text{df}} E \setminus (E_1 \cup E_2).$

Note that $A \subseteq V_1 \times V_2$.

The following observation will be helpful.

If a node requires q pebbles, then it has a predecessor requiring at least $q - d$ pebbles. (2)

Now we prove several auxiliary statements.

Claim 1. G_2 has a vertex requiring $\frac{n}{2} - d$ pebbles if the game is restricted to G_2.

Proof of Claim 1. If every vertex of G_2 requires at most $\frac{n}{2} - d - 1$ pebbles, then $n - 1$ pebbles are sufficient for G, as the following pebbling strategy for $v \in V$ shows: If $v \in V_1$, then $\frac{n}{2}$ pebbles are sufficient by definition of V_1. If $v \in V_2$, then v is pebbled using the strategy on G_2. By assumption $\frac{n}{2} - d - 1$ pebbles are sufficient. Whenever a vertex $v \in V_2$ having predecessors in V_1 must be pebbled (following the strategy on G_2), then first these predecessors of v must be pebbled (following the strategy on G_1). There are at most d predecessors $w_1, \ldots, w_d$ of v. Let $w_1, \ldots, w_{d'}$ $(d' \leqq d)$ belong to V_1. Each of these vertices can be pebbled using at most $\frac{n}{2}$ pebbles. The strategy is: Pebble first w_1. Remove all pebbles apart from that on w_1. Pebble w_2 etc. When $w_{d'}$ is pebbled, then at most $\frac{n}{2} + d - 1$ pebbles are on vertices of V_1. The set V_2 may contain simultaneously at most $\frac{n}{2} - d - 1$ pebbles, so that in this moment there are no more than $n - 2$ pebbles on V. The next moves are used to pebble v and to remove all pebbles from vertices in V_1. Thus $n - 1$ pebbles are enough to pebble an arbitrary vertex of V. □

As a consequence of Claim 1, G_2 has at least $R_d\left(\frac{n}{2} - d\right)$ edges.

Claim 2. G_1 has a vertex requiring at least $\frac{n}{2} - d$ pebbles if the game is restricted to G_1.

Proof of Claim 2. Take the end vertex of some edge in A. It requires at least $\frac{n}{2}$ pebbles. Apply observation (2) to get a vertex of V_1 which requires at least $\frac{n}{2} - d$ pebbles. □

From Claim 2 we conclude that G_1 has at least $R_d\left(\frac{n}{2} - d\right)$ edges.

Now we consider two cases

Case 1. card $A \geqq \frac{n}{4}$. Then

$$R_d(n) = \text{card } E \geqq 2R_d\left(\frac{n}{2} - d\right) + \frac{n}{4} \geqq 2R_d\left(\frac{n}{2} - d\right) + \frac{n}{4d}.$$

Hence in this case (1) is true.

Case 2. card $A < \frac{n}{4}$.

Claim 3. In this case G_2 has a vertex requiring $\frac{3n}{4}$ pebbles if the game is restricted to G_2.

Proof of Claim 3. If less than $\frac{3n}{4}$ pebbles would suffice for G_2, then G would not require n pebbles. To see this, pebble first all start vertices of edges in A and leave these pebbles in place. This are at most $\frac{n}{4}$ pebbles. Together with this initial pebbling every pebbling of G_2 becomes a legitimate one in G. □

Claim 4. In Case 2, G_2 has a subgraph $G_3 = (V_3, E_3)$ such that card $(E_2 \setminus E_3) \geqq \frac{n}{4d}$ and G_3 requires $\frac{n}{2}$ pebbles.

Proof of Claim 4. Take a node v requiring $\frac{3n}{4}$ pebbles and apply observation (2) to find a vertex v' requiring at least $\frac{3n}{4} - d$ pebbles. Delete v and all edges leading to v. If this procedure is applied $\frac{n}{4d}$ times, the remaining graph G_3 has at least $\frac{n}{4d}$ edges less than G_2 and still requires $\frac{n}{2}$ pebbles. □

Hence in Case 2 the inequality (1) is also true:

$$R_d(n) = \text{card } E \geqq R_d\left(\frac{n}{2} - d\right) + R_d\left(\frac{n}{2}\right) + \frac{n}{4d} \geqq 2R_d\left(\frac{n}{2} - d\right) + \frac{n}{4d}. \blacksquare$$

Remark. Another proof of 25.7 is given in [Lou 80].

In [PaTaCe 76] it is proved that for infinitely many n there are graphs of indegree 2 having n vertices and requiring $c_1 \cdot \frac{n}{\log n}$ pebbles. This shows the bound of 25.7 to be tight to within a constant factor:

25.8. Theorem. S-Comp $\mathscr{G}_d \asymp_{\text{io}} \frac{n}{\log n}$. ∎

Further restrictions of the graph classes may lead to lower estimations of the number of pebbles required, as the following theorem shows. Define

$$\text{TREE}_d = \{G\colon \text{indeg } G \leqq d \wedge G \text{ is a tree}\}$$

and let P_m denote the "pyramid" of $\frac{m(m+1)}{2}$ vertices of the form shown in Figure 25.5.

25.9. Theorem. [Coo 73a], [PaHe 70]

1. S-Comp(TREE$_d$) $\asymp \log n$.
2. S(P_m) $= m + 1$. ∎

Upper and lower **ST** bounds for graphs corresponding to computations of push-down automata are proved in [Ver 83].

In [EmLe 78] a modification of the pebble game is considered where move rule (R1) is replaced by

(R1′) Whenever all predecessors of a node v are pebbled, then one of these pebbles may be moved to v.

Let us denote the complexities for this pebble game by S′ and T′.

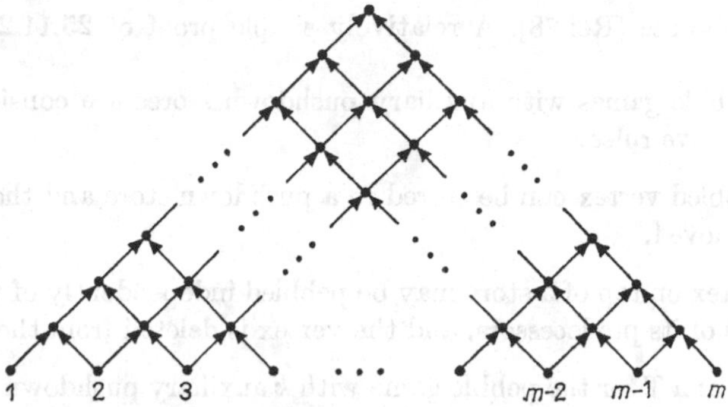

Figure 25.5. The pyramid P_m

25.10. Theorem. [EmLe 78]

$$S(s-1) \subseteq S'(s-1) \subseteq S(s),$$
$$T(t) \subseteq T'(\mathrm{Lin}\, t^2). \blacksquare$$

This means that (R1′) allows a saving of at most one single pebble, but can cause a quadratic blow-up in time (and this really occurs).

In [CoSe 76] pebble games with black and white pebbles are introduced, and, if bw-S denotes the space complexity for the pebble game with black and white pebbles, it is shown in [Meyh 79] that

$$\text{bw-S}(s) \subseteq S(s^2).$$

These two versions of the pebble game are compared for trees in [Lou 79] and [LeTa 80] and for a more general class of graphs including pyramids in [Kla 83].

25.3.2.1.2. Trade-Offs

So far we have considered only the space complexity of pebbling graph families. Theorem 25.8 together with the trivial observation $\mathscr{G}_d \in \mathrm{ST}(\mathrm{id}, \mathrm{id})$ leads to the question of what blow-up in time is required if the allowed space is reduced to some $s \geq \dfrac{n}{\log n}$.

We start with some upper bounds.

25.11. Theorem.

1. [Pau 78] $\mathscr{G}_d \in \mathbf{ST}\left(\mathrm{Lin}\,\frac{n}{\log n}, 2^n\right)$.
2. [LeTa 79]

$$\mathscr{G}_d \in \mathbf{ST}'\left(s(n), s(n) \cdot 2^{2^{\mathrm{Lin}\frac{n}{s(n)}}}\right), \text{ for } 300\frac{n}{\log n} \leqq s(n) \leqq n. \blacksquare$$

For earlier results see [Rei 78]. A relatively simple proof of 25.11.2 is given in [Lou 80].

In [Pip 80a] pebble games with auxiliary pushdown stores are considered. They have two further move rules:

(R3) A pebbled vertex can be stored in a pushdown store and the pebble can be removed.

(R4) A vertex on top of a store may be pebbled independently of the pebbling status of its predecessors, and the vertex is deleted from the store.

The measures S and T for the pebble game with k auxiliary pushdown storages are denoted by kauxPD-S and kauxPD-T.

It is interesting to compare 25.11.2 with the following result.

25.12. Theorem. [Pip 80a] For $s \geqq 3$,

1. $\mathscr{G}_d \in$ **1auxPD-ST**$\left(s(n), s(n)\, 2^{\mathrm{Lin}\left(\frac{n}{s(n)}\right)}\right)$.
2. $\mathscr{G}_d \in$ **2auxPD-ST**$\left(s(n), \mathrm{Lin}\left(\frac{n^2}{s(n)}\right)\right)$.
3. $\mathscr{G}_d \in$ **3auxPD-ST**$\left(s(n), \mathrm{Lin}\left(n \log \frac{n}{s(n)}\right)\right)$. $\blacksquare$

The next result describes trade-offs. The comparison with 25.11.2 and 25.12.1 shows that the upper bounds stated there are in a sense tight. 25.13.1 shows that superpolynomial time is required as soon as $s(n) \leqq c_0 \cdot \frac{n}{\log\log n}$ for sufficiently small $c_0 > 0$.

25.13. Theorem.

1. [LeTa 79] For sufficiently large d there is a $c > 0$ such that

$$\mathscr{G}_d \in \mathbf{ST}'(s, t) \text{ requires } t(n) \geqq_{\mathrm{io}} s(n) \cdot 2^{2^{c \cdot \frac{n}{s(n)}}} \text{ for } 300\,\frac{n}{\log n} \leqq s(n) \leqq n.$$

2. [Pip 80a] There are $c_1, c_2 > 0$ such that

$$\mathscr{G}_d \in \mathbf{1auxPD\text{-}ST}(s, t) \text{ requires } t(n) \geqq_{\mathrm{io}} c_1 \cdot s(n) \cdot 2^{c_2 \cdot \frac{n}{s(n)}} \text{ for } s \geqq 3. \blacksquare$$

Remark 1. The trade-offs of statement 1 remains valid also for the black-white pebble game.

Remark 2. For the proof of 25.13.1 a family of graphs called superconcentrators has been used. This family is constructible in the following sense (see [GaGa 79]): A family $\mathcal{G} = \{G_n: n \in \mathbb{N}\}$ is called constructive if card $V(G_n) = n$ for all n and the mapping $n \to \text{Code}(G_n)$ is recursive. Weaker results on time-space trade-offs for constructive graph families can be found in [Pip 78], [LeTa 79] and [CaSa 80].

Now we list some more dramatic space-time trade-offs for graph families where a time explosion below space $\prec \frac{n}{\log n}$ can be observed. On the other hand, these graph families can be pebbled simultaneously within space Lin $\frac{n}{\log n}$ and linear time.

25.14. Theorem. [PaTa 77] There is a family $\mathcal{G} \subseteq \mathcal{G}_2$ and constants c_1, c_2, c_3 such that $c_1 < c_2$ and

$$\mathcal{G} \in \mathbf{ST}\left(c_2 \sqrt{n}, n\right),$$

$$\mathcal{G} \in \mathbf{S}\left(c_1 \sqrt{n}\right),$$

$$\mathcal{G} \in \mathbf{ST}\left(c_1 \sqrt{n}, t\right) \to t \geqq_{\text{io}} 2^{c_3 \sqrt{n}}. \blacksquare$$

This shows that the time penalty that has to be paid for reducing the space by a constant factor is exponential. The next result shows that exponential time penalties can be caused by saving only one single pebble.

25.15. Theorem. [EmLe 78] There is a family $\mathcal{G} \subseteq \mathcal{G}_2$, a function $s \prec \frac{n}{\log n}$ and a constant $c > 0$ such that

$$\mathcal{G} \in \mathbf{ST}'(s + 1, n),$$

$$\mathcal{G} \in \mathbf{S}'(s) \setminus \mathbf{S}'(s - 1),$$

$$\mathcal{G} \in \mathbf{ST}'(s, t) \to t \geqq_{\text{io}} \left(\sqrt[4]{n}\right)! \blacksquare$$

Remark. The same family $\mathcal{G}$ can be used to prove a similar exponential blow-up for the original version of the pebble game. A similar result is proved in [Lin 78].

25.3.2.2. Further Results

A computation for which the notion of an intermediate result is defined gives rise to a graph which is defined in such a way that its nodes represent the intermediate results, and a node v has exactly those nodes as predecessors which correspond to the intermediate results required to compute the result belonging to v.

The applications of the pebble game for deriving space-time relations rest on the interpretation that a pebbled vertex indicates an intermediate result which is already computed and stored. Then the number of pebbles used to pebble the graph represents the number of storage units (registers, tape segments, ...) necessary for storing intermediate results. If straightline programs are considered, the computation time is the number of pebbling steps. Thus, trade-offs for the pebble game immediately provide us with space-time trade-offs for computations by straightline program or similarly simple types of algorithms.

A detailed review of the various results of this kind would go beyond the scope of our book. We only mention some relevant papers: [SaSw 78] (fast Fourier transform), [Tom 78] (polynomial multiplication and convolution, discrete Fourier transform, oblivious merging), [Tom 80] (transitive closure of Boolean matrices), [Jaj 80] (matrix multiplication and inversion), [Mis 79] (set operations).

There are also space-time trade-offs that are not based on the pebble game, for example [Gri 77] (matrix and polynomial multiplication), [DoMu 78] (finding the kth largest element), [SaSw 79] (integer multiplication), [Ruž 80] (producing partial order), [BoFiKiLyTo 79] and [ReSc 82] (sorting) or [BeBr 80]. Surveys on trade-offs are given in [Pip 80b] and [Sav 81].

§ 26. The General Space-Time Problem

In this section we once more take up the question of the connections between the basic measures TIME and SPACE. Here, however, we are interested only in those results for realistic machines which are invariant with respect to the choice of the machine type. This means that we confine ourselves to the consideration of complexity classes **XSPACE**(s) and **YTIME**(Pol t), X, Y $\in$ {D, N}. The results 25.1, 25.2, 25.3, and 25.4 are dependent on the machine models and hence not relevant for this section.

That the main problems are still open reflects our so far unsatisfactory knowledge of the proper nature of time and space. It is, however, worthwhile mentioning that in spite of the apparently completely different character of the space time problem its treatment is exactly parallel to that of the **P-NP** problem and the logspace problem. This will be discussed in the last subsection of this section.

26.1. The Problems

From § 5 we know the relationships **XSPACE**(s) $\subseteq$ **YTIME**($2^{\mathrm{Lin}\, s}$) for X, Y $\in$ {D, N} (for X = N and Y = D under certain honesty conditions, see 5.10 and 5.12). The question arises whether this result which was obtained by a very rough estimation can be improved or whether there is an s-space bounded machine such that every equivalent machine requires time $2^{cs(n)}$ for some $c > 0$. This is our first problem: For X, Y $\in$ {D, N} and for given s find t as small as possible such that **XSPACE**(s) $\subseteq$ **YTIME**(Pol t) holds.

The four special questions

NLINSPACE $\subseteq$ **P**?	**DLINSPACE** $\subseteq$ **P**?
NLINSPACE $\subseteq$ **NP**?	**DLINSPACE** $\subseteq$ **NP**?

are referred to as the *first space-time problem.*

It is not suprising that inclusion results may be extended to larger functions under certain honesty conditions.

26.1. Theorem. If $s \geqq \log$ and $t \geqq \text{id}$, $h \geqq \text{id}$, $s(|w|) \in \text{DSPACE}(s)$, $h(|w|) \in \text{DSPACE}(s \circ h) \cap \text{DTIME}(\text{Pol}\, t \circ h)$, then, for X, Y $\in$ {D, N},

$\mathbf{XSPACE}(\text{Pol}\, s) \subseteqq \mathbf{YTIME}(\text{Pol}\, t) \to \mathbf{XSPACE}(\text{Pol}\, s \circ h) \subseteqq \mathbf{YTIME}(\text{Pol}\, t \circ h)$,

$\mathbf{XSPACE}(s) \subseteqq \mathbf{YTIME}(\text{Pol}\, t) \to \mathbf{XSPACE}(s \circ h) \subseteqq \mathbf{YTIME}(\text{Pol}\, t \circ h)$.

Proof. The theorem is an immediate consequence of 6.17.1. ∎

Using 6.19, 6.16, and 6.15 the honesty conditions for h can be abandoned for X = D and Y = N.

26.2. Theorem. If $s \geqq \log$ and $t \geqq \text{id}$ are increasing, $s(|w|) \in \text{DSPACE}(s)$ and $h > \text{id}$, then

$\mathbf{DSPACE}(\text{Pol}\, s) \subseteqq \mathbf{NTIME}(\text{Pol}\, t) \to \mathbf{DSPACE}(\text{Pol}\, s \circ h) \subseteqq \mathbf{NTIME}(\text{Pol}\, t \circ h)$,

$\mathbf{DSPACE}(s) \subseteqq \mathbf{NTIME}(\text{Pol}\, t) \to \mathbf{DSPACE}(s \circ h) \subseteqq \mathbf{NTIME}(\text{Pol}\, t \circ h)$. ∎

Remark. In the preceding two theorems the case $s = t = \text{id}$ is of particular interest

26.3. Corollary.

1. The following statements are equivalent for X $\in$ {D, N} (where **DP** = **P**):

$$\mathbf{DLINSPACE} \subseteqq \mathbf{XP}, \quad \mathbf{NLINSPACE} \subseteqq \mathbf{XP}, \quad \mathbf{PSPACE} \subseteqq \mathbf{XP}.$$

2. If $h > \text{id}$ and $h(|w|) \in \text{DTIME}(\text{Pol}\, h)$, then

$$\mathbf{PSPACE} = \mathbf{NP} \to \mathbf{DSPACE}(\text{Pol}\, h) = \mathbf{NTIME}(\text{Pol}\, h),$$

$$\mathbf{PSPACE} = \mathbf{P} \to \mathbf{DSPACE}(\text{Pol}\, h) = \mathbf{DTIME}(\text{Pol}\, h).$$

Proof. Ad 1. We need only prove **DLINSPACE** $\subseteqq$ **(N)P** $\to$ **PSPACE** $\subseteqq$ **(N)P**. Every polynomial h satisfies the condition of 26.1 for $s = \text{id}$. Hence **DLINSPACE** $\subseteqq$ **(N)P** $\to$ $\mathbf{DSPACE}(h) \subseteqq \mathbf{(N)TIME}(\text{Pol}\, h) \subseteqq$ **(N)P**. This proves **PSPACE** $\subseteqq$ **(N)P**.
Ad 2. This statement is a special case of 26.1. ∎

Because of 26.3.1 we can also formulate the first space-time problem as "**PSPACE** $\subseteqq$ **(N)P**?". A special type of padding is the transition from the dyadic presentation of numbers to their unary presentation. The corresponding general translational results 6.21 and 6.22 yield as a special case

26.4. Theorem. For X $\in$ {D, N} (where **DP** = **P**),

$$\mathbf{PSPACE} \cap \mathfrak{P}(1^*) = \mathbf{XP} \cap \mathfrak{P}(1^*) \leftrightarrow \mathbf{EXPSPACE} = \mathbf{XEXPTIME}. \blacksquare$$

We now turn to the description of a second space time problem and start with the elementary fact $\mathbf{XTIME}(\text{Pol}\, t) \subseteqq \mathbf{YSPACE}(\text{Pol}\, t)$ (for X = N and Y = D under certain honesty conditions, see 5.10 and 5.12). Again we are interested in improvements of this trivial inclusion and pose the problem:

For X, Y $\in$ {D, N} and given t find an s as small as possible such that $\mathbf{XTIME}(\text{Pol}\, t) \subseteqq \mathbf{YSPACE}(s)$ holds.

The special case $t = \mathrm{id}$ and $s \in \mathrm{Pol}(\log)$ will be called the *second space-time problem*: Is there a $k \in \mathbb{N}$ such that

$$\mathbf{(N)P} \subsetneqq \mathbf{YSPACE}(\log^k)?$$

Again we have a translation result:

26.5. Theorem. If $h > \mathrm{id}$, $s \geqq \log$ and $t \geqq \mathrm{id}$, $t(|w|) \in \mathbf{DTIME}(\mathrm{Pol}\, t)$ and $h(|w|) \in \mathbf{DTIME}(\mathrm{Pol}\, t \circ h) \cap \mathbf{DSPACE}(s \circ h)$, then, for X, Y $\in$ {D, N},

$$\mathbf{XTIME}(\mathrm{Pol}\, t) \subsetneqq \mathbf{YSPACE}(s) \to \mathbf{XTIME}(\mathrm{Pol}\, t \circ h) \subsetneqq \mathbf{YSPACE}(s \circ h),$$

$$\mathbf{XTIME}(\mathrm{Pol}\, t) \subsetneqq \mathbf{YSPACE}(\mathrm{Pol}\, s) \to \mathbf{XTIME}(\mathrm{Pol}\, t \circ h) \subsetneqq \mathbf{YSPACE}(\mathrm{Pol}\, s \circ h).$$

Proof. Analogous to the proof of 26.1. ∎

Using 6.20, 6.16, and 6.15 the honesty conditions for h can be abandoned for X = D and Y = N.

26.6. Theorem. If $t \geqq \mathrm{id}$ and $s \geqq \log$ are increasing, $t(|w|) \in \mathbf{DTIME}(\mathrm{Pol}\, t)$ and $h > \mathrm{id}$, then

$$\mathbf{DTIME}(\mathrm{Pol}\, t) \subsetneqq \mathbf{NSPACE}(s) \to \mathbf{DTIME}(\mathrm{Pol}\, t \circ h) \subsetneqq \mathbf{NSPACE}(s \circ h),$$

$$\mathbf{DTIME}(\mathrm{Pol}\, t) \subsetneqq \mathbf{NSPACE}(\mathrm{Pol}\, s) \to \mathbf{DTIME}(\mathrm{Pol}\, t \circ h) \subsetneqq \mathbf{NSPACE}(\mathrm{Pol}\, s \circ h). \quad ∎$$

Remark. For $s \geqq \log$ and $h > \mathrm{id}$ the statement

$$\mathbf{NP} \subsetneqq \mathbf{NSPACE}(s) \to \mathbf{NTIME}(\mathrm{Pol}\, h) \subsetneqq \mathbf{NSPACE}(s \circ \mathrm{Pol}\, h)$$

can be proved by a method from [Coo 73a].

To the implications

$$\mathbf{XP} \subsetneqq \mathbf{YSPACE}(\log^k) \to \mathbf{XEXPTIME} \subsetneqq \mathbf{YSPACE}(n^k)$$

following from 26.5 we add the equivalences of the next theorem which are analogous to 26.4.

26.7. Theorem. For X, Y $\in$ {D, N},

$$\mathbf{XP} \cap \mathfrak{P}(1^*) \subsetneqq \mathbf{YSPACE}(\log^k) \cap \mathfrak{P}(1^*) \leftrightarrow \mathbf{XEXPTIME} \subsetneqq \mathbf{YSPACE}(n^k).$$

Proof. By 6.21 and 6.22. ∎

26.2. The First Space-Time Problem

By 3.2 we are able to reduce the space-time problem to questions about single sets.

26.8. Theorem. If X $\in$ {D, N} and A is $\leqq_m^P$-complete in **PSPACE**, then

$$\mathbf{PSPACE} \subsetneqq \mathbf{XP} \leftrightarrow A \in \mathbf{XP}. \quad ∎$$

Remark. If $X = D$, then $\leq_m^P$ can be replaced by $\leq_T^P$.

Theorem 14.14 presents a rich variety of problems from various fields of mathematics which can play the role of A in the foregoing theorem.

A sufficient condition for a $\leq_m^P$-complete set in **PSPACE** to belong to **P** can be derived from 24.14.

26.9. Theorem. If there is a sparse set which is $\leq_m^P$-hard for **PSPACE**, then
$$\mathbf{P} = \mathbf{PSPACE}.$$

Remark. In this connection see also 14.11.

Proof. Let B be a sparse set which is $\leq_m^P$-hard for **PSPACE**. Let B_ω play the role of A and let B play the role of B_1 and B_2 in 24.14. If

(1) B_ω is self-reducible (see p. 431) and

(2) $\mathrm{B}_\omega \leq_m^P B$ and $\overline{B}_\omega \leq_m^P B$,

then 24.14 yields $\mathrm{B}_\omega \in \mathbf{P}$. As B_ω is $\leq_m^P$-complete in **PSPACE** Theorem 26.8 implies $\mathbf{P} = \mathbf{PSPACE}$. Thus it remains to show (1) and (2). To prove (1) it suffices to write down the following equivalences

$$\bigwedge_p F(p) \in \mathrm{B}_\omega \leftrightarrow F(1) \in \mathrm{B}_\omega \wedge F(0) \in \mathrm{B}_\omega,$$

$$\bigvee_p F(p) \in \mathrm{B}_\omega \leftrightarrow F(1) \in \mathrm{B}_\omega \vee F(0) \in \mathrm{B}_\omega.$$

They show how the self-reduction works. The partial ordering on which the self-reducibility is based is defined by length ($x \prec y \leftrightarrow_{\mathrm{df}} |x| < |y|$).

Claim (2) follows from the fact that $\mathrm{B}_\omega \in \mathbf{PSPACE}$ (cf. 14.14). ∎

Next we extend Statement 1 of Corollary 26.3.

26.10. Theorem. [Boo 74b] For every rational $r > 0$ and $X, Y \in \{D, N\}$,

$$\mathbf{XSPACE}(n^r) \subseteq \mathbf{YP} \leftrightarrow \mathbf{PSPACE} \subseteq \mathbf{YP}.$$

Proof. We choose a set from $\mathbf{XSPACE}(n^r)$ according to 14.9.1, this set being $\leq_m^{\log}$-complete in **PSPACE**. Then 26.8 yields the nontrivial implication of our theorem. ∎

The following corollary reformulates the first space-time problem in terms of automata.

26.11. Corollary. For $X \in \{D, N\}$,

1. $\mathbf{2\text{-}DNESA} \subseteq \mathbf{XP} \leftrightarrow \mathbf{PSPACE} = \mathbf{XP}$,
2. $\mathbf{2\text{-}NCSA} \subseteq \mathbf{XP} \leftrightarrow \mathbf{PSPACE} = \mathbf{XP}$.

Proof. Statement 1 follows from the relationship $\mathbf{2\text{-}DNESA} = \mathbf{DSPACE}(n \log n)$ (13.29.1) and 26.10.

Statement 2 follows from 26.10 and the relationship $\mathbf{2\text{-}NCSA} = \mathbf{NLINSPACE}$ (13.40.3). ∎

For the next result recall the definition of f_k given before 13.11.

26.12. Theorem. [Gal 74] If A is $\leqq_m^{\log}$-complete in **PSPACE**, then

$$\mathbf{P} = \mathbf{PSPACE} \leftrightarrow f_k(A) \in \mathbf{2\text{-}DPDA} \quad \text{for some} \quad k \in \mathbb{N}.$$

Proof. Analogous to that for 24.34. ∎

From algebraic representations of **P**, **NP**, and **PSPACE** we can conclude some necessary and sufficient conditions for **(N)P** = **PSPACE** in terms of closure properties. As an immediate consequence of 11.20.1 and 11.23.2 we state

26.13. Theorem. [Boo 79c] $\mathbf{NP} = \mathbf{PSPACE} \leftrightarrow \mathbf{NP} = \Gamma_{\mathrm{wtc}}(\mathbf{NP})$. ∎

By 10.14.2 and 10.20.1

26.13. Theorem. $\mathrm{P} = \mathrm{PSPACE} \leftrightarrow \mathrm{P} = \Gamma_{\mathrm{BPR}}(\mathrm{P})$. ∎

From the relationships of time and space to size and depth of logical networks (cf. 21.13) we can conclude:

26.15. Theorem. $\mathbf{PSPACE} \subseteq \mathbf{P} \leftrightarrow \mathbf{DEPTH}(\mathrm{Pol}) \subseteq \mathbf{SIZE}(\mathrm{Pol})$. ∎

Adding parallelism or stronger instructions to realistic machine types enlarges their power if and only **P** ≠ **PSPACE**. More precisely

26.16. Theorem. The following statements are pairwise equivalent for $\mathrm{X} \in \{\mathrm{D}, \mathrm{N}\}$:
1. **P** = **PSPACE**,
2. **XTIME**(Pol) = **XPARTIME**(Pol),
3. **XTIME**(Pol) = **ATIME**(Pol),
4. **RAM-XTIME**(Pol) = **MRAM-XTIME**(Pol).

Proof. By 20.12, 20.35 and 20.38. ∎

Finally, we remember that the polynomial time hierarchy lies between **NP** and **PSPACE** (24.42):

$$\mathbf{NP} = \Sigma_1^{\mathrm{P}} \subseteq \Sigma_2^{\mathrm{P}} \subseteq \ldots \subseteq \bigcup_k \Sigma_k^{\mathrm{P}} \subseteq \mathbf{PSPACE}.$$

Furthermore, if $\bigcup_k \Sigma_k^{\mathrm{P}} = \mathbf{PSPACE}$ is known, then

$$\Sigma_1^{\mathrm{P}} \subseteq \ldots \subseteq \Sigma_k^{\mathrm{P}} = \Sigma_k^{\mathrm{P}} = \ldots = \mathbf{PSPACE} \quad \text{for some } k \text{ (24.45)}.$$

26.3. The Second Space-Time Problem

Our first step is again a transformation of the problem into a question about single complete languages with the help of 3.2.

26.17. Theorem. If $\mathrm{X}, \mathrm{Y} \in \{\mathrm{D}, \mathrm{N}\}$ and A is $\leqq_m^{\log}$-complete in **YP**, then

$$A \in \mathbf{XSPACE}(\log^k) \leftrightarrow \mathbf{YP} \subseteq \mathbf{XSPACE}(\log^k).\ ∎$$

Remark. If X = D, then $\leq_m^{\log}$ can be replaced by $\leq_T^{\log}$. We confine ourselves to the question **P** $\subseteq$ **XSPACE**($\log^k$) which seems less unlikely than **NP** $\subseteq$ **XSPACE**($\log^k$). Among the large number of languages which can play the role of A in 26.17 (see **14.19**) we particularly mention the solvable path problem SP which is known to be $\leq_m^{\log}$-complete in **P** (8.43.8).

26.18. Corollary. For X $\in$ {D, N}, SP $\in$ **XSPACE**($\log^k$) $\leftrightarrow$ **P** $\subseteq$ **XSPACE**($\log^k$). ∎

This shows that the main problem is to find space bounds for SP which are as low as possible. A first step in this direction makes use of the pebble game introduced in § 25. Every instance of SP can be considered as a path system on which the pebble game can be played successfully.

26.19. Theorem. If s is increasing and the pebble game can be played with $s(n)$ pebbles on path systems with n nodes, then SP $\in$ **NSPACE**$(s(n) \log n)$.

Proof. Let w be the code of a path system with N nodes. Then $n = |w| \geq N$. A 2-NTM M to accept SP within space $s(n) \log n$ works simply as follows. Simulating the play it keeps a list of the pebbled nodes. Each node can be coded by a word of length $\leq \log N$. If M simulates a successful sequence of moves reaching a source node with no more than $s(N)$ pebbles, then M needs no more than $s(N) \log N \leq s(n) \log n$ tape squares. ∎

The theorems 25.8 and 25.9.2 give evidence that SP $\notin$ NSPACE(Pol(log)). S. A. Cook and R. Sethi further support this conjecture by proving that "path machines" require space $c \cdot n^{1/4}$ for accepting SP (see [CoSe 76]). Path machines are not so powerful as Turing machines and therefore the lower bounds for path machines need not be valid for Turing machines. In [Koz 77a] and [Imm 79] there are results of the same kind. For instance, there are problems known to be $\leq_m^{\log}$-complete in P which require space $\frac{n}{\log n}$ for certain straightline programs [Koz 77a].

Theorem 26.17 can be further exploited by making use of the existence of specific languages which are complete in **P**.

26.20. Theorem. The following statements are mutually equivalent for X $\in$ {D, N}: For $k \geq 1$,

1. **P** $\subseteq$ **XSPACE**($\log^k$).
2. **2-DPDA** $\subseteq$ **XSPACE**($\log^k$).
3. **LINTIME** $\subseteq$ **XSPACE**($\log^k$).

Proof. To prove "2 → 1" one uses **P** = **2:multi-DPDA** (13.20.4), 13.13 and 26.17 (see also the proof of 24.34). "3 → 1" follows from 26.17 because **LINTIME** contains a set which is $\leq_m^{\log}$-complete in **P**. Such a set can be obtained by padding from a $\leq_m^{\log}$-complete set in **P** (14.9.2). ∎

26.21. Theorem. The following statements are mutually equivalent: For $k \geq 1$,

1. **NP** $\subseteq$ **XSPACE**($\log^k$).
2. **1-T-NTIME**(id) $\subseteq$ **XSPACE**($\log^k$).
3. **1-NCSA** $\subseteq$ **XSPACE**($\log^k$).

Proof. As the proof of 24.10, where 26.17 is used. ∎

From the relationship of time and space to size and depth of logical networks (cf. 21.13) we can conclude:

26.22. Theorem.

1. $\mathbf{SIZE}(\mathrm{Pol}) \subseteq \mathbf{DEPTH}(\mathrm{Lin}\log^k n) \to \mathbf{P} \subseteq \mathbf{DSPACE}(\log^k n)$.
2. $\mathbf{P} \subseteq \mathbf{NSPACE}(\log^k n) \to \mathbf{SIZE}(\mathrm{Pol}) \subseteq \mathbf{DEPTH}(\mathrm{Lin}\log^{2k} n)$. ∎

Compare this latter with the best known relationship between SIZE and DEPTH proved in [PaVa 76], namely $\mathbf{SIZE}(n) \subseteq \mathbf{DEPTH}\left(\mathrm{Lin}\,\frac{n}{\log n}\right)$.

Theorem 11.3 presents some conditions concerning algebraic closure properties of **L** and **NL** which are equivalent to **NP** = **L** and **NP** = **NL**, respectively. The following theorem extends 11.3 in some cases and can be proved as this theorem.

26.23. Theorem. For $X \in \{D, N\}$ and $k \geq 1$,

$$\mathbf{NP} \subseteq \mathbf{XSPACE}(\log^k) \leftrightarrow \Gamma_{\mathrm{eh}}(\mathbf{L}) \subseteq \mathbf{XSPACE}(\log^k). \blacksquare$$

The following theorem reformulates the problems **XP** = **YL** ($Y \in \{D, N\}$) in terms of automata defined classes of languages.

26.24. Theorem. The following statements are pairwise equivalent for $X \in \{D, N\}$:

1. **P** = **XL**.
2. **2-DPDA** ⊆ **2:multi-XCA**.
3. **2:2-AFA** ⊆ **2:multi-XFA**.

Proof. The proof is analogous to that of 26.20 where the relationships **2:multi-XCA** = **XL** (13.7), **2:multi-AFA** = **P** (13.3.2), **2:multi-XFA** = **XL** (13.2) and an analogue of 13.13 for 2:multi-AFA are used. ∎

From theorem 26.5 we know that **XP** ⊆ **L** implies $\mathbf{XTIME}(2^{\mathrm{Pol}}) \subseteq \mathbf{PSPACE}$. For the latter problem we have the following equivalent formulation which is mainly based on automata characterizations of complexity classes.

26.25. Theorem. $\mathbf{DTIME}(2^{\mathrm{Pol}}) = \mathbf{PSPACE} \leftrightarrow$ **2-DSA** ⊆ **2:multi-NNESA**.

Proof. Because of **PSPACE** = **2:multi-NNESA** (13.29.3) and **2-DSA** = $\mathbf{DTIME}(2^{\mathrm{Lin}(n\log n)})$ (13.35.1) the implication "→" is trivial. For the other implication it is sufficient to prove the inclusion $\mathbf{DTIME}(2^{\mathrm{Pol}}) \subseteq \mathbf{PSPACE}$. This follows from the existence of a $\leq_{\mathrm{m}}^{\log}$-complete set in $\mathbf{DTIME}(2^{\mathrm{Pol}})$ belonging to **DEXPTIME** ⊆ **2-DSA** (14.9.2). ∎

Finally, $\mathbf{PSPACE} = \mathbf{NTIME}(2^{\mathrm{Pol}})$ if and only if the deterministic and nondeterministic variants of the parallel machines in the sense of [FoWy 78] can do the same in polynomial time (cf. Remark 3 after 20.35).

26.4. The Interface between the Space-Time Problems, the Logspace Problem and the P-NP Problem

Comparing § 23, § 24 and § 26, one first of all will notice a strong parallelity in the treatment of the problems that leads to analogous theorems. Examples are 23.14, 24.9, 26.8, and 26.17 which are specific versions of a general fact about the reduction of inclusion problems between two classes to a membership problem of single languages (3.2). These theorems have in their turn very similar applications in these sections. This strong analogy of results convincingly demonstrates that complexity theory is a homogeneous theory far from being merely a collection of completely different results, but it cannot serve as an indication of a close relationship between our main problems.

There are, however, facts actually showing the interplay between the main problems:

1. Several results relate the problems directly, for instance theorem 23.3.

2. The hierarchies of complete problems first mentioned in [Gal 76a] reveal quite another connection between time and space. Complete sets in the classes **L**, **NL**, **P**, **NP**, **PSPACE**, **XTIME**(2^{Pol}) can be expressed in a uniform way. This makes explicit the difference between these sets in terms of that area from which the hierarchy is chosen (for example in terms of automata, graph accessibility etc.). An instance from the area of automata is given by Table 26.1. Similar interpretations can be given to all the other hierarchies known from the literature.

Table 26.1. The problem EW($\mathfrak{A}$) is $\leq_m^{\log}$-complete in $\mathscr{C}$ ($\leq_m^{1-\log}$-complete if $\mathscr{C}$ = **L**)

$\mathfrak{A}$	$\mathscr{C}$
1-DFA	**L**
1-NFA	**NL**
1-DPDA	**P**
1-DCSA	**NP**
1-DNESA	**PSPACE**
1-DSA	**DTIME**(2^{Pol})

Chapter IX
Relativized Computations

The relationships between measures relativized to different oracles are dealt with in an extra chapter.

§ 27 investigates the problem of diminishing the complexity by using oracles (helping). The various interconnections between relativized versions of the D-ND problems and the space-time problems suggest that we should treat them together. They form the contents of § 28.

§ 27. Oracles

Three trends in the theory of computing have accelerated interest in the study of relativized computations, i.e. computations which have a word input and a set input (the oracle), and corresponding relativized complexity measures. These trends are the development of the subroutine technique in programming theory, the intensive study of reducibilities among combinatorial problems and the method of giving evidence for an unresolved problem to be difficult by constructing positive and negative relativizations of this problem.

An axiomatic treatment of relativized complexity measures leads to relativizations of several results of the theory of "ordinary" complexity measures. Such relativizations, for example a relativized Compression Theorem and a relativized Union Theorem, can be found in [LyMeFi 76].

In this section we deal with the following question: How can the use of an oracle diminish the complexity of a given set, i.e. how can an oracle help the decision of a given set? There are two types of results: results on oracles which do help and results on oracles which cannot help.

27.1. Oracles that Do Help

In this subsection we present some results about oracles which can help the decision of a given set. In one case (27.2) the amount of help of a given oracle for deciding a certain set is determined. The remaining results concern the construction of oracles with a desired amount of help.

First, as a prelude, we show for some time and space measures that oracles whose symmetric difference is a regular set have the same power. In particular, regular oracles cannot help.

27.1. Theorem. Let $B \oplus C$ be regular and A recursive.

1. $\bigwedge_t$ (T-DTIME^B-Comp$(A) \leqq_{ae} t \to$ T-DTIME^C-Comp$(A) \leqq_{ae} t$).
2. $\bigwedge_t$ (multiT-DTIME^B-Comp$(A) \leqq_{ae} t \to$ multiT-DTIME^C-Comp$(A) \leqq_{ae} t$).
3. $\bigwedge_s$ (2-T-DSPACE^B-Comp$(A) \leqq_{ae} s \to$ 2-T-DSPACE^C-Comp$(A) \leqq_{ae} s$).

Proof. Ad 1. Let M be an oracle DTM which decides the set A using the oracle B, i.e. $\varphi_M^B = c_A$. An oracle DTM M' can decide the set A using oracle C as follows: M' works as M, but in addition it stores in its finite control whether the word on the oracle tape is a member of $B \oplus C$. If M asks whether the word u on the oracle tape is in B, then M' asks whether $u \in C$ and knows, because of $c_B(u) = c_B(u) \oplus c_{B \oplus C}(u)$, whether $u \in B$. Consequently, T-DTIME$_{M'}^C(u) =$ T-DTIME$_M^B(u)$ and $\varphi_M^C = c_A$.

The statements 2 and 3 can be proved in the same manner. ∎

The following theorem shows that the oracle $C = \{0^n 1^n : n \in \mathbb{N}\}$ really diminishes the complexity of the decision of the set S of symmetrical words by a certain amount. Remember that T-DTIME-Comp(S) $\asymp n^2$ (cf. 8.13.1).

27.2. Theorem. [Cho 69] T-DTIME^C-Comp(S) $\asymp \dfrac{n^2}{\log n}$.

Proof. First we show T-DTIME^C-Comp(S) $\preceq \dfrac{n^2}{\log n}$. We describe the work of an oracle DTM M using the oracle C on the input word w. Let $n =_{df} |w|$. In the first stage of its work M computes $\lfloor \log n \rfloor$, marks the middle of w and marks every $\lfloor \log n \rfloor$th symbol from the left to the middle and from the right to the middle. Thus w is subdivided into $w = u_1 u_2 \ldots u_r w' v_r \ldots v_2 v_1$ such that $|u_1| = \ldots = |u_r| = |v_r| = \ldots |v_1| = \lfloor \log n \rfloor$, $|w'| < \lfloor 2 \cdot \log n \rfloor$ and $r = \left\lfloor \dfrac{n}{2\lfloor \log n \rfloor} \right\rfloor$. This takes M at most $\preceq n \cdot \log n$ steps. In the second step of its work M checks whether $u_1 = v_1^{-1}$, $u_2 = v_2^{-1}$, ..., $u_r = v_r^{-1}$ and $w' = w'^{-1}$. In order to check $u_i = v_i^{-1}$, the machine M writes $0^{\mathrm{bin}^{-1}(u_i)} 1^{\mathrm{bin}^{-1}(v_i^{-1})}$ on the oracle tape and asks the oracle whether this is in C. Evidently, the answer is yes if and only if $u_i = v_i^{-1}$. Since $|u_i| = |v_i| = \lfloor \log n \rfloor$ implies $\mathrm{bin}^{-1}(u_i) < n$ and $\mathrm{bin}^{-1}(v_i^{-1}) < n$, this can be done within $\preceq n$ steps (see 7.13). The validity of $w' = w'^{-1}$ is checked in the ordinary way within $|w'|^2 \preceq \log^2 n$ steps. Altogether M works at most $\preceq n \cdot \log n + r \cdot n + \log^2 n \preceq \dfrac{n^2}{\log n}$ steps.

Now we show T-DTIME^C-comp(S) $\succeq \dfrac{n^2}{\log n}$. Assume that T-DTIME^C-Comp(S) $\preceq t(n) \prec_{io} \dfrac{n^2}{\log n}$. Let M be an oracle DTM which, using the oracle C, decides S within time t. We describe the work of an ordinary DTM M' which simulates the work of M. The tape of M' has two tracks. The upper track has the same contents as the tape of M. The lower track contains the binary presentation of the number $k - m$, where $0^k 1^m$ is the content of the oracle tape of M. This presentation is situated immediately to the left of the head of M'. If M asks the oracle, then M' checks whether the lower track contains the binary presentation of 0. Thus one step of M can be simulated by M' within $\lesssim \log t(n)$ steps. Altogether M' works $\lesssim t(n) \cdot \log t(n)$ steps. But $t(n) <_{io} \frac{n^2}{\log n}$ implies $t(n) \cdot \log t(n) <_{io} n^2$, contradicting the fact that T-DTIME-Comp(S) $\gtrsim n^2$ (cf. 8.13). ∎

Remark. In [Cho 69] a different model of an oracle DTM and an oracle different from C is used to establish the same result. There a machine has no extra oracle tape; an oracle question refers to the contents of the worktape. However, the proof is essentially the same.

Theorem 27.2 is only an example of the phenomenon that an oracle can diminish the complexity of a given set by a well-defined amount. This phenomenon leads us to the following question: Let $r_1 \lesssim r_2 \lesssim t_2$ and $t_1 \lesssim \Phi\text{-Comp}(A) \lesssim t_2$. Under which suppositions regarding r_1, r_2, t_1 and t_2 is there an oracle B such that $r_1 \lesssim \Phi^B\text{-Comp}(A) \lesssim r_2$ or even $\Phi^B\text{-Comp}(A) \asymp r_2$? The following theorems answer this question for the measures T-DTIME, multiT-DTIME and T-DSPACE.

For $r\colon \mathbb{N} \to \mathbb{N}$ such that $\lim_{n\to\infty} r(n) = \infty$ let $r^{-1}(n) =_{df} \max\{k : r(k) \leqq n\}$ and $\bar{r}(n) =_{df} \min\{k : k \cdot r^{-1}(k) \geqq n\}$. Note that

$$r(r^{-1}(n)) \leqq n,\ r^{-1}(r(n)) \geqq n \quad \text{and} \quad \bar{r}(n \cdot r^{-1}(n)) \leqq n.$$

Furthermore, we have $r^{-1}(r(n)) = n$ for strictly increasing r. For $A \subseteq \Sigma^*$ and $r \geqq \text{id}$ let $A_r =_{df} \{wv : r(|w|) = |wv| \wedge w \in A \wedge v \in \Sigma^*\}$.

27.3. Theorem. Let $r, t_1, t_2 \in \mathbb{R}_1$ such that

a) $\text{id} \leqq t_1 \lesssim \text{T-DTIME-Comp}(A) \lesssim t_2$ and $\text{id} \leqq r \lesssim t_2$,
b) r strictly increasing, t_2^{-1} increasing and $t_2 \circ r^{-1}$ superadditive,
c) $r^{-1}(|w|) \in \text{T-DTIME}(t_2 \circ r^{-1})$ and $r(|w|) \in \text{T-DTIME}(r)$.

Then there is a $c > 0$ such that

$$\min\left\{r\left(t_2^{-1}(t_1(n))\right), \bar{r}(c \cdot t_1(n)), \frac{t_1(n)}{\log t_1(n)}\right\} \lesssim \text{T-DTIME}^{A_r}\text{-Comp}(A) \lesssim r(n).$$

If in addition $t_1 = t_2$, t_2 strictly increasing, $r(n) < \frac{t_2(n)}{n}$ and $r \lesssim \frac{t_2(n)}{\log t_2(n)}$, then $\text{T-DTIME}^{A_r}\text{-Comp}(A) \asymp r(n)$.

Proof. Upper bound. We construct an oracle DTM M which, using the oracle A_r, decides the set A as follows: Let w be the input word and $n = |w|$. First M computes $r(n)$ (within time $r(n)$, because of supposition c), then M writes $w0^{r(n)-n}$ on the oracle tape (within time $\lesssim r(n)$ since counting down from m to 0 by successive binary subtraction of 1's can be done within time $\lesssim m$, see 7.13) and finally M asks whether $w0^{r(n)-n} \in A_r$ (within one step). Because of $w0^{r(n)-n} \in A_r \leftrightarrow w \in A$, the machine M decides A within time $r(n)$.

Lower bound. Let M be an oracle DTM such that $L(M^{A_r}) = A$ and let $t_3(n) =_{df} \text{T-DTIME}^{A_r}(n)$. We describe the work of a DTM M' (without oracle) which simulates M. Let w be an input word and $n =_{df} |w|$. Assume that M asks on this input the questions

$w_1, w_2, \ldots, w_m$. The work of M' between the questions w_i and w_{i+1} is said to be the ith stage of the work of M on w ($i = 0, \ldots, m$, where the question w_0 (w_{m+1}) is identified with the start (the end) of the computation). The tape of M' has three tracks. After the simulation of a step of the work of M, the first track of M' has the same contents as the tape of M and the head of M' has the same position as the head of M. The second and the third track are used as follows: At the beginning of the simulation of stage i of M these tracks are empty. The machine M' copies the contents of the first track onto the second track. Then all steps of stage i are simulated while on the third track immediately to the left of the actual head position the length of the word on the oracle tape is counted in binary notation (the word itself not being stored). On completion of the simulation of stage i, the number $|w_{i+1}|$ is stored on the third track. Now M' computes on the third track first $r^{-1}(|w_{i+1}|)$ (within time $\lesssim t_2(r^{-1}(|w_{i+1}|))$) and then the string $+^{r^{-1}(|w_{i+1}|)}$ (within time $\lesssim r^{-1}(|w_{i+1}|) \cdot \log r^{-1}(|w_{i+1}|)$). Then, using the contents of the second track, M' re-simulates the stage i of the work of M while on the third track immediately to the left of the actual head position, the initial part of length $r^{-1}(|w_{i+1}|)$ of the sequence $v+++\ldots$ is stored (v being the actual contents of the oracle tape). On completion of the re-simulation of stage i the number $|w_{i+1}|$ and the initial part u_{i+1} of w_{i+1} of length $r^{-1}(|w_{i+1}|)$ are stored on che third track. Because of $w_{i+1} \in A_r \leftrightarrow r(|u_{i+1}|) = |w_{i+1}| \wedge u_{i+1} \in A$ the machine M' tan decide whether $w_{i+1} \in A_r$ within time $\lesssim t_2(|u_{i+1}|) \lesssim t_2(r^{-1}(|w_{i+1}|))$. Consequently

$$
\begin{aligned}
t_1(n) &\lesssim \text{T-DTIME}_{M'}(n) \\
&\lesssim t_3(n) \cdot \log t_3(n) && \text{(first simulation of all steps)} \\
&\quad + t_3(n) \cdot r^{-1}(t_3(n)) && \text{(re-simulation of all steps)} \\
&\quad + \sum_{i=1}^{m} t_2(r^{-1}(|w_i|)) && \text{(computation of the } r^{-1}(|w_i|) \text{ and decision of all questions of } M) \\
&\lesssim t_3(n) \cdot \log t_3(n) + t_3(n) \cdot r^{-1}(t_3(n)) + t_2(r^{-1}(t_3(n))) && \text{(because of the superadditivity of } t_2 \circ r^{-1}).
\end{aligned}
$$

There is a $c > 0$ such that at least one of the inequalities $c \cdot t_1(n) \leqq t_3(n) \log t_3(n)$, $c \cdot t_1(n) \leqq t_3(n) \cdot r^{-1}(t_3(n))$ and $c \cdot t_1(n) \leqq t_2(r^{-1}(t_3(n)))$ holds true. Since r and t_2^{-1} are increasing functions we get

$$\min\left\{r\left(t_2^{-1}(t_1(n))\right), \bar{r}(c \cdot t_1(n)), \frac{t_1(n)}{\log t_1(n)}\right\} \lesssim t_3(n).$$

If in addition $t_1 = t_2$ and t_2 strictly increasing, then $r\left(t^{-1}(t_2(n))\right) = r(n)$. Furthermore, $r(n) < \frac{t_2(n)}{n}$ implies $r(n) \cdot r^{-1}(r(n)) = r(n) \cdot n < c \cdot t_1(n)$ for all $c > 0$. Consequently $r(n) < \min\{k : k \cdot r^{-1}(k) \geqq c \cdot t_1(n)\} = \bar{r}(c \cdot t_1(n))$ for all $c > 0$. Hence we get $r(n) \lesssim t_3(n)$. ∎

Now we give some applications of theorem 27.3.

27.4. Examples.

1. If T-DTIME-Comp$(A) \asymp n^2$ and $r_k(n) = n^{1+1/3^k}$ $(k = 1, 2, \ldots)$, then

$$\text{T-DTIME}^{A_{r_{k+1}}}\text{-Comp}(A) \lesssim n^{1+1/3^{k+1}} < n^{1+1/2\cdot 3^k+1}$$
$$\lesssim \text{T-DTIME}^{A_{r_k}}\text{-Comp}(A) \lesssim n^{1+1/3^k}.$$

Consequently, there is an infinite hierarchy of oracles with respect to their ability to help the decision of A. For example, take $A = \mathrm{S}$.

2. If T-DTIME-Comp$(A) \asymp n^\beta$, $\mathrm{r(n)} = \mathrm{n}^\alpha$ and $1 \leqq \alpha < \beta - 1$, then

$$\text{T-DTIME}^{A_r}\text{-Comp}(A) \asymp n^\alpha.$$

3. If $2^{d'\cdot n} \lesssim$ T-DTIME-Comp$(A) \lesssim 2^{d\cdot n}$ and $r(n) = n^\alpha$ $(d \geqq d' > 0, \alpha \geqq 1)$, then

$$\text{T-DTIME}^{A_r}\text{-Comp}(A) \asymp n^\alpha.$$

4. If $2^{d'\cdot n} \lesssim$ T-DTIME-Comp$(A) \lesssim 2^{d\cdot n}$ and $r(n) = 2^{d''\cdot n}$ $(d \geqq d' > 0, d \geqq d'' > 0)$, then

$$2^{\frac{d''\cdot d'}{d}\cdot n} \lesssim \text{T-DTIME}^{A_r}\text{-Comp}(A) \lesssim 2^{d''\cdot n}. \blacksquare$$

Remark. Theorem 27.3 remains valid if T-DTIME is replaced by kT-DTIME for $k > 1$.

Now we state analogous theorems for multiT-DTIME and T-DSPACE, which can be proved in the same manner.

27.5. Theorem. Let $r, t_1, t_2 \in \mathrm{R}_1$ such that

a) $\mathrm{id} \leqq t_1 \lesssim$ multiT-DTIME-Comp$(A) \lesssim t_2$ and $\mathrm{id} \leqq r \lesssim t_2$,

b) r strictly increasing, t_2^{-1} increasing and $t_2 \circ r^{-1}$ superadditive,

c) $r^{-1}(|w|) \in$ multiT-DTIME$(t_2 \circ r^{-1})$ and $r(|w|) \in$ multiT-DTIME(r).

Then

$$r\left(t_2^{-1}(t_1(n))\right) \lesssim \text{multiT-DTIME}^{A_r}\text{-Comp}(A) \lesssim r(n).$$

If in addition $t_1 = t_2$ and t_2 strictly increasing, then

$$\text{multiT-DTIME}^{A_r}\text{-Comp}(A) \asymp r(n). \blacksquare$$

27.6. Theorem. [WaWe 77] Let $r, s_1, s_2 \in \mathbb{R}_1$ such that

a) $\mathrm{id} \leqq s_1 \lesssim$ T-DSPACE-Comp$(A) \lesssim s_2$ and $\mathrm{id} \leqq r \lesssim s_2$,

b) r strictly increasing and s_2^{-1}, $s_2 \circ r^{-1}$ increasing,

c) $r^{-1}(|w|) \in$ T-DSPACE$(s_2 \circ r^{-1})$ and $r(|w|) \in$ T-DSPACE(r).

Then

$$r\left(s_2^{-}((s_1(n))\right) \lesssim \text{T-DSPACE}^{A_r}\text{-Comp}(A) \lesssim r(n).$$

If in addition $s_1 = s_2$ and s_2 strictly increasing, then

$$\text{T-DSPACE}^{A_r}\text{-Comp}(A) \asymp r(n). \blacksquare$$

Remark. The Examples 27.4.2, 27.4.3, and 27.4.4 remain valid if T-DTIME is replaced by multiT-DTIME or T-DSPACE.

It is an immediate consequence of the preceding theorem that there are arbitrarily complex recursive sets*) A such that for every "well-behaved" $r \leqq$ T-DSPACE-Comp(A) there is a recursive set B such that T-DSPACE^B-Comp$(A) \asymp r(n)$. This statement will be strengthened by the following theorem.

27.7. Theorem. [Dek 74]

$$\bigwedge_{s \in \mathbb{R}_1} \bigvee_{A \in \mathrm{REC}} \left(\text{T-DSPACE-Comp}(A) >_{\mathrm{ae}} s \right.$$
$$\left. \wedge \bigvee_{c>0} \bigwedge_{\substack{r \leqq s \\ r\,\text{T-DSPACE-} \\ \text{constructible}}} \bigvee_{B \in \mathrm{REC}} r(n) - c \leqq_{\mathrm{ae}} \text{T-DSPACE}^B\text{-Comp}(A) \leqq r(n) \right).$$

The theorems 27.3, 27.5, and 27.6 show that the amount by which the oracle A_r can diminish the complexity of A depends on the size of r, and consequently on the complexity of A_r. This observation will be generalized in the subsequent two theorems.

27.8. Theorem. For recursive sets A, B and for $t_1, t_2 \in \mathbb{R}_1$ such that $t_1 > \mathrm{id}$, $t_2 \geqq \mathrm{id}$ and t_2 is superadditive,

1. multiT-DTIME-Comp$(A) >_{\mathrm{ae}} t_2 \circ t_1 \wedge$ multiT-DTIME-Comp$(B) \leqq_{\mathrm{ae}} t_2$
$\rightarrow$ multiT-DTIME^B-Comp$(A) >_{\mathrm{ae}} t_1$.
2. multiT-DTIME-Comp$(A) >_{\mathrm{io}} t_2 \circ t_1 \wedge$ multiT-DTIME-Comp$(B) \leqq_{\mathrm{ae}} t_2$
$\rightarrow$ multiT-DTIME^B-Comp$(A) >_{\mathrm{io}} t_1$.

Proof. We prove Statement 1, Statement 2 can be proved in the same manner. Assume that there is an oracle kT-DM M which, using the oracle B, decides A within time $t_3 \leqq_{\mathrm{io}} t_1$. Then we can construct an lT-DM M' (for some $l > k$) without oracle which simulates the work of M within time

$$t_3(n) + \sum_{i=1}^{m} t_2(|w_i|) \leqq t_3(n) + t_2\left(\sum_{i=1}^{m} |w_i|\right) \leqq t_3(n) + t_2(t_3(n)) \leqq 2t_2(t_3(n)),$$

where $w_1, \ldots, w_m$ are the questions of M.

Now, t_2 is increasing because it is superadditive. Hence $t_3 \leqq_{\mathrm{io}} t_1$ implies $2 \cdot t_2(t_3(n)) \leqq_{\mathrm{io}} 2 \cdot t_2(t_1(n))$. Consequently, the inequality $t_2(t_1(n)) \geqq t_1(n) \geqq_{\mathrm{ae}} 2n$ and the linear speed-up for multiT-DTIME give multiT-DTIME-Comp$(A) \leqq_{\mathrm{io}} t_2 \circ t_1$, which contradicts the supposition. ∎

The next theorem can be proved in the same manner.

---

*) The expression "there are arbitrarily complex sets A such that ..." stands for

$$\bigwedge_{t \in \mathbb{R}_1} \bigvee_{A \in \mathrm{REC}} (\Phi\text{-Comp}(A) >_{\mathrm{ae}} t \wedge \ldots).$$

The expression "for all sufficiently complex sets B we have ..." stands for

$$\bigvee_{t \in \mathbb{R}_1} \bigwedge_{A \in \mathrm{REC}} (\Phi\text{-Comp}(A) \geqq_{\mathrm{ae}} t \rightarrow \ldots).$$

27.9. Theorem. For recursive sets A, B and for functions $s_1, s_2 \in \mathbb{R}_1$ such that $s_1 \geqq \mathrm{id}$, $s_2 \geqq \mathrm{id}$ and s_2 is strictly increasing,

1. $\text{T-DSPACE-Comp}(A) >_{\text{ae}} s_2 \circ s_1 \wedge \text{T-DSPACE-Comp}(B) \leqq_{\text{ae}} s_2 \rightarrow \text{T-DSPACE}^B\text{-Comp}(A) >_{\text{ae}} s_1$.
2. $\text{T-DSPACE-Comp}(A) >_{\text{io}} s_2 \circ s_1 \wedge \text{T-DSPACE-Comp}(B) \leqq_{\text{ae}} s_2 \rightarrow \text{T-DSPACE}^B\text{-Comp}(A) >_{\text{io}} s_1$. ∎

Statement 2 of the preceding theorem can be generalized to hold for arbitrary relative Blum measures.*)

27.10. Theorem. [LyMeFi 76] There is a $g \in \mathbb{R}_1$ such that for recursive sets A, B and for increasing unbounded $t_1, t_2 \in \mathbb{R}_1$,

$$\Phi\text{-Comp}(A) >_{\text{io}} g \circ t_2 \circ g \circ t_1 \wedge \Phi\text{-Comp}(B) \leqq t_2 \rightarrow \Phi^B\text{-Comp}(A) >_{\text{io}} t_1 . \blacksquare$$

The next theorem shows that for any sufficiently complex set A there exist arbitrarily complex sets B that do help the decision of A.

27.11. Theorem. [LyMeFi 76] There is an $h \in \mathbb{R}_2$ and a $g \in \mathbb{R}_1$ such that

$$\bigwedge_{A \in \mathbf{REC}} \bigwedge_{\substack{t \text{ incr.} \\ t \in \{\Phi_i : i \in \mathbb{N}\}}} \Big(\Phi\text{-Comp}(A) \leqq_{\text{ae}} t \rightarrow \bigvee_{B \in \mathbf{REC}} \big(t <_{\text{ae}} \Phi\text{-Comp}(B) \leqq_{\text{ae}} h \,\Box\, t \wedge \Phi^B\text{-Comp}(A) \leqq g\big)\Big) . \blacksquare$$

Finally we mention a result which states that there is a recursive set A whose decision is helped by all sets whose complexities are compressed around "well-behaved" bounds.

27.12. Theorem. [LyMeFi 76] For any $s \in \mathbb{R}_2$ there exists an $s' \in \mathbb{R}_2$ such that for all $h \in \mathbb{R}_2$,

$$\bigvee_{A} \bigwedge_{\substack{t \in \mathbb{R}_1 \\ t \in \{\Phi_i : i \in \mathbb{N}\}}} \bigwedge_{B} \Big(s' \,\Box\, t <_{\text{io}} \Phi\text{-Comp}(B) \leqq_{\text{ae}} h \,\Box\, t \rightarrow \bigvee_{t' \in \mathbb{R}_1} \big(\Phi^B\text{-Comp}(A) \leqq_{\text{ae}} t' \wedge s \,\Box\, t' \leqq_{\text{io}} \Phi\text{-Comp}(A)\big)\Big) . \blacksquare$$

In [Schg 84b] "robust" algorithms are considered. An oracle machine is called robust if the choice of the oracle can only influence the computation time but not the set accepted by the machine. For nondeterministic robust oracle machines the oracle can always be eliminated without loss of time. For deterministic robust oracle machines working in polynomial time the help of an oracle cannot always be eliminated without increasing the computation time, unless $\mathbf{P} = \mathbf{NP} \cap \mathbf{coNP}$.

27.2. Oracles that Do Not Help

It is a corollary of Theorem 27.1 that regular oracles cannot help the decision of any set. However, this is not surprising since regular sets have a very small complexity. Thus the question arises of whether there are arbitrarily complex oracles that cannot

*) In the following theorems the quantification "for all relative Blum measures $(\varphi^{(\,)}, \Phi^{(\,)})$" has generally been omitted. Furthermore, (φ, Φ) stands for $(\varphi^{\emptyset}, \Phi^{\emptyset})$.

help the decision of a given set. This question is answered in the affirmative by the following theorem. This and all subsequent theorems of this section are proved by diagonalization and (except for theorem 27.22) by priority arguments which, in general, are more complex than the classical priority constructions (cf. [Rog 67], Chapter 3). As an example of this kind of arguments we give the proof of Theorem 27.23.

27.13. Theorem. [LyMeFi 76], [Mac 72] There is a $g \in \mathbb{R}_2$ such that

$$\bigwedge_{A \in \mathrm{REC}} \bigwedge_{t \in \mathbb{R}_1} \left(\Phi\text{-}\mathrm{Comp}(A) >_{\mathrm{io}} g \square t \to \bigvee_{\mathrm{arb.compl.}\ B} \Phi^B\text{-}\mathrm{Comp}(A) >_{\mathrm{io}} t \right). \quad \blacksquare$$

Remark 1. For $\Phi = \mathrm{T}\text{-}\mathrm{DSPACE}$ we can choose $g(x, y) = y$. However, g cannot be omitted for all measures. This can be verified by the construction of certain pathological measures (cf. [LyMeFi 76]).

Remark 2. The complexity of B in Theorem 27.13 can be compressed between t' and $g \square t'$ if some additional assumptions for t' are fulfilled, namely that $t' \in \{\Phi_i : i \in \mathbb{N}\}$, and that t' is increasing and much larger than the complexities of t and A (cf. [LyMeFi 76]).

A variation of Theorem 27.13 for the space measure is given by the next theorem.

27.14. Theorem. [Cho 69] For every increasing $h \in \{\mathrm{T}\text{-}\mathrm{DSPACE}_i : i \in \mathbb{N}\}$,

$$\bigvee_A \bigvee_B \left(\mathrm{T}\text{-}\mathrm{DSPACE}\text{-}\mathrm{Comp}(A) \asymp \mathrm{T}\text{-}\mathrm{DSPACE}\text{-}\mathrm{Comp}(B) \asymp h \right.$$
$$\left. \wedge\ \mathrm{T}\text{-}\mathrm{DSPACE}\text{-}\mathrm{Comp}^B(A) \succeq_{\mathrm{io}} h \right). \quad \blacksquare$$

The next theorem strengthens theorem 27.13 in two ways: the set B can be chosen in such a way that its decision cannot be helped by A and the function t can be replaced by a recursive enumerable set of functions.

27.15. Theorem. [Dek 74] There is a $g \in \mathbb{R}_2$ such that

$$\bigwedge_{A \in \mathrm{REC}} \bigwedge_{T \subseteq \mathbb{R}_1} \left(\bigwedge_{t \in T} \Phi\text{-}\mathrm{Comp}(A) >_{\mathrm{io}} g \square t \to \bigvee_{\mathrm{arb.compl.}\ B} \left(\bigwedge_{t \in T} \Phi^B\text{-}\mathrm{Comp}(A) >_{\mathrm{io}} t \right.\right.$$
$$\left.\left. \wedge \bigwedge_{t'} \left(\Phi\text{-}\mathrm{Comp}(B) >_{\mathrm{io}} t' \to \Phi^A\text{-}\mathrm{Comp}(B) >_{\mathrm{io}} t' \right) \right) \right). \quad \blacksquare$$

Remark 1. For $\Phi = \mathrm{T}\text{-}\mathrm{DSPACE}$ we can choose $g(x, y) = y$ (cf. [Dek 74]).

Remark 2. The strongest result in this direction would be: There is a $g \in \mathbb{R}_2$ such that for any recursive A there exists an arbitrarily complex set B such that

$$\bigwedge_t \left(\Phi\text{-}\mathrm{Comp}(A) >_{\mathrm{io}} g \square t \to \Phi^B\text{-}\mathrm{Comp}(A) >_{\mathrm{io}} t \right)$$
$$\wedge \bigwedge_t \left(\Phi\text{-}\mathrm{Comp}(B) >_{\mathrm{io}} t \to \Phi^A\text{-}\mathrm{Comp}(B) >_{\mathrm{io}} t \right).$$

However, it is not known whether this statement holds true. It is only known that there are arbitrarily complex sets A and B with this property. This is the object of the next theorem. Furthermore, it is even not known whether

$$\bigvee_{g \in \mathbb{R}_2} \bigwedge_A \bigvee_{\mathrm{arb.compl.}\ B} \bigwedge_t \left(\Phi\text{-}\mathrm{Comp}(A) >_{\mathrm{io}} g \square t \to \Phi^B\text{-}\mathrm{Comp}(A) >_{\mathrm{io}} t \right)$$

holds true. Note that theorem 27.12 does not contradict this statement since the property $\bigvee\limits_{t'\in\mathbb{R}_1} \left(\Phi^B\text{-Comp}(A) \leq_{ae} t' \wedge s \,\square\, t' <_{io} \Phi\text{-Comp}(A)\right)$ cannot be guaranteed for all sufficiently complex sets B.

27.16. Corollary. (implicit in [Mac 72])

$$\bigwedge_{A\in\text{REC}\backslash\mathbb{P}r} \bigvee_{\text{arb.compl.}B} \bigwedge_{t\in\mathbb{P}r} \left(\Phi^B\text{-Comp}(A) >_{io} t \wedge \Phi^A\text{-Comp}(B) >_{io} t\right). \blacksquare$$

Remark. As a corollary of 27.16 we get the earliest result in the literature which can be stated in terms of the existence of two recursive sets not helping each other's decision: There are nonprimitive recursive sets A and B such that neither can make the other primitive recursive (cf. [Axt 59]).

The next theorem is due to M. K. VALIEV and to A. R. MEYER and M. J. FISCHER and includes a complexity-theoretic analogue to the Friedberg-Muchnik Theorem (cf. [Rog 67], § 10.2.).

27.17. Theorem. [MeFim 72], [LyMeFi 76]

$$\bigvee_{\text{arb.compl.}A} \bigvee_{\text{arb.compl.}B} \left(\bigwedge_t \left(\Phi\text{-Comp}(A) >_{io} t \to \Phi^B\text{-Comp}(A) >_{io} t\right)\right.$$
$$\left.\wedge \bigwedge_t \left(\Phi\text{-Comp}(B) >_{io} t \to \Phi^A\text{-Comp}(B) >_{io} t\right)\right). \blacksquare$$

A direct complexity-theoretic analogue of the theorem of FRIEDBERG and MUCHNIK is the following corollary of 24.31.

27.18. Theorem. $\bigvee\limits_{A\in\text{DEXPTIME}\backslash\mathbb{P}} \bigwedge\limits_{B\in\text{DEXPTIME}\backslash\mathbb{P}} (A \notin \mathbb{P}^B \wedge B \notin \mathbb{P}^A). \blacksquare$

In the remainder of this subsection we deal with results in which ae lower bounds are preserved (instead of io lower bounds as in the results above). First we give the ae analogue to Theorem 27.13.

27.19. Theorem. [LyMeFi 76] There is a $g \in \mathbb{R}_2$ such that

$$\bigwedge_{A\in\text{REC}} \bigwedge_{\substack{t\in\mathbb{R}_1 \\ t\geq\text{id} \\ t\in\{\Phi_i : i\in\mathbb{N}\}}} \left(\Phi\text{-Comp}(A) >_{ae} g \,\square\, t \to \bigvee_{\text{arb.compl.}B} \Phi^B\text{-Comp}(A) >_{ae} t\right). \blacksquare$$

Remark. A stronger result in this direction would be: there is a $g \in \mathbb{R}_2$ such that

$$\bigwedge_A \bigvee_{\text{arb.compl.}B} \bigwedge_t \left(\Phi\text{-Comp}(A) >_{ae} g \,\square\, t \to \Phi^B\text{-Comp}(A) >_{ae} t\right).$$

However, it is not known whether this holds true.

The complexity of the set B in Theorem 27.19 can be compressed:

27.20. Theorem. [LyMeFi 76] There is a $g \in \mathbb{R}_2$ such that

$$\bigwedge_{A\in\text{REC}} \bigwedge_{\substack{t\in\mathbb{R}_1 \\ t\geq\text{id} \\ t\in\{\Phi_i : i\in\mathbb{N}\}}} \left(\Phi\text{-Comp}(A) >_{ae} g \,\square\, t\right.$$
$$\left. \to \bigwedge_{\substack{t'\in\mathbb{R}_1 \\ t' \text{ incr.} \\ t'\in\{\Phi_i : i\in\mathbb{N}\}}} \bigvee_B \left(t' <_{ae} \Phi\text{-Comp}(B) \leq_{ae} g \,\square\, t' \wedge \Phi^B\text{-Comp}(A) >_{ae} t\right)\right). \blacksquare$$

Now we would like to present a result which strengthens Theorem 27.19 in such a way that the decision of B cannot be helped by A (i.e. an analogue to Theorem 27.15). Unfortunately only the following result is known, stating the existence of arbitrarily complex sets A and B which do not help each other very much.

27.21. Theorem. [LyMeFi 76] There is a $g \in \mathbb{R}_2$ such that for all $t, t' \in \{\Phi_i : i \in \mathbb{N}\} \cap \mathbb{R}_1$

$$\bigvee_{A \in \mathrm{REC}} \bigvee_{B \in \mathrm{REC}} \big(\Phi\text{-Comp}(A) \leqq_{\mathrm{ae}} g \circ t \wedge \Phi^B\text{-Comp}(A)\big) >_{\mathrm{ae}} t$$

$$\wedge\ \Phi\text{-Comp}(B) \leqq_{\mathrm{ae}} g \circ t' \wedge \Phi^A\text{-Comp}(B) >_{\mathrm{ae}} t'\big). \ \blacksquare$$

Remark. Theorem 27.21 can be shown using Theorem 27.23. The set A in Theorem 27.23 is split into two sets such that neither set can help the other's decision.

27.3. Autoreducibility

The following question is raised by B. A. Trachtenbrot in [Tra 70]: Can the computation of a (0-1 valued) function for a given input be helped by free access to other values of this function? There are two results concerning this question, stating that there are arbitrarily complex sets whose decision can (cannot) be helped in this way. For the formulation of these results we use the following notation: For a relativized Blum measure Φ, a set A and $g \in \mathbb{R}_1$,

$$\Phi^{[A]}\text{-Comp}(A) \leqq_{\mathrm{ae}} g \leftrightarrow_{\mathrm{df}} \bigvee_i \Big(\bigwedge_x \varphi_i^{A\setminus\{x\}}(x) = c_A(x) \wedge \bigwedge_x^{\mathrm{ae}} \Phi_i^{A\setminus\{x\}}(x) \leqq g(x)\Big),$$

$$\Phi^{[A]}\text{-Comp}(A) >_{\mathrm{ae}} g \leftrightarrow_{\mathrm{df}} \bigwedge_i \Big(\bigwedge_x \varphi_i^{A\setminus\{x\}}(x) = c_A(x) \rightarrow \bigwedge_x^{\mathrm{ae}} \Phi_i^{A\setminus\{x\}}(x) > g(x)\Big).$$

The next theorem is due to M. Paterson and can be found in [LyMeFi 76].

27.22. Theorem. There exist a $g \in \mathbb{R}_2$ and an $r \in \mathbb{R}_1$ such that

$$\bigwedge_{\substack{t \in \mathbb{R}_1 \\ t \text{ incr.} \\ t \in \{\Phi_i : i \in \mathbb{N}\}}} \bigvee_{A \in \mathrm{REC}} (t <_{\mathrm{ae}} \Phi\text{-Comp}(A) \leqq_{\mathrm{ae}} g \circ t \wedge \Phi^{[A]}\text{-Comp}(A) \leqq_{\mathrm{ae}} r). \ \blacksquare$$

27.23. Theorem. [Tra 70], [MeFim 72], [LyMeFi 76] There is a $g \in \mathbb{R}_2$ such that for any $t \in \mathbb{R}_1 \cap \{\Phi_i : i \in \mathbb{N}\}$,

$$\bigvee_{A \in \mathrm{REC}} (\Phi\text{-Comp}(A) \leqq_{\mathrm{ae}} g \circ t \wedge \Phi^{[A]}\text{-Comp}(A) >_{\mathrm{ae}} t). \ \blacksquare$$

Proof. The proof presented here is taken with slight modifications from [LyMeFi 76]. We would like to ensure

$$\bigwedge_i \Big(\Phi_i^{A\setminus\{x\}}(x) \leqq_{\mathrm{io}} t(x) \rightarrow \bigvee_z \varphi^{A\setminus\{z\}}(z) \neq c_A(z)\Big).$$

At any time during the construction some indices may be "cancelled". An index i is cancelled when the construction has ensured that

$$\bigvee_x \varphi_i^{A\setminus\{x\}}(x) \neq c_A(x).$$

In addition, at any time during the construction a single index may be "tentatively cancelled". An index i is tentatively cancelled when the process of attempting to cancel i by defining A in a suitable manner has not yet finished. If this process succeeds, then i will be cancelled; otherwise, the tentative cancellation of i will be removed.

The construction defines a function $\varphi(a, n)$ for a fixed number a. By the s-m-n-Theorem there is a recursive function s such that $\varphi_{s(a)}(n) = \varphi(a, n)$. For $\Phi_a = t$ we get the desired set A by $c_A = \varphi_{s(a)}$.

The construction is made in stages. If $m = \langle n, \Phi_a(n)\rangle$, then in stage m we will define $\varphi(a, n)$. For $m \geqq 0$ let

$$B_m = \{k: \varphi(a, k) \text{ has been defined in one of the stages } 0, 1, \ldots, m\}.$$

In the following algorithm we use the variable TC. If $TC = n$, then $i_n \in \mathbb{N}$ and a finite set $E_n \subseteqq \mathbb{N}$ are defined. $TC = n$ means that in stage $\langle n, \Phi_a(n)\rangle$ the index i_n is tentatively cancelled because E_n is tentatively different from the set A to be constructed. If TC is assigned the value $\emptyset$, then the previous tentative cancellation is removed. Initially, $TC = \emptyset$.

Stage m. If $m = \langle n, \Phi_a(n)\rangle$ for some n, then go to 1. Otherwise go to 4.

1. Find the smallest $i \leqq n$ that is not yet cancelled and such that
   a) if $TC = k$, then $i < i_k$,
   b) there exists an E such that
      α) $E \subseteqq \{0, \ldots, h(i, n, \Phi_a(n))\} \cup B_{m-1}$
      (h is chosen according to Theorem 5.31),
      β) $\bigwedge_{l \in B_{m-1}} c_E(l) = \varphi(a, l)$,
      γ) $n \notin E$,
      δ) $\Phi_i^E(n) \leqq \Phi_a(n)$.
2. If such an i exists then $TC = \emptyset$,
$$\varphi(a, n) = 1 \dot{-} \varphi_i^E(n),$$
$$TC = n,$$
$$i_n = i,$$
$$E_n = E.$$
3. If no such i exists then
   3.1. If $TC = k$ then $\varphi(a, n) = c_{E_k}(n)$
   if $\{0, \ldots, h(i_k, k, \Phi_a(k))\} \subseteqq B_m$ then $TC = \emptyset$ and cancel i_k.
   3.2 If $TC = \emptyset$ then $\varphi(a, n) = 0$.
4. Go to stage $m + 1$.

In order to show that $\Phi_i^{A\setminus\{x\}}(x) \leqq t = \Phi_a$ implies $\varphi_i^{A\setminus\{x\}}(x) \neq c_A(x) = \varphi_{s(a)}(x)$ we prove first two lemmas.

Lemma 1. If an index i is tentatively cancelled at some stage, then some index $j \leqq i$ will be cancelled later.

Proof. A tentative cancellation of i is removed if and only if i is cancelled or an index $j < i$ is tentatively cancelled. Furthermore, it is easy to see that every tentative cancellation will be removed if $\Phi_a \in \mathbb{R}_1$. □

Lemma 2. If an index i is cancelled, then $\varphi_i^{A\setminus\{x\}}(x) \neq c_A(x)$ for some x.

Proof. Let i be cancelled. Let stage $m = \langle k, \Phi_a(k)\rangle$ be the last stage at which i was tentatively cancelled. Thus $i_k = i$, $\varphi(a, k) = 1 \dot{-} \varphi_{i_k}^{E_k}(k)$, $c_{E_k}(l) = \varphi(a, l)$ for all $l \in B_{m-1}$ and $E_k \subseteq \{0, \ldots, h(i_k, k, \Phi_a(k))\} \cup B_{m-1}$ at stage m. Up to the stage $m' = \langle n, \Phi_a(n)\rangle$ at which i is finally cancelled, the construction gives $\varphi(a, l) = c_{E_k}(l)$ for all $l \in B_{m'} \setminus B_m$ and $B_{m'} \supseteq \{0, \ldots, h(i_k, k, \Phi_a(k))\}$. Consequently, we have $c_A(l) = \varphi(a, l) = c_{E_k}(l)$ for all $l \in \{0, \ldots, h(i_k, k, \Phi_a(k))\} \setminus \{k\}$. By theorem 5.31 and $\Phi_{i_k}^{E_k}(k) \leq \Phi_a(k)$ we get

$$\varphi_i^{A\setminus\{k\}}(k) = \varphi_{i_k}^{E_k}(k) \neq \varphi(a, k) = c_A(k). \ \square$$

Now assume that $\Phi_i^{A\setminus\{x\}}(x) \leq_{\text{io}} t(x) = \Phi_a(x)$. Let m_0 be the first stage at which all indices smaller than i that will ever be cancelled have already been cancelled. Choose $n \geq i$ such that $m_0 < m_1 =_{\text{df}} \langle n, \Phi_a(n)\rangle$ and $\Phi_i^{A\setminus\{n\}}(n) \leq \Phi_a(n)$. Assume that i has not been cancelled up to stage m_1.

Case 1. $TC = l$ and $i_l = i$ at stage m_1. Then by Lemma 1 the index i will be cancelled in this or a later stage.

Case 2. $TC = l$ and $i_l \neq i$ at stage m_1. Then by Lemma 1 we have $i < i_l$, and $i \leq n$ is the smallest index fulfilling 1a) and 1b) where $E = (A \setminus \{n\}) \cap \{0, \ldots, h(i, n, \Phi_a(n))\}$. Hence i is tentatively cancelled at this stage and, by Lemma 1, finally cancelled later.

Case 3. $TC = \emptyset$ at stage m_1. Here we proceed as in Case 2. In all cases i will be cancelled during the construction. By Lemma 2 we get $\varphi_i^{A\setminus\{x\}}(x) \neq c_A(x)$ for some x. Consequently, $\Phi^{[A]}$-Comp$(A) >_{\text{ae}} t$. Finally, since $\mathbf{D}_i \subseteq \mathbf{D}_{s(i)}$ for all i, the Combining Lemma (5.27) yields $\Phi_{s(i)} \leq_{\text{ae}} g \mathbin{\square} \Phi_i$ for all $i \in \mathbb{N}$ and some $g \in \mathbb{R}_2$. Consequently, $\Phi_{s(a)} \leq_{\text{ae}} g \mathbin{\square} t$, i.e. Φ-Comp$(A) \leq g \mathbin{\square} t$. ∎

§ 28. Relativization of the D-ND-Problems and the Space-Time Problems

In this section we deal with the behaviour of our main problems under relativization. It turns out that they become positively solvable for some oracles and negatively solvable for some other oracles. This suggests that it might be not so easy to solve the unrelativized problems, for instance to prove $\mathbf{P} \neq \mathbf{NP}$, since such a proof must not be relativizable. The intuitive belief that diagonalizations are always relativizable and that therefore $\mathbf{P} \neq \mathbf{NP}$ cannot be proved by diagonalization is, however, wrong. On the contrary, if $\mathbf{P} \neq \mathbf{NP}$, then this can be proved by diagonalization. This is a consequence of the following general result from [Koz 80b]: If $C = \{\varphi_{s(i)} : i \in \mathbb{N}\} \subseteq \mathbb{R}_1$ is a class of predicates, $s \in \mathbb{R}_1$, then for every $A \in \mathbf{REC}$ with $c_A \notin C$ there exists an $h \in \mathbb{R}_1$ such that $c_A(x) = 1 \dot{-} \varphi_{s(h(x))}(x)$ and for every $i \in \mathbb{N}$ there exists an x such that $\varphi_{s(i)} = \varphi_{s(h(x))}$. Thus A arises as a "diagonal over C" in a diagonalization process controlled by h. In particular, if $\mathbf{P} \neq \mathbf{NP}$, then SAT must be a diagonal over $\mathbf{P}$. For further investigations on this topic see [KoMt 80].

In studying relativizations we take into consideration that the proofs of the time and space hierarchy results in § 17 (including those by diagonalization) relativize without essential difficulties, and thus we can start with the situation represented in Figure 28.1. A justification for the position of $\mathbf{P}^A$ relative to $\mathbf{NL}_*^A$ in the figure, which does not correspond to the known fact $\mathbf{NL} \subseteq \mathbf{P}$, will be given in 28.11. (For relativized space classes see p. 83.)

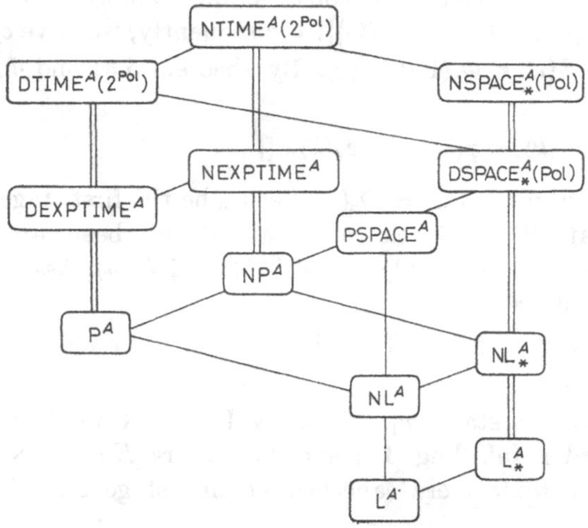

Figure 28.1. Double lines mean strict inclusion

28.1. Space

We start with the logspace problem. Our first result is stated in a broader context than relevant only for the logspace problem.

28.1. Theorem. If A is $\leqq_m^{\log}$-complete in **PSPACE**, then $\mathbf{L}_*^A = \mathbf{NL}_*^A = \mathbf{P}^A = \mathbf{NP}^A$.

Proof. Let $B \in \mathbf{NP}^A$. This together with $A \in \mathbf{PSPACE}$ implies $B \in \mathbf{PSPACE}$. But then $B \leqq_m^{\log} A$ because of the completeness of A, and this means $B \in \mathbf{L}_*^A$. ∎

For a suitable relativization a result contrasting to theorem 28.1 can be proved:

28.2. Theorem. [LaLy 76] For every $k \in \mathbb{N}$ there exists a recursive A such that $\mathbf{NL}_*^A \nsubseteq \mathbf{DSPACE}_*^A(n^k)$ and thus $\mathbf{L}_*^A \subset \mathbf{NL}_*^A$.

Remark. The proof of 28.2 goes through for: For every $k \in \mathbb{N}$ there exists a recursive A such that $\mathbf{NSPACE}_*^A(\mathrm{Pol}) \nsubseteq \mathbf{DSPACE}_*^A(2^{n^k})$ and thus $\mathbf{DSPACE}_*^A(\mathrm{Pol}) \subset \mathbf{NSPACE}_*^A(\mathrm{Pol})$. Consequently, W. J. SAVITCH's result 23.1 does not relativize to arbitrary oracles. See also p. 492. For $\mathbf{NSPACE}^A$, however, 23.1 relativizes.

Proof. We define an auxiliary function h by

$$h(0) = 1, \qquad h(n+1) = 2^{h(n)^{k+1}}$$

and define $H =_{\mathrm{df}} \{1^{h(n)} : n \in \mathbb{N}\}$. Obviously, $H \in \mathbf{L}$.

We fix upon an effective enumeration $\mathfrak{M} = \{M_0^{()}, M_1^{()}, \ldots\}$ of oracle machines working within space n^k and accepting all sets of $\mathbf{DSPACE}_*^Y(n^k)$ for all oracles Y. We construct A and B such that $B \in \mathbf{NL}_*^A \setminus \mathbf{DSPACE}_*^A(n^k)$. A and B are defined in stages by diagonalization over $\mathfrak{M}$. In stage i certain segments A_i and B_i of A and B are defined, and finally we set $A = \bigcup_{i \in \mathbb{N}} A_i$, $B = \bigcup_{i \in \mathbb{N}} B_i$.

Let M_j have q_j states and p_j symbols. Then on inputs of length n the machine M_j can make at most $(q_j \cdot n \cdot n^k \cdot p_j^{n^k})^2$ moves. Let $A_{i-1}, B_{i-1}, n_0, \ldots, n_{i-1}$ be already defined. To define A_i and B_i we choose $n_i > n_{i-1}$ in such a way that

$$1^{n_i} \in H \quad \text{and} \quad 2^{n_i^{k+1}} > \left(q_i n_i^{k+1} \cdot p_i^{n_i^k}\right)^2. \tag{1}$$

Thus, $2^{n_i^{k+1}}$ is an upper bound on the number of steps of M_i on input 1^{n_i}. Hence there must exist a word y of length n_i^{k+1} which is not used as a query of $M_i^{A_{i-1}}$ on input 1^{n_i}. Choose such a y and define

$$A_i = \begin{cases} A_{i-1} & \text{if } M_i^{A_{i-1}} \text{ accepts } 1^{n_i}, \\ A_{i-1} \cup \{y\} & \text{if } M_i^{A_{i-1}} \text{ rejects } 1^{n_i}, \end{cases}$$

$$B_i = \begin{cases} B_{i-1} & \text{if } M_i^{A_{i-1}} \text{ accepts } 1^{n_i}, \\ B_{i-1} \cup \{1^{n_i}\} & \text{if } M_i^{A_{i-1}} \text{ rejects } 1^{n_i}. \end{cases}$$

Thus, taking into account that y is not used by $M_i^{A_{i-1}}$ as a query on input 1^{n_i} we obtain

$$1^{n_i} \in B_i \leftrightarrow 1^{n_i} \notin \mathrm{L}(M_i^{A_i}). \tag{2}$$

From the choice of the length of y in the previous definition it immediately follows that later steps in the construction of A do not affect its power for earlier diagonalization points. More precisely

$$\bigwedge_i \left(1^{n_i} \in \mathrm{L}(M_i^{A_i}) \leftrightarrow 1^{n_i} \in \mathrm{L}(M_i^A)\right). \tag{3}$$

Similarly, later steps in the definition of B may add only larger elements to B:

$$\bigwedge_i (1^{n_i} \in B_i \leftrightarrow 1^{n_i} \in B). \tag{4}$$

We prove $B \notin \mathbf{DSPACE}_*^A(n^k)$. If this were not the case there would be a machine M_j^A accepting B in space n^k. Then by (4), (2) and (3) we obtain

$$1^{n_j} \in B \leftrightarrow 1^{n_j} \notin \mathrm{L}(M_j^A)$$

contradicting the assumption.

That B belongs to $\mathbf{NL}^A$ follows from the evident representation

$$B = \left\{1^n : 1^n \in H \wedge \bigvee_{w \in A} (|w| = n^{k+1})\right\}.$$

The decidability of A follows from its being enumerated by an increasing function. ∎

28.2. Time

We consider relativized versions of **P**, **NP** and **coNP**. The trivial implications for these classes are also valid for the relativized versions so that we have the situation shown in Figure 28.2.

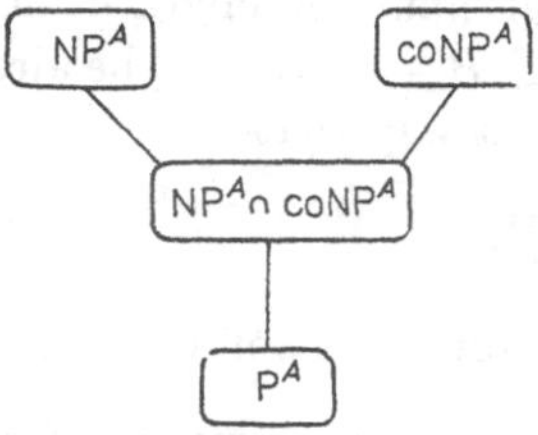

Figure 28.2

The following implications are straightforward:

28.3. Proposition.

1. $\mathbf{P}^A = \mathbf{NP}^A \to \mathbf{NP}^A = \mathbf{coNP}^A = \mathbf{NP}^A \cap \mathbf{coNP}^A$.
2. $\mathbf{NP}^A = \mathbf{coNP}^A \leftrightarrow \mathbf{NP}^A = \mathbf{NP}^A \cap \mathbf{coNP}^A$. ∎

All relations compatible with these implications may be realized by suitably constructed oracles:

28.4. Theorem. [BaGiSo 75] There are recursive oracles A, B, C, D such that

1. $\mathbf{P}^A = \mathbf{NP}^A$,
2. $\mathbf{P}^B \subset \mathbf{NP}^B \cap \mathbf{coNP}^B = \mathbf{NP}^B = \mathbf{coNP}^B$,
3. $\mathbf{P}^C = \mathbf{NP}^C \cap \mathbf{coNP}^C \subset \mathbf{NP}^C$,
4. $\mathbf{P}^D \subset \mathbf{NP}^D \cap \mathbf{coNP}^D \subset \mathbf{NP}^D$. ∎

Remark 1. Statement 2 has been discovered independently by A. Meyer and M. Fischer and by R. Ladner.

Remark 2. Statement 1 has been proved in 28.1. The remaining statements are proved by constructing oracles with the required properties using diagonalization. It is, however, not always obvious how to do this.

Remark 3. That there must exist an oracle A such that $\mathbf{P}^A \neq \mathbf{NP}^A$ is a simple consequence of 6.1.1 and 6.5.1. Such an $A \in$ **DEXPTIME** is constructed in [Lad 75].

Remark 4. In a sense there exist only "a few" oracles A such that $\mathbf{P}^A = \mathbf{NP}^A$. In [Meh 73] it is shown that $\{A : \mathbf{P}^A = \mathbf{NP}^A\}$ is a meagre set in a suitable defined topological space of recursive oracles, and in [BeGi 81] it is proved that, relative to a random oracle A, $\mathbf{P}^A \neq \mathbf{NP}^A \neq \mathbf{coNP}^A$ with probability 1. On the other hand, the set of all oracles B with $\mathbf{P}^B \neq \mathbf{NP}^B$ has a rich structure: In [Háj 77] it is proved that $\{i : \varphi_i = c_B \wedge \mathbf{P}^B \neq \mathbf{NP}^B\}$ is $\leqq_m$-complete in Π_2.

Remark 5. A completely different way to prove $\mathbf{P}^A \subset \mathbf{NP}^A$ for suitable oracle A is shown in [Mor 82]. In this paper, the notion of accepting density of an NTM M on input x is defined as

$$\frac{\text{number of accepting computations of } M \text{ on input } x}{\text{number of all computations of } M \text{ on input } x},$$

and for a suitable oracle A a strict ω-hierarchy in $\mathbf{NP}^A$ with respect to density is shown.

Remark 6. If $\mathbf{P}^A \neq \mathbf{NP}^A$ ($\mathbf{NP}^A \neq \mathbf{coNP}^A$) is true for an SLA-language A then $\mathbf{P} \neq \mathbf{NP}$ ($\mathbf{NP} \neq \mathbf{coNP}$) [BaBoLoScSe 84].

The next theorem extends the methods of [BaGiSo 75] to the classes introduced in [KiFi 77] (see before 24.37).

28.5. Theorem. [KiFi 77]

1. There is a recursive oracle A such that

$$\mathbf{P}^A \subset \mathbf{NP}^A_n \subset \mathbf{NP}^A_{n^2} \subset \ldots \subset \mathbf{NP}^A_{n^k} \ldots, \subset \mathbf{NP}^A.$$

2. For every $k \in \mathbb{N}$ there exists a recursive oracle B such that

$$\mathbf{P}^B \subset \mathbf{NP}^B_n \subset \ldots \subset \mathbf{NP}^B_{n^k} = \mathbf{NP}^B_{n^{k+1}} = \ldots = \mathbf{NP}^B. \blacksquare$$

Similar results are contained in [MeDoBo 81].

Concerning the relativized polynomial time hierarchy the following results are known in addition to 28.4 (note that $\mathbf{P}^A = (\Sigma_0^P)^A$, $\mathbf{NP}^A = (\Sigma_1^P)^A$, $\mathbf{coNP}^A = (\Pi_1^P)^A$):

28.6. Theorem.

1. [BaSe 76] There exists a recursive oracle A such that

$$(\Sigma_2^P)^A \neq (\Pi_2^P)^A.$$

2. [Hel 81] There exists a recursive oracle B such that

$$(\Sigma_1^P)^B \subset (\Sigma_2^P)^B = (\Pi_2^P)^B = \mathbf{DTIME}^B(2^{\mathrm{Pol}}) = \mathbf{NTIME}^B(2^{\mathrm{Pol}}).$$

3. [Ang 80b] There exists a recursive oracle C such that

$$\mathbf{RTIME}^C(\mathrm{Pol}) \setminus \left((\Sigma_2^P)^C \cup (\Pi_2^P)^C\right) \neq \emptyset.$$

4. [FuSaSi 81] If there exists a set in **PSPACE** which is not computable by non-uniform circuits of bounded depth and $2^{\mathrm{Pol(log)}}$ size, then there exists an orcale A such that

$$\bigcup_{i=1}^{\infty} (\Sigma_i^P)^A \subset \mathbf{PSPACE}^A. \blacksquare$$

Remark 1. It is not known whether there exists an oracle ensuring the properness of the relativized polynomial-time hierarchy.

Remark 2. For results similar to Theorem 28.6.3 see [Sto 83].

Remark 3. If there exists a sparse set relative to which the polynomial-time hierarchy collapses (does not collapse) then the unrelativized hierarchy collapses (does not collapse). This and similar results are proved in [BaBoLoScSe 84] and [BaBoSc 84b].

Now we extend our scope a little further and add some results about the time classes represented in Figure 28.3. The figure states all inclusions that are known to relativize (see [Dek 76]).

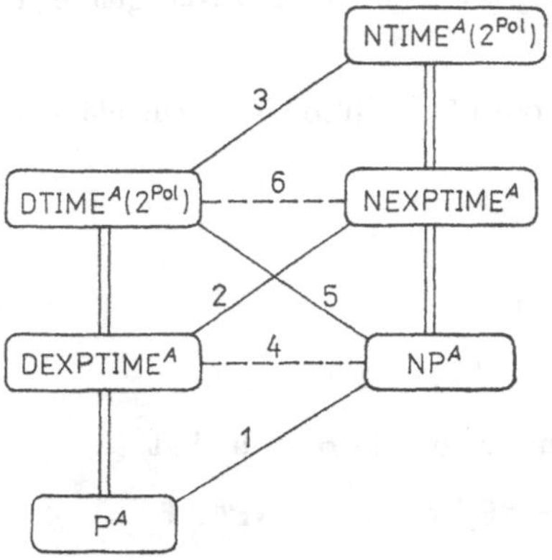

Figure 28.3. || indicates strict inclusion, | indicates $\subseteq$, --- indicates that $\subset$, $\supset$ and incomparability are possible

Furthermore, by relativizing the usual padding techniques it can be shown

28.7. Theorem. [Dek 76] For every oracle A,

1. $\mathbf{P}^A = \mathbf{NP}^A \to \mathbf{DEXPTIME}^A = \mathbf{NEXPTIME}^A$
2. $\mathbf{DEXPTIME}^A = \mathbf{NEXPTIME}^A \to \mathbf{DTIME}^A(2^{\mathrm{Pol}}) = \mathbf{NTIME}^A(2^{\mathrm{Pol}})$.
3. $\mathbf{DEXPTIME}^A \subseteq \mathbf{NP}^A \to \mathbf{DTIME}^A(2^{\mathrm{Pol}}) \subseteq \mathbf{NP}^A$.
4. $\mathbf{NEXPTIME}^A \subseteq \mathbf{DTIME}^A(2^{\mathrm{Pol}}) \to \mathbf{NTIME}^A(2^{\mathrm{Pol}}) \subseteq \mathbf{DTIME}^A(2^{\mathrm{Pol}})$. ∎

According to these results and to Figure 28.3, Table 28.1 shows the possibilities for the relationships represented by the labelled lines in this figure.

Table 28.1 (|| means incomparability)

1	2	3	4	5	6	this possibility has been proved to occur by
$=$	$=$	$=$	$\supset$	$\supset$	$\supset$	Theorem 28.4.1
$\subset$	$=$	$=$	$\supset$	$\supset$	$\supset$	Theorem 28.8.1
$\subset$	$\subset$	$=$	$\|$	$\supset$	$\supset$	[Lis 84]
$\subset$	$\subset$	$=$	$\supset$	$\supset$	$\supset$	[Lis 84]
$\subset$	$\subset$	$\subset$	$\subset$	$=$	$\subset$	Theorem 28.8.2
$\subset$	$\subset$	$\subset$	$\|$	$\supset$	$\|$	[Lis 84]
$\subset$	$\subset$	$\subset$	$\|$	$\supset$	$\subset$	[Lis 84]
$\subset$	$\subset$	$\subset$	$\supset$	$\supset$	$\|$	[Lis 84]
$\subset$	$\subset$	$\subset$	$\supset$	$\supset$	$\subset$	[Lis 84]

Remark. There exists an oracle A such that $\mathbf{NP}^A \subseteq \mathbf{DTIME}^A(n^{\log n})$ but $\mathbf{DEXPTIME}^A \neq \mathbf{NEXPTIME}^A$ (see [Sew 83]).

28.8. Theorem [Dek 76], [BoWiMe 81] There are recursive oracles A, B such that

1. $\mathbf{P}^A \neq \mathbf{NP}^A \wedge \mathbf{DEXPTIME}^A = \mathbf{NEXPTIME}^A$
2. $\mathbf{NP}^B = \mathbf{DTIME}^B(2^{\mathrm{Pol}})$. ∎

Remark. Since the statement of Remark 3 after Theorem 24.17 relativizes, Theorem 28.8.1 and the result in [Kut 82a] (that there exists a recursive oracle A such that $\mathbf{P}^A \subset \mathbf{NP}^A$ and $\mathbf{NP}^A \setminus \mathbf{P}^A$ does not contain sparse sets) are equivalent. In [Kut 83] it is shown that, for every subexponential increasing function f there exists an oracle A such that $\mathbf{P}^A \subset \mathbf{NP}^A$ but $\mathbf{NP}^A \setminus \mathbf{P}^A$ does not contain f-sparse sets. For related results see [HaSeIm 83].

28.9. Theorem. [Dek 76] There are recursive oracles A, B, C such that

$$\mathbf{DEXPTIME}^A \setminus \mathbf{NP}^A \neq \emptyset \wedge \mathbf{NP}^A \setminus \mathbf{DEXPTIME}^A \neq \emptyset. \;\blacksquare$$

The proofs of the preceding two theorems are by diagonalization.

Relativized results including probabilistic time classes can be found in [Rac 78b].

28.3. Space and Time

28.10. Theorem. [Dek 76] There exist recursive sets A, B, C such that

1. $\mathbf{P}^A = \mathbf{NP}^A = \mathbf{PSPACE}^A$
2. $\mathbf{NP}^B = \mathbf{PSPACE}^B = \mathbf{DTIME}^B(2^{\mathrm{Pol}})$
3. $\mathbf{PSPACE}^C = \mathbf{DTIME}^C(2^{\mathrm{Pol}}) = \mathbf{NTIME}^C(2^{\mathrm{Pol}})$. ∎

Remark 1. Compare statement 1 with 6.10 and the corresponding remark.

Remark 2. Figure 28.1 shows that the equality chains in theorem 28.10.2 and 3 cannot be extended. This implies the fact that there exist recursive sets B and C such that $\mathbf{P}^B \neq \mathbf{NP}^B$, $\mathbf{DTIME}^B(2^{\mathrm{Pol}}) \neq \mathbf{NTIME}^B(2^{\mathrm{Pol}})$, and $\mathbf{NP}^C \neq \mathbf{PSPACE}^C$. This latter statement has been proved independently in [SiGi 77] in the following stronger version: For every $X \notin \mathbf{NP}$ there exists a recursive C such that

$$X \in \mathbf{PSPACE}^C \setminus \mathbf{NP}^C.$$

Remark 3. Theorem 20.12 shows that deterministic and nondeterministic polynomial time for multiplication RAM are the same. The additional ability that multiplications can be done within one step has in connection with the Boolean operations the same effect as an oracle guaranteeing 28.10.1.

A table similar to Table 28.1 would contain 9 cases. Two of them are realized by 28.10.

In [LaLy 76] it has been conjectured that inclusion relations between complexity classes not provable by a step-by-step simulation do not relativize to arbitrary oracles. A first example is W. J. Savitch's result (see the remark after 28.2), and a second one is the relation $\mathbf{NL} \subseteq \mathbf{P}$:

28.11. Theorem. [LaLy 76] There exist recursive oracles A, B such that

1. $\mathbf{NL}_*^A \subset \mathbf{P}^A$,
2. $\mathbf{NL}_*^B \not\subseteq \mathbf{P}^B \not\subseteq \mathbf{NL}_*^B$. ∎

It is not known whether a recursive oracle A can be found such that $\mathbf{P}^A \subset \mathbf{NL}_*^A$. However, theorem 28.1 shows that there are oracles A such that $\mathbf{P}^A = \mathbf{NL}_*^A$.

Regarding the conjecture from [LaLy 76] further work has revealed that relativizability heavily depends on how relative space bounded TM's are defined. In particular, it depends on bounding the length of the queries. This is illustrated by the following result from [Ang 80a]. Let τ denote a machine type and let τ-$\mathbf{XSPACE}^A$-$\mathbf{Q}(s, t)$ denote for the moment the class

$$\{X : \bigvee_{M^{()}} (M^{()} \text{ is a } \tau\text{-XM with oracle} \wedge M^A \text{ accepts } X \text{ within space } s \text{ in such a way that all queries on the accepting paths are length bounded by } t(n) \text{ on inputs of length } n\}.$$

Note that in accordance with § 5.3 we have

$$\tau\text{-}\mathbf{XSPACE}_*^A(s) = \tau\text{-}\mathbf{XSPACE}^A\text{-}\mathbf{Q}(s, \mathbb{R}_1),$$
$$\tau\text{-}\mathbf{XSPACE}^A(s) = \tau\text{-}\mathbf{XSPACE}^A\text{-}\mathbf{Q}(s, s).$$

28.12. Theorem. [Ang 80a] For $s \geqq \log$,

1. $\bigwedge_A$ 2-auxPD-T-NSPACE$^A(s) \subseteq \mathbf{DTIME}^A(2^{\mathrm{Lin}\, s})$, and
   $\bigvee_A$ 2-auxPD-T-NSPACE$^A(s) \subset \mathbf{DTIME}^A(2^{\mathrm{Lin}\, s})$.
2. $\bigwedge_A \mathbf{DTIME}^A(2^{\mathrm{Lin}\, s}) \subseteq$ 2-auxPD-T-NSPACE$_*^A(s)$, but
   $\bigvee_A$ 2-auxPD-T-NSPACE$_*^A(s) \not\subseteq \mathbf{NTIME}^A(2^{\mathrm{Lin}\, s})$.
3. $\bigwedge_A \mathbf{DTIME}^A(2^{\mathrm{Lin}\, s}) =$ 2-auxPD-DSPACE^A-$\mathbf{Q}(s, 2^{\mathrm{Lin}\, s})$, but
   $\bigvee_A$ 2-auxPD-T-NSPACE^A-$\mathbf{Q}(s, 2^{\mathrm{Lin}\, s}) \not\subseteq \mathbf{NTIME}^A(2^{\mathrm{Lin}\, s})$. ∎

Remark 1. Theorem 28.12 shows that two relativized classes can be different although the unrelativized classes are known to be equal (see 20.13.1.). Two further results of this type: There exist recursive oracles A and B such that $\mathbf{ASPACE}^A(\mathrm{Pol}) \subset \mathbf{DTIME}^A(2^{\mathrm{Pol}})$ and $\mathbf{ATIME}^B(\mathrm{Pol}) \neq \mathbf{DSPACE}_*^B(\mathrm{Pol})$ ([Orp 83]).

Remark 2. In [Orp 83] it is moreover proved that $\mathbf{ASPACE}^A(\mathrm{Pol}) \neq \mathbf{DTIME}^A(2^{\mathrm{Pol}})$ is true with probability 1. This is not in accordance with the "random oracle hypothesis" of [BeGi 81] which, intuitively, says that relativized statements which are true with probability 1 are also true in the unrelativized case. The random oracle hypothesis is criticized also in [Kut 82b].

Now we restrict the use of the oracles in such a way that the number of queries is polynomially bounded. To make this more precise we define $\mathbf{NPSPACE}^{A,\mathrm{Pol}} =_{\mathrm{df}} \{L(M^{(A)}) : M^{(A)}$ is a nondeterministic oracle machine working in polynomial space in such a way that on every computation the number of queries is bounded by a polynomial depending only on $M^{(A)}$}. Note that we cannot simply omit the specification "N" because a deterministic simulation might lead to a number of queries that cannot be bounded by a polynomial.

28.13. Theorem. [Boo 79a] $\mathbf{NP} = \mathbf{PSPACE} \leftrightarrow \bigwedge_A (\mathbf{NP}^A = \mathbf{NPSPACE}^{A,\mathrm{Pol}})$. ∎

Remark. Theorem 28.13 is an example of a "positive relativization". A positive relativization of a problem "$\mathbf{C} = \mathbf{D}$?" where $\mathbf{C}$ and $\mathbf{D}$ are complexity classes is a statement of the form

$$\mathbf{C} = \mathbf{D} \leftrightarrow \bigwedge_A \mathbf{C}_R^A = \mathbf{D}_R^A,$$

where R is a restricted class of machines and $\mathbf{C}_R^A$ denotes the class of languages acceptable by oracle machines from R with oracle A and complexity restricted according to $\mathbf{C}$. Positive relativizations for $\mathbf{P} = \mathbf{NP}$, $\mathbf{P} = \mathbf{NP} \cap \mathbf{coNP}$ and further problems are given in [BoLoSe 82], [Long 83], [BoLoSe 84] and [Schg 84b].

Further investigations concerning bounded query machines can be found in [Boo 79a], so for example the equivalence (for every A)

$$\mathbf{NP}^A = \mathbf{NPSPACE}^{A,\mathrm{Pol}} \leftrightarrow \mathbf{DLINSPACE}^{A,\mathrm{Lin}} \subseteq \mathbf{NP}^A,$$

and in [Boo 81], [BoWr 81], [BoWiMe 81], [SeMeBo 81] and [BaBoSc 84b].

The relationships (cf. 11.20.1 and 11.23.2, for the relativized classes see [Boo 79b])

$$\mathbf{NP} = \Gamma_{\cap,\mathrm{hr}_{\mathrm{pol}}}(\mathbf{REG}),$$

$$\mathbf{PSPACE} = \Gamma_{\cap,\mathrm{hr}_{\mathrm{pol}},\mathrm{wtc}}(\mathbf{REG}),$$

$$\mathbf{NP}^A = \Gamma_{\cap,\mathrm{hr}_{\mathrm{pol}},\mathrm{h}^{-1}}\left(\mathbf{REG} \cup \{(A \oplus \bar{A})^*\}\right),$$

$$\mathbf{PSPACE}^A = \Gamma_{\cap,\mathrm{hr}_{\mathrm{pol}},\mathrm{h}^{-1},\mathrm{wtc}}\left(\mathbf{REG} \cup \{(A \oplus \bar{A})^*\}\right)$$

show that the difference between $\mathbf{NP}^A$ and $\mathbf{PSPACE}^A$ (if any) can be expressed exactly in algebraic terms in that $\mathbf{PSPACE}^A$ is additionally closed under weakly transitive closure. This observation immediately yields

28.14. Theorem. [Boo 79b] For every oracle set A, $\mathbf{NP}^A = \mathbf{PSPACE}^A$ if and only if $\mathbf{NP}^A$ is closed under wtc. ∎

Remark. In [Boo 79b] it is pointed out that 28.14 is true for all reasonably chosen complexity bounds.

Appendix

Bibliography

We use the following abbreviations.

Journals

AI	— Acta Informatica
AML	— Archiv für math. Logik und Grundlagenforschung
CACM	— Communications of the ACM
COMP	— Computing
DAN	— Doklady akademii nauk
EIK	— Elektronische Informationsverarbeitung und Kybernetik
I&C	— Information and Control
IJCM	— International Journal of Computer Mathematics
IPL	— Information Processing Letters
JACM	— Journal of the ACM
JSL	— Journal of Symbolic Logic
JCSS	— Journal of Computer and System Science
LNCS	— Lecture Notes in Computer Science
MST	— Mathematical Systems Theory
RAIRO	— Revue Française d'Automatique, Informatique et Recherche Opérationelle
SIAM-AMS	— SIAM-AMS Proceedings
SIAM JC	— SIAM Journal of Computing
TAMS	— Transactions of the AMS
TCS	— Theoretical Computer Science
ZML	— Zeitschrift für Mathematische Logik und Grundlagen der Mathematik

Conference Proceedings

FCT	— Conference on Fundamentals of Computation Theory
FOCS	— IEEE Annual Symposium on Foundation of Computer Science
GI	— Fachtagung der Gesellschaft für Informatik
ICALP	— International Colloquium on Automata, Languages and Programming
IFIP	— IFIP Conference
MFCS	— Conference on Mathematical Foundations of Computer Science
STOC	— Annual ACM Symposium on Theory of Computing
SWAT	— IEEE Annual Symposium on Switching and Automata Theory

The items are listed in alphabetic order with respect to their acronyms. Note that this is not the alphabetic order with respect to the full names of the authors.

Books are labelled by *. The numbers in brackets at the end of the quotations refer to the sections and (if without §) theorems where the papers and books are cited.

[Aan 73] Aanderaa, S. O., On k-tape versus $(k-1)$-tape real time computation, SIAM-AMS 7 (1974), 75–96. (8.37)

[AdIwKa 81] ADACHI, A., IWATA, S., KASAI, T., Low level complexity for combinatorial games, 13th STOC (1981), 228–237. (§ 8.2.3.3)

[Adl 78] ADLEMAN, L. M., Two theorems on random polynomial time, 19th FOCS (1978), 75–83. (§ 24.2.2.2.1)

[Adl 80] —, On distinguishing prime numbers from composite numbers, 21st FOCS (1980), 387–406. (§ 7.3.3)

[AdLo 80] ADLEMAN, L. M., LOUI, M. C., Space bounded simulation of multitape Turing machines, Lab. for Comp. Science MIT, TM-148 (Jan. 1980). See also: MST **14** (1981), 215–222. (25.1)

[AdMa 75] ADLEMAN, L. M., MANDERS, K., Computational complexity of decision procedures for polynomials, 16th FOCS (1975), 169–177. (§ 8.2.3.4)

[AdMa 76] —, Diophantine complexity, 17th FOCS (1976), 81–88. (§ 24.2.2.2.6)

[AdMa 77] —, Reducibility, randomness, and intractibility, 9th STOC (1977), 151–163. (§ 6.1.2, § 8.2.1, 14.21, 24.9, 24.17, 24.24)

[AdMa 79] —, Reductions that lie, 20th FOCS (1979), 397–410. (§ 8.2.1, 14.21, 24.14)

[AdMe 84] ADLER, I., MEGIDDO, N., A simplex algorithm whose average number of steps is bounded between two quadratic functions of the smaller dimension, 16th STOC (1984), 312–323. (7.22)

[Aga 75]* AGAFONOV, V. N., Complexity of Algorithms and Computations (in Russian), Novosibirsk State University, Novosibirsk 1975. (Intr.)

[AhCo 75] AHO, A. V., CORASICK, M., Efficient string matching: an aid to bibliographic search, CACM **18** (1975), 333–340. (§ 7.2.1)

[Aho 80] AHO, A. V., Pattern matching in strings, in: R. V. BOOK (ed.), Formal Language Theory. Perspectives and Open Problems, Academic Press, New York 1980, 325–347. (§ 7.2.1)

[AhHoUl 68] AHO, A. V., HOPCROFT, J. E., ULLMAN, J. T., Time and tape complexity of pushdown automaton languages, I&C **13** (1968), 186–206. (13.20)

[AhHoUl 74]* —, The Design and Analysis of Computer Algorithms, Addison-Wesley, Reading (Mass.) 1974. (Intr., § 7, § 7.2.2, § 8.2.3.1)

[AlKaLiLoRa 79] ALELIUNAS, R., KARP, R. M., LIPTON, R. J., LOVÁSZ, L., RACKOFF, C., Random walks, universal traversal sequences, and the complexity of maze problems, 20th FOCS (1979), 218–223. (§ 23.3)

[AlMe 75] ALT, H., MEHLHORN, K., A language over a one symbol alphabet requiring only $O(\log\log n)$ space, SIGACT News **7**, 4 (1975), 31–33. (5.6)

[AlMe 76] —, Lower bounds on space complexity for context-free recognition, 3rd ICALP (1976), Edinburgh Press. See also: AI **12** (1979), 33–61. (7.10, 8.31)

[Aln 73a] ALTON, D. A., Diversity of speed-up and embeddability in computational complexity, Univ. of Iowa, Dept. of Comp. Sci., TR 73-01. See also: JSL **41** (1976), 199–214. (17.21)

[Aln 73b] —, Non-existence of program-optimizers in an abstract setting, University of Iowa, Dep. of Comp. Sci., TR 73-08. See also: JCSS **12** (1976), 368–393. (§ 18.2)

[Aln 76] —, "Natural" complexity measures, subrecursive languages, and speed-up, Univ. of Iowa, Dept. of Comp. Sci., TR 76-05. (18.2)

[Aln 77] —, Natural complexity measures and a subrecursive speed-up theorem, IFIP 77, North-Holland 1977. (18.2)

[Alt 77] ALT, H., Eine untere Schranke für den Platzbedarf bei der Analyse beschränkter kontextfreier Sprachen, LNCS **48** (1977), 123–131. See also: AI **12** (1979), 33–61. (8.31)

[Amb 84] AMBOS-SPIES, K., On the structure of polynomial time degrees, Symposium on Theoretical Aspects of Theoretical Computer Science (Eds.: M. FONTET and K. MEHLHORN), LNCS **166** (1984), 198–201. (§ 24.2.1.2.3)

[Amb 85] —, On the relative complexity of subproblems of intractable problems, to appear in the Proceedings of STACS '85. (§ 24.2.1.2.3)

[AmFlHu 84] AMBOS-SPIES, K., FLEISCHHACK, H., HUWIG, H., Diagonalizations over poly-

nomial time computable sets, Forschungsbericht Nr. 177 (1984), Universität Dortmund, Abt. Informatik. (§ 24.2.1.2.3)
[Ang 80a] ANGLUIN, D., On relativizing auxiliary pushdown machines, MST 13 (1980), 283–299. (§ 5.3, 28.12)
[Ang 80b] —, On counting problems and the polynomial time hierarchy, TCS 12 (1980), 161–173. (28.6)
[AnVa 77] ANGLUIN, D., VALIANT, L. G., Fast probabilistic algorithms for Hamiltonian circuits and matchings, 9th STOC (1977), 30–41. See also: JCSS 18 (1979), 155–193. (§ 9.2.1, § 9.2.2)
[ApDi 82] APOLLONI, B., DIGREGORIO, S., A probabilistic analysis of a new satisfiability algorithm, RAIRO, Informat. Théor. 16 (1982), 201–223. (§ 9.2.1)
[ArSu 76] ARORA, A. K., SUDBOROUGH, I. H., On languages log-tape reducible to context-free languages, Proc. 1976 Conf. on Inf. Sc. and Systems, 10th Ann. Meeting, Baltimore 1976, 27–32. (14.3)
[Ass 56] ASSER, G., Das Repräsentantenproblem im Prädikatenkalkül der ersten Stufe mit Identität, ZML 1 (1956), 252–263. (§ 24.2.2.2.2)
[Ass 77] —, Bemerkungen zum Labyrinth-Problem, EIK 13 (1977), 203–216. (§ 23.3)
[AsSh 79] ASPVALL, B., SHILOACH, Y., A polynomial time algorithm for solving systems of linear inequalities with two variables per inequality, 20th FOCS (1979), 205–217. See also: SIAM JC 9 (1980), 827–845. (7.22)
[Asv 80] ASVELD, P. R. J., Time and space complexity of inside-out macro languages, Preprint IW 141/80, Math. Centr. Amsterdam. (14.3)
[AtPu 82] D'ATRI, G., PUECH, C., Probabilistic analysis of the subset sum problem, Discr. Appl. Math. 4 (1982), 329–334. (§ 9.2.1)
[Atr 65] ATRUBIN, A. J., A one-dimensional real-time iterative multiplier, IEEE Trans. on El. Comp. EC-14, 3 (1965), 394–399. (7.15)
[AuD'AGaPr 77] AUSIELLO, G., D'ATRI, A., GAUDIANO, M., PROTASI, M., Classes of structurally isomorphic NP-optimization problems, MFCS 1977, LNCS 53 (1977), 222–230. (§ 24.2.1.1)
[AuD'APr 76] AUSIELLO, G., D'ATRI, A., PROTASI, M., A characterization of reductions among combinatorial problems, Univ. di Roma, Inst. di Automatica, Rep. 76–23. (§ 24.2.1.1)
[AuD'APr 77] —, On the structure of combinatorial problems and structure preserving reductions, 4th ICALP (1977), LNCS 52, 45–60. See also: JCSS 21 (1980), 136–153. (§ 24.2.1.1)
[AuD'APr 78] —, Lattice theoretic ordering properties for NP-complete optimization problems. Univ. di Roma, Inst. di Automatica, preprint 1978. See also: Fund. Inform. 4 (1981), 83–94. (§ 24.2.1.1)
[Aus 71] AUSIELLO, G., Abstract computational complexity and cycling complexity, JCSS 5 (1971), 118–128. (§ 5.2)
[Aus 75]* —, Complessità di Calcolo delle Funzioni, Boringhieri, Torino, 1975. (Intr., 2.21, § 2.4.1)
[Aus 77] —, On the structure and properties of NP-complete problems and their associated optimization problems, MFCS 1977, LNCS 53 (1977), 1–16. (§ 24.2.1.1)
[AvMa 80] AVENHAUS, J., MADLENER, K., String matching and algorithmic problems in free groups, Rev. Colombiana Mat. 14 (1980), 1–15. (§ 7.2.1)
[Axt 59] AXT, P., On a subrecursive hierarchy and primitive recursive degrees, TAMS 92 (1959), 85–105. (27.16)
[Axt 65] —, Iteration of primitive recursion, ZML 11 (1965), 253–255. (2.18)
[Bab 81] BABAI, L., Moderately exponential bound for graph isomorphism, FCT '81, LNCS 117 (1981), 34–50. (§ 7.3.4)
[BaBo 74] BAKER, B. S., BOOK, R. V., Reversal-bounded multipushdown machines, JCSS 8 (1974), 315–332. (4.14, 11.20, 11.21, 21.7, 22.16, 24.33)
[BaBoLoScSe 84] BALCÁZAR, J. L., BOOK, R. V., LONG, T. J., SCHÖNING, U., SELMAN, A. L., Sparse oracles may as well be empty, 25th FOCS (1984), 308–311. (28.4, 28.6)

[BaBoSc 84a] BALCÁZAR, J. L., BOOK, R. V., SCHÖNING, U., Sparse oracles, lowness and highness, MFCS '84, LNCS **176** (1984), 185–193. (§ 24.3)

[BaBoSc 84b] —, A note on bounded query machines, preprint May 1984, Santa Barbara. (28.6, § 28.3)

[Bac 82] BACH, E., Fast algorithms under the extended Riemann hypothesis: a concrete estimate, 14th STOC (1982), 290–295. (7.29)

[BaDr 80] BAILEY, T. A., DROMEY, R. G., Fast string searching by finding subkeys in subtext, IPL **11** (1980), 130–133. (§ 7.2.1)

[BaErSe 80] BABAI, L., ERDŐS, P., SELKOW, S. M., Random graph isomorphisms, SIAM JC **9** (1980), 628–635. (§ 7.3.4)

[BaEv 82] BAR-YEHUDA, R., EVEN, S., On approximating a vertex cover for planar graphs, 14th STOC (1982), 303–309. (9.15)

[BaGiSo 75] BAKER, T. P., GILL, J., SOLOVAY, R., Relativizations of the P = NP question, SIAM JC **4** (1975), 431–442. (28.4)

[BaGrMo 82] BABAI, L., GRIGOREV, D. Ju., MOUNT, D. M., Isomorphism of graphs with bounded eigenvalue multiplicity, 14th STOC (1982), 310–324. (7.30)

[Bak 74] BAKER, T. P., Natural properties of flowchart complexity measures, 15th SWAT (1974), 178–184. See also: JCSS **16** (1978), 1–22. (§ 16.2, 16.15)

[Bak 79] —, On "provable" analogs of P and NP, MST **12** (1979), 213–218. (§ 24.2.2.2.3)

[BaLiRo 78]* BARAŠKO, A. S., LIPSKAJA, V. A., ROIZEN, S. J., Studies in Computational Complexity and Formal Languages (in Russian), Naukova Dumka, Kiev 1978. (Intr.)

[BaLu 83] BABAI, L., LUKS, E. M., Canonical labeling of graphs, 15th STOC (1983), 171 to 183. (§ 7.3.4)

[Bar 74] BARAŠKO, A. S., Tape complexity of languages that can be recognized by two-way finite-turn pushdown automata (in Russian), Kibernetika (Kiev) **1** (1974), 40–45. (12.14)

[Bar 77] —, Bikernels and complexity hierarchies (in Russian), DAN USSR, Seria A, Nr. 1 (1977), 3–5. (15.18, 15.20)

[BaRo 76] BARAŠKO, A. S., ROIZEN, S. J., Kernels of partial recursive functions naming complexity classes (in Russian), Kibernetika (Kiev) 1976, Nr. 5, 10–15. (15.25)

[Barz 65] BARZDIN, Ja. M., Complexity of the recognition of the symmetry predicate on Turing machines (in Russian), Problemy Kibernetiki **15** (1965), 245–248. (8.13, 8.14)

[Barz 69] —, On computability on probabilistic machines (in Russian), DAN SSSR **189** (1969), 699–702. (§ 9.2.2)

[Bas 72] BASS, L. J., A note on the intersection of complexity classes of functions, SIAM JC **1** (1972), 288–289. (15.20)

[BaSe 76] BAKER, T. P., SELMAN, A. L., A second step toward the polynomial hierarchy, 17th FOCS (1976), 71–75. See also: TCS **8** (1979), 177–187. (28.6)

[BaYo 73] BASS, L. J., YOUNG, P. R., Ordinal hierarchies and naming complexity classes, JACM **20** (1973), 668–686. (§ 15.3)

[BeBr 80] BENTLEY, J. L., BROWN, D. J., A general class of resource trade-offs, 21st FOCS (1980), 217–228. (§ 25.3.2.2)

[BeGi 81] BENNETT, C. H., GILL, J., Relative to a random oracle A, $P^A \neq NP^A \neq coNP^A$ with probability 1, SIAM JC **10** (1981), 96–113. (28.4, 28.12)

[BeHa 77] BERMAN, L., HARTMANIS, J., On isomorphisms and density of NP and other complete sets, SIAM JC **6** (1977), 305–322. (14.11)

[BeJoLeMcMc 84] BENTLEY, J. L., JOHNSON, D. S., LEIGHTON, F. D., MCGEOCH, C. C., MCGEOCH, L. A., Some unexpected expected behavior results for bin packing, 16th STOC (1984), 279–288. (§ 9.2.1)

[BeKoRe 84] BEN-OR, M., KOZEN, D., REIF, J., The complexity of elementary algebra and geometry, 16th STOC (1984), 457–464. (8.47)

[Bel 79] BELTJUKOV, A. P., A machine description and the hierarchy of initial Grzegorczyk classes (in Russian), Issledovanija po konstr. mat. i mat. logiki VIII, Zap. nauč. sem. 88 (1979), 30–44, Leningrad. (11.16)

[Ben 79] BENNISON, V. L., Information content characterizations of complexity theoretic properties, 4th GI (1979), LNCS 67 (1979), 58–66. (18.17, § 18.2)

[BeOtWi 83] BENTLEY, J. L., OTTMANN, T., WIDMEYER, P., The complexity of manipulating hierarchically defined sets of rectangles, Adv. in Comp. Res. 1 (1983), 127–158. (14.14, 14.25)

[Ber 50] BERECKI, I., Nem-elemi rekurziv függvény létezése C.R. du 1. Congr. des Math. Hongr. (1950), 409–417. (§ 2.4.1)

[Berl 76] BERMAN, L., On the structure of complete sets: almost everywhere complexity and infinitely often speed-up, 17th FOCS (1976), 76–80. (§ 18.2)

[Berl 77] —, Precise bounds for Presburger arithmetic and the reals with addition, 18th FOCS (1977), 95–99. (§ 17.2.6)

[Berl 78] —, The complexity of logical theories, IBM Research Report RC 7028 (1978). See also: TCS 11 (1980), 71–77. (8.47)

[Berp 78] BERMAN, P., Relationship between density and deterministic complexity of NP-complete languages, 5th ICALP (1978), LNCS 62 (1978), 63–71. (24.16)

[Berp 79] —, Complexity of the theory of atomless Boolean algebras, FCT 1979, 64–70. (8.45, 14.16)

[BeSi 83] BERMAN, P., SIMON, J., Lower bound on graph threading by probabilistic machines, 24th FOCS (1983), 304–311. (§ 8.2.1, § 23.3)

[BeSo 78] BENNISON, V. L., SOARE, R. I., Some lowness properties and computational complexity sequences, TCS 6 (1978), 233–254. (§ 18.2)

[Bis 78] BISKUP, J., The time measure of one-tape Turing machines does not have the parallel computation property, SIAM JC 7 (1978), 115–117. (16.14)

[BlKo 78] BLUM, M., KOZEN, D., On the power of the compass, 19th FOCS (1978), 132–142. (§ 23.3)

[BlMa 73] BLUM, M., MARQUES, J., On complexity properties of recursively enumerable sets, JSL 38 (1973), 579–593. (18.16–17, 18.23–24)

[Blo 80] BLONIARZ, P., A shortest path algorithm with expected time $O(n^2 \log n \log^* n)$, 12th STOC (1980), 378–384. See also: SIAM JC 12 (1983), 588–600. (§ 9.2.1)

[BlSa 77] BLUM, M., SAKODA, W., On the capability of finite automata in 2 and 3 dimensional space, 18th FOCS (1977), 147–161. (§ 23.3)

[Blu 67a] BLUM, M., Machine-independent theory of the complexity of recursive functions, JACM 14 (1967), 322–336. (§ 5, § 5.2, 17.8–9, 18.1–2)

[Blu 67b] —, On the size of machines, I&C 11 (1967), 257–265. (Intr.)

[Blu 71] —, On effective procedures for speeding up algorithms, JACM 18 (1971), 290–305. (18.3, 18.7, 18.22)

[BoBr 79] BOOK, R. V., BRANDENBURG, F.-J., Representing complexity classes by equality sets, 6th ICALP (1979), LNCS 71 (1979), 49–57. See also: SIAM JC 9 (1980), 729–743. (11.20, 11.21, 11.23)

[BoCb 79] BOOTH, K. S., COLBOURN, C. J., Problems polynomially equivalent to graph isomorphism, TR CS-77-04, University of Waterloo 1979. (§ 24.2.1.2.3)

[BoCoHo 69] BORODIN, A., CONSTABLE, R. L., HOPCROFT, J. E., Dense and nondense families of complexity classes, 10th SWAT (1969), 7–19. (16.20)

[BoCo 80] BORODIN, A., COOK, S. A., A time-space tradeoff for sorting on a general sequential model of computation, 12th STOC (1980), 294–301. See also: SIAM JC 11 (1982), 287–297. (§ 8, § 25.3.1)

[BoDe 76] BORODIN, A., DEMERS, A., Some comments on functional self-reducibility and the NP-hierarchy, Cornell Univ., Dept. Comp. Sci. TR 76-284, (1976). (24.47)

[Boe 83] BÖRGER, E., Spektralproblem and completeness, LNCS 171 (1983), 333–356. (8.42)

[BöKl 80] BÖRGER, E., KLEINE BÜNING, H., The reachability problem for Petri nets and decision problems for Skolem arithmetic, TCS 11 (1980), 207–220. (§ 8.2.3.4)

[BoFiKiLyTo 79] Borodin, A., Fischer, M. J., Kirkpatrick, D. G., Lynch, N. A., Tompa, M., A time-space trade-off for sorting on non-oblivious machines, 20th FOSC (1979), 319–327. See also: JCSS **22** (1981), 351–364. (§ 25.3.2.2)

[BoGr 70] Book, R. V., Greibach, S.A., Quasi-realtime languages, MST **4** (1970), 97 – 111. (11.9, 18.13, 22.2)

[BoGrIbWe 70] Book, R. V., Greibach, S. A., Ibarra, O. H., Wegbreit, B., Tape bounded Turing acceptors and principal AFL's, JCSS **4** (1970), 622–625. (11.5)

[BoGrWe 70] Book, R. V., Greibach, S. A., Wegbreit, B., Time- and tape-bounded Turing acceptors and AFL's, JCSS **4** (1970), 606–621. (11.1, 11.21, 11.22, 19.19)

[BoGrWr 78] Book, R. V., Greibach, S. A., Wrathall, C., Comparisons and reset machines, 5th ICALP (1978), LNCS **62** (1978), 113–124. See also: JCSS **19** (1979), 256–276. (11.10, 11.19, 22.8)

[BoLoSe 82] Book, R. V., Long, T. J., Selman, A. L., Controlled relativizations of complexity classes, 1982 Summer Institute on Recursion Theory, Cornell University. Special issue of Recursive Function Theory: Newsletters, 1982, 6–10. (28.13)

[BoLoSe 84] —, Qualitative realtivizations of complexity classes, preprint 1984, Santa Barbara. (28.13)

[BoMo 77] Boyer, R. S., Moore, J. S., A fast string searching algorithm, CACM **20** (1977), 262–272. (§ 7.2.1)

[BoMu 75]* Borodin, A., Munro, I., Computational Complexity of Algebraic and Numeric Problems, Amer., Elsevier, New York 1975. (Intr., § 7.2.2)

[BoNi 78] Book, R. V., Nivat, M., Linear languages and the intersection closure of languages, SIAM JC **7** (1978), 167–177. (11.18)

[BoNiPa 74] Book, R. V., Nivat, M., Paterson, M., Reversal-bounded acceptors and intersections of linear languages, SIAM JC **3** (1974), 283–295. (11.4, 11.10, 22.2, 22.8)

[Boo 74a] Book, R. V., Tally languages and complexity classes, I&C **26** (1974), 186–193. (24.4)

[Boo 74b] —, On the structure of complexity classes, LNCS **14** (1974), 437–445. (26.10)

[Boo 74c] —, Comparing complexity classes, JCSS **9** (1974), 213–229.

[Boo 76] —, Translational lemmas, polynomial time and $(\log n)^j$-space, TCS **1** (1976), 215–226.

[Boo 77] —, Language representation theorems. How to generate the recursively enumerable sets from the regular sets, 18th FOCS (1977), 58–61. (4.14)

[Boo 78] —, Simple representations of certain classes of languages, JACM **25** (1978), 23–31. (11.4, 11.19, 11.20)

[Boo 79a] —, On relativizations of the NP =? PSPACE question, Univ. of California, Santa Barbara, Dept. of Math. preprint May 1979. (28.13)

[Boo 79b] —, Complexity classes of formal languages, MFCS (1979), LNCS **74** (1979), 43–56. (11.19, 11.21, 11.22, 28.13, 28.14)

[Boo 79c] —, Polynomial space and transitive closure, SIAM JC **8** (1979), 434–439. (11.23, 26.13)

[Boo 79d] —, On languages accepted by space bounded oracle machines, AI **12** (1979), 177–185. (11.23)

[Boo 79e] —, A remark on tally languages and complexity classes, I&C **43** (1979), 198–201. (24.4)

[Boo 81] —, Bounded query machines: On NP and PSPACE, TCS **15** (1981), 27–39. (28.13)

[Boot 78] Booth, K. S., Isomorphism testing for graphs, semigroups and finite automata are polynomially equivalent problems, SIAM JC **7** (1978), 273–279. (§ 24.2.1.2.3)

[Bor 72] Borodin, A., Computational complexity and the existence of complexity gaps, JACM **19** (1972), 158–174. (15.3, 17.5, 17.14)

[Bor 77] —, On relating time and space to size and depth, SIAM JC **6** (1977), 733–744. (21.13)

[BoWiMe 81] Book, R. V., Wilson, C. B., Xu Mei-rui, Relativizing time and space, 22nd FOCS (1981), 254–259. See also: SIAM JC 11 (1982), 571–581. (28.8, § 28.3)

[BoWr 78] Book, R. V., Wrathall, C., On languages specified by relative acceptance, TCS 7 (1978), 185–195. (11.19)

[BoWr 81] –, Bounded query machines: On NP( ) and NPQUERY( ), TCS 15 (1981), 41–50. (28.13)

[BoWrSeDo 77] Book, R. V., Wrathall, C., Selman, A. L., Dobkin, D., Inclusion complete tally languages and the Hartmanis-Berman conjecture, MST 11 (1977), 1–8. (§ 24.2.1.2.1)

[BoYa 77] Book, R. V., Yap, C. K., On the computational power of reversal-bounded machines, 4th ICALP (1977), LNCS 52 (1977), 111–119. (22.10)

[Bra 77] Brandenburg, F.-J., On one-way auxiliary pushdown automata, LNCS 48 (1977), 132–144. (8.5, 8.6, 23.8)

[Bra 78] –, Die Zusammenhangskomplexität von nicht-kontextfreien Grammatiken, Diss., Bonn 1978. (12.17)

[Bra 79a] –, Multiple equality sets and Post machines, Univ. of California, Santa Barbara, Dept. of Math. TR (1979). See also: JCSS 21 (1980), 292–316. (§ 11.2.2.1)

[Bra 79b] –, Three write heads are as good as k, Univ. of California, Santa Barbara, Dept. of Math., TR (1979). See also: MST 14 (1981), 1–12. (13.4)

[Bra 81a] –, Oral communication. (§ 11.2.2.1)

[Bra 81b] –, Analogies of PAL and COPY, FCT '81, LNCS 117 (1981), 61–70. (11.10)

[Bra 82] –, Homomorphic equality operations on languages, Informatik-Berichte Universität Bonn, Nr. 36 (1982). (4.14)

[Bra 84] –, A truly morphic characterization of recursively enumerable sets, MFCS '84, LNCS 176 (1984), 205–213. (4.14, 11.10)

[Bran 80] Brandstädt, A., Space classes, bounded erasing homomorphisms and transductions and intersection of one-counter languages, Friedrich Schiller University Jena, TR N/80/40. (11.20, 11.22)

[BrCoMeVe 83] v. Braunmühl, B., Cook, S. A., Mehlhorn, K., Verbeek, R., The recognition of deterministic CFLs in small time and space, I&C 56 (1983), 34–51. (§ 25.3.1)

[Bre 77] Bremer, H., Ein vollständiges Problem auf der Baummaschine, 3rd GI (1977), LNCS 48 (1977), 391–406. (20.35)

[BrMe 78] Bruss, A., Meyer, A. R., On time-space classes and their relation to the theory of real addition, 10th STOC (1978), 233–239. See also: TCS 11 (1980), 59–69. (8.47)

[BrSa 77] Brandstädt, A., Saalfeld, D., Eine Hierarchie beschränkter Rückkehrberechnungen auf on-line Turingmaschinen, EIK 13 (1977), 571–583. (8.26, 17.48)

[BrVe 80] v. Braunmühl, B., Verbeek, R., A recognition algorithm for deterministic CFL's optimal in time and space, 21th FOCS (1980), 411–420. (25.6)

[BrVe 84] –, Input driven languages are recognized in log n space, preprint Bonn 1984. (12.14)

[BrVo 82] Brandstädt, A., Vogel, J., Kompliziertheitsbeschränkte Checking-Stack-Bänder und Raum-Zeit-Probleme, Wiss. Z. Friedrich-Schiller-Univ. Jena, Math.-Nat. Reihe, 31 (1982), 569–577. (22.3)

[BrWa 82] Brandstädt, A., Wagner, K., Reversal-bounded and visit-bounded realtime computations, Techn. Rep. N/82/54, Friedrich-Schiller-Univ. Jena, 1982. See also: Proc. FCT '83, LNCS 158 (1983), 26–39. (11.10, 22.8)

[Buc 74] Buchberger, B., On certain decompositions of Gödel numberings, AML 16 (1974), 85–96. (16.5)

[Bud 75] Budach, L., On the solution of the labyrinth problem for finite automata, EIK 11 (1975), 661–672. (§ 23.3)

[Bud 81] –, Two pebbles don't suffice, MFCS 1981, LNCS 118 (1981), 578–589. (§ 23.3)

[Büc 60] Büchi, J. R., Weak second order arithmethic and finite automata, ZML **6** (1960), 66–92. (8.46)

[Bul 72] Bulnes, J., On the speed of addition and multiplication on one-tape, off-line Turing machines, I&C **20** (1972), 415–431. (8.14)

[BuVa 71] Burkhard, W. A., Varaiya, P. P., Complexity problems in real time languages, Inf. Sci. **3** (1971), 87–100. (22.9)

[CaLiMe 76] Cardoza, E., Lipton, R., Meyer, A. R., Exponential space complete problems for Petri nets and commutative semigroups, 8th STOC (1976), 50–54. (§ 8.2.3.4)

[CaSa 80] Carlson, D. A., Savage, J. E., Graph pebbling with many free pebbles can be difficult, 12th STOC (1980), 326–332. See also: JCSS **26** (1983), 65–81. (§ 25.3.2.1.2)

[Chb 84] Chrobak, M., Nondeterminism is essential for two-way counter machines, 11th MFCS, LNCS **176** (1984), 240–244. (§ 13.10)

[Chb 85] —, Hierarchies of one-way multihead finite automata, to appear in TCS 1985. (13.24, 13.25)

[ChKoSt 78] Chandra, A. K., Kozen, D., Stockmeyer, L. J., Alternation, Research report RC 7489, IBM Thomas J. Watson Research Center 1978. See also: JACM **28** (1981), 114–133. (20.36, 20.37, 20.38)

[Chl 81] Chlebus, B. S., On the computational complexity of satisfiability in propositional logics of programs, TCS **21** (1982), 179–212. (§ 8.2.3.2)

[ChMa 80] Chew, P., Machtey, M., A note on structure and looking back applied to the relative complexity of computable functions, Purdue Univ., preprint 1980. See also: JCSS **22** (1981), 53–59. (§ 24.2.1.2.2)

[ChNiSa 82] Chiba, N., Nishizeki, T., Saito, N., An approximation algorithm for the maximum independent set problem on planar graphs, SIAM JC **11** (1982), 663–675. (§ 9.1.3.2)

[Cho 69] Chodshajev, D., Independent predicates and relative computations of the symmetry predicate, Vopr. kib. i vyč. mat., **33** (1969). (27.2, 27.14)

[Chom 62] Chomsky, N., Context-free grammars and pushdown storage, M.I.T. Res. Lab. Electron. Quart. Progr. Rep. 65 (1962). (12.4)

[Chr 75]* Christofides, N., Graph Theory. An Algorithmic Approach, Academic Press, New York 1975. (§ 7.2.2)

[Chr 76] —, Worst-case analysis of a new heuristic for the travelling salesman problem, CMU Symp. on New Directions and Recent Results in Algorithms and Complexity (1976). (9.14)

[ChSt 76] Chandra, A. K., Stockmeyer, L. J., Alternation, 17th FOCS (1976), 98–108. (13.3, 14.24, 17.34, 17.44, § 20.4.1)

[ChStVi 82] Chandra, A. K., Stockmeyer, L. J., Vishkin, U., A complexity theory for unbounded fan-in parallelism, 23rd FOCS (1982), 1–13. (§ 6.1.2)

[ChTa 76] Cheriton, D., Tarjan, R. E., Finding minimum spanning trees, SIAM JC **5** (1976), 724–742. (7.19)

[ChWa 84] Chytil, M. P., Wagner, K., Separation results for low nondeterministic auxiliary pushdown space complexity classes, preprint 1984. (§ 17.2.2)

[Chy 75] Chytil, M. P., On complexity of nondeterministic Turing machine computations, MFCS 1975, LNCS **32** (1975), 199–205. (21.2, 21.3, 21.6)

[Chy 76] —, Analysis of the non-context-free component of formal languages, LNCS **45** (1976), 230–236. (8.22)

[Chy 77] —, Comparison of the active visiting and the crossing complexities, MFCS 1977, LNCS **53** (1977), 272–281. (8.20, 21.8)

[Chy 82] —, The lower bound for context-sensitivity: A correction, Memorandum Nr. 382, Twente University of Technology, Enschede (1982). (8.7)

[Chy 84] —, Almost context-free languages, preprint 1984. (§ 8.1.1.2)

[Cle 63] Cleave, J. P., A hierarchy of primitive recursive functions, ZML **9** (1963), 331–346. (§ 2.4.2)

[CoAa 69] Cook, S.A., Aandera, S. O., On the minimum computation time of functions, TAMS 142 (1969), 291–314. (§ 8.1.4, 19.32)

[Cob 64] Cobham, A., The intrinsic computational complexity of functions, Proc. Intern. Congr. for Logic, Methodology and Philosophy of Science, North-Holland, Amsterdam 1964, 24–30. (10.14, 25.5)

[Cobn 80] Colbourn, C. J., Isomorphism complete problems on matrices, Proc. of the West Coast Conf. on Combinatorics, Graph Theory and Computing, Winnipeg (1980), 101–107. (§ 24.2.1.2.3)

[CoBo 72] Constable, R. L., Borodin, A. B., Subrecursive programming languages, part I: Efficiency and program structure, JACM 19 (1972), 526–568. (18.2)

[CoHa 71] Constable, R. L., Hartmanis, J., Complexity of formal translations and speed-up results, 3rd STOC (1971), 244–250. (18.5)

[CoHe 84] Compton, K., Henson, C. W., A simple method for proving lower bounds on the complexity of logical theories, preprint 1984. (§ 8.2.1)

[Col 69] Cole, S. N., Real-time computation by n-dimensional iterative arrays of finite state machines, IEEE Trans. on Comp. C-13 (1969), 349–365. (7.4, 7.8)

[Col 71] —, Deterministic pushdown store machines and realtime computations, JACM 18 (1971), 306–328. (13.17, 13.25)

[CoLe 77] Coffman, E. G. Jr., Leung, J. Y.-T., Combinatorial analysis of an efficient algorithm for processor and storage allocation, 18th FOCS (1977), 214–221. (§ 9.1.3.2)

[CoLeSl 78] Coffman, E. G. Jr., Leung, J. Y.-T., Slutz, D., On the optimality of fast heuristics for scheduling and storage allocation problems, Found. Contr. Engineering 3 (1978), 161–169. (§ 9.1.3.2)

[Com 79] Commentz-Walter, B., A string matching algorithm fast on the average, 6th ICALP (1979), LNCS 71 (1979), 118–132. (§ 7.2.1)

[Con 71a] Constable, R. L., Two types of hierarchy theorem for axiomatic complexity classes, Courant Computer Science Symposium 7 (1971), Computational Complexity, Ed. R. Rustin, 37–63. (17.12, 17.13, 17.14, 17.15)

[Con 71b] —, Subrecursive programming languages III, the multiple-recursive functions $\mathscr{R}^n$, Symp. on Computers and Automata, Polytechn. Inst. Brooklyn 1971, 393–410. (§ 2.4.1)

[Con 72] —, The operator gap, JACM 19 (1972), 175–183. (15.26, 17.8)

[Con 73] —, Type two computational complexity, 5th STOC (1973), 108–121. (§ 5.3, § 10.2.4)

[Coo 71a] Cook, S. A., The complexity of theorem proving procedures, 3rd STOC (1971), 151–158. (§ 6.1.1, 8.42, 14.21)

[Coo 71b] —, Characterizations of pushdown machines in terms of time-bounded computers, JACM 18 (1971), 4–18. (13.20, 13.34, 13.35, 20.13)

[Coo 72] —, Linear time simulation of deterministic two-way pushdown automata, Inf. Proc. 71 (Proc. IFIP Congr.) North-Holland, Amsterdam 1972, 75–80. (13.20)

[Coo 73a] —, An observation on time-storage trade off, 5th STOC (1973), 29–33. See also: JCSS 9 (1974), 308–316. (8.43, 14.19, 25.9, 26.6)

[Coo 73b] —, A hierarchy for nondeterministic time complexity, JCSS 7 (1973), 343–352. (17.43)

[Coo 79] —, Deterministic CFL's are accepted simultaneously in polynomial time and log squared space, 11th STOC (1979), 338–345. (§ 12.3, 13.15)

[Coo 81] —, Towards a complexity theory of synchronous parallel computation, L'énseignement mathématique, Série II, 27, Fasc. 1–2 (1981). (§ 21.2)

[Coo 83] —, The classification of problems which have fast parallel algorithms, FCT '83, LNCS 158 (1983), 78–93. (§ 21.2)

[CoRa 78] Cook, S. A., Rackoff, C. W., Space lower bounds for maze threadability on restricted machines, Univ. Toronto, Dep. Comp. Sci., TR No. 117 (1978). See also: SIAM JC 9 (1980), 636–652. (§ 8.2.1, § 23.3)

[CoRe 72] COOK, S. A., RECKHOW, R. A., Time-bounded random access machines, 4th STOC (1972), 73–80. See also: JCSS **7** (1973), 354–375. (17.42, 19.31)

[CoSe 76] COOK, S. A., SETHI, R., Storage requirements for deterministic polynomial time recognizable languages, JCSS **13** (1976), 25–37. (§ 8.2.1, § 25.3.2.1.1, § 26.3)

[CoWi 81] COPPERSMITH, D., WINOGRAD, S., On the asymptotic complexity of matrix multiplication. Extended summary. 22nd FOCS (1981), 82–90. See also: SIAM JC **11** (1982), 472–492. (§ 12.3)

[CrKi 80] CORNEIL, D. G., KIRKPATRICK, D. G., A theoretical analysis of various heuristics for the graph isomorphism problem, SIAM JC **9** (1980), 281–293. (§ 7.3.4)

[Čul 79] ČULIK, K., On the homomorphic characterizations of families of languages, 6th ICALP (1979), LNCS **71** (1979), 161–170. (4.14, 11.22)

[ČuDi 79] ČULIK, K., DIAMOND, N. D., A homomorphic characterization of time and space complexity classes of languages, Univ. of Waterloo, 1979, Res. Rep. CS-79-07 (1979). See also: IJCM **8** (1980), 207–222. (11.22)

[ČuYu 84] ČULIK, K., YU, S., Iterative tree automata, TCS **32** (1984), 227–247. (§ 20.3.2)

[Dah 84] DAHLHAUS, E., Reductions to NP-complete problems by interpretations, LNCS **171** (1984), 357–365. (24.9)

[Dal 71] VAN DALEN, D., A note on some systems of Lindenmayer, MST **5** (1971), 128–150. (12.19)

[Dek 69] DEKHT'AR, M. I., The impossibility of eliminating complete search in computing functions from their graphs (in Russian), DAN SSSR **189** (1969), 748–751. (7.13)

[Dek 74] —, On the complexity of relative computations (in Russian), DAN SSSR **214** (1974), 999–1001. (27.7, 27.15)

[Dek 76] —, On the relativization of deterministic and nondeterministic complexity classes, MFCS 1976, LNCS **45** (1976), 255–259. (28.7–28.10)

[Dek 79] —, Complexity spectra of recursive sets and approximability of initial segments of complete problems, EIK **15** (1979), 11–32. (§ 24.2.1.2.1)

[DeMe 77] DEUSSEN, P., MEHLHORN, K., Von Wijngaarden grammars and space complexity class EXSPACE, AI **8** (1977), 193–199. (§ 12.5)

[DiKa 78] DINIC, E. A., KARZANOV, A. B., A Boolean optimization problem under restrictions of one symbol, VNIISI Moskva, preprint 1978. (9.1)

[DoLiRe 79] DOBKIN, D. P., LIPTON, R. J., REISS, S. V., Linear programming is log-space hard for P, IPL **8** (1979), 96–97. (14.19)

[DoMu 78] DOBKIN, D., MUNRO, I., Time and space bounds for selection problems, LNCS **62** (1978), 192–204. (§ 25.3.2.2)

[Dow 72] DOWNEY, P. C., Undecidability of Presburger arithmetic with a single monadic letter, Center for Research in Comp. Technology, Harvard University 18–72 (1972). (8.47)

[DuGa 81] ĎURIŠ, P., GALIL, Z., Fooling a two way automaton or One pushdown is better than one counter for two way machines (preliminary version), 13th STOC (1981), 177–188. See also: TCS **21** (1982), 39–53. (13.27)

[DuGa 82] —, Two tapes are better than one for nondeterministic machines, 14th STOC (1982), 1–7. (§ 22.1, § 22.1)

[DuGaPaRe 83] ĎURIŠ, P., GALIL, Z., PAUL, W., REISCHUK, R., Two nonlinear lower bounds, 15th STOC (1983), 127–132. (8.37, § 22.1)

[DyCo 80] DYMOND, P. W., COOK, S. A., Hardware complexity and parallel computation, 21st FOCS (1980), 360–372. (§ 20.3.2, 21.7, 21.14, 21.16)

[DyTo 83] DYMOND, P. W., TOMPA, M., Speedups of deterministic machines by synchronous parallel machines, 15th STOC (1983), 336–343. (20.35, 20.42)

[ElRa 66] ELGOT, C. C., RABIN, M. O., Decidability and undecidability of extensions of second (first) order theory of (generalized) successor, JSL **31** (1966), 169–181. (8.46)

[Emd 74] VAN EMDE BOAS, P., Abstract resource bound classes, Diss., Math. Centrum Amsterdam 1974. (§ 17)

[Emd 75] —, Ten years of speed-up, MFCS (1975), LNCS 32 (1975), 13—29. (§ 18)
[Emd 78] —, Some applications of the McCreight-Meyer algorithm in abstract complexity theory, TCS 7 (1978), 79—98. (§ 15.2, 15.17)
[Emd 79] —, Complexity of linear problems, FCT 1979, 117—120. (§ 7.2.2.4)
[Emd 81] —, How good are the constants in Lenstra's algorithm for integer linear programming?, preprint 1981, Math. Inst. University of Amsterdam. (§ 7.2.2.4)
[EmLe 78] VAN EMDE BOAS, P., VAN LEEUWEN, J., Move rules and trade-offs in the pebble game, Rijksuniv. Utrecht, RUU-CS-78-4 (1978). See also: LNCS 67 (1979), 101—112. (25.10, 25.15)
[Eng 83] ENGELFRIET, J., Iterated pushdown automata and complexity classes, 15th STOC (1983), 365—373. (13.43)
[Ern 76] ERNI, W., Some further languages log-tape reducible to context-free languages, Univ. Karlsruhe, Bericht 45, 1976. (14.3)
[Eve 63] EVEY, R. J., The theory and application of pushdown store machines, Math. Linguistics and Automatic Translation, Harvard Univ., Computation Lab. Rep. NSF-IO (1963). (12.4)
[Even 79]* EVEN, S., Graph Algorithms, Computer Science Press, Inc., Woodland Hills (Calif.) 1979. (§ 7.2.2)
[EvItSh 75] EVEN, S., ITAI, A., SHAMIR, A., On the complexity of time table and multi-commodity flow problems, 16th FOCS (1975), 184—193. See also SIAM JC 5 (1976), 691—703. (14.21)
[EvTa 76] EVEN, S., TARJAN, R. E., A combinatorial problem which is complete in polynomial space, JACM 23 (1976), 710—719. (8.52, 14.14)
[Fag 73] FAGIN, R., Generalized first-order spectra and polynomial-time recognizable sets. In: Compl. of Comp. (Proc. Symp., New York, (1973), SIAM-AMS Proc. Vol. VII (1974), 43—73. (Ch. IV, § 24.2.2.2.2)
[FeLu 81] FERNANDEZ DE LA VEGA, W., LUEKER, G. S., Bin packing can be solved within $1 + \varepsilon$ in linear time, Combinatorica 1 (1981), 349—355. (§ 9.1.3.2)
[Fer 74] FERRANTE, J., Some upper and lower bounds on decision procedures in logic, MAC-TR-139, 1974. (8.45, 8.46, 8.48, 14.14, 14,25, 14.26)
[FeRa 75] FERRANTE, J., RACKOFF, C. W., A decision procedure for the first order theory of real addition with order, SIAM JC 4 (1975), 69—76. (8.47, 14.25)
[FeRa 79]* —, The Computational Complexity of Logical Theories, Lecture Notes in Mathematics, vol. 718, Springer-Verlag, Berlin—Heidelberg—New York 1979. (Intr., § 7.4)
[Fil 72] FILOTTI, I., On effectively levelable sets, Rec. Fctn. Newsletter No. 2 (1972), 12—13. (§ 18.2)
[FiLi 84] FINN, J., LIEBERHERR, K., Primality testing and factoring, TCS 23 (1983), 211—215. (9.24)
[Fim 69] FISCHER, M. J., Two characterizations of context-sensitive languages, 10th SWAT (1969), 149—156. (13.40)
[FiMa 80] FILOTTI, I. S., MAYER, J. N., A polynomial time algorithm for determining the isomorphism of graphs of fixed genus, 12th STOC (1980), 236—243. (7.30)
[FimLa 77] FISCHER, M. J., LADNER, R. E., Propositional modal logic and programs, 9th STOC (1977), 286—294. See also: JCSS 18 (1979), 194—211. (8.49)
[FimPa 74] FISCHER, M. J., PATERSON, M. S., String matching and other products, SIAM-AMS 7 (1974), 75—96. (§ 7.2.1)
[FimRa 74] FISCHER, M. J., RABIN, M. O., Super-exponential complexity of Presburger arithmetic, MAC Techn. Mem. 43. MIT 1974. (8.47, 14.26)
[FimRo 68] FISCHER, M. J., ROSENBERG, A. L., Real-time solutions of the origin-crossing problem, MST 2 (1968), 257—263. (7.9)
[FimSt 74] FISCHER, M. J., STOCKMEYER, L. J., Fast on-line integer multiplication, JCSS 9 (1974), 317—331. (7.16)
[Fip 65] FISCHER, P. C., Generation of primes by a one-dimensional real-time iterative array, JACM 12 (1965), 388—394. (7.27, 18.14)

[Fip 68] —, The reduction of tape-reversals for off-line one-tape Turing machines, JCSS **2** (1968), 136–146. (21.4)

[FipMeRo 68] FISCHER, P. C., MEYER, A. R., ROSENBERG, A. L., Counter machines and counter languages, MST **2** (1968), 265–283. (§ 8.1.1.1, 11.4, 13.6, 18.10, 19.11, § 20.1.1, 22.12, 22.13, 22.15)

[FlMaSi 76] FLEISCHMANN, K., MAHR, B., SIEFKES, D., Bounded concatenation theory as a uniform method for proving lower complexity bounds, Logic Coll. 76 (Oxford 76), Studies in Logic and Found. of Math. 87, North-Holland, Amsterdam 1977, 471–490. (8.51)

[FlSt 74a] FLAJOLET, P., STEYART, J. M., Complexity of classes of languages and operators, IRIA Res. Rep. 92 (1974). (11.3)

[FlSt 74b] —, On sets having only hard subsets, LNCS **14** (1974), 446–457. (17.12)

[FöMa 79] FÖRSTER, K. J., MAHR, B., Relating graphisomorphism and clique problems, Preprint TU Berlin (1979). (§ 24.2.1.2.3)

[FoHo 79] FORTUNE, S., HOPCROFT, J., A note on Rabin's nearest neighbour algorithm, IPL **8** (1979), 20–23. (§ 9.2.2)

[For 79] FORTUNE, S. A., A note on sparse complete sets, SIAM JC **8** (1979), 431–433. (24.16)

[FoWy 78] FORTUNE, S. A., WYLLIE, S., Parallelism in random access machines, 10th STOC (1978), 114–118. (20.35, § 26.3)

[Fre 79] FREDERICKSON, G. N., Approximation algorithms for some postman problems, JACM **26** (1979), 538–554. (§ 9.1.3.2)

[Fre 80] —, Probabilistic analysis for simple one- and twodimensional bin packing problems, IPL **11** (1980), 156–161. (§ 9.2.1)

[FrGaJoScYe 78] FRAENKEL, A. S., GAREY, M. R., JOHNSON, D. S., SCHAEFER, T., YESHA, Y., The complexity of checkers on an $N \times N$ board — prel. rep., 19th FOCS (1978), 55–64. (14.14)

[FrLi 81] FRAENKEL, A. S., LICHTENSTEIN, D., Computing a perfect strategy for $n \times n$ chess requires time exponential in n, Journ. Comb. Theory, Ser. A, **31** (1981), 199–213. (14.24)

[FrPa 83] FRANCO, J., PAULL, M., Probabilistic analysis of the Davis-Putnam procedure for solving the satisfiability problem, Discr. Appl. Math. **5** (1983), 77–87. (§ 9.2.1)

[Frv 65] FREIVALDS, R. V., Complexity of palindrome recognition by Turing machines with input (in Russian), Algebra i Logika **4**, 1 (1965), 47–58. (8.25)

[Frv 74] —, Fast probabilistic algorithms, MFCS (1979), LNCS **74** (1979), 57–69. (§ 9.2.2)

[Frv 75] —, Fast computations on probabilistic machines (in Russian), Teorija algoritmov i programm II, Riga 1975, 201–205. (9.20, 9.22)

[Frv 78] —, Recognition of languages with high probability on different classes of automata (in Russian), DAN SSSR **239** (1978), 60–62. (§ 9.2.2)

[Frv 79a] —, Speeding the recognition of certain sets by the use of a random number transducer (in Russian), Problemy Kibernetiki **36** (1979), 209–224. (9.20, 9.22)

[Frv 79b] —, Language recognition using finite probabilistic multitape and multihead automata (in Russian), Prob. Pered. Inf. **15**, 3 (1979), 99–106. (§ 9.2.2)

[Frv 79c] —, Fast probabilistic checking the correctness of integer multiplication (in Russian), Automatika i Vyč. Tekhnika **1** (1979), 40–42. (9.21)

[Frv 79d] —, Recognition of languages on probabilistic Turing machines in real-time and automata with pushdown storage (in Russian), Prob. Pered. Inf. **15**, 4 (1979), 97–101. (9.23)

[Frv 79e] —, On the worktime of deterministic and nondeterministic Turing machines (in Russian), Latviĭskiĭ matematiceskiĭ eshegodnik **23** (1979), 158–165. (24.6)

[Für 78] FÜRER, M., Simulation of multi-dimensional Turing machines, Bull. EATCS **5** (1978), 36–70. (19.22, 19.25)

[Für 80] —, The complexity of the inequivalence problem for regular expressions with intersection, 7th ICALP 1980, LNCS **85** (1980), 234–245. (8.44, 14.14, 14.16)

[Für 82] —, The tight deterministic time hierarchy, 14th STOC (1982), 8–16. (17.39, 17.51)

[Für 83] —, The computational complexity of the unconstrained limited domino problem (with implications for logical decision problems), LNCS **171** (1983), 312–319. (8.49)

[FüScSp 83] FÜRER, M., SCHNYDER, W., SPECKER, E., Normal forms for trivalent graphs and graphs of bounded valence, 15th STOC (1983), 161–170. (§ 7.3.4)

[FuSaSi 81] FURST, M., SAXE, J. B., SIPSER, M., Parity circuits and the polynomial time hierarchy, 22nd FOCS (1981), 260–270. (28.6)

[GaGa 79] GABBER, O., GALIL, Z., Explicit constructions of linear size concentrators and superconcentrators, 20th FOCS (1979), 364–370. See also: JCSS **22** (1981), 407–420. (§ 25.3.2.1.2)

[GaGrJoKn 78] GAREY, M. R., GRAHAM, R. L., JOHNSON, D. S., KNUTH, D. E., Complexity results for bandwidth minimization, SIAM J. Appl. Math. **34** (1978), 477–495. (14.21)

[GaHoLuScWe 82] GALIL, Z., HOFFMANN, C. M., LUKS, E. M., SCHNORR, C. P., WEBER, A., An $O(n^3 \log n)$ deterministic and an $O(n^3)$ probabilistic isomorphisms test for trivalent graphs, 23rd FOCS (1982), 118–125. (7.30)

[GaJo 75] GAREY, M. R., JOHNSON, D. S., Complexity results for multiprocessor scheduling under resource constraints, SIAM JC **4** (1975), 397–411. (§ 9.1.3.1.2, 14.21)

[GaJo 76] —, The complexity of near optimal graph coloring, JACM **23** (1976), 43–49. (9.11)

[GaJo 79]* —, Computers and Intractability: A Guide to the Theory of NP-Completeness, Freeman, San Francisco 1979. (Intr., § 7.2, § 7.3.2, § 9.1, 9.7, § 9.1.3.2, 14.21, 24.9)

[GaJoSt 74] GAREY, M. R., JOHNSON, D. S., STOCKMEYER, L. J., Some simplified NP-complete problems, 6th STOC (1974), 91–95. See also: TCS **1** (1976), 237–267. (14.21)

[GaJoTa 76] GAREY, M. R., JOHNSON, D. S., TARJAN, R. E., The planar Hamiltonian circuit problem is NP-complete, SIAM JC **5** (1976), 704–714. (14.21)

[Gal 74] GALIL, Z., Two way deterministic pushdown automata languages and some open problems in the theory of computation, 15th SWAT (1974), 170–177. See also: MST **10** (1976), 211–228. (13.20, 24.34, 26.12)

[Gal 76a] —, Hierarchies of complete problems, AI **6** (1976), 77–88. (§ 8.2.3.4, 14.12, 14.13, 14.14. 14.21, § 26.4)

[Gal 76b] —, Real-time algorithms for string matching and palindrome recognition, 8th STOC (1976), 161–173. See also: JCSS **16** (1978), 140–157. (7.5)

[Gal 76c] —, String-matching in real time, University of Tel-Aviv, Dep. Math. Sc., TR July 1976. See also: JACM **28** (1981), 134–149. (§ 7.2.1, 7.1)

[Gal 78] —, On improving the worst case running time of the Boyer-Moore string matching algorithm 5th ICALP (1978), LNCS **62** (1978), 241–250. See also: CACM **22** (1979), 505–508. (§ 7.2.1)

[Gal 84] —, Optimal parallel algorithms for string matching, 16th STOC (1984), 240–248. (7.1)

[Gall 69] GALLAIRE, H., Recognition time of context-free languages by on-line Turing machines, I&C **15** (1969), 288–295. (8.10)

[GaLo 79] GÁCS, P., LOVÁSZ, L., Khachian's algorithm for linear programming, TR Stanford University 1979. (7.22)

[GaNa 79] GALIL, Z., NAAMAD, A., Network flow and generalized path compression, 11th STOC (1979), 13–26. See also: JCSS **21** (1980), 203–217. (7.21)

[GaSe 77] GALIL, Z., SEIFERAS, J., Saving space in fast string matching, 18th FOCS (1977), 179–188. See also: SIAM JC **9** (1980), 417–438. (7.3)

[GaSe 78] —, A linear-time on-line recognition algorithm for "palstar", JACM **25** (1978), 102–111. (7.5)

[GaSe 81a] —, Linear time string matching using only a fixed number of local storage locations, TCS **13** (1981), 331–336. (§ 7.2.1)

[GaSe 81 b] —, Time-space-optimal string matching, 13th STOC (1981), 106—113. See also: JCSS **26** (1983), 280—294. (§ 7.2.1)

[GaWi 83] GALPERIN, H., WIGDERSON, A., Succinct representation of graphs, I&C **56** (1983), 183—198. (14.14, 14.25)

[GeLe 79] GENS, G. V., LEVNER, E. V., Computational complexity of approximation algorithms for combinatorial problems, MFCS 79, LNCS **74** (1979), 292—300. (9.5, 9.12)

[Geo 76] GEORGIEVA, N. V., Classes of one-argument recursive functions, ZML **22** (1976), 127—130. (2.21)

[GiGr 66] GINSBURG, S., GREIBACH, S., Deterministic contextfree languages, I&C **9** (1966), 620—648. (13.19)

[GiGrHa 67] GINSBURG, S., GREIBACH, S., HARRISON, M. A., Stack automata and compiling, JACM **14** (1967), 172—201. (13.33)

[Gil 74] GILL, J. T., Computational complexity of probabilistic Turing machines, 6th STOC (1974), 91—95. See also: SIAM JC **6** (1977), 675—695. (§ 9.2.2, 14.23, 20.43, 20.44, 20.46)

[GiLeTa 79] GILBERT, J. R., LENGAUER, T., TARJAN, R. E., The pebbling game is complete in polynomial space, 11th STOC (1979), 237—248. See also: SIAM JC **9** (1980), 513—524. (14.14)

[Gin 66]* GINSBURG, S., The Mathematical Theory of Context-Free Languages, McGraw-Hill, New York 1966. (§ 4, § 11)

[Gin 75]* —, Algebraic and Automata Theoretic Properties of Formal Languages, North-Holland, Amsterdam 1975. (Pref., § 4, § 11, 11.8)

[GiRo 70] GINSBURG, S., ROSE, G., On the existence of generators for certain AFL, Inf. Sci. **2** (1970), 431—446. (11.5)

[GiRo 74] —, The equivalence of stack-counter acceptors and quasi-realtime stack-counter acceptors, JCSS **8** (1974), 243—269. (13.33)

[GiSi 76] GILL, J., SIMON, I., Ink, dirty tape Turing machines, and quasicomplexity measures, 3rd ICALP (1976) Edinburgh University Press, 285—306. (§ 5.2)

[Gli 71] GLINERT, E. P., On restricted Turing computability, MST **5** (1971), 331—343. (17.50)

[Gob 80]* GOLUMBIC, M. C., Algorithmic Graph Theory and Perfect Graphs, Academic Press, New York 1980. (§ 7.2.2)

[Gol 77] GOLDBERG, A., Average case complexity of the satisfiability problem, Proc. 4th Workshop on Automated Deduction, Austin (Texas) (1977). (§ 9.2.1)

[Gols 78] GOLDSCHLAGER, L. M., A unified approach to models of synchronous parallel machines, 10th STOC (1978), 89—94. (§ 20.3.2)

[GoNe 78] GOETZE, B., NEHRLICH, W., Loop programs and classes of primitive recursive functions, MFCS 1978, LNCS **64** (1978), 232—237. See also: ZML **26** (1980), 255—278. (§ 2.4.2)

[GoPuBr 82] GOLDBERG, A., PUDOM, B., BROWN, C., Average-time analysis of simplified Davis-Putnam procedures, IPL **15**, 2 (1982), 72—75, Corrigendum, IPL **16**, 4 (1983), 213. (§ 9.2.1)

[Gra 77] GRANT, P. W., Recognition of EOL languages in less than quartic time, IPL **6** (1977), 174—175. (12.19)

[Gra 80] —, Some more independence results in complexity theory, TCS **12** (1980), 119—126. (§ 24.5)

[Graw 81] GRAW, B., Savitch-Labyrinthe und Speicherkomplexitätsklassen, Dissertation A, Humboldt-Univ. Berlin, 1981. (§ 23.3)

[Grd 83] GRANDJEAN, E., Universal quantifiers and time complexity of random access machines, LNCS **171** (1983), 366—379. (Ch. IV)

[GrDi 75] GRIMMET, G. R., MCDIARMID, C. J. H., On colouring random graphs, Math. Proc. Cambr. Phil. Soc. **77** (1975), 313—324. (9.16)

[Gre 73] GREIBACH, S. A., The hardest context-free language, SIAM JC **2** (1973), 304—310. (13.16, 14.2)

[Gre 75a] —, A note on the recognition of one-counter languages, RAIRO, Informat. Théor. 9 (1975), R-2, 5—12. (13.6)

[Gre 75b] —, Erasable context-free languages, I&C 29 (1975), 301—326. (13.6)

[Gre 76] —, Remarks on the complexity of nondeterministic counter languages, TCS 1 (1976), 269—288. (§ 20.1.1, 20.6, 22.13—22.15)

[Gre 77] —, A note on NSPACE ($\log_2 n$) and substitution, RAIRO Informat. Théor. 11 (1977), 127—132. (11.3)

[Gre 78a] —, One way finite visite automata, TCS 6 (1978), 175—221. (11.22)

[Gre 78b] —, Visits, crosses, and reversals for nondeterministic off-line machines, I&C 36 (1978), 174—216. (21.2, 21.3, 21.6)

[Gre 78c] —, Remarks on blind and partially blind one-way multi-counter machines, TCS 7 (1978), 311—324. (11.4, 11.12, 22.16)

[GrHaIb 67] Gray, J. N., Harrison, M. A., Ibarra, O. H., Two-way pushdown automata, I&C 11 (1967), 30—70. (13.20)

[GrHo 69] Greibach, S. A., Hopcroft, J. E., Scattered context grammars, JCSS 3 (1969), 233—247. (11.8)

[Gri 77] Grigorev, D. Ju., Embedding theorems for Turing machines of different dimensions and Kolmogorov's algorithms (in Russian), DAN SSSR 254 (1977), 15—18. (19.28, § 25.3.2.2)

[Gri 79a] —, Two reductions of the graph isomorphism to problems for polynomials (in Russian), Issled. po konstr. mat. i mat. log. VIII (1979), 47—54, Leningrad. (§ 7.3.4)

[Gri 79b] —, Time complexity of multidimensional Turing machines (in Russian), Zapiski naučnych seminarov, AN SSSR LOMI, 88 (1979), 47—55. (19.27, 19.31)

[Gru 76] Gruska, J., Descriptional complexity (of languages) — A short survey, MFCS '76, LNCS 45 (1976), 65—80. (Intr.)

[Grz 53] Grzegorczyk, A., Some classes of recursive functions, Rozprawy Matematiczne IV, Warszawa 1953. (§ 2.4.1, 2.19, 2.20)

[GuIb 79a] Gurari, I. M., Ibarra, O. H., On the space complexity of recursive algorithms, IPL 8 (1979), 267—271. (§ 20.3.1)

[GuIb 79b] —, Simple counter machines and number-theoretic problems, JCSS 19 (1979), 145—162. (22.16)

[GuIb 81] —, The complexity of the equivalence problem for two characterizations of Presburger sets, TCS 13 (1981), 295—314. (22.16)

[GuIb 82] —, (Semi) alternating stack automata, MST 15, 3 (1981/82), 211—224. (20.39)

[GuOd 80] Guibas, L. J., Odlyzko, A. M., A new proof of the linearity of the Boyer-Moore string searching algorithm, SIAM JC 9 (1980), 672—682. (§ 7.2.1)

[HaBa 75] Hartmanis, J., Baker, T. P., On simple Goedel numberings and translations, SIAM JC 4 (1975), 1—11. (§ 2.2)

[HaBa 79] —, Relative succinctness of representations of languages and separation of complexity classes, MFCS 1979, LNCS 74 (1979), 70—88. (24.39)

[HaBe 75] Hartmanis, J., Berman, L., A note on tape bounds for SLA language processing, 16th FOCS (1975), 65—70. See also: TCS 3 (1976), 213—224. (5.6, 8.32, 17.23, 23.12)

[HaBe 76] —, On isomorphism and density of NP and other complete sets, 8th STOC (1976), 30—40. See also: JCSS 16 (1978), 418—422. (24.9)

[HaHo 71a] Hartmanis, J., Hopcroft, J. E., An overview of the theory of computational complexity, JACM 18 (1971), 444—475. (§ 17.2.3, 18.1, 18.4)

[HaHo 71b] —, What makes some language theory problems undecidable?, JCSS 4 (1971), 368—376. (4.14)

[HaHo 76] —, Independence results in computer science, SIGACT News 8 (1976), 13—24. (§ 8, 24.49)

[HaHu 74] Hartmanis, J., Hunt III, H. B., The LBA problem and its importance in the theory of computing, SIAM-AMS 7 (1974), 1—26. (23.24)

[HaIb 68] Harrison, M. A., Ibarra, O. H., Multitape and multihead pushdown automata, I&C **13** (1968), 433–470. (13.19)

[HaImMa 78] Hartmanis, J., Immerman, N., Mahaney, S., One-way log-tape reductions, 19th FOCS (1978), 65–71. (§ 6.1.2, 8.43, 14.12)

[Háj 77] Hájek, P., Arithmetical complexity of some problems in computer science, MFCS 1977, LNCS **53** (1977), 282–287. (§ 24.5, 28.4)

[HaMa 80] Hartmanis, J., Mahaney, S., An essay about search on sparse NP-complete sets, MFCS 1980, LNCS **88** (1980), 40–57. (24.16)

[HaMa 81] —, Languages simultaneously complete for one-way and two-way log-tape automata, SIAM JC **10** (1981), 383–390. (14.13)

[Han 66] Hanák, J., On real-time Turing machines, Archivum mathematicum (Brno) **2** (1966), 79–92. (22.6)

[Har 75] Harrow, K., Small Grzegorczyk classes and limited minimum, ZML **21** (1975), 417–426. (§ 2.4.1)

[Har 78] —, The bounded arithmetic hierarchy, I&C **36** (1978), 102–117. (§ 2.4.1, 4.19)

[Har 79] —, Equivalence of some hierarchies of primitive recursive functions, ZML **25** (1979), 411–418. (§ 2.4.1)

[Harj 79] Harju, T., A simulation result for the auxiliary pushdown automata, JCSS **19** (1979), 119–132. (20.14)

[Hart 68a] Hartmanis, J., Computational complexity of one-tape Turing machine computations, JACM **15** (1968), 325–339. (8.15, 17.35)

[Hart 68b] —, Tape reversal bounded Turing machine computations, JCSS **2** (1968), 117 to 135. (8.27, 17.47, 21.5, 22.9)

[Hart 70] —, A note on one-way and two-way automata, MST **4** (1970), 24–28. (§ 5.1.1.2)

[Hart 71] —, Computational complexity of random access stored program machines, MST **5** (1971), 232–245. (17.42, 19.31)

[Hart 73] —, On the problem of finding natural complexity measures, Proc. of MFCS High Tatras, Czechoslovakia, 1973, 95–104. (15.1, § 16.2)

[Hart 74] —, Computational complexity of formal translations, MST 8 (1974/75), 156–166. (§ 2.2)

[Hart 77] —, Relations between diagonalization, proof systems and complexity gaps, 9th STOC (1977), 223–227. (§ 17.2.3)

[Hart 78] —, On log-tape isomorphisms of complete sets, TCS **7** (1978), 273–286. (14.11, 23.17)

[Hart 80] —, On the succinctness of different representations of languages, SIAM JC **9** (1980), 114–120. (24.39)

[Hart 83a] —, On sparse sets in NP ∖ P, IPL **16** (1983), 55–60. (§ 24.2.1.2.1, 24.17)

[Hart 83b] —, Generalized Kolmogorov complexity and the structure of feasible computations, 24th FOCS (1983), 439–445. (24.17)

[Hart 84] —, On non-isomorphic NP-complete sets. Bull. EATCS **24** (1984), 73–78. (24.9)

[HaSeIm 83] Hartmanis, J., Sewelson, V., Immerman, N., Sparse sets in NP-P; Exptime versus Nexptime, 15th STOC (1983), 382–391. (24.17, 28.8)

[HaSh 68] Hartmanis, J., Shank, H., On the recognition of primes by automata, JACM **15** (1968), 382–389.

[HaSi 76] Hartmanis, J., Simon, J., On the structure of feasible computations, In: Advances in Computers, vol. 14, Academic Press, New York 1976, 1–43. (20.12)

[HaSt 65] Hartmanis, J., Stearns, R. E., On the computational complexity of algorithms, TAMS May (1965), 285–306. (8.8, 15.15, 18.11, 18.12, 19.25)

[Hau 80] Haulitschke, M., Rückkehrmaße, Diplomarbeit, Friedrich-Schiller-Univ. Jena, 1980. (21.12)

[Hav 71] Havel, I. M., Weak complexity measures, SIGACT News **3** (1971), 21–30. (§ 5.2)

[HaYe 83] Hartmanis, J., Yesha, Y., Computation times of NP sets of different densities. ICALP '83, LNCS **154** (1983), 319–330. (24.17)

[Hec 75] Hecker, H. D., Eine Bemerkung zur Berechnungskompliziertheit auf Registermaschinen, EIK **11** (1975), 437–438. (16.5)

[HeKr 84] Hemmerling, A., Kriegel, K., On searching of special classes of mazes and finite embedded graphs, MFCS '84, LNCS **176** (1984), 291–300. (§ 23.3)

[Hel 81] Heller, H., Relativized polynomial hierarchies extending two levels, Thesis, München 1981. (28.6)

[Helm 71] Helm, J. P., On effectively computable operators ZML **17** (1971), 231–244. (§ 3.2)

[Hem 79a] Hemmerling, A., On the space complexity of multidimensional Turing automata, Ernst-Moritz-Arndt-Univ. Greifswald, preprint 2, 1979. (5.2)

[Hem 79b] —, Zur Raumkompliziertheit von Berechnungs-, Erkennungs- und Entscheidungsprozessen (mit einer Bemerkung zur P-NP-Problematik), Ernst-Moritz-Arndt-Univ. Greifswald, preprint, 1979. (§ 23.2)

[Hen 65] Hennie, F. C., One-tape off-line Turing machine computations, I&C **8** (1965), 553–578. (8.13, 17.38)

[Hen 66] —, On-line Turing machine computations, IEEE Trans. on Electr. Comp. **EC-15** (1966), 35–44. (8.9, 19.25)

[Her 71] Herman, G. T., The equivalence of different hierarchies of elementary functions, ZML **17** (1971), 219–224. (10.18)

[HeSt 66] Hennie, F. C., Stearns, R. E., Two-way simulation of multitape Turing machines, JACM **13** (1966), 533–546. (17.40, 19.16)

[HeYo 71] Helm, J., Young, P., On size vs. efficiency for programs admitting speed-ups, JSL **36** (1971), 21–27. (18.5)

[Hfm 82a]* Hoffmann, C. M., Group Theoretic Algorithms and Graph Isomorphism, LNCS **136** (1982). (§ 7.3.4)

[Hfm 82b] —, Subcomplete generalizations of graph isomorphism, JCSS **25** (1982), 332–359. (§ 7.3.4, § 24.2.1.2.3)

[Hib 67] Hibbard, T. N., A generalization of context-free determinism, I&C **11** (1967), 196–238. (12.5, 17.48)

[HiWo 76] Hirschberg, D. S., Wong, C. K., A polynomial-time algorithm for the knapsack problem with two variables, JACM **23** (1976), 147–154. (14.21)

[Hoc 82] Hochbaum, D. S., Approximation algorithms for the set covering and vertex cover problems, SIAM JC **11** (1982), 555–556. (9.15)

[Hof 80] Hofri, M., Two-dimensional packing: Expected performance of simple level algorithms, I&C **45** (1980), 1–17. (§ 9.2.1)

[HoKe 83] Hodgson, B. R., Kent, C. F., A normal form for arithmetical representations of NP-sets, JCSS **27** (1983), 378–388. (24.40)

[Hon 80a] Hong, Jia-wei, On some deterministic space complexity problem, 12th STOC (1980), 310–317. See also: SIAM JC **11** (1982), 591–601. (14.13)

[Hon 80b] —, On similarity and duality of computation, 21st FOCS (1980), 348–359. (20.35, 21.7)

[Hon 84] —, A trade-off theorem for space and reversal, TCS **32** (1984), 221–224. (§ 8.1.2.4)

[HoPaVa 75] Hopcroft, J. E., Paul, W., Valiant, L., On time versus space and related problems, 16th FOCS (1975), 57–64. See also: JACM **24** (1977), 332–337. (19.30, 19.31, 25.1, 25.7)

[HoSa 78]* Horowitz, E., Sahni, S., Fundamentals of Computer Algorithms, Comp. Sci. Press, INC. 1978. (Intr., § 7, § 7.2.2, § 9.1, 9.1, 9.6)

[HoSh 84] Hochbaum, D. S., Shmoys, D. B., Powers of graphs: A powerful approximation technique for bottleneck-problems, 16th STOC (1984), 324–333. (9.14)

[HoTa 74] Hopcroft, J. E., Tarjan, R., Efficient planarity testing, JACM **21** (1974), 549–568. (7.18)

[HoUl 67] Hopcroft, J. E., Ullman, J. D., Nonerasing stack automata, JCSS **1** (1967), 166–186. (13.29)

[HoUl 68a] —, Relations between time and tape complexities, JACM **15** (1968), 414–427. (25.3)

[HoUl 68b] —, Sets accepted by one-way stack automata are context-sensitive, I&C **13** (1968), 114–133. (13.33)

[HoUl 69a]* —, Formal Languages and Their Relation to Automata, Addison-Wesley, Reading (Mass.) 1969. (Intr.)

[HoUl 69b] —, Some results on tape bounded Turing machines, JACM **16** (1969), 168—177. (8.28, 8.30, 17.33)

[HoUl 79]* —, Introduction to Automata Theory, Languages and Computation, Addison-Wesley, Reading (Mass.) 1979. (Intr., Pref., § 17.2.3)

[Hro 84] HROMKOVIČ, J., Hierarchy of reversal and zerotesting bounded multicounter machines, 11th MFCS, LNCS **176** (1984), 212—321. (17.47, § 17.2.6, 22.16)

[HsYe 72] HSIA, P., YEH, R. T., Finite automata with markers, Aut., Languages and Programming, Roquencourt 1972, 443—451. (§ 7.3.3)

[Hul 79] HULL, B., Containment between intersection families of linear and reset languages, Ph. D. Thesis, University of California, Berkeley 1979. (11.10)

[Hun 73a] HUNT III, H. B., The equivalence problem for regular expressions with intersection is not polynomial in tape, Cornell Univ., Dep. Comp. Sci., TR 73-161, 1973. (8.44)

[Hun 73b] —, On the time and tape complexity of languages I, 5th STOC (1973), 10—19. (8.44)

[Hun 76] —, On the complexity of finite, pushdown and stack automata, MST **10** (1976), 33—52. (§ 8.2.3.4, 14.12—14.14, 14.19, 14.21, 14.24)

[Hun 77] —, A complexity theory for computation structures, Columbia University New York, Dep. of El. Eng. and Computer Sci., preprint 1977. (14.15)

[Hun 79] —, Observations on the complexity of regular expression problems, JCSS **19** (1979), 222—236. (8.44)

[HuRo 74] HUNT III, H. B., ROSENKRANTZ, D. J., Computational parallels between the regular and context-free languages, 6th STOC (1974), 64—74. See also: SIAM JC **7** (1978), 99—114. (8.44)

[HuRo 77] —, Complexity of grammatical similarity relations, Proc. Conf. on Theor. Comp. Sci., Waterloo, 1977, 139—145. (§ 8.2.3.4, 14.12, 14.14, § 24.2.1.2.3)

[HuRo 80] —, The complexity of recursion schemes and recursive programming languages, 21st FOCS (1980), 152—160. (§ 8.2.3.4)

[HuRoSz 76] HUNT III, H. B., ROSENKRANTZ, D. J., SZYMANSKI, T. G., On the equivalence, containment and covering problems for the regular and context-free languages, JCSS **12** (1976), 222—268. (§ 8.2.3.4, 14.12)

[HuSz 76] HUNT III, H. B., SZYMANSKI, T. G., Dichotomization, reachability and the forbidden subgraph problem, 8th STOC (1976), 126—134. (§ 8.2.3.4)

[HuSzUl 75] HUNT III, H. B., SZYMANSKI, T. G., ULLMAN, J. D., On the complexity of LR(k) testing, Proc. Conf. on Princ. of Comp. Lang., Palo Alto (1975), 130—136. (§ 8.2.3.4)

[Huy 80] HUYNH, T. D., The complexity of semilinear sets, 7th ICALP (1980), LNCS **85** (1980), 324—337. See also: EIK **18** (1982), 291—338. (14.22)

[Huy 84] —, Deciding the inequivalence of context-free grammars with 1-letter terminal alphabet is Σ_2^p-complete, 23rd FOCS (1982), 21—31. (14.22)

[Iba 71] IBARRA, O. H., Characterizations of some tape and time complexity classes of Turing machines in terms of multihead and auxiliary stack automata, JCSS **5** (1971), 88—117. (13.29, 13.35, 13.40, 20.19, 20.22, 20.26)

[Iba 72] —, A note concerning nondeterministic tape complexities, JACM **19** (1972), 608—612. (17.32)

[Iba 73a] —, On two-way multihead automata, JCSS **7** (1973), 28—36. (13.24)

[Iba 73b] —, Controlled pushdown automata, Inf. Sci. **6** (1973), 327—342. (§ 12.2)

[Iba 74] —, A hierarchy theorem for polynomial-space recognition, SIAM JC **3** (1974), 184—187. (19.1)

[IbKi 75a] IBARRA, O. H., KIM, C. E., Fast approximation algorithms for the knapsack and sum of subset problems, JACM **22** (1975), 465—468. (§ 9.1.3.1.1)

[IbKi 75b] —, On 3-head versus 2-head finite automata, AI **4** (1975), 193—200. (13.4)

[IbMo 83] Ibarra, O. H., Moran, S., Some time-space tradeoff results concerning single-tape and offline Turing machines, SIAM JC **12** (1983), 388—394. (25.4)

[IbRo 81] Ibarra, O. H., Rosier, L. E., On restricted one-counter machines, MST **14** (1981), 241—245. (22.16)

[IbSa 75] Ibarra, O. H., Sahni, S. K., Hierarchies of Turing machines with restricted tape alphabet size, JCSS **11** (1975), 56—67. (17.30, 19.1)

[Iga 77] Igarashi, Y., The tape complexity of some classes of Szilard languages, SIAM JC **6** (1977), 460—466. (12.18)

[IgHo 74] Igarashi, Y., Honda, N., Deterministic multitape automata computations, JCSS **8** (1974), 167—189. (12.6, 17.35, 19.17)

[Imm 79] Immerman, N., Length of predicate calculus formulas as a new complexity measure, 20th FOCS (1979), 337—347. See also: JCSS **22** (1981), 384—406. (§ 8.1.5, 8.43, § 8.2.1, 14.19, § 26.3)

[Imm 80] —, Upper and lower bounds for first order expressibility, 21st FOCS (1980), 74—82. See also: JCSS **25** (1982), 76—98. (§ 20.4.1, Ch. IV)

[Imm 83] —, Languages which capture complexity classes, 15th STOC (1983), 347—354. (Ch. IV)

[InTaTa 82] Inoue, K., Takanami, J., Taniguchi, H., A note on alternating on-line Turing machines, IPL **15** (1982), 164—168. (20.36)

[ItRo 77] Itai, A., Rodeh, M., Some matching problems, 4th ICALP (1977), LNCS **52** (1977), 258—268. (§ 7.2.2.3, 14.21)

[ItRo 78] —, Covering a graph by circuits, 5th ICALP (1978), LNCS **62** (1978), 289—299. (§ 9.2.2)

[ItRoTa 78] Itai, A., Rodeh, M., Tanimoto, S. L., Some matching problems for bipartite graphs, JACM **25** (1978), 517—525. (7.20)

[ItSh 79] Itai, A., Shiloach, Y., Maximum flow in planar networks, SIAM JC **8** (1979), 135—150. (§ 7.2.2.3)

[Jab 58] Jablonskii, S. V., On algorithmic difficulties of synthezising minimal contact networks (in Russian), Problemy Kibernetiki **2** (1958), 75—121. (§ 9.2.2)

[Jaj 80] Ja' Ja', J., Time-space tradeoffs for some algebraic problems, 12th STOC (1980), 339—350. See also: JACM **30** (1983), 657—667. (§ 25.3.2.2)

[Jan 79] Janiga, L., Real-time computations of two-way multihead finite automata, FCT 1979, 214—218. (13.4)

[JaSi 80] Ja' Ja', J., Simon, J., Parallel algorithms in graph theory: Planarity testing (preliminary version), MFCS 1980, LNCS **88** (1980), 305—319. See also: SIAM JC **11** (1982), 314—328. (§ 7.2.2.3)

[Jhn 83] John, T., A weak separation result between deterministic and nondeterministic log space, Conf. Comp. Complexity Theory, March 21—25, 1983, Santa Barbara, 128—139. (§ 23.3)

[Joh 74a] Johnson, D. S., Approximation algorithms for combinatorial problems, JCSS **9** (1974), 256—278. (9.15)

[Joh 74b] —, Fast algorithms for bin packing, JCSS **8** (1974), 272—314. (9.13)

[Joh 81-] —, The NP-completeness column: an ongoing guide, J. Algorithms, appears quarterly starting with **2** (1981), 393—405. (§ 8.2.3.3, 14.21, 24.2.1.1)

[JoLa 77] Jones, N. D., Laaser, W. T., Complete problems for deterministic polynomial time, 6th STOC (1974), 40—46. See also: TCS **3** (1977), 105—117. (8.52, 14.12, 14.19)

[JoLaLi 77] Jones, N. D., Landweber, L. H., Lien, Y. E., Complexity of some problems in Petri nets, TCS **4** (1977), 277—299. (§ 8.2.3.4)

[JoLiLa 76] Jones, N. D., Lien, Y. E., Laaser, W. T., New problems complete for nondeterministic log space, MST **10** (1976), 1—17. (14.13)

[JoMa 82] Jones, J. P., Matijasjevič, Ju., A new representation for the symmetric binomial coefficient and its applications, Ann. Sc. Math., Quebec, **6**, 1 (1982), 81—97. (§ 10.2.2)

[Jon 68] JONES, N. D., Classes of automata and transitive closure, I&C **13** (1968), 207–229. (11.23)

[Jon 75] —, Space-bounded reducibility among combinatorial problems, JCSS **11** (1975), 68–85. (8.43, 14.12, 14.13)

[JoSe 74] JONES, N. D., SELMAN, A. L., Turing machines and the spectra of first-order formulas, JSL **39** (1974), 139–150. (§ 24.2.2.2.2, Ch. IV)

[JoSk 77 a] JONES, N. D., SKYUM, S., Complexity of some problems concerning L-systems, 4th ICALP (1977), LNCS **52** (1977), 301–308. See also: MST **13** (1979/80), 29–43. (12.19)

[JoSk 77 b] —, Recognition of deterministic ETOL languages in logarithmic space, I&C **35** (1977), 177–181. (12.19, 14.3)

[JoVe 82] JOHNSON, D. B., VENKATESON, S. M., Parallel algorithms for minimum cut and maximum flows in planar networks, 23rd FOCS (1982), 244–254. (§ 7.2.2.3)

[JoYo 80] JOSEPH, D., YOUNG, P., Independence results in computer science?, 12th STOC (1980), 58–69. See also: JCSS **23** (1981), 205–222, and Corrigendum in JCSS **24** (1982), 378. (§ 24.5)

[JoYo 81] —, A survey of some recent results on computational complexity in weak theories of arithmetic, MFCS '81, LNCS **118** (1981), 46–60. (§ 24.5)

[Juk 82] JUKNA, S. P., Arithmetical representations of classes of machine complexity (in Russian), Matematičeskaja logika i jejo primenenija **2** (1982), 92–106. (24.40)

[Jun 81] JUNG, H., Relationships between probabilistic and deterministic tape complexity, MFCS '81, LNCS **118** (1981), 339–346. (20.44)

[Jun 84] —, On probabilistic tape complexity and fast circuits for matrix inversion problems, 11th ICALP, LNCS **172** (1984), 281–291. (23.1)

[KaAdIw 79] KASAI, T., ADACHI, A., IWATA, S., Classes of pebble games and complete problems, SIAM JC **8** (1979), 574–586. (14.14)

[Kal 43] KALMÁR, L., Egyszerü példa eldöntheretlen aritmetikai problémáre Matematikai és Fizikai Lapok 50 (1943), 1–23. (§ 2.4.1)

[KaLi 80] KARP, R. M., LIPTON, R. J., Some connections between nonuniform and uniform complexity classes, 12th STOC (1980), 302–309. (24.36)

[Kam 70] KAMEDA, T., Constant-tape-reversal bounded nondeterministic machine computations, Int. Comp. Symp., European Chapter ACM, 1970, 845–853. (12.2)

[Kam 72] —, Pushdown automata with counters, JCSS **6** (1972), 138–150. (13.21, § 20.2.1)

[Kan 81 a] KANNAN, R., Towards separating nondeterministic time from deterministic time, 22nd FOCS (1981), 235–243. (24.7)

[Kan 81 b] —, A circuit-size lower bound, 22nd FOCS (1981), 304–309. See also: I&C **55** (1982), 40–56. (24.36)

[Kan 83 a] —, Improved algorithm for integer programming and related lattice problems, 15th STOC (1983), 193–206. (7.22)

[Kan 83 b] —, Alternation and the power of nondeterminism, 15th STOC (1983), 344–346. (23.9)

[Kan 83 c] —, A method for proving the power of nondeterminism, Conf. Comp. Complexity Theory, March 21–25, 1983, Santa Barbara, 69–74. (24.6)

[Kano 84] KANOVIČ, M. J., A uniform independence of invariant sentences, MFCS '84, LNCS **176** (1984), 348–354. (§ 24.5)

[KaPa 80] KARP, R. M., PAPADIMITRIOU, C. H., On linear characterizations of combinatorial optimization problems, 21st FOCS (1980), 1–9. (§ 7.2.2.4)

[Kar 72] KARP, R. M., Reducibility among combinatorial problems, IBM Symp. 1972: Complexity of Computer Computations, Plenum Press, New York 1972, 85–103. (§ 6.1.2, 14.21, § 24.2.1.2.3)

[Kar 75] —, The fast approximate solution of hard combinatorial problems, 6th Southeastern Conf. on Combinatorics, Graph Theory and Computing (1975), 15–31. (9.16)

[Kar 76] —, The probabilistic analysis of some combinatorial search algorithms, in:

Algorithms and Complexity, New Directions and Recent Results, Academic Press, New York 1976, p. 1–19. (9.16, 9.17, 9.18)

[Karm 84] KARMARKAR, N., A new polynomial-time algorithm for linear programming, 16th STOC (1984), 302–310. (§ 7.2.2.4)

[Karz 74] KARZANOV, A. V., Determining the maximal flow in a network by the method of preflows, DAN **215** (1974), 434–437. (7.21)

[Kas 72] KASAMI, T., A note on computing time for recognition of languages generated by linear grammars, I&C **10** (1972), 209–214. (12.3)

[KaTa 80] KARP, R. M., TARJAN, R. E., Linear expected-time algorithms for connectivity problems, 12th STOC (1980), 368–377. See also: J. Algorithms **1** (1980), 374–393. (§ 9.2.1)

[KaVe 84] KARPINSKI, M., VERBEEK, R., On the Monte-Carlo space-constructible functions and separation results for probabilistic complexity classes, Inst. f. Informatik, Universität Bonn, Bericht I/3, August 1984. (§ 17.2.2)

[KaVo 70] KAMEDA, T., VOLLMAR, R., Note on tape reversal complexity of languages, I&C **17** (1970), 203–215. (13.18, 13.22, 21.5, 21.6)

[KeHo 80] KENT, C., HODGSON, B., An arithmetical characterization of NP, Lakehead Univ. Tech. Rep. 6–80 (1980) (Dept. of Mathematics, Thunder Bay, Ontario, Canada). See also: TCS **21** (1982), 255–267. (24.40)

[Kha 79] KHAČIAN, L. G., A polynomial time algorithm for linear programming (in Russian), DAN SSSR **244** (1979), 1093–1097. (7.22)

[KiFi 77] KINTALA, C. M. R., FISCHER, P. C., Computations with a restricted number of nondeterministic steps, 9th STOC (1977), 178–185. See also: MST **12** (1979), 219–231. (24.37, 24.38, 28.5)

[Kin 81a] KING, K. N., Alternating multihead finite automata, 8th ICALP, LNCS **115** (1981). (13.3, 13.4, 13.5, 13.25, 13.26)

[Kin 81b] —, Measures of parallelism in alternating computation trees, 13th STOC (1981), 189–201. (20.36)

[Kinb 75] KINBER, E. B., On frequency realtime computations, Uč. zap. Lat. gos. univ. im. P. Stucki **233** (1975), 174–182. (9.25)

[Kinb 76] —, Frequency computations on finite automata (in Russian), Kibernetika (Kiev) 1976, Nr. 2, 7–15. (§ 9.2.3)

[KiWr 78] KING, K. N., WRATHALL, C., Stack languages and log n space, JCSS **17** (1978), 281–299. (11.4, 11.13, 13.33)

[Kla 83] KLAWE, M. M., A tight bound for black and white pebbles on the pyramid, 24th FOCS (1983), 410–419. (§ 25.3.2.1)

[Kle 82] KLEINE BÜNING, H., Note on the E_1^*-E_2^* problem, ZML **28** (1982), 277–284. (§ 2.4.1)

[KlMi 72] KLEE, V., MINTY, G. J., How good is the simplex algorithm?, In: O. SHISHA (ed.), Inequalities III, Academic Press, New York 1972, 159–175. (§ 7.2.2.4)

[KnMoPr 77] KNUTH, D. E., MORRIS, J. H., PRATT, V. R., Fast pattern matching in strings, SIAM JC **6** (1977), 323–350. (§ 7.2.1)

[Knö 81] KNÖDEL, W., A bin packing algorithm with complexity $O(n \cdot \log n)$ and performance 1 in the stochastic limit, MFCS 1981, LNCS **118**, 369–378. (§ 9.2.1)

[Knu 69]* KNUTH, D. E., The Art of Programming, Vol. 2, Addison Wesley, Reading (Mass.) 1969. (§ 7.2.2, § 7.2.2.2)

[Knu 73]* —, The Art of Programming, Vol. 3, Addison Wesley, Reading (Mass.) 1973. (§ 7.2.2, § 8)

[Ko 82] KO, KER I, Some observations on the probabilistic algorithms and NP-hard problems, IPL **13** (1982), 39–43. (§ 9.2.2)

[Kob 83] KOBAYASHI, K., On the structure of one-tape nondeterministic Turing machine time hierarchy, TR C-56, 1983, Tokyo Institute of Technology. (17.43)

[KoKu 84] KORTAS, K., KUBIAK, W., Manuscript Gdansk University 1984. (14.20)

[Kol 65] KOLMOGOROV, A. N., Three approaches to the notion of "amount of information" (in Russian), Problemy peredači informacii **1** (1965), 3–11. (Intr.)

[KoMa 77] Kou, L. T., Markowsky, G., Multidimensional bin packing algorithms, IBM J. Res. Develop. **21** (1977), 443–448. (§ 9.1.3.2)

[KoMč 69] Kozmidiadi, V. A., Marčenkov, S. S., On multihead automata, Problemy Kibernetika **21** (1969), 127–158. (13.4)

[KoMt 80] Kozen, D., Machtey, M., On relative diagonals, Preprint 1980, IBM T. J. Watson Res. Center, Yorktown Heights. (§ 28)

[Kor 79] Kortas, K., Subgraph isomorphism is NQL-complete, FCT 1979, 559–563. (14.21)

[KoŘí 81] Koubek, V., Říha, A., The maximum k-flow in a network, MFCS 1981, LNCS **118** (1981), 389–397. (§ 7.2.2.3)

[Kos 75] Kosaraju, S. R., Speed of recognition of context-free languages by array automata, SIAM JC **4** (1975), 331–340. (12.13)

[KoSc 84] Ko, Ker I, Schöning, U., On circuit size complexity and the low hierarchy in NP, SIAM JC **13** (1984), to appear. (24.17)

[KoUs 58] Kolmogorov, A. N., Uspenskij, V. A., On the definition of an algorithm (in Russian), Usp. mat. nauk **13**: 4 (1958), 3–28. (19.29)

[Kow 84] Kowalczyk, W., Some connections between presentability of complexity classes and the power of formal systems of reasoning, MFCS '84, LNCS **176** (1984), 364–369. (§ 14.21, § 24.5)

[Koz 76] Kozen, D., On parallelism in Turing machines, 17th FOCS (1976), 85–97. (§ 20.4.1)

[Koz 77 a] —, Lower bounds for natural proof systems, 18th FOCS (1977), 254–266. (§ 8.2.1, § 8.2.3.4, 14.22, § 26.3)

[Koz 77 b] —, Complexity of finitely presented algebras, 9th STOC (1977), 164–177. (§ 8.2.3.4, § 24.2.1.2.3)

[Koz 78] —, A clique problem equivalent to graph isomorphism, SIGACT News **10** (1978), 50–52. (§ 24.2.1.2.3)

[Koz 79] —, Automata and planar graphs, FCT 1979, 243–254. (§ 23.3)

[Koz 80 a] —, Complexity of Boolean algebras, TCS **10** (1980), 221–247. (§ 8.2.3.2, 14.15)

[Koz 80 b] —, Indexing of subrecursive classes, TCS **11** (1980), 277–301. (§ 28)

[Kro 79]* Kronsjö, L. J., Algorithms: Their Complexity and Efficiency, Wiley Series-Computing, John Wiley & Sons, New York–London–Sidney–Toronto 1979. (Intr., § 7.2.2)

[Kuč 81] Kučera, L., Maximum flow in planar networks, MFCS 1981, LNCS **118** (1981), 418–422. (§ 7.2.2.3)

[Kuč 83]* —, Combinatorial Algorithms (in Czech.), SNTL Praha 1983. (§ 7.2.2)

[Kuč 84] —, Finding a maximum flow in (S,T)-planar network in linear expected time, MFCS '84, LNCS **176** (1984), 370–377. (7.21)

[Kui 71] Kuich, W., The complexity of skewlinear tuple languages and o-regular languages, I&C **19** (1971), 353–367. (12.2)

[KuMa 83] Kurtz, S., Maass, W., Some time, space and reversal tradeoffs, Conf. Comp. Complexity Theory, March 21–25, 1983, Santa Barbara, 30–40. (25.4)

[Kur 64] Kuroda, S. Y., Classes of languages and linear-bounded automata, I&C **7** (1964), 207–223. (12.15)

[Kut 82 a] Kurtz, S., An oracle relative to which there are no sparse sets in NP ∖ P. 1982 Institute on Recursive Function Theory held at Cornell University. Special issue of Recursive Function Theory: Newsletter, 79–82. (28.8)

[Kut 82 b] —, On the random oracle hypothesis, 14th STOC (1982), 224–230. (28.12)

[Kut 83] —, The fine structure of NP: Relativizations, Conf. Comp. Complexity Theory, March 21–25, 1983, Santa Barbara, 42–50. (28.8)

[KuTr 81] Kučera, L., Trnkova, V., Isomorphism completeness for some algebraic structures, FCT '81, LNCS **117** (1981), 218–225. (§ 24.2.1.2.3)

[Lad 75] Ladner, R. E., On the structure of polynomial time reducibility, JACM **22** (1975), 155–171. (6.13, 24.23, 24.27, 24.30, 28.4)

[Lad 77] —, The computational complexity of provability in systems of modal propositional logic, SIAM JC 6 (1977), 467–480. (8.49, 14.14)

[Lag 84] LANGE, K.-J., Nondeterministic logspace reductions, MFCS '84, LNCS 176 (1984), 378–388. (6.6, 6.9, 14.3)

[LaLiRo 78] LANDWEBER, L. H., LIPTON, R. J., ROBERTSON, E. L., On the structure of sets in NP and other complexity classes, Univ. of Wisconsin, Madison, TR 342 (1978). See also: TCS 15 (1981), 181–200. (24.20, § 24.2.1.2.3, 24.32)

[LaLiSt 78] LADNER, R. E., LIPTON, R. J., STOCKMEYER, L. J., Alternating pushdown automata, 19th FOCS (1978), 92–106. See also: SIAM JC 13 (1984), 135–155. (13.23, 13.30, 13.36, 20.39, 20.40)

[LaLy 76] LADNER, R. E., LYNCH, N. A., Relativization of questions about log space computability, MST 10 (1976), 19–32. (§ 6.1.1, 6.9, 6.11, 28.2, 28.11)

[LaLySe 75] LADNER, R. E., LYNCH, N. A., SELMAN, A. L., A comparison of polynomial time reducibilities, TCS 1 (1975), 103–123. (6.2, 6.5, 6.7, 6.9)

[Lan 63] LANDWEBER, L. M., Three theorems on phrase structure grammars of type 1, I&C 6 (1963), 131–137. (12.15)

[LaNo 83] LADNER, R. E., NORMAN, J. K., Solitaire automata, Conf. Comp. Complexity Theory, March 21–25, 1983, Santa Barbara, 21–26. (§ 20.4.1)

[LaRo 72] LANDWEBER, L. H., ROBERTSON, E. L., Recursive properties of abstract complexity classes, JACM 19 (1972), 296–308. (§ 15.1.1, 15.3, 15.5, 15.6, 15.7, 15.9, § 15.2, 15.19, 16.18)

[LaRo 78] —, Properties of conflict-free and persistent Petri nets, JACM 25 (1978), 352–364. (§ 8.2.3.4)

[Law 76] LAWLER, E. L., A note on the complexity of the chromatic number problem, IPL 5 (1976), 66–67. (§ 7.3.1)

[Law 77] —, Fast approximation algorithms for knapsack problems, 18th FOCS (1977), 206–213. (9.5)

[LaWi 83] LAKSHMIPATHY, N., WINKLMANN, K., Information flow properties of graph coloring problems, Conf. Comp. Complexity Theory, March 21–25, 1983, Santa Barbara, 140–144. (14.21)

[Lee 75a] VAN LEEUWEN, J., The membership question for ETOL-languages is polynomially complete, IPL 3 (1975), 138–143. (12.19, 14.3)

[Lee 75b] —, The tape complexity of context-independent developmental languages. JCSS 15 (1975), 203–211. (12.19)

[Lee 76] —, A study of complexity in hyper-algebraic families, in: Automata, Languages and Development, Ed. A. LINDENMAYER and G. ROZENBERG. North-Holland, Amsterdam 1976, 323–334. (12.19)

[Len 79] LENSTRA, H. W. Jr., Miller's primality test, IPL 8 (1979), 86–88. (7.28)

[Len 81] —, Integer programming with a fixed number of variables, Report 81-03, Math. Inst. Univ. of Amsterdam. (§ 7.2.2.4)

[LePa 80] LEWIS, H. R., PAPADIMITRIOU, CH., Symmetric space-bounded computation, 7th ICALP (1980), LNCS 85 (1980), 374–384. See also: TCS 19 (1982), 161–187. (8.43, 14.12)

[LeSe 77] LEONG, B., SEIFERAS, J., New real-time simulations of multihead tape units, 9th STOC (1977), 239–248. See also: JACM 28 (1981), 166–180. (19.11, 19.12)

[LeStHa 65] LEWIS II, P. M., STEARNS, R. E., HARTMANIS, J., Memory bounds for recognition of context-free and context-sensitive languages. IEEE Conf. Switch. Circuit Theory and Logic Design, 1965, 191–202. (12.14)

[LeTa 79] LENGAUER, T., TARJAN, R. E., Upper and lower bounds on time-space tradeoffs, 11th STOC (1979), 262–277. See also: JACM 29 (1982), 1087–1130. (25.11, 25.13)

[LeTa 80] —, The space complexity of pebble games on trees, IPL 10 (1980), 184–188. (§ 25.3.2.1.1)

[Lev 74] LEVIN, L. A., Step counting functions of computable functions (in Russian), in: Složn. vyčisl. i algor. Moscow 1974, 174–185. (16.1)

[Lewf 71] Lewis, F. O., The enumerability and invariance of complexity classes, JCSS **5** (1971), 286–303. (§ 15.1.1, 15.10, § 15.2, 16.6)

[Lewh 78a] Lewis, H. R., Complexity of solvable cases of the decision problem for the predicate calcalus, 19th FOCS (1978), 35–47. (7.32, 8.48, 8.49)

[Lewh 78b] —, Satisfiability problems for propositional caculi, 16th Ann. Allerton Conf. on Comm., Control and Computing, University of Illinois, Urbana-Champaign 1978, 513–520. (8.49)

[Lia 80] Liang, F. M., A lower bound for on-line bin packing, IPL **10** (1980), 76–79. (§ 9.1.3.2)

[Lic 82] Lichtenstein, D., Planar formulae and their uses, SIAM JC **11** (1982), 329–343. (14.21)

[Lin 78] Lingas, A., A PSPACE-complete problem related to a pebble game, 5th ICALP (1978), LNCS **62** (1978), 300–321. (14.14, 25.15)

[Lip 78] Lipton, R. J., Model theoretic aspects of computational complexity, 19th FOCS (1978), 193–200. (11.16, § 24.5)

[Lip 79] —, On the consistency of P = NP and fragments of arithmetic, FCT 1979, 269–278. (§ 24.5)

[Lis 75a] Lischke, G., Flußbildmaße — Ein Versuch zur Definition natürlicher Kompliziertheitsmaße, EIK **11** (1975), 423–436. (§ 16.2, 16.15)

[Lis 75b] —, Über die Erfüllung gewisser Erhaltungssätze durch Kompliziertheitsmaße, ZML **21** (1975), 159–166. (16.10)

[Lis 75c] —, Erhaltungssätze in der Theorie der Blumschen Kompliziertheitsmaße, Dissertation, Friedrich-Schiller-Univ. Jena 1975. (16.20)

[Lis 76] —, Natürliche Kompliziertheitsmaße und Erhaltungssätze I, ZML **22** (1976). 413–418. (16.11)

[Lis 77] —, Natürliche Kompliziertheitsmaße und Erhaltungssätze II, ZML **23** (1977), 193–200. (16.12, 16.13, 16.18)

[Lis 84] —, Oracle constructions to prove all possible relationships between relativizations of P, NP, EL, NEL, EP and NEP, to appear in ZML. (28.7)

[LiSi 78] Lichtenstein, D., Sipser, M., GO is PSPACE-hard, 19th FOCS (1978), 48–54, See also: JACM **27** (1980), 393–401. (14.14)

[Lon 79] Long, T. J., On γ-reducibility versus polynomially time many-one reducibility, 11th STOC (1979), 278–287. See also: TCS **14** (1981), 91–101. (§ 6.1.2)

[Lon 82a] —, Strong nondeterministic polynomial time reducibilities, TCS **21** (1982), 1–25. (§ 6.1.2, 6.8, 6.9, 6.10)

[Lon 82b] —, A note on sparse oracles for NP, JCSS **24** (1982), 224–232. (24.17)

[Lon 83] —, On relativizing complexity classes, Conf. Comp. Complexity Theory, March 21–25, 1983, Santa Barbara, 104–108. (28.13)

[Lou 79] Loui, M. C., The space complexity of two pebble games on trees, Laboratory for Computer Science, MIT, TM-133 (May 1979). (§ 25.3.2.1.1)

[Lou 80] —, A note on the pebble game, IPL **11** (1980), 24–26. (25.7, 25.11)

[Lou 81] —, Simulations among multidimensional Turing machines, 22nd FOCS (1981), 58–67. See also: TCS **21** (1982), 145–161. (19.18, 19.27)

[Lub 81] Lubiw, A., Some NP-complete problems similar to graph isomorphism, SIAM JC **10** (1981), 11–21. (§ 7.3.4)

[LuBo 79] Lueker, G. S., Booth, K. S., A linear time algorithm for deciding interval graph isomorphism, JACM **26** (1979), 183–195. (§ 7.3.4)

[Luk 80] Luks, E. M., Isomorphism of graphs with bounded valence can be tested in polynomial time, 21st FOCS (1980), 42–49. See also: JCSS **25** (1982), 42–65. (7.30)

[Lup 63] Lupanov, O. B., On the synthesis of some classes of control systems (in Russian), Problemi kibernetiki **10** (1963), 88–96. (§ 21.2)

[Lup 74] —, On methods for proving complexity estimations and computations of special functions (in Russian), Diskr. Analiz **25** (1974), 3–18. (Intr., § 8)

[LyMeFi 76] LYNCH, N. A., MEYER, A. R., FISCHER, M. J., Relativization of the theory of computational complexity, TAMS 220 (1976), 243–287. See also [Lyn 72]. (§ 5.3, 5.30, 5.31, § 27, 27.10–27.13, 27.17, 27.19–27.23)

[Lyn 72] LYNCH, N. A., Relativization of the theory of computational complexity, MIT Proj. MAC TR-99 (1972). See also [LyMeFi 76]. (6.12, 6.13)

[Lyn 75] —, On reducibility to complex or sparse sets, JACM 22 (1975), 341–345. (24.18)

[Lyn 76] —, Complexity-class-encoding sets, JCSS 13 (1976), 100–118. (8.41)

[Lyn 77] —, Log space recognition and translation of parenthesis languages, JACM 24 (1977), 583–590. (12.14)

[Lyn 78] —, Log space machines with multiple oracle tapes, TCS 6 (1978), 25–39. (§ 5.3, § 6.2)

[Maa 84] MAASS, W., Quadratic lower bounds for deterministic and nondeterministic one-tape Turing machines, 16th STOC (1984), 401–408. (§ 22.1)

[Mac 72] MACHTEY, M., Augmented loop languages and classes of computable functions, JCSS 6 (1972), 603–624. (§ 3.3, 27.13, 27.16)

[Mac 76] —, Minimal pairs of polynomial degrees with subexponential complexity, TCS 2 (1976), 73–76. (§ 24.2.1.2.3)

[Mah 80] MAHANEY, S. R., Sparse complete sets for NP: Solution of a conjecture of Berman and Hartmanis, 21st FOCS (1980), 54–60. See also: JCSS 25 (1982), 130–143. (24.16, 24.17)

[Mah 81] —, On the number of P-isomorphism classes of NP-complete sets, 22nd FOCS (1981), 271–278. (24.9)

[MaMe 81] MAYR, E. W., MEYER, A. R., The complexity of the finite containment problem for Petri nets, JACM 28 (1981), 561–576. (§ 8.2.3.4)

[MaMe 82] —, The complexity of the word problems for commutative semigroups and polynomial ideals, Adv. in Math. 46 (1982), 305–329. (§ 8.2.3.4)

[Mar 80] MARČENKOV, S. S., On a basis with respect to superposition in the class of functions which are elementary in the sense of Kalmár (in Russian), Mat. zametki 27, 3 (1980), 321–332. (§ 10.2.1)

[Mar 84] —, Oral communication. (§ 10.2.1)

[Marq 75] MARQUES, I., On speedability of recursively enumerable sets, ZML 21 (1975), 199–214. (18.19)

[Math 79] MATHON, R., A note on the graph isomorphism counting problem, IPL 8 (1979), 131–132. (§ 24.4)

[Mati 71] MATIJASJEVIČ, JU. V., The recognition in realtime of the occurence relation (in Russian), Zap. naučn. sem. LOMI 20 (1971), 104–114. (§ 7.2.1)

[Matr 76] MATROSOV, V. L., Complexity classes and classes of step counting functions of computable functions (in Russian), DAN SSSR 226 (1976), 513–515. (§ 10.2.1)

[MaWiYo 78] MACHTEY, M., WINKLMANN, K., YOUNG, P., Simple Gödel numberings, isomorphisms and programming properties, SIAM JC 7 (1978), 39–60. (§ 2.2)

[MaYo 78]* MACHTEY, M., YOUNG, P., An Introduction to General Theory of Algorithms, North-Holland, New York 1978. (Intr.)

[MaYo 81] —, Remarks on recursion versus diagonalization and exponentially difficult problems, JCSS 22 (1981), 442–453. (§ 8.2.1)

[McCo 83] MCKENZIE, P., COOK, S. A., The parallel complexity of the abelian permutation group membership problem, 24th FOCS (1983), 154–161. (§ 21.2)

[McCr 69] MCCREIGHT, E. M., Classes of computable functions defined by bounds on computation, Diss., Carnegie-Mellon Univ. Pittsburgh, 1969. See also [McMe 69]. (15.4, 17.20)

[McMe 69] MCCREIGHT, E. M., MEYER, A. R., Classes of computable functions defined by bounds on computation, 1st STOC (1969), 79–88. See also [Mccr 69]. (5.26, 15.15, 15.16, 15.21, 15.23, 16.19, 17.13, 17.16)

[McN 61] MCNAUGHTON, R., The theory of automata, A survey, in: Advances in Comp. (1961), vol. 2, Academic Press, New York–London, 379–241. (§ 9.2.3)

[MeDoBo 81] Mei-rui, Xu, Doner, J. E., Book, R. V., Refining nondeterminism in controlled relativizations of complexity classes, Res. Rep. Santa Barbara (1981). See also: JACM **30** (1983), 677–685. (28.5)

[MeFim 72] Meyer, A. R., Fischer, M. J., Relatively complex recursive sets, JSL **37** (1972), 55–68. (27.17, 27.23)

[MeFip 72] Meyer, A. R., Fischer, P. C., Computational speed-up by effective operators, JSL **37** (1972), 55–67. (18.1, 18.2, 18.5)

[Meh 73] Mehlhorn, K., On the size of sets of computable functions, 14th SWAT (1973), 190–196. (28.4)

[Meh 75] —, Bracketed languages are recognizable in logarithm space, Univ. d. Saarlandes, Saarbrücken 1975. See also: IPL **5** (1976), 168–170. (12.14)

[Meh 76] —, Polynomial and abstract subrecursive classes, JCSS **12** (1976), 147–178. (§ 6.2, § 6.3)

[Meh 77]* —, Effiziente Algorithmen, Teubner Studienbücher, Informatik, Stuttgart 1977. (Intr., § 3.3, § 7, § 7.2.2)

[Meh 80] —, Pebbling mountain ranges and its application to DCFL-recognition, 7th ICALP (1980), LNCS **85** (1980), 422–435. (§ 25.3.1)

[Mei 84 a] Meinhardt, D., Tape reversal bounded Turing machines with an auxiliary pushdown or an auxiliary counter, Proc. 2nd Frege Conf. Schwerin 1984, 338–344. (21.7)

[Mei 84 b] —, Umkehrbeschränkte und überquerungsbeschränkte Berechnungen auf zwei Turingbändern oder einem Turingband und subrekursivem Speicher, Dissertation, Friedrich-Schiller-Universität, Jena 1984. (21.3, 21.6, 21.7)

[MePa 79] Meyer, A. R., Paterson, M. S., With what frequency are apparently intractable problems difficult?, MIT, TR Feb. 1979. (24.14, 24.18)

[MeRi 67] Meyer, A. R., Ritchie, D. M., The complexity of Loop programs, Proc. 22nd National ACM Conference (1967), 465–470. (§ 2.4.1, 2.22, 18.2)

[MeSt 72] Meyer, A. R., Stockmeyer, L. J., The equivalence problem for regular expressions with squaring requires exponential space, 13th SWAT (1972), 125–129. (§ 6.1.2, 8.44, 14.14, 14.16, 14.17, § 24.3)

[MeWa 82] Meinhardt, D., Wagner, K., Eine Bemerkung zu einer Arbeit von Monien über Kopfzahl-Hierarchien für Zweiweg-Automaten, EIK **18** (1982), 69–74. (13.9, 13.24, 13.37, 13.47)

[MeWi 78] Meyer, A. R., Winklmann, K., The fundamental theorem of complexity theory, Found. of Comp. Sc., III (Third Advanced Course, Amsterdam 1978), Part I, 97–112. (16.3)

[Mey 73] Meyer, A. R., Weak monadic second order theory of successor is not elementary-recursive, Proj. MAC TM 38 (1973), MIT, Cambridge (Mass.). (8.46, 14.18)

[Mey 74] —, The inherent computational complexity of theories of ordered sets, Proc. Int. Congr. Math., Vancouver 1974, 477–482. (8.45, 8.48, 8.50, 14.17, 14.18, 14.25)

[Meyh 79] Meyer auf der Heide, F., A comparison between two variations of a pebble game on graphs, 6th ICALP (1979), LNCS **71** (1979), 411–421. See also: TCS **13** (1981), 315–322. (14.14, § 25.3.2.1.1)

[Mil 75] Miller, G. L., Riemann's hypothesis and tests for primality, 7th STOC (1975), 234–239. See also: JCSS **13** (1976), 300–317. (7.29)

[Mil 78] —, On the $n^{\log n}$ isomorphism technique, 10th STOC (1978), 51–58. (§ 7.3.4)

[Mil 79] —, Graph isomorphism, General remarks, JCSS **18** (1979), 128–142. (§ 24.2.1.2.3)

[Mil 80] —, Isomorphism testing for graphs of bounded genus, 12th STOC (1980), 225 to 235. (7.30)

[MiLi 79] De Millo, R. A., Lipton, R. J., Some connections between mathematical logic and complexity, 11th STOC (1979), 153–159. (§ 24.5)

[MiLi 80] —, The consistency of "P = NP" and related problems with fragments of number theory, 12th STOC (1980), 45–57. (§ 24.5)

[Mis 79] Misra, J., Space-time tradeoff in implementing certain set operations, IPL **8** (1979), 81–85. (§ 25.3.2.2)

[MiVa 80] Micali, S., Vazirani, V. V., An $O(\sqrt{|V|} \cdot |E|)$ algorithm for finding maximum matching in general graphs, 21st FOCS (1980), 17–27. (7.20)

[Miy 80] Miyano, S., $k+1$ heads are better than k for one-way multihead stack-counter automata, Res. Inst. Fund. Inform. Sci. Res. Rep. Nr. 98 (1980). See also: AI **17** (1982), 63–67. (13.37)

[Mol 76] Moll, R., An operator embedding theorem for complexity classes of recursive functions, TCS **1** (1976), 193–198. (17.21)

[MoMe 74] Moll, R., Meyer, A. R., Honest bounds for complexity classes of recursive functions, JSL **39** (1974), 127–138. (5.26, § 15.2, 15.23, 15.28–15.30)

[Mon 72] Monien, B., Relationships between pushdown automata and tape bounded Turing machines, ICALP (1972), Proc. Symp. Rocquencourt, North-Holland, Amsterdam 1973, 575–583. (23.2)

[Mon 74a] —, Komplexitätsklassen von Automatenmodellen und beschränkte Rekursion, Bericht Nr. 8, Institut für Informatik, Univ. Hamburg, 1974. (§ 10.2.4)

[Mon 74b] —, Characterization of time bounded computations by limited primitive recursion, 2nd ICALP (1974), LNCS **14** (1974), 280–293. (§ 20.2.1)

[Mon 74c] —, Beschreibung von Zeitkomplexitätsklassen bei Turingmaschinen durch andere Automatenmodelle, EIK **10** (1974), 37–51. (20.16)

[Mon 75a] —, About the simulation of nondeterministic log n-tape bounded Turing machines, LNCS **33** (1975), 118–126. (11.3)

[Mon 75b] —, Relationships between pushdown automata with counters and complexity classes, MST **9** (1975), 248–264. (§ 20.2.1)

[Mon 76] —, Transformational methods and their application to complexity problems, AI **6** (1976), 95–108. See also: Corrigenda, AI **8** (1977), 383–384.

[Mon 77a] —, A recursive and grammatical characterization of exponential time languages, TCS **3** (1977), 61–74. (10.22, 12.17)

[Mon 77b] —, The LBA-problem and the deterministic tape complexity of two-way one-counter languages over a one-letter alphabet, AI **8** (1977), 371–382. (23.24)

[Mon 77c] —, About the derivation languages of grammars and machines, 4th ICALP (1977), LNCS **52** (1977), 337–351. (19.26, 19.29)

[Mon 78a] —, Beziehungen zwischen dem LBA-Problem und dem Knapsackproblem, Preprint, Paderborn 1978. See also: Proc. Frege Conference 100 Jahre Begriffsschrift, Mai 1979, Jena. (23.18)

[Mon 78b] —, Two-way multihead automata over a one-letter alphabet, preprint Paderborn, 1978. See also: RAIRO, Informat. théor. **14** (1980), 67–82. (13.4, § 20.1.1, 22.15)

[Mon 80] —, On a subclass of pseudopolynomial problems, MFCS 1980, LNCS **88** (1980), 414–425. (§ 7.3.2)

[Mon 81] —, Personal communication 1981. (13.25)

[Mon 83] —, Deterministic two-way one-head pushdown automata are very powerful, preprint University of Paderborn 1983. (13.20)

[Monr 80] Monier, L., Evaluation and comparison of two efficient probabilistic primality testing algorithms, TCS **12** (1980), 97–108. (9.24)

[Mor 81] Moran, S., Some results on relativized deterministic and nondeterministic time hierarchies, JCSS **22** (1981), 1–8. (§ 17.2.4)

[Mor 82] —, On the accepting density hierarchy in NP, SIAM JC **11** (1982), 344–349. (28.4)

[MoSp 79] Monien, B., Speckenmeyer, E., 3-Satisfiability is testable in $O(1.62^r)$ steps, Bericht 3/1979, Universität Paderborn. See also: Discr. Appl. Math. **10** (1985), 287–295. (§ 7.3.1)

[MoSp 82] —, Some further approximation algorithms for the vertex cover problem, preprint, Paderborn 1983. See also: CAAP '83, LNCS **159** (1983), 341–349. (9.15)

[MoSp 84] —, Ramsey numbers and an approximation algorithm for the vertex cover problem, Proc. 2nd Frege-Conf. (1984), 345–353. (9.15)

[MoSpVo 80] Monien, B., Speckenmeyer, E., Vornberger, O., Upper bounds for covering problems, preprint University of Paderborn 1980. (§ 7.3.1)

[MoSu 79] Monien, B., Sudborough, I. H., On eliminating nondeterminism from Turing machines which use less than logarithmic worktape space, 6th ICALP (1979), LNCS **71** (1979), 431–445. See also: TCS **21** (1982), 237–253. (8.43, 23.1)

[MoSu 80] —, The interface between language theory and complexity theory, in: Formal Languages Theory, Perspectives and Open Problems, ed. R. V. Book, Academic Press, New York 1981, 287–323. (14.21)

[MoSu 81] —, Time and space bounded complexity classes and bandwidth constrained problems, MFCS '81, LNCS **118** (1981), 78–93. (14.21)

[Mül 73] Müller, H., Characterization of the elementary functions in terms of depth of nesting of primitive recursions, Rec. Fctn. Theory Newsletter **5** (1973), 14–15. (2.18)

[Muc 76] Muchnik, S. S., The vectorized Grzegorczyk hierarchy, ZML **22** (1976), 441–480. (§ 2.4.1)

[Myh 60] Myhill, J., Linear bounded automata, WADO Tech. Note 60–165, Rep. No. 60–22, Univ. of Pennsylvania, 1960. (11.16)

[Nek 72] Nekvinda, M., On a certain event recognizable in real time, Kybernetika (Prague) **8** (1972), 149–153. (7.2)

[Nek 73] —, On the complexity of events recognizable in real time, Kybernetika (Prague) **9** (1973), 1–10. (§ 17.2.6)

[Nep 66] Nepomnjaščij, V. A., Rudimentary predicates and Turing computations (in Russian), DAN SSSR **170** (1966), 1262–1264. See also: A redumentary interpretation of two-tape Turing computations (in Russian), Kibernetika (Kiev) 1970, Nr. 2, 29–35. (11.16)

[Nep 75] —, On the space complexity of the recognition of context-free languages (in Russian), Kibernetika (Kiev) 1975, Nr. 5, 64–68. (23.19)

[Nig 75] Nigmatullin, R. G., Complexity of the approximate solution of combinatorial problems (in Russian), DAN SSSR **224** (1975), 289–292. (9.3, 9.4)

[Nig 78] —, On approximate algorithms with restricted absolute error for discrete extremal problems (in Russian), Kibernetika (Kiev) 1978, Nr. 1, 95–101. (9.4)

[Odo 79] O'Donnell, M., A programming language theorem which is independent of Peano arithmetic, 11th STOC (1979), 176–188. (§ 24.5)

[Orp 83] Orponen, P., Complexity classes of alternating machines with oracles, ICALP '83, LNCS **154** (1983). (28.12)

[PaFiMe 74] Paterson, M. S., Fischer, M. J., Meyer, A. R., An improved overlap argument for on-line multiplication, SIAM-AMS **7** (1974), 97–111. (§ 8.1.4, 8.35, 8.36)

[PaHe 70] Paterson, M. S., Hewitt, C. M., Comperative schematology, Project MAC Conference on Concurrent Systems and Parallel Computations (1970), 119–128. (§ 25.3.2.1.1, 25.9)

[Paj 80] Pajunen, S., On two theorems of Lenstra, IPL **11** (1980), 224–228. (§ 7.3.3)

[Pak 79] Pakhomov, S. V., Maschine-independent description of some machine complexity classes (in Russian), Issledovanija po konstrukt. matemat. i mat. logike VIII, LOMI 1979, 176–185. (§ 10.2.1)

[PaMo 77] Paz, A., Moran, S., Nondeterministic polynomial optimization problems and their approximation, 4th ICALP (1977), LNCS **52** (1977), 370–379. See also: TCS **15** (1981), 251–277. (9.10)

[Pap 75] Papadimitriou, Ch. H., The euclidean travelling salesman problem is NP-complete, Princeton Univ. CSTR 191 (1975). See also: TCS **4** (1977), 237–244. (14.21)

[Pap 82] —, On the complexity of unique solutions, 23rd FOCS (1982), 14–20. (14.22)

[PaPiSzTr 83] Paul, W., Pippenger, N., Szemerédi, E., Trotter, W. T., On determinism versus non-determinism and related problems, 24th FOCS (1983), 429–438. (22.5, 24.6)

[PaPrRe 79] Paul, W. J., Prauss, E. J., Reischuk, R., On alternation I, preprint Bielefeld 1979. See also: AI **14** (1980), 243–255. (20.41, 20.42)

[PaRe 79] PAUL, W. J., REISCHUK, R., On time versus space II, 20th FOCS (1979), 298 to 306. See also: JCSS **22** (1981), 312–327. (25.2)

[PaSeSi 80] PAUL, W. J., SEIFERAS, J. I., SIMON, J., An information-theoretic approach to time bounds for on-line computation, 12th STOC (1980), 357–367. See also: JCSS **23** (1981), 108–126. (8.37, 19.18)

[PaSi 83] PATURI, R., SIMON, J., Lower bounds on the time of probabilistic on-line simulations, 24th FOCS (1983), 343–350. (8.37, § 19.2.2.2)

[Pat 72] PATERSON, M. S., Tape bounds for time bounded Turing machines, JCSS **6** (1972), 116–124. (25.3)

[PaTa 77] PAUL, W. J., TARJAN, R. E., Time-space trade-offs in a pebble game, 4th ICALP (1977), LNCS **52** (1977), 365–369. (25.14)

[PaTaCe 76] PAUL, W. J., TARJAN, R. E., CELONI, J. R., Space bounds for a pebble game on graphs, 8th STOC (1976), 149–160. See also: MST **10** (1977), 239–251. (25.7)

[Pau 77] PAUL, W. J., On time hierarchies, 9th STOC (1977), 218–222. See also: JCSS **19** (1979), 197–202. (17.39)

[Pau 78]* –, Komplexitätstheorie, Teubner Studienbücher, Informatik, Stuttgart 1978. (Intr., 25.5, 25.11)

[Pau 79] –, Kolmogorov complexity and lower bounds, FCT 1979, 325–334. (§ 8.1.5)

[Pau 81] –, On heads versus tapes, 22nd FOCS (1981), 68–73. (19.12)

[Pau 82] –, On-line simulation of $k+1$ tapes by k tapes requires nonlinear time, 23rd FOCS (1982), 53–56. See also: I&C **53** (1982), 1–8. (8.37)

[PaVa 76] PATERSON, M. S., VALIANT, L. G., Circuit size is nonlinear in depth, TCS **2** (1976), 397–400. (§ 26.3)

[PaYa 82] PAPADIMITRIOU, C. H., YANNAKAKIS, M., The complexity of facets (and some facets of complexity), 14th STOC (1982), 255–260. See also: JCSS **28** (1984), 244–259. (14.21)

[PaZa 82] PAPADIMITRIOU, CH., ZACHOS, S. K., Two remarks on the power of counting, preprint M.I.T. Cambridge, 1982. (§ 24.4)

[Pec 77] PECKEL, J., On a deterministic subclass of contextfree languages, MFCS 1977, LNCS **53** (1977), 430–434. (§ 8.1.3, 17.48)

[Pen 77] PENTTONEN, M., Szilard languages are log n tape recognizable, EIK **13** (1977), 595–602. (12.18)

[Per 79] PERRY, H. M., An improved proof of the Rabin-Hartmanis-Stearns conjecture, MIT Lab. f. Comp. Sci. TM-123 (1979). (8.37)

[PeRe 79] PETERSON, G. L., REIF, J. H., Multiple-person alternation, 20th FOCS (1979), 348–363. (§ 20.4.1)

[Pét 51]* PÉTER, R., Rekursive Funktionen, Akadémiai Kiadó, Budapest 1951. (§ 2.4)

[Pet 80] PETERSON, G. L., Succint representations, random strings, and complexity classes, 21st FOCS (1980), 86–95. (§ 20.4.1)

[PiFi 76] PIPPENGER, N., FISCHER, M. J., Relations among complexity measures, preprint 1976. See also: JACM **26** (1979), 361–381. (19.17, 19.25, 21.13)

[Pip 78] PIPPENGER, N., A time-space trade-off, JACM **25** (1978), 509–517. (25.13)

[Pip 79] –, On simultaneous resource bounds, 20th FOCS (1979), 307–311. (21.13, 21.14)

[Pip 80a] –, Comparative schematology and pebbling with auxiliary pushdowns, 12th STOC (1980), 351–356. See also: JCSS **23** (1981), 151–165. (25.12, 25.13)

[Pip 80b] –, Pebbling, Proc. 5th IBM Symposium on Math. Found of Comp. Science, Acad. and Scient. Programs, IBM Japan, (1980). (§ 25.3.2.2)

[Pip 82] –, Probabilistic simulations, 14th STOC (1982), 17–26. (19.27, 25.2)

[Pla 84] PLAISTED, D. A., Complete problems in the first-order predicate calculus, JCSS **29** (1984), 8–35. (§ 8.2.3.2)

[Ple 79] PLESNÍK, J., The NP completeness of the Hamiltonian cycle problem in planar digraphs with degree bound two, IPL **8** (1979), 199–201. (14.21)

[Pos 76] POSÁ, L., Hamiltonian circuits in random graphs, Discr. Math. **14** (1976), 359 to 364. (9.18)

[Pra 75] PRATT, V. R., Every prime has a succint certificate, SIAM JC 4 (1975), 214–220. (7.24)

[PrSt 76] PRATT, V. R., STOCKMEYER, L. J., A characterization of the power of vector machines, JCSS **12** (1976), 192–221. (20.12, § 20.3.2)

[Rab 60] RABIN, M. O., Degree of difficulty of computing a function and a partial ordering of recursive sets, Hebrew Univ. Jerusalem, TR 2 (1960). (Intr., § 5.2, 17.1)

[Rab 63] —, Real-time computation, Israel J. Math. **1** (1963), 203–211. (§ 8.1.2, 8.24, 22.9)

[Rab 76] —, Probabilistic algorithms, Symp. on New Directions and Recent Results in Algorithms and Complexity (1976), 21–39. (§ 9.2.2, 9.24)

[Rab 80] —, Probabilistic algorithm for testing primality, J. Number Theory **12** (1980), 128–138. (9.24)

[Rac 75] RACKOFF, C., The complexity of theories of the monadic predicate calculus, TR IRIA (8.48)

[Rac 76] —, On the complexity of the theories of weak direct powers, JSL **41** (1976), 561–573. (8.47)

[Rac 78a] —, The covering and boundedness problems for vector addition systems, TCS **6** (1978). (§ 8.2.3.4)

[Rac 78b] —, Relativized questions involving probilistic algorithms, 10th STOC (1978), 338–342. See also: JACM **29** (1982), 261–268. (§ 28.2)

[Ref 79] REIF, J. H., Universal games of incomplete information, 11th STOC (1979), 288–308. (§ 20.4.1)

[Ref 82a] —, Symmetric complementation, 14th STOC (1982), 201–214. (§ 20.4.1)

[Ref 82b] —, Parallel time $O(\log N)$ acceptance of deterministic CFL's, 23rd FOCS (1982), 290–296. (§ 12.3)

[Ref 83] —, Parallel algorithms for graph isomorphism, TR 14-83 Aiken Computation Lab., Harvard University, May 1983. (7.30)

[Reg 83] REGAN, K. W., Arithmetical degrees of index sets for complexity classes, LNCS **171** (1983), 118–130. (§ 24.5)

[Rei 78] REISCHUK, R., Improved bounds on the problem of time-space trade-off in a pebble game, 19th FOCS (1978), 84–91. See also: JACM **27** (1980), 839–849. (25.11)

[ReSc 82] REISCH, S., SCHNITGER, G., Three applications of Kolmogorov complexity, 23rd FOCS (1982), 45–52. (8.35, § 25.3.2.2)

[RiSp 72] RITCHIE, R. W., SPRINGSTEEL, F. N., Language recognition by marking automata, I&C **20** (1972), 313–330. (12.14)

[Ritd 66] RITCHIE, D. M., Program structure and computational complexity, Ph. D. Thesis, Harvard Univ. 1966. (§ 2.4.1)

[Ritr 63a] RITCHIE, R. W., Classes of predictably computable functions, TAMS **106** (1963), 139–173. (10.16, 10.18)

[Ritr 63b] —, Classes of recursive functions based on Ackermann's function. Pacific J. Math. **15** (1963), 1027–1044. (§ 2.4.1)

[Riv 77] RIVEST, R. L., On the worst-case behavior of string-searching algorithms, SIAM JC 6 (1977), 669–674. (§ 7.2.1)

[Rob 74a] ROBERTSON, E. L., Complexity classes of partial recursive functions, JCSS **9** (1974), 69–87. (15.11, § 15.2)

[Rob 74b] —, Structure of complexity in the weak monadic second order theories of the natural numbers, 6th STOC (1974), 161–171. (8.45)

[Rob 76] —, Honesty procedures and complexity classes of partial functions, preprint 1976, Indiana Univ., Bloomington (Ind.). (15.24)

[Robi 47] ROBINSON, R. M., Primitive recursive functions, Bull. Amer. Math. Soc. **53** (1947), 925–942. (2.3)

[Robs 83] ROBSON, J. M., The complexity of go, Information Processing 83, IFIP 1983 (R. E. A. MASON, ed.), 413–418. (14.24)

[Robs 84] —, $N \times N$ checkers is exptime complete, SIAM JC, to appear. (14.24)

[Rog 67]* ROGERS, H. Jr., Theory of Recursive Functions and Effective Computability, McGraw-Hill, New York 1967. (Pref., § 3, 18.16, § 27.2)

[Rol 79] ROLLIK, H. A., Automaten in planaren Graphen, 4th GI (1979), LNCS 67 (1979), 266–275. See also: AI 13 (1980), 287–298. (§ 23.3)

[Ros 67] ROSENBERG, A. L., Real-time definable languages, JACM 14 (1967), 645–662. (8.8, 11.2)

[Rou 75] ROUNDS, W. C., A grammatical characterization of exponential time languages, 16th FOCS (1975), 135–143. (12.17)

[RuFi 65] RUBY, S. S., FISCHER, P. C., Translational methods and computational complexity, 6th Ann. Symp. on Circuit Theory and Log. Design (1965), 173–178. (§ 6.4, 17.37)

[RuSiTo 82] RUZZO, W. L., SIMON, J., TOMPA, M., Space-bounded hierarchies and probabilistic computations, 14th STOC (1982), 215–223. (20.44)

[Ruz 78] RUZZO, W. L., General context-free language recognition, Ph. D. Diss., Univ. of Calif., Berkeley, Comp. Sci. Div., 1978. (20.36)

[Ruz 79a] —, On uniform circuit complexity, 20th FOCS (1979), 312–318. See also: JCSS 22 (1981), 365–383. (§ 12.3, § 20.3.2, 21.14, 21.15)

[Ruz 79b] —, Tree-size bounded alternation, 11th STOC (1979), 352–359. (20.36)

[Ruž 80] RUŽIČKA, P., Time and space bounds in producing certain partial orders, MFCS 1980, LNCS 88 (1980), 539–551. (§ 25.3.2.2)

[Ryt 84] RYTTER, W., Fast recognitions of pushdown automaton and context-free languages, MFCS 1984, LNCS 176 (1984), 507–515. (12.12, 13.20)

[SaGo 76] SAHNI, S., GONZALES, T., P-complete approximation problems, JACM 23 (1976), 555–565. (9.12, 9.15)

[Sah 75] SAHNI, S., Approximate algorithms for the 0/1 knapsack problem, JACM 22 (1975), 115–124. (§ 9.1.3.1)

[Sah 76] —, Algorithms for scheduling independent tasks, JACM 23 (1976), 116–127. (9.6)

[SaJa 81] SAVAGE, C., JA' JA', J., Fast efficient parallel algorithms for some graph problems, SIAM JC 10 (1981), 682–691. (§ 7.2.2.3)

[Sal 73]* SALOMAA, A., Formal Languages, Academic Press, New York and London 1973. (Pref., § 4, § 11)

[Sal 78] —, Equality sets for homomorphisms of free monoids. Acta Cybernetica 4 (1978), 127–139. (4.14)

[SaSt 76] SAVITCH, W. J., STIMSON, M. J., Time bounded random access machines with parallel processing, Mathematisch Centrum, Amsterdam, Dept. of Comp. Sci. TR, 1976. See also: JACM 26 (1979), 103–118. (20.35)

[SaSw 78] SAVAGE, J. E., SWAMY, S., Space-time trade-offs on the FFT algorithm, IEEE Trans. Inf. Theory IT-24 (1978), 563–568. (§ 25.3.2.2)

[SaSw 79] —, Space-time tradeoffs for oblivious integer multiplication, 6th ICALP (1979), LNCS 71 (1979), 498–504. (§ 25.3.2.2)

[Sav 76]* SAVAGE, J. E., The Complexity of Computing, John Wiley & Sons, New York—London 1976. (Intr., § 8)

[Sav 81] —, Space-time tradeoffs — A survey, 3rd Hungarian Computer Conference (1981), 93–104. (§ 25.3.2.2)

[Savi 70] SAVITCH, W. J., Relationships between nondeterministic and deterministic tape complexities, JCSS 4 (1970), 177–192. (23.1)

[Savi 73a] —, Maze recognizing automata and nondeterministic tape complexity, JCSS 7 (1973), 389–403. (8.43, 14.13, 23.15)

[Savi 73b] —, A note on multihead automata and context-sensitive languages, AI 2 (1973), 249–252. (§ 6.4)

[Savi 77] —, Recursive Turing machines, IJCM 6 (1977), 3–31. (§ 20.3.1, 20.33)

[Savi 78] —, Parallel and nondeterministic time complexity classes, 5th ICALP (1978), LNCS 62 (1978), 411–424. (20.35)

[SaVi 77] SAVITCH, W. J., VITÁNYI, P. M. B., Linear time simulation of multihead Turing machines with head-to-head jumps, 4th ICALP (1977), LNCS 52 (1977), 453–464. (§ 19.2.2.3)

[Saz 79] SAZONOW, V. JU., A theory in which no lower exponential bounds for NP complete problems are provable (in Russian), 5. Vsjesojusnaja konferencia po mat. logike, Novosibirsk 1979, 133. See also [Saz 80]. (§ 24.5)

[Saz 80] —, A logical approach to the problem "P = NP", MFCS 1980, LNCS 88 (1980), 562–575. (§ 24.5)

[Sca 83] SCARPELLINI, B., Second order spectra, LNCS 171 (1983), 380–389. (Ch. IV)

[Sch 76] SCHAEFER, T. J., Complexity of decision problems based on finite two-person perfect-information games, 8th STOC (1976), 41–49. See also: JCSS 16 (1978), 185–225. (8.52, 14.14)

[Scha 77] SCHAKUOV, S. N., Fast Turing calculations and their linear speed-up (in Russian), DAN 236 (1977), 556–557. (18.12)

[Schg 83] SCHÖNING, U., A low and high hierarchy within NP, JCSS 27 (1983), 14–28. (§ 8.2.1, § 24.3)

[Schg 84a] —, Minimal pairs for P, TCS 31 (1984), 41–48. (24.31, 24.32)

[Schg 84b] —, Robust algorithms: a different approach to oracles, 11th ICALP, LNCS 172 (1984), 448–453. (§ 27.1, § 28.3)

[Schi 83] SCHNITZLER, M., The isomorphism problem is polynomially solvable for certain graph languages, LNCS 153 (1983), 369–379. (7.30)

[Schn 73a] SCHNORR, C. P., Does the computational speed-up concern programming?, 1st ICALP (1972), 585–592. (18.6)

[Schn 74a]* —, Rekursive Funktionen und ihre Komplexität, Teubner, Stuttgart, 1974. (Intr., Pref., § 2.2, 18.1)

[Schn 74b] —, Optimal enumerations and optimal Goedel numberings, MST 8 (1974), 182 to 191. (§ 2.2)

[Schn 76] —, The network complexity and the Turing machine complexity of finite functions, AI 7 (1976), 95–107. (§ 21.2)

[Schn 78] —, Satisfiability is quasilinear complete in NQL, JACM 25 (1978), 136–145. (14.20)

[Schn 79] —, On self-transformable combinatorial problems, preprint 1979. (24.47)

[Schö 73] SCHÖNHAGE, A., Real-time simulation of multidimensional Turing machines by storage modification machines, MIT Technical Memo 37, 1973. (19.29)

[Schö 78] —, Storage modification machines, preprint 1978. See also: SIAM JC 9 (1980), 490–508. (7.17, 19.29)

[Schö 80] —, Random access machines and Presburger arithmetic, preprint 1980. (8.47)

[Scht 79] SCHWEIGGERT, F., On the relative complexity in deciding graph isomorphism, Discrete Structures and Algorithms, Proc. Conf. Berlin 1979, 65–76. (7.30)

[Schw 69] SCHWICHTENBERG, H., Rekursionstiefe und Grzegorczyk-Hierarchie, AML 12 (1969), 85–97. (2.18)

[ScKl 77] SCHNORR, C. P., KLUPP, H., A universally hard set of formulae with respect to nondeterministic Turing acceptors, IPL 6 (1977), 35–37. (8.47)

[ScSt 72] SCHNORR, C. P., STUMPF, G., A characterization of complexity sequences, Tagungsbericht 46/1972, Algorithmen und Komplexitätstheorie, Oberwolfach, 1972. See also: ZML 21 (1975), 47–56. (18.1)

[ScStr 71] SCHÖNHAGE, A., STRASSEN, V., Schnelle Multiplikation großer Zahlen, COMP 7 (1971), 281–292. (7.16)

[SeFiMe 77] SEIFERAS, J. I., FISCHER, M. J., MEYER, A. R., Seperating nondeterministic time complexity classes. Pennsylvania State Univ., Comp. Sci. Deptm., TR 205, 1977. See also: JACM 25 (1978), 146–167. (17.40, 17.43)

[SeGa 77] SEIFERAS, J. I., GALIL, Z., Real-time recognition of substring repetition and reversal, MST 11 (1977), 111–146. (7.6, 7.8)

[Sei 76] SEIFERAS, J. I., A note on notions of tape constructibilities, Pennsylvania State Univ., Comp. Sci. Deptm., TR 187, 1976. (5.6)

[Sei 77a] —, Techniques for separating space complexity classes, JCSS **14** (1977), 73–99. (8.30, 19.1, 19.6, 19.7)

[Sei 77b] —, Relating refined space complexity classes, JCSS **14** (1977), 100–129. (17.23, 17.32, 19.1, 19.6, 19.7)

[Sei 77c] —, Iterative arrays with direct central control, AI **8** (1977), 177–192. (19.32)

[Sel 78] Selman, A. L., Polynomial time enumeration reducibility, SIAM JC **7** (1978), 440–457. (§ 6.1.2)

[Sel 79] —, P-selective sets, tally languages, and the behavior of polynomial time reducibilities on NP, 6th ICALP (1979), LNCS **71** (1979), 546–555. See also: MST **13** (1979), 55–65. (§ 6.1.2)

[Sel 80] —, Reductions on NP and P-selective sets, Iowa State Univ., Ames, Comp. Sci. Deptm., TR Juni 1980. See also: TCS **19** (1982), 287–304. (§ 6.1.2)

[Sel 82] —, Analogues of semirecursive sets and effective reducibilities to the study of NP complexity, I&C **52** (1982), 36–51. (§ 6.2)

[SeMeBo 81] Selman, A. L., Xu Mei-rui, Book, R. V., Controlled relativizations of complexity classes, Res. Rep., Santa Barbara, Oct. 1981. See also: SIAM JC **12** (1983), 565–579. (28.13)

[Sew 83] Sewelson, V., The structure of NP under relativization, Conf. Comp. Complexity Theory, March 21–25, Santa Barbara, 1983, 94–103. (28.7)

[SeWe 76] Seiferas, J., Weicker, R., Linear-time simulation of an iterative array of finite-state machines by a Boolean RAM with uniform cost criterion, Pennsylvania State Univ., Comp. Sci. Deptm., TR 201, 1976. (19.34)

[Sha 77] Shapiro, S. D., Performance of heuristic bin packing algorithms with segments of random length, I&C **35** (1977), 146–158. (§ 9.2.1)

[ShBe 74] Shamir, E., Beeri, C., Checking stacks and context-free programmed grammars accept p-complete languages, 2nd ICALP (1974), LNCS **14** (1974), 27–33. (14.21)

[Sho 67]* Shoenfield, J. R., Mathematical Logic, Addison-Wesley, Reading (Mass.) 1967. (§ 8.2.3.2)

[SiGi 77] Simon, I., Gill, J., Polynomial reducibilities and upward diagonalizations, 9th STOC (1977), 186–194. (6.10, 28.10)

[SiGiHu 78] Simon, J., Gill, J., Hunt, J., On tape-bounded probabilistic Turing machine transducers, 19th FOCS (1978), 107–112. (20.44)

[Simi 76] Simon, I., Two results on polynomial time reducibility, SIGACT News **8** (1976), 33–37. (6.8)

[Simj 75] Simon, J., On some central problems in computational complexity, Cornell Univ., Ithaca, Dept. of Comp. Sci., TR 75–224, 1975. (14.23)

[Simj 76] —, Polynomially bounded quantification over higher types and a new hierarchy of the elementary sets, Nonclassical Logics, Model Theory, and Computability, 3rd Latin-Amer. Symp. Math. Logic (1976), in: Stud. Logic Found. Math. **89** (1977), 267–281. (§ 24.3)

[Simj 77] —, On the difference between one and many, 4th ICALP (1977), LNCS **52** (1977), 480–491. (20.44, § 20.4.2, § 24.4)

[Simj 79] —, Division is good, 20th FOCS (1979), 411–420. See also: JCSS **22** (1981), 421–441. (§ 20.1.3)

[Simj 81] —, Space-bounded probabilistic Turing machine complexity classes are closed under complement, 13th STOC (1981), 158–167. (20.44)

[Sip 78] Sipser, M., Halting space bounded computations, 19th FOCS (1978), 73–74. See also: TCS **10** (1980), 335–338. (17.22)

[Sip 83] —, A complexity theoretic approach to randomness, 15th STOC (1983), 330–335. (§ 24.4)

[Sir 71] Siromoney, R., Finite turn checking stack automata, JCSS **5** (1971), 549–559. (§ 12.2, 22.10)

[SlEm 83] Slot, C., van Emde Boas, P., On tape versus core: an application of space efficient perfect hash functions to the invariance of space, preprint 1983. (§ 5.1.1.2)

[Sli 77a] SLISENKO, A. O., A simplified proof of real-time recognizability of palindroms on TM's (in Russian), in: Teoret. primen. metodov mat. logiki II, Isd. Nauka, Leningradsk. otd., Leningrad 1977. (7.4)

[Sli 77b] —, Recognition of the substring predicate in realtime (in Russian), preprint LOMI P-7-77, 1977. (7.1)

[Sli 78] —, String matching in real-time: Some properties of the data structure, MFCS 1978, LNCS **64** (1978), 493–496. (7.8)

[Sli 79] —, Computational complexity of string and graph identification, MFCS 1979, LNCS **74** (1979), 182–190. (7.8)

[Sli 81] —, Search of periodicities and identification of subwords in realtime (in Russian), Zap. naučn. sem. LOMI **105** (1981), 62–173. (§ 7.2.1)

[Smi 72] SMITH III, A. R., Real-time language recognition by one-dimensional cellular automata, JCSS **6** (1972), 233–253. (12.13)

[Soa 77] SOARE, R. I., Computational complexity, speedable and levelable sets, JSL **42** (1977), 545–563. (18.17, § 18.2)

[SoSt 77] SOLOVAY, R., STRASSEN, V., Fast Monte Carlo test for primality, SIAM JC **6** (1977), 84–85. (9.24)

[Spr 76] SPRINGSTEEL, F. N., On the pre-AFL of [log *n*] space and related families of languages, TCS **2** (1976), 295–304. (11.13)

[SpSt 76]* SPECKER, E., STRASSEN, V., Komplexität von Entscheidungsproblemen, Ein Seminar, LNCS **43** (1976). (Intr.)

[Sta 79] STATMAN, R., Intuitionistic propositional calculus is PSPACE-complete, TCS **9** (1979), 67–72. (8.49, 14.14)

[StCh 79] STOCKMEYER, L. J., CHANDRA, A. K., Provably difficult combinatorial games, SIAM JC **8** (1979), 151–174. (8.52, 14.24)

[StHaLe 65] STEARNS, R. E., HARTMANIS, J., LEWIS II, P. M., Hierarchies of memory limited computations, IEEE Conf. on Switch. Circuit Theory and Logical Design 1965, 179–190. (5.6, 8.3, 8.28, 8.29, 17.23, 17.27, 18.11)

[StMe 73] STOCKMEYER, L. J., MEYER, A. R., Word problems requiring exponential time, 5th STOC (1973), 1–9. (8.44, 8.49, 14.14, 14.21, 14.22, 14.25, 24.46)

[Sto 74] STOCKMEYER, L. J., The complexity of decision problems in automata theory and logic, Thesis, M.I.T. Project MAC TR-133, 1974. (8.44, 8.45, 8.46, 14.17, 14.18, 18.25–18.27)

[Sto 77] —, The polynomial time hierarchy, TCS **3** (1977), 1–22. (8.48, 14.14, 24.43, § 24.3, Ch. IV)

[Sto 83] —, The complexity of approximate counting, 15th STOC (1983), 118–126. (§ 24.4, 28.6)

[Stor 83] STORER, J. A., On the complexity of chess, JCSS **27** (1983), 77–100. (14.14)

[Stoß 70] STOSS, H. J., *k*-Band-Simulation von *k*-Kopf-Turing-Maschinen, COMP **6** (1970), 309–317. (19.9)

[Stoß 71] —, Zwei-Band-Simulation von Turingmaschinen, COMP **7** (1971), 222–235. (19.23)

[Strn 68] STRNAD, P., On-line Turing machine recognition, I&C **12** (1968), 442–452. (§ 8.1.1.3, 17.40)

[StSa 79] STIMSON, M. J., SAVITCH, W. J., Hierarchies of recursive computations, IJCM **7** (1979), 271–286. (§ 20.3.1)

[Sud 73] SUDBOROUGH, I. H., On tape-bounded complexity classes and multihead finite automata, 14th SWAT (1973), 138–144. See also: JCSS **10** (1975), 62–76. (8.43, 14.13, 23.18, 23.19)

[Sud 75] —, A note on tape-bounded complexity classes and linear context-free languages, JACM **22** (1975), 499–500. (8.43, 14.3)

[Sud 76] —, On multihead writing automata, I&C **30** (1976), 1–20. (13.4)

[Sud 77a] —, Some remarks on multihead automata, RAIRO Inf. Théor. **11** (1977), 181 to 195. (13.11, 13.26)

[Sud 77b] —, On the time and tape complexity of developmental languages, 4th ICALP (1977), LNCS **52** (1977), 507—523. (12.19, 14.3)

[Sud 77c] —, Time and tape bounded auxiliary pushdown automata, MFCS 1977, LNCS **53** (1977), 493—503. (11.4, 14.3)

[Sud 77d] —, Separating tape bounded auxiliary pushdown automata classes, 9th STOC (1977), 208—217. (13.24, 13.25)

[Sud 78] —, On the tape complexity of deterministic context free languages, JACM **25** (1978), 405—414. (14.3)

[Sud 80] —, Efficient algorithms for path system problems and applications to alternating and time-space complexity classes, 21st FOCS (1980), 62—73. (14.21)

[Sud 81] —, Pebbling and bandwidth, Proc. FCT '81, LNCS **117** (1981), 373—383. (14.14)

[SuZa 73] Sudborough, I. H., Zalcberg, A., On families of languages defined by time-bounded random access machines, MFCS 1973, 333—338. See also: SIAM JC **5** (1976), 217—230. (17.42)

[Sym 71] Symes, D. M., The extension of machine independent computational complexity theory to oracle machine computation and the computation of finite functions, Res. rep. CSRR 2057, University of Waterloo, Oct 71. (§ 5.3)

[Sze 82] Szepietowski, A., A finite 5-pebble automaton can search every maze, IPL **15** (1982), 199—204. (§ 23.3)

[Tar 78] Tarjan, R. E., Complexity of combinatorial algorithms, SIAM Rev. **20** (1978), 457—491. (§ 7.2.2)

[Tar 80] Tarjan, R. E., Recent developments in the complexity of combinatorial algorithms, Proc. 5th IBM Symposium on Math. Found. of Comp. Science, Academic and Scientific Programs, IBM Japan, (1980). (§ 12.3)

[Tat 81] Tat-hung Chan, Reversal complexity of counter machines, 13th STOC (1981), 146—157. (21.7)

[TaTr 77] Tarjan, R. E., Trojanowski, A. E., Finding a maximum independent set, SIAM JC **6** (1977), 537—546. (§ 7.3.1)

[Tho 72] Thompson, D. B., Subrecursiveness: machine independent notions of computability in restricted time and storage, MST **6** (1972), 3—15. (10.10, 10.20)

[Tom 78] Tompa, M., Time-space tradeoffs for computing functions using connecting properties of their circuits, 10th STOC (1978), 196—204. (§ 25.3.2.2)

[Tom 80] —, Two familiar transitive closure algorithms which admit no polynomial time, sublinear space implementations, 12th STOC (1980), 333—338. (§ 25.3.2.2)

[Tom 81] —, An extension of Savitch's theorem to small space bounds, IPL **12** (1981), 106—108. (23.1)

[Tra 63] Trakhtenbrot, B. A., On the frequency computation of functions (in Russian), Algebra i logika **2**, 2 (1963), 18—26. (§ 9.2.3)

[Tra 64] —, Turing machine computations with logarithmic delay (in Russian), Algebra i logika **3**, 4 (1964), 33—48. (8.15)

[Tra 65] —, Optimal computations and the frequency phenomenon of Jablonskiĭ (in Russian), Algebra i logika **4**, 5 (1965), 79—93. (16.4, 16.20)

[Tra 67]* —, Complexity of Algorithms and Computations (lectures for students of the NGU) (in Russian), Novosibirsk 1967. (Intr., 17.5, § 18.2)

[Tra 70] —, On autoreducibility (in Russian), DAN SSSR **192** (1970), 1224—1227. (27.23)

[Tra 74] —, Notes on the complexity of computations on probabilistic machines (in Russian), Teorija algorifmov i mat. logika, Comp. Center ANSSSR, Moscow 1974, 159—176. (§ 9.2.2)

[Tra 77] —, Frequency algorithms and computations, MFCS 1977, LNCS **53** (1977), 148—161. (§ 9.2.3)

[Ukk 83a] Ukkonen, E., Two results on polynomial time truth-table reductions to sparse sets, SIAM JC **12** (1983), 580—587. (24.17)

[Ukk 83b] —, Exponential lower bounds for some NP-complete problems in a restricted linear decision tree model, BIT **23** (1983), 181—192. (§ 8.2.1)

[Ull 75] ULLMAN, J. D., NP complete scheduling problems, JCSS **10** (1975), 384–393. (14.21)

[Vai 76] VAISER, A. V., Stochastic languages and complexity of computations on probabilistic Turing machines (in Russian), Kibernetika (Kiev) 1976, Nr. 1, 21–25. (9.19)

[Val 75] VALIANT, L. G., General context-free recognition in less than cubic time, JCSS **10** (1975), 308–315. (12.7, 12.10)

[Val 77a] —, The complexity of computing the permanent, Univ. of Edinburgh, Dept. of Comp. Sci., CSR-14-77, 1977. See also: TCS **8** (1979), 189–201. (§ 24.4, § 20.4.2)

[Val 77b] —, The complexity of enumeration and reliability problems, Univ. of Edinburgh, Dept. of Comp. Sci., CSR-15-77, 1977. See also: SIAM JC **8** (1979), 410–421. (§ 24.4)

[Var 82] VARDI, M. Y., The complexity of relational query languages, 14th STOC (1982), 137–146. (14.14, 14.25)

[Vav 77] VALIEV, M. K., Realtime computations with restrictions on tape alphabet, MFCS 1977, LNCS **53** (1977), 532–536. (22.7)

[Vav 80] —, Decision complexity of variants of propositional dynamic logic, MFCS 1980, LNCS **88** (1980), 656–664. (8.49)

[VaVa 82] VAZIRANI, U. V., VAZIRANI, V. V., A natural encoding scheme proved probabilistic polynomial complete, 23rd FOCS (1982), 40–44. See also: TCS **24** (1983), 291–300. (14.21, § 24.2.1.2.3)

[Ver 81] VERBEEK, R., Time-space trade-offs for general recursion, 22nd FOCS (1981), 228–234. (25.6)

[Ver 83] —, Complexity of pushdown computations, Habilitationsschrift an der Rheinischen Friedrich-Wilhelms-Universität Bonn 1983. (§ 25.3.2.1.1)

[Vit 76] VITÁNYI, P. M. B., Deterministic Lindenmayer languages, nonterminals and homomorphisms, TCS **2** (1976), 49–71. (4.10, 12.19)

[Vit 77] —, Context sensitive table Lindenmayer languages and a relation to the LBA problem. I&C **33** (1977), 217–226. (4.10, 12.19)

[Vit 80] —, On the power of real-time Turing machines under varying specifications, 7th ICALP (1980), 658–671. (8.38)

[Vit 82] —, Real-time simulation of multicounters by oblivious one-tape Turing machines, 14th STOC (1982), 27–36. See also: I&C **55** (1982), 20–39. (20.2)

[Vit 84] —, The simple roots of real-time computation hierarchies, 11th ICALP (1984), LNCS **172** (1984), 486–489. (8.38)

[Vog 78] VOGEL, J., Bemerkungen zur Zeithierarchie bei einbändrigen deterministischen Turingmaschinen, Diplomarbeit, Jena 1978. (17.37)

[Vol 83a] VOLGER, H., Turing machines with linear alternation, theories of bounded concatenation and the decision problems of first order theories, TCS **23** (1983), 333–337. (§ 8.2.3.2)

[Vol 83b] —, Rudimentary relations and Turing machines with linear alternations, LNCS **171** (1983), 131–136. (11.16)

[VoWa 81] VOGEL, J., WAGNER, K., On a class of automata accepting exactly the languages which are elementary in the sense of Kalmár, Third Hungarian Computer Science Conference 1981, 289–299. (13.41–13.45, 20.29, 20.30)

[Waa 81] WAACK, S., Tape complexity of word problems, FCT '81, LNCS **117** (1981), 467–471. (14.13)

[Wag 79] WAGNER, K., Bounded recursion and complexity classes, MFCS (1979), LNCS **74** (1979), 492–498. (10.13, 10.19, 10.25)

[Wag 84a] —, The complexity of combinatorial problems with compactly described instances, preprint, Friedrich Schiller University Jena, N/84/23 (1984). (14.21, 14.22, § 24.3)

[Wag 84b] —, The complexity of problems concerning graphs with regularities, preprint, Friedrich Schiller University Jena, N/84/52. (14.14, 14.22, 14.25)

[WaWe 77] WAGNER, K., WECHSUNG, G., Complexity hierarchies of oracles, MFCS (1977), LNCS **53** (1977), 543–548. (27.6)

[WeBr 79] WECHSUNG, G., BRANDSTÄDT, A., A relation between space, return and dual return complexities, TCS **9** (1979), 127–140. (21.9, 21.12)

[Wec 76] WECHSUNG, G., Kompliziertheitstheoretische Charakterisierung der kontextfreien und linearen Sprachen, EIK **12** (1976), 289–300. (12.5, 21.9)

[Wec 79] —, A crossing measure for 2-tape Turing machines, MFCS (1979), LNCS **74** (1979), 508–516. (§ 17.2.5)

[Wec 85] —, On sparse complete sets, ZML **31** (1985), 281–287. (24.17)

[Weic 71] WEICKER, R., Tabulator-Turingmaschinen und Komplexität, COMP **7** (1971), 264–274. (§ 19.2.2.3)

[Weic 74] —, Turing machines with associative memory access. 2nd ICALP (1974), LNCS **14** (1974), 458–472. (19.35, 19.36)

[Weic 77] —, The influence of the machine model on the time complexity of context-free language recognition, MFCS (1977), LNCS **53** (1977), 560–569. (12.12)

[Weid 80] WEIDE, B. W., Random graphs and graph optimization problems, SIAM JC **9** (1980), 552–557. (§ 9.2.1)

[Weih 74] WEIHRAUCH, K., On the computational complexity of program schemata, GMD Bericht Nr. 70 (1974), See also: 2nd ICALP (1974), LNCS **14** (1974), 326–334, and JCSS **12** (1976), 80–107. (16.5)

[Wein 73] WEINER, P., Linear pattern matching algorithm, 14th SWAT (1973), 1–11. (§ 7.2.1)

[Wel 83] WELSH, D. J. A., Randomised algorithms, Discr. Appl. Math. **5** (1983), 133–145. (§ 9.2.1)

[Wer 71] WERNER, G., Propriété d'invariance des classes des fonctions de complexité bornée, C. R. Acad. Sci. Paris, Ser. A–B, **273** (1971), A133–A136. (15.27)

[Wer 74] —, Sous-classes récursivement énumerables d'une classe de complexité, C. R. des Journées Mathématiques de la Société Mathématique de France (Univ. Sci. Techn. Languedoc, Montpellier, 1974), 369–375. (15.8)

[Wie 83] WIEDERMAN, J., Deterministic and nondeterministic simulation of the RAM by the Turing machine, Information Processing 83 (R. E. A. MASON, ed.), Elsevier Science Publ. B.V. (North-Holland), IFIP (1983), 163–168. (19.31)

[Wig 82] WIGDERSON, A., A new approximate graph colouring algorithm, 14th STOC (1982), 325–329. (9.15)

[Wil 83] WILSON, C. B., Relativized circuit complexity, 24th FOCS (1983), 329–334. (§ 24.2.2.2.1, 24.36)

[Woo 75] WOOD, D., Iterated a-NGSM maps and L-systems, I&C **32** (1975), 1–26. (12.19)

[Wot 78] WOTSCHKE, D., Nondeterminism and boolean operations in pda's. JCSS **16** (1978), 456–461. (11.16)

[Wra 75] WRATHALL, C., Subrecursive predicates and automata, Ph. D. Diss., Harvard Univ., 1975. (11.20)

[Wra 77a] —, Complete sets and the polynomial-time hierarchy, TCS **3** (1977), 23–33. (24.45, 24.46)

[Wra 77b] —, Characterizations of the Dyck sets, RAIRO (1977). (11.15)

[Wra 78] —, Rudimentary predicates and relative computation, SIAM JC **7** (1978), 194–209. (11.4, 11.14)

[YaLe 82] YAMNITSKY, B., LEVIN, L. A., An old linear programming algorithm runs in polynomial time, 23rd FOCS (1982), 327–328. (§ 7.2.2.4)

[Yam 62] YAMADA, H., Real-time computation and recursive functions not real-time computable, IRE Trans. El. Comp. **EC-11** (1962), 753–760. (Intr.)

[Yaoa 75] YAO, A., An $O(|E| \cdot \log\log|V|)$ algorithm for finding minimum spanning trees, IPL **4** (1975), 21–23. (7.19)

[Yaoa 80] —, New algorithms for bin packing, JACM **27** (1980), 207–227. (§ 9.1.3.2)

[Yaoa 81] —, On the parallel computation for the knapsack problem, 13th STOC (1981), 123–127. See also: JACM **29** (1982), 898–903. (§ 20.3.2)

[Yaof 79] YAO, F. F., Graph 2-isomorphism is NP-complete, IPL **9** (1979), 68–72. (§ 7.3.4)

[Yap 83] YAP, C. K., Some consequences of nonuniform conditions on uniform classes, TCS **26** (1983), 287–300. (24.17)

[YaRi 76] YAO, A. C., RIVEST, R. L., $k+1$ heads are better than k, 17th FOCS (1976), 67–70. See also: JACM **25** (1978), 337–340. (13.4)

[Yes 83] YESHA, Y., On certain polynomial-time truth-table reducibilities of complete sets to sparse sets, SIAM JC **12** (1983), 411–425. (24.17)

[You 71] YOUNG, P. R., A note on dense and nondense families of complexity classes, MST **5** (1971), 66–70. (16.21)

[You 77] —, Optimization among provably equivalent programs, JACM **24** (1977), 693 to 700. (18.3)

[You 83] —, Some structural properties of polynomial reducibilities and sets in NP, 15th STOC (1983), 392–401. (24.9)

[Your 67] YOUNGER, D. H., Recognition and parsing of contextfree languages in time n^3, I&C **10** (1967), 189–208. (12.6)

[Yu 70] YU, Y. Y., Rudimentary relations and formal languages, Ph. D. Diss., Univ. of Calif., Berkeley, 1970. See also: MST **10** (1977), 337–343. (11.14)

[Zac 83] ZACHOS, S., Collapsing probabilistic polynomial hierarchies, Conf. Comp. Complexity Theory, March 21–25, 1983, Santa Barbara, 75–81. (§ 24.3. § 24.4)

[Žák 79] ŽÁK, S., A Turing machine space hierarchy, Kibernetika (Prague) **15** (1979), 100–121. (17.30)

[Zej 71 a] ZEJTIN, G. S., A normal form of normal algorithms and the theorem on linear speed-up (in Russian), Zap. naučn. sem. LOMI **20** (1971), 234–242. (18.13)

[Zej 71 b] —, A lower bound on the number of steps for reversing normal algorithms and other analogous algorithms (in Russian), Zap. naučn. sem. LOMI **20** (1971), 243–262. (§ 8.1.5)

[ZvLe 70] ZVONKIN, A. K., LEVIN, L. A., Complexity of finite objects and foundation of the notions information and randomness by the help of the theory of algorithms (in Russian), Usp. mat. nauk **25**: 6 (1970), 85–127. (Intr.)

Summary of the Notational System for TM's

We summarize all relevant definitions and conventions for TM's made in § 1.1.1, § 1.1.2 and § 1.1.3. Let $r \geqq 1$ and $X \in \{D, N\}$. A

$$Z_1\text{-}Z_2\text{-}\ldots\text{-}Z_r\text{-XM}$$

is a $\left\{\begin{array}{l}\text{deterministic if } X = D\\ \text{nondeterministic if } X = N\end{array}\right\}$ Turing machine having the storage media Z_1, Z_2, …, Z_r where $Z_1 \in \{2{:}h, 1{:}h, 1^*, T^d_m h, T^d h\}$ (this storage medium holds the input) and $Z_2, \ldots, Z_r \in \{T^d_m h, T^d h, S, NES, CS, PD, C\}$. Here the name

2:h denotes a two-way input tape with *h* heads,

1:h denotes a one-way input tape with *h* heads,

1* denotes the input tape of an on-line TM,

$T^d h(T^d_m h)$ denotes a *d*-dimensional Turing tape with *h* heads (and at most *m* work-tape symbols),

S denotes a stack,

NES denotes a nonerasing stack,

CS denotes a checking stack,

PD denotes a pushdown store,

C denotes a counter,

The following conventions are made:

1. 2- stands for 2:1-,
2. 1- stands for 1:1-,
3. 0- stands for no input tape, i.e. 0- can be omitted,
4. kZ stands for $\underbrace{\text{Z-Z-}\ldots\text{-Z}}_{k\,\text{times}}$, $Z \in \{T^d_m h, T^d h, S, NES, CS, PD, C\}$,
5. *d* and *h* can be omitted if they have the value 1,
6. i-XTM stands for i-T-XM ($i = 0, 1, 2$, $X \in \{D, N\}$),
7. i:h-XZA stands for i:h-Z-XM ($i = 1, 2$, $X \in \{D, N\}$, $Z \in \{S, NES, CS, PD, C\}$),
8. i:h-XFA stands for i:h-XM ($i = 1, 2$, $X \in \{D, N\}$).

Symbol Index

Author Index

Subject Index

9 781402 003134